CDX Learning Systems™

ELECTRICITY for the HVACR TECHNICIAN

HVAC Excellence™

JONES & BARTLETT
LEARNING

World Headquarters
Jones & Bartlett Learning
5 Wall Street
Burlington, MA 01803
978-443-5000
info@jblearning.com
www.jblearning.com

Jones & Bartlett Learning books and products are available through most bookstores and online booksellers. To contact Jones & Bartlett Learning directly, call 800-832-0034, fax 978-443-8000, or visit our website, www.jblearning.com.

Substantial discounts on bulk quantities of Jones & Bartlett Learning publications are available to corporations, professional associations, and other qualified organizations. For details and specific discount information, contact the special sales department at Jones & Bartlett Learning via the above contact information or send an email to specialsales@jblearning.com.

978-1-284-14477-2

Production Credits
General Manager: Kimberly Brophy
Product Manager: Jesse Mitchell
Product Owner: Christopher Benson
VMO Manager: Sara Kelly
Marketing Manager: Amanda Banner
VP, Manufacturing and Inventory Control: Therese Connell
Composition and Project Management: Integra Software Services Pvt. Ltd.
Cover Design: Scott Moden
Rights & Media Specialist: Robert Boder
Media Development Editor: Shannon Sheehan
Cover Image (Title Page): © InkkStudios/Getty images
Printing and Binding: LSC Communications
Cover Printing: LSC Communications

Library of Congress Cataloging-in-Publication Data unavailable at time of printing.

6048

Printed in the United States of America
22 21 20 19 18   10 9 8 7 6 5 4 3 2 1

BRIEF CONTENTS

CONTENTS

ACKNOWLEDGMENTS

Lead Writer

Troy Staton
Certified Master HVACR Craft Professional, Technical Writer, and Curriculum Design Consultant

Contributing Writers

Don Steeby
Associate HVAC Professor, Grand Rapids Community College

Scott Oakley
Certified Master HVACR Educator, NATE, Q Program Liaison, HVACRedu.net Faculty

Review Board

Herman Simien
Instructor, Nance HVAC Technical School

Steven Phyillaier
Associate HVAC Professor, Thaddeus Stevens College of Technology

Bob Feathers
Technical Trainer (Retired), Emerson Educational Services

Dick Wirtz
HVACR Consultant, Richard's Consulting Services

Don Steeby
Associate HVAC Professor, Grand Rapids Community College

CHAPTER 1

Electricity in HVACR

Knowledge Objectives

After reading this chapter, you will be able to:

- **K01001** Understand how electricity is generated and distributed.
- **K01002** Define common electrical terms.
- **K01003** Distinguish the difference between alternating current (AC) and direct current (DC).

Skill Objectives

There are no Skill Objectives for this chapter.

Introduction

The motors, controls, and many other components used in HVACR (heating, ventilating, air conditioning, and refrigeration) systems are electrical devices. Without electricity, the systems could not operate. Many of the problems that technicians encounter on service calls will be electrical in nature. Therefore, the HVACR technician requires a thorough knowledge of electricity.

Where Does Electricity Come From?

K01001 Understand how electricity is generated and distributed.

It is commonplace to insert a plug into an electrical outlet to obtain **power** for appliances and other electrical devices. The electricity used in homes and businesses originates at power plants that are often many miles away from the people who use the electricity. Power plants are traditionally classified based on their source of energy. Power plants dependent on a variety of energy sources will be introduced here.

Fossil-Fuel Power Plants

Many power plants are powered by burning coal, which is a **fossil fuel** used in the generation of electricity. Natural gas and oil are also fossil fuels, although oil is not a popular fuel for generating power. Natural gas, however, has grown in popularity for smaller plants that can be started quickly to provide additional power when the demand is high.

There are two concerns related to the burning of fossil fuels. First, it is a limited resource that cannot be replaced. Second, the burning of fossil fuels (especially coal) produces harmful gases and waste by-products. Efforts are being made to increase the use of alternative energy sources, such as solar power, wind power, and **biomass** materials, which do not have these disadvantages. On the other hand, coal and natural gas–fired power plants are very reliable and produce power at a rate that can easily be regulated based on demand. There are a number of reasons why they are still in operation across the world.

A coal-fired power plant burns coal in a huge boiler to produce steam. The steam is in turn used to drive a steam turbine generator, which produces electricity. **FIGURE 1-1** shows the basic arrangement of a coal-fired power plant.

Burning coal produces smoke that carries harmful pollutants into the air. The various anti-pollution devices used in coal-fired plants reduce these emissions but do not eliminate them. In addition, the burning of coal produces a large amount of ash, called fly ash, which must be disposed of. The ash also contains substances harmful to humans and the environment. This means that it must be disposed of in special containment facilities that prevent these substances from leaking into groundwater.

Natural gas is also used to power turbines, but on a smaller scale than coal-fired plants. Natural-gas plants continue to grow in popularity for a variety of reasons. The projections for supply are very good, and these plants can be started up and shut down in response to peak power demands without a great deal of effort. They are generally smaller in scale than plants based on coal, often used to supplement both coal and nuclear power plants.

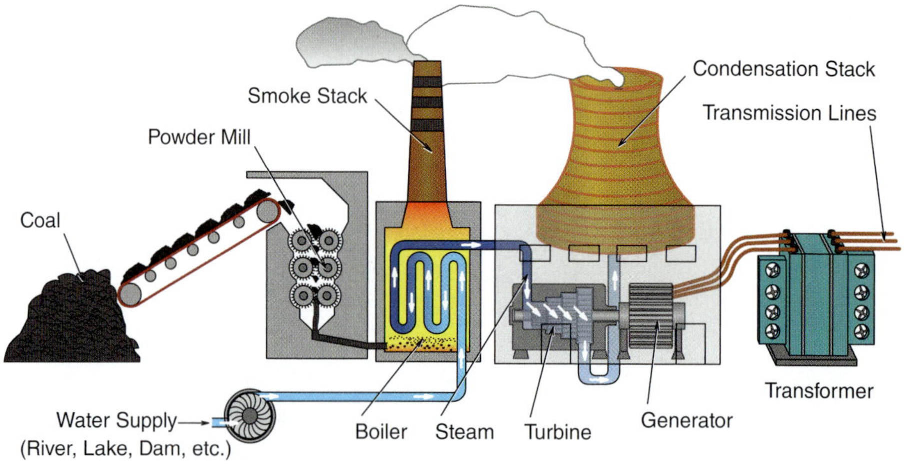

FIGURE 1-1 Coal-fired power plant.

▶ CORE CONNECTION

Reviewing the structure of the atom is necessary to fully understand electric current. All forms of matter are composed of atoms. It is the smallest component of an element that can exist on its own and retain the element's properties.

Three even smaller particles—the proton, electron, and neutron—complete the structure of a typical atom (**FIGURE 1-2**). Two of the three particles possess a natural electric charge. The proton has a positive charge, while the electron has a negative charge. The neutron has no electrical charge. The protons and neutrons form the nucleus, or center, of the atom, while the electrons orbit around it, much like planets orbit around the sun. The mass of an atom is concentrated in its nucleus.

Unlike orbiting planets, the negatively charged electrons of some atoms can leave their orbits easily and travel to another atom. The atoms of materials such as copper fall into this category, and these materials are a good conductor of electricity. The electrons of other atoms, like rubber, cannot readily leave their orbits. These materials are not good conductors; instead, they fall into the category of insulators.

What is considered to be a neutral atom is one with a neutral charge, possessing an equal number of positive protons and negative electrons. As electrons come and go in a specific atom, the charge of the atom changes. When there are more electrons than protons, the atom takes on a negative charge. When there are fewer electrons than protons, the atom takes on a positive charge.

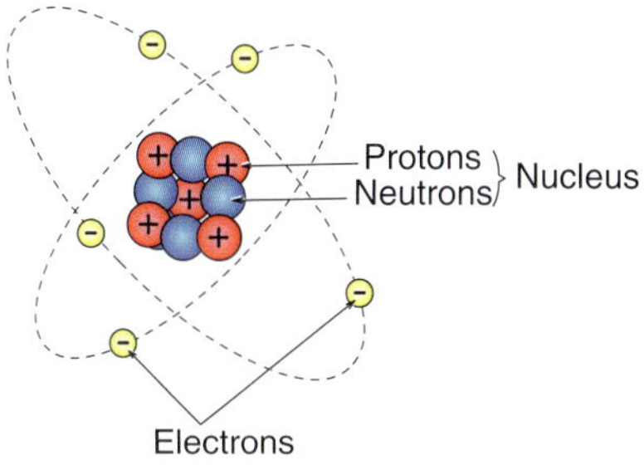

FIGURE 1-2 The structure of an atom.

Nuclear Power Plants

Nuclear power plants (**FIGURE 1-3**) are an alternative to fossil-fuel plants for large-scale needs. In a nuclear power plant, radioactive material is used to heat water and turn it into steam that drives a steam turbine generator. These plants do not produce toxic emissions, but the spent nuclear fuel rods remain radioactive and therefore present a disposal problem. Still, there are many active nuclear plants in the United States.

Controlling the output of nuclear plants is not as simple as it is for fossil-fuel plants. They typically are designed to produce power at a given level around the clock. As a result, they are a valued source of power to serve the **base load**, which is the minimum amount of power needed by the power grid it serves. Nuclear power plants are able to consistently serve users with stable power for years. When power demands begin to rise, exceeding the capacity of the nuclear plant, their output is supplemented by fossil-fuel plants or by alternative energy sources where available.

FIGURE 1-3 A nuclear power plant.
Kodda/Shutterstock

▶ CORE CONNECTION

For many years, fly ash was simply released to the atmosphere, but no longer. Fly ash must be collected, which has required the installation of special equipment at coal-fired power plants to capture it at the source. Most of the fly ash produced by coal-fired power plants ends up in special retention sites. It cannot be placed in standard landfills, due to its toxic ingredients. Roughly 40% of the fly ash collected in the United States, over 40 million tons (36.3 metric tonnes), is repurposed as an additive in Portland cement mixtures and hydraulic plasters. The use of recycled fly ash reduces the amount of Portland cement used in the mix and therefore results in lower-cost concrete while making use of an otherwise useless and hazardous material.

Other Power Generation Sources

Scientists and engineers are continuously searching for methods of generating electricity that do not rely on fossil fuels. One common approach is hydroelectric power, where the power of flowing water is used to drive a turbine generator. Hundreds of U.S. rivers have been dammed over the years to harness the energy produced by flowing water. **FIGURE 1-4** shows how a typical hydroelectric power plant works.

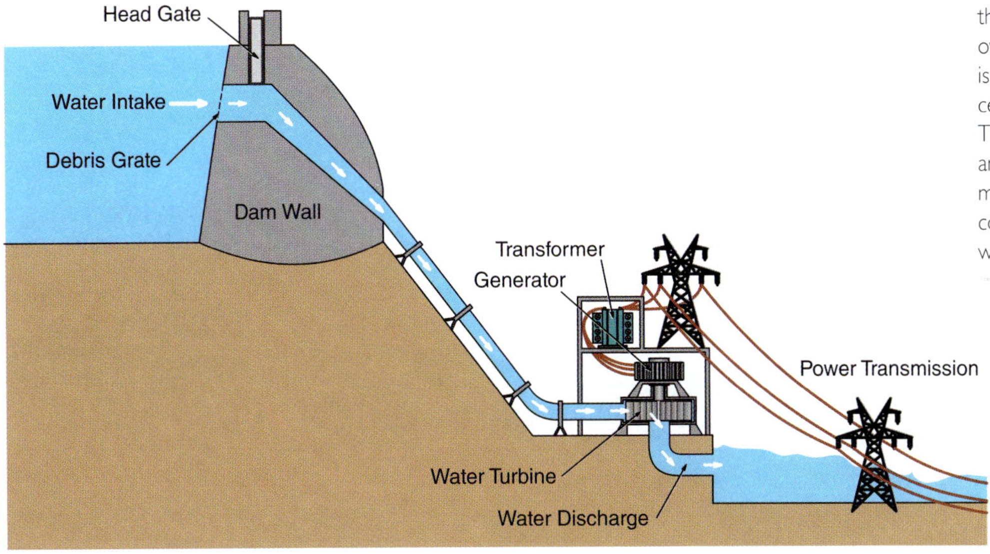

FIGURE 1-4 Hydroelectric power generation.

FIGURE 1-5 Wind farm.

Andrew Zarivny/Shutterstock

FIGURE 1-6 Solar photovoltaic power installation.

It has become increasingly common to see fields containing large numbers of wind turbines (**FIGURE 1-5**) and solar panels. Improvements in these technologies have made it possible to use these methods on a larger scale. Wind turbines rely on the power of the wind to turn a turbine generator, which produces electricity. Solar panels consist of **photovoltaic cells** (**FIGURE 1-6**) that convert the energy from the sun into electricity. Solar power is also used to generate steam directly at concentrated solar-thermal facilities. Instead of photovoltaic cells, a solar-thermal facility uses masses of mirrors to focus the sun's energy on a single elevated point. This point acts as a boiler, where the concentration of heat from the reflected solar rays creates steam to operate a turbine.

Another alternative power source is biomass material (**FIGURE 1-7**), including material such as plants, trees, and even garbage. These materials can be burned to produce the necessary steam to drive a turbine generator.

While the power output from biomass plants can be easily controlled, power generated by wind and solar sources is not as reliable and consistent. Power generated but not consumed is power lost, regardless of the energy source, since the large-scale storage of electrical power is not yet practical. Since the demand for power does not necessarily coincide with peak output from an alternative energy source, the power may be wasted. Demand can also outpace the available power generation, especially when the sun isn't

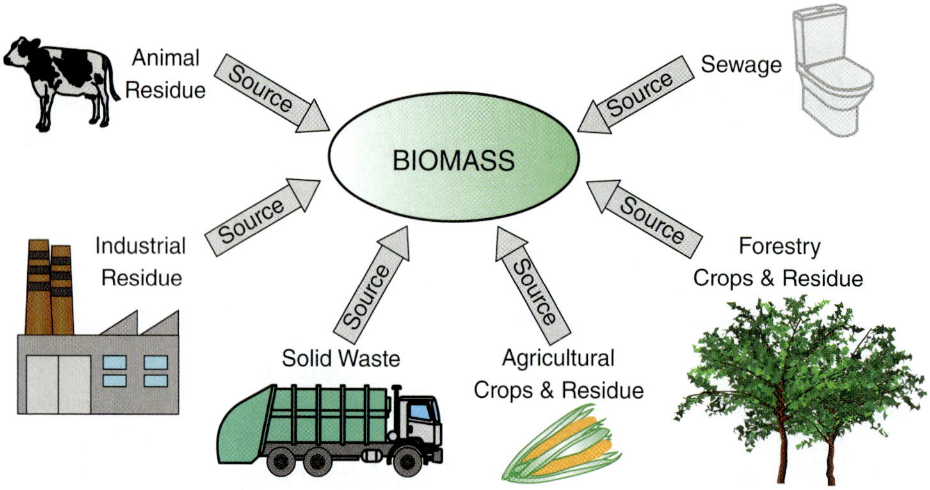

FIGURE 1-7 Sources of biomass materials.

▶ CORE CONNECTION

Since air conditioning systems consume a significant amount of the power serving many homes, making them more efficient has been a work in progress for many years. Common air conditioning units now far exceed the electrical efficiency of their peers that were on the market 25 years ago, and progress continues. Some manufacturers have also teamed up with specialists in the field of solar photovoltaic systems to offer packages that include both air conditioning equipment and solar panels (**FIGURE 1-8**). Although it takes a considerable number of functional solar panels to fully power a hungry cooling system, just a few panels tied into a home's electrical system can help offset the energy usage of the cooling system and serve other electrical needs when the cooling system isn't operating.

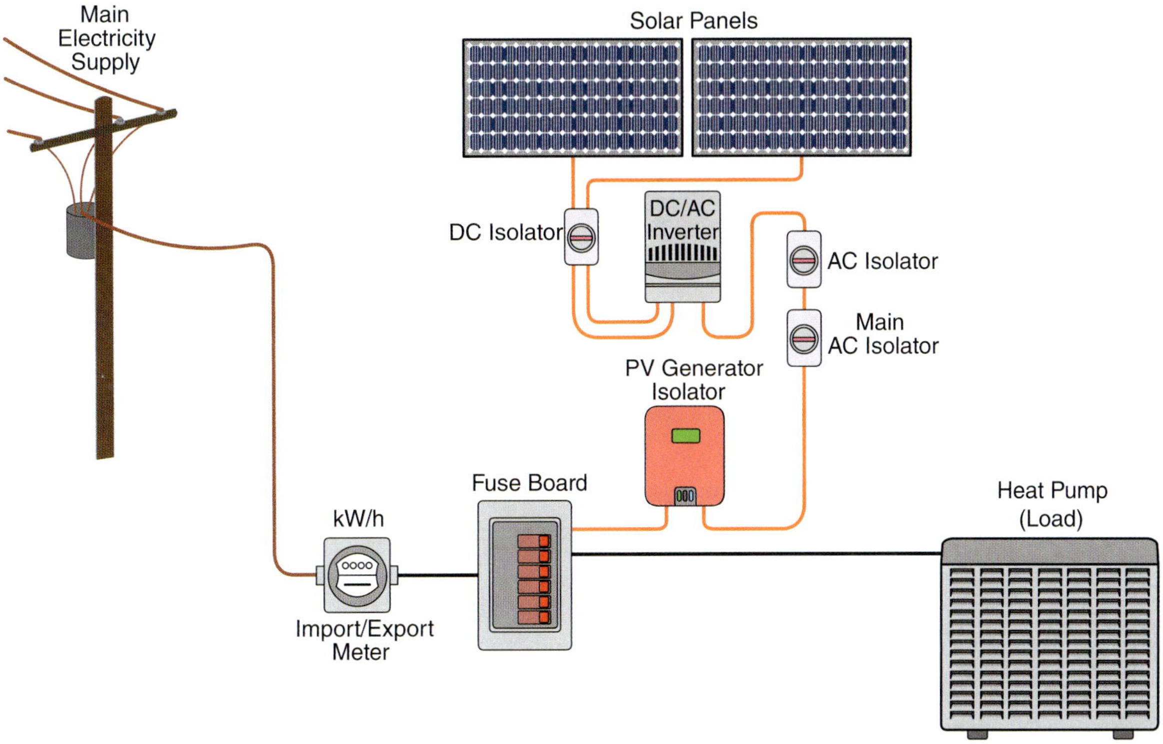

FIGURE 1-8 An air conditioning unit and a solar panel.

shining brightly or wind speeds are low. Consequently, wind and the sun cannot yet be relied on as primary energy sources for large-scale power generation in most cases. Future advances in power-storage technology will hopefully improve our ability to integrate these sources and make full use of their output.

Power Transmission and Distribution

The output of a power generation facility is typically **alternating current** (AC) in the range of 2,500 to 25,000 **volts**. This voltage is supplied to a **transformer** (**FIGURE 1-9**). Transformers are used throughout the power transmission and distribution grid to increase (step up) or decrease (step down) AC voltage levels. At the power station, the generator output is stepped up by the transformer to a very high voltage level, as high as 765,000 volts, before it is placed on the transmission lines (**FIGURE 1-10**). Using high AC transmission voltages reduces **current** flow in the lines; therefore, the higher voltage allows smaller wire sizes to be used to transmit a great deal of power.

FIGURE 1-9 Transmission-level step-up transformer.

sunipix55/Shutterstock

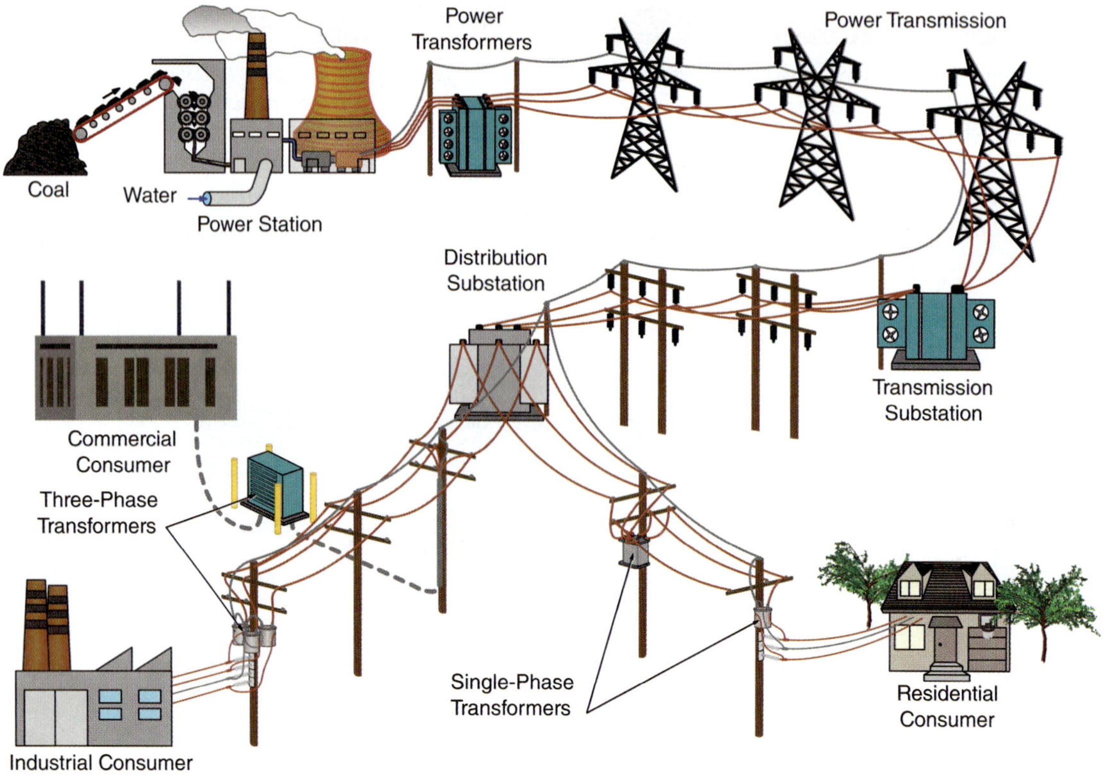

FIGURE 1-10 A power transmission and distribution system.

FIGURE 1-11 A local power substation.

Troy Staton

FIGURE 1-12 A pole-mounted power distribution transformer.

This power travels for many miles along high-voltage transmission lines to a transmission **substation**. Transformers at this substation step down the voltage in preparation for transmission to local substations (**FIGURE 1-11**). Localized substations further reduce the voltage and distribute the power to a variety of residential, commercial, and industrial users.

A four-wire system is used in power transmission and distribution. Three wires are hot (energized), each carrying one phase of a **three-phase power** system. (This topic will be discussed in more detail in a later chapter.) The fourth wire is a neutral wire that acts as a return path for each of the three phases of voltage. For residential use, a single-phase of the three-phase voltage, along with the neutral line, is connected to a distribution transformer. This transformer may be mounted on a pole (**FIGURE 1-12**) or on a pad at ground level. The transformer steps the voltage down to roughly 208–240 VAC, which is the range of voltage typically supplied to residential and light-commercial customers. This level of voltage is needed to power items such as centralized air conditioning equipment, electric clothes dryers, and electric ranges—the appliances that consume the most electricity in an average household. Lights, TVs, computers, and small appliances only need 120 VAC, which is the voltage available at electrical wall outlets in the United States. The 120 VAC power is obtained by splitting the 240 VAC supply and pairing each energized **conductor** with a neutral wire to complete the **circuit**. This topic will be covered in a later chapter.

Stores, offices, and other small businesses generally use the same voltages as residential users. Larger commercial businesses and industrial firms need more power and are thus supplied with all three phases of the distributed voltage. The most common three-phase voltages used by industrial firms are 460 VAC and 575 VAC. Some large industrial facilities have transmission-level voltage fed directly into their own dedicated substation, where the voltage is stepped down to a usable level.

▶ Important Electrical Terms

Electricity has a language of its own, so one of the first steps in learning about electricity is to understand the key terms used to describe it.

K01002 Define common electrical terms.

Voltage, Current, and Resistance

Voltage, measured in volts, is the difference in electrical potential that causes an electrical current to flow. Voltage is the electrical pressure that pushes electron flow. Volts are usually represented by a capital V.

An electrical current is the movement of electrically charged particles from one point to another through a conductive material. An electrical current flowing through a device with any level of **resistance** to the flow will cause that device to perform some type of work. The work performed may be as simple as producing heat. For example, current flowing through a light bulb causes the filament to glow extremely hot, producing light (**FIGURE 1-13**). Any device that performs work in an electrical circuit, such as a motor, light bulb, and appliance, acts as a **load** on the circuit in that it resists the flow of current and consumes energy in the process. Resistance in electrical circuits is measured in **ohms** and is represented by the Greek letter omega (Ω).

Current is measured in **amperes**, shortened to amps for almost all purposes. A current value would be represented by a capital A, such as 5 A or 5 amps. Currents smaller than one amp are generally measured in milliamps, which is one-thousandth (1/1,000th) of an amp or 0.001 A. Milliamps are represented by the letters mA, properly written as shown here. An example would be 325 mA, which is equal to 0.325 amps. In HVACR systems, technicians will encounter very small currents measured in milliamps, as well as much larger currents. Motors and electric heaters are the two loads in HVACR systems that traditionally use the most current.

▶ CORE CONNECTION

Many of the terms used to designate electrical values have their origin in the names of the people who are credited with their discovery. The term "voltage," for example, is derived from the name of Alessandro Volta, an eighteenth-century physicist who observed that an electric current can be made to flow by immersing silver discs in a chemical bath and connecting the discs with a wire.

Power

Power is the measure of the amount of electrical energy consumed by a load device in an electrical circuit. Power, which is represented by a capital P, is measured in **watts** (W) in honor of James Watt, who invented the steam engine. There is a direct relationship between voltage, current, and power, which can be expressed by the following formula:

$$P \text{ (power)} = E \text{ (voltage)} \times I \text{ (current)}.$$

This formula is sometimes referred to as **Watt's law**. The capital I is used to represent current because it originally was identified as "intensity" back in the days when electrical phenomena were first being studied. Voltage is represented by a capital E in the formula since it is an electromotive force. If two of the three variables in the equation are known, the value of the third can be calculated.

The power formula, like most equations, can be rewritten in a variety of ways to find an unknown value. Refer to **FIGURE 1-14**. Each segment on the wheel has a formula for determining power in watts, voltage in volts, current in amps, or resistance in ohms. This wheel is known as the power wheel, and it is based on **Ohm's law** and Watt's law. The application of the formula in DC circuits will be presented in another chapter.

Understanding power is important because many electrical devices are rated in terms of their power consumption in watts, represented by a capital W.

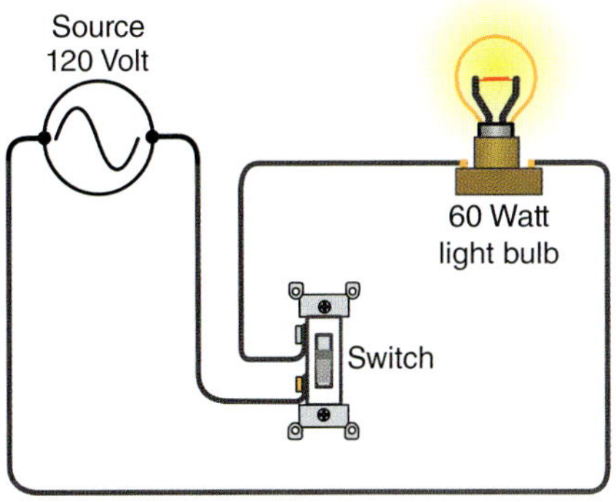

FIGURE 1-13 A basic circuit with a power source and a load.

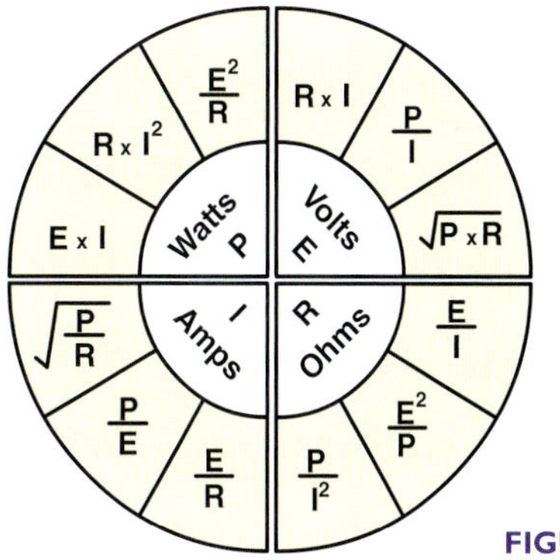

FIGURE 1-14 The power wheel.

▶ CORE CONNECTION

How much current will a specific electrical device, such as an electric heating element, consume? Electric heating elements are rated in watts. Many heating elements used in HVACR systems are rated at 4,500 W with a 230 VAC power source applied.

To determine the current through an electrical load, use this version of the power formula (Watt's law):

$$I = P \div E$$

where I = current in amps; P = power in watts; and E = voltage in volts.

Substitute the two known variables—power and voltage—and then solve the equation for I, as follows:

$$I = 4{,}500 \text{ watts} \div 230 \text{ VAC} = 19.6 \text{ amps}$$

So this would be the value seen when measuring the current flow to the electric heater at the given voltage. However, if the voltage provided to the heater is not 230 VAC, then the heater will produce a lower wattage and consume less current.

If the voltage provided is only 208 VAC, for example, the current will fall to roughly 17.7 amps. The resulting wattage is then calculated using this form of the power formula:

$$E \times I = P$$

$$208 \text{ VAC} \times 17.7 \text{ amps} = 3{,}682 \text{ watts}$$

Although the difference of two amps in the current may seem insignificant at this point, it can affect the required overcurrent protection provided for the circuit (fuse or circuit breaker selection), as well as the size of the wire required to provide the unit with power. The lower voltage also produces less heat, which may not be enough to satisfy the need. These concepts are important for both HVACR designers and electricians to understand and consider.

SAFETY TIP

There is much to learn about electrical safety. It is very important to understand early on that voltage has a great deal to do with the level of hazard. For example, the 120 VAC power in a home is certainly powerful enough to electrocute someone, but physical contact with an energized conductor is required to receive a shock. However, contact is not required to receive a deadly shock from a transmission-level conductor. Although there are a number of factors involved, an air gap of 1 inch (2.5 cm) can generally be crossed by 75,000 volts. In other words, if a person is standing on the ground or is in contact with anything that is grounded and then reaches out toward a 75,000-volt conductor, an electrical arc from the conductor to the hand will form while the hand is still 1 inch away, delivering a deadly shock (electrocution). The process of welding is based on maintaining an appropriate gap between a high-voltage electrode and a grounded surface to keep an electrical arc actively melting metal.

Conductors and Insulators

As previously mentioned, wires are the conductors that usually carry current to the loads and control devices in a circuit. Technically, however, a conductor is any substance that will carry an electrical current. Most good electrical conductors are also good conductors of heat. Most, if not all, wiring used in HVACR equipment is made of copper, which is a very good conductor of electricity. Gold and silver are better conductors, but they are too expensive to use for common wiring.

An **insulator** is the opposite of a conductor. It opposes or prevents the flow of electrical current or heat. Wire conductors are typically wrapped with an insulating material (**FIGURE 1-15**). Rubber, fiberglass, plastic, and ceramics are common insulating materials. Insulation protects people from electrical shock and prevents **short circuits**, which occur if an energized conductor touches another energized conductor or conductive surface. The higher the voltage, the more substantial the insulation must be to prevent injury and short circuits.

FIGURE 1-15 Insulated copper conductors.

Electromagnetism and Induction

Magnetism plays a very important role in electricity. A magnet is generally a piece of steel or iron that has been exposed to an intense magnetic field. As a result, the magnet will attract or repel steel or iron objects. If an AC current passes through a bare wire, that wire will become magnetized and develop a magnetic field around it as long as power is applied. If many turns of wire are wrapped around an iron core, a more intense magnetic field can be created. This principle—called electromagnetism—is the basis for the operation of relays and contactors, which are electrical switching devices used extensively in HVACR control circuits.

Another important principle of electromagnetism is **induction**. If a coil of wire is wrapped around an energized electromagnet, a voltage will be induced into that coil and current will flow. This is the principle on which a transformer operates. The active coil, called the primary winding, induces a voltage into the passive coil, called the secondary winding (Figure 1-15). The amount of voltage transferred is determined by the number of wraps or passes, in each of the two windings. For example, if 120 VAC is applied to the primary winding and the secondary winding has the same number of windings as the primary, 120 VAC will be induced into the secondary. To increase the voltage output, more windings must be wrapped on the secondary side.

We saw an example of the need to increase the voltage in the discussion of power transmission, where a transformer is used to raise the output from the power plant by thousands of volts. In order to decrease, or step down, the voltage, the secondary winding is created with fewer wraps. A good example of this is found in typical HVACR control circuits. While the compressor and other devices require line voltage to operate, most HVACR control circuits are designed to operate at 24 VAC. A step-down transformer is therefore used to reduce the line voltage. In **FIGURE 1-16**, there are more windings (wraps of wire) around the primary side of the transformer core than there are on the secondary side. Therefore, the voltage on the secondary side will be lower than the voltage feeding the primary side. In reality, there are far more passes on each side than are shown in the figure, but the ratio of wraps shown is correct.

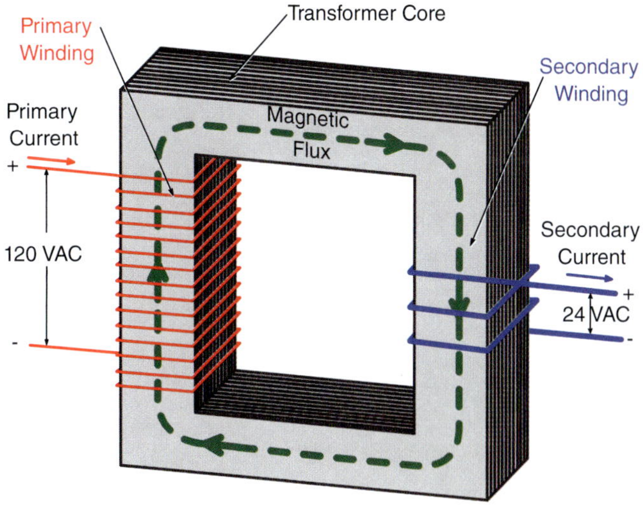

FIGURE 1-16 Step-down transformers have fewer windings on the secondary side than on the primary side.

Although step-up transformers have been mentioned, the clear majority of the transformers encountered in the HVACR field are used to reduce voltage for control circuits.

Impedance

Impedance and resistance have similar definitions. To a certain degree, they mean the same thing. However, impedance applies only to AC circuits.

A very simple AC circuit may contain only a resistive load. An electric heating element is an example of purely resistive load. In this case, the current flow can be calculated in the same way as it is for a DC circuit. This will be covered in the application of Ohm's law in Chapter 4. But AC circuits may also contain inductive loads, such as motors or relay coils. Inductive loads create **reactance** rather than resistance. Inductive reactance opposes a change in AC flow.

To complicate matters further, there are also **capacitors** in AC circuits, primarily those that support motor operation. Capacitors also cause their own form of reactance (capacitive reactance), because they oppose a change in voltage across a capacitive device.

Impedance, therefore, is the sum of the resistance and reactance in an AC circuit. Calculating impedance in an AC circuit requires more complex math than calculating simple resistance in a DC circuit. Although the concept of impedance may be difficult to fully understand at this level (and rightfully so), understand that it is the sum of an AC circuit's resistance and reactance. Although the term will come up, such as in the future discussion of impedance relays, it is rarely a factor that must be considered or calculated in the course of common HVACR service tasks.

Ground

Grounds are an important part of electrical systems. Electrical circuits are connected to earth ground, which is typically a copper rod driven into the earth near the structure. A ground serves several important purposes. As a safety measure, it provides a low-resistance path through which fault currents can flow if an energized conductor contacts the conductive surface of a system component, such as a furnace cabinet. This is known as a short circuit, as previously described.

Assume, for example, that the insulation on a 240 VAC conductor is damaged and that the live conductor touches the metal cabinet of an air conditioning unit. Without the cabinet being grounded, the current could pass through to a person touching the cabinet.

SAFETY TIP

The existence of a ground connection does not automatically mean the equipment has a good and safe ground to protect the technician and the equipment. The ground wire could have a loose connection or be broken somewhere along its length. Use a voltmeter to check the equipment ground. There should be an applied voltage reading from an energized power lead to the case ground wire.

Grounding the cabinet provides a low-resistance path to earth for the current. In such a case, the high current flowing through the low-resistance path would exceed the rating of the fuse or circuit breaker protecting the circuit, and the protective device would open the circuit.

A ground also serves as a common reference for measuring voltage, so that all voltages in the electrical systems are measured from the same zero-potential reference point. Another important purpose of grounding is to provide a discharge path for tiny stray currents, referred to by the term "noise," in an electronic system. Electronic control devices and networked connections are common in modern HVACR systems. Noise can have a negative effect on electronic signals unless it is suppressed.

Alternating Current and Direct Current in Heating, Ventilating, Air Conditioning, and Refrigeration Systems

K01003 Distinguish the difference between alternating current (AC) and direct current (DC).

AC is the type of power used across the world to power homes and businesses. DC is also used, but in very limited ways and rarely as a primary power supply. This section provides some insight on how these two power supplies differ.

CORE CONNECTION

Back in the 1870s, when electric lighting was coming into popular use, Thomas Edison and George Westinghouse were engaged in what became known as the "current wars." Edison favored a DC system, whereas Westinghouse was an AC man. AC won out because it could be transmitted over long distances and the voltage could be raised and lowered easily using transformers. Using AC, many cities could be powered from a single power plant, while DC required local power plants. Edison eventually came to terms with the AC approach and joined forces with another contemporary to form the General Electric Company.

Alternating Current

Alternating current is produced by AC generators—devices that convert mechanical energy to electrical energy. The operation of a generator is based on the principle that moving a conductor through a magnetic field will induce an electric current in the conductor. This principle is shown in **FIGURE 1-17**. The magnetic poles are permanent magnets that emit magnetic lines of flux, through which the rotating conductor passes, inducing a voltage into the moving conductor. The rotating portion of the generator is called the armature. The permanent magnet that produces the lines of flux is called the stator because it is stationary.

The induced voltage is extracted from the moving conductor at the slip rings. The amount of voltage induced into the armature at any time is a function of the position of the rotating conductor in relation to the magnetic lines of flux. For that reason, the output voltage varies as the conductor rotates.

In most power plants, the armature is mechanically connected to a gas or steam turbine, which provides the power to turn the armature. In a hydroelectric facility, the armature is turned by a turbine powered by the flowing water. Generators used in power plants are very large. **FIGURE 1-18** shows an example of a power plant turbine and generator set.

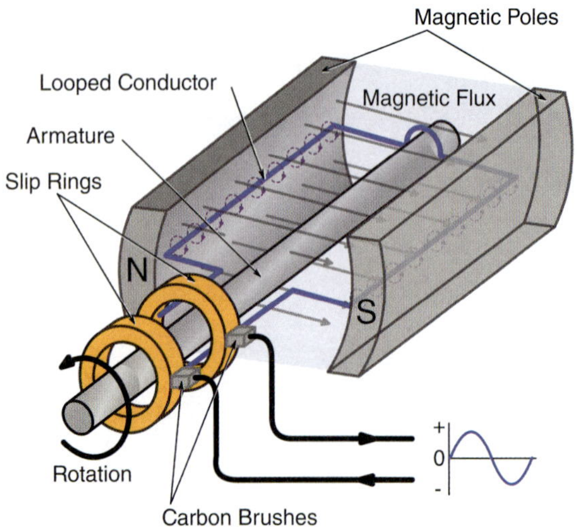

FIGURE 1-17 AC generator operating principle.

FIGURE 1-18 A power plant turbine and generator set.

Theodore Clutter/Getty Images

When plotted on a graph, the generator's electrical output takes on the shape shown in **FIGURE 1-19**: This is known as a sine wave and is universally used to represent AC voltage. It starts at zero volts, representing the armature at a position where the conductors are not cutting through the lines of flux. It then moves in a positive direction until it reaches the peak voltage output, at which point the armature is cutting through the lines of flux at its maximum. As it continues to rotate and complete half of one full rotation, the voltage returns to zero. At that point, the armature has rotated 180 degrees. As rotation continues, the armature breaks the lines of flux in a negative direction. This causes the line of the sine wave to dip down below the base line (0). The voltage again peaks at the 270-degree mark. Once a full 360-degree turn is complete, the voltage returns to zero.

The speed at which this cycle repeats is known as its **frequency**, determined by the rotation speed of the generator armature. The frequency standard in the United States is 60 cycles per second, which is stated as 60 Hertz (Hz). This means that the armature of a generator with one set of magnetic poles would go through 60 full rotations in one second. In many other countries, the power is generated at a frequency of 50 Hz.

Because AC voltage varies between zero and maximum (peak), the amount of voltage available to perform work—the effective voltage—is less than the peak voltage. The effective voltage is determined by multiplying the peak voltage by 0.707. In the case of standard 120 VAC, the peak voltage is 169.1 VAC (169.1 × 0.707 = 120).

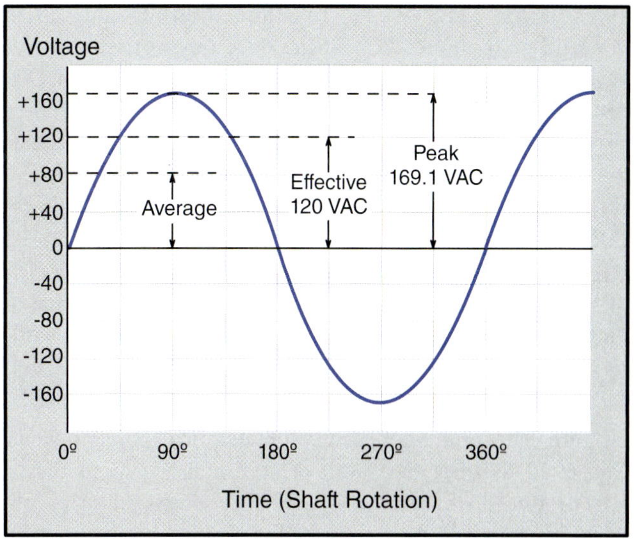

FIGURE 1-19 An AC sine wave.

Direct Current

AC voltage alternates in relation to the position of the generator armature windings, as represented by the sine wave. When DC voltage is plotted on a graph, a straight line is seen instead of a wave—as shown at the top of **FIGURE 1-20**. The most familiar source of DC power is the common battery. However, DC can also be derived from an AC power source. Microprocessor-based and electronic control devices are common in modern HVACR systems. Many of these devices are indirectly powered by AC, but they incorporate electronic devices known as **rectifiers** on board to change the voltage from AC to DC. In the process, the voltage is also reduced. For example, a control board in a furnace is powered by 120 VAC, but it has a built-in rectifier that changes the power supply to 5 VDC or a similar value to power its electronic components.

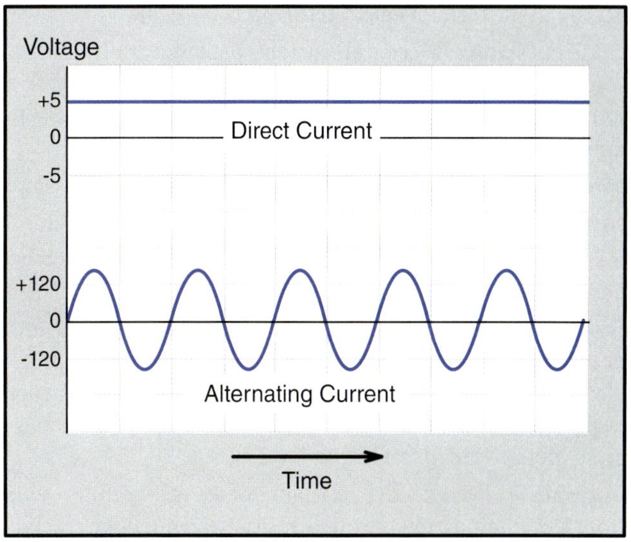

FIGURE 1-20 A comparison of DC and AC wave forms.

▶ Wrap-Up

Summary

HVACR systems are powered by electricity, so a thorough knowledge of electrical principles is essential for HVACR technicians. The electrical power used in homes and businesses originates at distant power plants. Many power plants are coal-fired plants, but other fossil fuels are used as well. Nuclear power and hydroelectric power are other sources that provide utility-scale electrical power. Alternative fuel sources such as wind, the sun, and biomass materials are used as well, but usually on a smaller scale to augment other, more reliable sources.

The electricity generated by power plants is alternating current, which is produced by an AC generator. The output of the power plant is stepped up to a higher voltage by a large transformer and fed through high-voltage transmission lines to substations. Substations traditionally reduce the voltage to a safer, more usable level and distribute it to utility customers.

Transformers play a key role in the transmission, distribution, and application of AC voltages since they are used to increase or reduce the voltage to the level needed for consumption. Direct current (DC) voltage, which is produced by batteries or rectifying circuits, is used on a limited basis in HVACR systems. Its use in the trade is generally restricted to electronic devices.

Key Terms

alternating current (AC) Electric current that reverses its direction many times per second at a consistent interval. The interval is referred to as its frequency, which is typically 50 or 60 times per second. AC frequency is generally standardized by each country.

ampere The unit of measure for electrical current, usually shortened to amp. One amp of current is equal to the current produced when 1 volt is applied to a resistance of 1 ohm. Amperes are typically represented in electrical formulas by the letter I, but by the capital letter A in other contexts.

base load The minimum level of electrical power demand required from a power system over a 24-hour period.

biomass Plant materials, such as trees, vegetation, and animal waste, used as a fuel source.

capacitor An electrical device designed to store electrical energy in the form of an electric field.

circuit A closed electrical current path generally including components that consume electric power, as well as often the source of the electrical energy. It is a complete path for electron flow.

conductor A material that readily transfers electricity and heat.

current The flow of electrically charged electron particles.

fossil fuels Fuels such as coal, oil, and natural gas, which are derived from the decomposition of plants and/or animals.

frequency In power generation, the frequency refers to the number of times per second that a generator's armature completes one full revolution. The unit of measure for frequency is hertz (Hz). In the United States, the standard frequency is 60 cycles per second, or 60 Hz.

impedance The effective resistance of an AC circuit, resulting from the combination of simple resistance, inductive reactance, and capacitive reactance.

induction The process in which a conductor becomes electrically charged by its proximity to a magnetic field or another electrically charged body.

insulator A material that resists the flow of electricity and/or heat.

load Any electrical device that performs work and consumes energy in the process.

ohm The unit of measure for electrical resistance. One ohm is equal to the resistance of a component or circuit when 1 volt is applied and result in 1 amp of current flow. Ohms are represented by the Greek letter omega (Ω) or by the capital letter R in formulas, such as Ohm's law.

Ohm's law The common name for an electrical concept that states that the strength of a direct current is directly proportional to the difference in electrical potential, and inversely proportional to the resistance of the circuit. It is written mathematically as E (volts) = I (amps) × R (resistance).

photovoltaic cell An electronic device that converts sunlight (radiant energy) into electrical energy.

power The energy, measured in watts, consumed by a load or loads in an electrical circuit.

reactance The component(s) of impedance in an AC circuit that are not resistive in nature, resulting from the effect of inductance and/or capacitance.

rectifier An electronic component used to convert AC to DC.

resistance Opposition to the flow of electrical current.

short circuit An accidental condition in which a low-resistance path around the intended load is created by a live conductor coming into contact with another conductive object.

substation A facility in which power from the power transmission system is reduced to voltage levels required by utility customers. Substations also serve as a point of protection for the downstream power distribution system, using circuit breakers, fuses, and power monitoring systems.

three-phase power An electrical distribution system in which the three legs of the power supply are electrically out of phase with each other by one-third of a cycle.

transformer An electrical device consisting of two or more coils of wire (windings) wrapped around an iron core in which voltage is transferred from one set of windings to the other(s) by induction.

volt The unit of measure for electrical potential difference or electromotive force. One volt is equal to the electrical potential required to generate a current of 1 amp in a circuit with a resistance of 1 ohm. Voltage is electrical pressure used to push electrons in circuit. Voltage is represented by the capital letter V in most contexts but often by the capital letter E in power formulas.

watt The unit of measure for power. One watt represents the power produced by a current of one amp with a potential difference of 1 volt.

Watt's law A common name for the electrical formulas that state the relationship between power, voltage, and current. The various versions of the formula are also known as the power formulas.

Review Questions

1. Which of the following is *not* a fossil fuel?
   a. coal
   b. oil
   c. nuclear fuel
   d. natural gas
2. An atom takes on a negative electrical charge when ________.
   a. there are more neutrons than electrons
   b. there are more electrons than protons
   c. all of its electrons are lost
   d. all of its neutrons become protons
3. Dams are used in the generation of what type of power?

4. The voltage output from a power plant is _____ before it is placed on transmission lines.
   a. stepped up
   b. stepped down
5. The voltage typically supplied to residential customers from the power distribution system is _____.
   a. 120 VAC
   b. 7,200 VAC
   c. 208–240 VAC
   d. 12 VDC
6. The operation of a transformer is based on the principle of _____.
   a. rectification
   b. transmission
   c. infusion
   d. induction
7. The flow of electrically charged particles in an electrical circuit is known as _____.
   a. voltage
   b. current
   c. power
   d. resistance
8. A device that performs work and consumes power in an electrical circuit is the _____.
   a. load
   b. battery
   c. switch
   d. conductor
9. A 1,000-watt electric heating element is in a circuit supplied by a 230 VAC power supply. The current in this circuit would be _____ amps.
10. The unit of measure for resistance is _____.
   a. the watt
   b. the volt
   c. the amp
   d. the ohm
11. Which of the following is *not* considered a good conductor?
   a. Gold
   b. Silver
   c. Rubber
   d. Copper
12. In a step-down transformer, the number of wraps of wire on the secondary winding is _____ the number of turns on the primary winding.
   a. equal to
   b. less than
   c. greater than
13. If an active conductor comes into unintended contact with another conductor or a conductive surface, the result is called a _____.
14. In a hydroelectric facility, the armature is turned by a turbine powered by __________.
15. When the armature of a generator has reached 180 degrees of rotation, the voltage is _____.
16. An AC generator has fixed magnets (a stator) and a(n) ______ in order to produce electricity.
   a. rotating armature
   b. split commutator
   c. transformer
   d. insulator
17. Microprocessor-based controls in HVACR systems obtain DC operating voltages from __________.
   a. transformers
   b. batteries
   c. rectifiers
   d. motors
18. In power generation, the term "Hertz" refers to the ________.
19. Which of the following values is greater output from an AC generator?
   a. the peak voltage
   b. the effective voltage
20. The amount of energy consumed by an electrical load is known as _____.
   a. voltage
   b. current
   c. resistance
   d. power

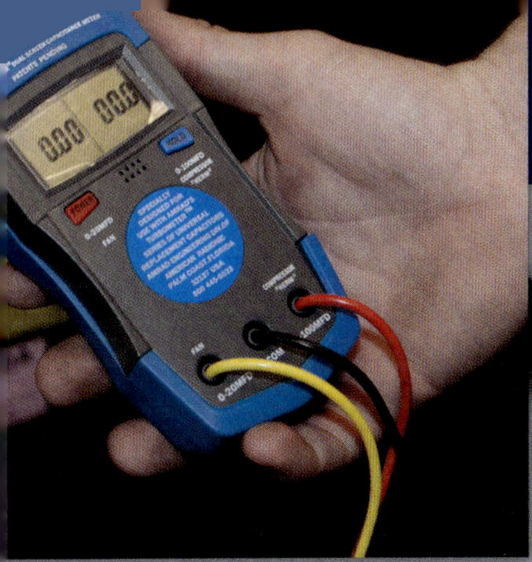

CHAPTER 2

Electrical Test Instruments

Knowledge Objectives

After reading this chapter, you will be able to:

- **K02001** Select, test, and safely use a multimeter.
- **K02002** Select, test, and safely use a clamp-on ammeter.
- **K02003** Select, test, and safely use a capacitor tester.
- **K02004** Select, test, and safely use a non-contact voltage tester.

Skill Objectives

Skills required to meet the objectives of this chapter:

- **S02001** Use a multimeter to measure voltage.
- **S02002** Use a multimeter to measure resistance.
- **S02003** Use a multimeter to measure current values less than 2 amps.
- **S02004** Use a clamp-on ammeter to measure current values greater than 2 amps.
- **S02005** Use a capacitor tester to measure capacitance.
- **S02006** Use a non-contact voltage tester to detect voltage.

▶ Introduction

HVACR technicians use electrical test instruments to measure voltage, resistance, current, and capacitance when testing and troubleshooting HVACR systems. To do the job safely and efficiently, technicians must properly use the appropriate test instrument for the task being performed. This chapter presents how to select, test, and use instruments such as multimeters, ammeters, capacitor testers, and non-contact voltage testers.

SAFETY TIP

The Occupational Safety and Health Administration (OSHA) has established electrical safety guidelines to prevent electrical arc-flash injuries and property damage. The guidelines are found in the *Code of Federal Regulations* (*CFR*), Title 29, Sections 1910.333–1910.335 of the OSHA Standards. The guidelines are comprehensive and result from the analysis of a history of injury and incident reports.

Of course, there are many additional OSHA guidelines to which HVACR technicians must adhere. Before working with energized systems, the above sections of the OSHA standards should be reviewed in their entirety. There are many other publications that address electrical safety as well. It is a topic that deserves extensive study.

High-potential electrical energy can be deadly, especially when handled carelessly or without specific training. HVACR technicians do not typically possess the training and credentials that qualify them as electricians, but they must recognize that significant danger is present when near energized equipment or power sources. Be advised that OSHA has established minimum clearances to electrical energy sources, called the Minimum Approach Distance (MAD):

- For voltages under 300 volts, do not contact any energized surface.
- For voltages between 300 and 750 volts, a 12-inch (30 cm) clearance to the energized surface is required.
- For voltages between 750 volts and 2,000 volts, an 18-inch (46 cm) clearance is required.
- An HVACR technician should never approach equipment energized with a voltage exceeding 2,000 volts.

Maintaining these minimum approach distances increases the safety margin but does not guarantee that accidents won't happen. Always be aware of your surroundings and the unseen hazard of electrical energy.

▶ Selecting and Using Multimeters

K02001 Select, test, and safely use a multimeter.

S02001 Use a multimeter to measure voltage.

S02002 Use a multimeter to measure resistance.

S02003 Use a multimeter to measure current values less than 2 amps.

One type of electrical test equipment that HVACR technicians routinely use is a volt-ohm-milliammeter (VOM), commonly referred to as a multimeter. A multimeter combines the voltmeter, ohmmeter, and milliammeter into a single instrument. In the field, this one instrument can be used to measure (alternating current) AC and (direct current) DC voltages, AC and DC current flow, and electrical resistance.

Although **analog multimeters** have long been available, most multimeters used today are **digital multimeters (DMMs)**. Both types are shown in **FIGURE 2-1**.

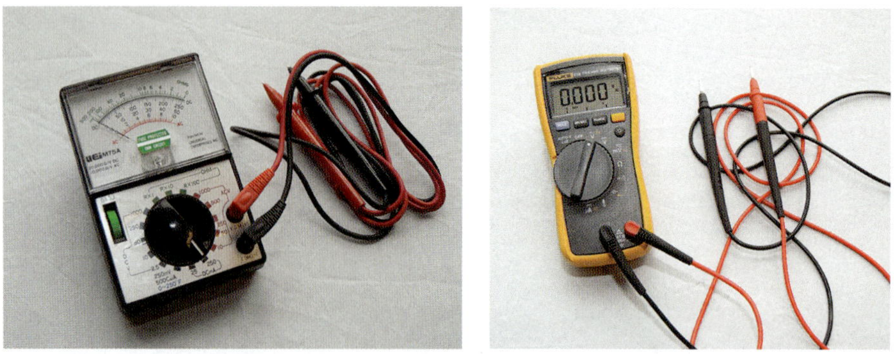

FIGURE 2-1 Analog and digital multimeters.

Multimeters have a selector switch for selecting the desired function and possibly for selecting the range within the selected function. Many multimeters and other digital electrical test instruments are auto-ranging. In other words, they automatically adjust to the proper range based on the value being measured at the test point. If the meter does not have an auto-ranging feature and the range for the value being tested is unknown, set the range at the highest setting. Then work down to a more appropriate range for the measured value after a reading is made.

Many DMMs also have a capacitor-testing feature. When the meter's probes are connected to a capacitor, the meter gives an estimate of the actual capacitance value. The measured value is then compared to the value marked on the capacitor label. Typically, a multimeter's capacitor-testing feature also checks for a shorted or open capacitor.

Most multimeters can measure low levels of current without special accessories. However, in order to measure current, the circuit must be opened and the meter test leads must be wired in series with the other circuit components. A common maximum value for multimeters, without accessories, is 10 amps (A) with the meter wired into the circuit. Some are limited to an even lower current value, due to the fragile nature of the internal circuitry.

In HVACR work, significantly higher current measurements are often required. To accommodate higher current measurements, a clamp-on ammeter accessory is available for many multimeters. A multimeter with a clamp-on ammeter accessory is shown in **FIGURE 2-2**. Such an accessory significantly extends the value and flexibility of the meter. They can measure much larger values since the current is not passing directly through the meter. Very low current values are still best measured by wiring the meter in series with the load.

Other common multimeter functions include diode and transistor testing. Temperature measurement is another feature that many meters offer, using an accessory sensing probe. Some meters offer a minimum/maximum memory function for recording the highest and lowest readings taken over a selected time period. Always refer to and follow the manufacturer's instructions for the specific multimeter being used, to make the best use of its features. The manual that accompanies each meter is a valuable resource.

Instrument Safety Categories

Electrical test instruments, such as multimeters and similar devices, must be safe for technician use in the event that they are exposed to voltages beyond the advertised range. The International Electrotechnical Commission (IEC) established *IEC Standard 61010*, which was also adopted by Underwriters Laboratories (UL) as *UL Standard 61010-1*. The standard defines four different overvoltage categories, identified as Category I through Category IV. They are typically written as CAT I through CAT IV. CAT I represents the least hazardous voltage category, while CAT IV represents the most hazardous relative to the potential for voltage surges that can destroy the meter and injure the user.

> **▶ TECHNICIAN TIP**
>
> The requirement for using true root-mean-square (RMS) test instruments has increased as the complexity of HVACR control systems has also increased. Without a working knowledge of the differences in test devices, the technician cannot accurately determine when an RMS meter is needed.
>
> Solid-state variable and adjustable speed motors contain semiconductors called nonlinear loads. A nonlinear load is one in which current is drawn in short pulses. By comparison, a linear load, such as an induction motor, draws current consistently rather than in pulses.
>
> A linear load returns a smoothly rounded sine wave when graphed, whereas a nonlinear load returns at least three varying types of sine waves that cannot be accommodated by the average-responding-RMS (standard) meter.
>
> The true RMS meter addresses these variances, while a standard meter responds correctly only to a continuously rounded sine wave. A standard meter erroneously returns a measured value when testing nonlinear loads. Error rates of 10–40% occur, depending on the type of load measured. Note, however, that this is an issue only with AC loads; DC measurements are not affected.

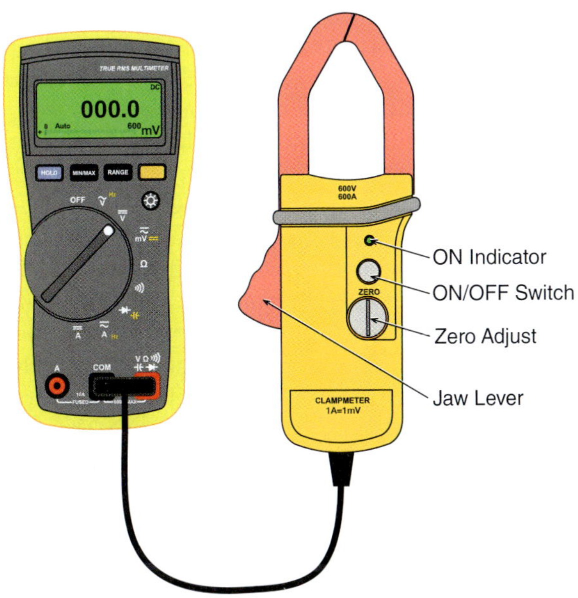

FIGURE 2-2 DMM with a clamp-on ammeter accessory.

TABLE 2-1 Overvoltage Categories for Test Instruments

Overvoltage Category	In Brief	Examples
CAT IV	Three-phase at utility connection, any outdoor conductors	• Refers to the "origin of installation"; i.e., where low-voltage connection is made to utility power. • Electricity meters, primary overcurrent protection equipment. • Outside and service entrance, service drop from pole to building, run between meter and panel. • Overhead line to detached building, underground line to well pump.
CAT III	Three-phase distribution, including single-phase commercial lighting	• Equipment in fixed installation, such as switchgear and polyphase motor. • Bus and feeder in industrial plants. • Feeders and short branch circuits, distribution panel devices. • Lighting system in larger buildings. • Appliance outlets with short connection to service entrance.
CAT II	Single-phase receptacle connected loads	• Appliance, portable tool and other household and similar loads. • Outlet and long branch circuits. • Outlets at more than 10 meters (30 feet) from CAT III source • Outlets at more than 20 meters (60 feet) from CAT IV source
CAT I	Electronic	• Protected electronic equipment. • Equipment connected to (Source)circuits in which measures are taken to limit transient overvoltages to an appropriately low level. • Any high-voltage, low-energy source derived from a highwinding resistance transformer, such as the high-voltage section of a copier.

TABLE 2-1 provides examples of electrical installations that fall into each category. Since a higher category number indicates a higher level of hazard, a meter with a CAT III rating is a safer instrument than one rated as CAT II. Ratings found on instruments may include a voltage, such as CAT III-600 V and CAT III-1000 V. The latter rating is better than the former because it indicates that the meter was tested at a higher voltage and offers superior protection.

The category rating is generally shown on multimeters in a very conspicuous location, such as near the test-probe jacks. Note the labels near the jacks on the meters shown in Figure 2-1. As a general rule, HVACR technicians are well protected by CAT III ratings since they are not typically required to work on CAT IV installations, where the voltage often exceeds standard consumer levels. However, a higher level of protection always works in the technician's favor.

Selecting an Appropriate Multimeter

When selecting a multimeter for a specific application, choose a meter that is properly rated for the circuit or component to be tested. Technician should always have a general idea of the voltage or current expected before they measure either value. Ensure that both the meter and the leads have the correct category rating for each task. Also, check that the ratings of the leads or accessories meet or exceed the rating of the multimeter.

Always inspect a multimeter and its accessories before using it. Ensure that the meter case is not wet, oily, or cracked; that the input jacks are not broken; and that there are no other obvious signs of damage.

Check the test leads carefully. Ensure that the insulation is not cut, cracked, or melted and that the tips are not loose. The best test leads have double insulation, shrouded input connectors, and probes with finger guards (**FIGURE 2-3**). These features reduce the risk of shock and **arc flash**. Since there are times when a measurement must be made while a technician's hands are free to do something else, accessory alligator clips are designed to slide onto the probe tips. They can then be used to grasp the test point securely, freeing the hands.

Meter test leads are generally color-coded for insertion into the matching jacks of the meter. The leads are not different, but the colors help identify the leads when **polarity** is important (most often with DC circuits).

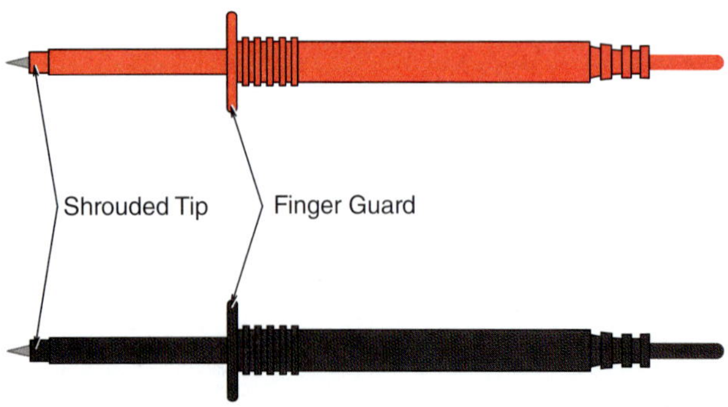

FIGURE 2-3 Multimeter test leads.

Testing a Multimeter

Always verify that the selected multimeter is working properly before it is used on the job. This ensures that the test instrument will provide accurate measurements. It also helps ensure a technician's safety. If the meter is not working properly, it may not indicate the presence of voltage at all. That is how electrical incidents often begin. Return a multimeter to the manufacturer for an annual calibration to ensure its accuracy.

To test the meter, verify that the meter responds as expected to resistance and/or voltage. To check its ability to measure **continuity** (a simple indication of whether a circuit is closed or open), as shown in **FIGURE 2-4**, set the multimeter to the lowest setting for resistance. Then touch the tips of the two probes together. The display should show 0 ohms (Ω) since there should be continuity and virtually no resistance between the two probes. A DMM will typically display OL (overload or out of limits) when a circuit is open (lacks continuity). If OL is displayed during this test, either the meter is defective or the test leads are defective.

To check the meter's ability to measure resistance, use a resistor with a known value. Set the multimeter to measure resistance. Place one probe at each end of the resistor, as shown in **FIGURE 2-5**. The value in the display should be very close to the known value of the resistor being used for the test. If the multimeter does not measure the resistance correctly, double-check it by repeating the test with a different resistor of a known value.

To check the DC voltage function, use a new battery (**FIGURE 2-6**). Set the multimeter to measure DC voltage. Place the red probe on the positive terminal of the battery and the black probe on the negative terminal. The multimeter should display a reading of, or close to, the battery's labeled voltage. Reversing the leads on a digital meter will display a negative (-) reading. It will still be an accurate reading, but the display will appear with a negative sign followed by the DC voltage of the connected source. For example, a 9-volt battery would display a reading of –9.2 VDC.

To check the AC-voltage function as shown in **FIGURE 2-7**, use a nearby 120 VAC receptacle known to be energized. Plug in and turn on a lamp or another simple device to ensure that the receptacle is energized. Set the multimeter to measure AC voltage. Place the red

SAFETY TIP

Never attempt to measure resistance on an energized circuit. Before checking a circuit for continuity or measuring resistance, use the voltmeter function to ensure the circuit is deenergized. Testing for continuity or resistance on an energized circuit can damage the meter. The meter's protective fuse will likely blow, at a minimum.

FIGURE 2-4 Testing the continuity function of a meter.

▶ TECHNICIAN TIP

Many HVACR technicians keep more than one multimeter on hand. This helps to ensure that a working meter is always available; a meter that fails suddenly can prevent service tasks from being performed until the problem has been resolved. Consider purchasing a good DMM with a clamp-on ammeter accessory, as well as a good clamp-on ammeter that offers the same functions as a DMM. This combination provides full redundancy and flexibility in the field.

FIGURE 2-5 Testing the resistance function.

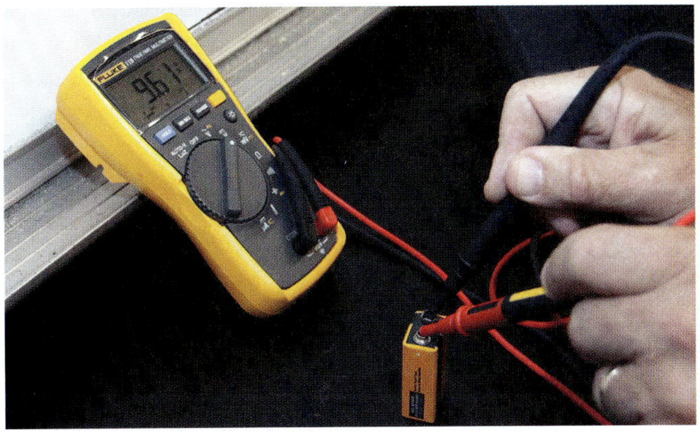

FIGURE 2-6 Testing the DC voltage function.

FIGURE 2-7 Testing the AC voltage function.

SAFETY TIP

Never use any electrical test instrument that is not in good working order. If problems are found when an instrument is inspected and tested, follow the employer's policy for tagging out, repairing, or discarding unserviceable instruments. Failing to remedy problems found during inspection and testing can lead to an electrical incident later if a technician happens to use a defective instrument on the job.

probe into the energized side—the smaller slot of the receptacle—and the black probe into the other side (neutral). The multimeter should display a reading of, or close to, 120 VAC.

If the multimeter fails any one of these tests, it is best to return it to the manufacturer for service. Technicians should never attempt to open and repair the meter's electronics or movement. If the multimeter will not perform any functions or provide a display when tested, the problem may be weak or dead batteries. DMMs have a battery indicator that will display when the battery is weak but still functional. Always replace the battery as soon as the indication is displayed, to ensure continued reliable operation.

If the batteries are not the problem, determine if there is an accessible fuse inside the battery compartment. If there is, check to see if it is open or not; another meter may be needed to check the continuity of the fuse if its condition cannot be determined visually. Replace an open fuse with one of the proper current and voltage rating. Using the wrong fuse, especially one that has a higher current rating than it should, can result in the meter being damaged beyond repair.

Using a Multimeter

When testing circuits with a meter that does not have an auto-ranging feature, test at a high range first, then set the meter to the appropriate range if necessary. If an analog meter is being used, set a range that allows the measured value to fall in the middle third of the scale. All meters have a stated accuracy, such as 1% of scale. If an analog meter is set to a voltage range of 1,000 VAC, the reading could be as far off as 10 VAC, since 1% of 1,000 is 10. However, this performance is still within the meter's stated specifications As a result, reading 24 VAC on a 1,000 VAC scale results in questionable accuracy, if accuracy is important at the time.

In addition, the stated accuracy often does not hold true at the extreme ends of the scale. Limited accuracy is available in those areas of an analog multimeter scale. For example, the least accurate areas for measurement on a 1,000 VAC scale are below 300 VAC and above 700 VAC. In the example above, when the measured voltage is roughly 24 VAC, switching from the 1,000 VAC range to a 60 VAC range will provide a more accurate reading. This is because the reading will be near the center of the scale and because an accuracy of 1% of scale means the reading should not be off by more than 0.6 VAC.

Note that DMMs work in an entirely different manner, and their accuracy is more about the components and methodology used in the construction of the meter. Few DMMs come without an auto-ranging feature, so range selection is rarely an issue.

When taking electrical measurements with a multimeter, always make contact with the neutral (black) lead first and then with the hot (red) lead second. After taking a reading, disconnect the hot lead first and the neutral lead second, reversing the order of contact.

Hang or rest the meter at eye level on a stand whenever possible, rather than holding it (**FIGURE 2-8**). This reduces the user's exposure to transient currents. Using a stand or hanger also frees up a technician's hands to concentrate on positioning the test probes. Do not try to watch the meter while making contact with the test probes; concentrate on the test probes until good contact is made. Try to keep the meter display where it can be viewed without turning significantly in one direction or another.

FIGURE 2-8 There are various stands, hangers, and magnets available to position a meter.

One common error that users make is plugging the test leads into the incorrect jack on the meter. The black lead should go to the black port, which is usually labeled COM (common). The red lead is connected to one of a few possible jacks, depending on the measurement being made. Although the two leads are identical in construction, they can be misleading when it is important to pay attention to polarity. Get in the habit of inserting the test leads in the correct jack.

▶ TECHNICIAN TIP

Be aware of phantom, or ghost, readings when using DMMs. These are low readings that can sometimes occur due to the multimeter's sensitivity. Phantom readings are often caused by **induction** that occurs around energized conductors and surfaces in close proximity to the meter.

Using a Multimeter to Measure Voltage

HVACR technicians measure both AC and DC voltages with multimeters, but AC voltage measurements are far more common. Voltage can only be detected if the power source is energized. Energized systems represent a constant shock hazard. Take all available and necessary precautions to avoid contact with energized surfaces, and ensure that the probe of the test lead does not accidentally bridge two points with different electrical potentials. Use test leads with finger guards, and always keep fingers behind them.

Try to avoid the following common errors when using a multimeter to measure voltage:

- *Measuring voltage while the red test lead is in the wrong jack.* The test-lead jack for the red lead is usually marked with the V symbol and is often marked with the Ω symbol as well (**FIGURE 2-9**). This jack is usually separate from the jack used for current measurements. Note that applying voltage to the meter while the red lead is plugged into the wrong jack may damage the meter.
- *Measuring AC voltage on a DC setting.* The dial symbol usually associated with measuring AC voltage is a V with a wavy line above it, or an mV with the same wavy line above. The latter is for measuring in the millivolt range.
- *Keeping the test probe in contact with an energized surface longer than necessary.* The longer it is in contact, the more time there is for an accident to occur.
- *Using the meter above its rated voltage.* As mentioned earlier, technicians should have a general idea of the expected voltage or current to be measured and ensure that the meter is capable of safely making such readings.

To use a multimeter to measure voltage, follow the steps in **SKILL DRILL 2-1**.

FIGURE 2-9 Test-lead jacks.

▶ TECHNICIAN TIP

There are three steps to ensure that an accurate voltage measurement is being made. Technicians in the power industry are often required to follow these steps:

1. Measure a known voltage source to ensure the meter functions and it accurately reads the voltage present.
2. Use the same voltmeter as planned to measure the voltage from the unknown source.
3. Repeat the measurement on the original, known voltage source to verify that the meter continues to work properly.

SKILL DRILL 2-1 Using a Multimeter to Measure Voltage

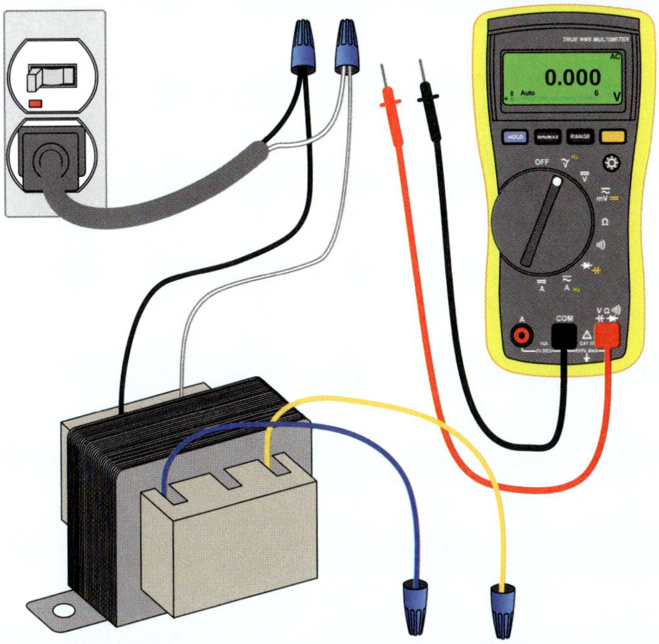

1. Prepare to measure the voltages associated with a 120 VAC/24 VAC control transformer found in a typical warm-air furnace. Select an appropriate multimeter for the job. Visually inspect it and the test leads for any signs of damage. Before using the multimeter on the job, always test it on a known voltage source to verify that it is functioning properly.

This drill can be done on the bench or with the transformer installed in the furnace.

SAFETY TIP

Working with electrically energized circuits is hazardous. Avoid contact with any surface that could potentially be energized. When working inside energized enclosures, maintain a stable position and, whenever possible, avoid placing both hands inside the enclosure. All work on energized equipment by unqualified individuals should be supervised by a qualified instructor or technician.

Continued

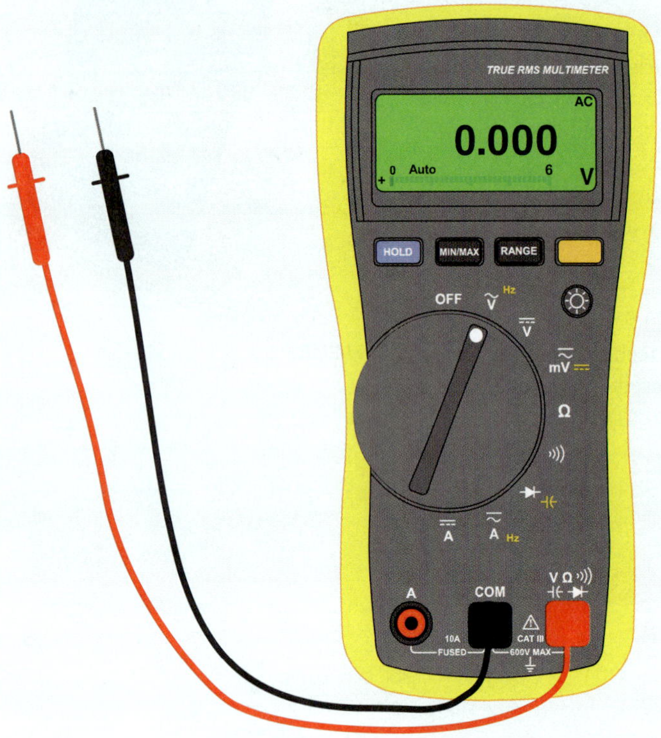

2. Insert the black test lead into the common input jack (COM) and the red test lead into the input jack for AC volts (V or VΩ). Take care to insert the test leads into the correct jacks. Select AC as the type of voltage to be measured. If the meter has no auto-ranging feature, select the voltage range.

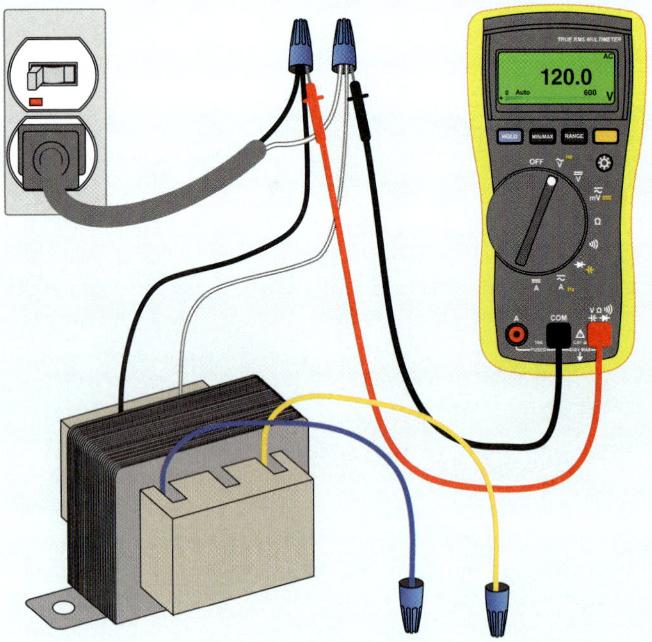

3. To measure the voltage of the incoming power supply to the transformer, insert the black test lead into one of the wire nuts or touch the appropriate exposed terminal on the terminal board. Next, connect the red test lead to the other wire or terminal. Read the voltage displayed on the meter while both probes make good electrical contact with their individual targets.

Continued

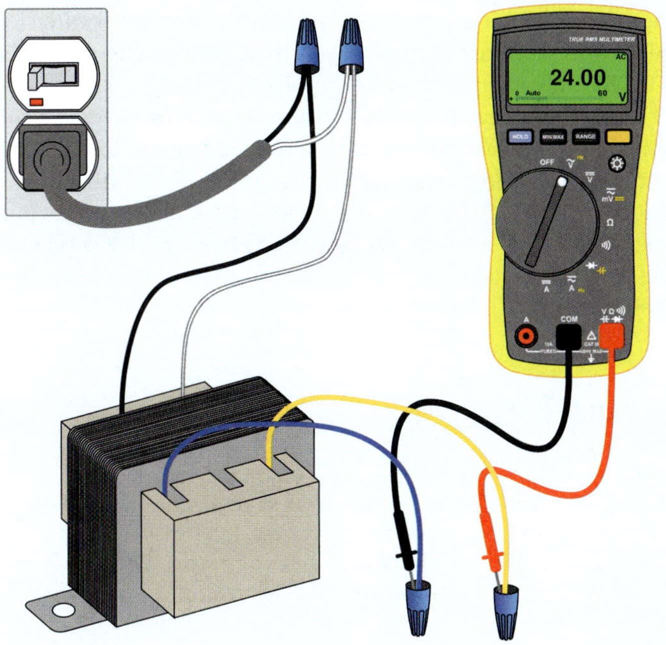

4. Disconnect the test leads from the incoming power supply once a voltage value has been recorded. Then locate the low-voltage wire leads emerging from the transformer. Connect the meter test leads to the wire nuts or terminals, and check the voltage displayed on the meter. The voltage should match the transformer secondary-voltage rating. Once both readings have been taken, carefully deenergize the transformer setup for bench measurement or deenergize the furnace and reinstall the access door(s).

Using a Multimeter to Measure Resistance

An HVACR technician often needs to measure resistance when troubleshooting components such as motors. This ohmmeter function is also used to determine whether a circuit has continuity. The only difference in the two tests is that when testing for continuity, the only interest is whether the circuit is intact or not. For example, it may be important to know whether the cooling circuit is complete through the wall thermostat, so that it can energize a cooling circuit when power is applied. In this case, a technician is interested only in whether the circuit is complete or not. The specific resistance value of the circuit in this case is essentially meaningless. A specific resistance value is important, however, when determining the condition of motor windings and similar components.

▶ TROUBLESHOOTING TIP

Even though there is much to learn about motors at this point, it might be interesting to show how using the measured resistance of the windings in a simple single-phase motor can help identify the individual windings. **FIGURE 2-10** shows a schematic common to many single-phase AC motors. There are three leads at the motor, identified as common (C), start (S), and run (R). The common lead is connected where the windings in the motor meet. If there is a question as to which lead is which, the following tests will reveal the answer:

- Label or identify the motor leads as 1, 2, and 3.
- Use a multimeter to take resistance measurements from one lead to another. Measure the resistance between 1 to 2 and record the value. Then test 1 to 3 and record that value. Finally, measure the resistance between 2 and 3 and record it.
- Now apply the following rules:
  - The resistance between S and R should be the highest value.
  - The resistance between C and R should be the lowest value.
  - The resistance between C and S should be some value between the other two.

Now the windings can be identified. This is important information when trying to properly connect capacitors and motor control devices.

▶ TECHNICIAN TIP

Most modern DMMs automatically detect the polarity of DC circuits. If the test probes are connected by using incorrect polarity, the multimeter screen will simply display a negative symbol. With an analog multimeter, however, connecting the red lead to a DC negative source and the black lead to a DC positive source could damage the meter's movement.

FIGURE 2-10 Single-phase motor windings.

When using a multimeter to measure an accurate resistance value, connect the meter in parallel, or across, the component or circuit to be tested. If the tested component is wired into an electrical circuit, fully disconnect it from the circuit to ensure reliable results. Otherwise, the reading could be affected by other components that are wired in parallel with the tested component. The accuracy of resistance measurements can also be affected by foreign substances such as dirt, soldering flux, or oil interfering with the test-lead connection.

When the test leads are connected to a component or circuit, a DMM with an auto-ranging feature automatically adjusts to the best resistance range. An analog meter must be set to the appropriate range. Remember to select a range that places the displayed value near the center of the scale for the most accurate reading. To use a multimeter to measure resistance, follow the steps in **SKILL DRILL 2-2**.

SKILL DRILL 2-2 Using a Multimeter to Measure Resistance

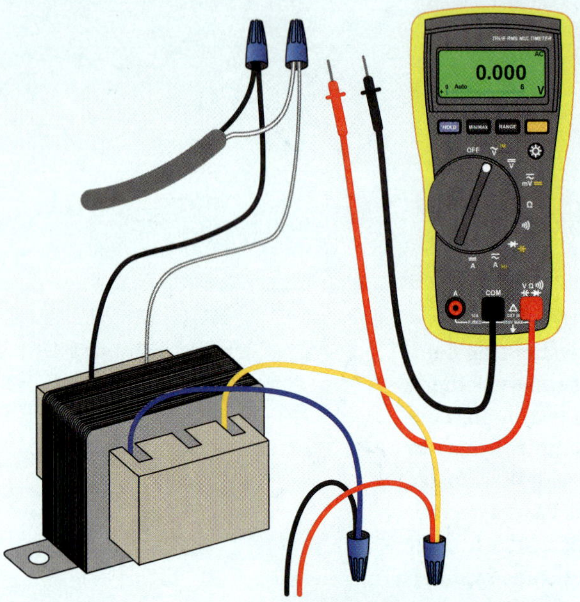

1. Prepare to measure the resistance of the windings in a 120 VAC/24 VAC control transformer, found in a typical warm-air furnace. First, select an appropriate multimeter for the job. Visually inspect it and the test leads for any signs of damage. Before using the multimeter on the job, always test it to verify that it is functioning properly and able to measure continuity and resistance. A small resistor of a known value can be used to verify meter operation.

SAFETY TIP

Working with electrically energized circuits is hazardous. Avoid contact with any surface that could potentially be energized. When working inside energized enclosures, maintain a stable position and, whenever possible, avoid placing both hands inside the enclosure. All work on energized equipment by unqualified individuals should be supervised by a qualified instructor or technician.

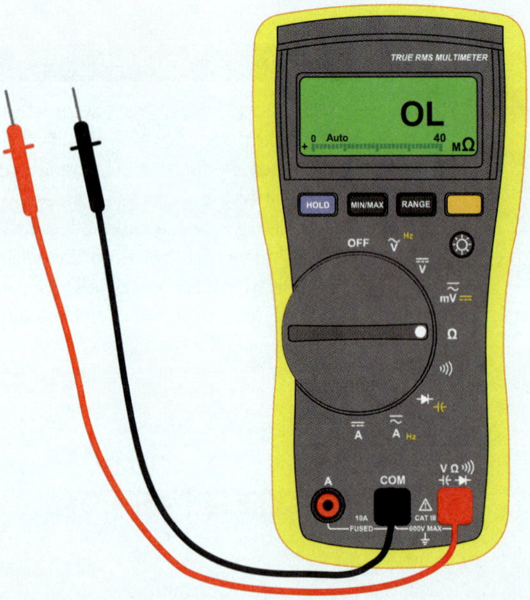

2. Insert the black test lead into the common input jack (COM) and the red test lead into the input jack for VΩ or Ω. If the meter has no auto-ranging feature, then select the voltage range, starting with the highest.

Continued

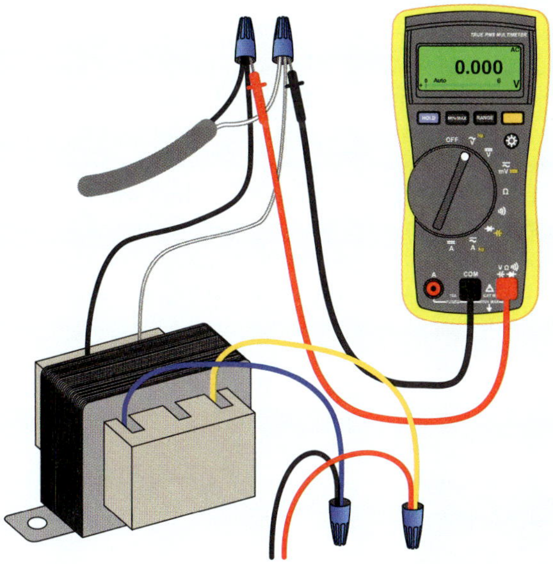

3. Ensure the circuit is deenergized. If a transformer that is installed or has a power source connected to the power leads is being worked on, set the meter appropriately and conduct a voltage test of the power source to be sure the power is off. Then electrically isolate the transformer by physically disconnecting all the leads. Mark the wires as necessary to ensure proper reconnection at the end of the skill drill.

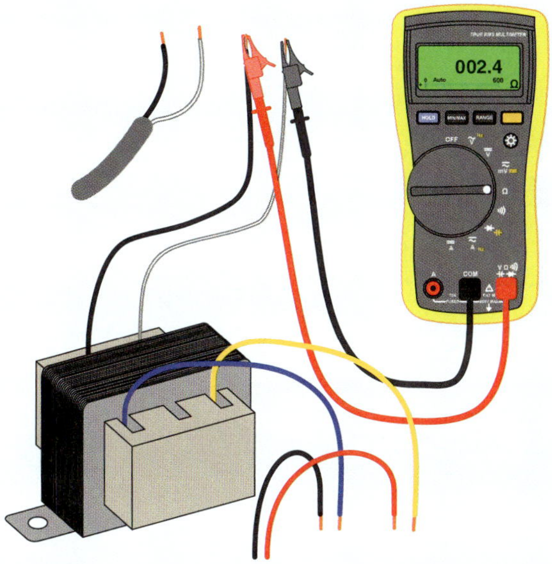

4. Set the meter to measure resistance, starting at the highest range if necessary. Touch the test leads to the leads for incoming power (the primary winding leads). Ensure that the test-lead probes make good contact with the conductors. Read and record the resistance value displayed on the meter.

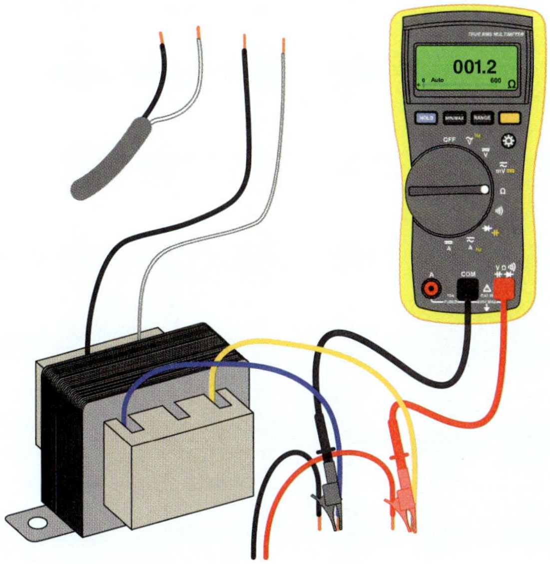

5. Touch the test leads to the secondary-winding terminals or leads of the transformer, ensuring good contact. Read and record the resistance value displayed on the meter. Then carefully reconnect the transformer primary and secondary wiring, and reinstall any removed access doors.

Using a Multimeter to Measure Current

An HVACR technician will frequently use a meter to measure AC current. In some cases, AC current values that are in the milliamp, or even the **microamp (μA)**, range may need to be measured. Most multimeters can measure these small current values by placing the meter in series with the circuit and then energizing the circuit to read the current flow (**FIGURE 2-11**). Multimeters typically have a maximum current limit of 10 A when using this method of measurement. Remember that exceeding the current limit with the meter in series with the load can result in significant damage to the meter. To measure current values that exceed the meter's limit, use either a clamp-on ammeter or a clamp-on ammeter accessory designed for the multimeter in use. To use a multimeter to measure low current values, follow the steps in **SKILL DRILL 2-3**.

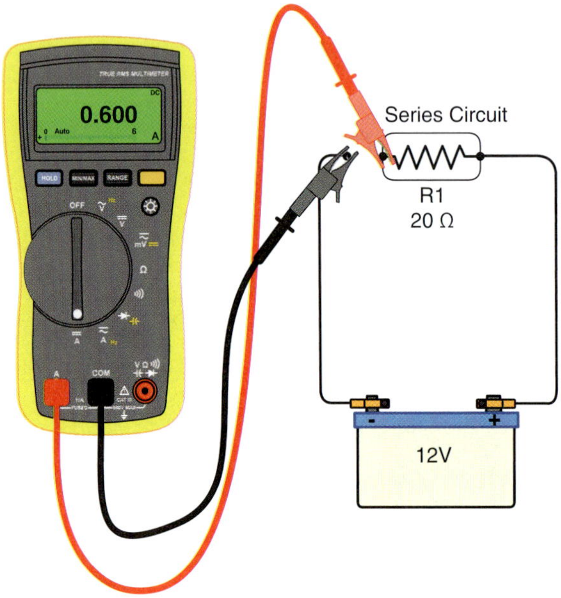

FIGURE 2-11 Placing a multimeter in series with a circuit for current measurement.

SKILL DRILL 2-3 Using a Multimeter to Measure Low Current Values

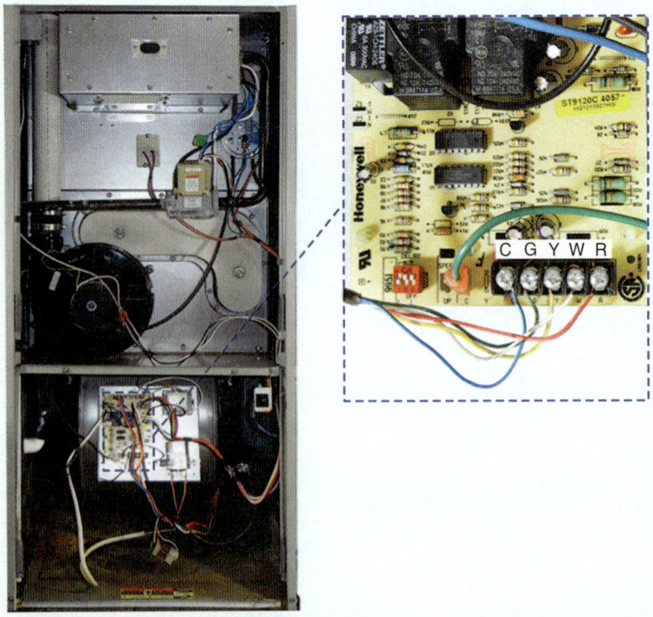

1. Prepare to measure the current flow in a typical 24 VAC gas- or oil-furnace control circuit. Select an appropriate multimeter for the job. The measured current in these circuits should be less than 2 amps. Ensure the meter is designed to measure AC current in this range. Visually inspect it and the test leads for any signs of damage. Before using the multimeter on the job, always test it to verify that it is functioning properly and able to measure continuity, voltage, and resistance.

Remove the furnace access door, and locate the low-voltage terminal board inside the furnace. These terminals are used to connect the furnace to the thermostat and to a remote air conditioning unit, if one is installed. The terminals are typically identified as shown in the photo, but they may differ by manufacturer.

SAFETY TIP

Working with electrically energized circuits is hazardous. Avoid contact with any surface that could potentially be energized. When working inside energized enclosures, maintain a stable position and, whenever possible, avoid placing both hands inside the enclosure. All work on energized equipment by unqualified individuals should be supervised by a qualified instructor or technician.

Continued

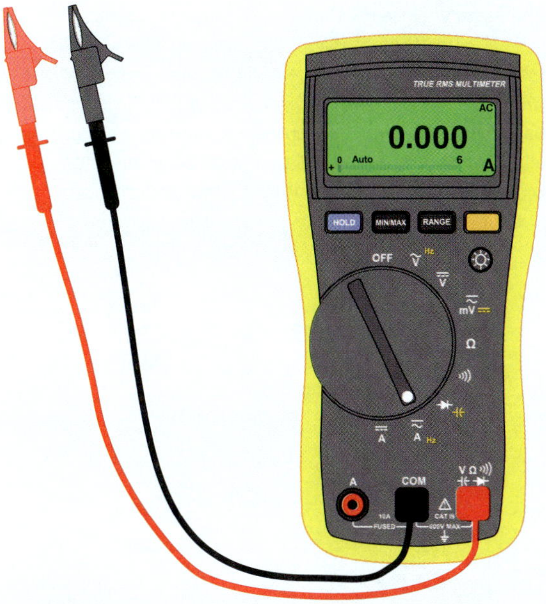

2. Turn off the power to the furnace. Use the voltmeter function where the power wiring connects to the furnace to confirm the power is off. Once the test is complete, move the red test lead to the jack marked for current measurement. This jack will typically be marked with the A and/or mA symbol. For this test, alligator-clip accessories for the test leads are very helpful. Set the meter dial to the proper position for current measurement, usually identified by an A or an mA, like the test-lead jack.

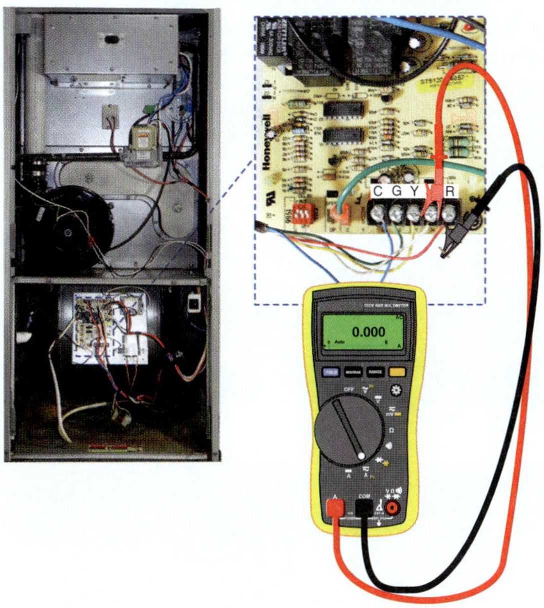

3. Disconnect the control wire from the thermostat connected to the W terminal of the furnace terminal board. This terminal enables the heating mode of the furnace. The wire may be white to match the standard color code, but not necessarily. Place the meter in series with the heating control circuit by connecting one of the alligator clips to the wire from the thermostat. The second alligator clip is then connected to the terminal where the wire was previously connected. Be sure the alligator clips are not touching any other wires, terminals, or conductive surfaces. Allow the meter to function in the auto-ranging mode, or set the range as needed. Position the meter and the leads securely so that both hands are free and the meter is not placing any tension on the leads or interfering with any other wiring or controls.

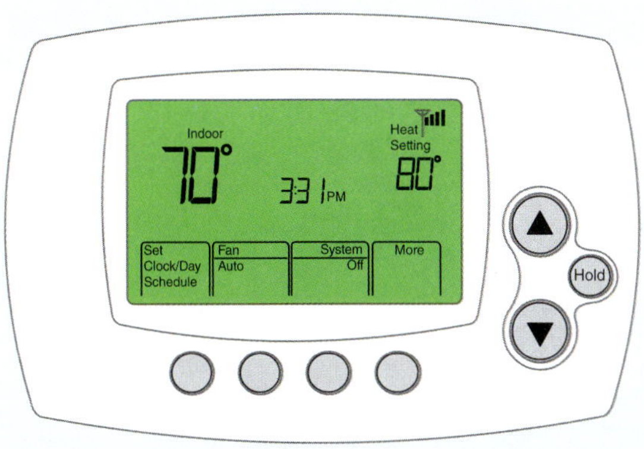

4. Set the system thermostat to the Heat mode, and change the heating set point to a value well above the room temperature.

Continued

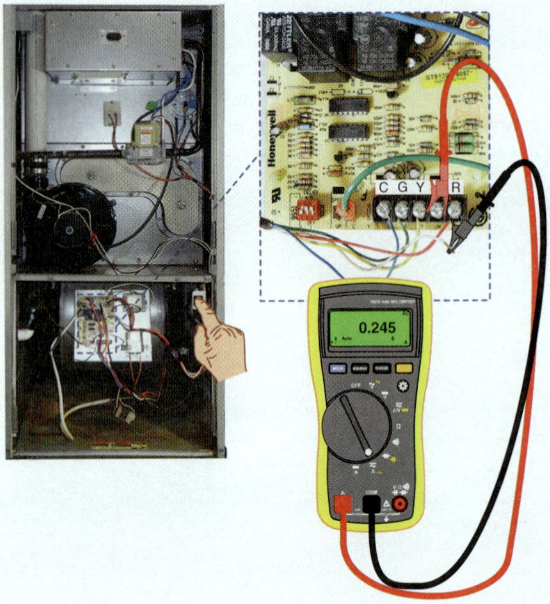

5. Restore power to the furnace. Since most warm-air furnaces have a switch that opens the power circuit when the access door is removed, the switch must be manually pressed to energize the furnace and complete the test. While depressing the door switch, be sure to keep all fingers away from the wiring and terminals on the back. Once the switch is depressed, power is restored to the furnace and the heating control circuit will be energized immediately or after a short delay while the thermostat is initialized. Clicks and/or motor operation are likely to be heard from the furnace as the heating cycle is initiated. Read and record the AC current value displayed on the meter.

6. Release the furnace door switch. All furnace activity should stop when the door switch is released; there is no need to allow the heating cycle to continue. Turn off the power supply to the furnace. Disconnect the meter leads, and carefully reconnect the heating control wire from the thermostat to the furnace terminal board. Reset the wall thermostat back to its previous setting. Reinstall the access door before restoring power to the unit.

▶ Selecting and Using Clamp-On Ammeters

K02002 Select, test, and safely use a clamp-on ammeter.

S02004 Use a clamp-on ammeter to measure current values greater than 2 amps.

Many clamp-on ammeters today are designed to provide most or all the functions of a DMM, with the added convenience of a built-in ammeter that can measure a significant amount of current. Therefore, they may be used to measure voltage, resistance, and current like any DMM. HVACR technicians often use clamp-on ammeters interchangeably with DMMs.

Most clamp-on ammeters today are digital models (**FIGURE 2-12**), although older analog instruments are still in use. A major distinguishing feature of a clamp-on ammeter is that it has a **current transformer** built into the jaws that can be opened and closed around a conductor. Closing the jaws around a conductor enables the current flow to be measured without having to handle the conductor or disrupt the operation of the component or system being tested. Through induction, a small current is induced in the jaws by the current in the conductor.

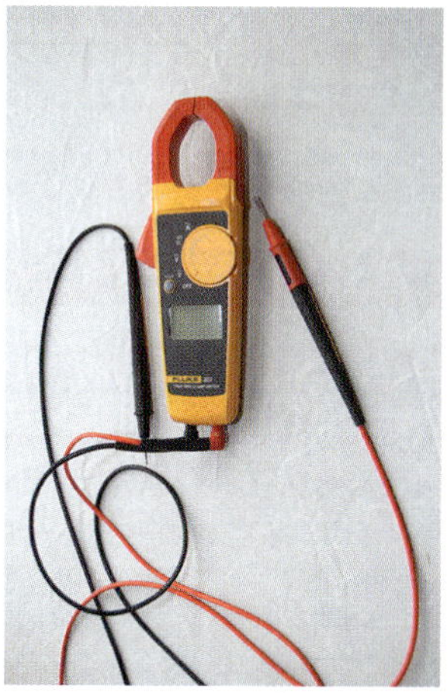

FIGURE 2-12 Clamp-on ammeter.

Typical clamp-on ammeter features and functions include but are not limited to the following:

- a selector switch for selecting the desired test function (voltage, current, or resistance)
- an auto-ranging feature to automatically select the proper measurement range
- a HOLD function to freeze the reading shown on the display
- a minimum/maximum memory function to determine the highest and lowest reading over the course of a test (the maximum function is valuable to have in order to measure the **inrush current** when a motor starts)
- a capacitor-testing feature that measures capacitance and checks for shorts and opens
- a continuity beeper that is activated when continuity through a circuit is detected (the beeper is an advantage for troubleshooting, particularly in tight spaces, because attention can remain focused on the meter leads—listening for the beeper instead of diverting attention from the leads to check the readings)
- overload protection to prevent damage to the meter and to protect the user.

Always read and follow the manufacturer's instructions for the meter, in order to make the best use of its features.

▶ TECHNICIAN TIP

Measuring current values in the microamp range is not a common task for HVACR technicians. One example, though, is measuring current for a gas-fired furnace that uses a flame-proving circuit generating 2-5 μA of DC current. Measurements in this range usually require the use of an in-line ammeter since the minute current flow is not sufficient to create a significant electromagnetic field. Clamp-on ammeters cannot effectively measure tiny current values using the jaws.

Selecting an Appropriate Clamp-On Ammeter

Carefully review an ammeter's specifications, features, and functions. Be sure that it has overload protection in order to protect the meter and the user. Choose an ammeter that can provide the degree of accuracy required. Most common HVACR applications do not require extreme precision in current measurements.

Ensure the ammeter is properly rated for the expected current. **Inductive loads**, such as electric motors, experience a significant inrush of current when they start. This surge occurs when the motor is first energized and must be brought up to speed from a complete stop. Although the surge may last less than one second, the current can be many times higher than the current shown on the motor data plate, especially if the motor has mechanical or electrical defects. Remember to consider the possible inrush current when selecting an ammeter and the range. The range can be changed, if needed, to a more appropriate level while the ammeter is actively measuring current. Meters with an auto-ranging feature eliminate this concern.

Ensure that the ammeter has the correct category rating for each task. Also, be sure that the ratings of the leads or other accessories meet or exceed the rating of the ammeter. When making a current measurement, ensure that the clamp is firmly and completely closed to get an accurate reading (**FIGURE 2-13**).

FIGURE 2-13 Inspecting the jaws of a clamp-on ammeter.

Testing a Clamp-On Ammeter

Verify that the clamp-on ammeter is working properly before using it. The procedures for testing a clamp-on ammeter are the same as those used to test the functions of a multimeter.

Always inspect a clamp-on ammeter before using it. Ensure the case is not cracked or greasy and that there are no obvious signs of damage. Be sure the jaw tips are not dirty and that they meet and interlock properly. Proper jaw alignment at the tips is essential for an accurate measurement. If the jaws are dirty or misaligned, the meter will not read correctly.

Test a clamp-on ammeter to verify that it can measure continuity/resistance and voltage. Do not use the meter if it fails any one of the tests. If the ammeter will not perform any functions at all when it is tested, check for a blown fuse and/or replace the battery.

> **▶ TECHNICIAN TIP**
>
> If a digital clamp-on ammeter is sensing current that is outside of its range, the display will read OL (overload). If an analog ammeter is overloaded, the indicating needle will strike the stop at the far end of the scale, possibly damaging the meter movement.

Using a Clamp-On Ammeter

Most digital clamp-on ammeters have an auto-ranging feature. If, however, the range needs to be set manually, always start at the highest range and adjust the range as needed to obtain an accurate reading.

When using a clamp-on ammeter to measure current, place the jaws of the ammeter around only one conductor at a time (**FIGURE 2-14**). Placing the jaws around two conductors at the same time will produce an inaccurate reading.

Using a Clamp-On Ammeter to Measure Current

Measuring current with a clamp-on ammeter is quite simple, since there are no test leads involved. It is important to recognize that whenever a current measurement is being made, the circuit is obviously energized. These tasks require complete attention, to avoid shock hazards and/or damage to the system. There are times when the ammeter jaws need to be placed around a specific conductor within a bundle of conductors. While the system is deenergized, select the location for the current measurement and separate the target conductor from others so that the jaws can safely and easily be snapped closed around the conductor when the power is on. Any preparations for measurement that can be made while the power is off to significantly reduce exposure to serious injury. Note that wire insulation is not a factor; there is no need to place the jaws around bare conductors. In fact, doing so can be dangerous, and it is rarely, if ever, required in the HVACR field.

If the current reading is expected to be at the low end of the lowest ammeter range, a more accurate reading can be obtained by wrapping the single conductor wire multiple times around one jaw of the ammeter (**FIGURE 2-15**). It is not often necessary for digital meters, but it is a significant help to read low current values on an analog model. Each time the wire passes through the jaws, the current it carries is sensed by the meter. For example, if the same conductor passes through the jaws five times, the current measured will be five times the actual current flow. Be sure, however, to divide the reading by the number of times the wire passes through the jaws.

While wrapping the conductor, do not allow the wraps to cross over each other. This can affect the accuracy. Note that this works equally well with a clamp-on ammeter accessory used with multimeters. To measure current with a clamp-on ammeter, follow the steps in **SKILL DRILL 2-4**.

> **▶ CORE CONNECTION**
>
> The watt was introduced before this chapter as a unit of measure for power. Technically, the watt is a unit of measure for what is known as true power, which represents the actual amount of power (P) used in a circuit. Another type of power, such as that used by capacitors, is referred to as reactive power. Combining reactive power and true power results in the apparent power, which is measured in units of volt-amps (VA). Of course, there is far more to this topic than these simple facts.
>
> A wattmeter is an instrument that directly measures the watts of power consumed by a circuit. Although a few multimeters are available that offer this feature, most do not. The meter's circuitry must be designed to handle the electrical load, since the meter must be wired in series with the load circuit to make the reading. Although there are times when measuring true power directly might be of value, it is rarely needed for typical HVACR service activities.

FIGURE 2-14 Clamp-on ammeter jaws placed around a single conductor.

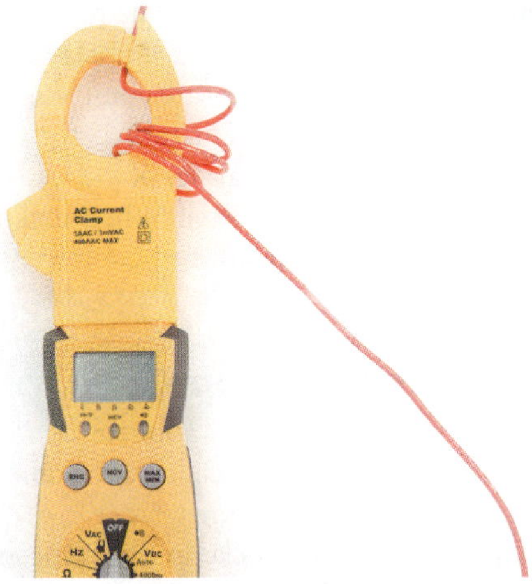

FIGURE 2-15 Conductor wrapped around the ammeter jaws.

SKILL DRILL 2-4 Measuring Current with a Clamp-On Ammeter

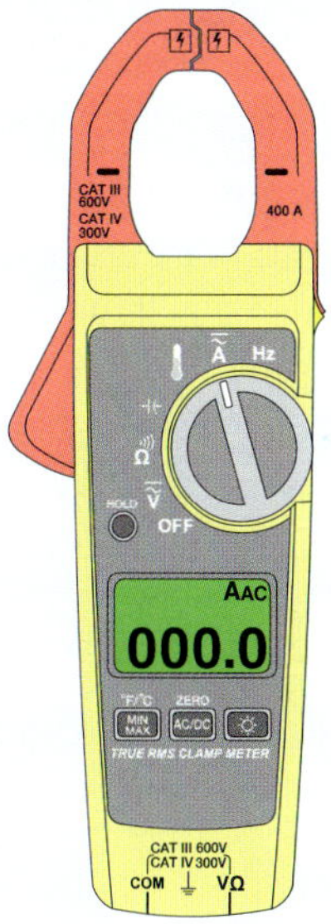

1. Prepare to measure the total current flow to an operating condensing unit. Select an appropriate clamp-on ammeter for the job, based on the expected current flow found on the unit data plate. Adding the current flow shown for the compressor(s) and any motors together produces a rough estimate of the unit's current flow, since these are the primary electrical loads found in the unit. Remember that these are inductive loads, so the inrush current will be several times higher than the current values shown. Inspect the ammeter for any obvious signs of damage. Before using the meter on the job, always test it to verify that it is functioning properly.

Select current as the value to be measured. The dial position is generally labeled with an A.

SAFETY TIP

Working with electrically energized circuits is hazardous. Avoid contact with any surface that could potentially be energized. When working inside energized enclosures, maintain a stable position and, whenever possible, avoid placing both hands inside the enclosure. All work on energized equipment by unqualified individuals should be supervised by a qualified instructor or technician.

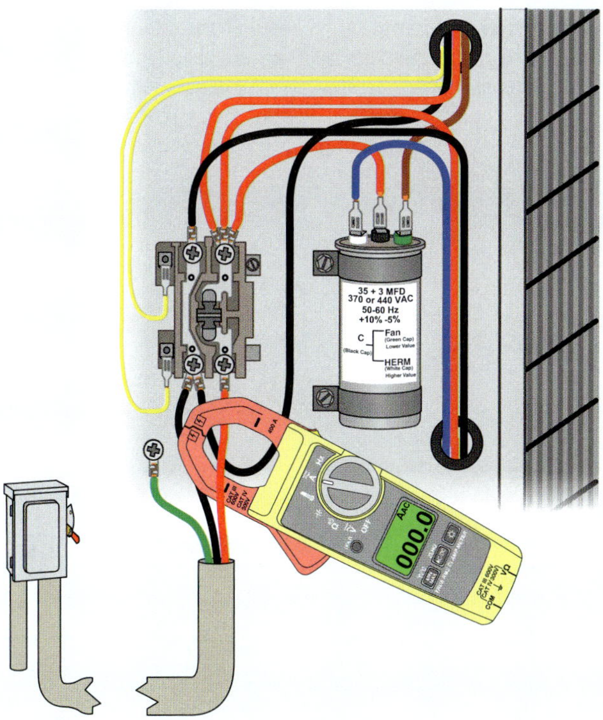

2. Turn off the power to the unit. Note that the 24 VAC control circuit might remain active because its power source is typically located in the indoor unit of split systems. Remove the access door, and locate where the power wiring enters the unit. There should be sufficient slack in the wires to easily work the jaws of the ammeter into position around a single conductor. For a residential condensing unit operating at 208–240 VAC, the conductors will consist of two power wires and a ground wire, which should be green and connected to the metal frame of the unit or control cabinet. Either of the two power wires can be chosen for this test. Open the jaws of the ammeter and close them around one of the conductors. Ensure that the ammeter is set properly.

Continued

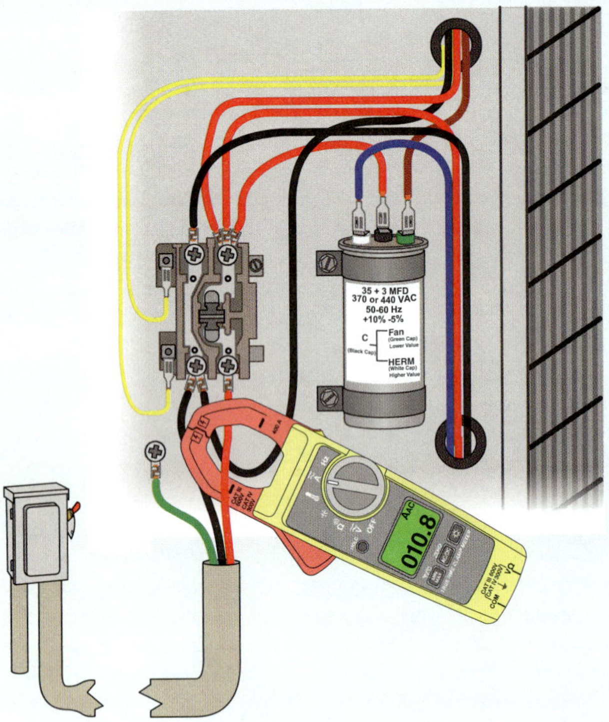

3. Set the room thermostat to Cool and move the set point well below room temperature. To observe the inrush current firsthand, set the thermostat first, and then return outdoors to restore power to the unit while watching the ammeter display. However, the inrush current is very short-lived and may not display accurately. Therefore, ..., it is best to engage the minimum/maximum feature of the ammeter (if equipped) to accurately record the highest reading. Restore power while watching the ammeter display. If the power disconnect switch position or location does not allow this, the minimum/maximum feature must be used to document the inrush.

Note: Do not allow the outdoor unit to run for an extended period with the access door removed. On most residential condensing units, this allows airflow through the unit to bypass the condenser coil, eventually causing the refrigerant discharge pressure to exceed the setting of the high-pressure switch or a relief valve. Once a stable current reading has been observed, discontinue testing.

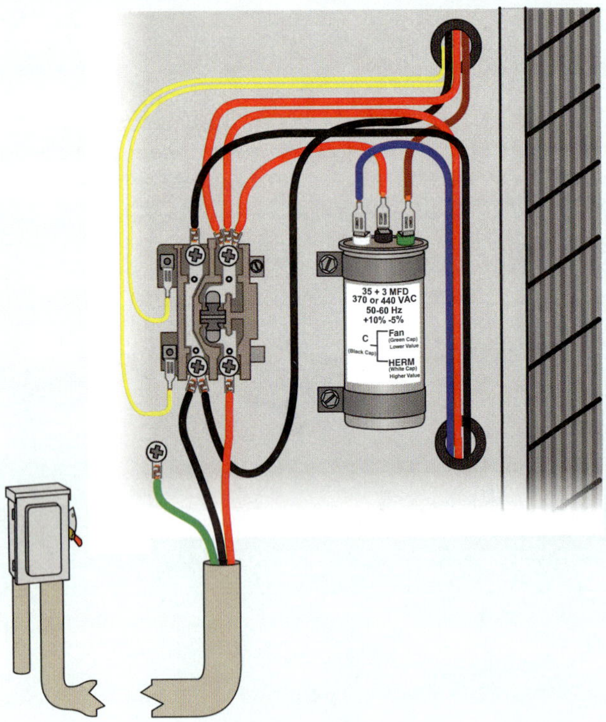

4. Turn off power to the unit; the unit should cease operation immediately. Remember that the 24 VAC control circuit might remain active. Remove the ammeter and reinstall the access door. Return the thermostat to the desired setting and restore power. If the inrush current was not observed, check the maximum value that was recorded on the meter. This indicates the current flow of the unit at start-up, when the inrush of current flow was at its highest. Compare the maximum value to the value recorded after the current measurement stabilized. This will demonstrate the relationship of the inrush current to the operating current.

▶ Selecting and Using a Capacitor Tester

K02003 Select, test, and safely use a capacitor tester.

S02005 Use a capacitor tester to measure capacitance.

Many DMMs have a capacitor-testing feature, but dedicated capacitor testers (**FIGURE 2-16**) are also popular. Capacitor testers can also check a capacitor for current leakage, insulation breakdown, shorts, and opens. HVACR technicians frequently use capacitor testers to test motor start and run capacitors.

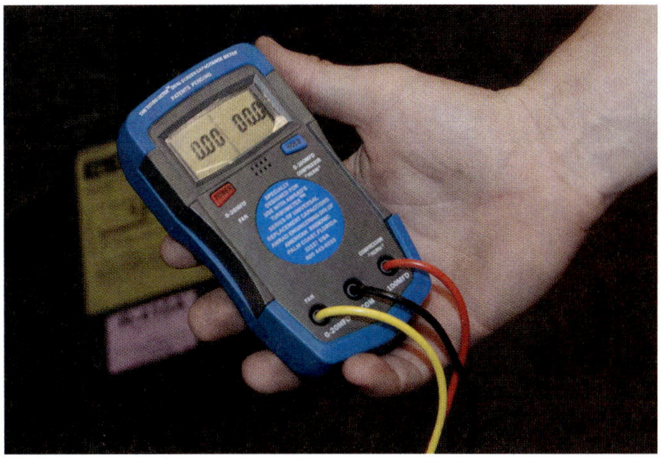

FIGURE 2-16 Capacitor tester.

A capacitor tester may have a manual scale selector for selecting the expected capacitance range. Others automatically select the range once a capacitance value is detected. If a scale selector is present, use the highest scale first if the capacitance expected is unknown. Common accessories for a capacitor tester include a set of leads, a carrying pouch, and replacement fuses.

Selecting an Appropriate Capacitor Tester

When selecting a capacitor tester, choose one that is properly rated for use in the environment where the measurement will be made. Carefully review its specifications, features, and functions. Ensure that the ratings of both the capacitor tester and its leads are adequate for the intended application. There are few, if any, capacitor testers marketed to HVACR technicians that are not suited for the purpose.

Always inspect the capacitor tester before using it. Ensure that the case is in good condition, that the input jacks are not broken, and that there are no other obvious signs of damage.

Check the test leads carefully. Ensure that the insulation is not cut, cracked, or melted and the tips (often alligator clips, for convenience) are not loose. Look for leads that have double insulation and probes/clips with finger guards. These features reduce the risk of shock and arc flash.

Testing a Capacitor Tester

To ensure the capacitor tester is functioning properly, use it to test the capacitance of a known good capacitor and the capacitance of a known defective capacitor. Follow the guidance provided in the next section to complete the tests. Ideally, for safety reasons, test capacitors that have not been recently connected to a power source. Do not use the capacitor tester if it fails to function properly during the tests.

Using a Capacitor Tester

Before connecting a capacitor tester to a capacitor, turn off the power source. Use a multimeter to check for voltage at the unit power-wiring connections to ensure power has been disabled. However, do not immediately handle the capacitor. Capacitors are designed to store a significant electrical charge, which is then used to assist in starting and/or running an electric motor. This stored charge can be hazardous, so do not touch the capacitor terminals or any of the associated wire leads until the charge has been dissipated. Before handling or testing any capacitor, always ensure that any remaining charge has been eliminated.

Some motor start capacitors are equipped at the factory with a bleed resistor installed across the terminals (**FIGURE 2-17**). This resistor bleeds off the capacitor charge once it is deenergized.

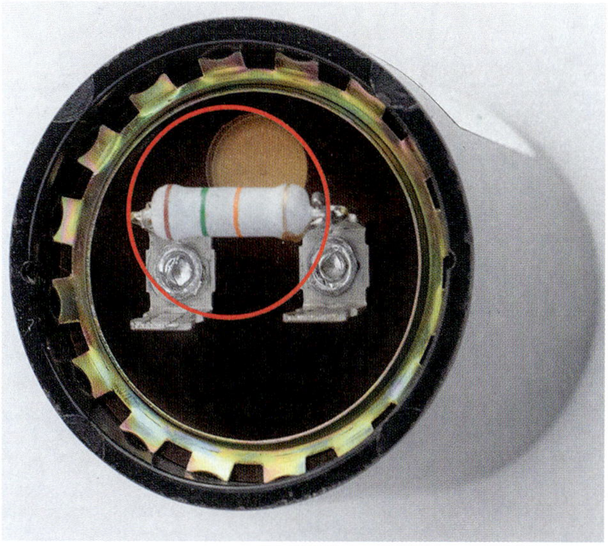

FIGURE 2-17 A motor start capacitor with a permanent bleed resistor.

▶ TECHNICIAN TIP

Typically, the measured microfarad (μF) value for a start capacitor should be ±20% of the value shown on the capacitor label. For a run capacitor, the measured value should be ±10% of the value shown on the capacitor. If the measured value is near or outside the range, replace the capacitor. Some capacitor manufacturers require a lower percentage tolerance to be considered good. Some are as low as ±5%.

If there is no permanent bleed resistor installed, carefully place a 1,000 Ω (minimum) resistor across the capacitor terminals for a few seconds. Use insulated needle-nose pliers to hold the resistor body. Permanent bleed resistors are usually 10,000–20,000 Ω, so they bleed more slowly and produce less heat. Resistors in this range can also be used to bleed the capacitor before handling and testing, but they require a little extra time in contact with the terminals. In any case, an undischarged capacitor can cause the capacitor tester's protective fuse to open, as well as deliver a significant shock. It is not wise to place a simple conductor, like a screwdriver, across the terminals to discharge capacitors, especially start capacitors. Using a proper resistor that bleeds the charge more slowly is by far the best and safest practice.

Run capacitors are not shipped with bleed resistors, because they remain in the motor operating circuit as long as the motor is running. They also must have their charge bled off using a resistor for safety's sake.

Follow the capacitor-tester manufacturer's specific instructions to operate the instrument. This usually involves attaching the leads of the tester to the capacitor terminals, turning on the tester, and viewing the capacitance reading displayed on the screen. To measure capacitance with a capacitor tester, follow the steps in **SKILL DRILL 2-5**.

SKILL DRILL 2-5 Measuring Capacitance with a Capacitor Tester

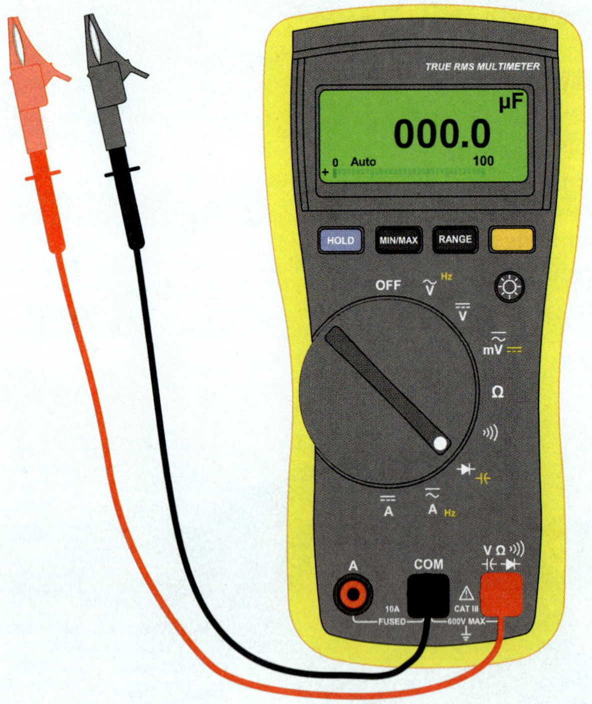

1. Prepare to test the capacitance of a motor start or run capacitor. Select an appropriate capacitor tester for the job. Inspect it for any obvious signs of damage. Before using a capacitor tester on the job, always test it to verify that it is functioning properly.

Run capacitors in common HVACR applications generally range from 5 to 40 **microfarad (μF)**; a run capacitor in this range is fine for this drill. A bleed resistor in a range from 1,000 to 10,000 ohms is also needed.

SAFETY TIP

Working with electrically energized circuits is hazardous. Avoid contact with any surface that could potentially be energized. When working inside energized enclosures, maintain a stable position and, whenever possible, avoid placing both hands inside the enclosure. All work on energized equipment by unqualified individuals should be supervised by a qualified instructor or technician.

Continued

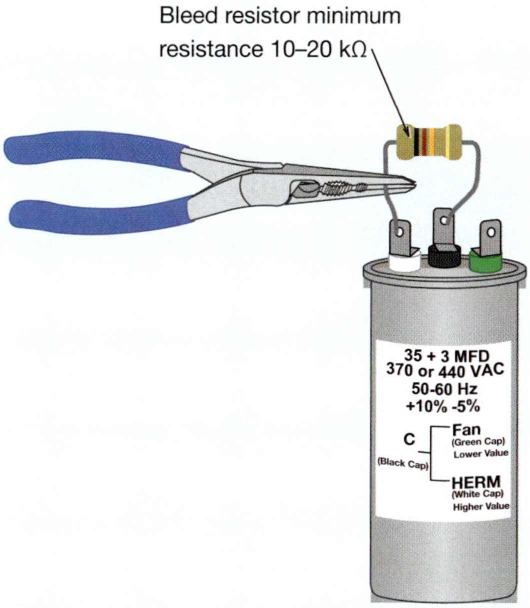

2. If the capacitor being tested is in service, first disconnect all power sources and verify with a voltmeter. Then bleed off any electrical charge by touching the bleed resistor across the capacitor terminals. Use a pair of needle-nose pliers to pick up the resistor and allow the ends to make contact with the capacitor terminals for 5–10 seconds. Once the charge has been bled off, carefully remove any wires attached to the capacitor terminals. Mark the wires as necessary before removing them to ensure that they are correctly reattached.

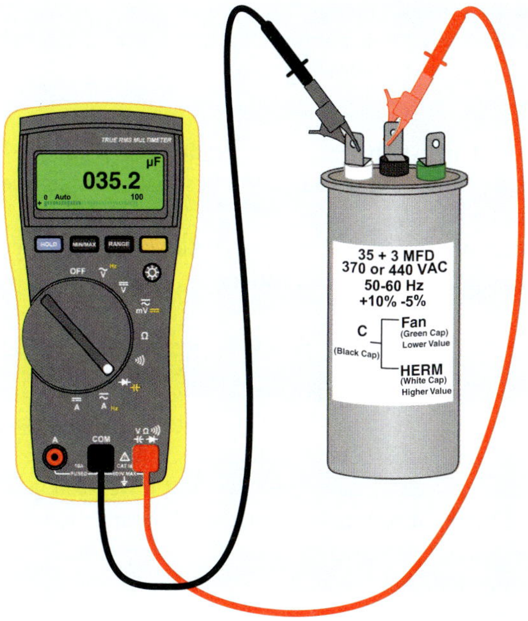

3. Attach the leads of the tester to the capacitor terminals. If it is a dual capacitor, as shown in the figure, make sure that the Common terminal is used to test each of the remaining two terminals. They are typically marked Herm to represent to higher capacitance for the compressor and Fan to represent the lower capacitance for the condenser fan motor.

Ensure both leads make good contact with the capacitor terminals. Using alligator clips simplifies the task. Set the scale to the highest range until an initial value is displayed. Consider changing to a lower range for greater accuracy if the value falls into a lower available range. Turn the tester on and read the capacitance. Complete this drill by determining whether the value displayed is within the acceptable range shown on the label. Note that open or shorted capacitors will typically cause all 8s to appear in the display (open) or zeros that may blink or flicker to other values (internally shorted to the case). Responses vary from one instrument model to another. If the tested capacitor is installed in an active unit, then carefully reconnect the wiring, reinstall the access door, and restore power.

▶ TECHNICIAN TIP

When test results indicate a shorted capacitor using a low range on the tester, test the capacitor again at a higher range. Good capacitors above 10 µF can sometimes indicate a shorted capacitor (displaying 0.000) when they are tested on the low range. On the high range, the correct reading should be displayed if the capacitor is good.

▶ Selecting and Using a Non-Contact Voltage Tester

Although a non-contact voltage tester (**FIGURE 2-18**) is not suitable for work that requires any level of accuracy, it is useful for quickly verifying that a circuit is energized or deenergized. HVACR technicians routinely use non-contact voltage testers to check for the presence of power. However, it is important to note that it should not replace voltmeter testing to prove that a circuit is deenergized for safety reasons. Electrical safety demands that a more dependable test device be used.

K02004 Select, test, and safely use a non-contact voltage tester.

S02006 Use a non-contact voltage tester to detect voltage.

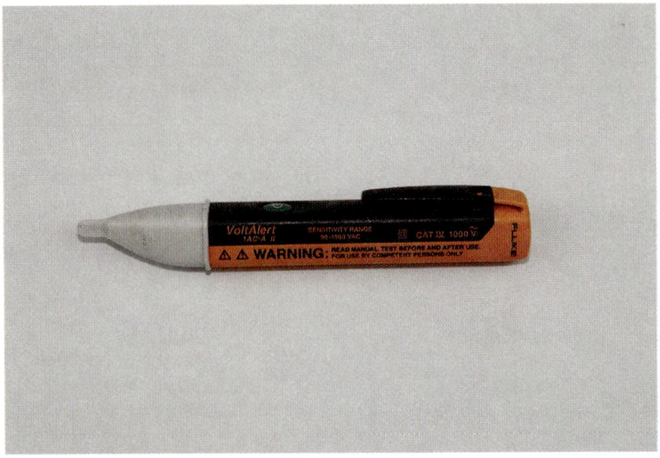

FIGURE 2-18 Non-contact voltage tester.

If voltage is present when a non-contact voltage tester is placed close to a source—such as an electrical outlet, lamp, or light switch—a light in the tester will flash and/or the tester will produce a warning tone, such as a chirp or beep. The amount of voltage present can sometimes be roughly estimated based on the intensity of the flashing light and/or warning sound being emitted by the tester, depending on the model.

Non-contact voltage testers are relatively small, compact, battery-powered instruments that fit in a shirt pocket. They typically have the following features:

- an On/Off power button
- a flashlight or indicator light that shows the tester is operational and the flashlight may help illuminate the workplace (in some models, the flashlight can be operated independently of the voltage tester)
- a high-intensity LED light and/or a warning tone to notify when voltage is detected
- an automatic shutoff to conserve power and extend battery life
- a low-battery indicator
- a built-in self-test feature.

Non-contact voltage testers are available in several voltage ranges. A standard model may be sensitive only to voltages above a range of 90–100 VAC. Note that this is not a low enough range to detect the traditional HVACR control voltage of 24 VAC. Dual-voltage models can register and differentiate between standard voltages and lower voltages. This feature makes these models more suitable for testing low-voltage AC circuits for the presence of power. Note that these tools are not designed to sense the presence of DC voltage, since they rely on the presence of a magnetic field generated by AC. They may occasionally respond to the presence of DC voltage, but not reliably.

Selecting a Non-Contact Voltage Tester

When selecting a non-contact voltage tester, choose one with a range suitable for the expected voltage. Be sure the instrument is sensitive enough in terms of the minimum voltage that it can detect. Inspect the tester for any obvious signs of damage. Ensure the batteries for the tester are working. Many testers offer a battery test button to ensure the device is working properly.

Testing a Non-Contact Voltage Tester

Check the tester on a known, energized voltage source prior to using the instrument in the field. Use the built-in self-test feature, if the tester is so equipped. Otherwise, a simple way to test it is to place the tip of the tester near the line-voltage side (smaller slot) of an electrical outlet known to be energized. If the tester is working properly, the tester should give a clear audible and/or visual indication that voltage is present.

Using a Non-Contact Voltage Tester

Although non-contact voltage testers may detect voltages up to 1,000 VAC or more, working near voltages this high can be extremely hazardous. They are best used to detect voltages well below that value. When safety is at stake and the intention is to make contact with conductors and/or electrical parts, use a multimeter to conduct the test to ensure that the circuit has been deenergized. Non-contact testers are best reserved for quick testing while troubleshooting or monitoring a system's sequence of operation. Likewise, it is possible in some rare instances for a component to be energized without showing any voltage. If the test is being done in the pursuit of safety before contact is made with electrical components, use a multimeter rather than a non-contact device. To use a non-contact voltage tester to detect voltage, follow the steps in **SKILL DRILL 2-6**.

SKILL DRILL 2-6 Using a Non-contact Voltage Tester to Detect Voltage

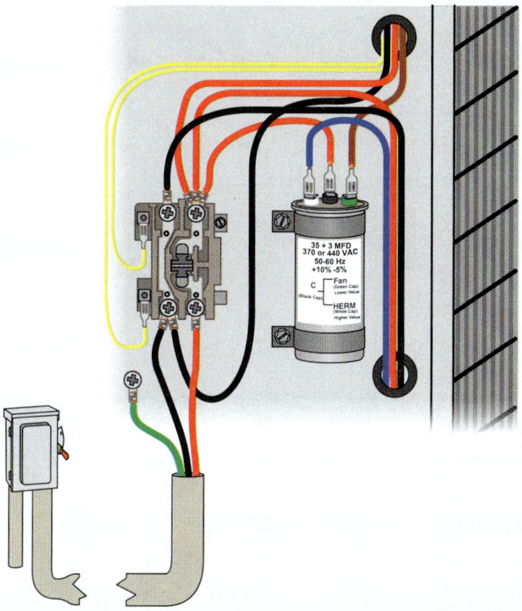

1. Prepare to check for the presence of voltage inside an outdoor condensing unit. Select an appropriate non-contact voltage tester for the job. Inspect it for any obvious damage. Ensure the batteries are working. If the instrument is equipped with a self-test feature, use it to ensure the instrument is functioning properly. Before using the voltage tester on the job, always test it to verify that it is functioning properly.

With the power to the condensing unit shut off, remove the control panel cover.

SAFETY TIP

Working with electrically energized circuits is hazardous. Avoid contact with any surface that could potentially be energized. When working inside energized enclosures, maintain a stable position and, whenever possible, avoid placing both hands inside the enclosure. All work on energized equipment by unqualified individuals should be supervised by a qualified instructor or technician.

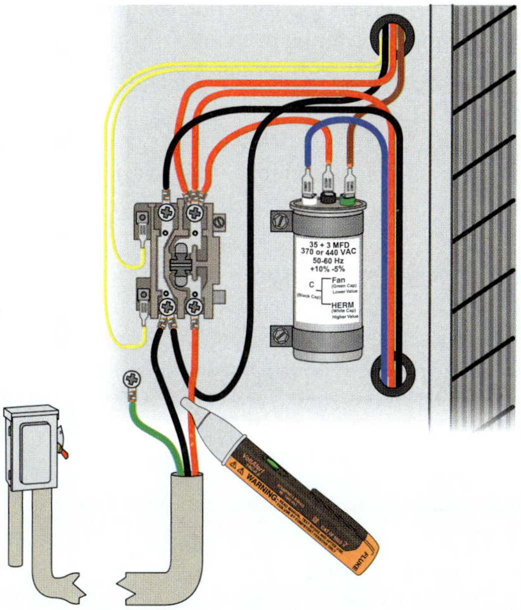

2. Gradually place the tester as close as possible to the power wires entering the unit. There should be no voltage present and no response from the tester. Move from one power wire to the other to be sure.

Continued

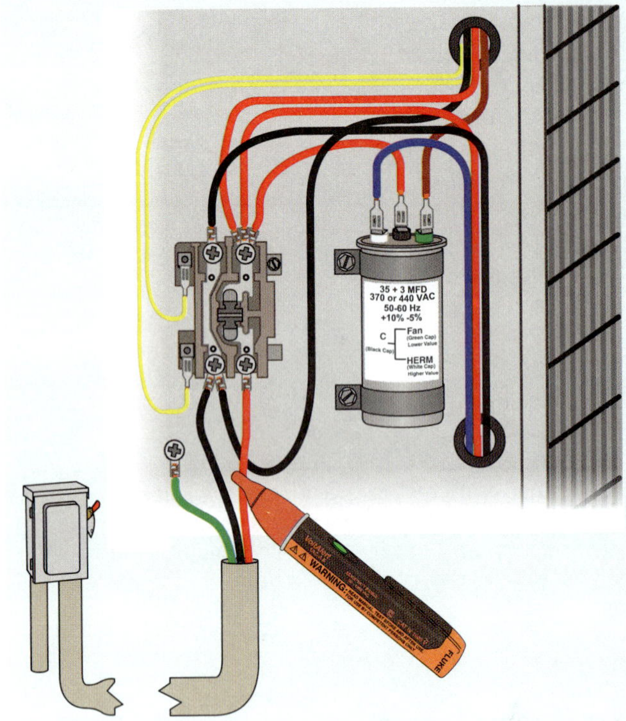

3. Turn the power to the unit back on. Remember that the unit is now energized and thus represents a shock hazard. Use the tester once again to check for the presence of voltage by placing the tip near the wiring. The tester should provide a visual, and possibly an audible, indication that voltage is present. Then cycle the power off and watch for the indication that voltage is no longer present. Once it is verified that no voltage is present, reinstall the unit access door and return the unit to service.

▶ Wrap-Up

Summary

When working with electricity, safety is always the first priority. All technicians share a responsibility to work safely for their own protection and for the protection of their coworkers.

Take care to select, test, and safely use the proper test instrument for the task at hand. First, determine the electrical value needed to measure or detect—voltage, resistance, current, or capacitance. Next, select the appropriate test instrument for the specific purpose, such as a multimeter, a clamp-on ammeter, a capacitor tester, or a non-contact voltage tester. Then inspect and test the selected instrument to ensure that it is in good working order. Finally, follow the proper procedures for using the instrument so that the test is performed safely and accurately in the field.

Although this chapter has introduced common electrical test instruments in the HVACR trade, it takes practice to apply learned skills effectively and become proficient in test instrument use. However, developing this skill is essential to the success of an HVACR technician. The instruments presented here are invaluable tools that an HVACR technician must master. Study the manufacturer's documentation for each owned instrument to take full advantage of the instruments' features and properly care for them.

Key Terms

analog multimeters Multimeters that use a moving pointer or needle to display electrical values such as volts, amps, and ohms.

arc flash An event where electric current leaves its intended path and travels across an air gap from one conductive surface to another, or to ground, resulting in a violent, bright flash. The energy released from an arc flash is often significant enough to cause serious personal injury and/or death.

continuity As it pertains to electrical circuits, a state of continuity indicates a complete, unbroken electrical path. However, some amount of resistance to electrical current is typically present.

current transformer A special type of transformer often used for metering and current measurement. It produces an alternating current in its secondary winding that is precisely proportional to the current in the primary winding. In the case of a clamp-on ammeter, the secondary winding is formed by the jaws of the instrument, and the conductor serves as the primary winding.

digital multimeters (DMMs) Multimeters that use a digital display of numbers and symbols to present electrical values such as volts, amps, and ohms.

induction The process through which one electrical conductor becomes energized when near another electrically charged source.

inductive loads Electrical loads that operate based on the principles of electromagnetism. An electric motor, for example, rotates due to the formation of magnetic fields by energized electrical windings. Other examples of inductive loads include transformers and the operating coils of relays and similar control devices.

inrush current The instantaneous, high-volume current drawn by many electrical devices, especially those that operate through induction, such as motors and transformers, when they are first energized.

microamp (μA) 1 microamp equals 1/1,000,000th of an ampere.

microfarad (μF) A unit of measure for capacitance, equal to 1/1,000,000th of a farad.

polarity A property of various points in an electrical circuit, related to their electrical potential. A negative pole has more electrons than a positive pole, and electrons tend to flow from negative poles to positive poles. Current, however, flows from a positive pole to a negative pole. For DC circuits, current flows in only one direction; one pole is always positive, and the other is always negative. In AC circuits, the polarity constantly changes, with current flowing in one direction and then the other and changing direction at a consistent frequency.

Review Questions

1. A single instrument that is used to measure electrical values such as AC and DC voltage, AC and DC current, and resistance is a _____.
   a. variable analyzer
   b. auto-ranging tester
   c. multimeter
   d. megohmmeter
2. If a DMM that does not have an auto-ranging feature is used and the expected voltage range is unknown, start the range at the _____ voltage setting.
   a. highest
   b. lowest
   c. middle
   d. overload
3. Before using a voltmeter or multimeter to ensure a circuit is safely deenergized, test it on a ____.
   a. known voltage source
   b. rectified continuity tester
   c. calibrated ohmmeter
   d. parallel circuit
4. When using a multimeter to measure resistance, it is necessary to connect the meter test probes in ______ with the circuit.
   a. series
   b. parallel
5. Before taking a resistance measurement in a circuit with a multimeter, _____.
   a. charge any capacitors in the circuit
   b. deactivate the multimeter's capacitance mode
   c. set the range manually
   d. ensure that the circuit is deenergized
6. To measure an AC current that is expected to be higher than 10 A, use a digital multimeter with _____.
   a. a clamp-on ammeter accessory
   b. a non-contact voltage tester
   c. an auto-ranging feature
   d. a capacitor-testing feature
7. Inrush current is associated with _____.
   a. components with a low resistance value
   b. inductive loads
   c. overload protection
   d. non-contact voltage testers
8. To accurately measure current, a clamp-on ammeter must have _____.
   a. silicone-coated jaws
   b. test leads with finger guards
   c. an analog pointer movement
   d. jaws that properly interlock
9. To get an accurate current measurement with a clamp-on ammeter, always place the jaws of the ammeter around _____ conductor(s) at a time.
10. How can a more accurate reading be obtained when the expected current reading is at the low end of a clamp-on ammeter's accurate range?
11. To capture the inrush current drawn by a motor during start-up, select a clamp-on ammeter capable of ________.
    a. measuring crest factors
    b. measuring microamperes
    c. recording minimum/maximum values
    d. identifying transient capacitance
12. Which of these instruments has a current transformer built into it?
    a. capacitor tester
    b. analog voltmeter
    c. clamp-on ammeter
    d. non-contact voltage tester
13. What should you do if a multimeter or clamp-on ammeter does not perform any functions at all when you attempt to test its functionality?
14. Which type of capacitor is less likely to have a permanent bleed resistor installed?
    a. a motor start capacitor
    b. a run capacitor
15. Before connecting a capacitor tester to a capacitor, confirm that the power is off and _______ the capacitor.
    a. charge
    b. bleed off
16. As a general rule, the measured capacitance value of a run capacitor should be within what percentage of the labeled value?

17. Which of the following pocket-sized instruments is often used for quickly verifying that a circuit is energized?
- **a.** A clamp-on ammeter
- **b.** A voltmeter with an auto-ranging feature
- **c.** An analog multimeter
- **d.** A non-contact voltage detector

18. Before using a non-contact voltage tester, _____.
- **a.** test it to determine if it responds to voltage
- **b.** connect the test leads to it
- **c.** ensure the jaws are not misaligned
- **d.** ensure the input jacks are not recessed

19. One potential disadvantage of a non-contact voltage detector is that some do not detect _____.
- **a.** voltages greater than 250 VAC
- **b.** voltages less than 90–100 VAC
- **c.** voltage in a circuit with an active load
- **d.** voltage in a parallel circuit

20. Which of the following overvoltage categories defined by the IEC is considered the most hazardous, and therefore a meter marked with this category is considered safer?
- **a.** CAT IV
- **b.** CAT III
- **c.** CAT II
- **d.** CAT I

CHAPTER 3
Power-Distribution Systems

Knowledge Objectives

After reading this chapter, you will be able to:

- **K03001** Identify the different types of overload protection devices used in electrical distribution circuits.
- **K03002** Explain how single-phase power is developed and distributed.
- **K03003** Explain how three-phase power is distributed.

Skill Objectives

Skills required to meet the objectives of this chapter:

- **S03001** Replace a cartridge fuse in a fused disconnect.

▶ Introduction

When HVACR equipment is installed, the required power supply must be provided to each independent unit. The wiring and circuit protection devices used to supply power are usually installed by licensed electricians. In a typical new installation, the electrician would route the wiring from the main distribution panel to the **disconnect** location that will serve an HVACR unit. This would happen during the electrical **rough-in** of the structure. When it is time to install and connect the HVACR units, either the HVACR installer will make the final electrical connection or an electrician will be called in.

All electrical circuits in a home or business must be protected to ensure the wiring and panels are not loaded beyond their safe capacity. HVACR technicians need to have a working knowledge of the protective devices involved.

▶ Distribution Circuit Protective Devices

K03001 Identify the different types of overload protection devices used in electrical distribution circuits.

S03001 Replace a cartridge fuse in a fused disconnect.

Electrical circuits used in homes and businesses must be protected against current overloads that could damage equipment or conductors and cause a fire. Circuit breakers and fuses are the protective devices commonly used for this purpose. These devices are designed to limit the amount of electrical current that can be allowed to flow in a circuit. If the current rating of the protective device is exceeded, the device opens and interrupts the circuit, disconnecting the load from the power source.

Fuses

A fuse contains a metal link designed to melt at a specific temperature or current. It is installed in series with an electrical circuit so that the current flows through the link. Fuses are rated for a specific current level, such as 30 or 60 amps. When the current through the fuse exceeds that level, the heat generated by the current will cause the link to melt, opening the circuit. The greater the overcurrent level, the faster the link will open. However, there are other fuse characteristics that affect how quickly and effectively a fuse responds to an overcurrent condition, depending on the level of protection needed.

The majority of fuses encountered by HVACR technicians are cartridge fuses (**FIGURE 3-1**). A great deal of the discussion in this chapter will be devoted to them. Plug-type fuses, such as the ones shown in **FIGURE 3-2**, are also used, although they represent an older design and are becoming rare.

Plug fuses are designed to screw into their sockets. There are two types of plug fuses: the Edison base and the Type S. The Edison base is an older type that can still be used today, but only to replace existing Edison-base fuses. The bases of these fuses are all the same size, regardless of current rating. If a person had only a 30-amp fuse on hand, they could use it to replace a 15-amp fuse, thereby creating a fire hazard; there is no mechanical method of preventing it. Type S fuses were designed to eliminate that problem and thus are required in new installations that use plug fuses. The sockets of Type S fuses are sized to match the fuse rating. This prevents a fuse with a higher current value from being installed in place of a fuse with a lower value.

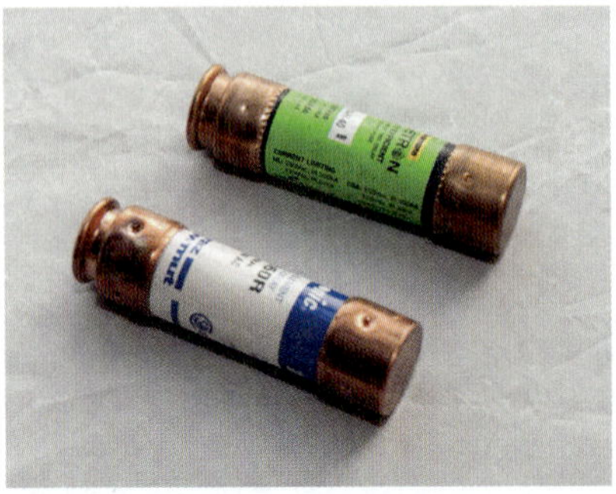

FIGURE 3-1 Cartridge fuses.

FIGURE 3-2 Plug-type fuses.

Some fuses are designed to hesitate before opening when an overload current occurs. These are called time-delay or slow-blow fuses. Their purpose is to prevent unnecessary circuit interruption when a brief yet acceptable high-current condition occurs.

One common use for time-delay fuses is in circuits serving AC (alternating current) induction motors, such as the type commonly used in HVACR applications. Induction motors experience a significant current surge in the first moment of start-up—up to eight times a motor's normal running current. This is normal operation for AC motors. If a standard fuse—one that opens as soon as its current rating is briefly exceeded—were used in that situation, it would open during every motor start-up. The alternative to that would be to size the fuse for the start-up current. In that case, the fuse would not react properly when the circuit is actually overloaded while running. This would create a hazard to equipment and conductors in the event of a sustained current overload. The solution to that issue is to use a properly rated time-delay fuse.

Fuses are often found in in the power disconnect switches that are located on or within sight of HVACR equipment. **FIGURE 3-3** shows a disconnect switch used to control power applied to a split-system condensing unit. This disconnect shown is designed to house cartridge fuses. The red-tipped handle on the left is used as an On/Off switch, controlling power to the load.

The disconnect switch shown in Figure 3-3 is a popular choice for both single-phase and three-phase electrical systems. A somewhat simpler, and less expensive, version of a fused disconnect is shown in **FIGURE 3-4**. The fuses are installed in a block that fits into the disconnect assembly. There is no external handle to turn power on and off; opening the door and pulling out the fuse block deenergizes the circuit.

Fuse Ratings

Fuses are primarily rated in terms of current and voltage. The current rating is the maximum current the fuse can generally tolerate without melting the internal link. The voltage rating of a fuse must be greater than that of the voltage source it would interrupt. According to electrical codes, a fuse selected for a standard electrical circuit must have a current rating that is at least 125% of the continuous current load, plus 100% of any non-continuous load. Circuits containing motors require a more complex calculation.

Plug fuses have voltage ratings up to 125 volts and current ratings up to 30 amps. Their low voltage capacity is why they are rarely seen outside of the residential environment and are rarely used today. Standard cartridge fuses have ratings of 250 volts and 600 volts, with current ratings up to 600 amps. A ferrule-type cartridge fuse is used in applications up to 600 volts and currents up to 60 amps. Knife-blade cartridge fuses are used for currents above 60 amps. See **FIGURE 3-5** for a visual comparison of these two types.

SAFETY TIP

Fused and non-fused disconnects have knife-like busbars that are forced between two spring-steel fingers when the switch is turned on, ensuring good electrical contact. On rare occasions, mechanical problems occur, and the components do not separate as they should when the switch is turned off. When a disconnect switch is placed in the Off position, open the disconnect cover and examine the switch, ensuring that the circuits have opened properly. This is one reason why every deenergized circuit should be carefully tested with a voltmeter to ensure it is indeed deenergized.

FIGURE 3-3 A fused disconnect.

FIGURE 3-4 Fused disconnect with a pullout block.

FIGURE 3-5 Knife-blade and ferrule-type cartridge fuses.

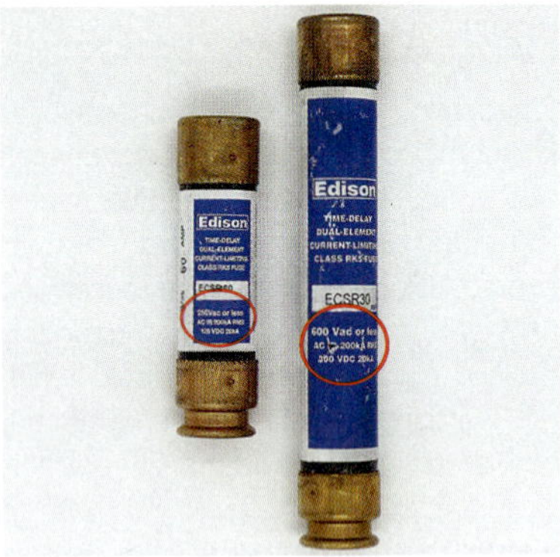

FIGURE 3-6 250- and 600-volt cartridge fuses.

▶ **TECHNICIAN TIP**

All fuse manufacturers provide cross-reference charts to assist in matching the characteristics of fuses from other manufacturers. They can be accessed in catalogs or via the Internet. In addition, some companies have developed apps that can be used to cross-reference fuses on a computer or mobile device.

▶ **TECHNICIAN TIP**

Large cartridge fuses can be quite expensive, especially when unique characteristics are required for the application. Some cartridge fuses are designed for disassembly and can be fitted with a new element (melting link). This type of serviceable fuse is often found in industrial settings where they are more likely to require frequent replacement and have many fuses in service.

The mounting style of cartridge fuses divides those with higher current ratings from those that are lower. To ensure that cartridge fuses are used with the correct voltage, 250-volt and 600-volt fuses differ in length. **FIGURE 3-6** shows two cartridge fuses of the same current value but used with different voltages. Note that the 250-volt fuse is significantly shorter in length and will not fit a fuse block rated for higher voltages. The dimensions of fuses vary dramatically, making it more difficult to use the incorrect type and create a hazardous situation.

Replacing Fuses

One essential point is this: If it is necessary to replace a fuse, always replace it with one of the same type that has the same voltage rating and the same, or a lower, current rating. The same brand may not be readily available, but cross-referencing from one brand to another is commonplace. Never oversize the fuse in terms of current; doing so can result in damaging the conductors and possibly cause a fire. The capacity of a fuse to withstand current cannot be greater than that of the conductors it protects. In some cases, a fuse with a higher voltage rating can be used, but never try to use one with a lower voltage rating. If there is any doubt about whether a replacement fuse is appropriate, contact the equipment manufacturer or an electrician qualified to answer the question.

To replace a cartridge fuse, turn off the power and use a fuse puller (**FIGURE 3-7**) to remove the fuse. Note that if a fuse blows repeatedly, this indicates a fault in the circuit, or a component that must be isolated and repaired before operation can continue. A number of things beyond a component fault can cause a fuse to blow, including loose or poorly connected wiring terminations.

Fuses are also installed with the fuse puller tool. It is best to avoid grasping the body of a long fuse in the center, because the fuse might be damaged from the stress of pulling or pushing on the center of the barrel. On longer fuses, apply installation pressure close to one end first, then snap the opposite end into position.

Testing Fuses

Some very small glass fuses are fully transparent, allowing a technician to see whether the thermal element inside is intact (**FIGURE 3-8**). However, failure can also occur at the extreme ends of the fuse, where the failure cannot be seen clearly. There are also some larger cartridge fuses that have a window in the barrel for visual inspection of the thermal element. Unless the condition of a fuse is completely obvious, such as the blown fuse shown in Figure 3-8, the best way to determine its condition is by testing it for continuity with an ohmmeter or multimeter (**FIGURE 3-9**). Fuses should not be tested while lying on,

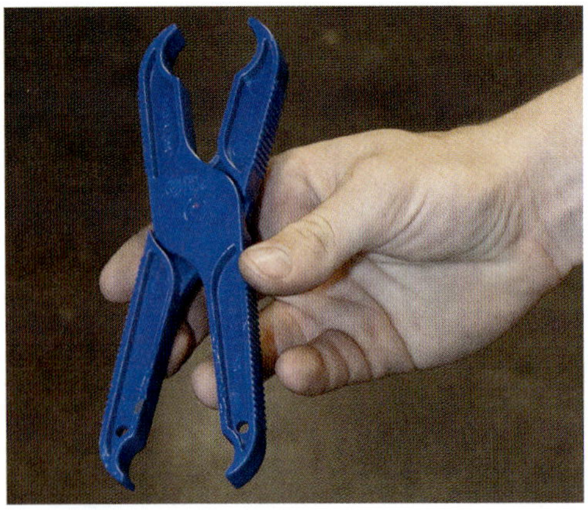

FIGURE 3-7 Fuse pullers.

FIGURE 3-8 Glass fuses.

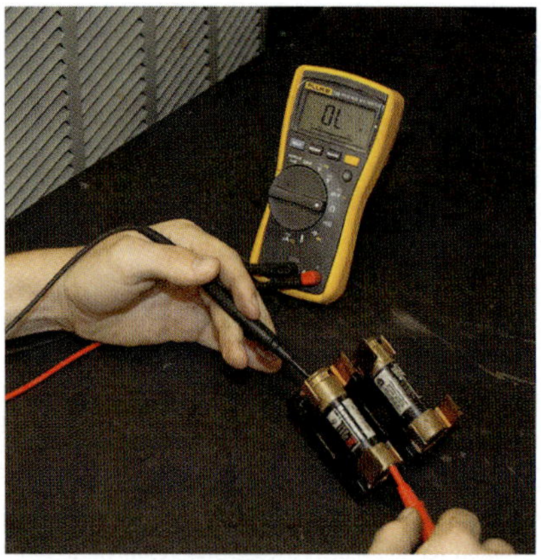

FIGURE 3-9 Testing a cartridge fuse for continuity (this one is open; no continuity).

or being in contact with, a metal surface, since continuity will be read through the surface. Set the meter to measure continuity/resistance. A blown fuse will be open: no continuity, no resistance value displayed, and no audible indication of continuity. A good fuse will have continuity and a very low resistance. To replace a cartridge fuse, follow the steps in **SKILL DRILL 3-1**.

SAFETY TIP

Remember that even when the power at a disconnect switch is turned off, there is still voltage on the **line side** (incoming side) inside the switch enclosure. Power generally enters the switch mechanism at the top, so the terminals at the top are always energized unless its source of power, perhaps a circuit breaker, is turned off. Never replace a fuse while the disconnect switch contacts remain closed. Doing so can cause a significant and hazardous **arc**. Many disconnect switches are designed so that the door cannot be opened unless the disconnect switch handle is in the Off position, which helps to prevent this dangerous error. However, it is not foolproof.

SKILL DRILL 3-1 Replacing a Cartridge Fuse

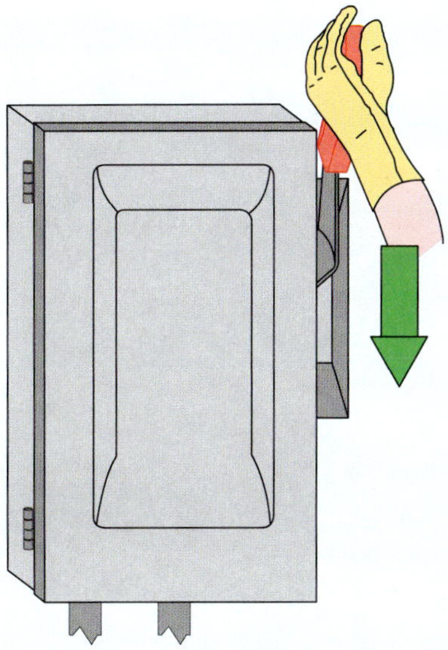

1. It is assumed that a fuse will be replaced inside of an equipment disconnect switch. Begin by turning the disconnect switch off (switch arm down). If there is no switch arm on the outside, open the door and pull the fuse block straight out to deenergize the circuit. Fuse blocks have a metal bail or molded handle so that they can be securely grasped.

SAFETY TIP

Working with electrically energized circuits is hazardous. Avoid contact with any surface that could potentially be energized. When working inside energized enclosures, maintain a stable position and, whenever possible, avoid placing both hands inside the enclosure. All work on energized equipment by unqualified individuals should be supervised by a qualified instructor or technician.

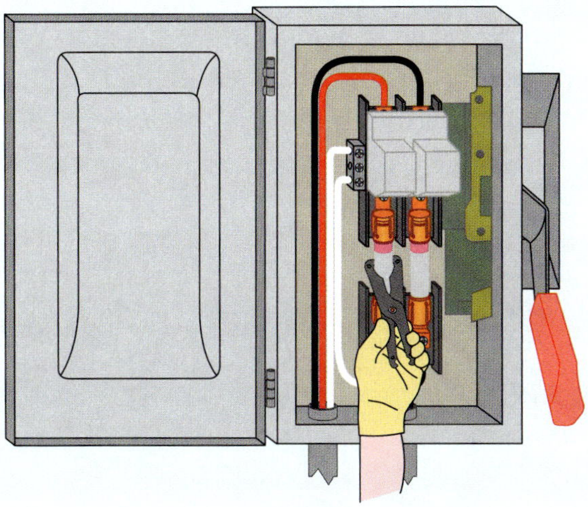

2. If working with a fuse mounted on the switch mechanism, exercise extreme caution. A hazardous level of voltage remains present at the line side (top) of the fuse holder assembly where the incoming conductors are attached. Using a fuse puller appropriate for the fuse size, grasp a fuse near one end or the other, and pull sharply. The fuse may separate from its mount at only one end. If so, use the fuse puller to grasp the fuse body close to the remaining end and pull sharply. Do not, under any circumstances, reach in and grab the fuse to pull it free by hand. Fuses mounted in a pullout block are handled in the same manner, although there is little safety concern since the assembly is completely free from the disconnect-switch mechanism.

If the fuses breaks apart or is too difficult to remove, the power supply feeding the disconnect (generally a circuit breaker in another location) must be turned off and verified to be off with a voltmeter. The disconnect switch can then be serviced safely.

Continued

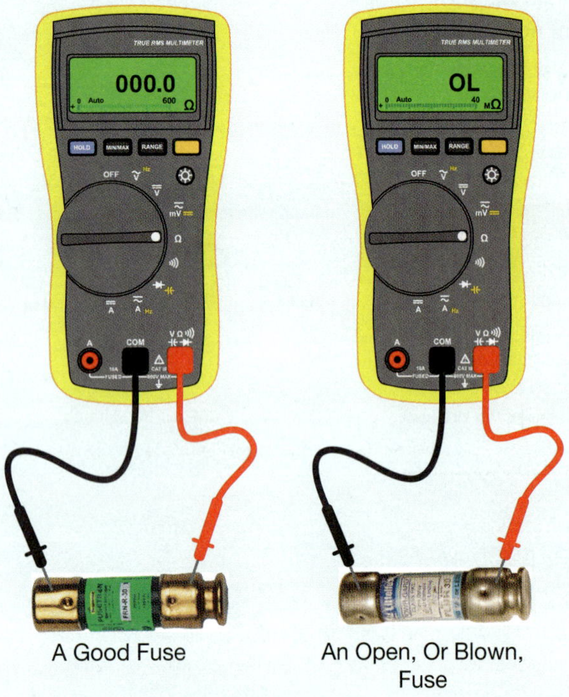

3. Test each fuse for continuity by using a multimeter.

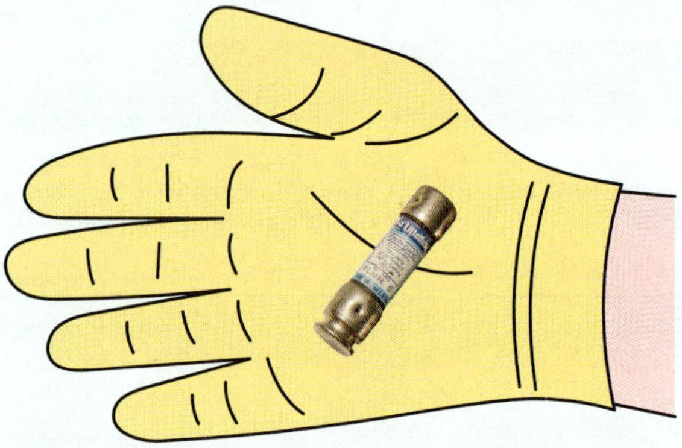

4. Check the current and voltage characteristics of the old fuse. Also record the manufacturer, manufacturer's part number, and any information regarding time-delay characteristics.

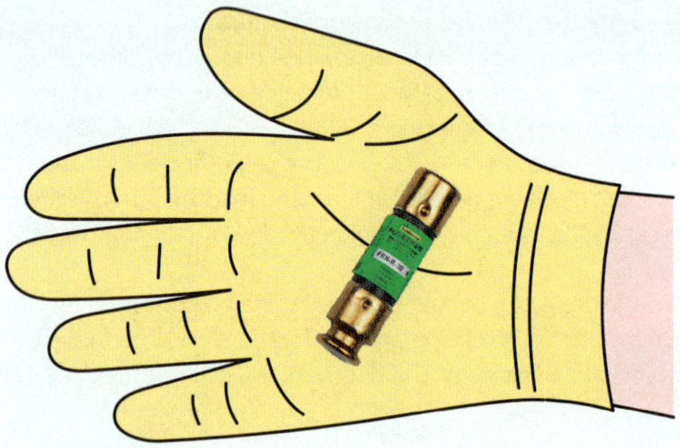

5. If necessary, cross-reference the fuse to a product from another fuse manufacturer. Obtain a fuse with the same current, voltage, and time-delay (if any) ratings.

Continued

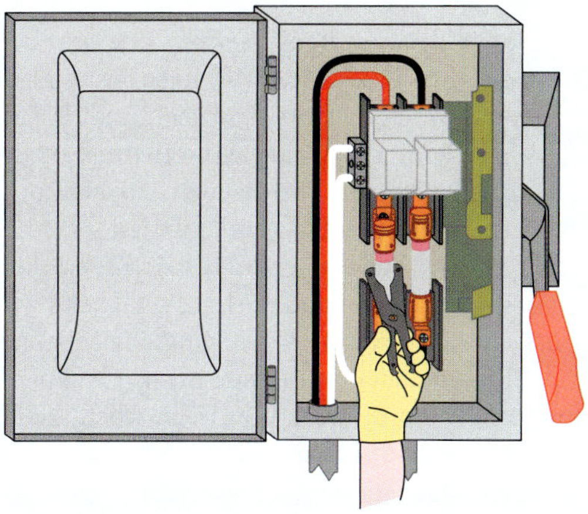

6. Insert the new fuse(s) into the fuse holder, using the same tool.

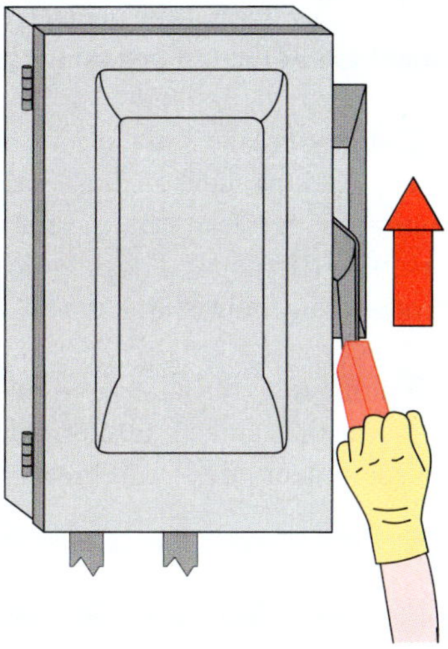

7. Once the fuses are installed, close the door of the disconnect completely and raise the switch handle to restore power. Alternatively, firmly push the fuse pullout assembly into its receptacle to restore power, then close the door.

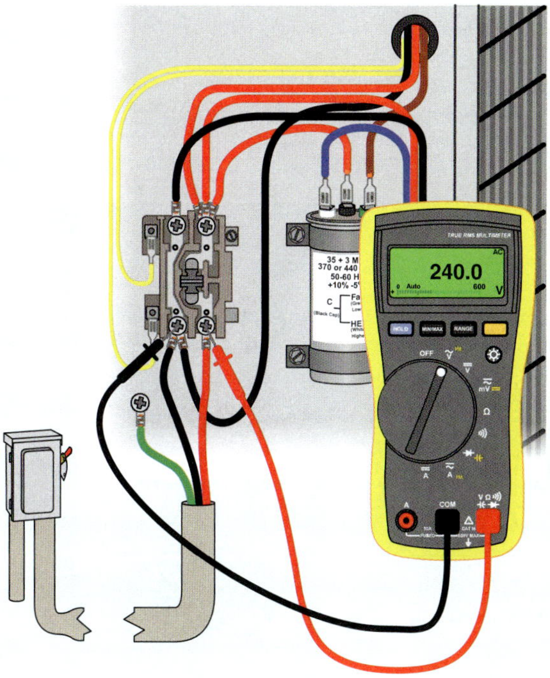

8. Remove the access door to the equipment and use a multimeter to confirm that power has been restored to the unit.

CIRCUIT BREAKERS

A circuit breaker is a switchable overcurrent safety device. Circuit breakers are most commonly found in power panels, such as the one shown in **FIGURE 3-10**. Keep in mind that induction motors experience a brief current surge when they are first energized. For that reason, circuit breakers used in these circuits have both a magnetic and a thermal method of interrupting the circuit in the event of a current overload. The magnetic element of the breaker is designed to instantly interrupt the circuit in the event of a **fault current** spike, such as one that would be caused by a short circuit. The thermal element in the circuit breaker is a **bimetal strip** made of two dissimilar metals, such as copper and steel, that will bend when heated by the current that flows through it. One side of a set of switching contacts inside the breaker is connected to the bimetal strip. This segment of the circuit breaker will permit smaller overloads to persist for a longer time and will only break the circuit when the heat builds up enough to bend the bimetal strip away from the switch contacts. **FIGURE 3-11** shows the symbol used on diagrams to represent thermal-magnetic circuit breakers.

The external switch handle on a circuit breaker (**FIGURE 3-12**) has three positions: On (closed), Off (open), and tripped. When the breaker opens because of a current overload, the switch handle moves to the center, or tripped, position. In order to reset the breaker, the switch handle must first be moved to the Off position, and then to the On position. They cannot be switched directly from their tripped position to On.

Heating, air conditioning, and refrigeration (HACR) circuit breakers are specifically listed for an electrical group that includes motor-related applications, such as those found in HVACR equipment. These breakers incorporate a built-in time delay that permits a higher current to momentarily flow through the breaker without tripping it. The time-delay period is long enough to accommodate the current surge of starting motors. In Figure 3-12, note the label information that shows the breaker is an HACR type.

Although most disconnect switches are either fused or non-fused (the latter used solely as a point of disconnect, without overcurrent protection), some disconnects accommodate circuit breakers instead (**FIGURE 3-13**). To deenergize the equipment, the circuit breaker is simply switched to OFF.

SAFETY TIP

Breakers trip and fuses blow due to an electrical overload caused by excessive current in the protected circuit. Before restoring power to the circuit, the circuit and the HVACR unit must be tested for obvious short circuits and grounds. Otherwise, if power is prematurely restored to a shorted or grounded system, serious personal injury could result due to electrical arcing or shocks. Of course, the breaker or fuse will also open again if the condition still exists.

FIGURE 3-10 Example of a distribution panel with circuit breakers.

▶ TECHNICIAN TIP

Circuit breakers have different structural characteristics. Aside from electrical characteristics, there are breakers that snap in place, whereas others require a mounting screw; breakers that are half size, full size, and have one, two, or three poles; and breakers that fit only specific panels made by a specific manufacturer. If a circuit breaker needs to be replaced, record all the information from the panel and the breaker before shopping for a new one.

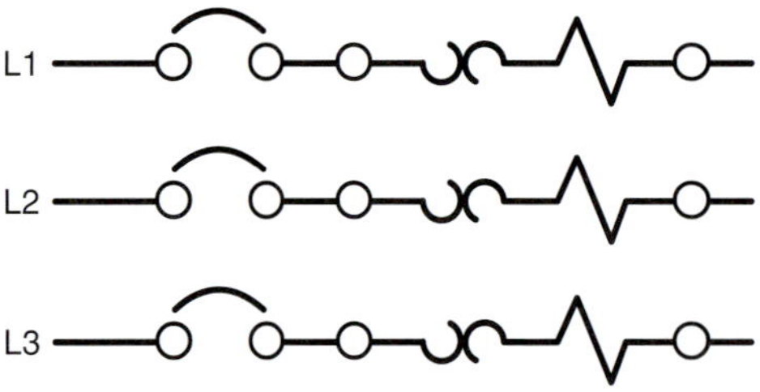

FIGURE 3-11 Symbol for a three-pole thermal-magnetic circuit breaker.

FIGURE 3-12 A double-pole circuit breaker, shown in the OFF position.

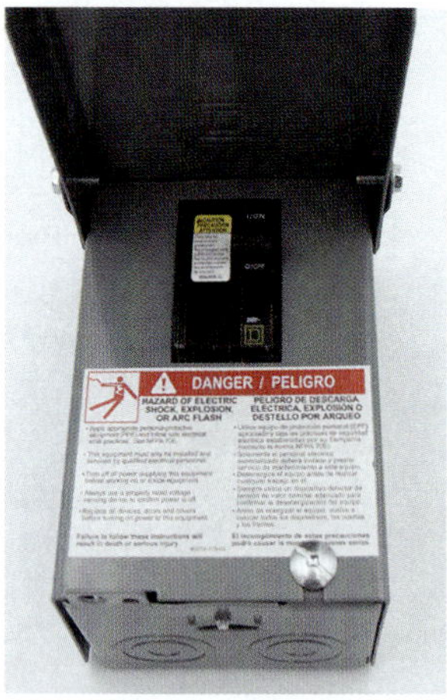

FIGURE 3-13 A disconnect switch with a circuit breaker.

▶ Single-Phase Power Distribution

K03002 Explain how single-phase power is developed and distributed.

As discussed in Chapter 1, the power distributed on the electrical grid consists of three phases of power and a neutral conductor. A pole- or pad-mounted transformer reduces the high voltage on the distribution lines to the 240 VAC single-phase voltage needed by residences and businesses. **FIGURE 3-14** shows how the 120 VAC and 240 VAC voltages are developed by the secondary side of a power transformer. These transformers are generally located close to the point of service and may also be placed underground in some locations.

The amount of current available to operate electrical loads is directly proportional to the amount of voltage available. Electrical loads such as compressor motors as well as the electric heaters used as backup and supplemental heat for heat pumps, typically require more current than a 120 VAC source can provide. Therefore, these components require the 240 VAC supply from the distribution transformer. The reason is rather simple: half the voltage would mean twice as much current for the same load, leading to much larger wire sizes.

The indoor air handling unit for a split-system heat pump or condensing unit usually requires 240 VAC, since optional electric heating elements may be added. Gas furnaces, on the other hand, generally require only 120 VAC because they do not contain any high-current loads.

The voltage supply from the distribution transformer is fed through overhead or underground lines to a utility meter and then into a structure's power panel. The meter records power usage for billing purposes. At the power panel, the incoming voltage is divided into branch circuits that supply either 120 or 240 VAC to power outlets, junction boxes, and equipment disconnects to support the building's needs. The voltage coming from the utility transformer is transferred to the structure by a three-wire system. Two of the wires are hot (energized), connected to the transformer secondary winding. Note that the third wire, a center-tapped neutral, is also grounded at the transformer. Therefore, a voltmeter will not show a voltage potential between ground and the neutral conductor, because they both have the same electrical potential. It serves as a reference point for the voltage, as well as the return path for current back to the transformer.

At the power panel (**FIGURE 3-15**), the two hot wires are connected to a **bus** that runs vertically through the panel. Each circuit breaker that provides 120 VAC is connected to one bus or the other. Double-pole breakers, which provide 240 VAC (or a nearly equal voltage, depending on the utility in a given area), are connected to both buses. Power to the buses is controlled by a main breaker, through which all the power for the structure must flow.

▶ CORE CONNECTION

Doubling the voltage provided to a load generally reduces the required current by half. From the power wheel, we know that P (power) = I (current) × E (volts). If a load such as an electric heater operates at 120 VAC with a current of 26 amps, the resulting power consumed would be 3,120 watts (3.12 kW). Based on the formula, the same amount of power is consumed (an equal amount of work is done) when the voltage is 240 VAC and the current is only 13 amps (240 × 13 = 3,120). As will be discussed in Chapter 6, the current passing through a conductor is the primary consideration in wire sizing; larger currents require larger conductors at a greater cost for both materials and labor.

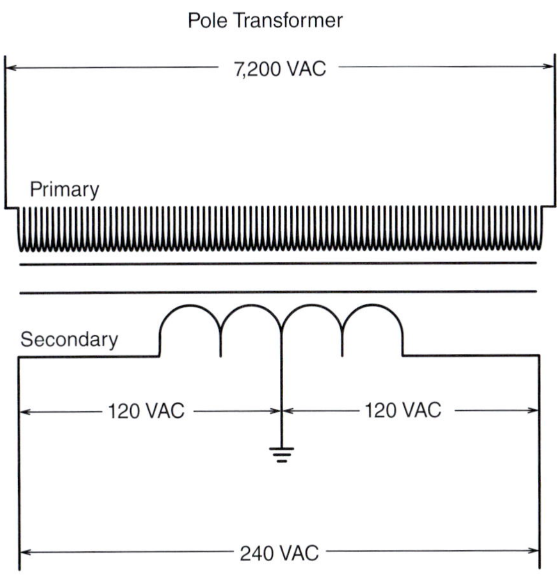

FIGURE 3-14 Single-phase power from the secondary of a transformer.

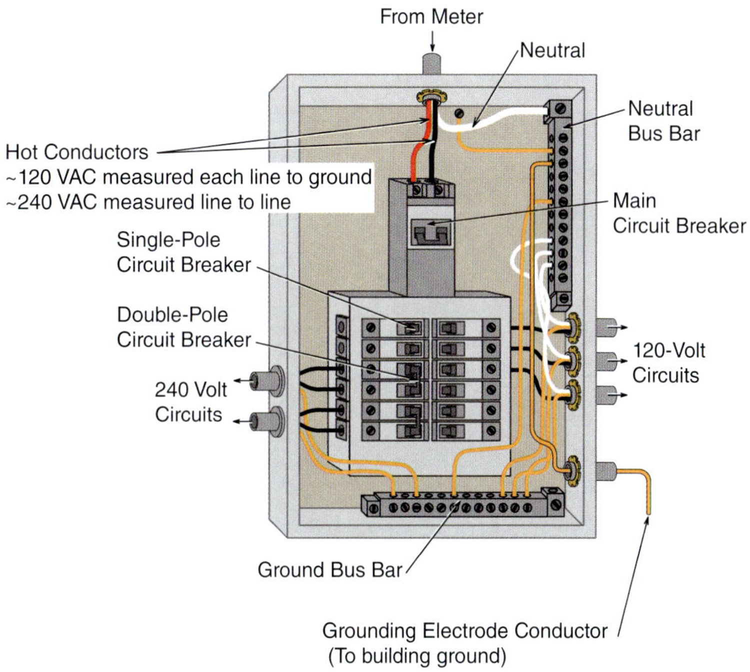

FIGURE 3-15 Interior of a power panel.

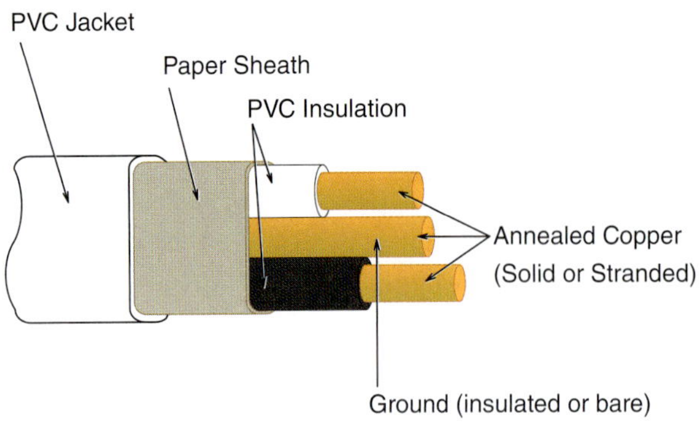

FIGURE 3-16 One example of NM cable used in residential construction.

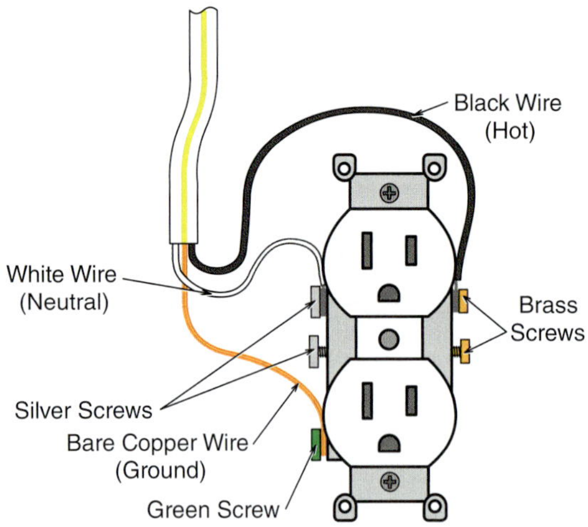

FIGURE 3-17 Standard 120-VAC receptacle wiring.

> **▶ CORE CONNECTION**
>
> A number of older homes were built without air conditioning systems, and people purchasing these homes often want the added comfort. However, many older homes were equipped with a 100-amp electrical service, which is not generally adequate to support the needs of the added air conditioning equipment. A 200-amp service is today's standard for residential construction. Larger homes often have larger services and may even be served by a three-phase power supply. Estimators need to be sure that the existing electrical service of a home will support the addition of air conditioning equipment. If not, an upgraded electrical service will be required, adding significantly to the cost of the system.

> **▶ TECHNICIAN TIP**
>
> Some power tools have a two-prong plug instead of a three-wire plug with ground. They are not required to have a ground connection, because the tools are double-insulated. That is, they have two layers of insulation between the electrical connections and the case of the tool. The two-prong plug can still be plugged into the three-prong 120-VAC receptacle.

K03003 Explain how three-phase power is distributed.

The incoming neutral wire from the distribution transformer is connected to the neutral bus in the panel. For 120 VAC branch circuits, a single hot wire, usually black, is connected to the **load side** of a single circuit breaker. It is combined with a neutral wire, usually white, from the neutral bus bar. The third wire is a ground wire, which is connected to the ground bus in the panel. The ground wire is often bare (uninsulated). In most electrical systems, the neutral bus and ground bus are actually connected together to prevent hazardous differences in potential from occurring.

Residential line-voltage wiring in the United States is often done with three-conductor, plastic-sheathed non-metallic (NM) cable, commonly referred to as Romex (**FIGURE 3-16**). The conductors in NM cable are covered with plastic insulation individually; the ground wire is usually bare, depending on the specific type and application.

The ground wire is a safety feature to direct current harmlessly to the earth in case there is a short circuit or similar event. When connected to a three-prong grounded receptacle (**FIGURE 3-17**), the neutral is connected to the silver-colored terminals and the hot wire is connected to the brass-colored terminals. The bare copper ground wire is connected to the grounding lug, which is usually green.

Notice in Figure 3-17 that the neutral prong slot on the left is larger than the hot on the right. This is a polarized receptacle. The male plug on most devices can be inserted only one way, which ensures that the grounded side of an AC supply is always connected to the correct terminal on the equipment. Of course, the receptacle must first be properly wired as shown. However, some devices have two equally sized prongs, and those devices can be plugged in without any concern for polarity.

▶ Three-Phase Power Distribution

Commercial and industrial facilities require larger HVACR equipment. These larger loads generally require three-phase power due to their significant power demands. Single-phase motors are typically limited to about 2.5 hp (horsepower) in size, although they are available up to 5 hp. Three-phase motors, on the other hand, can develop hundreds of horsepower. The most common three-phase voltages are 208–240 VAC and 480 VAC. The next step up from 480 VAC is typically 575 VAC, which is reserved primarily for industrial applications.

Three-phase power offers advantages. Three-phase induction motors do not require start assistance, which is often needed by single-phase motors. In addition, smaller, less expensive conductors can be used with three-phase power, due to the reduction in current. Finally, three-phase motors are less likely to experience significant vibration due to their construction, resulting in longer life.

Three-Phase Power Sources

Recall from Chapter 1 that the voltage produced by a power plant comes from a generator in which the poles are physically and electrically separated by 120 degrees. This produces three

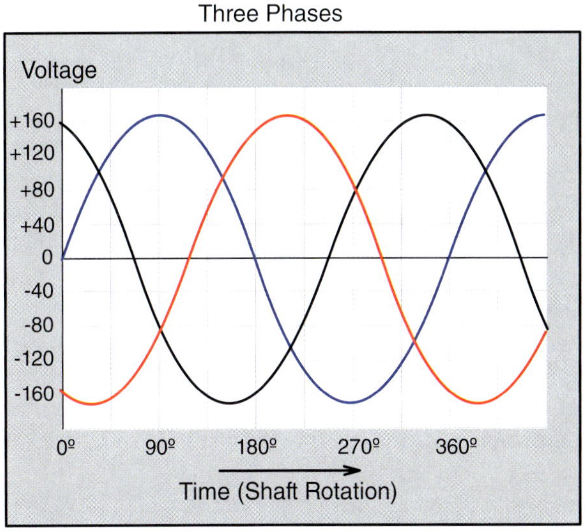

FIGURE 3-18 Three-phase sine waves.

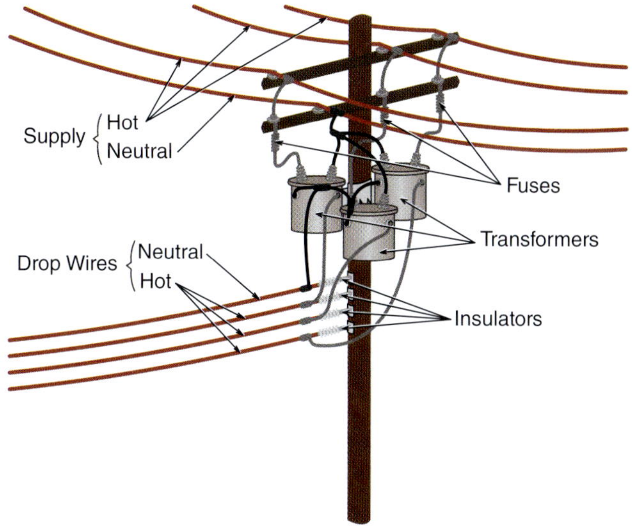

FIGURE 3-19 Three-phase pole-mounted transformers.

lines of voltage that are each 120 degrees out of phase with each other. **FIGURE 3-18** shows the associated sine-wave relationships. The step-up transformer at the power generation facility and the transformers at the substations along the way are configured the same way.

One common way to deliver three-phase voltage to utility customers is by using a three-transformer arrangement, such as the one shown in **FIGURE 3-19**. Each of the transformers provides one phase of the three-phase power supply. The transformer output can be wired in a wye (star) or delta arrangement, as shown in **FIGURE 3-20**. The four-wire wye arrangement includes a neutral. It produces a 120/208 VAC leg-to-neutral or 277/480 VAC leg-to-leg output. The 277 VAC output is commonly used for fluorescent lighting in larger commercial buildings. The delta arrangement produces 240/480 VAC, measured leg-to-leg. The wye configuration produces 208 VAC rather than 240 VAC because of the phase relationships between windings. Three-phase voltage distribution will be presented in greater detail with the discussion of transformers in Chapter 7.

Distribution of the three-phase voltage coming from the utility transformers is different from that of residential single-phase applications. **FIGURE 3-21** shows a typical arrangement for a commercial building. Power is delivered through buried cables to the bused gutter. The gutter contains bus bars that are used to distribute the voltage to the various power disconnects and utility meters. Each individual service in this example has a meter, since each individual office or store will be responsible for their own power costs. The power wiring is then routed from the meter to the circuit breaker panels indoors, where the power is distributed throughout the facility.

▶ TECHNICIAN TIP

If the term "low voltage" is used when talking to power line workers, they will think of voltages of less than 600 volts. In the world of the electrical utility worker, any voltage up to 600 volts is classified as low voltage. 600 to 1,000 volts (1 kV) is considered medium voltage. Values below 50 volts are classified as extra-low voltage. Among HVACR technicians, however, 24 VAC circuits are generally considered to be low-voltage circuits and most values above this to be high voltage. For these reasons, it is important to understand these technical differences. For example, if working near what is classified as low-voltage switchgear, it could actually be handling up to 600 volts, and perhaps higher depending on what standard is used for the classification, so special safety precautions are usually required.

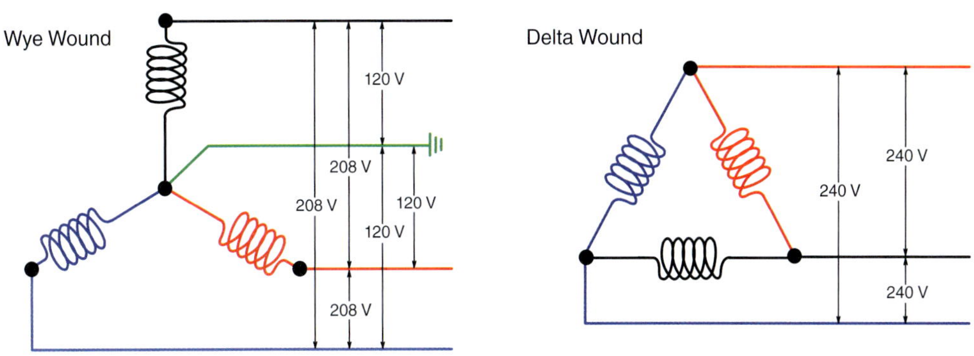

FIGURE 3-20 Three-phase transformer output arrangements.

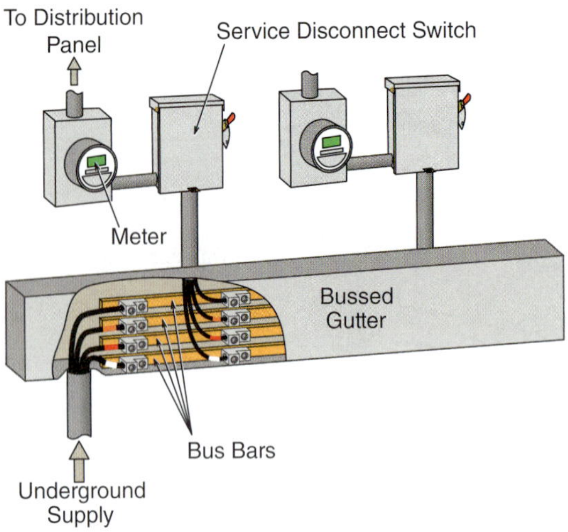

FIGURE 3-21 Typical commercial (retail) power-distribution arrangement.

FIGURE 3-22 Low-voltage switchgear.
ETAJOE/Shutterstock

Switchgear

Panelboards for lighting and other single-phase loads in commercial and industrial applications are similar to the power panels used in residential systems. They contain circuit breakers used to protect the branch circuits. However, in heavy commercial and industrial applications where high-voltage and high-current systems are common, simple circuit breakers are not up to the task of protection and control. More elaborate circuit breakers and motor-control devices capable of handling high levels of voltage and current are needed.

FIGURE 3-22 shows an example of low-voltage switchgear assemblies. Units like this are designed to house controls and circuit breakers designed to interrupt very large currents and high voltages. These low-voltage units are designed to handle voltages up to 600 VAC and 600 amps. Each cubicle can contain a circuit breaker, transformers, and a variety of control devices.

Heavy-duty circuit breakers are designed to suppress the arc that occurs between the contact points when the breaker is switched on or reset. In 120 VAC work, this arc is negligible, but in higher-voltage circuits, it is substantial. If not suppressed, the arc from a 600 VAC circuit breaker can build to an explosive level and create an **arc flash** and an **arc blast** (**FIGURE 3-23**) that can kill or injure anyone in the immediate vicinity. For that reason, circuit breakers and other components used in these circuits are housed in heavy-duty cabinets designed to withstand the blast that can come with a high-voltage arc. Special circuit breakers that use gas, air, or a vacuum to suppress the arc are often used in this type of switchgear assembly. These circuit breakers are much larger than those used in single-phase circuits.

Anyone working around switchgear like this must conform to National Fire Protection Association (NFPA) Standard 70E®, *Standard for Electrical Safety in the Workplace*, which specifies minimum approach boundaries for electrical equipment, depending on the voltage. This standard also specifies the protective gear that must be worn when working on or near the equipment, as well as warning labels that must be placed on the equipment.

SAFETY TIP

Circuit breakers that operate at high voltages and control high currents can be extremely dangerous. Large circuit breakers that roll or slide into racks are common in industrial facilities and utility substations. To reduce the hazard of arc blasts to personnel, simple remote-controlled robots (**FIGURE 3-24**) can be used to rack and set circuit breakers.

FIGURE 3-23 An arc blast.

FIGURE 3-24 A robotic helper setting a circuit breaker.

Motor-Control Centers

Motor-control centers (MCCs), such as the one shown in **FIGURE 3-25**, are used to supply voltage to motors. An MCC typically contains contactors or solid-state motor starters, along with fuses and circuit breakers, to provide short-circuit protection and overload relays for greater motor protection. On large assemblies that house a number of individual MCCs, bus bars located at the rear of the assembly are used to distribute the voltage to each MCC. The outside of the MCC door typically contains switches that can be used to shut off or manually activate a motor. In most cases, a power disconnect switch will be installed in view of the motor that the MCC supports. If so, it can be used to isolate the motor for service. When the motor is within sight of the MCC, the MCC itself can serve as the disconnecting means.

If a motor has stopped running and it is not receiving voltage from the MCC, it can be started using the manual switch on the front panel of the MCC. If the motor runs in a manual mode, then the problem is likely in the automatic control circuit.

SAFETY TIP

Keep in mind that lethal voltages exist in MCCs, panelboards, and switchgear. Minimum approach distance and PPE (personal protection equipment) requirements must be observed when working around this equipment.

FIGURE 3-25 Example of a motor-control center.

enigma/Shutterstock

▶ Wrap-Up

Summary

All electrical circuits must have overcurrent protection that will automatically open the circuit in the event that the current flowing through the circuit exceeds a safe level. Circuit breakers are a common form of overcurrent protection, but fuses are also used. Plug (screw-in) fuses are rated for currents up to 30 amps. There are many types and sizes of cartridge fuses. Whenever it is necessary to replace a fuse, it must be replaced with one that has the same characteristics. If a fuse blows more than once, this indicates a circuit or load problem that must be isolated and repaired before operation can continue.

Electrical codes require that a power disconnect be installed on or in sight of HVACR equipment, such as heat pumps, condensing units, fan coils, chillers, and pumps. The disconnect provides a positive electrical shutoff for equipment, for safety purposes. There are several types of disconnect switches commonly used. Some are simple mechanisms with conductive prongs that complete the circuit when they are installed. Others have pullout blocks that contain fuse holders. Still others have fixed fuse holders incorporated in the switch mechanism. There are also disconnects that contain circuit breakers.

Residential and light commercial operations typically require a single-phase voltage source of 120/240 VAC. The 120 VAC voltage is obtained by tapping one line and a neutral line from the three-phase voltage supplied by the electric utility. Most HVACR equipment, such as condensing units and electric heaters, require 240 VAC.

Heavy commercial and industrial facilities require a three-phase voltage source. An arrangement of three pole-mounted transformers is often used to distribute this voltage. The transformer output may be wired in a wye (star) or delta configuration, depending on the application. Because of the high voltages and current involved, the distribution equipment at an industrial site is significantly different from that used in residential and light commercial work. This equipment includes switchgear and MCCs. The equipment used in these applications must be able to withstand faults that can produce an arc blast.

Key Terms

arc An electric current that flows across an open space between two points, releasing heat and bright light in the process.

arc blast An intense explosion that occurs during an arc-flash incident.

arc flash A dangerous flash of light caused by the enormous release of thermal energy in an electric arc.

bimetal strip A narrow strip composed of two layered dissimilar metals. When heated, the metals expand at different rates, so the strip will bend or straighten as the temperature changes.

bus A thick strip of metal, often copper, that serves as a common distribution point for electricity; a bus is also referred to as a bus bar.

disconnect A safety switching device used to enable and disable power to a load, usually located adjacent to the equipment. A disconnect means must be provided on, or within sight of, HVACR equipment.

fault current Any abnormal electrical current, such as that developed in a short circuit.

line side The terminals of a device, such as a circuit breaker, connected to the voltage source.

load side The terminals of a device, such as a circuit breaker, connected to the load being powered.

rough-in The phase of construction in which electrical wiring, terminal boxes, and air conditioning ductwork are installed. Rough-in is done before walls and ceilings are covered with sheet rock or other finishing material.

Review Questions

1. The wiring and circuit protection devices used to supply power to HVACR equipment are generally installed by _____.
   - a. HVACR technicians
   - b. HVACR installers
   - c. licensed electricians
   - d. homeowners
2. The circuit protection devices used to protect against current overloads are _____.
   - a. resistors and capacitors
   - b. fuses and circuit breakers
   - c. resistors and disconnects
   - d. transformers and fuses
3. The type of plug fuse that has a socket size matched to the fuse rating is type _____.
4. A time-delay fuse is designed to delay circuit interruption to allow for _____.
   - a. momentary current surges
   - b. lightning strikes
   - c. short circuits
   - d. fault currents
5. Fuses are rated primarily in terms of voltage and _____.
6. A fuse used in a standard electrical circuit must be able to withstand _____ of the continuous current load.
   - a. 100%
   - b. 125%
   - c. 150%
   - d. 200%
7. Knife-blade cartridge fuses are available for 250 or 600 VAC applications when the current rating needed exceeds _____ amps.
8. The magnetic element in a thermal-magnetic circuit breaker is designed to react to _____ _____.
9. In order to reset a tripped circuit breaker, _____.
   - a. move the switch handle from the tripped position to the On position
   - b. move the switch handle from the tripped position to the Off position, then to the On position
   - c. replace it, since tripped circuit breakers cannot be reset safely
   - d. first remove and inspect the bimetal strip inside the circuit breaker
10. The typical allowed tolerance for voltage values is _____.
11. Which of the following statements about disconnect switches is true?
    - a. Disconnect switches never contain fuses.
    - b. All disconnect switches have external shutoff switches.
    - c. Disconnect switches never contain circuit breakers.
    - d. Some disconnect switches have removable plugs that contain fuses.
12. The leg-to-leg voltage in a single-phase power source is _____.
    - a. 120 VAC
    - b. 240 VAC
    - c. 480 VAC
    - d. 575 VAC
13. The indoor unit for a split-system heat pump or condensing unit typically requires a 240 VAC power supply because they are equipped to receive _____.
14. On a standard 120 VAC receptacle, the silver-colored terminal on the left is connected to _____.
    - a. neutral
    - b. 120 VAC
    - c. ground
    - d. three-phase
15. The gold-colored terminal on a 120 VAC receptacle is connected to the _____.
    - a. ground wire
    - b. neutral wire
    - c. hot wire
    - d. current wire
16. The three-phase voltage arrangement that produces a 240/480 VAC output is the _____ configuration.
17. A wye-configured transformer arrangement produces a leg-to-leg voltage of _____.
    - a. 120 VAC
    - b. 208 VAC
    - c. 240 VAC
    - d. 277 VAC

18. Bus bars in large motor-control center (MCC) assemblies are used to ______.
 a. serve as protective grounds
 b. distribute voltage to the individual MCCs
 c. act as overload protection devices
 d. provide control power

19. Switchgear cabinets serve as housings for ______.
 a. motors
 b. large transformers
 c. circuit breakers
 d. conduit

20. A bused gutter is an electrical assembly used to ______.
 a. distribute power
 b. drain moisture from MCCs
 c. turn power on and off
 d. house circuit breakers

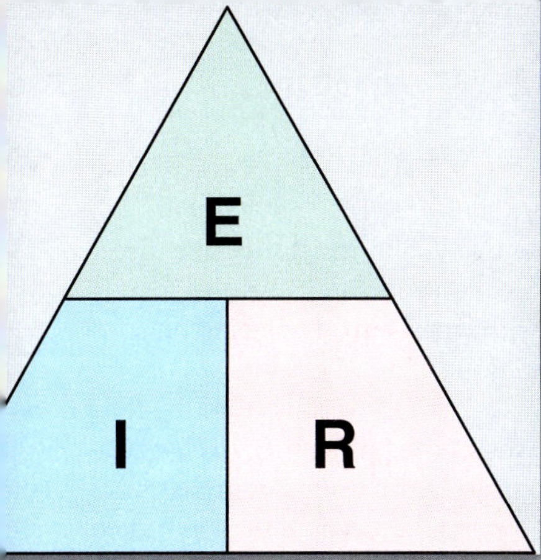

CHAPTER 4

Ohm's Law and DC Circuits

Knowledge Objectives

After reading this chapter, you will be able to:

- **K04001** Define and describe the significance of Ohm's law.
- **K04002** Define and describe a series electrical circuit.
- **K04003** Explain how Ohm's law applies to series circuits.
- **K04004** Define and describe a parallel electrical circuit.
- **K04005** Explain how Ohm's law applies to parallel circuits.
- **K04006** Explain how Ohm's law applies to series-parallel electrical circuits.

Skill Objectives

Skills required to meet the objectives of this chapter:

- **S04001** Apply Ohm's law to calculate the voltage applied to a series DC circuit.
- **S04002** Apply Ohm's law to calculate the current in a series DC circuit.
- **S04003** Apply Ohm's law to calculate the total resistance in a series DC circuit.
- **S04004** Apply Ohm's law to calculate the total resistance in a parallel DC circuit.
- **S04005** Apply Ohm's law to calculate the current in a parallel DC circuit.
- **S04006** Apply Ohm's law to calculate the voltage applied to a parallel DC circuit.

▶ Introduction

This chapter explores the relationships between voltage, current, and **resistance** in a direct current (DC) **circuit**. Whereas modern HVACR power and control circuits involve mainly alternating current (AC), DC circuits are easier to work with when introducing circuit concepts. This is because AC circuits involve complexities not found in DC circuits.

An electric current, measured in units called amperes (or amps), results from the flow of an electrical charge in a circuit. Voltage, measured in volts, is the difference in electrical potential that causes current to flow in a circuit. Current flows from points of higher electrical potential, indicated by a plus sign (+), to points of lower electrical potential, indicated by a – sign. The relationship between voltage and current is somewhat like the relationship between the pressure difference provided by a pump and the resulting flow of liquid. A pump creates a low pressure at its inlet and a higher pressure at its outlet, causing fluid to flow toward the area with the lesser pressure.

A device that uses current is known as an electrical **load**. Every load has a resistance value, measured in ohms, that affects how much current will flow in the circuit. A resistor is an example of a simple load in a DC circuit.

In 1827, Georg Ohm, a German physicist, identified the relationship between voltage, direct current, and resistance for electrical components and documented it in a simple formula known as Ohm's law. Ohm's law is based on the following relationship between the three units of measure: 1 volt of difference in potential will cause 1 amp of current to flow through a 1 ohm resistor. Mathematically, it states that voltage (E) equals current (I) multiplied by resistance (R), or:

$$E = I \times R, \text{ or } E = IR$$

▶ CORE CONNECTION

Resistance values shown on electrical diagrams commonly use the symbol Ω—the Greek capital letter omega (equivalent to the letter O in English)—to represent ohms. This was done to honor Georg Ohm. In Ohm's day, the letter E (used in formulas in this text), rather than V, was used to represent voltage. The E represented the electromotive force that caused current to flow.

Learning the concepts in this chapter will become easier as the symbols used for the different electrical quantities become more familiar. Refer to **TABLE 4-1** as needed.

This chapter is about working with electrical circuits, so we will begin with the term "circuit." An electrical circuit is a closed path for electrical current that includes a source of electrical potential (voltage), conductors to carry the current, and one or more loads that use the current to perform work. A key thing to remember is that, for an electrical circuit to function, it must start and end at the voltage source without breaks in the path. Nearly all circuits are purposely designed with controllable breaks (e.g., switches) to control the current flow as needed.

▶ Understanding Ohm's Law

K04001 Define and describe the significance of Ohm's law.

Ohm's law establishes a mathematical relationship between voltage, current, and resistance. Variations of the formula for Ohm's law can be used to ascertain one of the three values when the other two values are known. Like most equations, the arrangement can be changed to place the unknown information on one side of the equation.

There are three primary arrangements. The first form is used to ascertain the voltage by using current and resistance values, as shown here:

$$E = I \times R$$

To solve for current, the formula must be rearranged so that the I is alone on one side of the equation. In algebra, that is done by dividing both sides of the equation by R. The two Rs cancel each other, leaving the I alone on one side of the equation:

TABLE 4-1 Common Symbols for Electrical Units

Quantity	Unit (Abbreviation)	Equation Symbol
Current	Ampere (A)	I
Voltage	Volt (V)	E
Resistance	Ohm (Ω)	R

$$E \div R = I \times R \div R$$

or

$$E \div R = I, \text{ generally written as } I = E \div R$$

In algebra, the unknown quantity in an equation is typically listed before the equal sign (=)—hence the rearrangement to place the I at the beginning of the equation.

To solve for resistance, the equation is manipulated to get the R alone on one side:

$$E \div I = I \div I \times R$$

$$E \div I = R, \text{ generally written as } R = E \div I$$

Now that the three primary forms of Ohm's law have been discussed, some applications can be presented.

Determining Current

Current can be ascertained when the voltage and resistance are known. To calculate current, we use this arrangement of Ohm's law:

$$I = E \div R$$

Example: How much current will flow in a circuit with an applied-voltage source of 12 volts DC (12 VDC) and a load resistance of 200 ohms (200 Ω)?

$$I = E \div R$$

$$I = 12 \text{ VDC} \div 200\ \Omega = 0.06 \text{ A (or 60 mA)}$$

When a current value is less than 1 amp, it is common to convert it to **milliamps (mA)**, which equals one-thousandth (1/1,000th) of an amp. This is done by simply moving the decimal point three places to the right; this is the same as multiplying the number by 1,000. To convert milliamps to amps, move the decimal point three places to the left. This is important because the current value placed into the standard Ohm's law formula must be in amps, not milliamps.

Determining Voltage

The voltage applied to a circuit can be ascertained when the current and resistance are known. To ascertain voltage, use the following arrangement of Ohm's law:

$$E = I \times R$$

Example: A DC circuit with a total resistance of 50 Ω is consuming 5 A of current. What voltage is being applied to the circuit?

$$E = I \times R$$

$$E = 5 \text{ A} \times 50\ \Omega = 250 \text{ VDC}$$

Determining Resistance

The resistance of an individual circuit component can be ascertained when the **voltage drop** across it and the current flow through it are both known. The resistance of the entire circuit can be ascertained when the voltage applied to the circuit and the current are both known. To find resistance, Ohm's law is used in this arrangement:

$$R = E \div I$$

Example: What is the resistance of a load in a DC circuit that has an applied voltage of 12 VDC and draws a current of 4 A?

$$R = E \div I$$

$$R = 12 \text{ VDC} \div 4 \text{ A} = 3\ \Omega$$

▶ TECHNICIAN TIP

Resistance in AC circuits is not the same as it is in DC circuits. The AC creates an effective resistance in circuit components, other than resistors, that varies with frequency. This affects the total amount of resistance in the circuit. In AC circuits, resistance is called impedance.

▶ CORE CONNECTION

One method technicians use to remember the three primary arrangements of Ohm's law is by using a simple triangle with the three variables positioned in it, as shown in **FIGURE 4-1**. Their position identifies their relationship; the voltage (E) must be positioned at the top, above the other two. When solving for E, the I and R are multiplied, indicated by their side-by-side position in the triangle. To find I, E is divided by R, also as indicated in the E-over-R position. Finally, to solve for R, the E is divided by the I, as pictured.

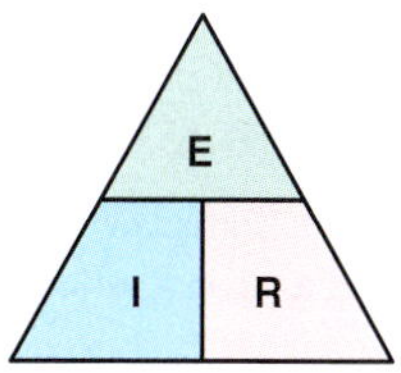

FIGURE 4-1 The Ohm's law triangle.

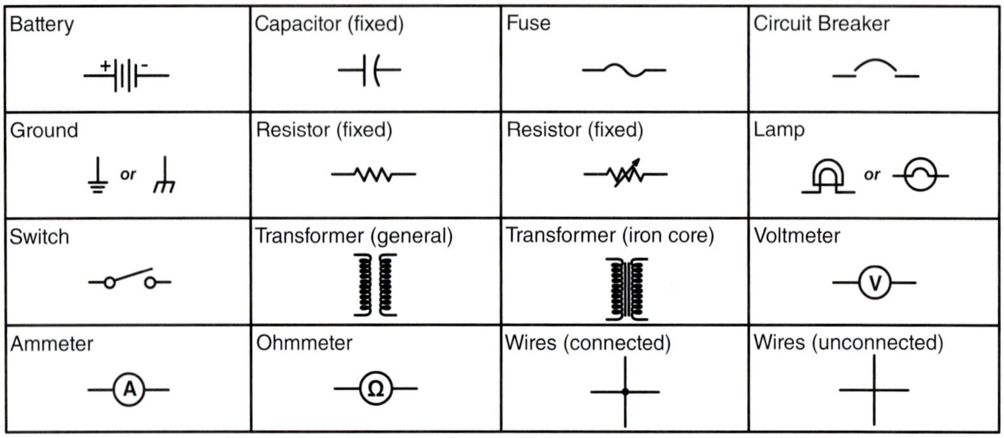

FIGURE 4-2 Common electrical circuit symbols.

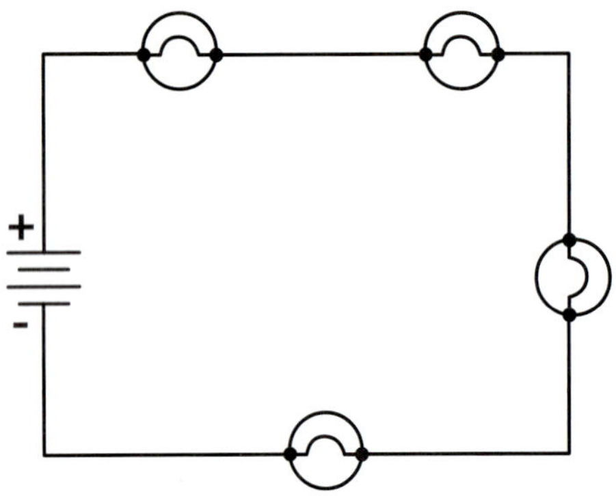

FIGURE 4-3 Example of a series circuit.

As the discussion progresses and circuit drawings are needed to present information, specific symbols will be used to represent electrical components. These symbols and more are defined in **FIGURE 4-2**.

▶ Series Circuits

K04002 Define and describe a series electrical circuit.

FIGURE 4-3 shows an example of a series DC circuit. It is called a series circuit because the loads are directly in line with each other; that is, there is only one path through which current can flow. The voltage source shown is a battery, which produces direct current. The other symbols shown are simple lamps, each of which has a resistance and represents a load.

Here are some important characteristics to remember about series circuits:

- Each load (resistance) in a series circuit creates a voltage drop proportional to its resistance. The greater the resistance, the greater the voltage drop.
- The sum of all the voltage drops across loads in the circuit equals the voltage of the source connected to the circuit.
- The current flow in a series circuit is the same through all the loads.
- The total resistance (R_T) in a series circuit is the sum of the individual resistances:

$$R_T = R_1 + R_2 + R_3 + \ldots$$

- If one of the lamps shown in the figure burns out, the circuit becomes open and all the lamps will turn off; current will cease to flow.

▶ Applying Ohm's Law to Series Circuits

K04003 Explain how Ohm's law applies to series circuits.

S04001 Apply Ohm's law to calculate the voltage applied to a series DC circuit.

S04002 Apply Ohm's law to calculate the current in a series DC circuit.

S04003 Apply Ohm's law to calculate the total resistance in a series DC circuit.

Calculating voltage, current, and resistance for series DC circuits using Ohm's law requires relatively simple math. The three forms of Ohm's law previously presented are used to find the unknown value.

Calculating Voltage in a Series Direct Current Circuit

When voltage is the unknown value, add the known values of the individual resistances in a series circuit to ascertain total resistance. The current must be known as well. Remember that this arrangement of Ohm's law is used to ascertain voltage:

$$E = I \times R$$

For example, the series circuit shown in **FIGURE 4-4** has three resistances, each with a different value. Note that the resistors are identified as R1, R2, etc. Before using Ohm's law to ascertain the voltage, calculate the total resistance of the circuit as follows:

$$R_T = R_1 + R_2 + R_3$$

$$R_T = 20\ \Omega + 30\ \Omega + 60\ \Omega = 110\ \Omega$$

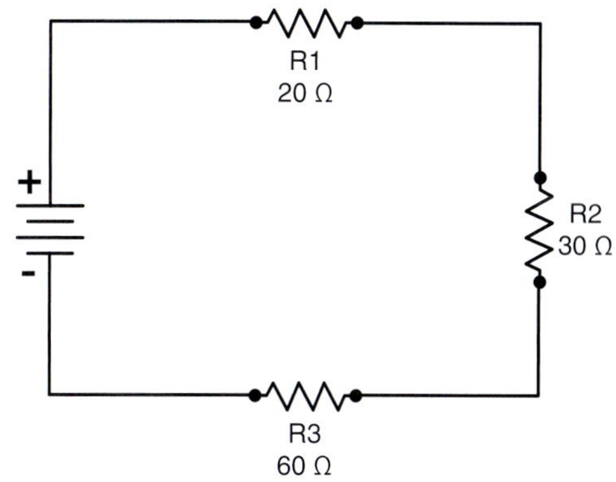

FIGURE 4-4 A series DC circuit with the resistances identified.

Assume that the current flow in the circuit has been measured at 200 mA with an ammeter. With both the total resistance and the current established, Ohm's law can now be applied to ascertain the voltage applied to the circuit. Remember that the 200 mA must be changed to the equivalent value in amps before inserting the value into the equation. Moving the decimal point three digits to the left (same as dividing the number by 1,000) results in 0.2 A, as shown in the equation below:

$$E = I \times R_T$$

$$E = 0.2\ A \times 110\ \Omega = 22\ VDC$$

To summarize, a series circuit with these resistance and current characteristics would be operating at 22 VDC. To calculate the voltage in a series circuit, follow the steps in **SKILL DRILL 4-1**.

Calculating Current in a Series Direct Current Circuit

To calculate the current in a series circuit, the total circuit resistance and the applied voltage must be known. The series circuit in **FIGURE 4-5** shows an applied voltage of 18 VDC. Again, the first step is to ascertain the total resistance, using the values shown in Figure 4-5:

$$R_T = R_1 + R_2 + R_3 + R_4$$

$$R_T = 22\ \Omega + 13\ \Omega + 20\ \Omega + 15\ \Omega = 70\ \Omega$$

With the total resistance and applied voltage established, Ohm's law can be used to calculate the current flow through the circuit:

$$I = E \div R_T$$

$$I = 18\ VDC \div 70\ \Omega = 0.257\ A, \text{ or } 257\ mA$$

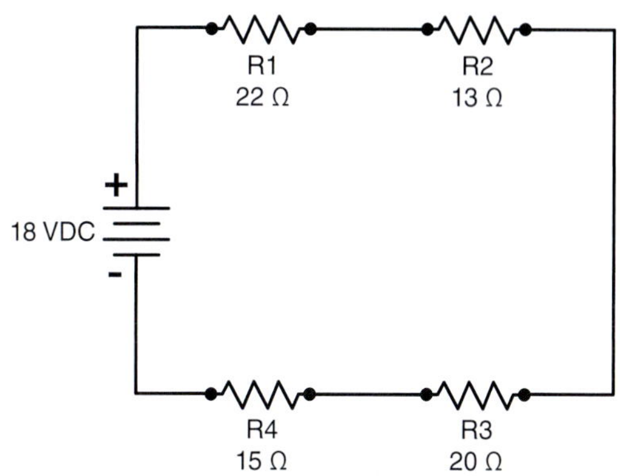

FIGURE 4-5 A series DC circuit with four resistances.

SKILL DRILL 4-1 Calculate Voltage in a Series Circuit

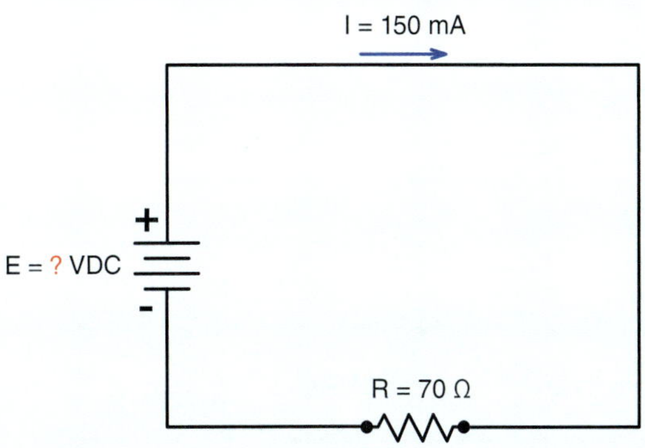

A series DC circuit is drawing a current of 150 mA and has a total resistance of 70 Ω. What voltage is being applied to the circuit?

1. Use the proper arrangement of Ohm's law, with voltage alone on one side of the equation:

$$E = I \times R$$

2. Convert milliamps to amps, and substitute the current and resistance values into the equation:

$$E = 0.150\ A \times 70\ \Omega$$

3. Multiply the two values to calculate the voltage applied to the circuit:

$$E = 10.5\ VDC$$

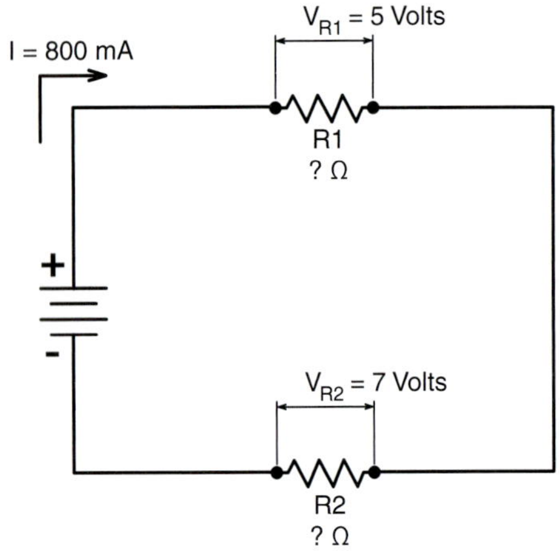

FIGURE 4-6 A series DC circuit with two loads of unknown resistance.

In this example, with the resistance and voltage characteristics presented, the current flow through the circuit would be 257 mA. To calculate the current in a series circuit, follow the steps in **SKILL DRILL 4-2**.

Calculating Resistance in a Series DC Circuit

If there is a single load in a DC circuit, its resistance can be calculated if the applied voltage and the circuit current are known. If there are multiple loads in the circuit, this method can be used to ascertain the total resistance of all the loads. However, if there is more than one load in the circuit and the resistance of an individual load needs to be ascertained, measure the voltage drop across that load as well as the current flow, and then use Ohm's law to calculate its resistance. Remember that the current flow through a series circuit is the same through each device.

In the circuit shown in **FIGURE 4-6**, assume that the voltage drop measured across R1 (shown as V_{R1}) is 5 volts, that the voltage drop across R2 (shown as V_{R2}) is 7 volts, and that the current draw of the circuit is 800 mA.

Again, remember that the milliamp value must be converted to amps before entering the current value into the equation. Otherwise, the result will be wrong. Using the resistance form of Ohm's law to ascertain the resistance of R1 yields the following:

$$R_1 = E_{R1} \div I$$

$$R_1 = 5 \text{ VDC} \div 0.8 \text{ A} = 6.25\ \Omega$$

Applying the same equation to ascertain the resistance of R2 yields the following:

$$R_2 = E_{R2} \div I$$

$$R_2 = 7 \text{ VDC} \div 0.8 \text{ A} = 8.75\ \Omega$$

Adding the two resistance values together results in the total resistance of the circuit:

$$R_T = R_1 + R_2$$

$$R_T = 6.25\ \Omega + 8.75\ \Omega = 15\ \Omega$$

To test the accuracy of the calculation, we can also use Ohm's law to ascertain the voltage applied to this circuit. This arrangement of Ohm's law is used to ascertain voltage:

$$E = I \times R_T$$

$$E = 0.8 \text{ A} \times 15\ \Omega = 12 \text{ VDC}$$

To summarize, this is a series DC circuit with two loads. One has a resistance of 6.25 Ω, and the other has a resistance of 8.75 Ω. The voltage applied to the circuit was calculated to be 12 VDC. Note that the voltage applied to the circuit, 12 VDC, is the same as the total

SKILL DRILL 4-2 Calculate Current in a Series Circuit

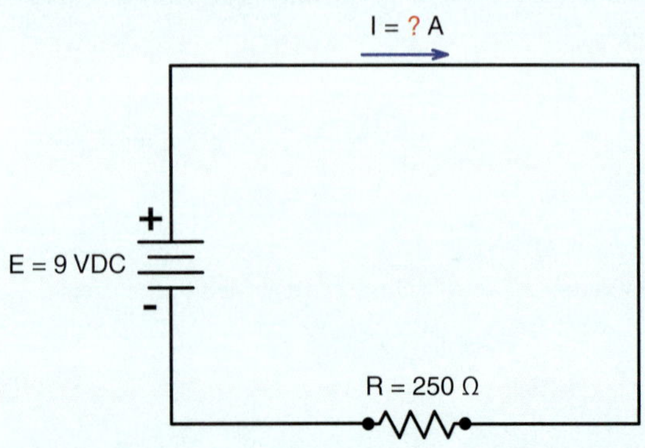

How much current will flow in a series DC circuit powered by a 9 VDC battery with one or more loads totaling 250 Ω of resistance?

1. Use the proper arrangement of Ohm's law, with current alone on one side of the equation:

$$I = E \div R$$

2. Substitute the voltage and resistance values into the equation:

$$I = 9 \text{ VDC} \div 250\ \Omega$$

3. Divide 9 by 250 to calculate the current flowing through the circuit:

$$I = 9 \text{ VDC} \div 250\ \Omega = 0.036 \text{ A, or } 36 \text{ mA}$$

voltage drop measured through the two loads individually. This confirms the accuracy of the work. All the unknown information was ascertained by measuring the voltage drop across each load and by measuring the current flow in the circuit.

An alternate way to find the resistance for this series circuit is to measure the supply voltage, which was 12 VDC, and then calculate the total current, which is 0.8 A. Use the formula for Ohm's Law to find the resistance.

$$R_T = \frac{E}{I}$$

$$R_T = \frac{12}{0.8}$$

$$R_T = 15\,\Omega$$

To calculate resistance in a series circuit, follow the steps in **SKILL DRILL 4-3**.

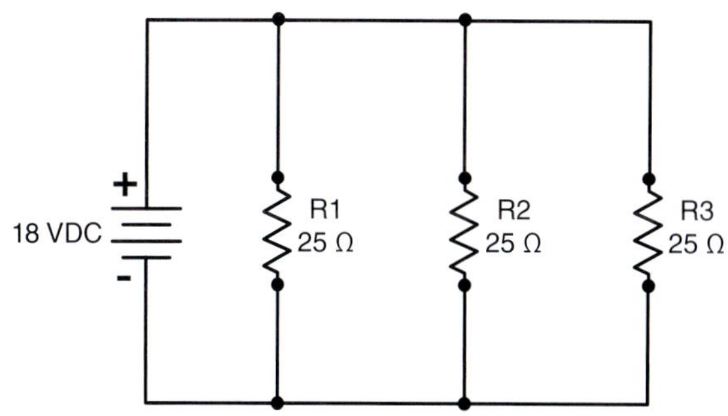

FIGURE 4-7 A parallel DC circuit.

▶ Parallel Circuits

A parallel circuit has two or more branches, each of which contains an electrical load. **FIGURE 4-7** is an example of a parallel DC circuit. The current in the circuit has multiple paths through which to flow. Unlike a series circuit, if one of the loads or load circuits opens, the remaining loads can continue to operate since there are multiple current paths. The amount of current flow through each branch of the circuit is determined by the amount of resistance of the load in the branch.

Here are some important characteristics to remember about parallel circuits:

- The voltage drop across any branch of a parallel circuit is the same as the voltage applied to the circuit. If a voltmeter is connected across any of the loads in a parallel circuit, it would read the same as the voltage applied to the entire circuit.
- The total current flowing through a parallel circuit is equal to the sum of the currents flowing through the branch circuits.
- The total resistance of a parallel circuit is less than the smallest resistance of any load in a branch, which will be proven mathematically later in the chapter.

K04004 Define and describe a parallel electrical circuit.

K04005 Explain how Ohm's law applies to parallel circuits.

S04004 Apply Ohm's law to calculate the total resistance in a parallel DC circuit.

S04005 Apply Ohm's law to calculate the current in a parallel DC circuit.

S04006 Apply Ohm's law to calculate the voltage applied to a parallel DC circuit.

▶ Applying Ohm's Law to Parallel Circuits

Ohm's law applies to parallel DC circuits as well as series circuits, but there are some differences. The primary difference is in the relationship between the individual resistance values and the total resistance of the circuit.

SKILL DRILL 4-3 Calculate Resistance in a Series Circuit

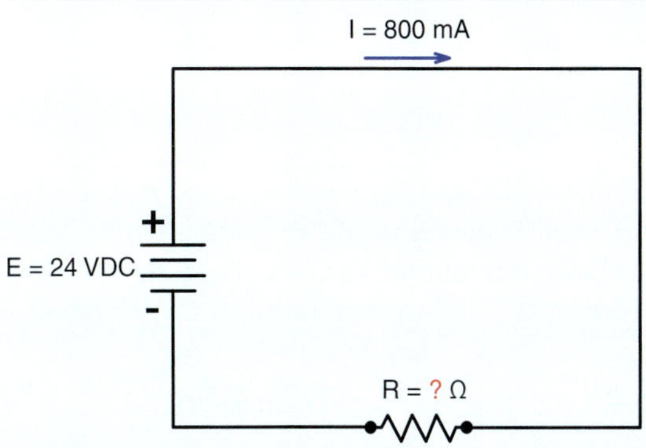

What is the resistance of a load in a series DC circuit if the voltage is 24 VDC and the current is 800 mA?

1. Use the proper arrangement of Ohm's law, with resistance alone on one side of the equation:

$$R = E \div I$$

2. Convert milliamps to amps, and substitute the voltage and current values into the equation:

$$R = 24\,\text{VDC} \div 0.8\,\text{A}$$

3. Divide 24 by 0.8 to calculate the resistance of the load in the circuit:

$$R = 24\,\text{VDC} \div 0.8\,\text{A} = 30\,\Omega$$

Calculating Total Circuit Resistance with Multiple Loads Having the Same Resistance Values

It may be challenging to grasp that the total resistance of loads in a parallel circuit is less than the smallest resistance of any load in a branch, but it will be explained in this section. Figure 4-7 contains three loads, each with the same resistance of 25 Ω. The circuit has 18 VDC applied to it, so the load in each branch will draw 0.72 A (720 mA) as demonstrated here by using Ohm's law:

$$I = E \div R$$

$$I = 18\ \text{VDC} \div 25\ \Omega = 0.72\ \text{A}$$

The total current (I_T) flowing through this entire parallel circuit can be ascertained by simply adding the three current values flowing through each branch together:

$$I_T = I_1 + I_2 + I_3$$

$$I_T = 0.72\ \text{A} + 0.72\ \text{A} + 0.72\ \text{A} = 2.16\ \text{A}$$

With the voltage and total current flow through the circuit known, the total resistance of the circuit can be calculated by using Ohm's law:

$$R_T = E \div I_T$$

$$R_T = 18\ \text{VDC} \div 2.16\ \text{A} = 8.33\ \Omega$$

The total circuit resistance in this case is one-third of the resistance of any of the three loads, demonstrating the principle that the total resistance in a parallel circuit is less than the resistance in the branch with the least resistance.

Alternate Formula for Calculating Parallel Resistance of the Same Value

Two or more parallel resistors that have the same value can be easily calculated; simply divide the number of resistors into the common resistance value. For example, two 10 Ω parallel resistors will have a total resistance of 5 Ω. Three 10 Ω parallel resistors will be 3.33 Ω. This formula, however, works only with parallel resistors that have the same resistance.

Calculating Total Circuit Resistance with Multiple Loads of Different Resistance Values

As demonstrated above, the total resistance in a parallel circuit is easy to ascertain if all the branch circuit resistances are the same. There is a little more work to do, however, when they are different. **FIGURE 4-8** shows an example of such a circuit. A more complex formula is needed to ascertain the total resistance in parallel circuits that have loads with different resistance values. Of course, it works equally well when the resistance values are the same:

$$R = \frac{1}{\frac{1}{R_1} + \frac{1}{R_2} + \frac{1}{R_3}}$$

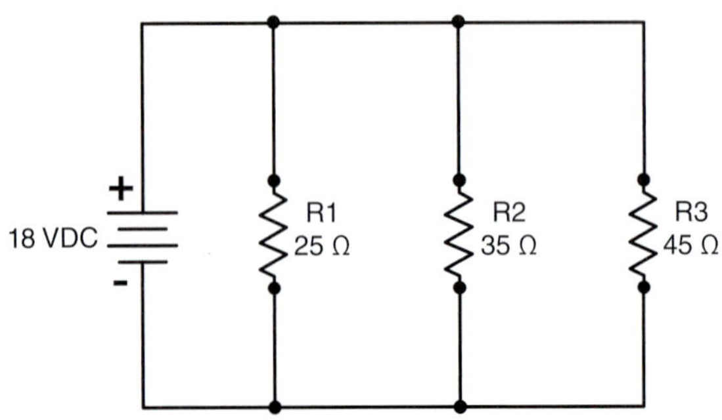

FIGURE 4-8 A parallel circuit with three loads that have different resistance values.

It might be helpful to see this equation written in text form: the total resistance is equal to 1 divided by the sum of the reciprocals of the individual resistance values. To work this unusual equation, begin by finding the **reciprocal** for each different resistance value shown in the denominator above as follows:

Reciprocal of R_1 $(1/R_1) = 1 \div 25\ \Omega = 0.040\ \Omega$

Reciprocal of R_2 $(1/R_2) = 1 \div 35\ \Omega = 0.029\ \Omega$

Reciprocal of R_3 $(1/R_3) = 1 \div 45\ \Omega = 0.022\ \Omega$

The three values are then added together, forming the denominator of the original equation above:

$$0.040\ \Omega + 0.029\ \Omega + 0.022\ \Omega = 0.091\ \Omega$$

The equation can now be rewritten in a much simpler form and the total resistance calculated:

$$R_T = 1 \div 0.091\ \Omega = 10.99\ \Omega$$

Note that the result is smaller than the smallest resistance value found in the individual loads. As noted earlier, the total resistance will always be smaller than the smallest resistance value in the circuit.

Calculating Voltage and Current in Parallel Circuits

Using Ohm's law to calculate current and voltage in parallel DC circuits is essentially the same as its use in series DC circuits. The current flow through each load can easily be calculated by applying Ohm's law, using the circuit voltage and the resistance of the single load in that branch. Further, the total current flow in the circuit can be calculated by first calculating the current flow through each individual load and then adding them together.

However, to directly ascertain the total current flow in a circuit, remember to work with the total resistance of the loads in the circuit. This slightly changes the appearance of the Ohm's law equation to show the resistance represents the total resistance:

$$I_T = E \div R_T$$

Calculating the voltage applied to a parallel DC circuit also relies on the use of the total current flow in the circuit, as shown here:

$$E = I_T \times R_T$$

To calculate the total resistance of loads in a parallel circuit, follow the steps in **SKILL DRILL 4-4**.

SKILL DRILL 4-4 Calculate the Total Resistance of Loads in a Parallel Circuit

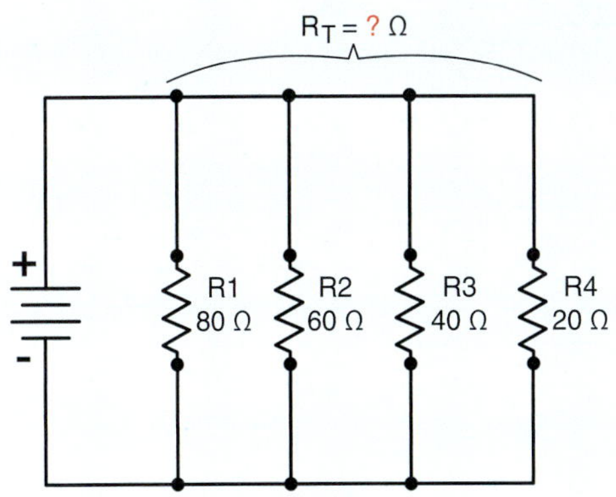

Refer to the parallel DC circuit shown in the figure. There are four loads in this circuit, each with a different resistance value. Calculate the total resistance of the loads in this parallel circuit.

1. Remember that this is not a calculation that uses Ohm's law; the equation shown here is used to calculate the total resistance of parallel circuits (R_T):

$$R_T = \frac{1}{\frac{1}{R_1} + \frac{1}{R_2} + \frac{1}{R_3} + \frac{1}{R_4}}$$

2. Ascertain the reciprocal of each resistance value.

Reciprocal of $R_1 = 1 \div 80 = 0.0125\ \Omega$
Reciprocal of $R_2 = 1 \div 60 = 0.0167\ \Omega$
Reciprocal of $R_3 = 1 \div 40 = 0.0250\ \Omega$
Reciprocal of $R_4 = 1 \div 20 = 0.0500\ \Omega$

3. Add the reciprocals together:

$$0.0125\ \Omega + 0.0167\ \Omega + 0.0250\ \Omega + 0.050\ \Omega = 0.1042\ \Omega$$

4. Rewrite the equation for clarity:

$$R_T = 1 \div 0.1042\ \Omega$$

5. Complete the division to calculate the total resistance of the circuit:

$$R_T = 1 \div 0.1042\ \Omega = 9.597\ \Omega$$

Unless extreme precision is necessary (which is unlikely in HVACR applications for field technicians), the result can be rounded to 9.6 Ω.

To calculate the total current flow in a parallel circuit, follow the steps in **SKILL DRILL 4-5**. To calculate the voltage in a parallel circuit, follow the steps in **SKILL DRILL 4-6**.

▶ Series-Parallel Circuits

K04006 Explain how Ohm's law applies to series-parallel electrical circuits.

A series-parallel circuit, also referred to as a combination circuit, is one that contains loads in series as well as loads in parallel. **FIGURE 4-9** shows an example of such a circuit. Notice that R2 and R3 form a parallel circuit in series with R1 and R4. Solving for total resistance in such a circuit requires first reducing the two parallel resistances to a single value—the same process used to ascertain total resistance in parallel circuits in this chapter.

SKILL DRILL 4-5 Calculate the Total Current Flow in a Parallel Circuit

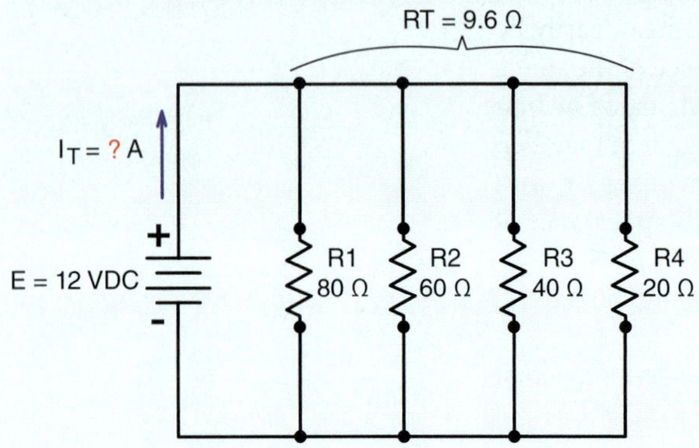

This exercise is based on the same figure used in the previous exercise: Figure SD4-4-1. The voltage applied to the circuit is 12 VDC, and the total resistance of the circuit was found to be 9.6 Ω. Calculate the total current flow in the circuit.

1. Use the proper form of Ohm's law, with current alone on one side of the equation. Remember to work with the total resistance of the circuit to find the total current flow in the circuit directly:

$$I_T = E \div R_T$$

2. Substitute the known values of voltage and total resistance into the equation:

$$I_T = 12\text{ VDC} \div 9.6\ \Omega$$

3. Complete the division problem to ascertain the current flow:

$$I_T = 12\text{ VDC} \div 9.6\ \Omega = 1.25\text{ A}$$

SKILL DRILL 4-6 Calculating the Voltage in a Parallel Circuit

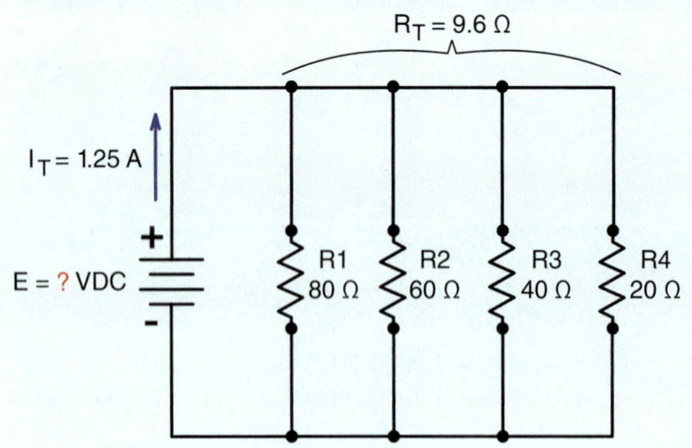

This exercise will provide the opportunity to complete a voltage calculation by using Ohm's law and will serve to confirm the accuracy of the math in previous Skill Drills. Based on a circuit that has a total current flow of 1.25 A and a total resistance of 9.6 Ω, what is the applied voltage?

1. Use the proper form of Ohm's law, with voltage alone on one side of the equation. Remember to work with the total resistance and the total current flow of the circuit in order to find the voltage applied to the circuit:

$$E = I_T \times R_T$$

2. Substitute the known values of total current flow and total resistance into the equation:

$$E = 1.25\text{ A} \times 9.6\ \Omega$$

3. Complete the multiplication problem to ascertain the applied voltage:

$$E = 1.25\text{ A} \times 9.6\ \Omega = 12\text{ VDC}$$

This matches the applied-voltage value shown in Figure SD4-1-1, confirming the accuracy of the math in the previous Skill Drills.

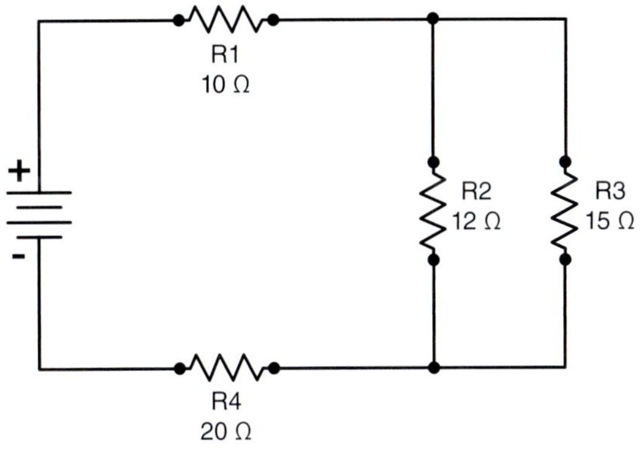

FIGURE 4-9 A series-parallel circuit.

This value can then be added to the resistance value of the other loads in the series portion of the circuit, R1 and R4. Stated another way, the parallel portion of the circuit is treated as a parallel circuit should be in terms of calculating total resistance and current flow. The resulting resistance is then handled as another load value in the series circuit when adding those values together.

▶ Wrap-Up

Summary

Ohm's law provides a simple formula designed to answer questions about the operating characteristics of an electrical circuit. This chapter has focused on the application of Ohm's law in DC circuits. Although it also applies to AC circuits, it is applied differently, and therefore, the two applications are best presented separately.

When testing or troubleshooting electrical circuits, it is sometimes necessary to ascertain how much current should be flowing through a circuit or how much resistance a circuit offers to the flow of current. Loads can be connected in series or in parallel arrangements. Ohm's law is applied to both types of DC circuits, but they behave differently, and therefore, Ohm's law is applied a bit differently. For example, the applied voltage across the load in any branch of a parallel circuit is equal to the applied voltage. In a series circuit, the applied voltage equals the voltage drop across each of the loads. Although HVACR technicians typically spend most of their time working with AC power, it is also important to understand the basic principles of DC power to prepare for advanced work with electronic controls and building management systems.

Key Terms

circuit A closed electrical path that includes one or more components or a portion of the circuit that consumes electric power.

load Any electrical device that performs work and consumes energy in the process. Common loads in DC circuits include resistors and capacitors.

milliamps (mA) A unit of measure for current equaling 1/1,000 of an amp or 0.001 A.

reciprocal A mathematical value obtained by dividing the number one by another value.

resistance The characteristic of an electrical device that causes it to impede the flow of electrical current.

voltage drop The amount of voltage loss that occurs in all or part of a circuit due to resistance.

Review Questions

1. Does the following statement apply to series or parallel circuits? "The total resistance in the circuit is the sum of the individual resistances."
2. Current is expressed in _____.
   a. volts
   b. amps
   c. ohms
   d. megawatts
3. Convert 450 milliamps to amps.
   a. 0.00450 A
   b. 0.0450 A
   c. 0.450 A
   d. 4.50 A
4. In a series circuit containing resistances totaling 90 Ω and an applied voltage of 12 VDC, the current would be _____ milliamps.

5. A series circuit contains resistances of 30 Ω and 45 Ω. The current flow in the circuit is measured at 120 mA. The voltage applied to the circuit is _____ VDC.

6. You are shown a parallel DC circuit, and there are three loads shown in the circuit. Which of the following statements related to the total circuit resistance is true?

a. The total circuit resistance is larger than the sum of their resistance.

b. The total circuit resistance is less than the smallest of the three loads.

c. The total circuit resistance is larger than the average resistance of the three loads.

d. The total circuit resistance lies between the smallest resistance and largest resistance.

7. Resistance is measured in _____.

Questions 8–11 rely on **FIGURE 4-10.**

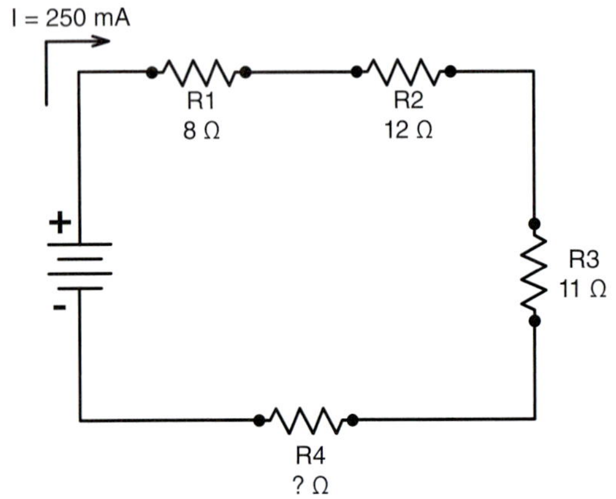

FIGURE 4-10 Example of a series DC circuit.

8. If 18 VDC is applied to this series circuit and the current draw is 250 mA, what is the resistance value of the R4 resistor?

9. What is the voltage drop across the R2 resistor if the voltage remains at 18 VDC? Round your answer to the nearest volt.

10. What would the current flow through the circuit be if R4 were 30 Ω?

11. What voltage is being applied to the circuit if the current is measured at 300 mA and the circuit has a total resistance of 40 Ω?

12. In a parallel DC circuit, the voltage drop across any branch is _____.

a. dependent on the resistance in each branch

b. dependent on the current flow through each branch

c. the same as the voltage applied to the circuit

d. greater than the total resistance of the circuit

13. Which of the following is a characteristic of a parallel DC circuit?

a. The total current is equal to the sum of the currents flowing through each branch.

b. Voltage can be ascertained by adding the voltage drops across each branch together.

c. The applied voltage from the source changes if a resistor is changed.

d. Total circuit resistance is the sum of all the branch resistances.

14. Which of these two formulas is used to calculate total resistance in a parallel DC circuit with three resistances?

a. $R_T = R_1 + R_2 + R_3$

b.

$$R_T = \frac{1}{\frac{1}{R_1} + \frac{1}{R_2} + \frac{1}{R_3}}$$

Questions 15–19 rely on **FIGURE 4-11.**

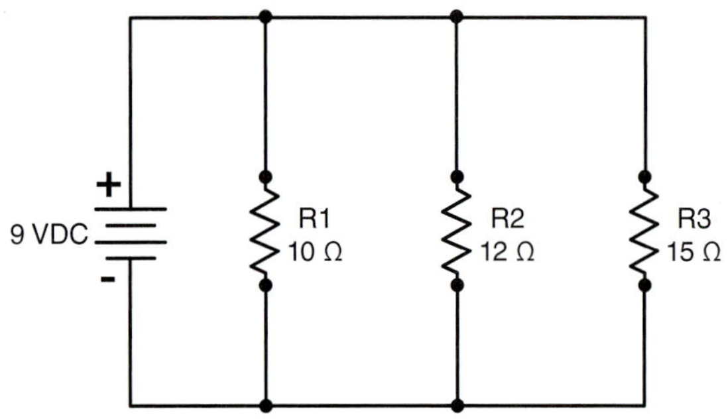

FIGURE 4-11 Example of a parallel DC circuit.

15. The total resistance in this circuit is _____. Round your interim results to the thousands position (0.000) and round your final answer to the nearest ohm.

16. The total current flowing through the circuit is _____.

17. The voltage drop across R3 is 5 volts.

a. True

b. False

18. The current flowing through the branch containing R2 is _____.

19. If a 12 VDC battery is supplying the circuit, the current would _____.

a. increase

b. decrease

20. A series-parallel circuit may also be referred to as a(n) _____.

a. reciprocal circuit

b. combination circuit

c. unbalanced circuit

d. impractical circuit

CHAPTER 5

Electrical Safety

Knowledge Objectives

After reading this chapter, you will be able to:

- **K05001** Describe the hazard of electric shock and identify related factors.
- **K05002** Identify OSHA's general electrical safety requirements.
- **K05003** Describe the NFPA and the contents of NFPA 70E.
- **K05004** Identify general electrical emergency responses and procedures.

Skill Objectives

Skills required to meet the objectives of this chapter:

- **S05001** Follow an established set of procedures to lock out and tag an HVACR unit for service and then remove the lock and restore power.

▶ Introduction

Electrocution is a leading cause of death in the workplace. Likewise, electrical shocks and severe burns that occur from making contact with live electrical circuits are among the most serious hazards to which HVACR technicians are regularly exposed. However, the presence of electricity is often not apparent. Knowledge and common sense are essential to recognize the presence of electricity in a work environment. Technicians must follow safe work practices and stay alert for potential hazards every moment they are on the job.

This chapter describes the nature and severity of electric shock hazards, reviews the general electrical safety regulations of the Occupational Safety and Health Administration (OSHA), and describes National Fire Protection Association (NFPA) standards for electrical installations and electrical work practices. It also discusses how to respond to electrical incidents.

In no way, however, does this chapter represent all the knowledge that a technician needs regarding electrical safety. This is a topic that should be studied throughout a technician's career. All users are encouraged to study the electrical and other safety requirements found in the OSHA standards, as well as other industry publications, at length. Although technical knowledge is required to be a successful HVACR technician, it is secondary to a technician's knowledge of safety. No technician can demonstrate technical prowess while injured.

▶ The Electrical Shock Hazard

K05001 Describe the hazard of electric shock and identify related factors.

Electrical current ideally flows in a closed circuit along a defined route through a **conductor**, such as copper, steel, silver, or aluminum. However, the current will travel along any path in which the voltage can overcome the resistance.

The human body is also a conductor of electricity, and the body's conductivity greatly increases when the skin is wet or broken open. If a human body in contact with the ground or another point in an electrical circuit contacts an electrically energized point, the body becomes a part of the circuit and the person sustains an electrical shock.

Electrical Shock and the Body

Any time contact between energized components carrying a load is made or broken, an **arc**, or spark, is produced. An electric arc between components at different electrical potentials can also occur due to poor electrical contact or failed wire insulation. However, electric arcs are also part of the normal operation of nearly all electrical equipment. For example, when electrical contacts come together to energize a significant load such as a compressor, an arc occurs (**FIGURE 5-1**). How long the arc lasts, the voltage, and the amount of current involved determine the intensity of the arc and the damage that may ensue.

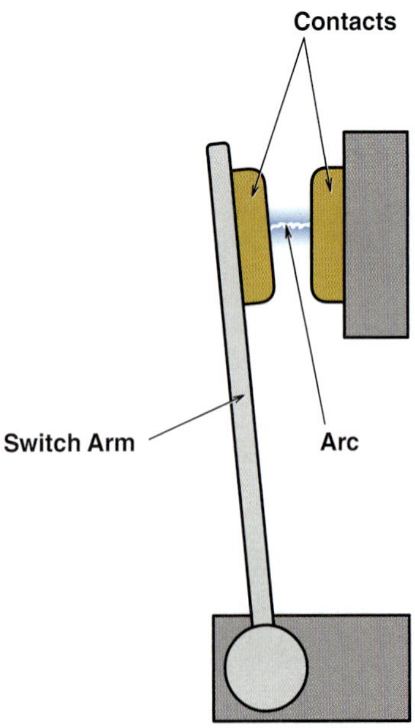

FIGURE 5-1 Normal arcing occurs when load or switch contacts come together.

Significant electrical hazards in the HVACR work environment include electrical shock; an **arc fault**, which may produce an **arc flash** and an **arc blast**; and burns caused by electrical shocks and arc faults (**FIGURE 5-2**). Assume that all electrical equipment presents potential shock and arc-fault hazards unless it has been verified otherwise.

An arc fault is a high-energy discharge of electricity between two or more conductors. It can happen when an electrical current crosses the air gap between two energized sources that have different potentials. It can also occur between an energized electrical circuit and ground. The resulting current is called arc-fault current.

An arc flash is created when a small arc is not interrupted and grows into a larger arc. The events that accompany an arc flash typically include intense light, fire, an explosion (arc blast), temperatures up to 35,000°F (19,427°C), toxic fumes, and shockwaves that damage anything near the center of the arc flash. A worker can be blown of their feet, even sent head over heels, by the pressure of an arc blast—enough pressure to blow structures apart and to fling equipment and shrapnel over great distances. The intense pressure wave of an arc blast can also deafen people and create vaporized metal that can be inhaled by those nearby.

Electrical Burns

Burns are a common shock-related injury. They are serious injuries that require immediate attention. Burns are classified as first, second, or third degree, based on the extent of cell damage. First-degree burns are superficial, damaging only the outer layer of the skin. Second-degree burns penetrate beyond the outer layer of skin to the inner layer. Third-degree burns, the most serious, penetrate all the skin layers to reach the muscles and fatty tissues inside, causing cell death. The three categories of burns associated with electrical incidents are electrical burns, arc burns, and thermal contact burns. Each of them can result in first-, second-, or third-degree burns.

Electrical burns are caused by electric current flowing through a body's tissues or bones. The heat generated by the current is very damaging. What may initially appear to be a small surface wound could actually be associated with significant internal damage. Tissue inside the body that is in the current path may be damaged or destroyed, which causes the skin to die from the inside out over time.

FIGURE 5-2 An arc blast.

Arc burns, as the name suggests, can result when electrical arcs occur. Even people who are several feet from the source of a major arc can receive severe or even fatal burns. Thermal contact burns are often caused by contact with objects thrown during an arc blast. In addition, molten metal droplets from an arc blast can cause thermal contact burns as they land. More hospital admissions resulting from electrical accidents are for burns than for injuries directly related to the electrical shock itself.

Electrocution results when a person dies from the current that passes through the body when the person receives an electric shock. Statistics issued by the National Institute of Safety and Health (NIOSH) consistently report electrocution as one of the most common fatal workplace injuries.

Factors Related to Shock Severity

If the body is in contact with the ground or grounded objects and contact with an energized part is made, an electrical shock will result. One way an electrical shock can be sustained is if an internal fault causes the cases or enclosures of tools, appliances, or electrical equipment to become energized. These items are not expected to be energized, so physical contact is made with them without thought. The greatest damage is caused when current flows through or near the body's nerve centers or a vital organ, such as the heart.

A severe electrical shock can cause much more damage than can be seen externally. Shock victims may suffer internal injuries, such as hemorrhages and tissue, nerve, and muscle damage. In addition, electrical shock is often merely the first in a chain of destructive events. A shock victim's most serious injury may actually be from a fall, cuts, burns, or broken bones that follow the shock. For all these reasons, it is critical to seek medical attention when even a minor electrical shock is sustained.

Voltage (volts), body resistance (ohms), and current (amps) are key factors in the severity of an electrical shock. Simply put, the amount of electrical current that flows through the body goes up when resistance goes down.

> **SAFETY TIP**
>
> To prevent electric shock, stay well insulated from the ground and never form part of a closed circuit. For instance, to prevent a shock when working on an electrical circuit, place one end of a conductor on the part of the circuit being worked on and the other end of the conductor onto a ground point. Because the conductor has virtually no resistance, current would then flow through the conductor to ground if something goes wrong. This is also the final step in assuring that a circuit is 100% deenergized—by shorting it to ground.

Voltage

The higher the voltage, the greater the current, and the greater the chance for electrocution to occur. High voltage (above 480 VAC) is nearly 10 times as likely to kill a person as lower voltages. However, HVACR technicians spend most of their time working on or near voltages that range from 120 VAC to 240 VAC. It is not surprising then, in the HVACR field, shocks and even electrocutions most frequently occur at these voltages. Clearly, the voltages that HVACR technicians work with are not harmless.

Body Resistance

During an electrical shock, current enters the body at one point and exits at another, often leaving corresponding entry and exit wounds. **FIGURE 5-3** illustrates resistance values of various current paths through the human body.

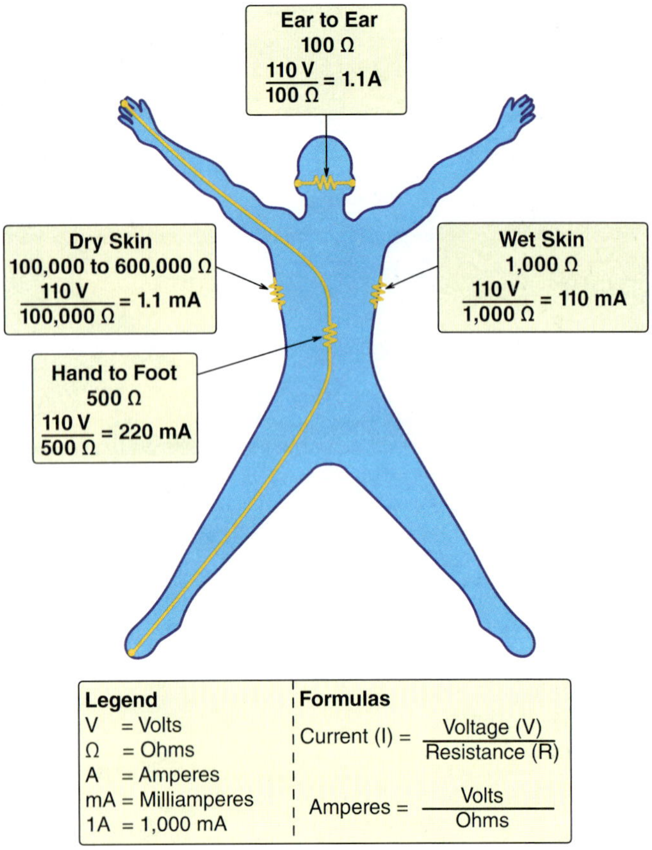

FIGURE 5-3 Resistance values of various paths through the human body.

Electrical shock is damaging to the skin, the heart, and other muscles. For the skin, the damage caused varies depending on the skin's resistance, which varies from one individual to another. The outer layer of dry skin may have a resistance of several thousand ohms. Resistance greatly decreases if the skin becomes moist or if the outer layer of the skin is pierced by a cut or abrasion. Resistance also varies significantly at different parts of the body. For instance, an individual may have high-resistance skin on the hands but low-resistance skin on the backs of the legs.

Muscles throughout the body are controlled by electrical impulses delivered by nerves. Electric shock can cause a person to lose muscular control, often leaving the person unable to let go of an electrical conductor. Electric shock can also produce involuntary muscle reactions that can cause bruises, fractures, and even death as the result of collisions or falls.

Current

Normally, the heart muscle contracts and pumps blood in response to a very low-level electrical signal that the heart receives from the body. If an abnormal electrical signal, such as current from an electrical shock, reaches the heart, the low-level heartbeat signals from the body can become ineffective. The heart then begins twitching irregularly. When the normal rhythm of the heart is disrupted, it is referred to as **fibrillation**.

Cardiopulmonary resuscitation (CPR) can be used to keep blood and oxygen flowing to the body, but unless the normal heartbeat is restored by using **defibrillation** equipment (paddles), a shock victim can die. In some cases, electrical shock can immediately stop the heart altogether.

The severity of the shock sustained depends primarily on the following:

- how much current flows through the body
- the path that the current takes through the body
- how long the body remains in the circuit.

The general effects of electric current passing through the human body are summarized in **TABLE 5-1**.

TABLE 5-1 Effects of Current Passing Through the Human Body

Current	Reaction
Below 1 Milliampere	Generally not perceptible
1 Milliampere	Faint Tingle
5 Milliampere	Slight shock felt. Not painful but disturbing. Average individual can let go. Strong involuntary reactions can lead to other injuries.
6 to 25 Milliampere (Women)	Painful shocks. Loss of muscle control.
9 to 30 Milliampere (Men)	The freezing current or "Let go" range. If extensor muscles are excited by shock, the person may be thrown away form the power source. Individuals cannot let go. Strong involuntary reaction can lead to other injuries.
50 to 150 Milliampere	Extreme pain, respiratory arrest, severe muscle reaction. Death is possible.
1.0 to 4.3 Amperes	Rhythmic pumping action of the heart ceases. Muscular contraction and never damage occur; death is likely.
10 Amperes	Cardiac arrest, severe burns, death is probable.

▶ Occupational Safety and Health Administration and Electrical Safety

The law that established OSHA specifies the duties of both employers and employees related to workplace safety. OSHA is authorized to enforce standards, and to assist and encourage the states in their efforts to ensure safe and healthful working conditions for all workers. To help accomplish these goals, OSHA provides research, information, education, and training related to occupational safety and health.

K05002 Identify OSHA's general electrical safety requirements.

S05001 Follow an established set of procedures to lock out and tag an HVACR unit for service and then remove the lock and restore power.

General Requirements

The electrical safety regulations that OSHA publishes in the *Code of Federal Regulations* (*CFR*) must be followed. *OSHA Standard 29, Part 1910, Subpart S, Sections 331–335* addresses electrical safe work practices for the general industry, including the deenergization of equipment and the use of personal protective equipment (PPE) appropriate to the hazard. *Section 1910.332* specifies the type and degree of training that all employees who are exposed to electrical shock must receive. *Section 1910.333* specifies a **minimum approach distance (MAD)** of 10 feet (305 cm) to energized high-power voltages at 50 kilovolts (kV) or below for unqualified employees; for voltages over 50 kV, the MAD is set at 10 feet (305 cm) plus an additional 4 inches (10 cm) for every 10 kV over 50 kV. *Section 1910.333* also provides a table of MADs for qualified employees, shown in **TABLE 5-2**.

Never take any action that could cause bodily contact with live circuits. Never touch an electrical conductor or component without first making absolutely sure that it is not energized. Always follow the employer's safety rules, policies, and procedures. These include the lockout/tag-out (LOTO) procedures specified by the employer or jobsite to deenergize equipment and prevent anyone from energizing the circuit. The employer's requirements on the job may exceed those of OSHA. All work on electrical equipment should be done with circuits deenergized and cleared or grounded. Avoid working on energized electrical equipment if at all possible.

Ground-Fault Circuit Interrupters

Most electrical incidents can be prevented simply by thinking ahead and avoiding unsafe actions. Follow proper safety practices, including using properly maintained grounded or double-insulated/ungrounded tools and using **ground-fault circuit interrupter (GFCI)** protection. HVACR technicians are often required to use extension cords on the job for power tools, vacuum pumps, and similar equipment. A GFCI is a form of circuit breaker that senses small imbalances in a circuit caused by current leakage to ground. This current leakage can happen, for example, when the insulation inside a power tool fails. The failure can result in a very high resistance path to ground, yet a path that still permits current flow. A catastrophic failure of the insulation, of course, would result in a very low resistance path to ground. The GFCI can respond in a fraction of a second to such a condition,

TABLE 5-2 Approach Distances for Qualified Employees—Alternating Current

Voltage Range (phase to phase)	Minimum Approach Distance
300 V and less	Avoid contact.
Over 300 V, not over 750 V	1 ft 0 in. (30.5 cm)
Over 750 V, not over 2 kV	1 ft 6 in. (46 cm)
Over 2 kV, not over 15 kV	2 ft 0 in. (61 cm)
Over 15 kV, not over 37 kV	3 ft 0 in. (91 cm)
Over 37 kV, not over 87.5 kV	3 ft 6 in. (107 cm)
Over 87.5 kV, not over 121 kV	4 ft 0 in. (122 cm)
Over 121 kV, not over 140 kV	4 ft 6 in. (137 cm)

opening the power circuit and shutting down the tool. A GFCI monitors the amount of current going to a load and compares it to the current returning from the device on the neutral line. If the difference reaches roughly 5 milliamps (mA), the GFCI responds. Again, the time it takes for a GFCI to sense a current imbalance and respond is a fraction of a second.

A GFCI, however, is not designed to protect workers from line-to-line shock hazards. If a worker accidently grabs energized conductors and a circuit path is completed, an electric shock will result. But it does protect against ground faults, which are the largest source of jobsite shock hazards.

OSHA requires employers to provide worker protection through the use of GFCI devices on temporary circuits. This can be accomplished by using GFCI receptacles or portable devices that can be plugged into an electrical receptacle **FIGURE 5-4**. One visible characteristic on both is the presence of Test and Reset buttons. The Test button is used to ensure the device will open as intended. The Reset button is used to reclose the power circuit following a test or a ground-fault incident.

Every HVACR technician should possess a GFCI-protected extension cord accessory and use it any time that a GFCI-protected receptacle is not available.

The following are some general guidelines for avoiding electrical hazards on the job:

- When working with electrical equipment or circuits, assume that the equipment or circuit is energized until it is personally verified to be deenergized.
- Remove metal jewelry, such as rings and watches, and do not wear wire-rimmed glasses.
- Never switch electrically powered equipment on or off while touching a wet surface.
- Always unplug electrical equipment before working on it. Unplug cords at their plugs; do not pull on the cords themselves.
- Short components to ground before touching deenergized wires.
- Do not kneel on the ground when making voltage measurements. If kneeling down is necessary for a task, do so on a piece of dry, electrically insulating material.
- Inspect all tools, PPE, and other equipment before every use. Never use damaged or unsafe equipment. Follow the employer's policy for tagging out, repairing, or discarding tools and equipment that are no longer in good working order.

▶ TECHNICIAN TIP

OSHA Standard 29, Part 1910.333(a)(1) states that "live parts that operate at less than 50 volts to ground need not be deenergized if there will be no increased exposure to electrical burns or to explosion due to electric arcs." Common 24 VAC control circuits in HVACR systems fall into this category.

SAFETY TIP

If it is not possible to avoid working near energized circuits, try to use one hand only near the hazard area. That way, if a shock is sustained, current is most likely to flow through one hand and arm, and then through the feet to the ground. A shock conducted through both hands and arms is more likely to follow a path through the heart and lungs, which could more easily prove fatal.

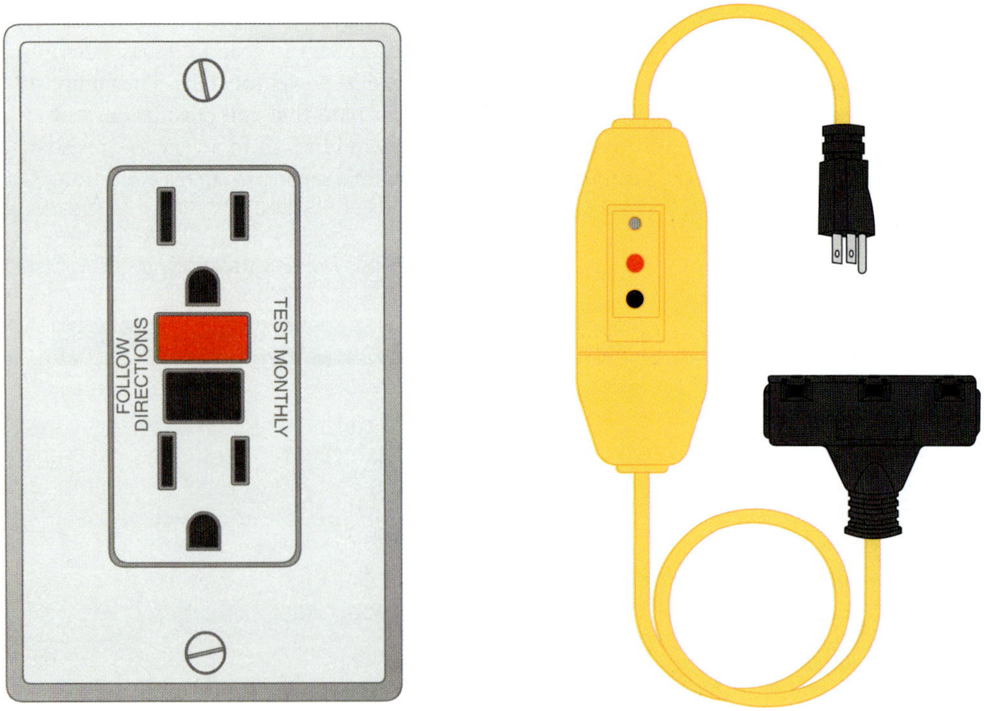

FIGURE 5-4 Examples of GFCI devices.

Lockout/Tag-Out Requirements

LOTO of electrical equipment is an orderly procedure for making equipment safe by removing the potential for the release of electrical energy. *NFPA 70E* requires that energized conductors and circuit parts to which an employee may be exposed must be put into an **electrically safe work condition (ESWC)** before an employee works within the **limited approach boundary**. Establishing an ESWC applies specifically to isolating the sources of electrical energy. Other sources of energy—including static or induced voltages, gravity, tension, thermal, chemical, hydraulic, and pneumatic pressures—must also be made safe, by verifying the absence of energy and by taking measures to prevent inadvertently reenergizing the normal source. **FIGURE 5-5** shows two types of LOTO devices.

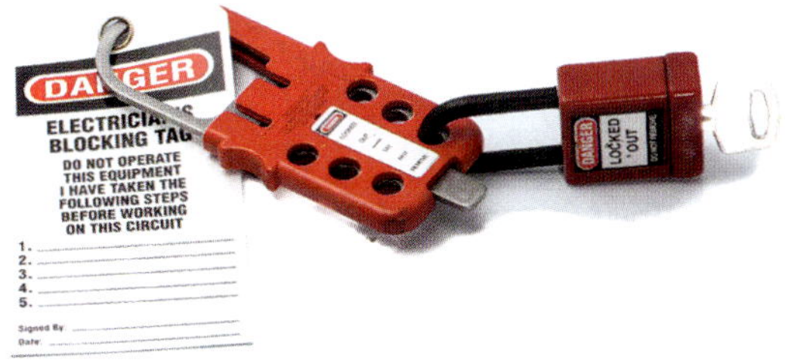

FIGURE 5-5 LOTO devices.

Lemau Studio/Shutterstock

NFPA 70E outlines two basic electrical energy control procedures: the simple LOTO procedure and the complex LOTO procedure.

Any lockout in which a **qualified person** (or persons) deenergizes a single set of conductors or electrical source for the sole purpose of safeguarding employees from exposure to electrical hazards is considered a simple LOTO device. Note that unqualified personnel may not carry out a simple LOTO procedure.

The complex LOTO procedure is required when any of the following conditions are present:

- multiple energy sources
- multiple crews or multiple crafts
- multiple locations
- multi-shift work.

The complex LOTO procedure requires a written execution plan to ensure that the multiple sources will be safely locked out. In addition, the multiple employees and crews who must place personal locks under the policy are clearly defined. The plan includes specific procedures that must be followed for group lockout devices or the use of lockboxes.

The basic procedure for an electrical lockout is as follows:

Step 1: Communicate with affected personnel and prepare for shutdown. Wear the appropriate PPE, and identify all sources of electrical power being supplied to the equipment.

Step 2: Shut down the equipment by normal means, and open the disconnecting device(s) to isolate the equipment from its hazardous energy source(s).

Step 3: Operate other switches, valves, and other energy-isolating devices so that the equipment is isolated from all its energy sources. Visually verify the opening of switch assembly blades or contacts when possible.

Step 4: Discharge any induced voltage or stored energy, such as that in capacitors or pneumatic lines.

Step 5: Verify the absence of voltage by using a multimeter or voltmeter. To further verify that the equipment is isolated from its energy source(s), turn the equipment on to determine whether it operates. Do this by operating the normal operating control(s) to ensure that the equipment will not operate or is in a zero-energy state. If the equipment does not respond, this confirms that the energy sources have been disconnected and that the equipment is now locked and tagged. Return the operating control(s) that were activated during the isolation verification step to the OFF or neutral position. Otherwise, retrace the previous steps and repeat them until the equipment no longer responds when turned on.

Step 6: Apply LOTO devices according to the established procedures and policies for the given jobsite or employer. One rule common to all LOTO procedures is that each authorized person must install and remove their own separate lock and tag. If the deenergized circuit parts could be reenergized, apply temporary grounds rated for the available fault current. With the lockout completed, proceed to service or repair the equipment as necessary.

To lock out an HVACR unit, follow the steps in **SKILL DRILL 5-1**.

SAFETY TIP

A three-step procedure is recommended whenever using a meter to verify the absence of electrical power, as follows:

Step 1: Always begin by testing the meter on a known energized source to prove that the meter is working and senses the presence of voltage.

Step 2: Test the circuit needed to test for the absence of power.

Step 3: Finally, return immediately to the same known energized source and test the meter once again to prove that it still does properly sense voltage.

This procedure is used on a daily basis by technicians and electricians across the world. Note that this procedure is a requirement when working on circuits of 600 VAC or higher, but it is also recommended for all voltages.

SAFETY TIP

If the work area is left temporarily unattended, retest the equipment upon return, to ensure that the equipment is still deenergized.

SKILL DRILL 5-1 Locking Out an HVACR Unit

Follow an established set of procedures to lock out and tag an HVACR unit for service, then remove the lock and restore power.

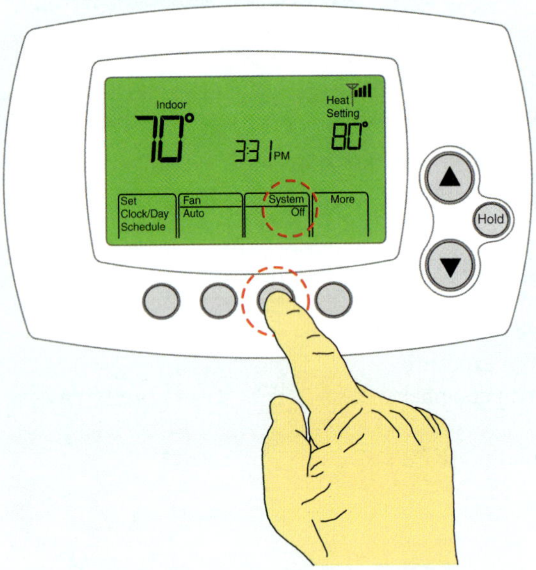

1. Begin by notifying all affected personnel that a LOTO is to be applied and explain the reason why. Identify all electrical energy sources that must be controlled. Don the appropriate PPE. Shut down the unit through its normal controls by using the OFF or STOP procedures. In this case, the system thermostat is likely the primary control.

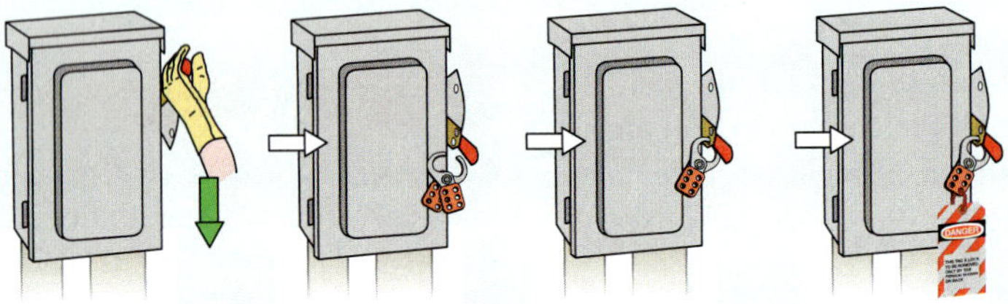

2. Operate the unit disconnect switch to deenergize the equipment. Lock the disconnect switch in the OFF position, or lock the disconnect door closed if no external handle is present.

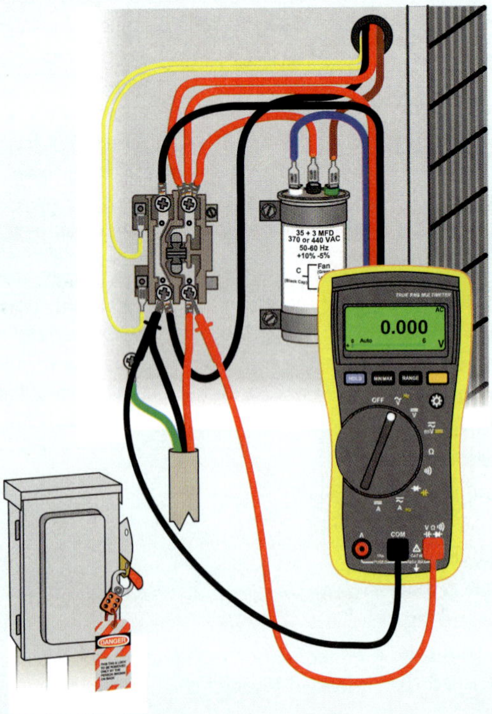

3. Use a voltmeter to ensure there is no voltage present in the condensing unit.

Continued

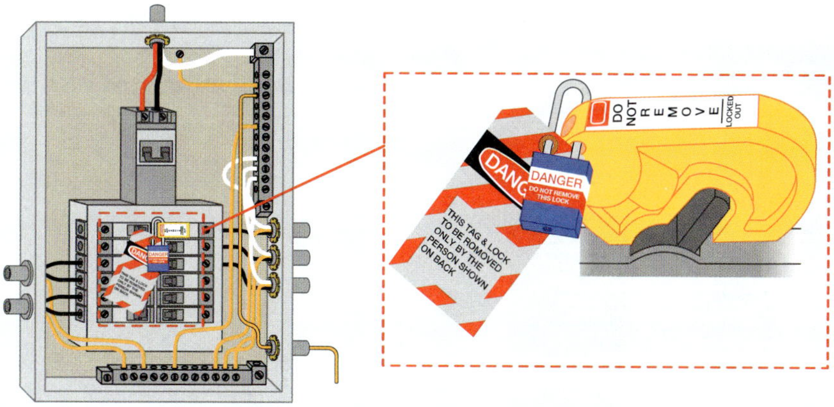

4. Although the control voltage provided to the unit is less than 50 VAC, it should also be deenergized to protect the circuit itself from damage while working. Small control transformers can easily be destroyed by a brief short circuit. The control voltage typically originates at the indoor unit of a split heating/cooling system (furnace, fan coil, or air handling unit). Locate the indoor unit power source, turn it off, and lock/tag the switch.

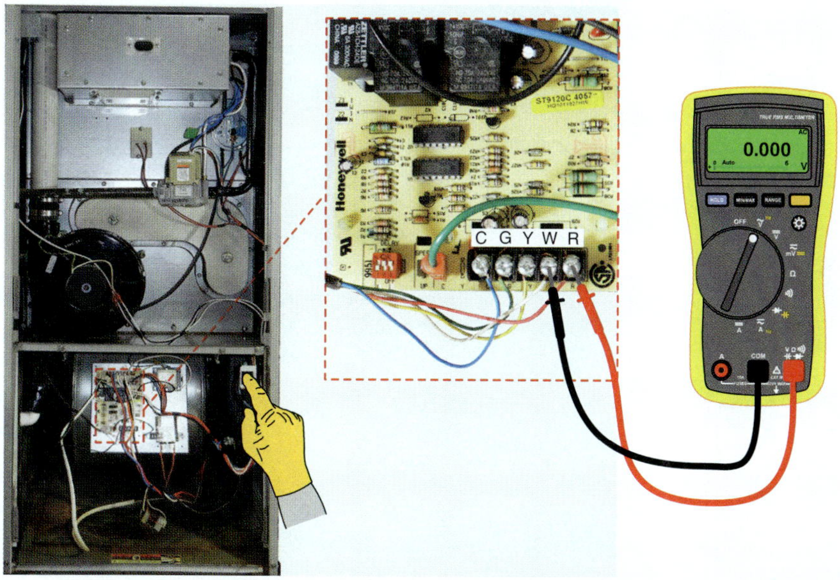

5. Use a voltmeter to ensure there is no voltage present in the indoor unit. If not, the system overall is now electrically safe.

6. Once ready to restore electrical power to the system, first confirm that all tools are accounted for and have been removed from the equipment. Notify affected personnel in the area that power is being restored. Verify that all controls, such as the thermostat, are in a safe (OFF) or neutral position.

Continued

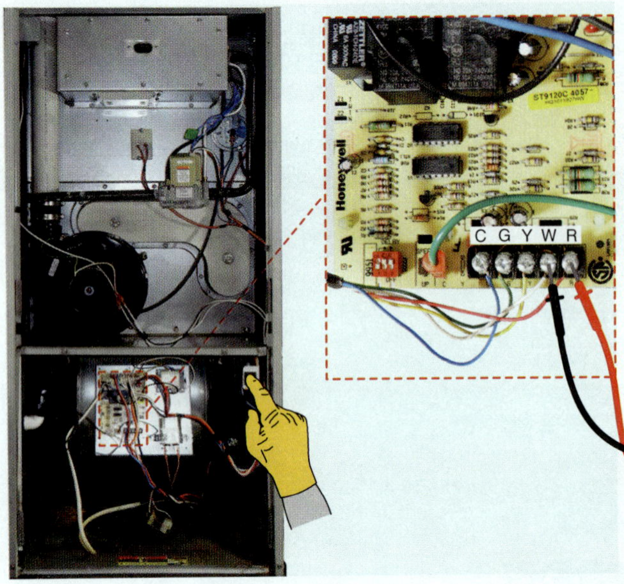

7. Remove the lock and tag from the indoor unit power source. Restore power, and verify the presence of the proper voltage at the indoor unit terminal block or other point of connection.

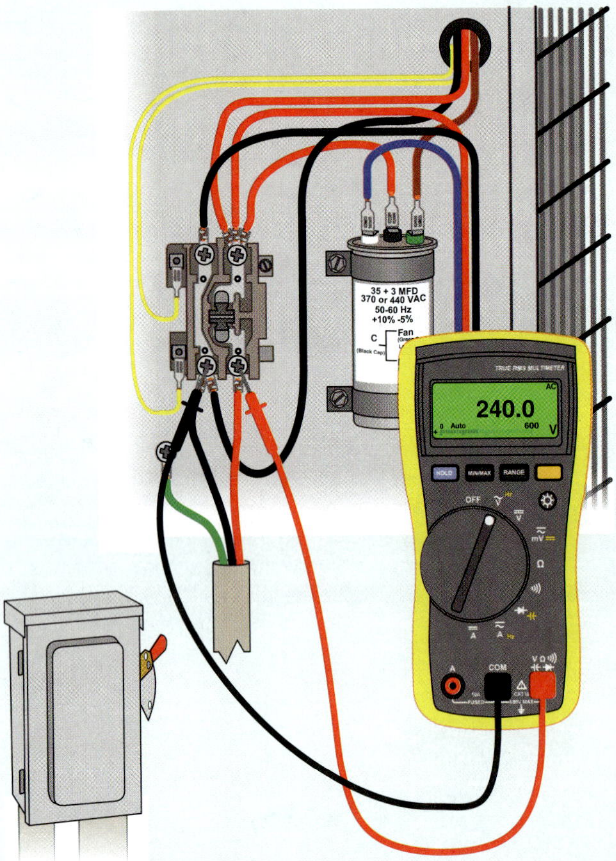

8. Remove the lock and tag from the outdoor unit power source. Restore power, and verify the presence of the proper voltage at the condensing unit contactor or other point of connection.

Continued

9. Reinstall all access doors and panels on both the indoor unit and outdoor unit. Ensure that the area is clear before restarting the unit to ensure that it operates normally in response to the thermostat.

Restoring electrical energy is essentially a matter of reversing the steps performed for the lockout procedure, as outlined below.

Step 1: Let affected personnel know that the equipment is going to be energized. Remove all tools from the equipment and reinstall any safety guards.

Step 2: Remove the LOTO devices that were recently installed. Any other individuals who had installed locks and tags must also personally remove the devices that they installed.

Step 3: Once all LOTO devices have been removed, operate the energy-isolating devices to restore energy to the equipment. Notify all affected personnel that the equipment is going to be energized, and again when it is energized.

▶ NFPA 70E

The NFPA publishes the following OSHA-recognized standards for electrical safety:

- *NFPA 70, National Electrical Code (NEC)*
- *NFPA 70E, Standard for Electrical Safety in the Workplace.*

K05003 Describe the NFPA and the contents of NFPA 70E.

The *NEC* specifies the minimum installation provisions necessary for protecting people and property from electrical hazards. *NFPA 70E* specifies requirements related to the hazards that arise during the installation, operation, and maintenance of electrical equipment. In short, the *NEC* applies to installations, and *NFPA 70E* applies to worksites and work practices.

▶ TECHNICIAN TIP

An electrician generally installs the circuit from a building's electrical service to the disconnecting device for the HVACR equipment. When installing an HVACR system, the HVACR technician may be required to connect the power from the disconnecting device to the unit. When doing so, be careful to follow the *NEC* code requirements for grounding and proper protection of the wiring.

NFPA 70E Content

NFPA 70E provides practical guidance in hazard assessment, training of qualified persons, and management of electrical hazards. The content of *NFPA 70E* is divided into three chapters, which are further subdivided into articles and sections. Additional annexes and supplements provide calculation methods, forms, and other references. The content is arranged as follows:

- *Chapter 1, Safety-Related Work Practices*—This chapter focuses on the basic safety requirements for working on or near electrical equipment. It covers safety training programs, explains the difference between qualified and unqualified individuals, and specifies requirements for an analysis of both the electrical shock hazard and the electrical arc-flash/arc-blast hazard for a given task. The analysis includes assessing the presence and extent of hazardous voltage and the available fault current for arc flash. The result of each hazard analysis is the selection of safe work practices and appropriate PPE to be used when performing the task.

- *Chapter 2, Safety-Related Maintenance Requirements*—This chapter covers the maintenance requirements for various types of electrical equipment. This includes premises wiring, various types of electric equipment and devices, and PPE.
- *Chapter 3, Safety Requirements for Special Equipment*—This chapter covers electrolytic cells, batteries, lasers, and various types of electronic equipment, such as radio and television transmitters, uninterruptible power supply (UPS) systems, arc welding equipment, and lighting controllers.
- *Annexes A through P*—The annexes provide various technical resources. These resources include safe approach distances, determination of flash protection boundaries, and incident energy exposure from arc flash, a sample electrical LOTO procedure, and other useful information.
- *Part Two*—This part contains supplements that cover the *NEC* requirements associated with safe work practices and electrical preventive maintenance programs. It also presents a case history of an arc-flash incident.

Electrical Boundaries

Distance is an effective means of protection from exposure to energized parts. Industry standards include specific electrical hazard boundaries for shock protection and arc-flash protection (the **arc-flash boundary**), as illustrated in **FIGURE 5-6**. Only qualified persons who are adequately trained and use all the required PPE are allowed to enter and work inside these boundaries.

Most electrical safety guidelines recommend safe working distances based on shock considerations. The three boundaries for shock protection are called approach boundaries—the limited approach boundary, the **restricted approach boundary** and the **prohibited approach boundary**. Note that these boundaries are subject to change because standards are revised and updated. There are also other factors that must be considered that are part of the *NEC*.

A limited approach boundary is set at a specified distance from an exposed energized part. Only qualified persons are allowed to cross this boundary. The one exception is that unqualified persons may cross a limited approach boundary for on-the-job training if they are escorted by a qualified person.

A restricted approach boundary is set in closer proximity to exposed energized parts. This boundary may be crossed only by qualified persons who are authorized through an energized electrical work permit, follow shock-protection techniques, and use specified PPE.

▶ TECHNICIAN TIP

The restricted approach boundaries, include added safety margins to compensate for inadvertent movement of the worker. If such movement allows exposed energized parts to make contact with test equipment or hand tools, the tool or test lead could become part of the circuit path. This frequently results in shock and arc-flash incidents. In fact, it is estimated that 75% or more of all arc-flash incidents begin in this way.

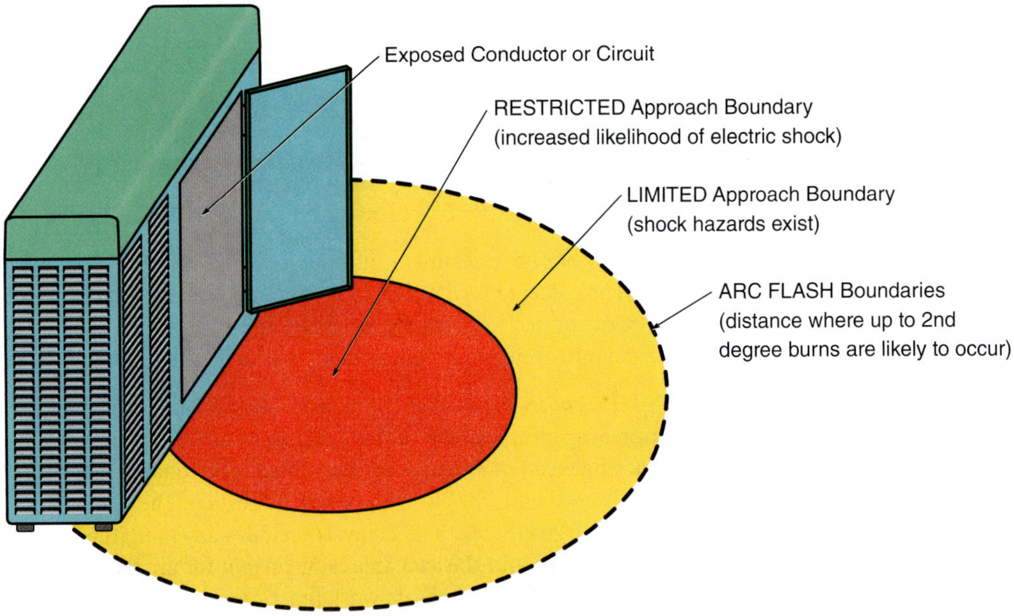

FIGURE 5-6 Shock and arc-flash protection boundaries.

A prohibited approach boundary defines an area within which workers will make contact with an electrically energized component or the work being performed is considered equivalent to contacting an energized component. Essentially, if a technician is crossing the restricted approach boundary, a prohibited area has been entered. Any part of a worker's body that crosses a prohibited approach boundary must be suitably insulated and protected from the hazard.

When an arc-flash hazard is present, an arc-flash boundary must be established. This boundary defines how far away a person needs to be so that they will avoid sustaining serious burns in the event that an arc flash were to occur. Note that an arc-flash boundary is determined independently from, and may be at a greater or a lesser distance than, any of the shock-protection boundaries. *NFPA 70E* provides direction for establishing flash protection boundaries and the levels of protection required when qualified persons must work within the flash protection boundary.

SAFETY TIP

Electrical safety guidelines generally recommend safe working distances based on shock considerations, not arc-fault hazards. So, even when the safety guidelines for electrical shock are being followed, there is still risk of arc and thermal contact burns.

Shock Hazard and Arc-Flash Hazard Analyses

Hazard analyses are performed to ensure that work can be performed safely when shock and/or arc-flash hazards exist. During an analysis, specially trained workers identify, analyze, and document possible electrical hazards within the work area and establish protective hazard boundaries. The area is then clearly marked with signs that identify the hazard(s) and provide information about the PPE that must be worn in the area.

At its simplest level, a **shock hazard analysis** identifies the presence of exposed energized parts, determines safe limits of approach for both qualified and unqualified personnel, and specifies appropriate PPE requirements. **FIGURE 5-7** shows a sample flow chart for the process developed by the Center for Construction Research and Training. The analysis should also identify concealed, covered, or guarded energized parts that would be exposed when a boundary or barrier is bypassed. In addition, it should specify methods to be used for preventing both deliberate and inadvertent contact with energized parts, and it should identify parts that are currently deenergized but that could either intentionally or unintentionally become energized.

The primary objectives of an **arc-flash hazard analysis** (or arc-flash risk assessment) are to identify when an arc-flash hazard is expected to exist, and to estimate the extent of the hazard. The analysis establishes the arc-flash boundary for a given electrical component or work area and identifies the PPE needed to protect workers within that boundary against second-degree burns.

Arc-flash hazard analysis is more complicated than analyzing shock hazards. Typically, with a shock hazard, there are known voltage levels in the environment and a chart to refer to that identifies approach boundaries. With an arc flash, numerous variables that control the intensity of the arc flash must be taken into account. They include available fault current, time required to detect and clear an arcing flash, arc length, and voltage and distance from the fault. *NFPA 70E Section 130.5* explains when an arc-flash hazard analysis is required, describes how to determine an arc-flash boundary, and covers how to select the required PPE. *NFPA 70E Informative Annex D* provides formulas for calculating flash protection boundaries and estimating incident energy exposure.

▶ TECHNICIAN TIP

An arc-flash boundary is determined independently from, and may be at a greater or a lesser distance than, any of the shock-protection boundaries. Many electrical safety programs set both the arc-flash boundary and the outer shock-protection boundary at whichever distance is greater, as determined by an arc-flash hazard analysis.

▶ Responding to Emergencies

K05004 Identify general electrical emergency responses and procedures.

Responding promptly and appropriately is essential when emergencies occur. The response to an electrical incident depends on the type and severity of injury that the victim has suffered. Victims of an electrical incident may be injured by any combination of electrical shock, arc flash, and arc blast.

In any case, being prepared saves precious time during an emergency. If there is an emergency response plan in place, be sure to understand it before starting work. At the outset of a job, take a moment to think about how to describe the jobsite location to a 911 operator, should that be necessary.

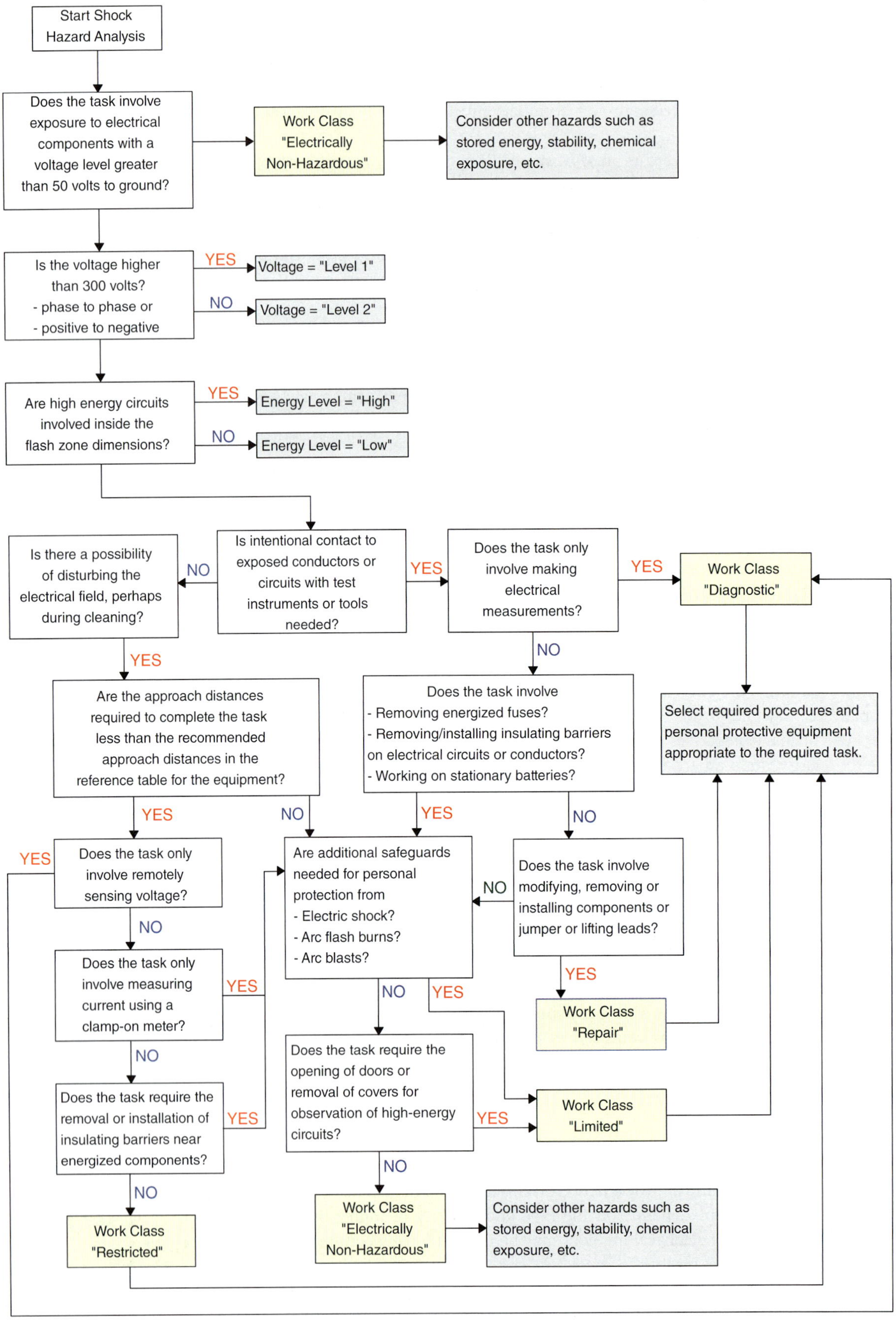

FIGURE 5-7 Shock hazard analysis flow chart.

Electrical Shock

A victim of electrical shock may not be able to let go of the object(s) that is shocking them. To avoid becoming part of the same circuit, do not touch the victim. If possible, immediately turn off the source of the electricity. If that is not possible, use an insulated rescue hook (**FIGURE 5-8**) to remove the victim from the electrical source. After the victim has been removed, ensure the area is safe for rescuers to begin first aid.

An electric shock victim may go into physical shock, so be prepared to deal with this as well. For this reason, companies often require workers to be trained in first aid and CPR. If the victim's heart has stopped or lost its normal rhythm (entered a state of fibrillation), use CPR to maintain the victim's breathing and blood flow until help arrives.

A machine called an AED (automated external **defibrillator**) may be available at the site of the incident. **FIGURE 5-9** shows one example of an AED. With this machine, a victim's heart rhythm can be checked, and the paddles can then be used if necessary to restore normal rhythm. Minimal training is needed to use an AED effectively. The machine essentially functions automatically, guiding its operator through the procedure. Unless fibrillation is present, it will not deliver a shock through the paddles.

Arc Flash and Arc Blast

Arc-flash and arc-blast victims are often injured more seriously than is apparent; they may require treatment at a special burn center. The initial response should be to douse the victim with water and remove him or her to a safe location. Also be aware that burns to the neck can make it difficult for the victim to breathe. Provide first aid and CPR as needed while waiting for help to arrive.

FIGURE 5-8 Insulated rescue hook.

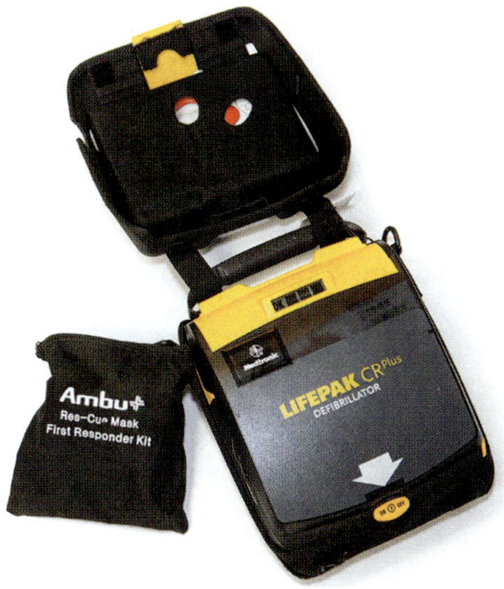

FIGURE 5-9 An AED.

▶ Wrap-Up

Summary

One of the basic responsibilities as an HVACR technician is to maintain a safe workplace for all staff. Because HVACR systems are powered by electricity, recognizing and avoiding electrical hazards are critical skills for performing jobs safely and efficiently. Be aware of how electricity can hurt people, follow safe work practices, and be prepared to respond quickly and appropriately if an electrical incident occurs.

Electrical injuries most often occur when individuals fail to follow safe work practices and/or they disregard or are not aware of hazards. Before beginning a job, take a moment to critically assess the electrical hazards that may exist. Ask the following questions:

- What are the electrical hazards involved?
- What must be done to avoid or minimize these hazards?
- What should be done if an electrical incident occurs?

Be concerned with safety at all times, to prevent being the victim or the cause of an electrical incident. Remember that learning to work safely in the presence of electricity is not a single event but rather a journey that continues throughout a career. There is far more to learn about electrical safety than can be presented in this single chapter.

Key Terms

arc An electric current that flows across an open space between two points, releasing heat and bright light in the process.

arc blast An intense explosion that occurs during an arc-flash incident.

arc fault A high-energy electrical discharge between two or more conductors.

arc flash A dangerous flash of light caused by the enormous release of thermal energy in an electric arc.

arc-flash boundary A boundary set at a distance from exposed energized electrical conductors or circuit parts; persons within the boundary could receive second-degree burns if an electrical arc flash were to occur.

arc-flash hazard analysis Investigation and documentation of a worker's potential exposure to arc-flash energy, which is conducted to prevent injury by determining safe work practices and appropriate levels of PPE.

conductor A material through which it is relatively easy to maintain an electric current.

defibrillator A device used to help the heart return to its normal rhythm by applying a controlled electric shock to the chest wall or heart.

defibrillation The process of administering a controlled electrical shock to stop fibrillation and restore the normal heart rhythm.

electrically safe work condition (ESWC) A state in which the conductor or circuit part to be worked on or near has been disconnected from energized parts, locked/tagged in accordance with established standards, tested to ensure the absence of voltage, and grounded if necessary.

electrocution Death that results from being exposed to a lethal amount of electrical energy.

fibrillation Very rapid irregular contractions of the muscle fibers of the heart that result in the heartbeat and pulse going out of rhythm with each other.

ground fault circuit interrupter (GFCI) A safety device, often in the form of an electrical receptacle, that disables power flow when it detects that current is flowing along an unintended path, such as ground.

limited approach boundary A shock-protection boundary to be crossed only by qualified persons (at a distance from a live part); it is not to be crossed by electrically unqualified persons unless escorted by a qualified person, such as in a training situation.

minimum approach distance (MAD) The closest distance at which a qualified worker is permitted to approach an exposed energized conductor or equipment. The MAD is the combination of two elements: the minimum air insulation distance (MAID) and the ergonomic factor.

prohibited approach boundary A shock-protection boundary to be crossed only by qualified persons. Once a person crosses this boundary with a part of their body or with an object, the same PPE and equipment is required as if direct contact is to be made with a live part.

qualified person An individual who has demonstrated the skills and knowledge related to the construction and operation of electrical equipment and installations and who has received safety training to recognize and avoid the hazards involved. See *OSHA Standard 1910.332(b)(3)* for details on the required minimal training and skills.

restricted approach boundary A shock-protection boundary set from an exposed energized electrical conductor or circuit part within which specific PPE and insulated equipment must be used. An electrical work permit is required prior to entry.

shock hazard analysis Investigation and documentation of a worker's potential exposure to voltages from energized parts that pose electrical shock hazards, which is conducted to prevent injury by determining safe work practices and appropriate levels of personal protective equipment (PPE).

Review Questions

1. A human body that is in contact with the ground or another point in an electrical circuit can become a part of the circuit if the body contacts ______.
   a. a resistance path
   b. a ground fault circuit interrupter (GFCI)
   c. an electrically energized point
   d. defibrillation paddles
2. What is the most common injury associated with electrical shock?
   a. Defibrillation
   b. Sprains
   c. Paralysis
   d. Burns
3. Which class (degree) of burns reaches the muscles and fatty tissues inside the body, causing cell death?
4. The irregular twitching of the heart muscle that occurs when an abnormal electrical signal reaches the heart is called ______.
   a. cardiac resuscitation
   b. fibrillation
   c. cardiac arrest
   d. defibrillation
5. What are the three key factors that affect the severity of shock that a person receives?
6. The minimum approach distance (MAD) for voltages less than or equal to 50 kV as specified in *OSHA Standard 1910.333* is ______.
   a. 4 feet (122 cm)
   b. 6 feet (183 cm)
   c. 8 feet (244 cm)
   d. 10 feet (305 cm)

7. Whenever possible, all work on electrical equipment should be done with circuits deenergized and ______.
   a. rectified or reconfigured
   b. recalibrated or certified
   c. cleared or grounded
   d. shunted or bypassed
8. Which of the following is a safe work practice to protect the body from electrical shock hazards?
   a. Use a GFCI only when using double-insulated power tools.
   b. Do not switch electrically powered equipment on or off while touching a wet surface.
   c. Use only two-prong adapter plugs with GFCIs.
   d. Always kneel on the ground when making voltage measurements.
9. What type of lock out/tag-out (LOTO) procedure is required when multiple energy sources and/or multiple crafts are involved?
10. When applying LOTO devices, each authorized person involved in the LOTO procedure must always ____.
   a. sign the execution plan
   b. deactivate the circuit testers
   c. use explosion-proof test instruments
   d. install a separate lock
11. During a LOTO procedure, to verify that a piece of equipment has been deenergized and there is no power, a technician should_____.
12. Before restoring energy to an HVACR unit that has been locked out and tagged, _________.
   a. drain the refrigerant lines
   b. ensure all tools are accounted for
   c. purge the compressor
   d. backwash the condenser coils
13. Which of these standards specifies requirements related to the hazards of operating and maintaining electrical equipment?
   a. *ANSI Standard C12.1*
   b. *29 CFR, Part 30*
   c. *NEC*
   d. *NFPA 70E*
14. Which chapter of *NFPA 70E* focuses on safety-related work practices?
15. Which approach boundary defines an area within which workers will contact an electrically energized component or within which the work being performed is considered the equivalent of contacting an energized component?
16. Which of the following boundaries is determined on the basis of how far away workers must be from an energized electrical component to avoid receiving second-degree burns?
   a. Arc-flash boundary
   b. Limited approach boundary
   c. Restricted approach boundary
   d. Prohibited approach boundary
17. Why is performing an arc-flash hazard analysis considered more complicated than performing a shock hazard analysis?
18. What is the first thing to do when responding to an electrical incident in which the victim has sustained an electrical shock?
   a. Douse the victim with water.
   b. Begin administering CPR.
   c. Transport the victim to the nearest burn center.
   d. Deenergize the electrical source.
19. When responding to an electrical shock victim, be prepared for the victim to ________.
   a. succumb to an infection
   b. become irradiated
   c. go into physical shock
   d. leave the scene of the incident
20. Arc-flash and arc-blast victims usually require ________.
   a. being wrapped in reflective blankets
   b. relatively minor first aid procedures
   c. treatment at a special burn center
   d. injections of anti-seizure drugs

CHAPTER 6

Conductors, Insulators, and Terminations

Knowledge Objectives

After reading this chapter, you will be able to:

- **K06001** Compare and contrast electrical conductors and insulators.
- **K06002** Identify the criteria for selecting conductors.
- **K06003** Explain how to identify a damaged conductor.
- **K06004** Describe the tools, equipment, and methods used to terminate wiring.

Skill Objectives

Skills required to meet the objectives of this chapter:

- **S06001** Select the appropriate wire size for a condensing-unit whip, based on ampacity.
- **S06002** Terminate wiring by using wire nuts.
- **S06003** Terminate wiring by using common crimp connectors.
- **S06004** Install a properly sized condensing-unit whip.

▶ Introduction

Technicians are often called upon to select and connect wiring while installing, servicing, or repairing HVACR equipment. It is vital to understand the relationship between wire size and type, the operating environment, and the required current-carrying capacity. It is also important to be aware of the applications of wire **conductors** and the insulating materials that protect both the user and the wire.

There are several methods used to terminate conductors. HVACR technicians and installers must know how to select and properly use conductor-termination hardware. Poor wiring connections can create problems that are challenging to identify and locate.

▶ Conductors and Insulators

K06001 Compare and contrast electrical conductors and insulators.

Conductors are the wires that carry current to the loads and control devices in a circuit. Technically, however, a conductor is any material that will carry an electrical current. The most common wire conductors used in HVACR are made of copper, which is a very good conductor of electricity. Gold and silver are better conductors, but they are prohibitively expensive to use for wiring. Aluminum is also a good conductor, but copper is the preferred conductor for wiring because a given size of copper wire has a higher current-carrying capacity than an equal size of aluminum wire. Aluminum wire is also more fragile and prone to damaging corrosion and degradation. It was once widely used in homes due to its low cost before the extent of its weaknesses were widely recognized.

An **insulator** is the opposite of a conductor. It opposes the flow of electrical current. Current-carrying conductors must be insulated to ensure the safety of people and the equipment (**FIGURE 6-1**). Insulation prevents arcing and short circuits that could occur when an energized conductor contacts another, or a nearby, conductive surface such as a metal equipment housing. If that happens and a current overload device does not interrupt the circuit, anyone touching the metal enclosure could be shocked or electrocuted.

> **SAFETY TIP**
>
> **Water**
>
> Water is thought to be an excellent conductor of electricity. Most everyone understands that water and electricity make a dangerous combination. Technically, however, water isn't the problem. Indeed, pure water that is free of minerals and similar materials is a good insulator rather than a conductor.
>
> Distilled water, when distilled to a high standard, is considered pure. During distillation, all minerals are removed and only neutral water molecules are left. Neutral molecules without any electrical charge aren't conductors.
>
> Of course, water that is not pure is often loaded with other molecules that do possess an electrical charge. The result is that the conductivity of water goes up sharply, resulting in a significant hazard.

The specific type of insulation on wires and cables is often determined by building codes. For example, low-voltage thermostat wire running unprotected in the space above a suspended ceiling must be **plenum-rated** wire if that space is used as the return-air plenum for the HVACR system. This is a common situation in commercial installations. Some insulation materials will release toxic fumes when they burn. In a return-air plenum, these fumes would be drawn into the return-air system and distributed to the conditioned space. For that reason, an insulating material that is **flame-retardant** and produces a limited amount of smoke must be used in these applications.

In addition to providing protection, insulation color is often used to help identify conductors. For example, as discussed in Chapter 3, Type NM cable used in 120 VAC, single-phase applications has a black wire that is designated as the energized conductor, and a white or gray wire that is the neutral conductor. A cable designed for 240 VAC applications often has red and black conductors instead, both of which are energized. Ground wires will have green insulation or no insulation at all.

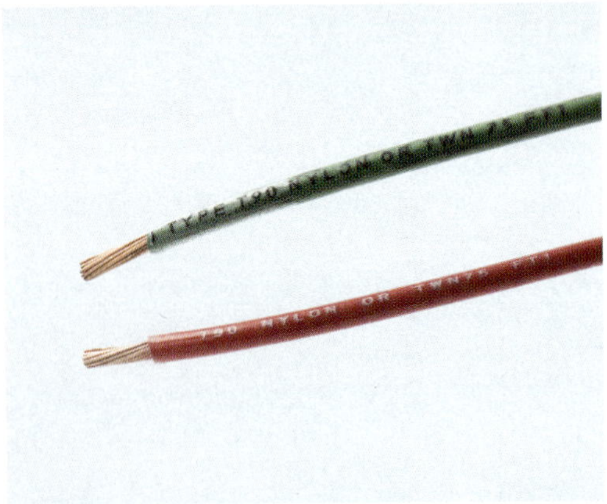

FIGURE 6-1 Insulated conductors.

Manufacturers and installers often follow standard color-coding schemes for conductor insulation in their equipment. Never take that for granted, however. For example, green is the standard color for a ground wire, but it's possible that someone doing a repair had only green wire available and used it to replace a damaged current-carrying conductor. Always verify the conductor status using a multimeter and/or by tracing its path.

Thermostat wire colors are designed to correspond to terminal markings on thermostats and terminal boards, such as R for red and G for green. However, installers do not always adhere to the color code for a variety of reasons. Make no assumptions when approaching an unfamiliar system.

When the control section of a unit is opened, what appears to be a tangle of wiring may be observed. Color-coding using abbreviations may be used on the wiring diagram to make it easier for technicians to trace circuits through the

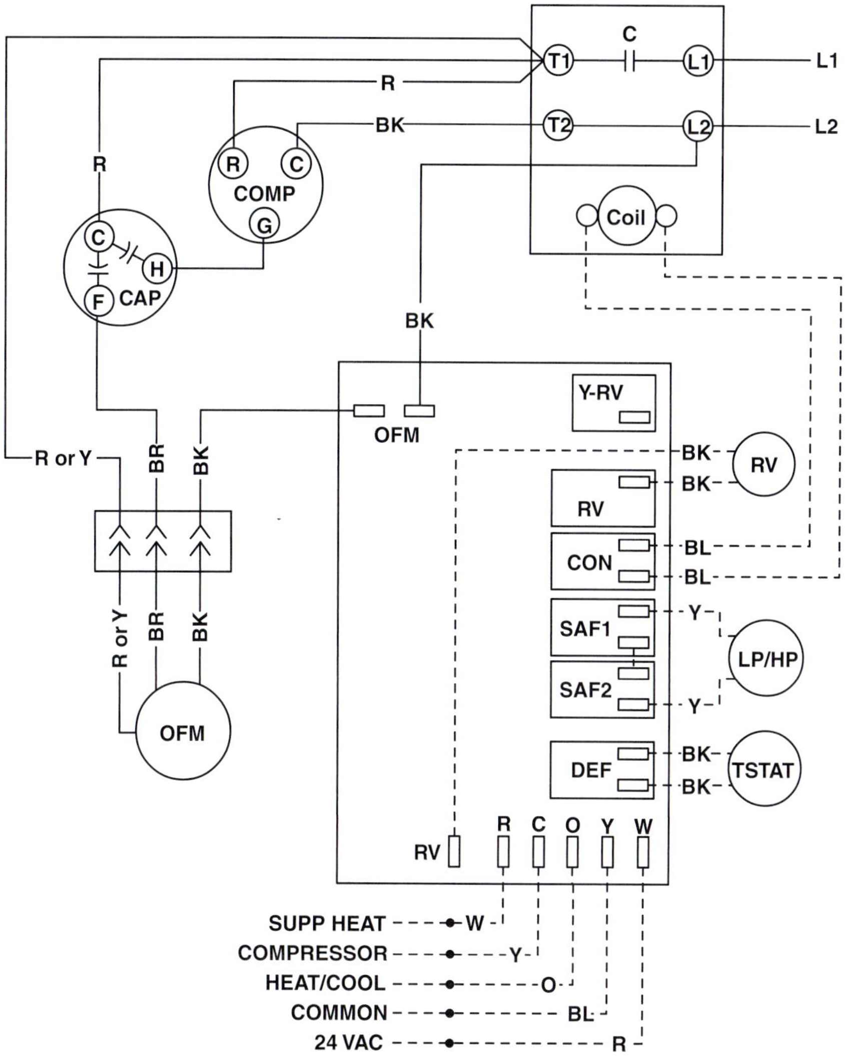

FIGURE 6-2 HVACR wiring diagram with color-coded wires.

maze. For that reason, if it is necessary to replace a conductor, it is always a good idea to use the same color insulation as the original wire. **FIGURE 6-2** shows a wiring diagram with each wire identified by a color. The colors are represented by abbreviations: BL for blue, BK for black, R for red, etc.

▶ Identifying and Selecting Conductors

Technicians and installers often need to replace damaged conductors or to install new conductors as part of an installation. The proper selection of conductors is crucial to safety and reliability, and it must be done in accordance with national and local codes.

K06002 Identify the criteria for selecting conductors.

S06001 Select the appropriate wire size for a condensing-unit whip, based on ampacity.

Wire Gauge and Ampacity

There are different conductor sizes due to the need to handle different levels of current. Wire sizes are stated as wire gauges under the **American Wire Gauge (AWG) system**. Note that a wire size has nothing at all to do with the thickness of the insulation.

Only the size of the conductor is considered. Wires with the greatest cross-sectional area have the lowest gauge number; gauge numbers increase as the wire size decreases. The wire gauge tool in **FIGURE 6-3** will give an approximation of the range of common wire sizes. In practice, HVACR installers and technicians regularly encounter wire sizes ranging from #6 AWG to #20 AWG. Wire that is larger than #16 AWG has its size, insulation type, and voltage rating printed on its insulating covering.

Wire conductors can be either stranded or solid (**FIGURE 6-4**). Wire that is #6 AWG and larger is generally stranded, while wire that is #18 AWG and smaller is typically solid. However, there are exceptions. Between these sizes, both stranded and solid conductors are readily available.

FIGURE 6-3 Standard wire gauge tool.

FIGURE 6-4 Stranded and solid conductors.
vldkont/Shutterstock

Ampacity

Ampacity is way of stating the maximum current that a conductor of a given size can continuously carry without risking damage or failure. The National Electrical Code (*NEC*®) contains tables that specify the ampacities of conductors under a variety of conditions and applications.

Voltage Drop

All common conductors offer some resistance to current flow. Therefore, some amount of voltage drop occurs along the length of a conductor. Four primary factors affect the resistance of a given wire, and this resistance leads to voltage drop when a load is applied:

- *Temperature*—The higher the temperature, the greater the resistance.
- *Cross-sectional area*—Resistance decreases as the wire size increases.
- *Length of wire*—The longer a wire is, the more total resistance it offers to the flow of current.
- *Conductor material*—The better the conductor, the lower the resistance. Copper has lower resistance than aluminum, for example.

Voltage drop is a significant concern for loads, especially inductive loads like motors, since the current will rise as the voltage drops. For example, motors may be unable to start or the running current at the reduced voltage may become too high, causing an overload condition. Resistive loads, such as an electric heating element, will usually operate but at a reduced capacity. Voltage drop also affects control wiring, even though control-system loads are relatively small. Since thermostat wiring is typically #18 AWG, its current-carrying capacity is already very limited. A significant voltage drop often results in relays and contactors that won't change position when energized.

Wiring charts provide information on the resistance per unit length for all types of wire. If the voltage drop in a length of wire is determined to be more than 3%, the conductor must be **derated**. That is, the amount of current it is rated to carry must be reduced. As a result, the wire size for long runs often needs to be increased. Remember that resistance decreases as the wire size increases. Therefore, a conductor with a larger cross-sectional area will have less voltage drop per unit length.

To measure the voltage drop in a conductor, first measure the voltage at the source, then measure the voltage at the input to the energized load; the load must be energized when measuring voltage drop. When the voltage is measured without a load, it will be the same at both ends of the conductor. To measure voltage drop in a simple cooling circuit, start the system using the thermostat, measure the voltage at the control transformer, and then measure it at the compressor contactor coil terminals.

After making both measurements, subtract the destination voltage from the source voltage to determine the voltage drop. Then calculate the percentage of voltage drop by dividing the source voltage by the measured voltage drop. For example, assume that the transformer output measures 24.4 VAC, and that the voltage measured at the energized contactor is 21.2 VAC. The difference is 3.2 volts. Dividing 3.2 by 24.4 yields 0.131, or 13% (moving the decimal point two places to the right yields percentage). This is a significant amount of voltage drop, although the voltage may still be within the specifications of the contactor coil for operation. However, if anything occurs that creates additional voltage drop, such as a wiring termination that becomes loose, the reduced voltage will probably not be sufficient to energize the contactor. This much voltage drop is a problem in waiting, if not an immediate problem.

▶ CORE CONNECTION

Wiring and Piping

As strange as it may seem, wiring and piping have some things in common. Think of pressure in a pipe as being the same as voltage: both represent the difference in potential that causes the movement of liquid or power. In piping, volume is measured in some unit over time, such as gallons per minute. The volume of electricity over time is measured in watts. The walls and changes of direction in piping present a resistance to the flow of liquid, just as a conductor resists the flow of electricity. Undersizing both pipe and wire in their respective applications causes operational problems that can only be effectively solved by increasing the size. Note that when there is no flow in a pipe, the pressure stabilizes, but the pressure typically falls at the far end of the piping run when flow begins. In wiring, the voltage is stable when there is no demand, but when a demand is placed on a circuit, the voltage drops. This is why no voltage drop is detected until a load places a demand on the circuit.

Selecting a Replacement Wire Size

As a general rule, HVACR technicians are not responsible for determining the size and type of conductors providing power to a piece of equipment. This chore typically falls to electricians in new installations. HVACR technicians are most often concerned with replacing wires inside a unit cabinet, or for selecting and installing what is known as a **whip** (**FIGURE 6-5**). The wiring that brings power from the disconnect switch to an air-cooled condensing unit is located outdoors and is therefore exposed to the elements. To protect this wiring, it is encased in flexible, weathertight **conduit**. This wiring assembly, with fittings on each end for connection to the electrical enclosures, is available as a prefabricated whip.

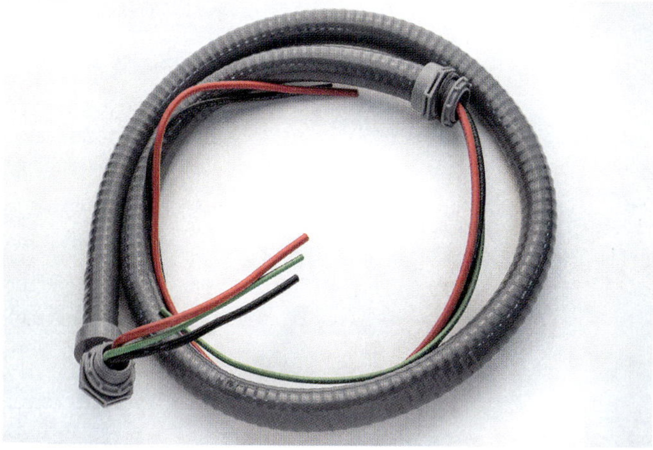

FIGURE 6-5 A typical HVACR unit whip.

It is easy to assume that an existing wire should be replaced with one that has the same characteristics as the damaged one. Determining the characteristics of the existing wire is the correct place to begin, but it is always best to confirm that it is the proper size for the load imposed on it. There is a certain risk in simply matching the wire size, because it assumes the installed wire was correctly sized.

For wiring outside of the equipment, confirm the correct wire size by referring to the manufacturer's installation, operation, and maintenance (IOM) manual and the appropriate tables in the *NEC*®. As a general rule, when replacing a factory-installed wire inside a unit cabinet, the wire size should be correct. HVACR equipment is built to comply with other standards, such as those put forth by Underwriters Laboratories (UL), rather than to *NEC*® standards.

FIGURE 6-6 shows a general guide for wire ampacity ranges, but it is important to remember that these values are very general. There are many *NEC*® tables, all of which specify the allowable ampacity of wire under a variety of specific conditions. There is a reason why electricians are licensed and require specific training in the use of the *NEC*® and the information it contains; ensuring that the correct table is being consulted can be a complex act in itself. In summary, the ampacity values shown in the figure will not apply in every case. Whenever there is doubt about the proper wire size, consult a licensed electrician. Although it is important for a technician to understand the practice of wire sizing, the guidance provided here is in no way sufficient to reach a level of proficiency.

HVACR equipment manufacturers provide information about the minimum circuit ampacity (MCA) for a unit, which is used to determine the correct wire size. This information is usually included on the equipment data plate, as well as in the engineering data.

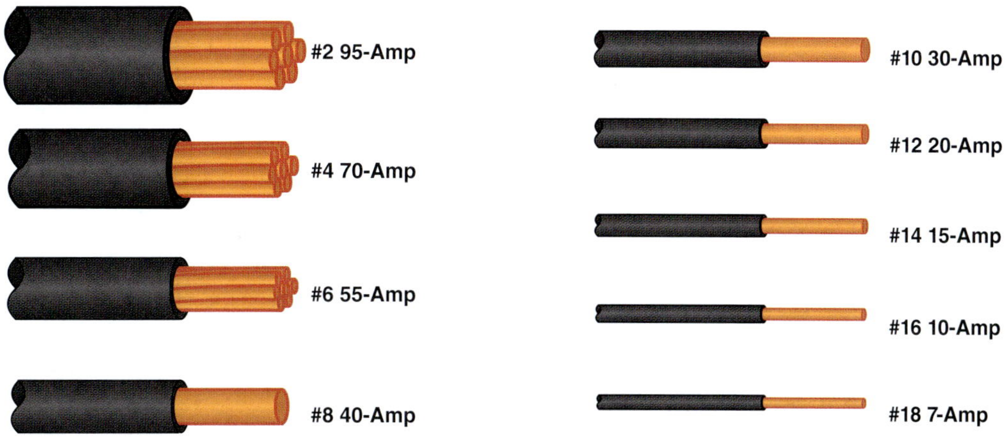

FIGURE 6-6 Wire gauge current ratings.

▶ TECHNICIAN TIP

Wire-Size Calculators

Some wire manufacturers have web-based calculators that can be used to determine the correct wire size. One such calculator is available at www.paigewire.com. To use the calculator, the technician must provide inputs, including but not limited to the following:

- voltage
- number of phases
- amperes
- length of run
- allowable voltage drop (%).

Using the 11.2 MCA shown in Figure 6-7, and selecting a 240 VAC, single-phase application with a maximum 3% voltage drop and 100 feet (30 meters) of wire, the resulting wire size is #14 AWG. However, note that if the length is increased by just 20 feet (6 meters), to 120 feet (37 meters), the wire size increases to #12 AWG due to the increased voltage drop at the new length.

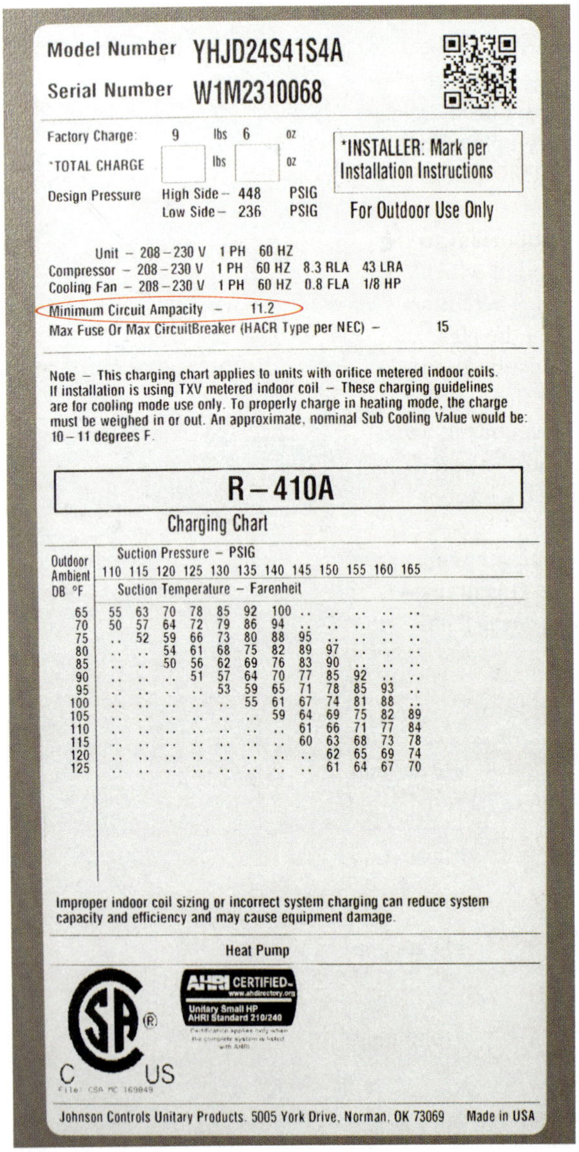

FIGURE 6-7 A condensing-unit data plate showing the ampacity rating.

FIGURE 6-7 shows the minimum ampacity rating on the data plate for a condensing unit used in a residential split system. Also note that the rated load amp (RLA) rating shown for the compressor is 8.3 amps, while the locked rotor amp (LRA) rating is 43 amps. The RLA represents the normal running current of the compressor under maximum load, whereas the LRA represents the instantaneous current surge that occurs in motors at start-up or when the rotor is locked in place and cannot turn. The ampacity rating and maximum circuit-breaker rating are based on the RLA rather than the LRA. This is because the LRA is so brief that its effect on the circuit is negligible, except in the selection of fuse or circuit breaker protection.

Some equipment manufacturers may provide helpful wire selection guides—such as the one shown in **FIGURE 6-8**, which is based on the length of the wiring for an induction motor of a specific horsepower rating. Notice that the wire size for a motor of a given horsepower increases as the length of the conductor increases. Keep in mind, however, that the manufacturers provide such tables and calculators as guides only, and that they point out *NEC*® and local codes must always prevail.

Selecting Conductor Insulation

In addition to its size and voltage rating, wiring labels include the insulation designation. There are many types of insulation. Each is rated for its ability to withstand heat and the environmental conditions for which it is suitable. For example, the designation THHN (**FIGURE 6-9**) stands for thermoplastic, heat-resistant, nylon-coated insulation. This type of insulation is commonly used on both stranded and solid wire in electrical distribution systems with voltages up to 600 VAC. If the wire marking includes THWN, then the wire is designated as thermoplastic, heat- and water-resistant, nylon-coated insulation. THHN insulation is rated for 194°F (90°C), while THWN insulation is rated for 167°F (75°C). For high-temperature uses, XHHW (cross-linked polyethylene, high-heat-resistant, water-resistant insulation) might be specified. It is made of a thermoset material that will not deform when heated. For wiring repairs inside a unit cabinet, it is generally acceptable to use wire with the same type of insulation found on the other wires in the cabinet.

To select the appropriate wire size, based on ampacity, for a condensing-unit whip, follow the steps is **SKILL DRILL 6-1**.

Three Phase (208 Volt) – Wire-Size Selection Guide*					
	Cable Length				
HP	Up to 30"	31"–50"	51"–75"	76"–100"	101"–150"
½	14	14	14	14	14
¾	14	14	14	14	14
1	14	14	14	14	14
1½	14	14	14	14	14
2	14	14	14	14	14
3	14	14	14	14	12
5	14	14	12	12	10
7½	14	12	10	10	8
10	14	12	10	8	6
15	12	10	8	6	6
20	12	8	8	6	4
25	10	8	6	4	4

FIGURE 6-8 Example of a wire-size selection guide.

> ▶ **TECHNICIAN TIP**
>
> **THHN Wire**
>
> THHN wire is popular in construction due to its durability and higher capacity for current and heat than some others. Many runs of wire in a building are quite long. The higher ratings of THHN can help prevent costly increases in wire size. Using larger wire can also cause an increase in the size of conduit to contain it. Inside HVACR equipment, however, very little THHN will be found. The wires used in these applications are sized to different standards and are typically very short in length, making voltage drop due to wire size and length insignificant. The wire used inside equipment cabinets and control panels is also softer and easier to manipulate than THHN.

FIGURE 6-9 Example of wire markings.

SKILL DRILL 6-1 Selecting the Appropriate Wire Size, Based on Ampacity, for a Condensing-Unit Whip

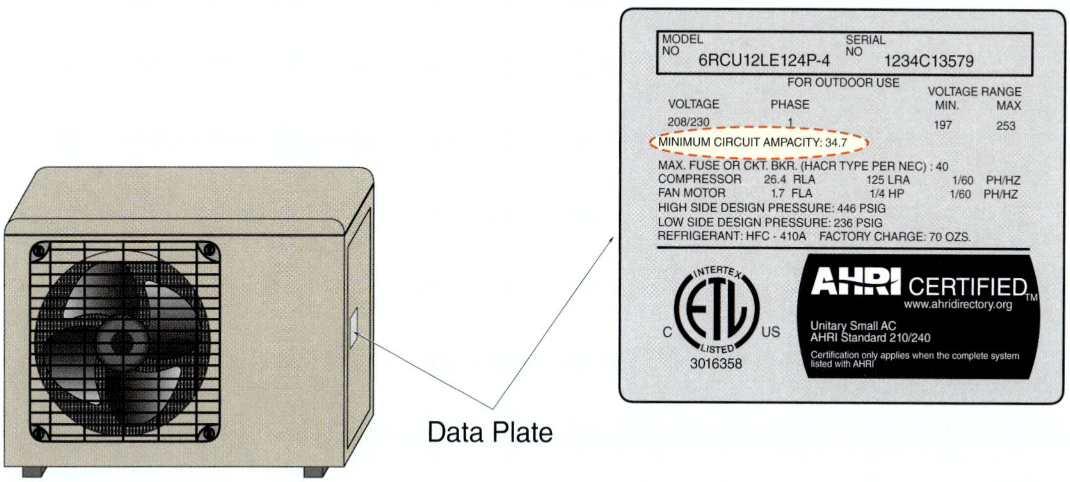

1. Locate the data plate for the condensing unit. Examine the information and look for an entry related to the terms "circuit ampacity" and "minimum circuit ampacity" (or "MCA"). Depending on the manufacturer and vintage of the unit, the information may or may not be on the data plate. If the information is not provided, the IOM or specifications for the specific unit model must be acquired. In the example, shown here, the circuit ampacity for wire sizing is 34.7 amps.

Continued

310.60 ARTICLE 310 - CONDUCTORS FOR GENERAL WIRING

Table 310.15(B)(16) (formerly Table 310.16) Allowable Ampacities of Insulated Conductors Rated Up to and Including 2000 Volts, 60°C Through to 90°C (140°F Through 194°F), Not More Than Three Current-Carrying Conductors in Raceway, Cable, or Earth (Directly Buried), Based on Ambient Temperatures of 30°C (86°F)*

	Temperature Rating of Conductor [See Table 310.104(A).]						
	60°C (140°F)	75°C (167°F)	90°C (194°F)	60°C (140°F)	75°C (167°F)	90°C (194°F)	
Size AWG or kcmil	Types TW, UF	Types RHW, THHW, THW, THWN, XHHW, USE, ZW	Types TBS, SA, SIS, FEP, FEPB, MI, RHH, RHW-2, THHN, THHW, THW-2, THWN-2, USE-2, XHN, XHHW, XHHW-2, ZW-2	Types TW, UF	Types RHW, THHW, THW, THWN, XHHW, USE, ZW	Types TBS, SA, SIS, FEP, FEPB, MI, RHH, RHW-2, THHN, THHW, THW-2, THWN-2, USE-2, XHN, XHHW, XHHW-2, ZW-2	Size AWG or kcmil
	COPPER			ALUMINUM OR COPPER-CLAD ALUMINUM			
18**	---	---	14	---	---	---	18**
16**	---	---	18	---	---	---	16**
14**	15	20	25	---	---	---	14**
12**	20	25	30	15	20	25	12**
10**	30	35	40	25	30	35	10**
8	40	50	55	35	40	45	8
6	55	65	75	40	50	55	6
4	70	85	95	55	65	75	4
3	85	100	115	65	75	85	3
2	95	115	130	75	90	100	2
1	110	130	145	85	100	115	1
1/0	125	150	170	100	120	135	1/0
2/0	145	175	195	115	135	150	2/0
3/0	165	200	225	130	155	175	3/0
4/0	195	230	260	150	180	205	4/0
250	215	255	290	170	205	230	250
300	240	285	320	195	230	260	300
350	260	310	350	210	250	280	350
400	280	335	380	225	270	305	400
500	320	380	430	260	310	350	500
600	350	420	475	285	340	385	600
700	385	460	520	315	375	425	700
750	400	475	535	230	385	435	750
800	410	490	555	330	395	445	800
900	435	520	585	355	425	480	900
1000	455	545	615	375	445	500	1000
1250	495	590	665	405	485	545	1250
1500	525	625	705	435	520	585	1500
1750	545	650	735	455	545	615	1750
2000	555	665	750	470	560	630	2000

* Refer to 310.15(B)(2) for the ampacity correction factors where the ambient temperature is other than 30°C (85°F).
Refer to 310.15(B)(3)(a) for more than three current-carrying conductors.
** Refer to 240,4(D) for conductor overcurrent protection limitations.

2. Refer to Table 310.15(B)(16) in the current *NEC*® handbook. Note that the table has separate sections within the columns, based on the conductor material: copper, copper/aluminum alloys, and aluminum. Since copper conductors are what will be worked with, locate that section of the chart. A selection must come from this area. Note that the current *NEC*® table may differ from the example here, due to the consistent update and revision process.

Continued

310.60 ARTICLE 310 - CONDUCTORS FOR GENERAL WIRING

Table 310.15(B)(16) (formerly Table 310.16) Allowable Ampacities of Insulated Conductors Rated Up to and Including 2000 Volts, 60°C Through to 90°C (140°F Through 194°F), Not More Than Three Current-Carrying Conductors in Raceway, Cable, or Earth (Directly Buried), Based on Ambient Temperatures of 30°C (86°F)*

	Temperature Rating of Conductor [See Table 310.104(A).]						
	60°C (140°F)	75°C (167°F)	90°C (194°F)	60°C (140°F)	75°C (167°F)	90°C (194°F)	
Size AWG or kcmil	Types TW, UF	Types RHW, THHW, THW, THWN, XHHW, USE, ZW	Types TBS, SA, SIS, FEP, FEPB, MI, RHH, RHW-2, THHN, THHW, THW-2, THWN-2, USE-2, XHN, XHHW, XHHW-2, ZW-2	Types TW, UF	Types RHW, THHW, THW, THWN, XHHW, USE, ZW	Types TBS, SA, SIS, FEP, FEPB, MI, RHH, RHW-2, THHN, THHW, THW-2, THWN-2, USE-2, XHN, XHHW, XHHW-2, ZW-2	Size AWG or kcmil
	COPPER			ALUMINIUM OR COPPER-CLAD ALUMINIUM			
18**	---	---	14	---	---	---	18**
16**	---	---	18	---	---	---	16**
14**	15	20	25	---	---	---	14**
12**	20	25	30	15	20	25	12**
10**	30	35	40	25	30	35	10**
8	40	50	55	35	40	45	8
6	55	65	75	40	50	55	6
4	70	85	95	55	65	75	4
3	85	100	115	65	75	85	3
2	95	115	130	75	90	100	2
1	110	130	145	85	100	115	1
1/0	125	150	170	100	120	135	1/0
2/0	145	175	195	115	135	150	2/0
3/0	165	200	225	130	155	175	3/0
4/0	195	230	260	150	180	205	4/0
250	215	255	290	170	205	230	250
300	240	285	320	195	230	260	300
350	260	310	350	210	250	280	350
400	280	335	380	225	270	305	400
500	320	380	430	260	310	350	500
600	350	420	475	285	340	385	600
700	385	460	520	315	375	425	700
750	400	475	535	230	385	435	750
800	410	490	555	330	395	445	800
900	435	520	585	355	425	480	900
1000	455	545	615	375	445	500	1000
1250	495	590	665	405	485	545	1250
1500	525	625	705	435	520	585	1500
1750	545	650	735	455	545	615	1750
2000	555	665	750	470	560	630	2000

* Refer to 310.15(B)(2) for the ampacity correction factors where the ambient temperature is other than 30°C (85°F).
Refer to 310.15(B)(3)(a) for more than three current-carrying conductors.

** Refer to 240,4(D) for conductor overcurrent protection limitations.

3. Above the ampacity tables listed for copper conductors, note that there are separate columns for various types of wire insulation, since the integrity and characteristics of the insulation affects safe current-carrying capacity. Whips manufactured for HVACR and similar outdoor applications are typically assembled with wire that is rated THHN. Locate this column on the table in the left section related to copper conductors. Note that the wire ratings for THHN and THWN are in different columns. If a given wire is clearly marked with both ratings, it is OK to prefer the column that shows the most favorable ampacity. The current-carrying capacity of THHN is higher than that of THWN, so the ampacity column for THHN may be consulted. If in doubt about the wire's rating, consult catalog data via the Internet or a distributor.

Continued

310.60 ARTICLE 310 - CONDUCTORS FOR GENERAL WIRING

Table 310.15(B)(16) (formerly Table 310.16) Allowable Ampacities of Insulated Conductors Rated Up to and Including 2000 Volts, 60°C Through to 90°C (140°F Through 194°F), Not More Than Three Current-Carrying Conductors in Raceway, Cable, or Earth (Directly Buried), Based on Ambient Temperatures of 30°C (86°F)*

	Temperature Rating of Conductor [See Table 310.104(A).]						
	60°C (140°F)	75°C (167°F)	90°C (194°F)	60°C (140°F)	75°C (167°F)	90°C (194°F)	
Size AWG or kcmil	Types TW, UF	Types RHW, THHW, THW, THWN, XHHW, USE, ZW	Types TBS, SA, SIS, FEP, FEPB, MI, RHH, RHW-2, THHN, THHW, THW-2, THWN-2, USE-2, XHN, XHHW, XHHW-2, ZW-2	Types TW, UF	Types RHW, THHW, THW, THWN, XHHW, USE, ZW	Types TBS, SA, SIS, FEP, FEPB, MI, RHH, RHW-2, THHN, THHW, THW-2, THWN-2, USE-2, XHN, XHHW, XHHW-2, ZW-2	Size AWG or kcmil
	COPPER			ALUMINIUM OR COPPER-CLAD ALUMINIUM			
18**	---	---	14	---	---	---	18**
16**	---	---	18	---	---	---	16**
14**	15	20	25	---	---	---	14**
12**	20	25	30	15	20	25	12**
10**	30	35	40	25	30	35	10**
8	40	50	55	35	40	45	8
6	55	65	75	40	50	55	6
4	70	85	95	55	65	75	4
3	85	100	115	65	75	85	3
2	95	115	130	75	90	100	2
1	110	130	145	85	100	115	1
1/0	125	150	170	100	120	135	1/0
2/0	145	175	195	115	135	150	2/0
3/0	165	200	225	130	155	175	3/0
4/0	195	230	260	150	180	205	4/0
250	215	255	290	170	205	230	250
300	240	285	320	195	230	260	300
350	260	310	350	210	250	280	350
400	280	335	380	225	270	305	400
500	320	380	430	260	310	350	500
600	350	420	475	285	340	385	600
700	385	460	520	315	375	425	700
750	400	475	535	230	385	435	750
800	410	490	555	330	395	445	800
900	435	520	585	355	425	480	900
1000	455	545	615	375	445	500	1000
1250	495	590	665	405	485	545	1250
1500	525	625	705	435	520	585	1500
1750	545	650	735	455	545	615	1750
2000	555	665	750	470	560	630	2000

* Refer to 310.15(B)(2) for the ampacity correction factors where the ambient temperature is other than 30°C (85°F).
Refer to 310.15(B)(3)(a) for more than three current-carrying conductors.

** Refer to 240,4(D) for conductor overcurrent protection limitations.

4. With the appropriate column of values identified, read down the column until a current value greater than 34.7 amps is encountered, which is the MCA for the unit in this example. Then follow the row to the left to identify the needed wire size. The first value encountered that is greater than 34.7 is 40 amps. Reading left indicates that the wire size is #10 AWG. #12 AWG wire, just above it on the table, can handle only 30 amps at these conditions and is therefore not acceptable.

Continued

5. Measure the distance from the condensing unit to the disconnect to determine the length of the whip. Remember to add length for turns and bends; the whip should not be taut once installed. Whips are typically sold in four-, six-, and eight-foot (1.2 m, 1.8 m, and 2.4 m) lengths, but the flexible conduit material and conductors can be cut if needed.

SAFETY TIP

This drill is intended to help demonstrate and practice the process of selecting appropriate wire sizes for a condensing unit. However, the selection of conductors and the installation of power wiring should only be attempted and completed by appropriately qualified individuals.

6. With the whip length and wire size determined, a product from a physical or digital catalog can be selected. Use the Internet or a catalog sheet to locate condensing-unit whips. Select a whip that matches the needed wire size and length. Depending on local codes and/or job specifications, a choice will need to be made between whips with metal box connectors and non-metallic box connectors. Note that the manufacturer determines the diameter of the flexible conduit, based on wire size. Both disconnect switch enclosures and HVACR units are designed to accommodate more than one size of connecting hardware.

▶ Identifying Damaged Conductors

K06003 Explain how to identify a damaged conductor.

An important part of troubleshooting an electrical problem is using system-specific knowledge to detect trends and eliminate possibilities based on a its behavior. For example, if a circuit breaker trips repeatedly or a fuse blows more than once, there can be several potential causes. If the problem always occurs when the compressor is trying to start, the first thing to test is the compressor starting sequence and the starting components. But when these events occur at random times while the compressor is running, the compressor starting components are far less likely to be the culprit. There is a chance that a conductor with damaged insulation is intermittently coming into contact with a conductive surface. Another possibility is a poor wiring termination that gets progressively hotter and causes high voltage drop and increased current under load.

Over time, vibration and expansion/contraction processes can cause conductors to loosen at their termination points. Insulated wiring that is not properly secured can also rub against sharp metal edges and be damaged to the point where the conductor is exposed and touches the metal. Before inspecting for wiring damage, always lock out and tag the system power sources first.

FIGURE 6-10 Example of a scorched wire.

If the problem is caused by a damaged conductor and/or a poor termination, the distinctive odor of scorched insulation might be noticeable. During a visual inspection, charring can often be seen at the point where a damaged conductor has touched another conductive surface, or where a loose connection has caused overheating (**FIGURE 6-10**).

When searching for wiring damage during a periodic inspection or troubleshooting process, focus on areas where wiring passes through or near metal cabinet components. Carefully running the fingers along a section of deenergized conductor can help to locate nicks and cuts in the insulation that may be difficult to see otherwise. Some damage may seem minimal, but if it exposes the copper wiring beneath, the door is open for a short circuit. A drop of water that falls between the nick and the metal cabinet can trip a breaker or blow a fuse. Once the water drop goes away, the system will operate again with no obvious problems.

Also look for areas where wiring runs near refrigerant piping. This is especially true of the hot-gas refrigerant line, where the surface temperature can approach 200°F (93°C). Continuous contact with the hot-gas line, which also vibrates, can damage the insulation. But a wire can rub against any refrigerant line and eventually expose the conductor inside over time. Heat simply accelerates the process.

The insulation is far more fragile on traditional thermostat wires than on products designed for higher voltages. Thermostat wiring routed to an outdoor unit is often given limited attention and is wrapped around piping and power wiring. It is also subject to damage from grass trimmers and similar equipment, since it is not usually inside a protective conduit. Wiring damage is often found in thermostat wiring between the structure wall and the unit cabinet.

▶ Wire Terminations

K06004 Describe the tools, equipment, and methods used to terminate wiring.

S06002 Terminate wiring by using wire nuts.

S06003 Terminate wiring by using common crimp connectors.

S06004 Install a properly sized condensing-unit whip.

A number of methods are used to connect wires to their source and destination points. It is very important for technicians and installers to be able to make these connections correctly and securely. There are three basic types of conductor terminations:

1. wire-to-wire, where wire nuts or butt connectors are used to make the connection
2. wire-to-lug terminal, where a stripped conductor end is inserted directly under a screw terminal or into a lug and secured
3. wire-to-connector, where a stripped conductor end is crimped into a connector and the connector is then attached to a screw terminal.

Some examples are shown in **FIGURE 6-11**.

Tools Used to Terminate Conductors

In order to terminate conductors, the wire must be cut to the correct length and the insulation must be stripped back to expose the conductor. **FIGURE 6-12** shows diagonal cutting pliers that work very well. Also shown is a tool that cuts smaller wire and can also strip the wire and crimp terminals on the end. Such a tool is handy since it can prepare and terminate a wire without changing tools, as long as the wire is not too large for the cutting jaws. **FIGURE 6-13** shows a wire stripper designed for production work where many wire ends need to be stripped. For a large control-wiring job, this tool would be very handy. The wire is simply placed in the jaws, and the tool holds the wire firmly as it strips off the insulation on the end. With common wire strippers, as shown in Figure 6-12, the user must hold the wire, squeeze the handles, and physically pull on the tool to remove the insulation.

FIGURE 6-11 Examples of wiring connections and terminations.

There are also tools that cut and crimp but that do not strip the wire (**FIGURE 6-14**). These tools might be considered better for crimping than the tool shown in Figure 6-12. Their construction provides a more comfortable grip and more leverage for crimping.

FIGURE 6-12 Wire cutters and strippers.

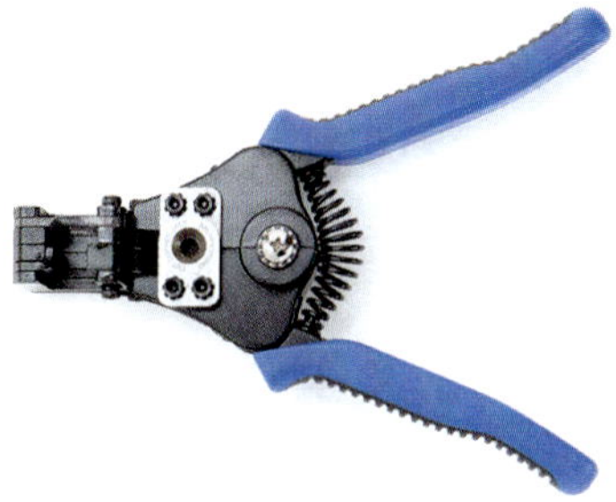

FIGURE 6-13 Production-quality wire strippers.

When using wire nuts to connect two conductors, side-cutting pliers (FIGURE 6-15) might be used to twist the two conductor ends together before applying the wire nut. This is commonly done with solid conductors, but stranded conductors are not generally twisted before they are inserted into the wire nut. These pliers, also known as linesman pliers or electrician's pliers, are used for a number of tasks, including wire cutting. Some versions also have limited crimping capabilities. Also shown are, needle-nose, or long-nose, pliers that are also valuable for many wiring tasks.

FIGURE 6-14 Crimping tool.

Wire Nuts

Wire nuts (FIGURE 6-16) are commonly used to connect two conductors. Some electrical components come with wires attached to them, referred to as pigtails. Control transformers and fan motors are examples. Although some of these pigtails are equipped with termination connectors, many are not. Wire nuts are often used to connect them to the electrical circuits. As shown in TABLE 6-1, wire nuts are color-coded to indicate the conductor sizes for which they

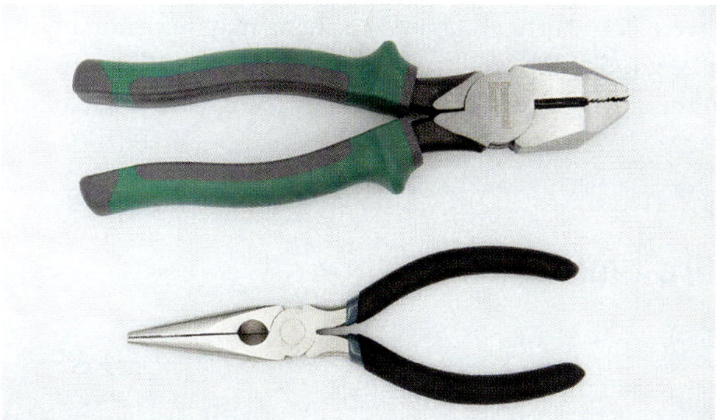

FIGURE 6-15 Side-cutting and needle-nose pliers.

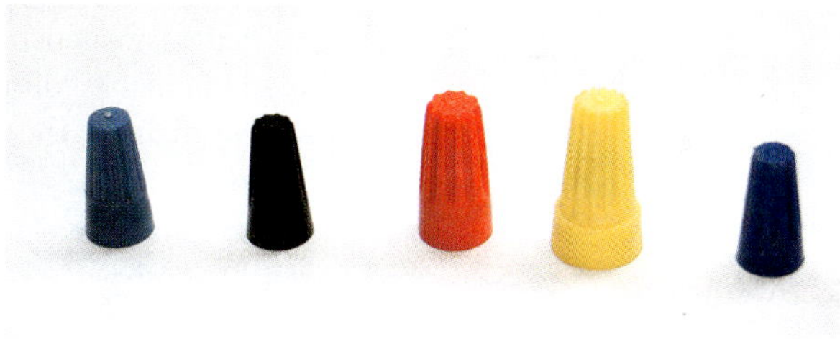

FIGURE 6-16 Examples of wire nuts.

TABLE 6-1 Wire-Nut Color Codes

Color	Type	Spring Type	Range (AWG)	Wire Min.	Max.
Gray	Non-Winged	Fixed	22-16	2#22	2#16
Blue	Non-Winged	Fixed	22-14	3#22	3#16
Orange	Non-Winged	Fixed	22-14	3#22	2#14w/ 1#18
Yellow	Non-Winged	Fixed	18-10	1#14w/ 1#18	1#10w/ 1#14
Red	Non-Winged	Fixed	18-10	2#14	4#12
Yellow	Winged	Freespring	22-10	2#18	3#12Sol
Red	Winged	Freespring	18-8	2#16	5#12 sol
Blue	Winged	Freespring	14-6	1#10w/ 1#12	1#6w/ 2#8

are intended. Note that the information shown in this table does not apply to all products. The appropriate data for each manufacturer and product line must be consulted.

Wire nuts can be either the free-spring or the fixed-spring (standard) type. Free-spring wire nuts are often winged, as shown in Figure 6-16. Some job or project specifications may require one type or the other.

Installing Wire Nuts

It is common to pre-twist solid wires together before applying the wire nut; stranded wires are simply placed side by side into the wire nut, and the wire nut twists the conductors together as it is installed. However, some wire-nut manufacturers may specify a different approach. The topic of pre-twisting wires is one that has stirred debate in the electrical community for years, and many electricians have their own thoughts on the matter. In general, it is best to follow the manufacturer's guidance, although sometimes that guidance does not specifically address twisting.

When pre-twisting solid wires, always twist in a clockwise direction as viewed from the tip of the wires. This ensures that the twist is in the same direction as the threads of the wire nut. If the wires are twisted in the opposite direction, they will try to untwist as the wire nut is applied. If pre-twisting is planned, strip the ends of the wires longer than usual, then cut the end of the twisted assembly off to the desired length. This ensures that the twisted splice is flush on the end.

When connecting a stranded wire to a solid wire, pre-twisting is not generally recommended. However, because the two wires are placed side by side, allow the stranded wire to extend slightly beyond the end of the solid wire by approximately ⅛" (3 mm).

The recommended strip length for the wires generally ranges from ¼" to as much as ¾" (6–19 mm), depending on the wire size and the wire-nut type and brand. The recommended strip length is generally printed on the exterior of the packaging or on one of the box-closure flaps; this offers the best guidance. As a general rule, the stripped length should be half to two-thirds the length of the wire nut. Over-stripping the wire can lead to exposed conductors and is cause for an electrical inspector to reject work. No bare wiring should be visible below the skirt of the wire nut.

Wire-nut connections that have electrical tape applied to them may occasionally be seen. This is not generally accepted as good practice. If tape is needed to secure the wire nut, it is an unreliable connection from the start. To terminate wiring by using wire nuts, follow the steps in **SKILL DRILL 6-2**.

SKILL DRILL 6-2 Terminating Wiring by Using Wire Nuts

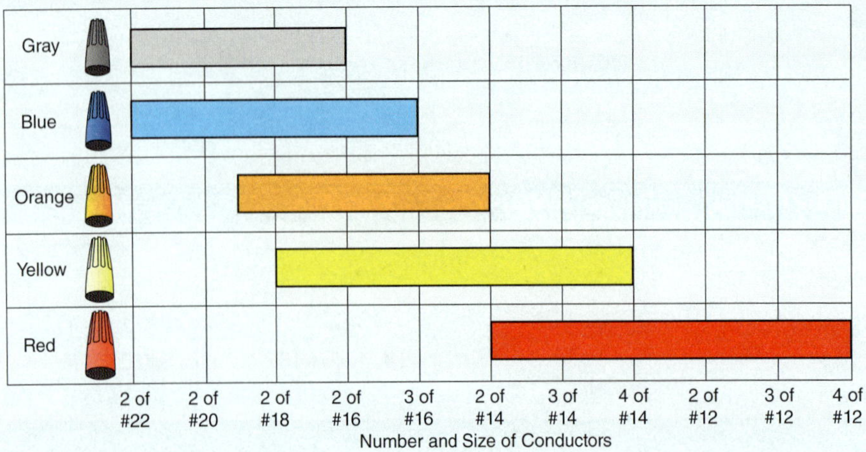

1. Select two or more pieces of insulated wire for practice. A wire size of #18 to #12 AWG is recommended for this exercise. Once the wire size has been chosen, select the correct size of wire nut from a manufacturer's chart, which is often shown on the box. The chart shown here is only an example and may not be appropriate for the product in use.

Continued

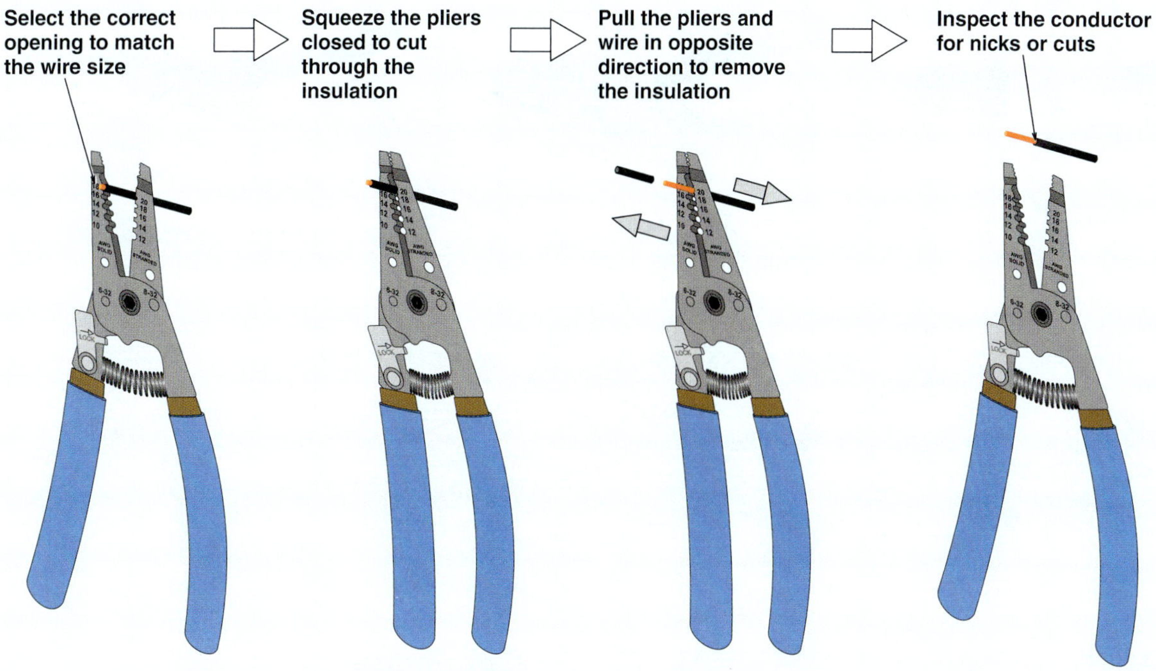

2. Use wire strippers to strip the insulation from one end of each of the two wires. Be sure to select the opening in the wire stripper that matches the size of the wire. Consult the connector box or manufacturer's literature for guidance on strip length. Check the stripped area carefully to ensure that a solid conductor has not been nicked, or that the strands of a stranded conductor have not been severed or significantly nicked. This happens easily if the correct wire-stripper opening is not used. If the conductor is damaged, remove the damaged portion and strip the wire again with greater care. Remember that solid conductors intended to be pre-twisted should be stripped longer and then cut back to the appropriate length after they are twisted.

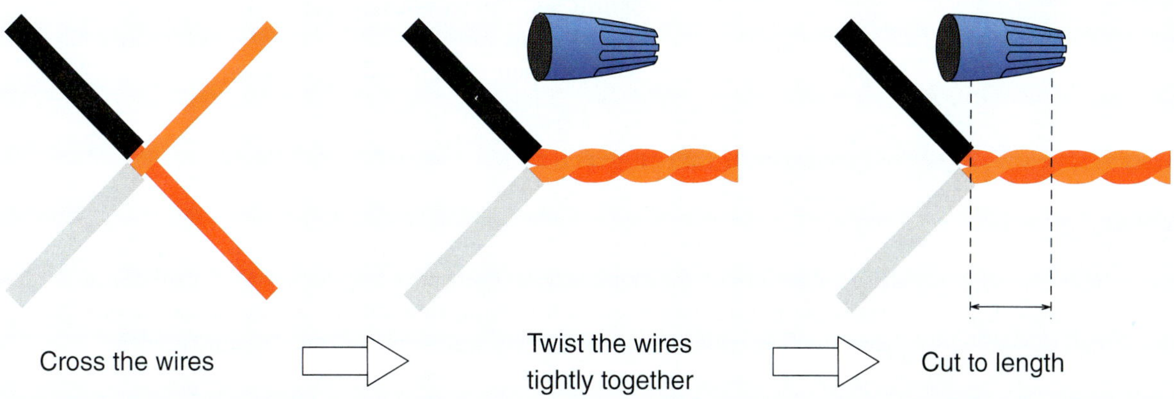

3. If stranded wire is being used, twist the strands of each individual conductor to ensure the strands remain tight to each other; do not twist the ends of the two conductors together, however. Solid conductors can be pre-twisted together if desired or recommended by the wire-nut manufacturer. Twist the solid-conductor ends until the conductors are twisted together roughly twice the length of the wire nut to be used. Ensure the bare portions are twisted tightly together, without intermittent gaps. Then cut the end of the twisted pair to make them even and the proper stripped length for the wire nut.

Continued

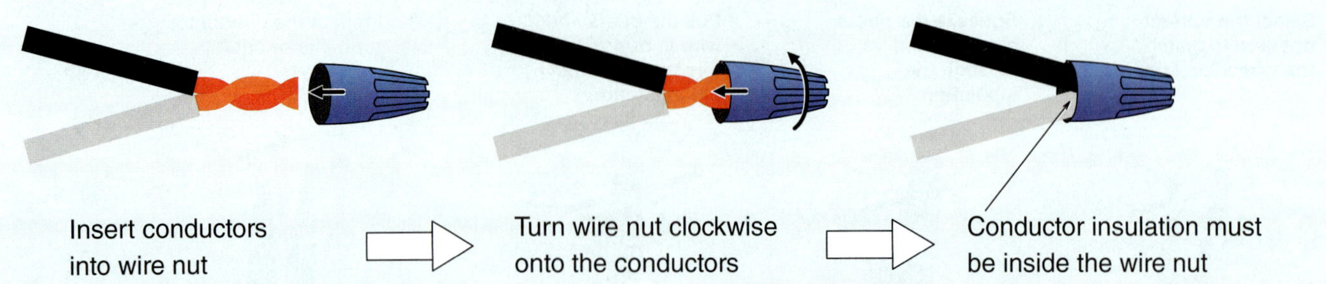

4. Place the chosen wire nut over the ends and push down as you thread the nut over the exposed conductors, turning clockwise. The ends of the wires should be flush on the end as they are inserted, unless stranded wire is being connected to solid wire. In that case, allow the stranded wire to extend $\frac{1}{8}$" (3 mm) past the solid wire. The nut should be good and tight, but do not allow the wires to bottom out in the nut while twisting continues. This will strip the threads that have been created in the wires. If the nut does not tighten firmly on the connection, then the joint isn't fitting together as it should; another wire-nut size might be more appropriate. Give the wire nut a slight tug to ensure it is firmly attached.

FIGURE 6-17 Crimp connectors.

Crimp Connectors

There are many types of crimp connectors (**FIGURE 6-17**) used to terminate conductors. What they have in common is that they are designed to be attached to a bare conductor using a crimping tool.

Notice that most of the terminals pictured throughout this section are insulated. The use of insulated terminals is highly recommended. The terminals crimped onto wires in an HVACR unit at the factory will often be uninsulated. However, the connectors used are also designed for crimping by machinery or tools that are not available in the field. The manufactured crimps are very sound, and the insulator would interfere with the crimping process. If some of the factory-made terminations are inspected closely, the difference in their appearance versus the appearance of field-installed terminations can be seen.

Like wire nuts, insulated crimp connectors are color-coded to match the wire size for which they are recommended, as follows:

- red—for #22 up to #18 AWG
- blue—for #16 or #14 AWG
- yellow—for #12 or #10 AWG

It is important to note that crimp connectors that HVACR technicians work with are designed exclusively for stranded wire. There are crimp connectors for solid wire, but solid wire is rarely used inside an HVACR unit. In addition, due to the nature of solid wire, crimping them with a hand tool is rarely successful; different tools are needed.

Blade Terminals

Blade terminals (**FIGURE 6-18**) consist of male and female parts. Blade terminals are very common in HVACR equipment, especially for low-voltage wiring. Most general-purpose relays used in HVACR equipment are equipped with male blade terminals. The connectors are also used extensively for line-voltage terminations that serve low-current loads. A piggyback version of the connector is shown in **FIGURE 6-19**. These extend the number of wires that can be connected to a single terminal. Blade terminals are often incorrectly referred to as spade terminals, which are presented in the next section.

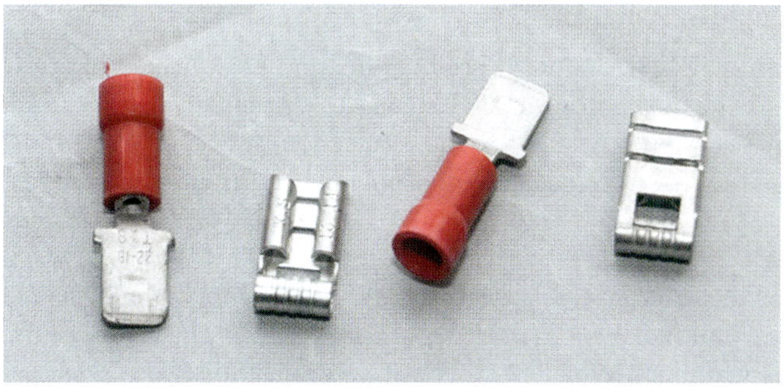

FIGURE 6-18 Male and female blade terminals.

FIGURE 6-19 Piggyback blade terminal.

▶ TECHNICIAN TIP

Compressor Terminals

Compressors are often fitted with electrical connections that have a special type of male blade terminal, as shown in **FIGURE 6-20**. Some compressor terminals may even be a simple round pin. These terminals are sealed using special glass materials that can withstand refrigerant pressures while electrically insulating the terminal. On the blade terminals, note that the blade terminal is welded to a slender pin.

Due to the presence of this pin in the center of the male terminal, standard blade terminals should not be used to make this connection. In addition, field-crimped connections at this high-current connection may not be very reliable. If compressor-terminal connectors must be replaced in the field, it is best to use a product manufactured for this purpose, such as the one shown in **FIGURE 6-21**. There are also a number of special connectors on the market that can be used to connect to a damaged compressor terminal or a round pin terminal, such as the product shown in **FIGURE 6-22**. Any of these products are likely a better choice than attempting to use a standard blade terminal and a field-crimped connection at this critical, hot, high-current location. Note that working at these compressor terminals can also be dangerous. The terminal can blow out due to the refrigerant pressure on the opposite side and cause a serious injury. When working on the compressor terminals, recover the refrigerant, or pump down the system down, so that the pressure is very low and not likely to cause injury.

FIGURE 6-20 Examples of male compressor terminals.

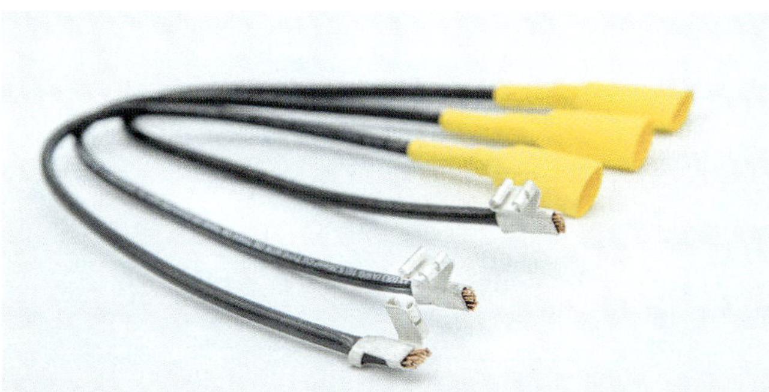

FIGURE 6-21 Female compressor terminals.

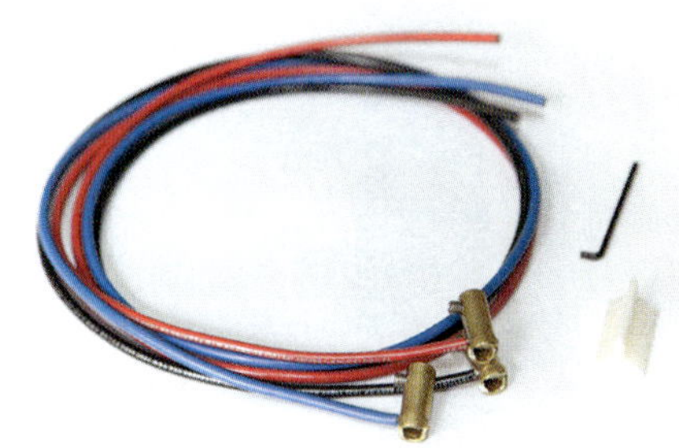

FIGURE 6-22 Compressor-terminal repair kit.

FIGURE 6-23 Spade and ring connectors.

Spade and Ring Terminals

Spade and ring terminals (**FIGURE 6-23**) are designed to be connected to screw terminals. The spade, or fork, terminal is designed to slip under a screw head, which is then tightened down to complete the connection. A bolt or screw must be inserted through the ring terminal, since it is not open at the end.

Butt Connectors

The butt connector (**FIGURE 6-24**) is another type of crimp connector, designed to receive a stripped conductor on each end to create a splice. The center of the connector contains a conductive material that presses against the two conductors when the body of the connector is crimped. The two conductors are inserted separately and crimped individually.

To terminate wiring by using common crimp connectors, follow the steps in **SKILL DRILL 6-3**.

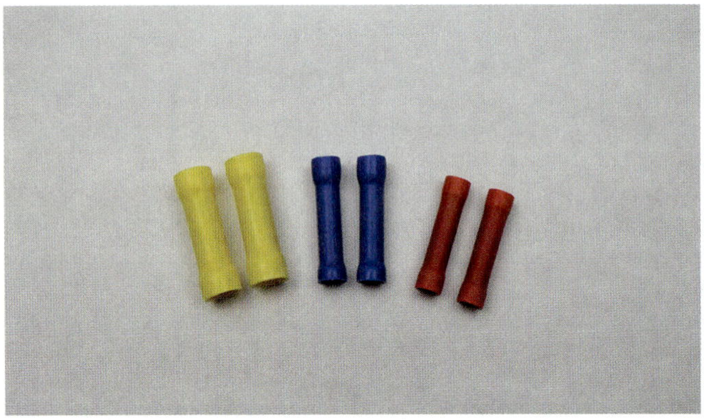

FIGURE 6-24 Butt connectors.

SKILL DRILL 6-3 Terminating Wiring by Using Common Crimp Connectors

1. Choose a piece of insulated, stranded wire and determine its wire size. Wire sizes from #16 to #12 AWG are recommended for practice. Select the appropriate size of crimp connector for the wire size. The crimp connector can be any of the types previously discussed.

Continued

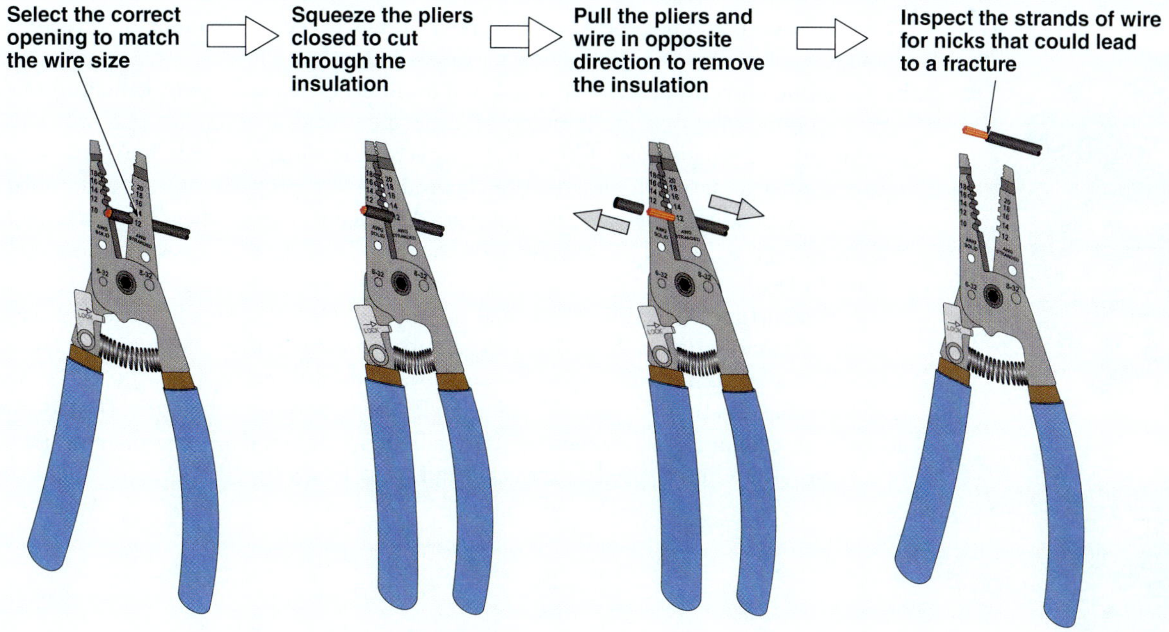

2. Strip the wire the appropriate distance from the tip. When pushed into the connector, the wire strands should not extend far enough forward to interfere with any termination hardware. The stripped end of the wire should extend only slightly past the metal portion of the crimp sleeve. Inspect the stripped conductor for damaged or nicked strands. Test-fit the stripped wire, with the wire insulation butted against the metal portion of the connector and trim the length of the stripped end if necessary.

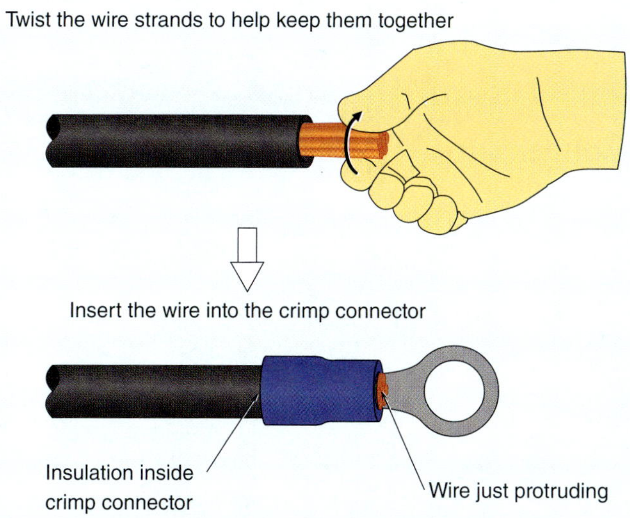

3. Slightly twist the strands together at the end, leaving no strands protruding so that the end slips cleanly into the connector. Then insert the wire into the connector.

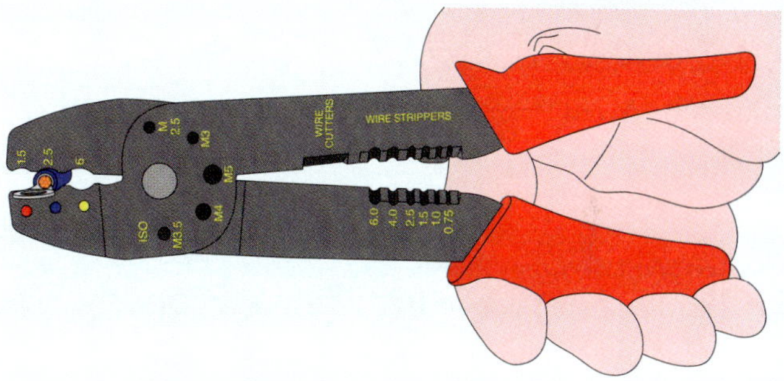

4. While holding the junction securely, place the jaws of the crimping tool over the barrel of the connector and squeeze. Be sure to use the jaws of the tool that match the size of the connector. It is also important to ensure the jaws of the tool are in the right place—surrounding the insulated portion of the connector that covers the crimp sleeve.

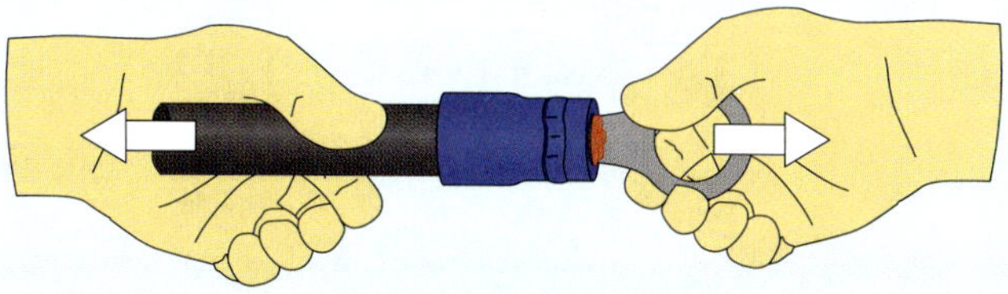

5. Inspect the crimp to ensure that it is uniform. Give the wire a slight tug to ensure that it is firmly seated in the connector.

Screw and Lug Terminations

There are several components that use screws or lugs to secure a conductor. **FIGURE 6-25** shows two contactors—one with screw terminals and one with lugs. When a screw terminal has no washer directly underneath it, it is best to connect stranded wire by installing a spade terminal on the wire, then slip it underneath the screw head. The contactor shown can also accommodate a female blade terminal. Many screw terminals, though, have an indented square washer underneath, designed to clamp down and properly capture a stranded or solid wire with no terminal attached. Lugs are designed to capture the wire in an opening, and the screw is tightened directly down on the wire. They can be used with both solid and stranded wire.

It is sometimes necessary to connect a solid wire to a screw terminal that has no indented washer underneath, in order to capture the wire. In that case, a loop is typically created in the end of the wire so that it wraps around the screw shank (**FIGURE 6-26**). When the loop is placed around the screw, place it with the open side to the right. The clockwise action of tightening the screw then tends to pull the wire around it rather than push it out.

To install a properly sized condensing-unit whip, follow the steps in **SKILL DRILL 6-4**.

FIGURE 6-25 **A.** Lug Terminals **B.** Screw Terminals.

FIGURE 6-26 Making a loop in a solid conductor.

SKILL DRILL 6-4 Installing a Properly Sized Condensing-Unit Whip

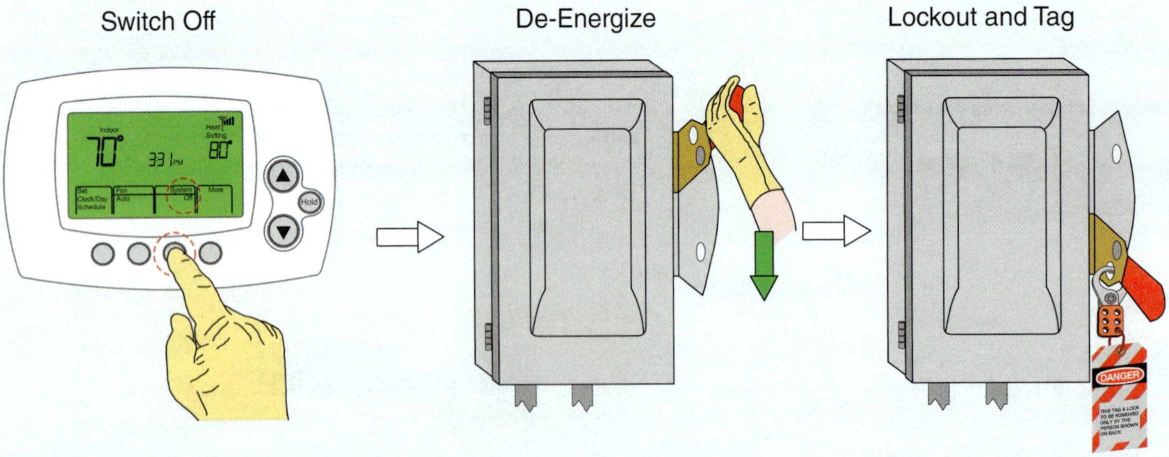

1. Ensure that all controls, including the thermostat, are set to the Off position. Deenergize, lock out, and tag the equipment to be serviced. Ensure that the process deenergizes not only the line voltage feeding the equipment but also the low-voltage control power source that may be located in a separate piece of equipment.

Continued

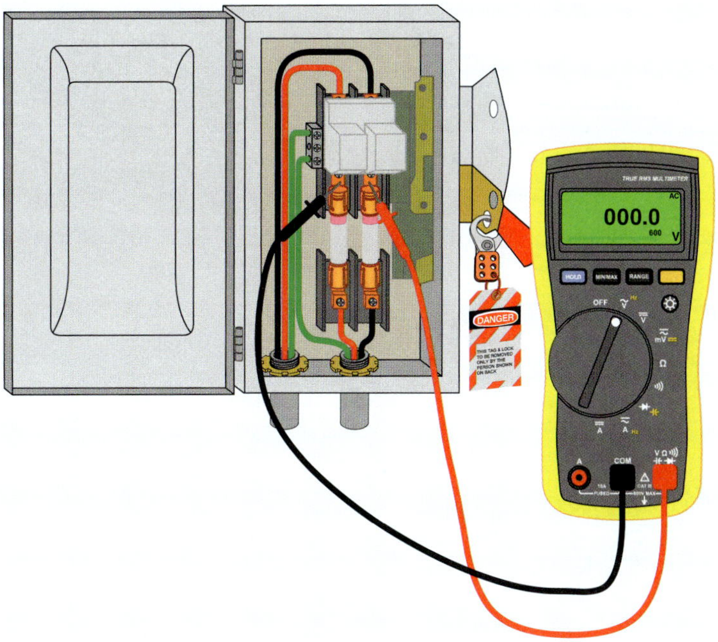

2. Next, ensure the power source serving the disconnect switch is switched off, locked out, and tagged. Then use a voltmeter to test the line-side terminals to confirm the power is off.

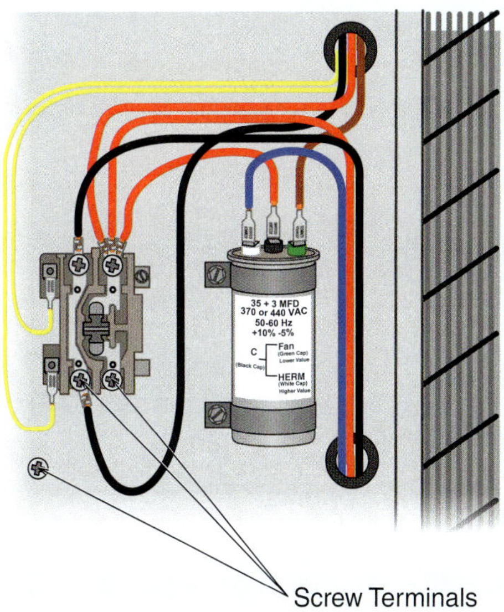

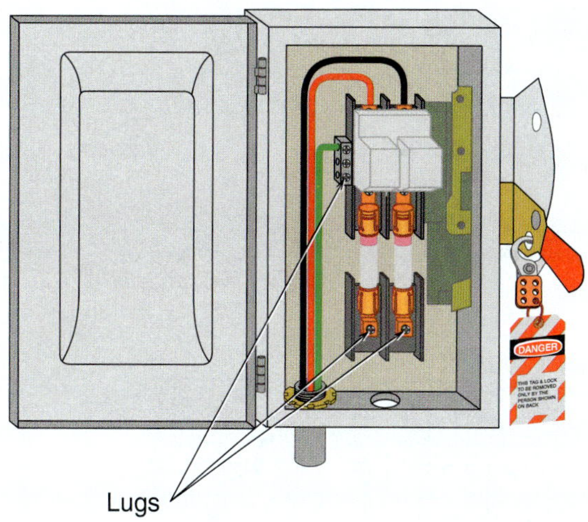

3. Examine the terminals inside the disconnect switch and inside the condensing unit where the whip terminations must be made. Most disconnect switches will be provided with lugs into which the bare conductor can be inserted and connected. In the condensing unit, you may find the wires go directly to the contactor, which may have lugs or screw terminals. If the screw terminals are not equipped with washers under the screw heads to capture the wire, use an appropriately sized spade terminal on the end of the conductors. Remember that the wire in the whip will be stranded rather than solid. Other units may have leads that are factory installed from the contactor to a small junction box where the incoming power connections are made. If so, those connections are usually made using wire nuts. Note that larger equipment often has a terminal board with lugs where the incoming power is connected. The ground wire will usually be connected at a green screw terminal or lug attached to the control cabinet wall.

Continued

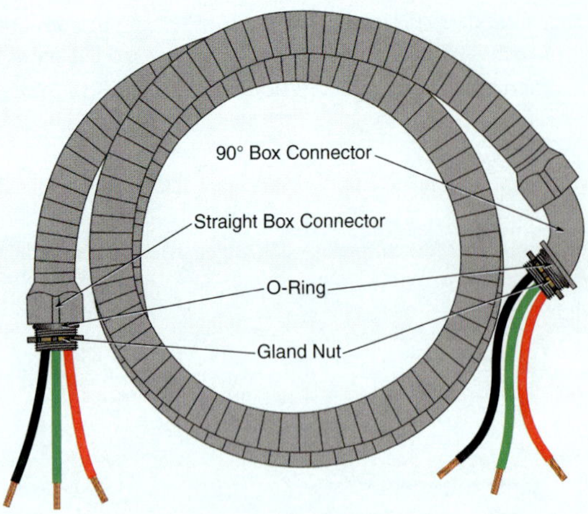

4. Whips are usually equipped with one straight box connector and one 90-degree box connector. Decide which end is best connected to the disconnect switch and which end is best connected to the unit.

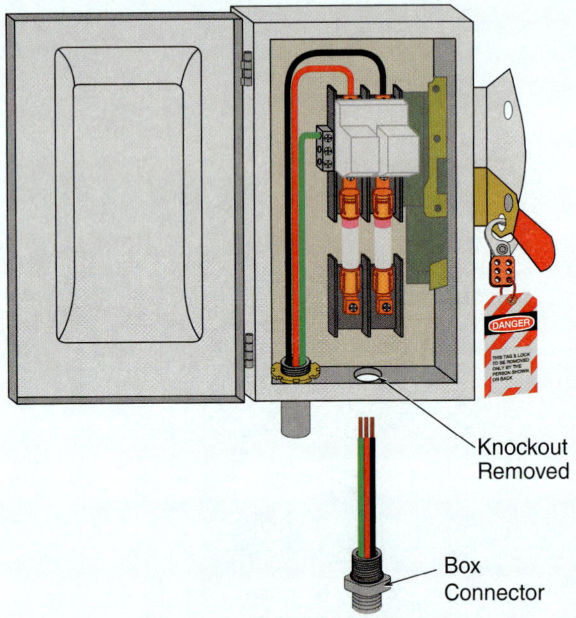

5. Most whips will be constructed with the use of ¾" box connectors. Remove the appropriate knockouts from the points of conduit connection on both the disconnect and unit. Note that a technician occasionally makes an error: a larger knockout than desired is removed. In such a case, special washers are available to install on both sides of the box connection as a method of repair.

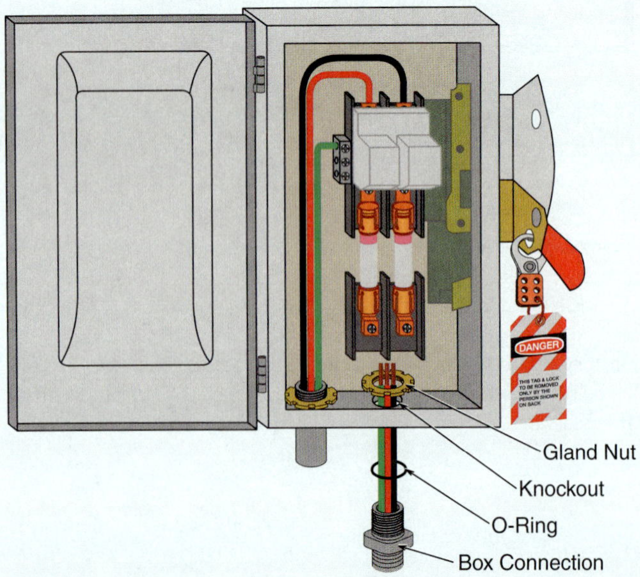

6. Route the conductors through the respective box openings, and connect the conduit fittings to each enclosure by using the nuts provided. There should also be O-rings inserted over the threads and positioned on the exterior of the box. Tighten each nut securely.

Continued

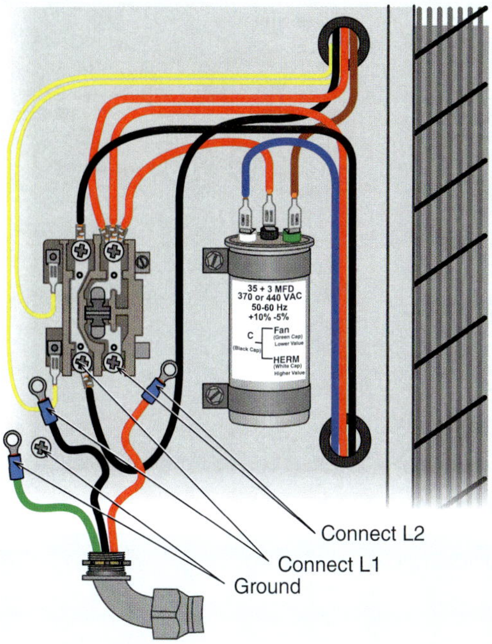

7. Locate the input voltage terminals on the condensing unit. As noted previously, the incoming power is often terminated at the compressor contactor, but it may be terminated in a small electrical enclosure instead. Also locate the lug or ground lead to connect the ground from the whip. Terminate the wires as needed, always ensuring that there is a little slack in the wiring; in the future, repairs may need to be made that require a little extra wire. The black wire from the whip should be connected to the contactor terminal labeled L1 or to the black lead provided.

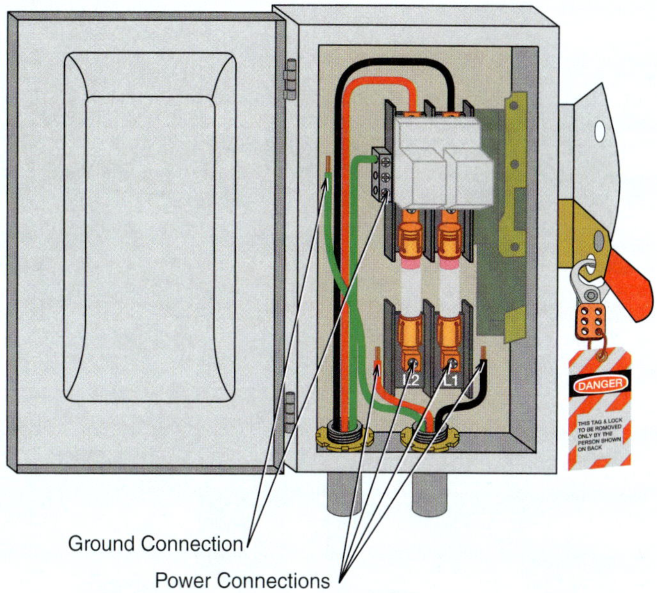

8. Now return to the disconnect switch and terminate the wires at the lugs provided, including the ground wire. If necessary, cut the conductors so that there is no excess wire rolled up inside the box, but some slack is needed and acceptable.

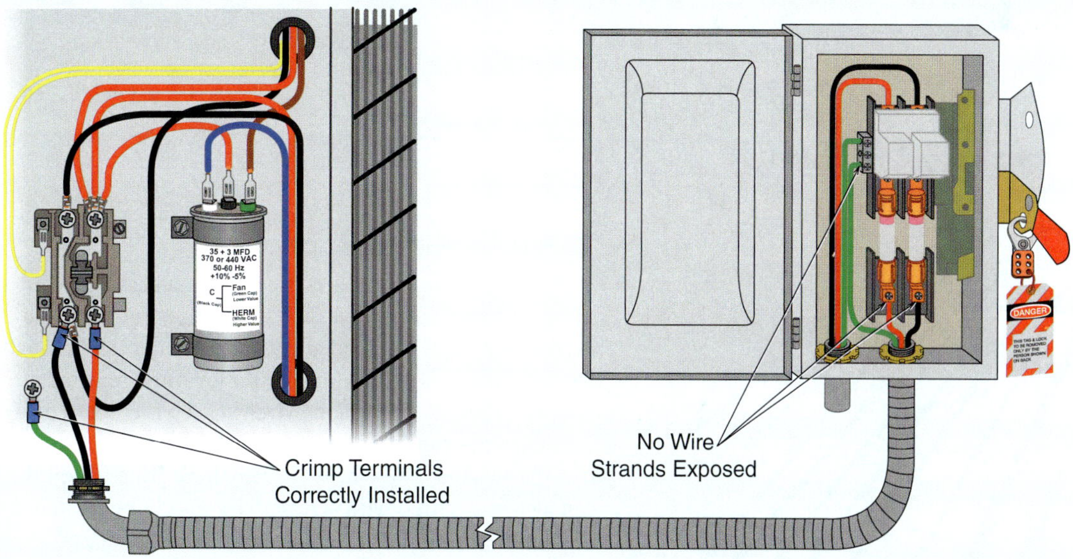

9. Examine the results at both ends, ensuring that each connection is sound and neatly done. Ensure that no wire strands from stranded conductors are protruding from their terminations.

Continued

10. Ensure that all controls, including the thermostat, remain set to the Off position. Remove the locks and tags from the power source(s). Restore power to the system.

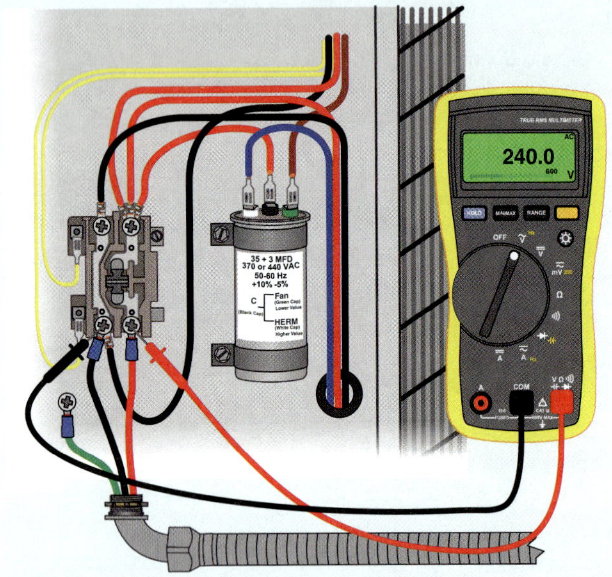

11. Use a voltmeter to check for the proper voltage at the condensing-unit terminations.

12. Carefully reinstall control panel covers and equipment access doors, and return the system to service.

Wire Ties and Tape

Wire ties, or cable ties, are strapping materials used to secure bundles of wire. These ties are available in a variety of sizes and colors (**FIGURE 6-27**). Longer ties are generally wider as well. Most are made of nylon, but some are made of Velcro for easy removal and reinstallation. Wire ties are wrapped around a bundle of wires to keep them neat and out of the way.

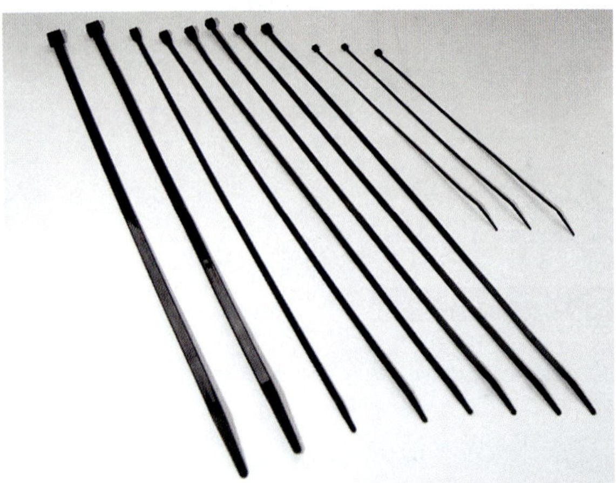

FIGURE 6-27 Wire ties.

Wire ties can be secured to small square bases to anchor the bundle to a flat surface. These bases usually have adhesive on the back, but they can also be secured with screws. Ties should be tightened snugly around the bundle, but not so tight that they pinch the wires.

Electrical tape is often used to mark conductors for identification. Electrical tape comes in a variety of colors (**FIGURE 6-28**) and can be made of vinyl, rubber, and a variety of plastics, including PVC. Vinyl tape is the most common. Black electrical tape is used for insulating purposes with certain splices and seals against moisture. Colored electrical tapes are often used to indicate their function or the phase of power on a three-phase system. For example, black tape, red tape, and blue tape is used to identify phases A, B, and C, respectively. Green tape is used to identify earth ground, and white identifies a single-phase neutral. Note that these color codes are common in the United States, but other countries may use different colors.

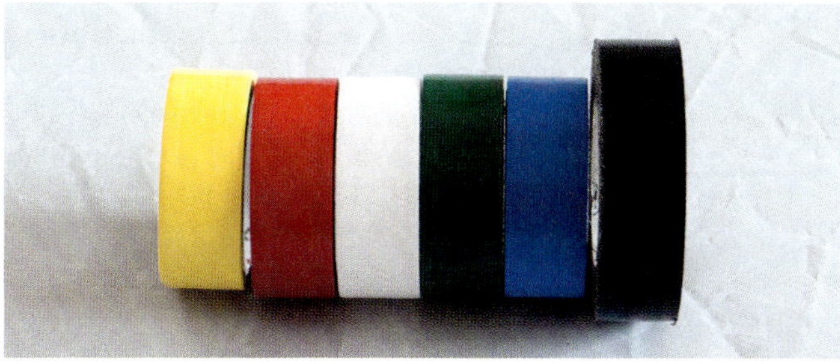

FIGURE 6-28 Electrical tape.

▶ Wrap-Up

Summary

There are occasions when wire will need to be selected and connected when installing, servicing, or repairing HVACR equipment. It is vital to understand the relationships between the type and size of wire and the current it can continuously carry, referred to as its ampacity. When installing or replacing wiring, it is essential to select wire with the correct ampacity and voltage rating. The insulation must also be rated for the conditions in which the conductor will be used.

The selection of wire must be based on the maximum ampacity as identified on a unit's data plate or in the equipment manufacturer's literature. Wire tables provided by the *NEC®* must be used to select the correct wire size. However, it is important to note that selecting the proper wire is an important task and should be done only by appropriately qualified individuals in strict accordance with the *NEC®* and other code requirements.

There are several methods for terminating conductors, including connecting solid wire directly under a screw terminal and using wire nuts to connect two wires together. HVACR technicians and installers must know how to select and properly use all types of conductor-termination hardware to ensure the safety and reliability of a system.

Key Terms

American Wire Gauge (AWG) system A system that assigns sizes to electrical wire, based on the cross-sectional area of the wire.

ampacity A measure of the maximum current that a conductor can carry without risking its immediate or progressive damage or failure.

conductors Materials that allow the flow of electrical current without substantial resistance.

conduit A plastic or metal tube used to protect wiring.

derated The quality of having the current-carrying capacity of a wire or wiring device reduced due to unusual circumstances.

insulator A material that resists the flow of electrical current.

flame-retardant Substances that use a chemical reaction to resist or inhibit the spread of fire.

plenum-rated Wire and cable insulation that is approved for unprotected use in spaces used as a return-air path (plenums) for conditioned air. Such material must be flame-retardant and exhibit limited smoke production.

whip An assembly of conductors and flexible conduit with connection fittings designed to connect a piece of equipment to a nearby source of power.

Review Questions

1. The most common conductors used in HVACR applications are made of _____.
   a. gold
   b. silver
   c. copper
   d. aluminum
2. List two reasons why aluminum conductors are not preferred.
3. The neutral wire of a single-phase power supply is typically _____.
   a. black
   b. white
   c. red
   d. green
4. Which of these conductors has the smallest cross-sectional area?
   a. #6 AWG
   b. #10 AWG
   c. #14 AWG
   d. #18 AWG

5. The ampacity rating of a conductor represents _____.
   a. the instantaneous current surge it can carry for 5 seconds
   b. the maximum current it can carry continuously without damage
   c. the minimum temperature at which it can be used
   d. the maximum current it can carry at 600 VAC
6. The four factors that affect the resistance of a conductor are temperature, cross-sectional area, the conductor material, and the _____.
7. When a conductor needs to be derated, it means that _____.
   a. it must be removed from service
   b. its capacity must be reduced
   c. its capacity must be increased
   d. it has been damaged
8. Which of the following conductors has the highest voltage drop per unit length?
   a. #10 AWG
   b. #12 AWG
   c. #16 AWG
   d. #20 AWG
9. If the voltage at the service panel is 120.4 VAC and the voltage at the load is 119.1 VAC, the percentage of voltage drop is _____.
10. As a general rule, if you are replacing a factory-installed wire inside a unit cabinet, _____.
    a. the wire size is probably correct
    b. the new wire should be a size larger
    c. the new wire must be sized according to the *NEC*® tables
    d. the new wire must be acquired from the manufacturer
11. The ampacity rating for an HVACR unit, such as a condensing unit, is often found on the ____.
    a. inside of the disconnect
    b. contactor
    c. equipment data plate
    d. compressor
12. The minimum circuit ampacity of a condensing unit includes the load of the compressor, using the ________ of the compressor.
    a. RLA
    b. LRA
13. Which of these types of insulation is *not* rated for wet environments?
    a. THWN
    b. THHN
    c. XHHW
14. Which of these is *not* a type of crimp connector?
    a. Wire nut
    b. Blade terminal
    c. Butt connector
    d. Spade terminal
15. Which of the following statements about wire nuts is true?
    a. All conductors should be pre-twisted together before installing the wire nut.
    b. While solid conductors may be twisted before installing a wire nut, stranded conductors should not be twisted together.
    c. Never use a wire nut to connect a stranded wire to a solid wire.
    d. Every wire-nut connection should be wrapped in electrical tape.
16. Which of the following crimp connectors has male and female components?
    a. Spade terminals
    b. Ring terminals
    c. Butt connectors
    d. Blade terminals
17. The type of wire termination that would generally be expected to be found in a disconnect switch is a _____.
    a. ring terminal
    b. lug
    c. blade terminal
    d. wire nut
18. Crimp connectors that HVACR technicians work with are designed exclusively for use with ______ wire.
    a. solid
    b. stranded
19. If red electrical tape is on one conductor of a three-phase power source, it indicates that the conductor is ______.
    a. phase A
    b. phase B
    c. phase C
    d. earth ground
20. A screw terminal equipped with a special square washer underneath is designed to accept _____.
    a. a ring terminal
    b. a blade terminal
    c. bare wire
    d. solid wire only

CHAPTER 7
Transformers

Knowledge Objectives

After reading this chapter, you will be able to:

- **K07001** Describe the basic construction and operation of a transformer.
- **K07002** Identify common types of single- and three-phase transformers.
- **K07003** Describe the features and characteristics of HVACR-related control transformers.
- **K07004** Discuss the process of testing and troubleshooting a control transformer.
- **K07005** Describe how to select and replace a control transformer.

Skill Objectives

Skills required to meet the objectives of this chapter:

- **S07001** Identify the primary and secondary sides of a control transformer by using a meter.
- **S07002** Troubleshoot a control transformer.
- **S07003** Remove and replace a multi-tap, 24VAC control transformer.

▶ Introduction

Most modern refrigeration and air conditioning equipment uses electrical energy to operate a refrigerant evaporation and condensation cycle to transfer heat. The major electrical loads required for the task are designed to operate on readily available, utility-provided specific voltages.

However, many other electrical components in systems do not require utility-level voltages to perform their jobs. In fact, they would have to be constructed of heavier materials and would cost significantly more if they did. Operating control systems at a reduced voltage is also safer. In addition, specific electrical licensing is not required for working with low-voltage wiring (generally considered to be 50 volts or lower), reducing installation and maintenance expenses. However, relying on the utility to provide a variety of voltages would be impractical and very costly.

This chapter will discuss the purpose, operation, maintenance, and replacement of devices called transformers that make other voltages available. Although not every transformer is designed to reduce a voltage to a lower value, the vast majority of them are used for that purpose.

▶ Transformer Construction and Operation

K07001 Describe the basic construction and operation of a transformer.

Simply stated, a transformer is an electrical device that can receive an input voltage and produce a different output voltage. They are based on the principle of electromagnetic **induction**, which was first discovered in the early nineteenth century. The concept of induction is based on the fact that an electrical current flowing through a conductor generates a magnetic field around it. As the current changes, the size and intensity of the magnetic field also changes. **FIGURE 7-1** shows the relationship between current in a conductor and the size of the magnetic field it generates.

If another conductor lies within the magnetic field of the first conductor, the magnetic field will induce a current in the second conductor. The faster the magnetic field expands and contracts around the first wire, the larger the induced current in the second wire (**FIGURE 7-2**). Since the induced current is generated only while the current changes in the first wire, the currents need to be continuously changing to make such a process useful. This is why transformers are found only in alternating current (AC) systems.

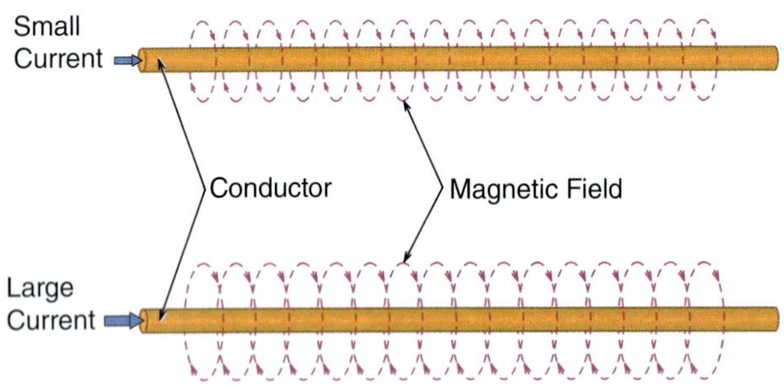

FIGURE 7-1 Current in a conductor and its magnetic field.

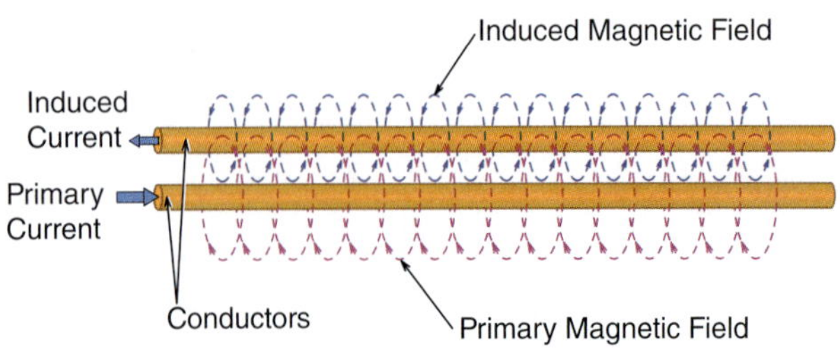

FIGURE 7-2 Magnetic induction of an electric current in a second conductor.

Transformer Construction

The strength of the induced magnetic field is proportional to several variables. If the first wire is wrapped into a stack of loops or coils, the magnetic field around the wire in each loop adds to the adjacent fields, creating a much stronger field. An example is a bar magnet, shown in **FIGURE 7-3A**. If these coils are wrapped around a **ferromagnetic metal** bar, or **core**, the presence of the metal concentrates the magnetic field even more, greatly increasing its intensity at a given current value (**FIGURE 7-3B**). The magnetic field is also proportional to the number of wraps, or turns, of a wire around the core.

Finally, the induced current in a second wire wrapped around the same core is proportional to the number of its turns around the core for a given intensity of magnetic field. The more turns, the stronger the induced current.

In a transformer, the first wire, which is energized by an existing power source, is called the **primary winding**. The second wire, which provides voltage through induction, is called the **secondary winding**. The conductors are insulated so that the currents are confined within the wires. **FIGURE 7-4** shows a very simple transformer.

Transformers may also have various materials for their cores, including metals, nonmetals, and even air. The core materials control the way the transformer works. Ferromagnetic metal cores produce the most efficient transformers. All HVACR control transformers are of the

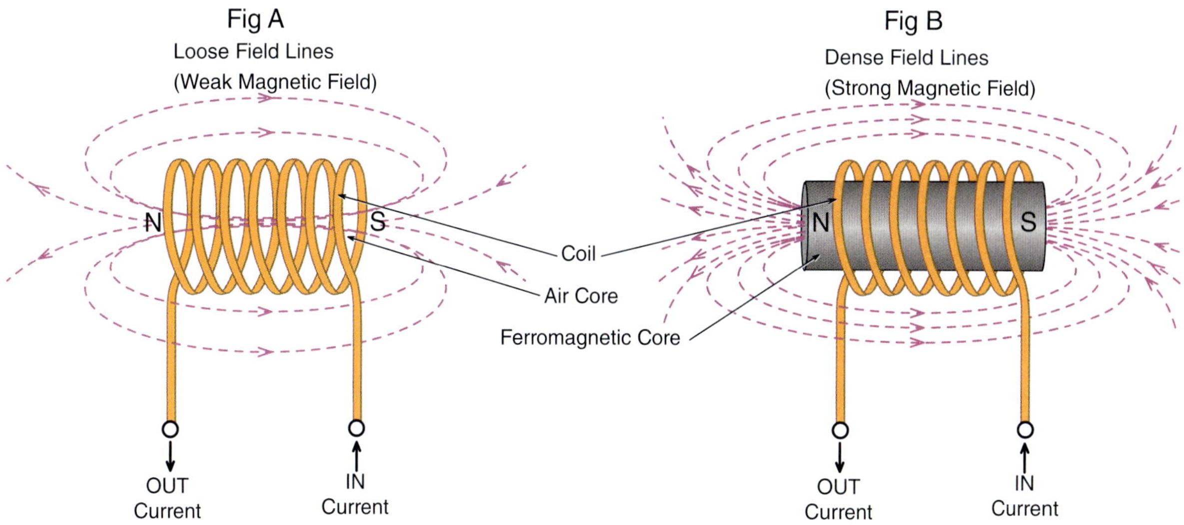

FIGURE 7-3 Coiling a current-carrying wire and including a metal core together increases the strength of a magnetic field.

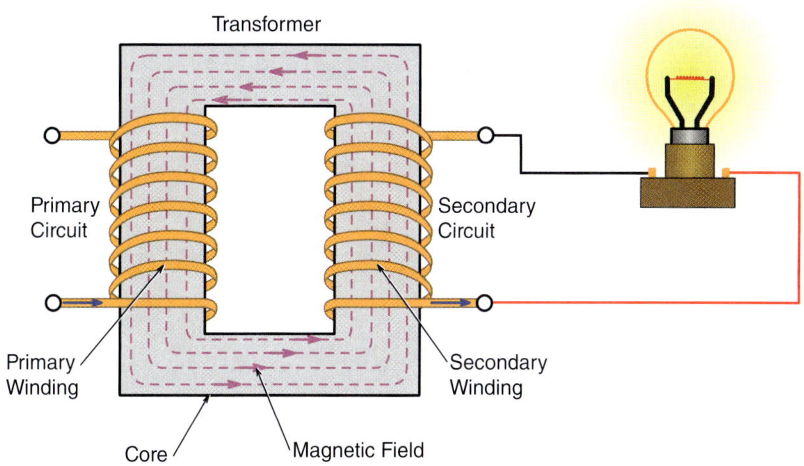

FIGURE 7-4 A simple transformer schematic showing the primary and secondary windings.

ferromagnetic metal-core type. The rectangular transformer core itself is formed of iron **laminations**. This feature minimizes the loss of electrical energy by the generation of **eddy currents** inside the core. Every metal-core transformer has them.

FIGURE 7-5 shows several examples of simple, low-voltage transformers that a technician will find in modern HVACR equipment. The primary and secondary windings wrap around the common core. A protective casing may cover the windings, as shown in the figure. An epoxy resin is generally applied for electrical insulation beneath the casing. The coating may also be covered with a paper wrapper.

The internal construction of a simple transformer is similar to what is shown in **FIGURE 7-6**. The primary windings may be wrapped inside or outside the secondary windings, or even side by side on the same core. The rapidly expanding and contracting magnetic field in the core generated by the current in the primary

FIGURE 7-5 Typical transformers used in HVACR equipment.

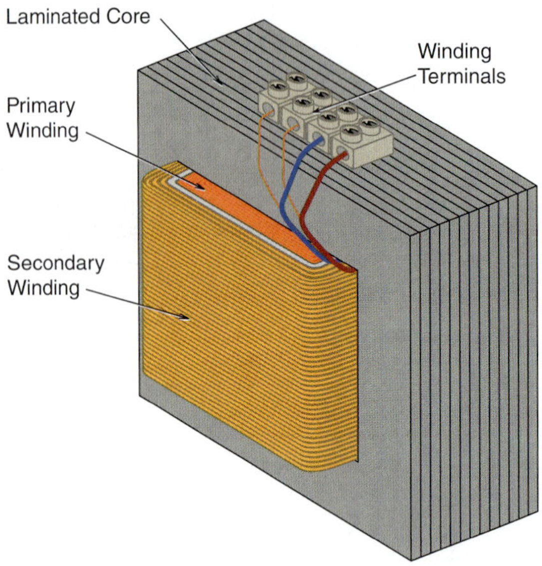

FIGURE 7-6 Construction of a simple transformer.

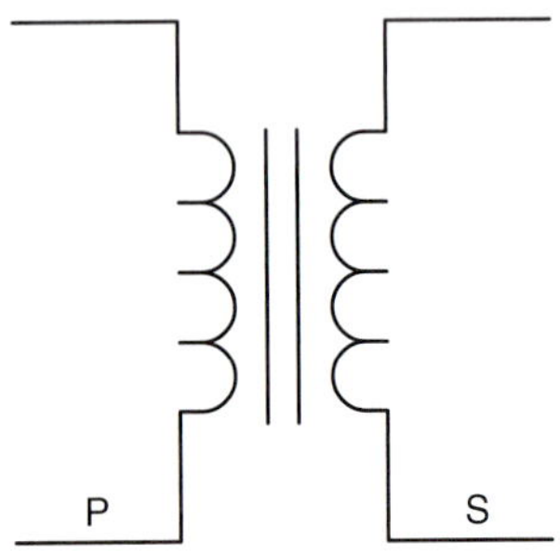

FIGURE 7-7 Transformer schematic symbol.

winding induces both the AC frequency and voltage in the secondary winding. A 60 Hz input frequency in the primary winding will produce a 60 Hz frequency in the secondary winding.

In electrical schematics, a symbol similar to the one shown in **FIGURE 7-7** identifies a transformer. The central pair of lines represents a ferromagnetic metal core. The lack of lines in a similar symbol means an air or other type of core. The scalloped lines represent the windings, which are often identified by a "P" or "S" for primary and secondary, respectively. However, in many cases, the letters are not shown on the symbol. The primary electrical source is connected to the leads of the primary windings; the downstream load circuit(s) are connected to the leads of the secondary windings.

Transformer Operation

The operation of a simple transformer was discussed along with its construction characteristics; they are difficult topics to separate. To summarize, AC voltage and current are supplied to the primary windings during operation. A rapidly alternating magnetic field is created and concentrated in the core of the transformer. The changing field in the core induces an AC voltage in the secondary winding. If the secondary winding is connected to a complete circuit, then AC also flows through that circuit.

▶ Single- and Three-Phase Transformers

K07002 Identify common types of single- and three-phase transformers.

As discussed in Chapter 3, AC electrical power is distributed as either single-phase or three-phase power. Not surprisingly, transformers can be designed to provide both single-phase and three-phase power outputs.

Single-Phase Transformers

A single-phase transformer usually consists of a single primary winding and a separate, single secondary winding wrapped around a common core. Recall from the previous discussion that the intensity of the magnetic field in the transformer core is directly proportional to the number of primary-winding turns around the core. Similarly, the induced voltage in the secondary winding is directly proportional to the number of its turns. Stated another way, the ratio of turns in the primary winding to the turns in the secondary winding determines the desired output voltage.

The sections that follow will introduce three basic types of single-phase transformers: isolation, step up, and step down.

Isolation Transformer

When designing complex electrical systems, it is sometimes necessary to ensure that a ground fault or short circuit in one part of the system will not affect the rest of the system.

Systems subject to environmental or combat damage (e.g., warship subsystems) will likely be supplied by isolation transformers. The numbers of primary and secondary windings of an isolation transformer are equal so that input and output voltages are the same. In this case, the transformer is exclusively used to isolate a circuit, and no change in voltage is desired.

Engineers call the ratio of primary windings to secondary windings a transformer's **turns ratio**. In an isolation transformer, the turns ratio is 1:1. This produces a secondary voltage that is equal to the voltage applied to the primary winding. Again, an isolation transformer serves no other function than to electrically isolate one portion of the system from another. **FIGURE 7-8** shows the schematic of an isolation transformer.

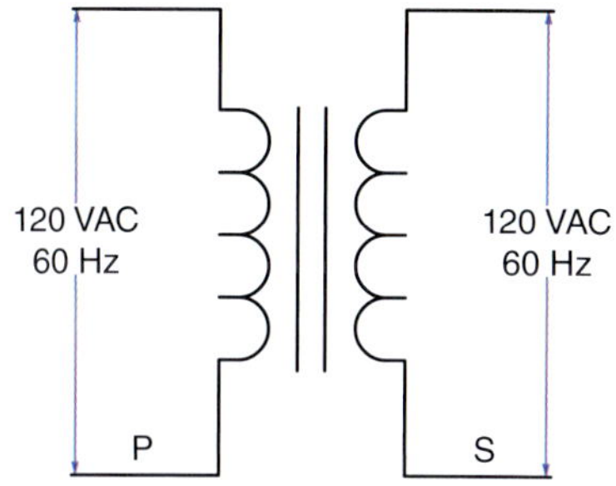

FIGURE 7-8 Schematic of an isolation transformer.

Step-Down Transformer

The most common transformer that an HVACR technician will encounter is the step-down transformer. These transformers commonly convert 240 VAC, 208 VAC, or 120 VAC line voltage provided by most residential and commercial power supplies to the 24 VAC secondary-winding control voltage used by typical heating and cooling systems. In the case of a 240/24 VAC transformer, the 24 VAC output voltage is one-tenth the input voltage. Thus, the turns ratio is 10:1. For a 120/24 VAC transformer, the output voltage is one-fifth the input voltage, requiring a turns ratio of 5:1. **FIGURE 7-9** shows the schematic for a step-down transformer.

Step-Up Transformer

In some electrical applications, it may be necessary to raise the AC voltage feeding a piece of equipment or process above the voltage supplied to the facility itself. More commonly, electrical utilities do this for economic reasons. A transformer can be used to do this as well.

Line voltage is raised by using step-up transformers. The higher output voltage is achieved by making more turns in the secondary winding than in the primary winding. The turns ratio in a step-up transformer is less than 1, which may seem unusual. However, if the primary-winding voltage is *divided* by a turns ratio of less than one, this provides the higher secondary voltage. **FIGURE 7-10** shows the schematic for a step-up transformer.

The disadvantage to using step-up transformers is that they can be quite large and are limited in the amount of current they can provide from the secondary winding. They are rarely encountered in HVACR applications.

Three-Phase Transformers

Although working directly with three-phase transformers is rarely required in the HVACR field, technicians should be aware of their characteristics and how they operate. They are used primarily at the substation level of the electrical distribution system, including substations and power-control rooms that are found at manufacturing or industrial facilities.

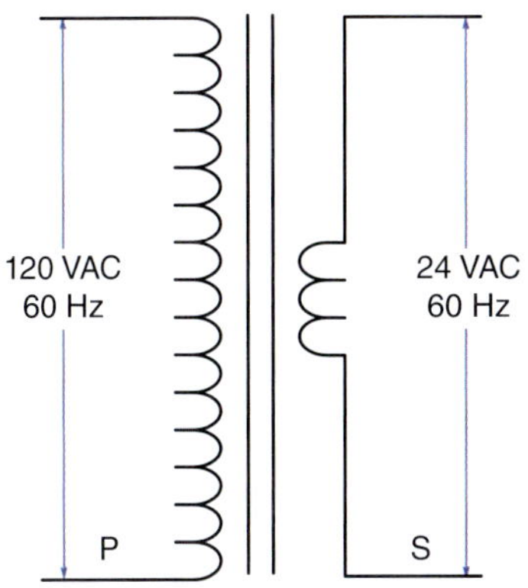

FIGURE 7-9 Schematic for a step-down transformer.

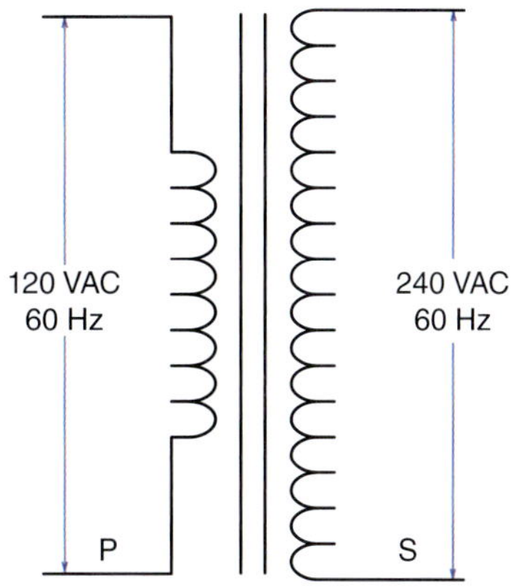

FIGURE 7-10 Schematic for a step-up transformer.

▶ TECHNICIAN TIP

Ignition Transformers

Transformers are also used in HVACR systems to generate the necessary spark to ignite fossil-fuel burners (**FIGURE 7-11**). This is one of the applications for step-up transformers in the HVACR trade.

The challenge is that a substantial voltage must be provided to cause an electric arc to develop across an air gap. A common ignition transformer for an oil burner, for example, has a secondary-winding voltage of 10,000 volts, developed from 120VAC applied to the primary winding. This allows an arc to establish across a gap of roughly 3⁄16" (4.8 mm) or less. Even though arcing across an air gap this wide is possible, a smaller gap reduces the stress on the secondary winding and still provides reliable ignition.

The current required is relatively low, yet a common oil-burner ignition transformer is a cube 4"–6" (10–15 cm) in size and weighs 4–6 lb (1.8–2.7 kg). In other words, they are not petite, but they generally produce less than 1 amp of current.

FIGURE 7-11 Ignition transformer for an oil burner.

Three-phase electrical distribution systems supply a given amount of electrical power more efficiently and at a lower cost than a single-phase system.

As the name suggests, a three-phase transformer contains three primary windings and three secondary windings on three sections of the same core or on three separate cores. Otherwise, these transformers operate using the same concepts as their single-phase counterparts. Each set of windings can be configured in one of two patterns, referred to as delta-connected configurations and wye-connected configurations.

Delta-Connected Windings

FIGURE 7-12 shows the arrangement of a 480 VAC delta-connected winding in a three-phase transformer. The connection pattern appears to be in the shape of the capital Greek letter delta (Δ). Each winding phase is connected between two conductors of the three-phase power supply, and the AC voltage applied to each winding is the phase voltage of the supply.

Note that in this configuration, there is no neutral wire.

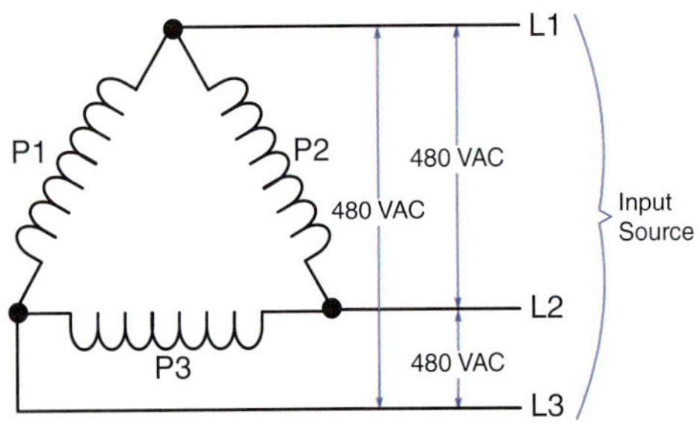

FIGURE 7-12 Arrangement of three-phase, delta-connected windings in a transformer.

Wye-Connected Windings

In the wye configuration, each winding is connected between a common neutral wire and a conductor, as shown in **FIGURE 7-13**. This connection pattern gets its name from the capital letter Y. While the phase voltage between lines in a wye-connected transformer is one value, the line-to-neutral AC voltage is smaller. Wye-connected transformer windings, especially on the secondary side, permit supplying one, two, or three single-phase circuits from the output of a three-phase transformer, which is a very common application.

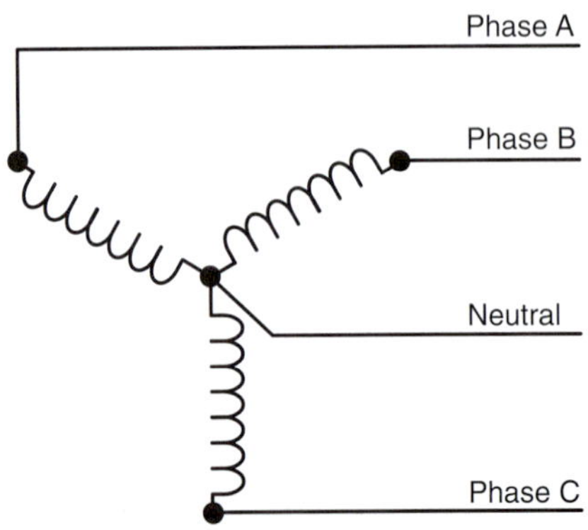

FIGURE 7-13 Arrangement of three-phase, wye-connected windings in a transformer.

Three-Phase Transformer Winding Combinations

Three-phase transformers contain two pairs of three windings, giving a total of six windings. Each of the sets of windings in a transformer can be configured at the factory in either the delta or wye pattern. This feature can result in one of four standard arrangements of primary and secondary windings: delta-delta, delta-wye, wye-delta, or wye-wye. However, for practical reasons, delta-delta and delta-wye configurations are the most common.

The delta-delta type transformer (**FIGURE 7-14**) permits dropping the one three-phase voltage to a lower three-phase voltage for further distribution within a facility. For example, a factory might receive three-phase, 4,160 VAC power from the grid and transform it to three-phase, 480 VAC or 240 VAC to supply its three-phase motors and other major electrical loads.

The delta-wye transformer is commonly used to distribute lower-voltage, single-phase power for smaller loads from a higher-voltage three-phase supply. For example, **FIGURE 7-15** shows a 240 VAC to 208 VAC delta-wye transformer. The wye line-to-line voltage is 208 VAC, but the line-to-neutral voltage is 120 VAC, which is the standard lighting- and service-circuit voltage.

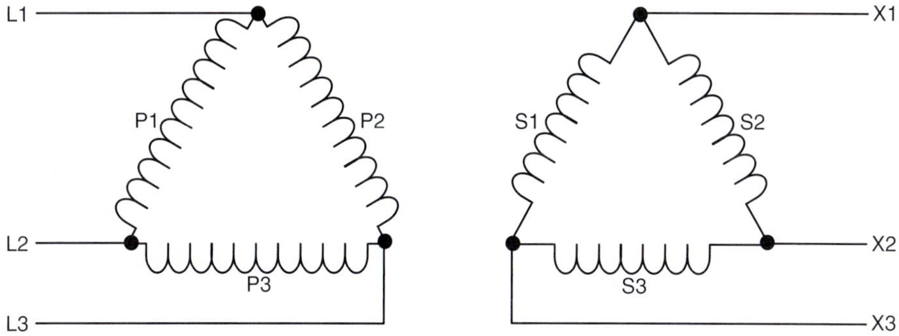

FIGURE 7-14 A delta-delta three-phase transformer schematic.

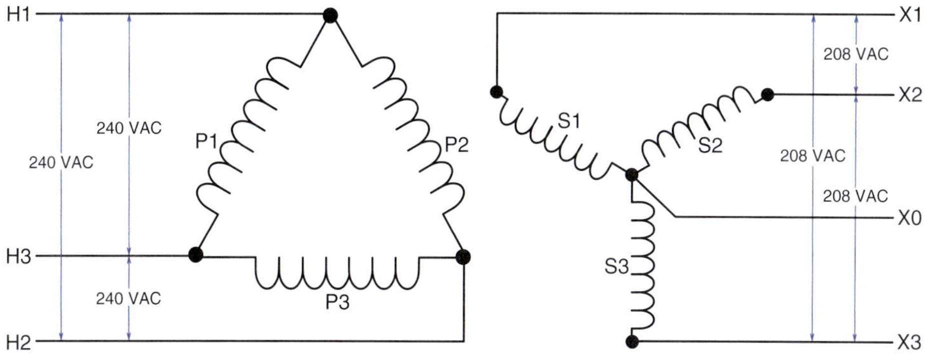

FIGURE 7-15 A delta-wye three-phase transformer schematic.

> **▶ CORE CONNECTION**
>
> **Autotransformers**
>
> There are transformers that are constructed by using only one winding, though they are rarely used in HVACR systems. Known as autotransformers, the single winding serves as both the primary and secondary of the transformer. They do have the disadvantage of lacking the electrical isolation between the two windings that a typical transformer provides. However, they are more compact, lighter, and less expensive than typical transformers with separate windings. Autotransformers are typically used to power household devices like hair dryers and laptop computers. This allows a device that operates on 120 VAC, 60 Hz power in the United States to also operate at 220 VAC, 50 Hz power commonly found in Europe and Asia.

The other configurations of three-phase transformers have their applications, but their electrical characteristics are similar to those discussed here. From this point, the chapter will concentrate on the operation and maintenance of HVACR-related control transformers.

▶ HVACR Control Transformers

K07003 Describe the features and characteristics of HVACR-related control transformers.

S07001 Identify the primary and secondary sides of a control transformer by using a meter.

The typical single-phase, step-down control transformers used in heating and cooling equipment are designed to convert the AC voltage supplied to the unit to the standard 24 VAC used for the system's control circuits. Control devices include thermostatic controls, relays, contactors, and safety devices. Control transformers also supply 24 VAC power to integrated furnace control (IFC) boards and similar printed-circuit (PC) boards. Electronic devices like these are usually equipped with a **rectifier** that converts the 24 VAC power to direct current (DC) for use by these solid-state devices.

Primary Versus Secondary Voltages

Supply voltages to a typical residential heating or cooling system in the United States will be either 240 VAC, 230 VAC, 208 VAC, or 120 VAC. The higher input voltages are usually found in packaged comfort-cooling systems, while the 120 VAC input voltage would be typical for transformers found in furnaces and small boilers. Manufacturers of HVACR control transformers will often provide multiple connections to the primary winding to allow use of a given transformer for different input voltages. This is made available through multiple wires extending from the primary winding, each of which is connected to a different point in the winding. These points of connection at the winding are referred to as taps. For example, if the 240 VAC tap is used, it produces a turns ratio of 10:1. Using the 120 VAC tap decreases the number of primary-winding coils in the circuit to provide a turns ratio of 5:1. **FIGURE 7-16** shows how the taps work to accept different supply voltages.

Since there is a predictable ratio between the turns of the two windings, the windings of a simple step-down control transformer can be identified based on their resistance. Since

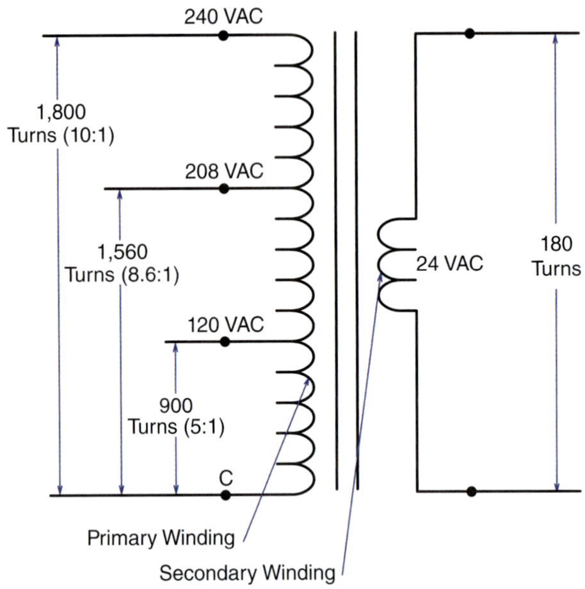

FIGURE 7-16 Multi-tap control transformer schematic.

there are more turns of wire in the primary winding, it should always have a higher resistance than the secondary winding. Note that this is reversed for a step-up transformer, which again, HVACR technicians will rarely encounter. To identify the primary and secondary windings of a control transformer by using a meter, follow the steps in **SKILL DRILL 7-1**.

SKILL DRILL 7-1 Identifying the Primary and Secondary Windings of a Control Transformer by Using a Meter

A suggestion to instructors, to make this exercise more interesting, is to add wire to the primary leads that match the leads of the secondary winding. Then conceal the transformer body so that only the two pairs of leads are visible to the tester, with both pairs now being the same color. This removes all visual clues from the exercise.

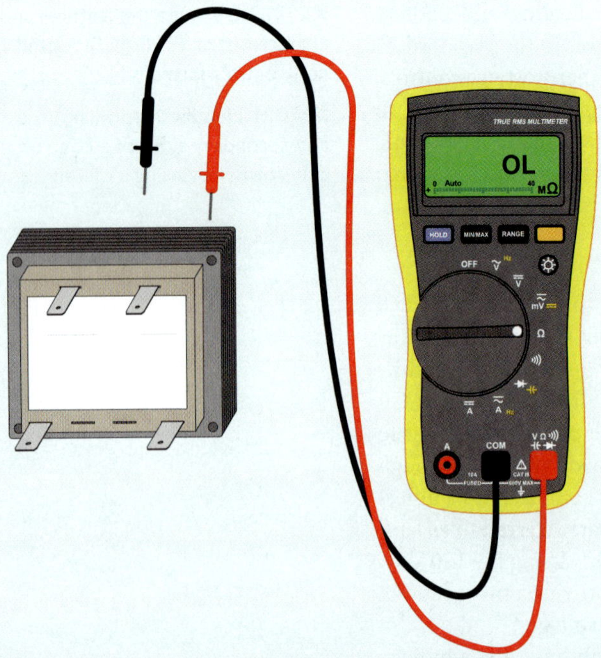

1. Acquire a 24 VAC step-down control transformer in good working order for bench testing.

Continued

2. Prepare a multimeter for continuity/resistance testing. Connect the meter probes to the two leads from one side of the transformer and record the resistance value.

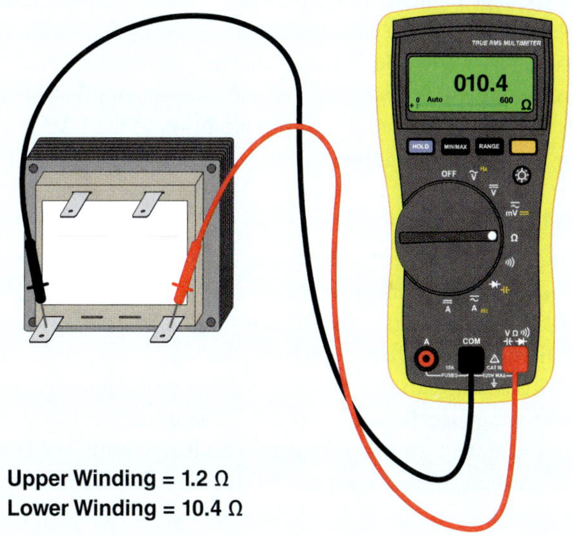

3. Now connect the meter probes to the two leads from the opposite side of the transformer and record the resistance value.

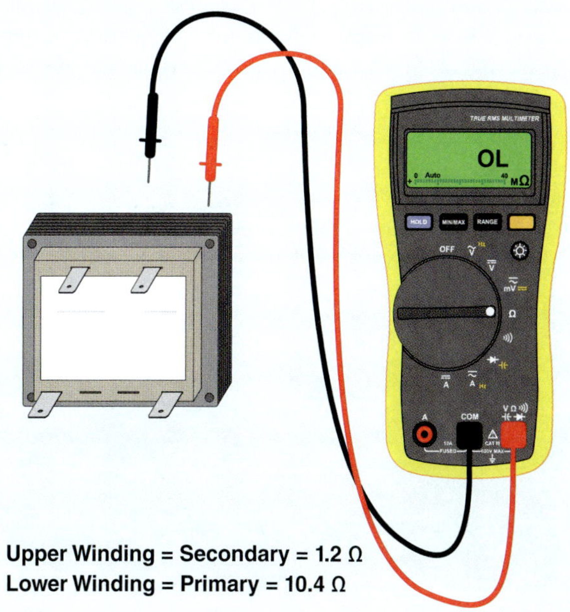

4. Evaluate the resistance values. The primary winding is the winding with the higher resistance value.

VA Ratings

A control transformer is designed to support a certain amount of current. If too much demand is placed on it, the heat generated can break down winding insulation or even create a fire hazard. The power rating of a transformer is expressed in terms of the product of the secondary voltage (V) and the maximum current (A) that the transformer can safely deliver. In the trade, this rating is known as its volt-amp, or **VA**, rating. **FIGURE 7-17** shows the nameplate data for a typical 24 VAC control transformer, which includes the transformer's VA rating.

FIGURE 7-17 Control transformer nameplate, displaying the VA rating and other characteristics.

A common HVACR control transformer rating is 40 VA at 24 VAC. The maximum current that the transformer can support can be determined with this information. The following equation shows that, by dividing the VA rating (40 VA) by the voltage of the secondary winding (24 volts), the current the transformer can support can be determined. Remember that voltage is identified by the letter E, current by the letter I, and power by the letter P:

$$I = P \div E$$

$$I = 40 \text{ VA} \div 24 \text{ volts}$$

$$I = 1.67 \text{ amps}$$

HVACR equipment design engineers must perform calculations to determine the maximum possible current flowing through the circuits supplied by the control transformer. A transformer with a VA and current rating that somewhat exceeds the design requirement is then selected for the system. Technicians should know that this limitation exists and thus pay attention to the VA rating of the transformer when replacement is needed.

FIGURE 7-18 Control transformer with a built-in circuit breaker.

Overload Protection

The secondary windings of a transformer are made of thin wire and have very little resistance. If a short circuit develops in the control circuits it powers, very high currents can quickly be generated in the transformer. This can easily lead to overheating and permanent damage. To prevent this from happening, overcurrent protection is needed. This can take the form of either a fuse or a miniature circuit breaker. Both of these may be incorporated into the transformer itself.

FIGURE 7-18 shows a control transformer with a built-in circuit breaker for overload protection. Few 40 VA transformers use this method; this feature is more likely to be found on transformers rated at 50 VA and higher. More expensive and elaborate machine-tool transformers with multiple primary and secondary taps often have fuse holders built in (**FIGURE 7-19**). However, most small control transformers used in HVACR have fuses soldered into the windings, buried beneath the exterior insulation (**FIGURE 7-20**). A technician may not even know

FIGURE 7-19 Machine-tool transformer with fuse holders.

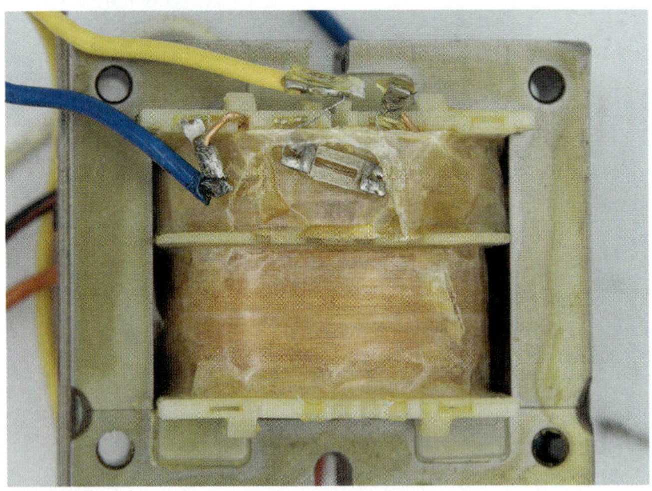

FIGURE 7-20 Fuse inside a control transformer.

that it is there, since that information is not frequently added to the label information. If the fuse blows, it will simply appear as if the winding is open (and it is, thanks to the open fuse) on an ohmmeter. Although the fuse can be accessed and replaced, it is usually impractical to do so, given the cost of labor and low-current fuses compared to the cost of a new transformer.

When there is concern about the health of a control circuit and/or transformers are failing, it is wise to install a more accessible fuse by using an inline fuse holder (**FIGURE 7-21**). Ideally, the chosen fuse will have a slightly lower current rating than the fuse embedded in the transformer, so that the accessible fuse opens first.

FIGURE 7-21 Inline fuse holders.

To select the fuse, determine the maximum current rating for the transformer, then select a fuse rating equal to or smaller than that current. For example, a 24 VAC, 40 VA transformer should have a maximum design continuous-current rating of 1.67 A. Therefore, a fuse with a slightly lower rating, such as 1.5 A, would be appropriate to protect this transformer. However, since the loads imposed on control transformers are often inductive, there is an inrush of current each time a relay or contactor coil is energized. The inrush can easily exceed the fuse rating. For this reason, always choose a fuse with a time-delay feature that can accommodate the short-lived inrush of current.

The above practice for sizing transformer-secondary fuses is a conservative approach. It is important to note that the National Electric Code (NEC) and United Laboratories (UL) allows the secondary-winding fuse to be sized as high as 167% of the secondary current rating in some cases and at 125% in other cases. Taking a conservative approach is best, however, to protect the transformer and control-circuit wiring. Refer to UL Standard 508 and NEC Standard 450.3 for additional information if necessary. Using a fuse with a lower rating than outlined by UL or the NEC is acceptable. In the long run, being conservative in transformer sizing is the best approach. This eliminates the need to use fuses that exceed the transformer rating.

Transformer Mounting

HVACR control transformers can be mounted in three primary ways. Some replacement transformers can be mounted all three ways, while others offer more limited options.

FIGURE 7-22 shows the different types of mounts. Foot-mounted transformers are very common in HVACR equipment. Other transformers have threaded hubs that allows it to be mounted directly into the knock-out opening of an electrical junction box. Some have a threaded hub on each end, while others have a threaded hub only on the primary side of the transformer. These transformers can also be mounted to a junction box cover with an appropriate hole in the center. Control transformers are also available permanently assembled to a cover plate for installation on a junction box. This type is rarely seen inside HVACR equipment and is more likely to be found powering a doorbell or similar load.

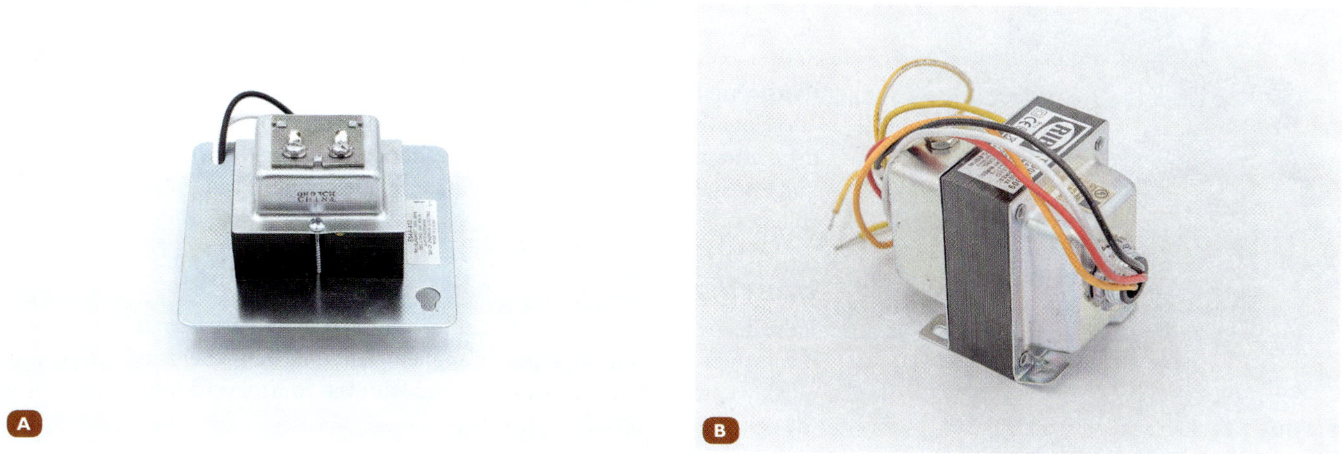

FIGURE 7-22 Transformer mounting methods. **A.** Box-mounted. **B.** Foot-mounted.

▶ Testing and Troubleshooting Control Transformers

K07004 Discuss the process of testing and troubleshooting a control transformer.

S07002 Troubleshoot a control transformer.

Testing a control transformer is fairly straightforward and can be performed with basic tools and a multimeter. Since a control transformer is the power source for the 24 VAC control system, a total loss of power to the control circuit sends a knowledgeable technician directly to the transformer for testing. However, it is important to point out that failed transformers or blown fuses in the control circuit are most often symptoms of a problem elsewhere in the circuit. When operated within their environmental and current limitations, transformers are extremely reliable.

Deenergized Inspection and Testing

Initial testing and troubleshooting begins with deenergizing the unit and the primary side of the transformer. A visual inspection is a good place to start. Look for discolored or charred windings, as shown in **FIGURE 7-23**. However, not every failed transformer will show visible signs of failure or damage. Also check for any burned or acrid odors, which if present would indicate an overheated component or wiring. The insulation coating applied to transformer and motor windings have a very strong smell when burned. Check the control-circuit fuse if one is present. A blown fuse suggests that there is a problem in the control circuit that will have to be addressed before the unit is restarted.

Before disconnecting the transformer for testing, perform a voltage check to verify that the equipment is deenergized. Then disconnect the transformer at its terminals or connecting wire nuts. Use an ohmmeter to check for continuity between the winding conductors and the transformer case. Continuity here would indicate that the windings are now grounded. Any resistance reading other than infinite/OL between the case or ground and any lead indicates that a winding is shorted to ground.

If there is no indication that the windings are grounded, check to find out whether the primary and secondary windings have become shorted to each other. Attach one meter probe to a primary-winding lead and the other to a secondary-winding lead. Any continuity or resistance reading other than infinity/OL here indicates a short circuit between the two windings. Note that when this type of failure occurs (which is rare), it is likely that the voltage feeding the primary winding (240 VAC, 120 VAC, etc.) was briefly applied to the control circuit of the unit. As a result, controls beyond the transformer may be damaged as well.

The next test ensures that the primary and secondary windings are intact. The primary windings will often have several leads for various input voltages. With one probe connected to the common lead, connect the other probe to each of the remaining primary-winding leads, one at a time. The resistance readings should be small and decrease in proportion to their indicated voltage tap label. For example, the resistance reading between the 120 VAC tap-to-common should be about half the 240 VAC tap-to-common resistance. Then check the resistance of the secondary winding by connecting the meter probes to the two secondary winding leads. The resistance reading of the secondary winding should be smaller than the resistance reading of the primary winding(s). An open winding is a common reason to replace a control transformer.

FIGURE 7-23 A failed control transformer.

Note that there is little need to focus on the precise resistance of the windings. The most important conclusion here is whether the windings have continuity or not. Energized testing will confirm whether it can provide the desired voltage or not. If the transformer passes all the tests above, it should be operable.

Energized Testing

If the deenergized electrical tests of the transformer indicate that it is functional, it can then be tested with power applied. The following energized tests can be quickly completed with basic tools and with the control transformer in place in the equipment.

No-Load Tests

A no-load test of a transformer simply verifies that it is producing the correct output voltage while the load is disconnected from the secondary winding. Using appropriate electrical safety precautions, deenergize the transformer and disconnect the 24 VAC connections to the control circuit. Pull the circuit leads away from the transformer, and ensure that the transformer leads are not in contact with other wires. Place a wire nut on each lead for protection. Then restore power to the transformer. Using a voltmeter, measure the voltage applied to the primary winding first, to verify that it is as expected. The displayed voltage value should be close to the connected voltage tap label.

Note that transformers that are to the power source using the wrong tap will still operate, but they will not produce the proper secondary voltage. For example, if the voltage applied to the primary is 208 volts, but the connection was made to the 240 VAC lead, the transformer will produce something less than 24 VAC. The output will likely be closer to 20 VAC. This can cause a lot of problems in the control circuit.

If the voltage applied to the primary winding is correct, measure the secondary voltage. It should be the stated voltage, plus-or-minus the manufacturer's voltage tolerance. Note that there are many HVACR control transformers on the market that are labeled to produce as much as 27 VAC. Be sure to check the transformer label and confirm that its design voltage is being produced.

> ▶ **TECHNICIAN TIP**
>
> **Start-Ups**
>
> New equipment designed to operate at 200+ VAC is designed to function at voltages ranging from 208 to 240 volts. If the equipment, such as a packaged heating/cooling unit, has a control transformer in it, the manufacturer can connect the primary winding only one way. For this reason, technicians must address the wiring to the primary side of the transformer before attempting a start-up. The voltage provided to the unit must be checked and the wiring examined to ensure that the correct transformer primary lead is in use. Otherwise, the secondary voltage could be incorrect and create problems from the start.

Tests Under Load

If the tests indicate that the transformer is operating normally, it can be tested under load. Deenergize the transformer primary and reconnect the control-circuit leads to the secondary winding. Ensure that all thermostats and control switches are in the OFF position. Restore power to the unit and transformer. Before starting the unit, check the secondary voltage again to ensure it is as expected. If it is good, place a clamp-on ammeter on one of the leads from the transformer secondary. Remember that the current should be low; if the ammeter used cannot accurately measure current under 10 amps, then a standard ammeter must be placed in series with the control circuit. There may be a small amount of current displayed immediately, as there may be some minor continuous loads on the transformer, such as a digital thermostat (**FIGURE 7-24**).

With an ammeter around one of the secondary leads, start the equipment while observing the ammeter display. The current should rise immediately and then fall to a stable level once the inrush load of the relay and contactor coils has passed. Ensure that the steady current reading under load is within the transformer's capabilities. For example, for a 40 VA transformer, the continuous current should be less than 1.6 amps.

The secondary voltage should now be checked once more. There will likely be some voltage drop while the transformer is under load, but it must remain above the minimum value needed by the control devices. Relay and contactor coils generally require the voltage to be within +10% or −15% of the rated voltage. For a 24 VAC coil, the minimum would then be 20.4 volts. For some relays and contactors, however, this may be insufficient. Operating too close to the minimum invites problems. For example, if the utility voltage drops during peak loads in the summer, the transformer secondary voltage will also drop slightly, and thus the unit controls may not operate reliably.

Higher-than-expected current is likely the result of a problem with a control-circuit device or poor wiring connections. A significant voltage drop is likely the result of similar problems. More tests may be required to identify problems with control devices. To troubleshoot a control transformer, follow the steps in **SKILL DRILL 7-2**.

FIGURE 7-24 Secondary current reading on an idle unit.

SKILL DRILL 7-2 Troubleshooting a Control Transformer

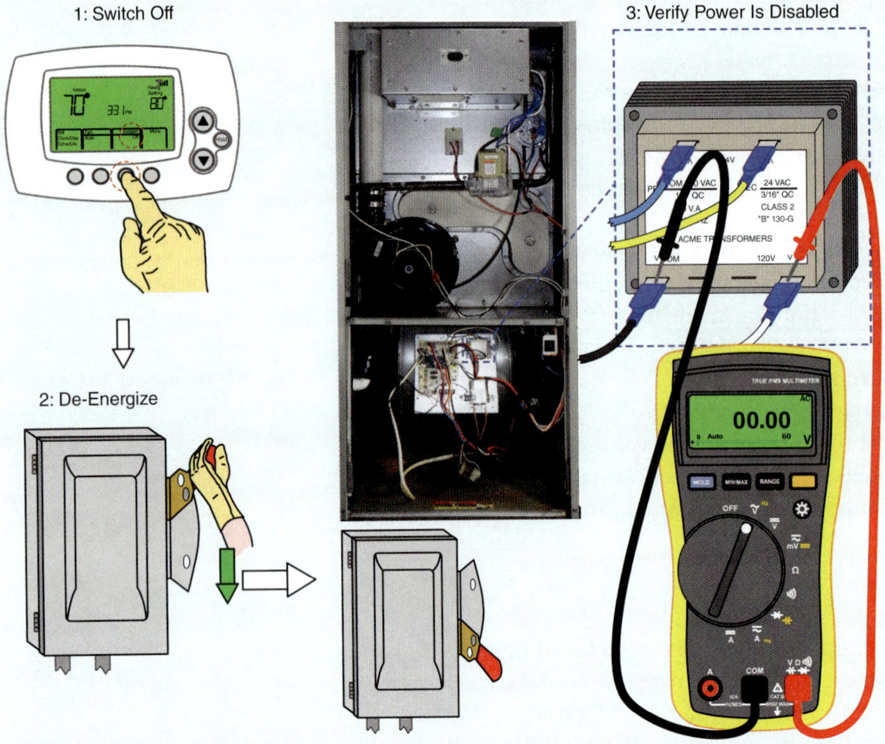

1. Ensure that all controls, including the thermostat, are set to the OFF position. Deenergize the equipment to be serviced by using the disconnect switch, but do not lock and tag, since the unit must operate for the testing. Verify with a voltmeter that power has been disabled.

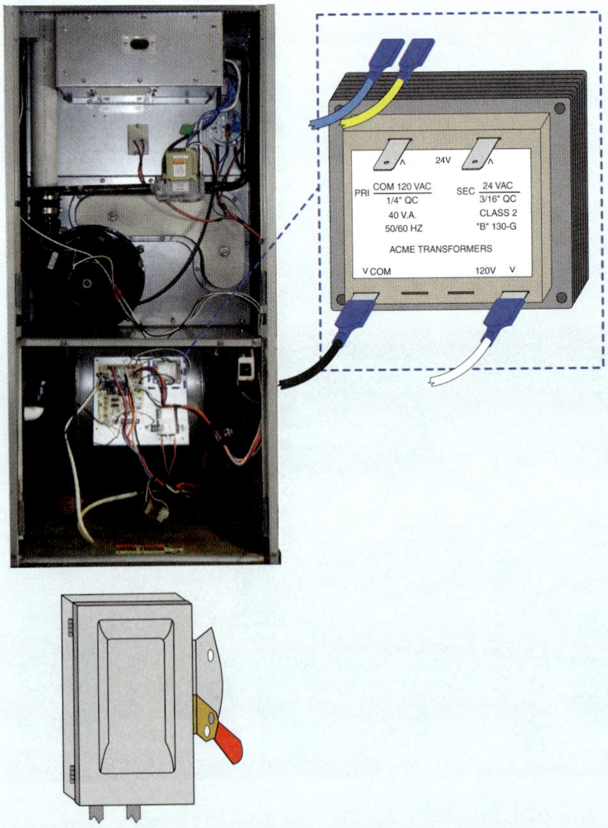

2. Disconnect the secondary leads of the transformer from the control circuit, and ensure that the leads are controlled and not in contact with anything.

Continued

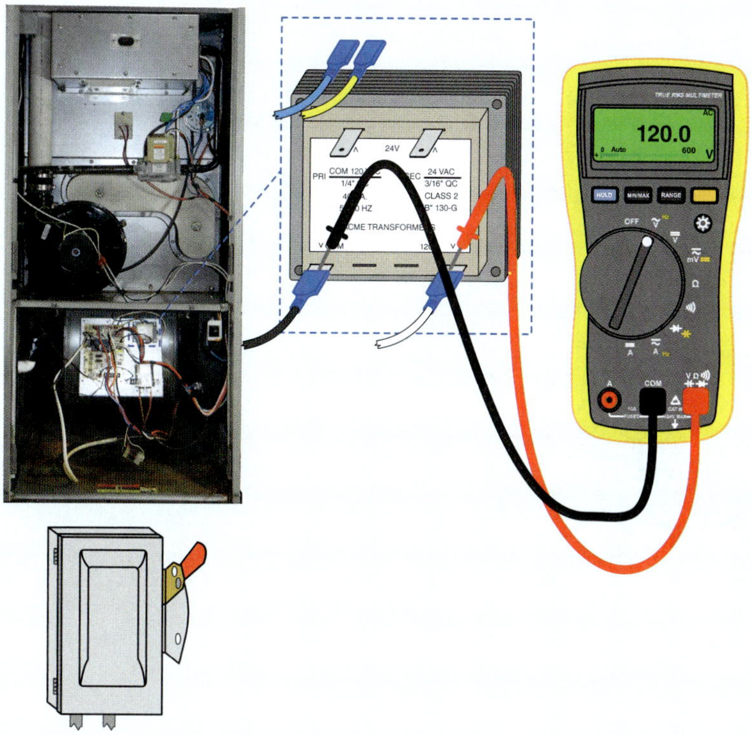

3. Restore power to the unit. Use a voltmeter to check the voltage supplied to the primary winding, and ensure that power is connected to the appropriate primary-winding lead of the transformer.

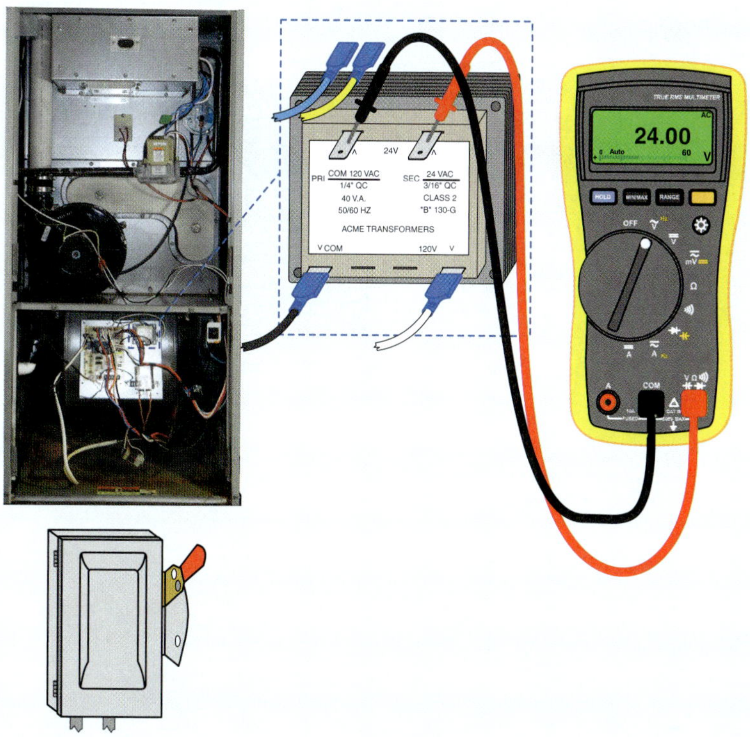

4. If the input voltage is correct, measure the voltage at the secondary-winding leads. The voltage should read very close to the design voltage rating of the transformer. Remember that not every transformer is rated to produce exactly 24 VAC.

Continued

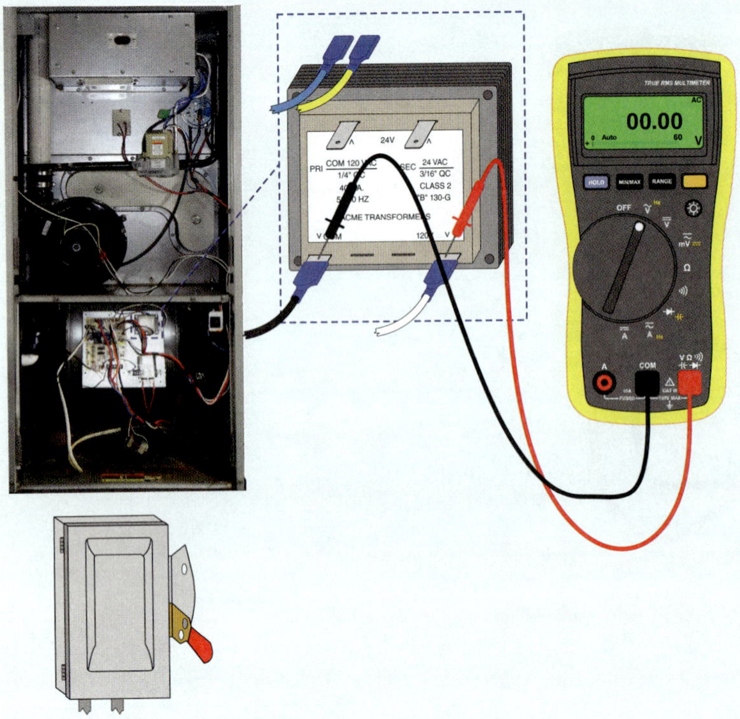

5. If the secondary voltage is appropriate, deenergize the system. Use a voltmeter to confirm that power has been disabled. Reconnect the transformer secondary-winding leads to the unit control circuit.

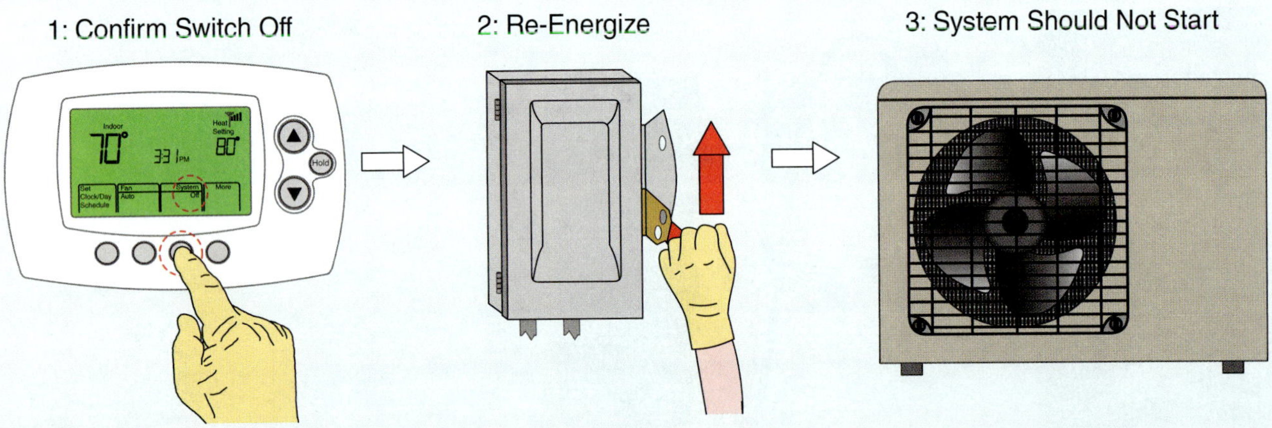

6. Make a final check to ensure that all control switches and/or thermostats for the unit remain off. Restore power to the unit; it should not start.

Continued

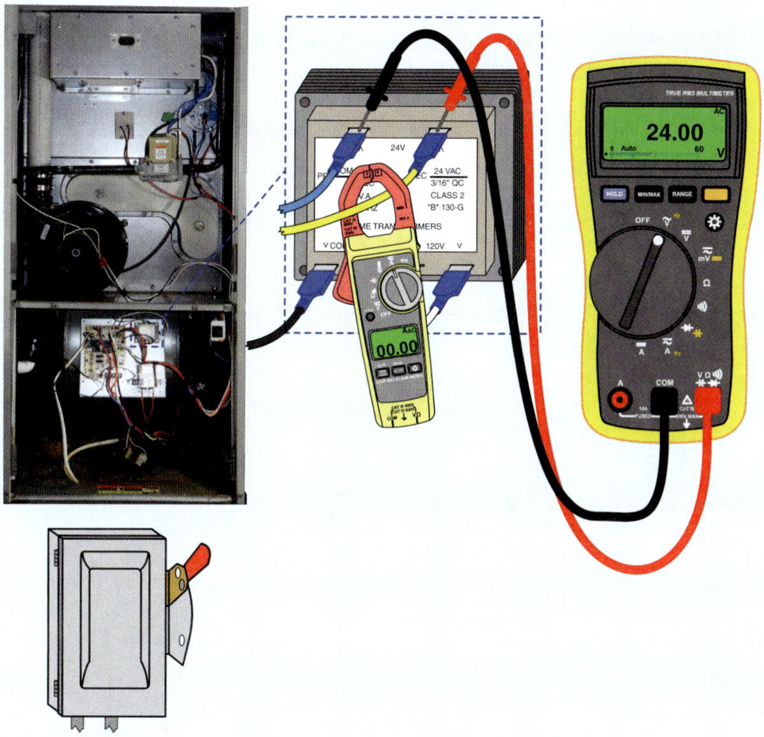

7. Before starting the unit, check the transformer-secondary voltage again to ensure that it is as expected. Then attach a clamp-on ammeter to one of the transformer-secondary leads to prepare for start-up. If it is not auto-ranging, ensure that it is set to read currents below 10 amps. Remember that there may be a small amount of current displayed, even before the unit starts.

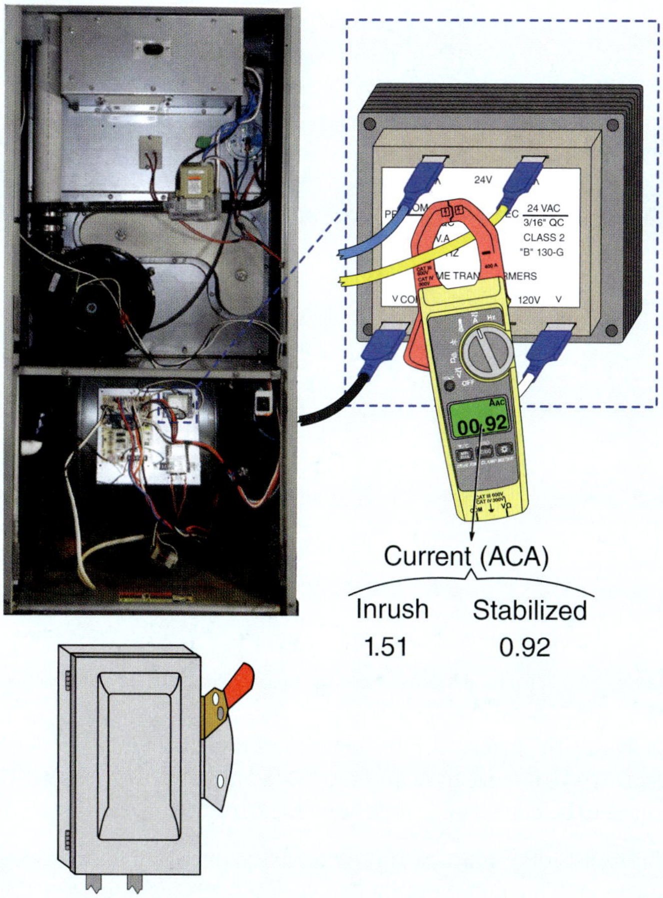

8. Start the unit while observing the ammeter display. The current should momentarily rise to a relatively significant level during the inrush of current to the relay and contactor coils, then fall to a normal and stable level. The current must stabilize below the maximum current capacity of the transformer.

Continued

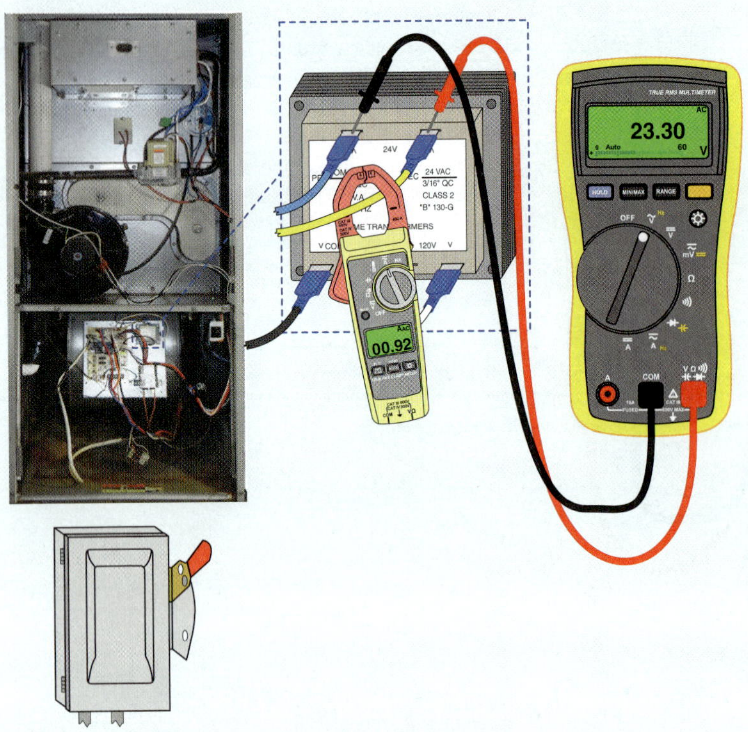

9. If the current is acceptable, make a final test of the secondary-winding voltage while the transformer is under load. Remember that many relay and contactor solenoid coils require the incoming voltage to be within +10% or −15% of the rated voltage.

10. If the transformer passes all the tests, it can be considered fully functional. Remove the ammeter, reinstall the equipment access doors, and return the unit to service.

K07005 Describe how to select and replace a control transformer.

S07003 Remove and replace a multi-tap, 24 VAC control transformer.

▶ Transformer Replacement

Transformers do fail, usually due to problems in the circuit rather than an internal defect. Fortunately, control transformers of the size used in HVACR systems are not expensive. The following characteristics need to be considered when selecting a replacement:

- *Primary voltage*—The transformer must have a compatible primary tap for the voltage supplied to the equipment. This will typically be 240 VAC, 208 VAC, or 120 VAC. Some, but not all, transformers sold for HVACR applications can accommodate all of these voltages with multi-tapped primary windings. Others may have a single-voltage primary winding.

- *Secondary voltage*—Most replacement transformers are rated to provide 24 VAC output. However, there are a number of transformers available that are rated for slightly more, as much as 27 VAC. These higher-voltage transformers are a good choice if the voltage drop measured in an existing circuit is significant. They have proved to be valuable in systems that have a longer run of thermostat wiring than is desirable, which is one of the primary causes of voltage drop.
- *VA rating*—The VA rating of the transformer must be large enough to support the loads. Matching the capacity of the failed transformer is generally a safe choice, unless there is evidence that it is insufficient. A transformer with greater capacity than needed is acceptable, but it will cost more and may be a poor fit in the space provided.
- *Mounting means*—Remember that transformers can be mounted several different ways. The most popular mount in HVACR equipment is a simple foot-mount.
- *Circuit protection*—If the failed transformer has a built-in circuit breaker, select a model with equal protection. Also consider the need for an inline fuse holder and fuse if there is concern about the health of the control circuit but the problem has not yet been identified.

To remove and replace a multi-tap 24 VAC control transformer, follow the steps in **SKILL DRILL 7-3**.

SKILL DRILL 7-3 Removing and Replacing a Multi-tap 24 VAC Control Transformer

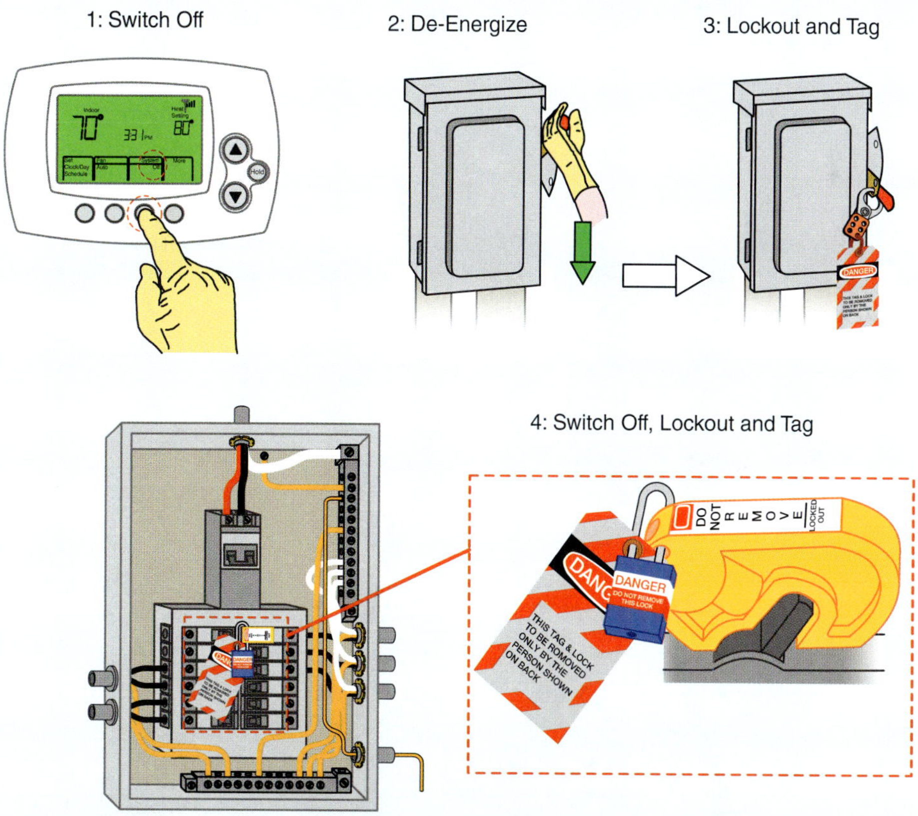

1. Ensure that all controls, including the thermostat, are set to the OFF position. Deenergize, lock out, and tag the equipment to be serviced. Be sure that the process deenergizes not only the line voltage feeding the equipment but also the transformer to be replaced.

Continued

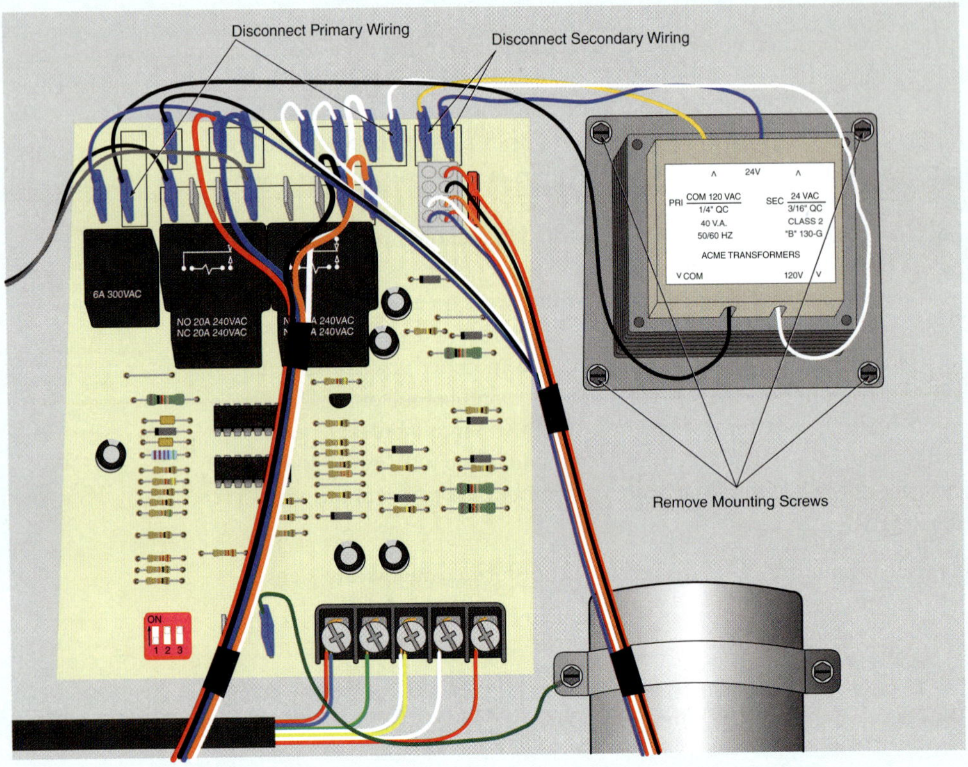

2. Disconnect the primary and secondary leads of the transformer from the unit wiring. Label the leads if necessary, and remove the transformer from its mount.

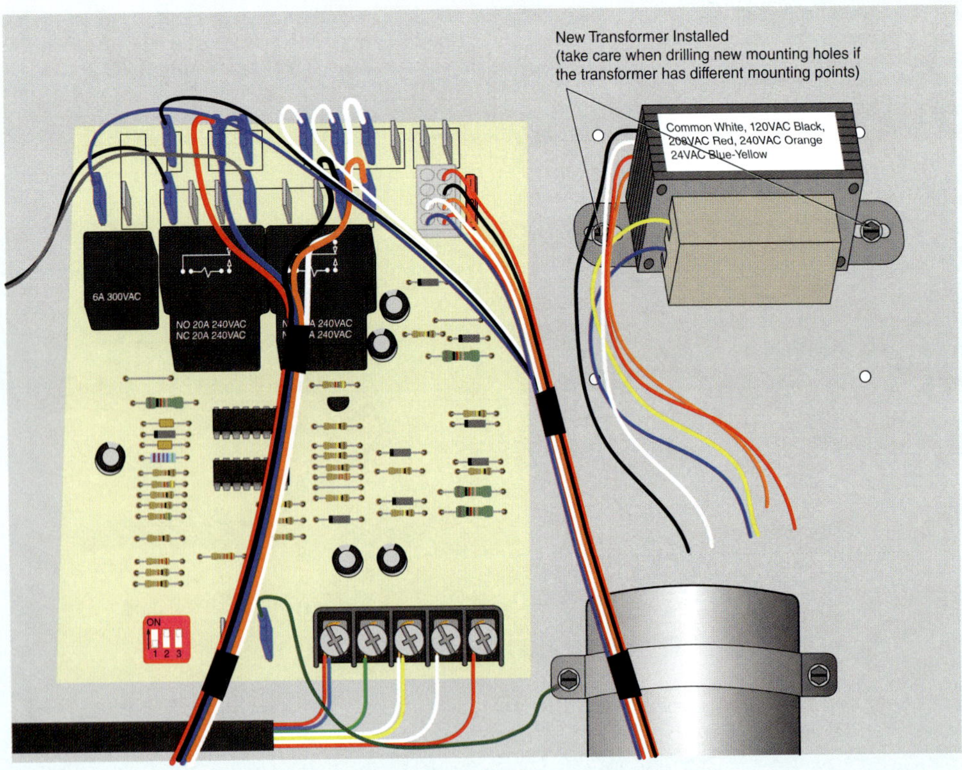

3. Set the new transformer in place and install the necessary hardware.

Continued

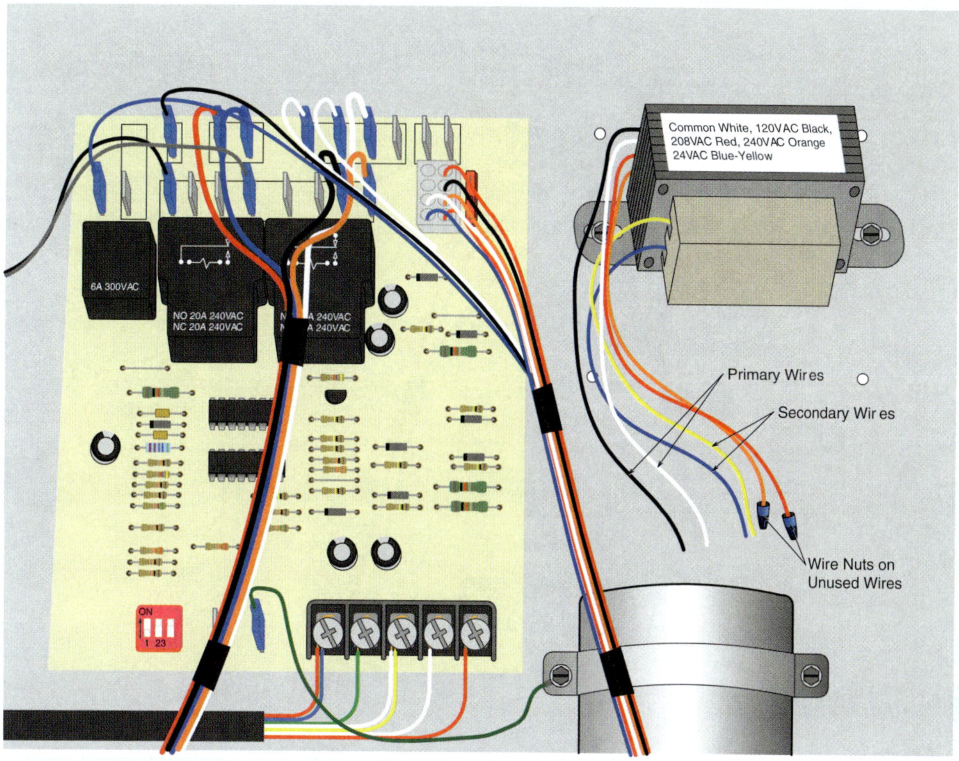

4. Examine the leads of the new transformer. Identify the common lead of the primary winding, as well as the primary lead that matches the available input voltage. Also identify the two secondary leads that will be connected to the control circuit. Apply a wire nut to each individual lead that will not be used; do not connect any of these unused leads together, however.

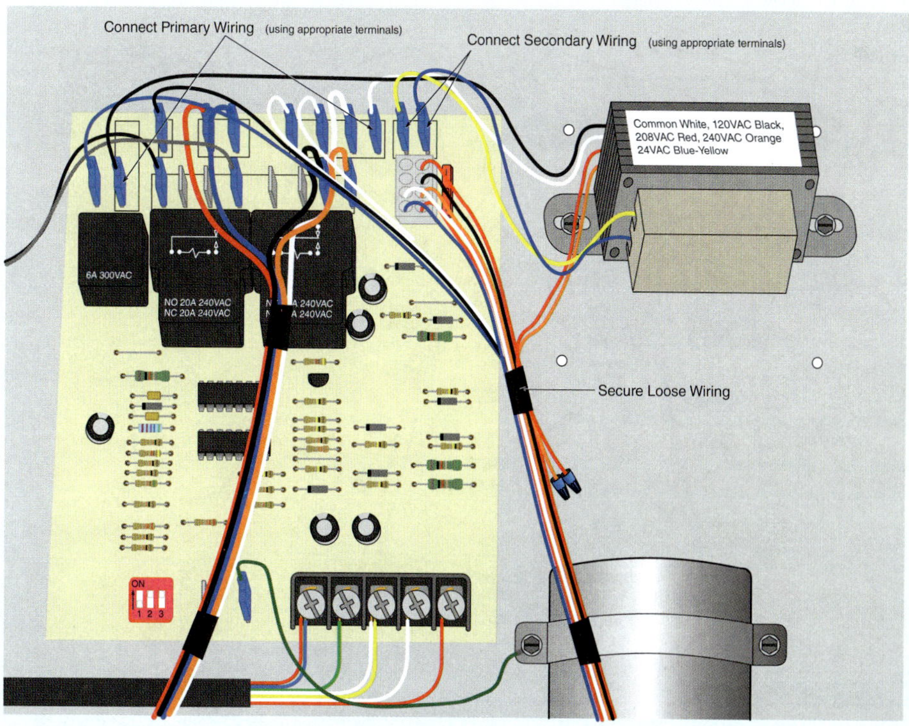

5. Now connect the primary and secondary leads to their respective circuits. This may require wire nuts or another type of termination hardware, such as spade terminals. Secure loose wiring as necessary, using wire ties.

Continued

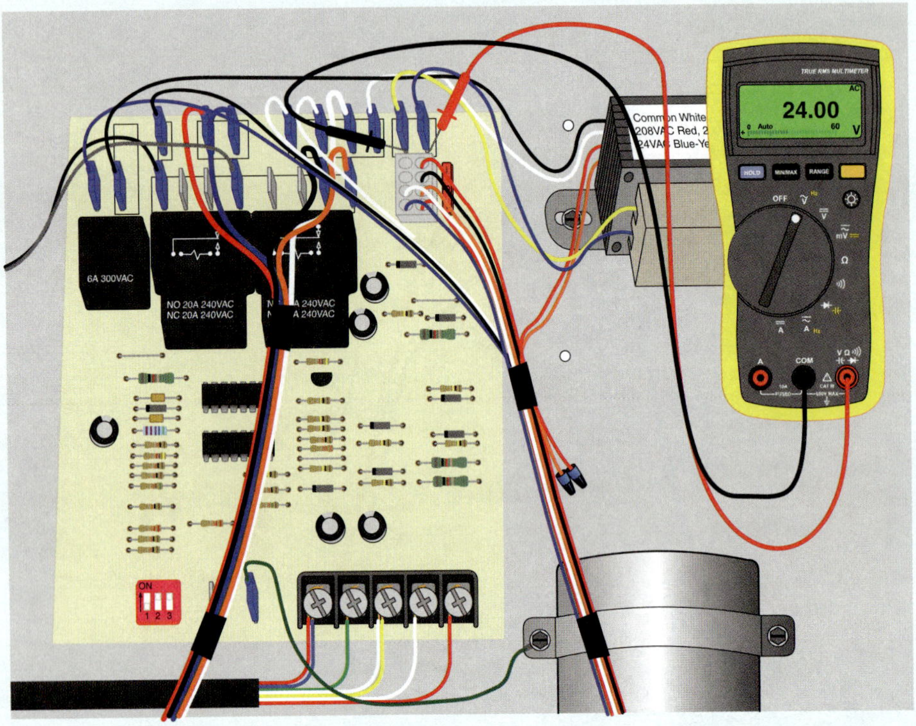

6. Ensure that all control switches and thermostats remain in the OFF position. Restore power to the unit. Use a voltmeter to check the primary and secondary voltage to ensure that it meets expectations. If the secondary voltage is higher or lower than expected, recheck the connections to the primary winding to ensure the proper lead was selected.

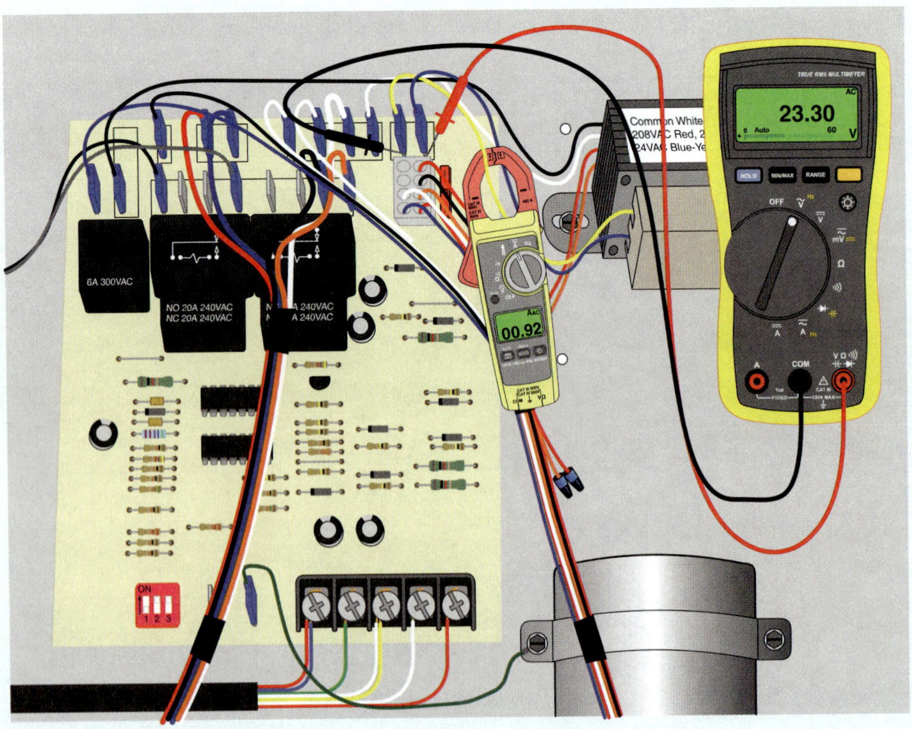

7. If the voltage tests are satisfactory, attach a clamp-on ammeter to one of the secondary-winding leads, start the unit, and repeat the tests outlined in the previous drill.

▶ Wrap-Up

Summary

HVACR control transformers are the source of electrical power to the control, monitoring, and logic circuits in most heating and air conditioning systems. Although they are generally very reliable, when they do fail, everything comes to a stop.

Understanding the construction and operation of transformers in general is essential to being able to intelligently test them. Due to their simplicity and lack of moving parts, they rarely fail due to an internal defect. More often than not, a problem in the control circuit causes their failure.

Transformer operating characteristics are easily checked with a multimeter. Their replacement involves the use of a few basic tools. If the technician understands the task, replacing a transformer is straightforward. However, improperly wiring a transformer can not only damage the new one, but also cause significant damage to other control-circuit components. It never hurts to double-check work before energizing the system.

Key Terms

core In a transformer, the material around which a winding is wrapped. Most transformers contain laminated metal cores, but some use air or other materials.

eddy currents Circular electrical currents generated inside conductors due to the action of changing magnetic fields surrounding them. These currents sap energy from the magnetic field, reducing its strength.

ferromagnetic metal Any metal that can be easily magnetized, such as iron, due to the presence of unpaired electrons. Naturally ferromagnetic elements include iron, cobalt, and nickel.

induction Refers to the process of a changing magnetic field generating (inducing) an electrical current in a conductor.

laminations The thin metal plates that make up the metal core of a transformer. They are required to minimize energy losses due to electrical eddy currents created within the metal by the alternating magnetic field.

primary winding The coiled wires in a transformer that are connected to the supplied voltage source.

secondary winding The set of coiled wires in which the transformer's output voltage is electromagnetically induced.

rectifier An electrical device that converts AC power to DC power.

turns ratio The ratio of the number of primary-winding loops and the number of secondary-winding loops. The ratio determines the value of the secondary voltage in relation to the primary voltage.

VA An abbreviation for volt-amps, which serves as a common method of rating the power output of a transformer that is equivalent to watts. It is the product of the output voltage and maximum continuous design current for a transformer.

Review Questions

1. The concept of induction is based on the fact that an electrical current flowing through a conductor generates a __________.
2. On which of the following physics principles does transformer operation depend?
   a. Nuclear transmutation
   b. Electromagnetic induction
   c. Electronegativity
   d. Electroencephalography
3. The most efficient transformers have a core comprised of __________.
   a. air
   b. copper alloys
   c. polyethylene
   d. ferromagnetic metals
4. Metal transformer cores are laminated to minimize the occurrence of __________.
5. When the primary and secondary windings of a transformer have a different number of turns, the one electrical property that remains the same in both windings is the __________.
6. The ratio of wraps, or turns, in the primary winding to the turns in the secondary winding determines the __________.
   a. output voltage
   b. output frequency
   c. input current
   d. VA or wattage
7. Except for autotransformers, how many separate windings are found in a typical single-phase transformer?
8. To calculate the secondary voltage of a transformer based on its primary voltage and the turns ratio, __________.
   a. add the turns ratio to the primary voltage
   b. subtract the turns ratio from the primary voltage
   c. divide the primary voltage by the turns ratio
   d. multiply the primary voltage by the turns ratio
9. A 240 VAC/24 VAC control transformer requires a turns ratio of __________.
10. An isolation transformer has a turns ratio of __________.
11. Which of the following would be the most likely use for a delta-wye, three-phase transformer?
    a. Supplying a three-phase circuit from a single-phase source.
    b. Supplying a single-phase circuit from a three-phase source.
    c. Stepping up three-phase voltage to a three-phase circuit.
    d. Stepping down three-phase voltage to a three-phase circuit.
12. Which of the following is *not* an accurate description of a typical HVACR control transformer?

a. Single phase
b. 24 VAC voltage output
c. Step down
d. Step up

13. The power capacity of a control transformer is expressed as its ______.

14. Which of the following fuse ratings would be the most reasonable choice to protect a 40 VA, 24 VAC control transformer?
a. 5.0 A
b. 3.0 A
c. 1.5 A
d. 1.0 A

15. Since DC power is not readily available, PC boards and similar electronic controls that require DC power must be equipped with a __________.

16. Which of the following can *not* be determined during deenergized testing of a control transformer?
a. Open primary winding.
b. Grounded secondary winding.
c. Primary-to-secondary winding short circuit.
d. The secondary-winding output voltage.

17. Which of the two windings in a typical HVACR control transformer should have the highest resistance—the primary winding or the secondary winding?

18. When checking for continuity between the secondary winding and the case of the transformer, any reading other than infinite/OL on the meter indicates that the winding is __________.
a. grounded
b. open
c. OK
d. fused

19. Which of the following is *not* an important consideration when selecting a replacement control transformer?
a. Output voltage
b. Brand
c. Mounting means
d. VA rating

20. When testing an energized and connected control transformer while the unit is not running, should a measurable current in the secondary winding be present? Explain why or why not.

CHAPTER 8

Switches and Relays

Knowledge Objectives

After reading this chapter, you will be able to:

- **K08001** Identify and describe various types of switches.
- **K08002** Identify and describe the operation of various control relays.
- **K08003** Explain how to select and install replacement relays.
- **K08004** Explain how to test and troubleshoot common relays.

Skill Objectives

Skills required to meet the objectives of this chapter:

- **S08001** Select and install a replacement control relay.
- **S08002** Troubleshoot a control relay.

FIGURE 8-1 A common HVACR relay.

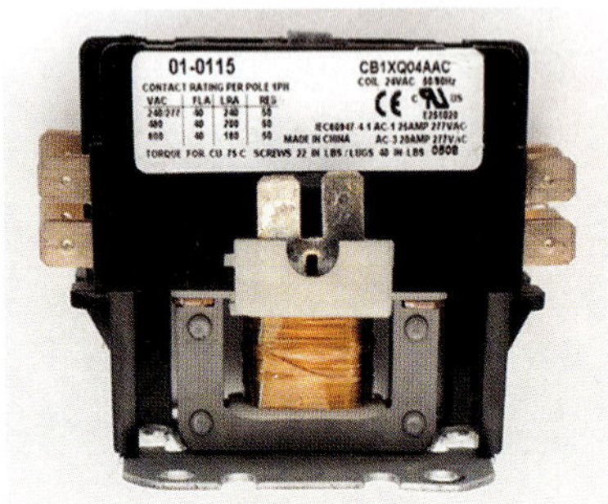

FIGURE 8-2 Compressor contactor.

K08001 Identify and describe various types of switches.

▶ Introduction

The operation of nearly all HVACR systems is automated to some degree. They are very rarely started and allowed to run indefinitely, without cycling off or changing some aspect of the equipment operating status. In addition, for a system to start and operate properly, a specific sequence of events must take place. Switching devices are needed to control a system and automate this sequence of events, providing the foundation of HVACR system control.

Simple, manually operated switches provide the most basic level of control. Some sort of manual switch operation is often required to enable or disable system operation, which is often fully automated once it has been enabled. The action of any manual switch can be duplicated using an automated device called a **relay** (**FIGURE 8-1**). Relays provide a system with what could be considered a primitive means of communication. The control of the relays themselves is where more advanced devices often enter the picture.

Relays generally control relatively small electrical loads. Larger electrical loads, such as compressors, are often controlled by a heavy-duty style of relay called a **contactor** (**FIGURE 8-2**). When motors require added protection from overloads, protective devices can be added to a contactor. The resulting assembly is known as a **motor starter** (**FIGURE 8-3**). Contactors and motor starters will be discussed in Chapter 9.

This chapter will describe the operation of switches and relays and explain how they operate and how they are applied. In addition, it will show how to apply basic relay testing and troubleshooting practices.

▶ Switches

Switches are electrical components used to control the flow of electricity to other devices and systems. A simple wall-mounted light switch is a perfect example. The toggle is raised to power a light fixture and lowered to disable it. Inside, a set of electrical contacts are coming together (making) or being separated (breaking) to complete or open the electrical circuit. Note that the other wire needed to power the light—the neutral wire for standard 120 VAC lighting—bypasses the switch altogether. The neutral circuit is continuous and unbroken from the distribution panel to the 120 VAC loads.

The simplest form of light switch has one set of contacts inside. A source of power is attached to one terminal, and the contacts inside can apply power to another terminal that is wired to the light assembly. This describes the action of a single-pole, single-throw switch (SPST), as shown in **FIGURE 8-4**. It is considered single-pole because there is only one power source terminal. It is single-throw because there is only one switch position that closes a circuit; the other toggle position is OFF (no circuit is closed). If a switch can close another circuit when the toggle is moved from one position to another, it is called a single-pole, double-throw (SPDT) switch.

FIGURE 8-3 Example of a motor starter.

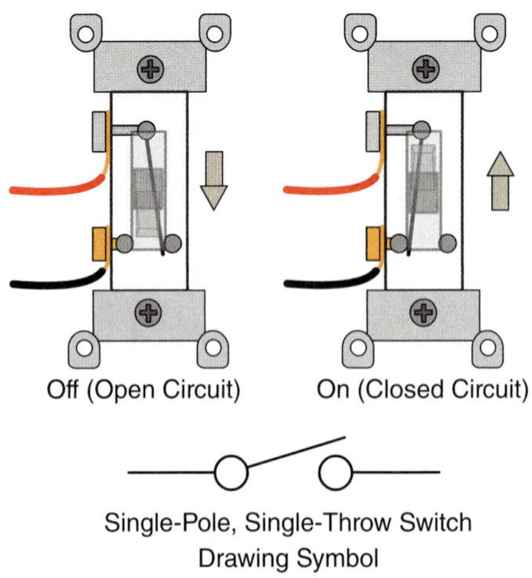

FIGURE 8-4 Action of a simple switch.

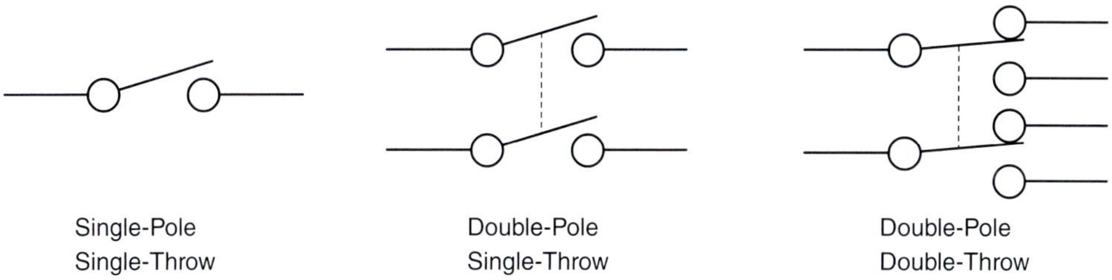

FIGURE 8-5 Common switch arrangements.

Switches with more complex contact arrangements than SPST are needed for many applications. **FIGURE 8-5** shows three common switch actions available. The dashed line connecting the switch arms in the center and lower drawings indicates that a single toggle or other switch actuator controls both of the arms; they move together rather than independently. Switches with multiple poles are used when there are two or more separate power sources to be controlled.

A multi-throw switch completes a circuit to terminals in more than one toggle position. As shown in the lower portion of Figure 8-5, when this double-throw switch toggle changes position, the circuits are broken to two terminals and completed to two other terminals. Multi-throw switches may also have an OFF position, where no circuit is completed. Although these three configurations represent the majority of switch types used, switches with more poles and/or throws are available for unique applications.

Manually Operated Switches

There are many different types and styles of switches. Some of the most common manually operated types found in and around HVACR installations are described below (**FIGURE 8-6**).

- *Toggle switches*—Toggle switches are actuated by a manually operated lever or bat. The common wall-mounted light switch is one example of a toggle switch. The position of the lever provides the user with a visual indication of switch position.
- *Rocker switches*—Rocker switches are similar to toggle switches in their operation. The switch lever rocks back and forth on an axis—hence the name. Also, as in a toggle switch, the position of the rocker visually tells the user the position of the switch.
- *Rotary switches*—Rotary switches can offer a user many different positions, allowing power from the source to be supplied to a number of terminals individually. This type of switch is often found on the front of motor-starter enclosures, providing

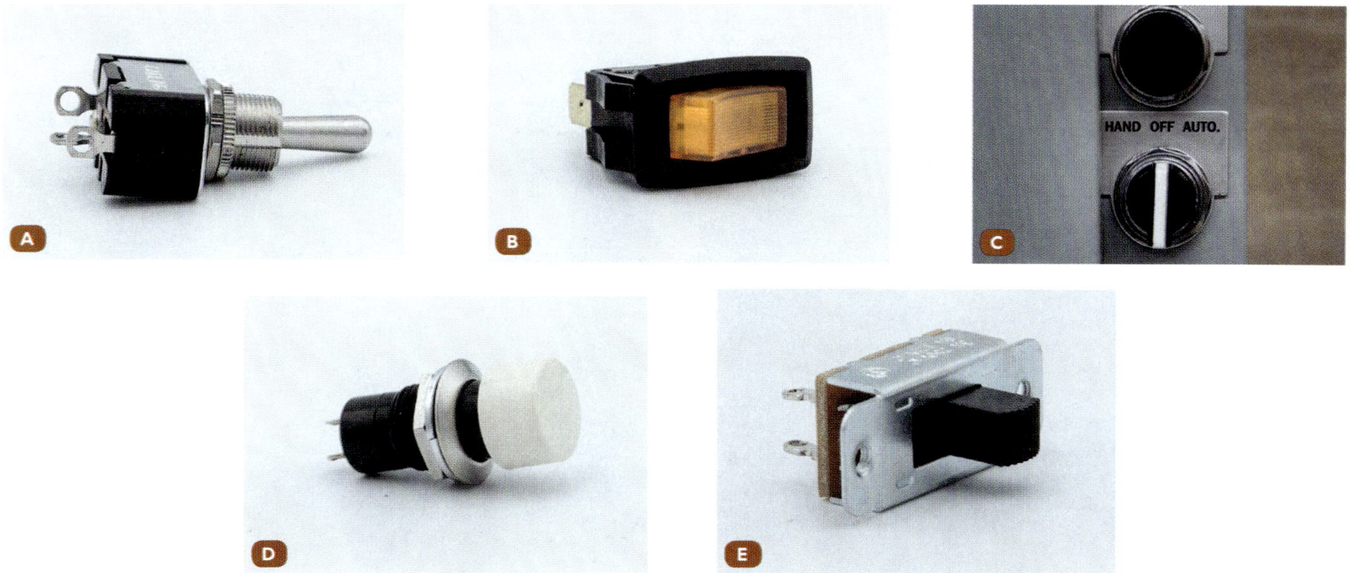

FIGURE 8-6 Manually operated switches. **A.** Toggle switch. **B.** Rocker switch. **C.** Rotary switch. **D.** Push-Button switch. **E.** Slide switch.

a hand-off-automatic (HOA) switching function for the motor. The Hand position immediately energizes the motor (manual operation). In the Automatic position, motor operation is enabled by the control circuit tied to the starter solenoid coil, responding to external control inputs as needed.

- *Pushbutton switches*—Pushbutton switches are generally simple, in that they are two-position switches, each with a limited contact arrangement. Some are designed so that the button remains depressed when the circuit is closed, providing a visual indication of position. A pushbutton switch may also be designed only for momentary contact, where the circuit is complete only as long as the button is depressed. In this case, the switch may be initiating an event, and a **holding circuit** is used to keep the circuit closed after the pushbutton has been released by the user. For motors that are powered by a motor starter but need no automated control circuit, a pushbutton switch can be used on the starter enclosure as a start-stop station.
- *Slide switches*—Slide switches are usually economical choices. Like pushbuttons, they typically have a simple, limited contact arrangement. Slide switches generally have only two or three positions.

Some of these switches are not routinely serviced or replaced by HVACR technicians, since they may be found on industrial control panels and even on printed-circuit (PC) boards in miniature form.

There are few HVACR systems that are fully controlled by a manually operated switch. However, simple switches to enable or disable the operation of a system are frequently encountered. A power disconnect switch is one example. Single-phase disconnect switches are double-pole, single-throw (DPST) switches, and three-phase models are three-pole, single throw (3PST) switches.

The switch that enables heating and air conditioning system operation is usually incorporated into a wall thermostat. A ventilation system, however, may be fully controlled by a simple two-position toggle switch for remote control, or even its power disconnect switch, unless the automatic control of temperature or some other air property is part of its design.

Special Switches

There are many switches that are operated neither manually nor by an external control circuit. HVACR systems must respond to a variety of system or environmental conditions, most notably temperature and pressure (**FIGURE 8-7**). These special switches will be covered in chapters devoted specifically to their operation and function. Temperature and pressure switches, for example, are a vital part of most HVACR systems, used for both safety and common control purposes.

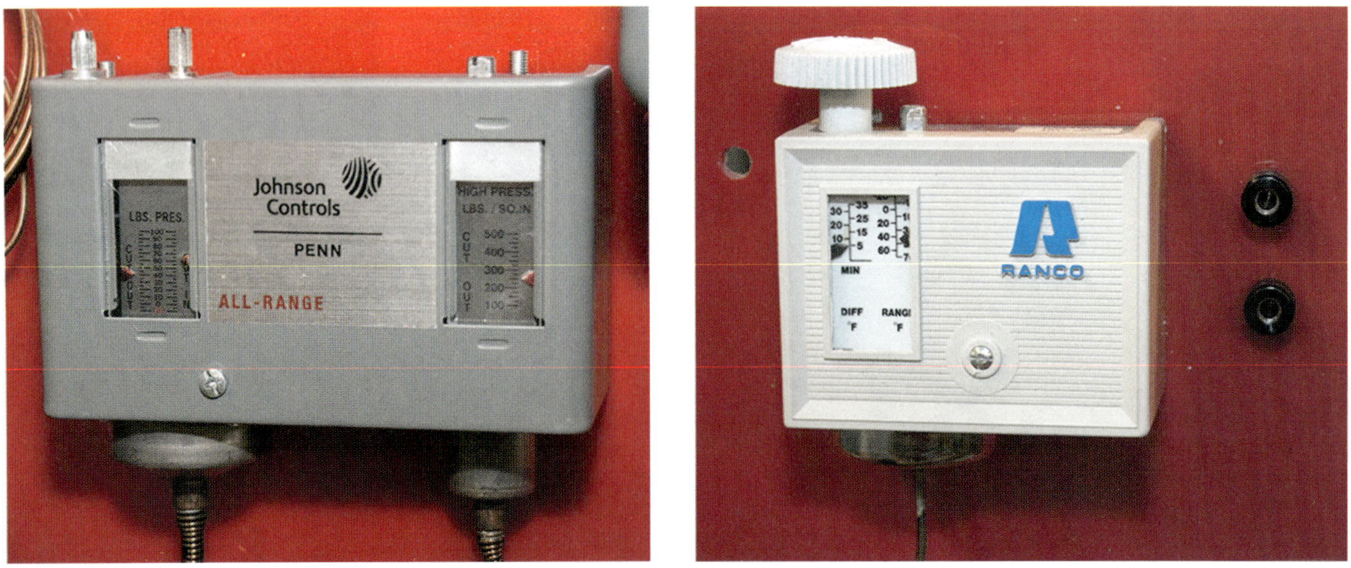

FIGURE 8-7 Examples of pressure and temperature switches.

Testing and Troubleshooting Switches

Testing manually operated switches is relatively simple with a multimeter. Testing to determine whether the contacts are open or closed can be done with power off, using the ohmmeter function to test for continuity. The meter probes are placed on the two-switch terminals in question, and the meter display should indicate whether there is or isn't continuity through the contacts. Testing can also be done with power on, using the voltmeter function. With power applied to a switch and the probes placed across the contacts, closed contacts will show no voltage present between the two terminals. This is because there is no difference in electrical potential between those two points, even though voltage is in fact present. If the contacts are open, the voltage being switched will be displayed. This process will be presented in greater detail and practiced in the next section related to relays.

▶ Relays

There are very few HVACR control circuits that do not contain one or more relays. Think of a relay as a magnetic switch; a switch that is operated electrically instead of manually. Like switches, relays can have multiple poles and throws. As a result, when a relay is energized, multiple circuits can be energized and/or deenergized.

At the heart of an electromechanical relay is the **solenoid coil**. A solenoid coil uses electrical energy to develop a magnetic field in a coil of wire wrapped around a core to produce movement (**FIGURE 8-8**). When a solenoid coil is energized, a magnetic field quickly develops around the core, drawing the armature toward it. The armature, of course, must be made of magnetic material. When it is deenergized, a spring typically pushes or pulls the armature toward the default (normal) position. The armature in turn changes the position of the switch contacts, and a circuit(s) is either opened or closed as a result.

FIGURE 8-9 shows an example of a relay wiring diagram. This example shows the contact structure of an SPDT relay. Wiring diagrams and relay specifications always show the position of the contacts when the relay is deenergized. In this case, there is a normally closed contact between terminals 4 and 5 and a normally open contact between terminals 4 and 2. In this case, terminal 4 is called the common terminal since it is common to both contact sets. When the relay solenoid is energized through terminals 1 and 3, the armature inside moves and closes the circuit between terminals 4 and 2, simultaneously opening the circuit between terminals 4 and 5.

K08002 Identify and describe the operation of various control relays.

▶ CORE CONNECTION

Solid-State Relays

Solid-state relays, unlike electromechanical relays, do not have a physical set of contacts. In fact, there are no moving parts at all. They operate exclusively by using electronic components to allow or prevent the flow of electricity. Solid-state relays are small and operate quietly and quickly, and there are no internal arcs at contacts to prevent using them in explosive atmospheres. They are well suited to applications that require a relay to survive thousands and possibly millions of cycles. However, there are disadvantages as well. Solid-state relays will be covered in greater detail in Chapter 11.

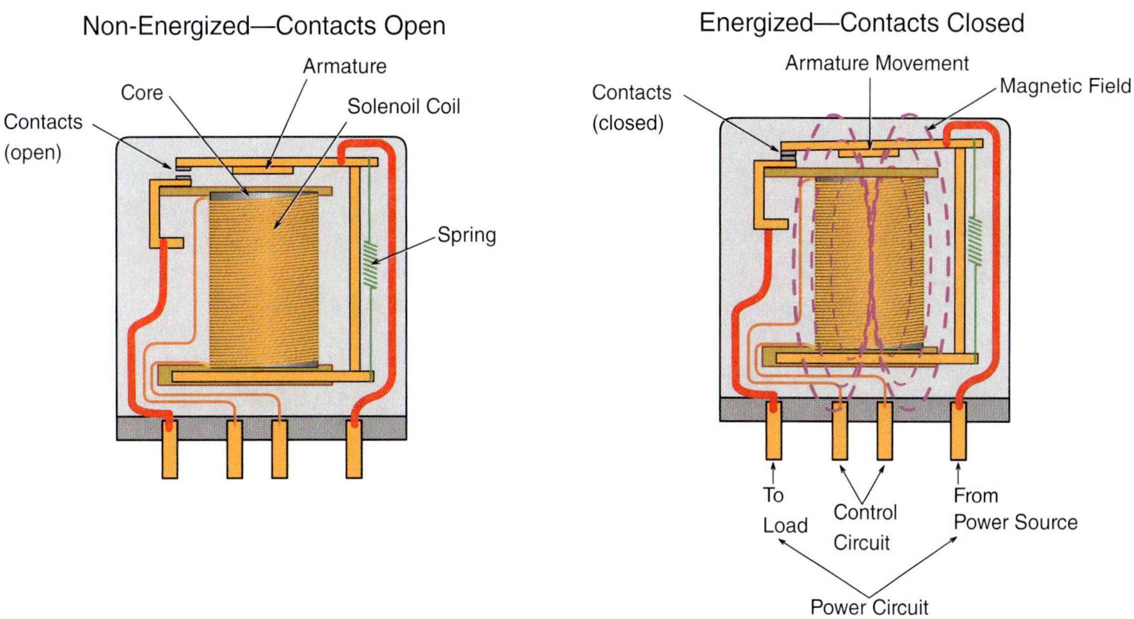

FIGURE 8-8 Solenoid action of a relay.

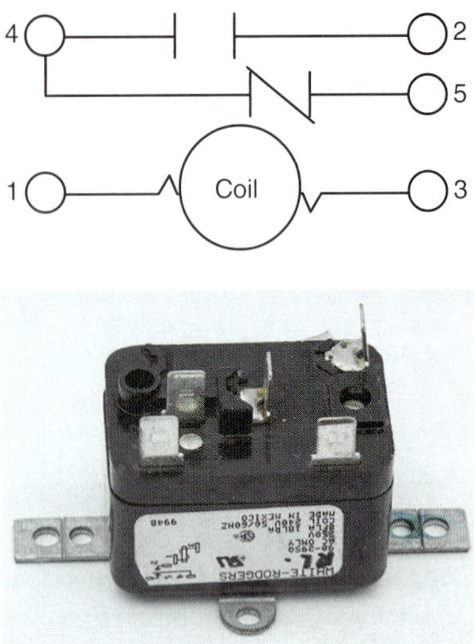

FIGURE 8-9 SPDT relay.

FIGURE 8-10 A compressor anti-short-cycle timer.

Relay Applications

Beyond their electrical characteristics, relays are also categorized based on their application or specific function. The following list identifies some of the common relay types used in HVACR applications:

- *Electromagnetic relay*—This is the most common type of HVACR relay, which uses the previously described solenoid coil to create a magnetic field to move an armature. A spring is usually provided to ensure the contacts return to their normal state when the coil is deenergized. Many of the relays listed here are electromagnetic relays with unique characteristics that set them apart.
- *Solid-state (electronic) relay*—This type uses solid-state devices internally to open and close electronic gates rather than physical contacts.
- *Timing relay*—Timing relays have a variety of applications and can differ significantly in their appearance and features. A delay-on-make relay begins the time-delay period when it is energized, then changes its position after the delay period has elapsed. This may be used to delay the start of a second compressor until another compressor has started, eliminating the inrush current associated with starting two compressors at once. A delay-on-break timer begins its timing period after it is deenergized. There are a number of other timing variations available. One type of delay-on-break timer, for example, is used as a compressor anti-short-cycle timer (**FIGURE 8-10**). This relay has a normally closed contact arrangement. When the control circuit is energized by a thermostat and the compressor starts, the relay does nothing. Once the circuit is deenergized, the relay then opens its contacts and will not reclose them until the time period has elapsed. This prevents the compressor from being restarted too quickly, especially when the thermostat is rapidly cycled OFF and ON or the temperature setting is changed multiple times in a short period. Rapid on/off cycling of any compressor can lead to motor failure.
- *Thermal relay*—Thermal relays are generally considered a type of timing relay. A small resistance heater (instead of a solenoid assembly) is used to position the relay. When power is applied, a tiny heating element (resistor) warms a bimetal element—which bends in response, closing or opening the contacts after it reaches a sufficient temperature. As a result, there is a delay in the relay's activity. One subtype is called an electric heat sequencer (**FIGURE 8-11**), which is used to control multiple electric heating elements. Do not confuse them with thermal overload relays, which have a completely different function in the operation of a motor starter.
- *Potential and current relays*—These two relay types are associated with motor-starting circuits. A potential relay opens to disconnect the start windings of a high-torque motor when starting. Current relays provide the same function on motors with low starting torque requirements. Potential and current relays respond to changes in the voltage or the current flow through the coil to support the start-up of single-phase motors that require a start capacitor. These relays will be discussed in greater detail in Chapter 10 and Chapter 14.
- *Overload relay*—Overload relays are used as motor-protection devices. They can be separated from the contactor or be a component part of a motor-starter assembly. Sensors may also be buried within the motor windings, with wiring terminals provided alongside the motor-wiring terminals. The most basic overload relays are thermal. They respond to heat generated by a thermal element in the motor circuit that grows hotter as current increases. The temperature change is very precise: if the motor current exceeds the value of the thermal elements by a small amount, the excess heat will cause the overload relay to open its contacts. The contacts of an overload relay are wired in series with the solenoid coil of the starter contactor; when the contacts open, the control circuit is broken and the motor shuts down. Another contact may be used to sound an alarm or illuminate a trouble light to get the attention of an operator or technician.

FIGURE 8-11 An electric-heat sequencer.

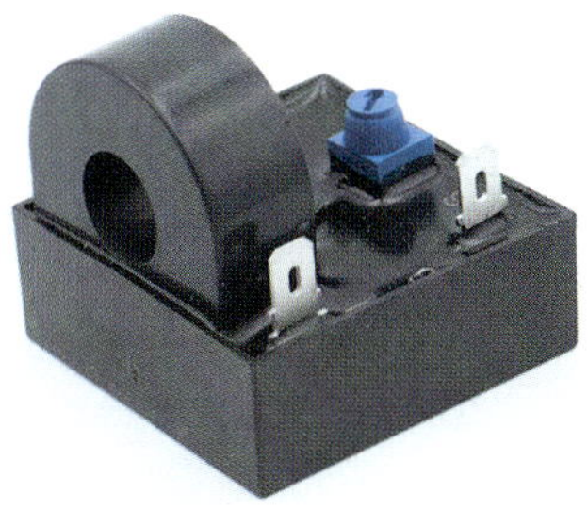

FIGURE 8-12 A current-sensing relay with a built-in CT.

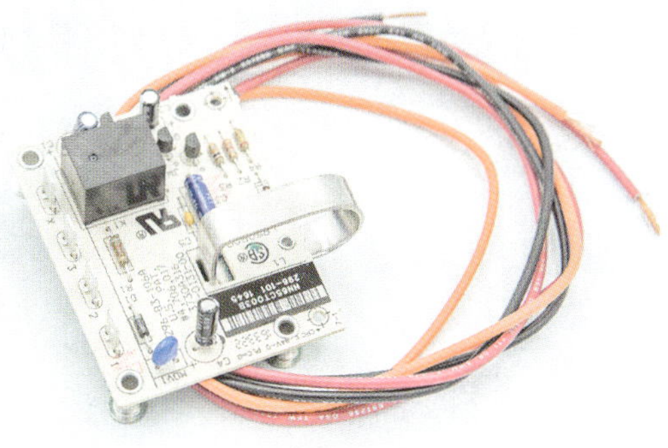

FIGURE 8-13 Compressor lockout relay.

- *Current-sensing relay*—A current-sensing relay (**FIGURE 8-12**) is most often used as a motor-protection device in HVACR applications, but it can also serve other purposes. They incorporate a **current transformer (CT)** that monitors motor current. The CT functions using the same principles as a clamp-on ammeter to sense the presence and level of current. When the relay detects continuous current values beyond the set point, the contacts open to stop the motor and prevent it from being damaged. They can also be used to sense the flow of current in one circuit and respond by opening/closing their contacts to control other circuits. Current-sensing relays are a specific type of overload relay.
- *Compressor lockout relay*—This relay prevents the compressor from operating after a safety device has opened the circuit. For example, a low-pressure switch may activate, indicating an excessive drop in refrigerant pressure and open the control circuit of the compressor contactor. The lockout relay (**FIGURE 8-13**) opens its contacts in response, and as long as power is applied to the unit, the circuit remains open. Note that there is a CT on this relay, but in this case, it serves to detect only the presence of current rather than the volume of current flow. The presence or lack of current flow tells the relay whether or not the compressor is actually running. This way, if the controls have directed the compressor to start and the relay senses that it did not start or that it stopped unexpectedly, it will lock out the circuit and prevent future start attempts until it is reset. They are reset by cycling the control power source off and then on again. As a result, the compressor remains off until there is human intervention to investigate and correct the underlying problem.

There are many ways that relays contribute to the operation of HVACR equipment. Although a number of common applications are listed here, there are many other possibilities. The fact that there are possibilities is why a technician must understand the basic operation of all types of relays and consult technical data for the specific part, when clarification is needed.

▶ Selecting and Installing Relays

Relays are available in a wide variety of shapes and sizes (**FIGURE 8-14**). Shape and size may even be a factor in choosing a proper replacement or when a new control arrangement is constructed in the confined space of a control panel. This section will address important factors in relay selection and the installation process.

K08003 Explain how to select and install replacement relays.

S08001 Select and install a replacement control relay.

Relay Selection

In general, when replacing a relay, it is easier to use a relay of the same style originally selected by the equipment manufacturer. The wiring has been installed to fit that particular shape and size of relay, and the mounting holes will match. So the task is generally without

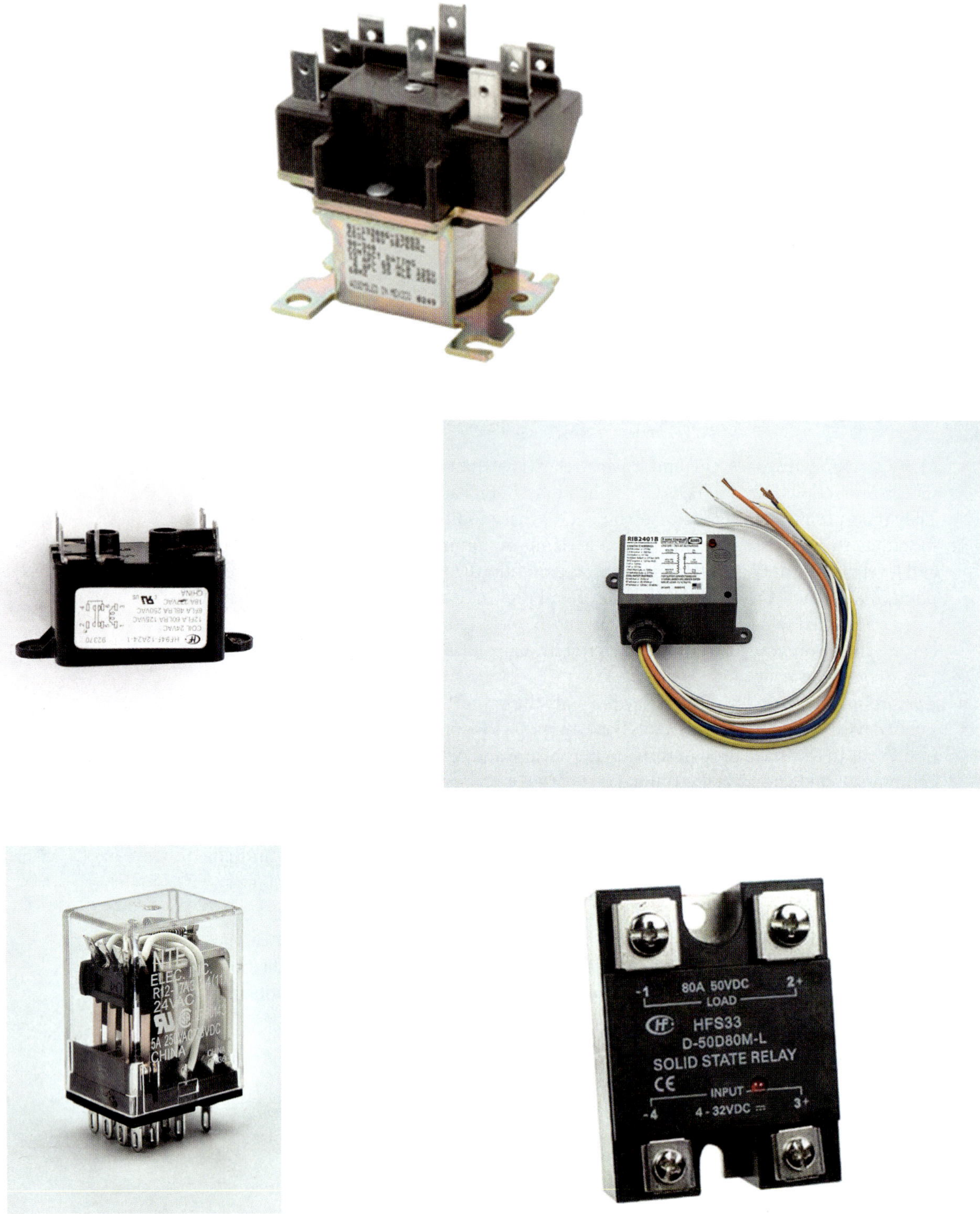

FIGURE 8-14 Relays found in HVACR applications.

complication when the same style is chosen. But there are certainly other factors involved in relay selection, including the electrical characteristics and the application.

Contact Structure

A replacement relay must first be able to provide the same switching functions as the one it replaces. Many possible contact configurations are available. **FIGURE 8-15** shows the electrical diagram for a double-pole, double-throw relay (DPDT), as one example. If SPST operation is needed, an SPDT switch can be used; the extra terminal simply remains unused. Substitutions like this can be made if there is no significant difference in cost and if the relay fits comfortably in the space provided.

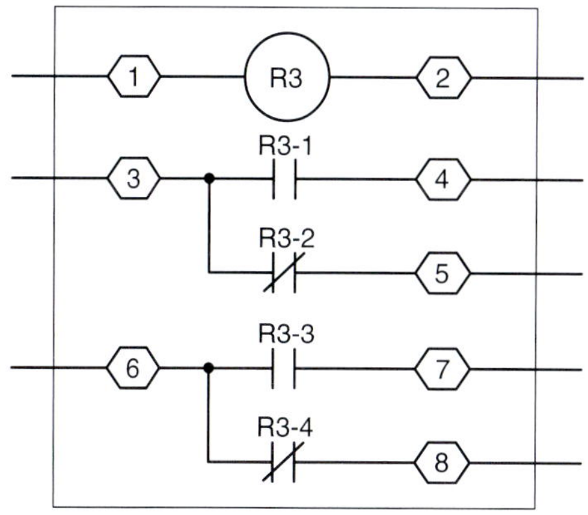

FIGURE 8-15 A DPDT relay diagram.

Coil Voltage

The voltage of the relay solenoid coil is the next factor to consider. Although the most common coil voltage for HVACR applications is 24 VAC, some relays operate at higher voltages. Large commercial and industrial equipment may make use of relays with 230 VAC coils, for example.

Although the information will not likely be found on the relay labeling, relays have a **pickup voltage** and a **dropout voltage**. The pickup voltage, sometimes called the must-operate voltage, is the lowest voltage at which the relay solenoid can operate. The dropout voltage defines the opposite condition—the voltage at which the relay solenoid must release the armature and return to its deenergized condition. If the voltage applied to a relay is not equal to or greater than the pickup voltage, the relay will fail to operate. Remember that the transformers used in many HVACR control circuits can produce only a small amount of current (a 40 VA, 24 VAC transformer can support only about 1.6 amps, for example). If a number of loads are energized at the same time and the transformer is overloaded, the voltage can drop below the pickup voltage and the relay will fail to operate.

When selecting a relay, be sure that the coil voltage meets the requirement. Note that this voltage has nothing to do with the voltage switched through the relay contacts.

> ▶ **TROUBLESHOOTING TIP**
>
> **Low-Voltage Conditions**
>
> Most relay coils require the incoming voltage to be within +10% or −15% of the rated voltage. For a 24 VAC coil, this range means the operating voltage should be 20.4–26.4 VAC. However, it is important to note that actual results may vary. A 24 VAC relay may actually operate with only 19 volts applied, but not reliably. If the voltage applied to the coil is too low, it will not develop a magnetic field strong enough to move the armature. The relay may hum or even chatter as it attempts to move. Applying a higher voltage than required will likely operate the relay but shorten its life significantly. Before deciding that the solenoid coil of a relay or contactor has failed when power is applied and the relay doesn't change position, ensure that the voltage applied is appropriate. Measure the voltage at the coil terminals while it is energized. If there is a voltage drop in the circuit causing the problem, that is when it will be apparent in the voltage displayed. The circuit may have 24 VAC available when there is no load imposed, but the voltage can drop significantly when components are energized. A poor wiring connection in the control circuit can also create this type of problem.

Switched Voltage

The voltage that a relay can safely switch is related to the air gap between the contacts and other internal insulating properties. The information may be reported on the label in several different ways. A label from a common relay is shown in **FIGURE 8-16**. The electrical data indicates that the relay contacts can handle 12 FLA (full load amps) at 125 VAC and an LRA (locked rotor amps) of 60 amps. Without directly stating what these values refer to, it is understood that they refer primarily to inductive loads such as motors, since inductive loads have an inrush current when they are first energized. As a result, they momentarily draw far more current than their normal running current. The LRA refers to the amount of current a motor might reach when it is first energized, or the continuous current experienced if the motor is mechanically seized and refuses to rotate. This relay can accommodate up to 60 amps of inrush. However, if the 60-amp load were continuous, the contacts would quickly burn off and/or melt together.

Note that there are other ratings listed for 250 VAC. Since this is the highest voltage value shown, it can be assumed to be the maximum voltage that should be applied to the contacts. Most HVACR relays are rated for a maximum of 250 VAC. To switch higher voltages, a higher quality and more expensive relay is required. Internally, the contacts will be larger and set farther apart to prevent arcing between the contacts when they are separated.

Although there are many exceptions, most HVACR relays are switching either 125 VAC in a standard furnace or 208/240 VAC in air conditioning equipment. Of course, relays can be used to switch any voltage below their maximum rating. Many relays switch 24 VAC circuits, for example, although they have a much higher voltage rating.

Switched Current

Refer again to Figure 8-16. The label indicates that it can switch up to 12 amps at 125 VAC or 8 amps at 250 VAC. As discussed previously, these values are for inductive loads. Resistive loads, like an electric heating element, do not require an inrush of current when they are energized—the current simply rises to a point and dwells at or near a given value. As noted on the label, the relay can switch up to 18 amps of current at voltages up to 277 VAC. This value is known, by the letters "RES," to be related to a resistive load in the specification.

FIGURE 8-16 Relay labeling.

To select the correct relay, it is important to know the current that will flow through the relay contacts during normal operation. It is also important to know whether the load

FIGURE 8-17 A properly mounted relay.

is inductive (motor, transformer, etc.) or resistive. Although the example relay can handle 8 amps at 250 VAC, it would not be the best choice for a steady 8-amp inductive load. Consistently placing the maximum load on the relay will likely lead to a short life. For a steady 8-amp inductive load, a relay with a higher contact rating would be more dependable. On the other hand, applying an 8-amp resistive load to the contacts would be well within the relay's maximum resistive capacity. It is common for a relay to have a higher current capacity for resistive loads than for inductive loads.

Installing a Relay

Once a proper relay has been selected, it can be installed. The three primary steps in the installation are proper mounting, correct wiring terminations, and testing.

Most relays are not position-sensitive; they can be mounted in any position. However, although many of today's potential relays used for compressor starting are not position-sensitive, a few still are. Be sure to review the installation instructions for the chosen relay and ensure that the mounting provided is appropriate.

The mount should be firm and dependable. Most relays provide a variety of mounting holes from which to choose. A good location provides clear access to the relay terminals and does not place the terminals too close to other component terminals or conductive enclosures (**FIGURE 8-17**). A relay should never be left dangling from, or be supported by, its wiring connections.

Plug-In and DIN-Mounted Relays

Although most of the relays that a technician will likely replace require that wiring connections be made, replacing one type of relay eliminates the need to do so. A plug-in relay (**FIGURE 8-18**) has electrical terminals built into a base that also serves as its mount. A socket base is first installed, and all wiring terminations are made at the base. The relay then plugs into the socket. Replacement is a simple matter of pulling the relay out and inserting a new one into the socket. Not only does this eliminate the wiring task, but it also eliminates the mounting task once the socket has been installed. However, this style of relay is typically found only in built-up control panels and heavy commercial/industrial HVACR equipment.

FIGURE 8-18 A four-pole, double-throw (4PDT), 24 VAC plug-in relay.

There are only a few styles of terminals used on plug-in relays, such as spade and round pins. The terminal style and their arrangement must match the socket provided. The relay and socket also have keyed elements to ensure that the relay can be pushed into the socket oriented only one way. This guarantees proper alignment and orientation and ensures that the relay and socket are compatible.

In Figure 8-18, note that the socket can be mounted to a panel or to a **DIN rail**. A DIN (Deutsches Institut für Normung) rail (**FIGURE 8-19**) is a simple stamped-metal bar used to mount a variety of electrical devices inside control enclosures in an organized and standardized manner. Switches, relays, contactors, circuit breakers, and many other components are available with DIN-rail mounting features. The relay socket in the figure has a wide groove on the bottom that is designed to lock on to the rail.

There are also relays that are designed for direct mounting to the DIN rail, without a socket. Wiring terminations are made where shown in **FIGURE 8-20**. Although not many plug-in and DIN-mounted relays will be found in residential or commercial equipment, they will be found in control systems and heavy commercial/industrial HVACR systems.

▶ CORE CONNECTION

What Does DIN Mean?

DIN is an abbreviation of Deutsches Institut für Normung, the name of the German organization that originally created the mounting rail's manufacturing specifications. Those specifications have since been accepted internationally.

Wiring Terminations

Correctly terminating the wiring is crucial to success. Since many relays are controlled by a low-voltage power source and switch circuits operating at a higher voltage, a wiring error can instantly destroy the relay. Applying 240 VAC to a 24 VAC solenoid coil will usually cause quick coil failure. Mixing up the wiring connections accidentally can also lead to line voltage being applied to a low-voltage control circuit or component, spreading the damage further. Follow the tips below for a successful installation:

FIGURE 8-19 A DIN rail.

FIGURE 8-20 DIN-mounted relays.

- Before disconnecting the defective relay, mark the individual wires relative to their purpose or terminal (**FIGURE 8-21**). If the new relay is an exact replacement, transferring the wiring should be a relatively simple matter. If there is working room, dismount the old relay with the wiring connections intact, mount the new relay, and then transfer the wires to the same terminals on the new relay. If the relay is a different style, be aware that the terminal numbering will likely be different. That can be more complicated than it sounds because manufacturers do not number terminals in a specific manner. Many HVACR relays, for example, use the numbers 1 and 3 to designate the terminals for the coil. One might expect them to be terminals 1 and 2, but that is not the case.
- Knowledge always contributes to a better job. Before beginning, know where each wire leads and what its function is.
- Inspect the work after it is done. Take a step back, think through the connections made, and ensure that it is wired correctly before applying power.

FIGURE 8-21 Marking wires for reconnection.

To select and install a replacement control relay, follow the steps in **SKILL DRILL 8-1**.

SKILL DRILL 8-1 Selecting and Installing a Replacement Control Relay

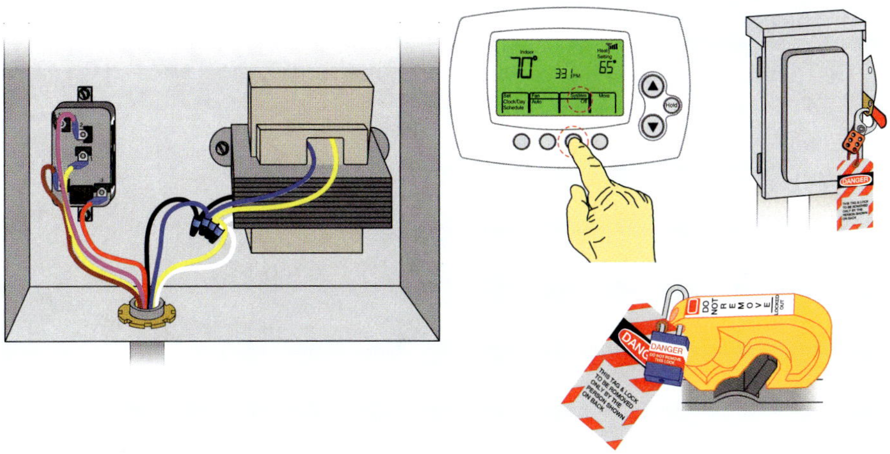

1. Ensure that all controls, including the thermostat, are set to the OFF position. Deenergize, lock out and tag the equipment to be serviced. Be sure that the process deenergizes not only the line voltage feeding the equipment but also the low-voltage control power source, which may be located in a separate piece of equipment.

Continued

SAFETY TIP

Working with electrically energized circuits is hazardous. Avoid contact with any surface that could potentially be energized. When working inside energized enclosures, maintain a stable position and, whenever possible, avoid placing both hands inside the enclosure. All work on energized equipment by unqualified individuals should be supervised by a qualified instructor or technician.

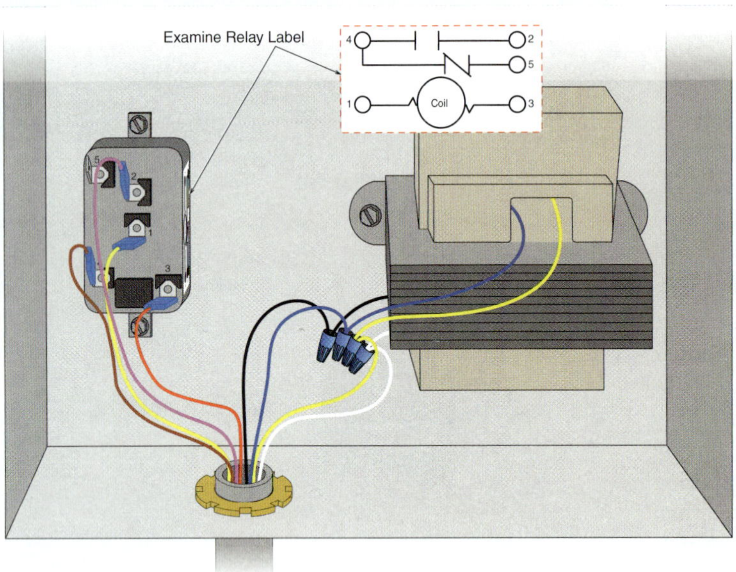

2. Examine the relay labeling carefully to determine its characteristics. In addition, consider the function of the relay and both the current and voltage of the circuit it is switching. Determine whether the characteristics of the existing relay are sufficient for the voltage and current load imposed on it. If so, an exact duplicate of the relay can be used as a replacement if desired.

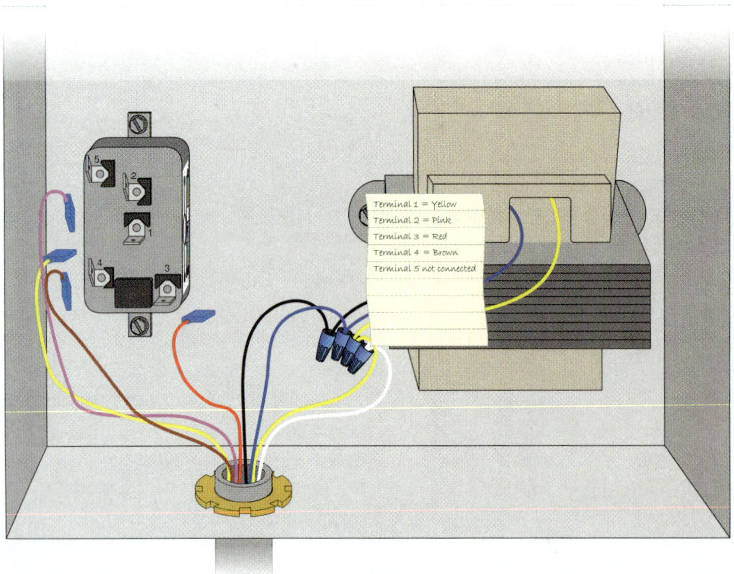

3. Mark or label the wires before removing them from the relay terminals, to ensure that they will be properly reconnected. Then remove the wires from each terminal.

Continued

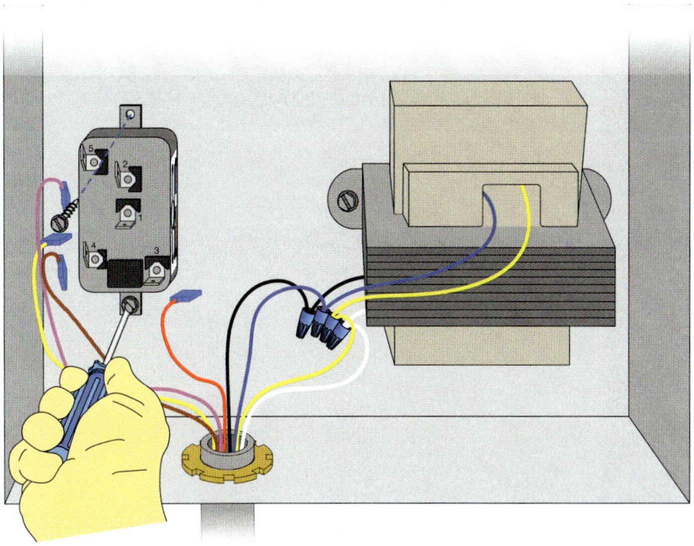

4. Remove the mounting screws, remove the old relay, and mount the new relay in a secure manner. A different style of relay may require drilling one or more new mounting holes. Always take great care when drilling in the control cabinet to ensure that the drill does not damage other wiring, control components, or refrigerant tubing behind the panel.

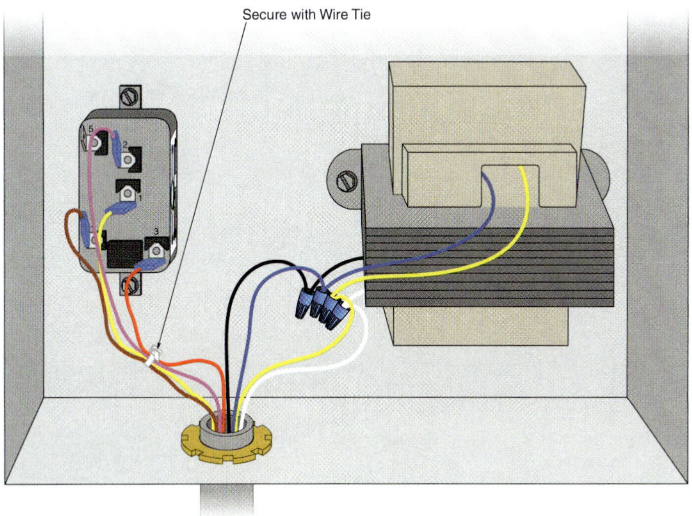

5. Reconnect the wires to their appropriate terminals. Inspect the condition of the termination hardware before making the connection, to ensure a good electrical connection exists. If it is questionable, cut and terminate the wire with a new connector. Recheck completed work, and ensure that the wires are properly connected and that no wire or termination hardware is in danger of making contact with another conductor or the cabinet. Use wire ties as necessary to neatly control wires.

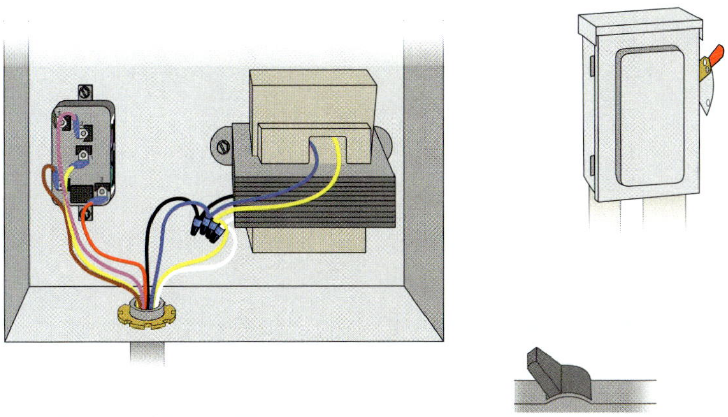

6. Ensure that all controls, including the thermostat, remain set to the OFF position. Remove the locks and tags from the power source(s). Restore power to the system.

Continued

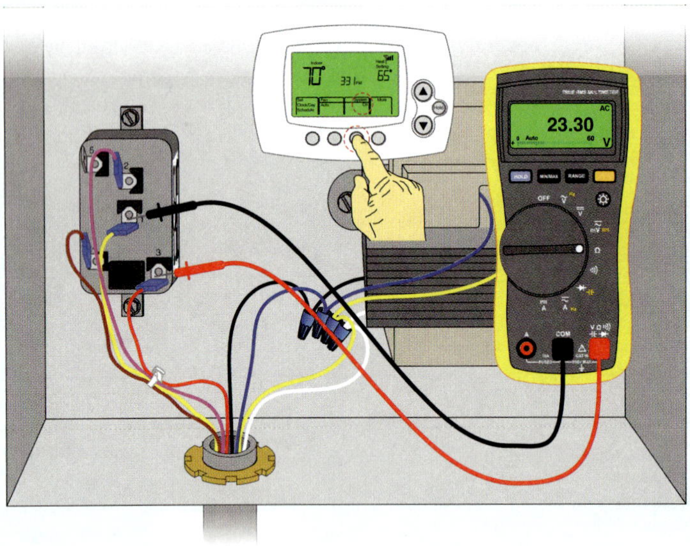

7. Start the equipment and ensure that the relay participates in the sequence of operation as expected. Use a voltmeter to check the voltage applied to the relay coil to confirm that it is steady and sufficient to reliably energize the relay.

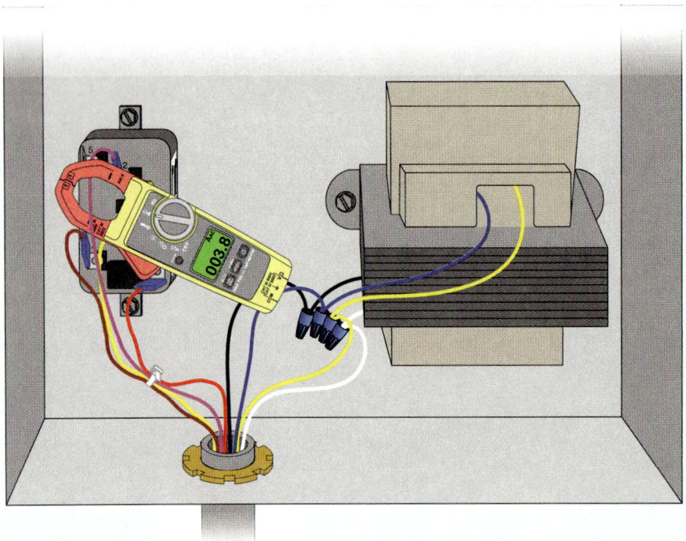

8. Use a clamp-on ammeter to measure the current passing through the relay contacts to confirm that it is within the capabilities of the relay.

9. Carefully reinstall the control panel covers and equipment access doors, and return the system to service.

▶ Testing and Troubleshooting Relays

K08004 Explain how to test and troubleshoot common relays.

S08002 Troubleshoot a control relay.

Do not assume an installed relay will work as intended. Operate the equipment while observing the relay to ensure that it responds as expected. In this case, observing the relay means monitoring it with a voltmeter and/or ammeter to determine when it is energized and whether the contacts have responded. Few relays have any visual method built in to determine its position or status.

Testing most relays is not a complex activity. It is not unusual for a relay to initially appear to be the source of a problem, but troubleshooting may reveal a different source. For example, if relay contacts are not opening/closing, disrupting operation of the system, begin troubleshooting by determining whether the relay is receiving voltage to the coil. If the closure of a system thermostat should result in 24 VAC being applied to a relay coil but there is no voltage present at the coil terminals, the relay is not the problem. The problem is obviously upstream from the relay.

Two types of failures are normally associated with relays. The internal coil that develops the electromagnetic force to move the relay can open or be shorted internally. The other type of failure is related to the contacts. The contact assembly can fail mechanically, resulting in the contacts failing to move even when the coil is energized, or the contacts can be burned or damaged by excessive current. In cases of current overload, the contacts may permanently weld themselves together and will not separate.

Testing a Relay Coil

A multimeter is first used to check for voltage at the coil terminals. If the proper voltage is applied to the coil but there is no response, deenergize the system (including the source of the control voltage), disconnect the leads from the coil, and use a multimeter to check for continuity through it. If there is no continuity through the coil, it has failed and must be replaced. If there is continuity, there may be mechanical or electrical damage to the contact mechanism that is preventing it from working.

Testing Relay Contacts

SAFETY TIP

Troubleshooting completed with power applied must be done with the greatest care. Working with energized electrical circuits represents a major shock hazard. In addition, if the meter probes cross multiple wires or terminals, arcing and possible damage to system components may result.

The contacts of a relay can be checked for their position (open or closed) in one of two ways. The task can be done by using a voltmeter with power applied or by using an ohmmeter to check for continuity through the contacts without power applied.

To test the contacts with power applied, use the voltmeter function of a multimeter. Refer to **FIGURE 8-22**. The voltmeter probes are placed on the two terminals that represent the set of contacts to be tested. All wiring can remain connected to the relay.

When testing normally open contacts, begin with the relay deenergized but with power applied to the system. Touch the meter probes to the two terminals representing the contact

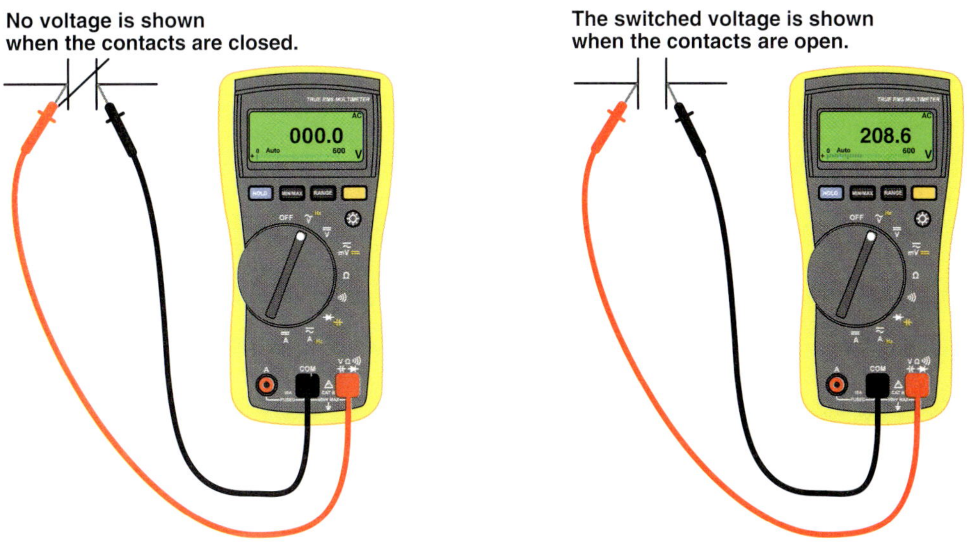

FIGURE 8-22 Voltmeter response to open and closed contacts, with power applied.

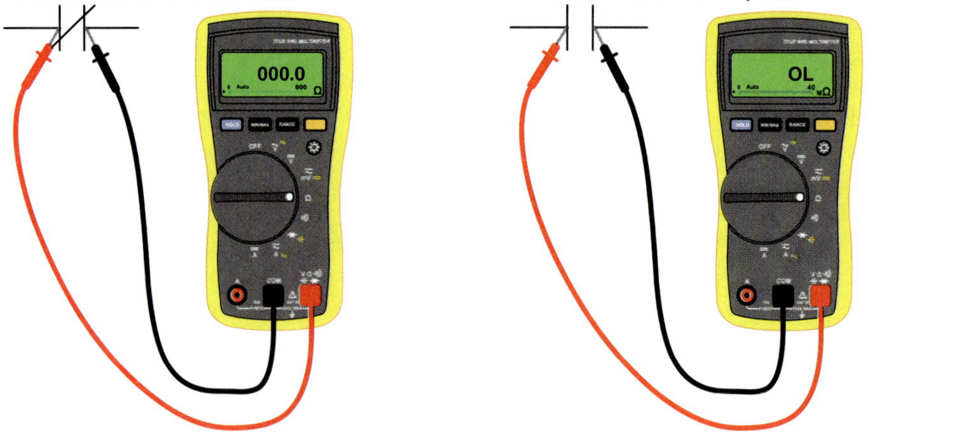

FIGURE 8-23 Ohmmeter response to open and closed contacts, with no power applied.

set to be tested. The voltmeter should read the switched voltage when the contacts are open. Next, energize the relay coil. The contacts should now close, indicated by no voltage being displayed on the meter across the same two terminals. Normally closed contacts are checked the same way. No voltage displayed indicates that the contacts are closed.

The contacts can also be tested with an ohmmeter. Remember that working on a deenergized system is always safer, and it can often yield the same information. Refer to **FIGURE 8-23**. Deenergize the system as well as the source of control power, mark the wires as necessary, and disconnect them from the relay. With the meter probes touched to the proper terminals, continuity is displayed when a set of contacts is closed. When they are open, there will be no continuity (infinite resistance). Continuity with any substantial resistance usually indicates that the relay contacts have been damaged and will likely cause an unacceptable voltage drop under load.

When testing relays such as timing relays, remember that timing periods have to elapse before the relay responds as expected. You must know the relay's timing and sequence of operation to test them effectively.

To troubleshoot a control relay, follow the steps in **SKILL DRILL 8-2**.

SKILL DRILL 8-2 Troubleshooting a Control Relay

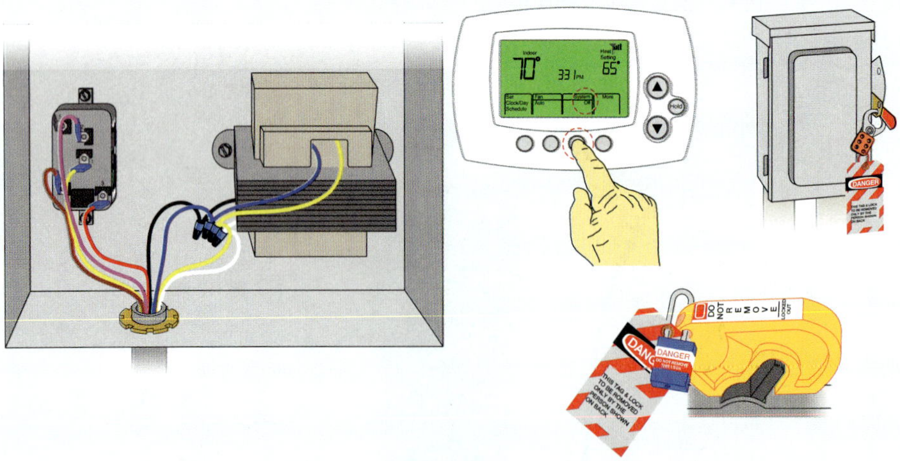

SAFETY TIP

Working with electrically energized circuits is hazardous. Avoid contact with any surface that could potentially be energized. When working inside energized enclosures, maintain a stable position, and, whenever possible, avoid placing both hands inside the enclosure. All work on energized equipment by unqualified individuals should be supervised by a qualified instructor or technician.

1. Ensure that all controls, including the thermostat, are set to the OFF position. Deenergize, lock out, and tag the equipment to be serviced. Be sure that the process deenergizes not only the line voltage feeding the equipment, but also the low-voltage control power source that may be located in a separate piece of equipment.

Continued

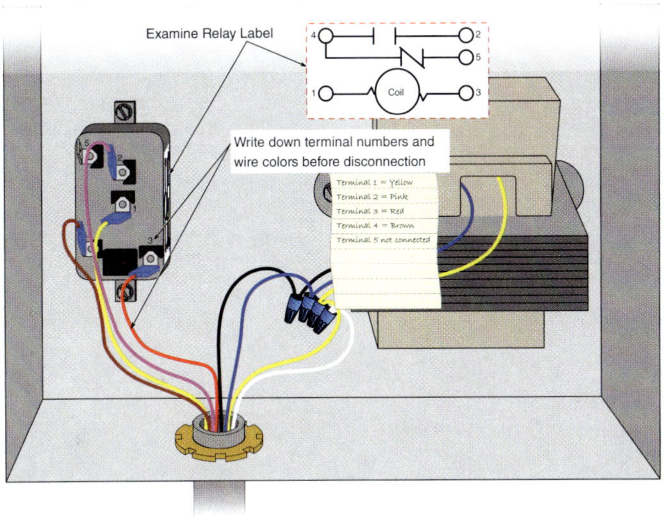

2. Locate the relay to be tested, then examine the relay label and equipment wiring diagram to identify the terminals. Determine the function and destination of each wire. Then mark the wires as they are disconnected from the terminals. Disconnect all the wires.

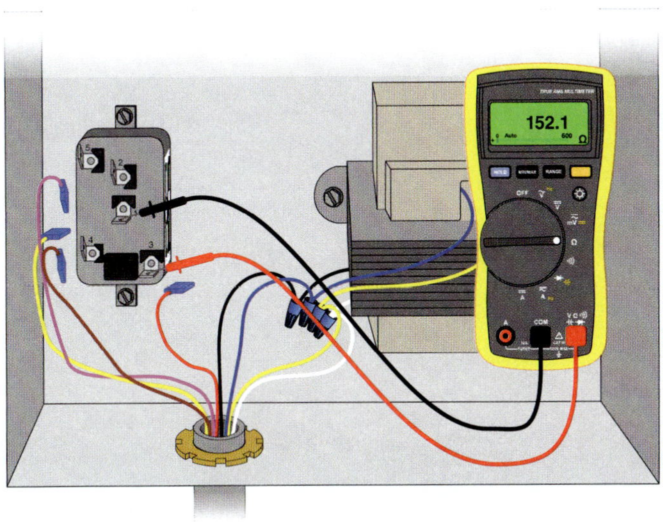

3. Set the multimeter up for continuity testing. Test the meter by touching the probes together and confirming its response, then touch the probes to the two coil terminals. Continuity with resistance should be displayed on the meter. An open circuit (infinite resistance) is an indication of coil failure.

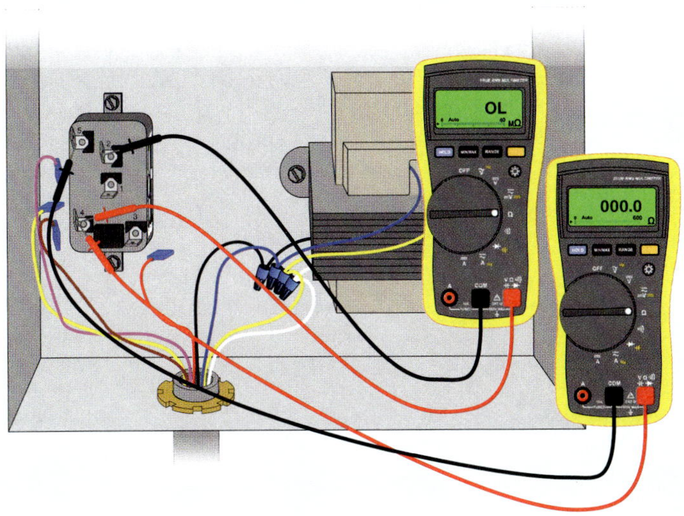

4. Locate the terminals associated with the contacts, and determine which contacts are normally open and/or normally closed. Place the meter probes on the two terminals representing one set of contacts. There should be continuity with no resistance across closed contacts, and no continuity (infinite resistance) through open contacts. Contacts that are closed when they should be open, or open when they should be closed, indicate a defective relay. Repeat the test for each set of contacts.

Continued

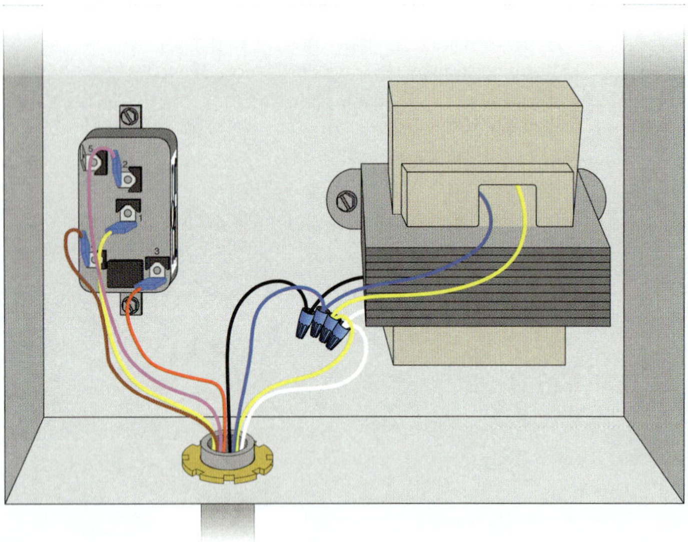

5. If the relay passes deenergized testing, reconnect all relay wiring.

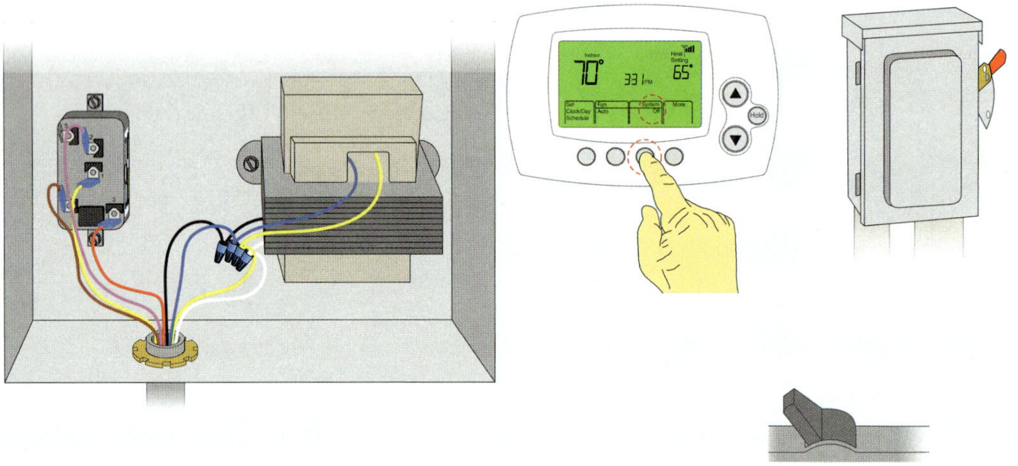

6. Ensure that all controls, including the thermostat, remain set to the OFF position. Remove the locks and tags from the power source(s), and restore power to the system.

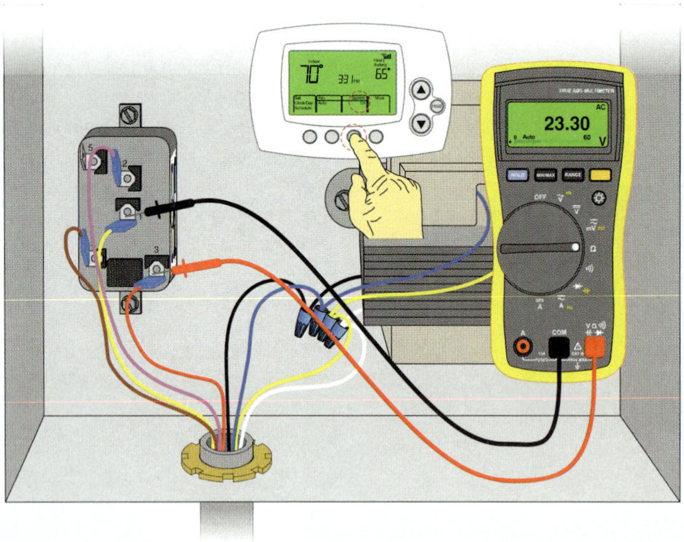

7. Manipulate the controls as necessary to energize the relay coil. Set the multimeter to test voltage. Test the meter on a known source to confirm it is functional, then touch the probes to the terminals of the relay coil. The meter should read the appropriate voltage for the control circuit, indicating the relay coil is energized and that the contacts should have now changed their position.

Continued

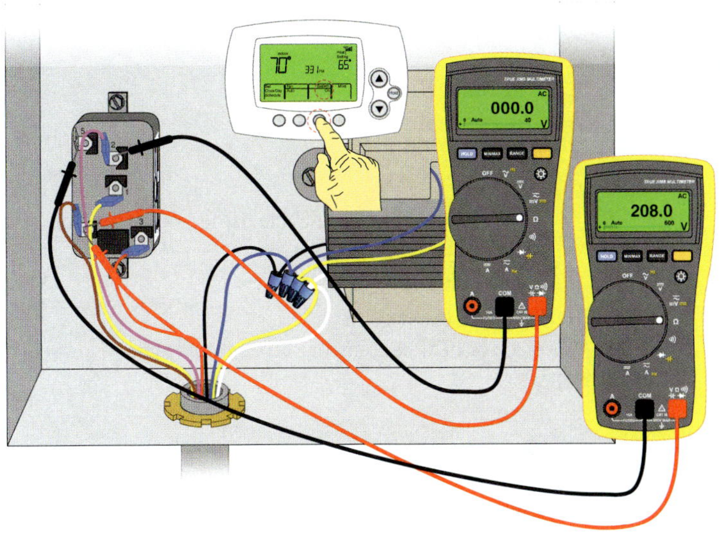

8. With the relay energized, carefully use the voltmeter to test each set of contacts. Contacts that are normally open should now be closed, and normally closed contacts should now be open. Remember that the switched voltage will be displayed when testing across contacts that are open, whereas no voltage is displayed when testing across contacts that are closed. If a set of contacts does not show the expected response, the relay is likely defective. Before condemning a relay, however, always recheck the results.

9. Carefully reinstall control panel covers and equipment access doors, then return the system to service.

▶ Wrap-Up

Summary

Switches and relays form the backbone of HVACR system control. While manually operated switches provide human control of systems and the ability to intervene in the operation, relays automate many control actions.

Both switches and relays are available in a wide variety of styles. But beneath the exterior, the same function is taking place—pathways are created for power to move from one component to another. There is no mystery to what happens inside a relay, no matter how unique its form. Sophisticated multi-pole timing relays may be the most difficult type to work with, but with the correct information, they can be tested easily as well.

Troubleshooting electrical control circuits is a large part of technician's life. Since these components represent the backbone of a control system, it is essential to understand how they operate and how to test them effectively. It is a skill that will be used often.

Key Terms

contactor An electrically controlled switch to control line voltage power flow to devices that represent a significant electrical load.

current transformer (CT) A device that produces alternating current in a secondary winding that is proportional to the current in a primary winding.

DIN rail A metal rail constructed to one of several standards for size, designed to accommodate the mounting of electrical devices in an organized manner.

dropout voltage The voltage value applied to a relay coil at which the contacts must return to their deenergized position.

holding circuit A circuit designed to allow an energized relay to remain energized until the power supply is deenergized, even though the original circuit/device that energized the relay has since returned to its normal position.

motor starter The combination of a contactor and one or more overload relays that protect the motor from damaging overload conditions.

pickup voltage The voltage value applied to a relay coil at which the contacts of a relay must function and move to their energized position.

relay An electrical device, often equipped with an electromagnetic solenoid coil, which is activated by one circuit to enable or disable another circuit(s).

solenoid coil An electrical mechanism that changes an electrical input into linear movement through magnetism.

Review Questions

1. What switch configuration is shown in the figure below?

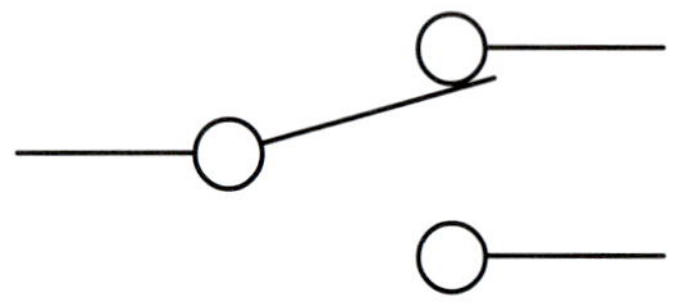

   a. SPST
   b. SPDT
   c. DPST
   d. 3PDT
2. An HOA switch is an example of a _____ switch.
3. When a pushbutton switch only closes a circuit while it is being physically pressed, a _____ circuit is likely required to keep the circuit closed, allowing an operation to continue when you release the switch.
4. When testing switch contacts while they are energized, closed contacts are indicated by a voltmeter display of _____.
   a. continuity without resistance
   b. no continuity
   c. the voltage applied to the contacts
   d. no voltage
5. The solenoid coil of a common electromechanical relay must overcome _____ to make the armature move.
   a. another, weaker magnetic field
   b. spring pressure or tension
   c. a contact latch
   d. gravity
6. Wiring diagrams and relay specifications always refer to the position of the contacts when the relay is _______________.
   a. energized
   b. deenergized
7. The type of relay that creates no electric arcs and therefore is a good choice for explosive environments is the _______________ relay.
   a. electromagnetic
   b. solid-state
   c. thermal
   d. current
8. A compressor anti-short-cycle timer is an example of a _______________ timing relay.
   a. delay-on break
   b. delay-on-make
9. An electric heat sequencer represents a type of _________ relay.
   a. thermal
   b. current
   c. potential
   d. overload
10. Basic overload relays respond to the amount of ___________ generated as the current passes through to a motor.
    a. voltage
    b. electromagnetic force
    c. heat
    d. resistance
11. The operation of a current transformer (CT) can be compared to the operation of a _____________.
12. A motor starter consists of an overload relay matched with a _____.
    a. control relay
    b. solenoid coil
    c. current transformer
    d. contactor
13. Inrush current is associated only with _______________ loads.
    a. inductive
    b. resistive
14. An electric heating element is an example of a(n) _______________ load.
    a. inductive
    b. resistive
15. It is common for a relay's contacts to have a higher current rating for _______________ loads.
    a. inductive
    b. resistive
16. Which of the following is an accurate statement about control relays?
    a. An SPDT relay can be used to replace an SPST relay, without any problem.
    b. If you select a relay with a 240 VAC solenoid coil, it will operate reliably at any voltage equal to or less than 240 VAC.
    c. Most control relays are rated to safely switch voltages up to 1,000 VAC.
    d. A relay with a 24 VAC coil should be expected to operate reliably at 19.2 VAC.

17. A simple stamped-metal component used as standardized mounting hardware for electrical components in control panels is called ______________.
18. If a relay coil is energized with the proper voltage but the relay does not respond, what is the next logical step?
    a. Replace the relay.
    b. Apply a higher voltage to see if the response changes.
    c. Change the orientation of the relay and try again.
    d. Deenergize the circuit and check for continuity through the coil.
19. The safest approach to electrical troubleshooting, especially when it can yield the same information, is ______________.
20. To effectively test a timing relay, you will need to know ____.

CHAPTER 9
Contactors and Motor Starters

Knowledge Objectives

After reading this chapter, you will be able to:

- **K09001** Identify and describe the operation of contactors.
- **K09002** Explain how to select and install a replacement contactor.
- **K09003** Explain how to troubleshoot a contactor.
- **K09004** Identify and describe the operation of motor starters.
- **K09005** Explain how to troubleshoot a motor starter.

Skill Objectives

Skills required to meet the objectives of this chapter:

- **S09001** Select and install a replacement contactor.
- **S09002** Troubleshoot a contactor.

Introduction

In the HVACR environment, manual switches and relays serve well in low-current applications. These devices, however, are not designed to interrupt the significant current of major electrical loads, such as motors. Those tasks are handled by contactors and motor starters that have more robust and durable contacts.

Contactors

K09001 Identify and describe the operation of contactors.

> **TECHNICIAN TIP**
>
> **Definite-Purpose Contactors**
>
> In the HVACR industry, the term **definite-purpose contactor** is often found in product information. Definite-purpose contactors are specifically designed for the HVACR industry. Although HVACR contactors must repeatedly switch heavy electrical loads, they are designed only to survive the intended life cycle of the equipment they serve. This designation came about in the 1950s at the request of major manufacturers that wanted a cheaper alternative to National Electric Manufacturers Association (NEMA)–standard contactors, which are more robust and could outlast the equipment in which they were installed. The development of definite-purpose contactors was the result of the manufacturers' efforts. However, definite-purpose contactors tend to hum when energized, creating more noise than their NEMA-standard counterparts.

Think of a contactor (**FIGURE 9-1**) as a heavy-duty relay. While small enclosed relays do not have the structure to switch loads that carry a significant amount of current, contactors are specifically designed for this function. Contactors are used to start and stop major inductive and resistive, high-current loads, such as compressors and large motors. Resistive heating elements in heat pumps and similar systems also draw large currents and may be switched by using contactors. Beyond a contactor's ability to switch higher currents, it functions in much the same way as a relay.

One of the differences between contactors and most relays is the fact that the contacts of a contactor are often visible or at least accessible, rather than being fully enclosed. This allows for an easy inspection of the contacts. An inspection cover must first be removed from larger contactors while the contacts of smaller models are fully visible. Note, though, that exposed contacts allow dust and debris to collect around them.

Another difference is serviceability. Although relays and smaller contactors are not generally designed to be repaired, larger contactors can be serviced in the field. The solenoid coils can be replaced or changed to another voltage, and contact sets can be replaced. These features are attractive since large contactors can be quite expensive, and the ability to change coils makes them more flexible in their application. When contactors are the focus of discussion, the descriptors "larger" and "smaller" refer to the device's current-carrying capacity. The higher the capacity, the heavier and more robust the contactor needs to be to handle the current and arcing as a load is energized and deenergized. The major components of a typical contactor are shown in **FIGURE 9-2**.

In general, contactors operate in the same manner as relays. A solenoid coil is energized, which then develops an electromagnetic field to pull an armature down and bring the electrical contacts together. When the solenoid coil is deenergized, springs pushing in the opposite direction separate the contacts. The coils, however, must be more robust and produce a stronger magnetic field than those of a relay.

FIGURE 9-1 A common compressor contactor.

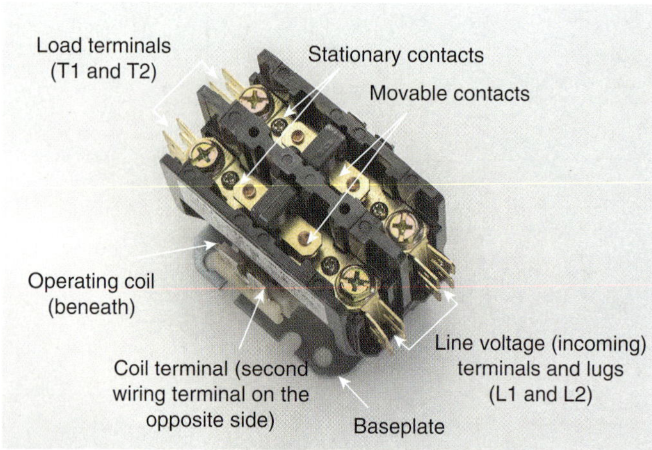

FIGURE 9-2 Contactor components.

▶ Contactor Selection and Installation

K09002 Explain how to select and install a replacement contactor.

S09001 Select and install a replacement contactor.

There are various contactor designs available, resulting from the fact that there are various standards organizations developing requirements for their construction. NEMA was established in 1926 in order to promote the voluntary standardization of electrical equipment. This work pertained mainly to the commercial, industrial and military sectors within the United States. The International Electrotechnical Commission (IEC), headed by British scientists, engineers, and manufacturers, was established two decades before NEMA, serving the same purpose in international markets. With respect to applied or built-up HVACR systems, electrical devices manufactured to NEMA and IEC standards are both in use. There are some differences in their construction and appearance. As would be expected, IEC devices are more likely to be found in equipment assembled outside of the United States.

NEMA standards promote devices that can be broadly applied, are easily serviceable, and are repairable with replacement parts. They tend to be bulkier and more expensive than IEC devices as a result. NEMA-standard devices are so well built that they often exceed the life expectancy of the equipment in which they serve. This was the motivation for the development of the definite-purpose contactor. NEMA-standard contactors are commonly found in motor-control centers and large packaged HVACR units built in the United States.

NEMA-standard contactors are numbered according to their current-carrying capacity, as shown in **TABLE 9-1**. Note that this is a simple, abbreviated table used as an example only. Current ratings are increased when the contactor is not in an enclosure, since they cool more easily that way. This table also does not include three-phase applications. Do not size a contactor based solely on this data; again, it is merely an example of the relationship between NEMA numbers, current capacity, and horsepower.

IEC-standard devices are manufactured in a wider selection because of their application-driven design philosophy, which results in devices that are built to meet specific needs. **TABLE 9-2** shows the designations and applications for IEC contactors. IEC Class AC3 is the most popular model for HVACR applications. IEC contactors and motor starters are much smaller than a NEMA-standard device with an equal current rating, up to those that can handle 100 amps or more. They are also flexible in their means of mounting, often incorporating the snap-in, DIN (Deutsches Institut für Normung)-rail mounting system. Their main disadvantage is that a given contactor type can be used in only a very narrow range of voltage and current conditions. IEC contactors are less expensive than NEMA devices, but they are still more expensive than the popular definite-purpose contactors that continue to dominate the HVACR unitary-equipment market.

TABLE 9-1 NEMA Contactor Ratings

NEMA Size	Continuous Amp Rating	HP 230 VAC	HP 460 VAC
00	9	1	2
0	18	3	5
1	27	5	10
2	45	15	25
3	90	30	50
4	135	50	100
5	270	100	200
6	540	200	400
7	810	300	600
8	1,215	450	900
9	2,250	800	1,600

Reprinted with permission from National Electrical Manufacturers Association.

TABLE 9-2 IEC Contactor Designations and Applications

Utilization Category	IEC Category Description
AC1	Non-inductive or slightly inductive rows
AC2	Starting of slip-ring motors
AC3	Starting of squirrel-cage motors and switching off only after the motor is up to speed. (Make LRA, Break FLA)
AC4	Starting of squirrel-cage motors with inching and plugging duty. Rapid Start/Stop. (Make and Break LRA)
AC11	Auxiliary (control) circuits

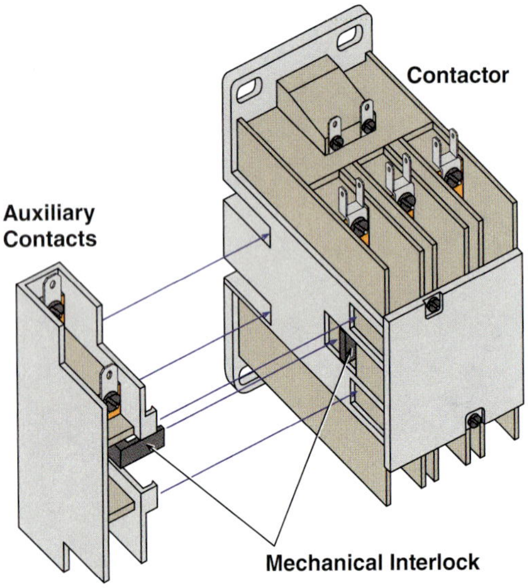

FIGURE 9-3 A contactor with auxiliary contacts.

Like relays and switches, contactors have specific characteristics that identify them. Each characteristic must match the requirements of the application. These characteristics include the physical structure of the contactor, the voltages and currents it is designed to handle, and the solenoid coil voltage.

Contact Structure

For the most part, contactors are simpler than relays in terms of contact structure. Common definite-purpose contactors have one, two, or three poles, all of which are normally open. When energized, all sets of contacts close simultaneously. Single- and two-pole contactors are used for single-phase loads, and three-pole contactors are used to serve three-phase loads. Technicians may occasionally encounter a four-pole contactor in heavy-commercial or industrial applications. The extra pole can be used in a variety of ways.

Some contactors can also be fitted with **auxiliary contacts** (see **FIGURE 9-3**). Auxiliary contacts are added on to the frame of a contactor, allowing it to operate as both a contactor and a relay. When the contactor is energized, the auxiliary contacts change position through a mechanical interlock. Auxiliary-contact sets that are normally closed or normally open are available, as well as sets that are single-pole, double-throw (SPDT). An auxiliary contact might be used to energize another contactor controlling a condenser water pump, for example. The auxiliary-contact assembly is a customization to support control functions that are related to contactor activity and may be installed in the field or at the factory.

Single-pole contactors (**FIGURE 9-4**) are common in residential and light-commercial, single-phase equipment. Some manufacturers refer to them as 1.5-pole contactors. The primary pole has a set of contacts, but the other side is simply a busbar that passes power continuously. Some manufacturers use this arrangement to maintain a continuous trickle of current through the run capacitor and the start winding of the compressor, causing the winding to warm slightly and act as a crankcase heater. In other cases, it simply serves to reduce cost. This is considered acceptable for smaller single-phase systems, but it does present an increased safety hazard. For instance, with the contactor in a small condensing unit deenergized, touching one of the leads at the compressor or run capacitor can result in a shock. However, allowing one leg of power to remain constantly energized on three-phase equipment is never advised. The safety hazard it represents is unacceptable.

FIGURE 9-4 A typical single-pole contactor.

Solenoid Coils

Contactors operate by using a solenoid action like that of common relays. Since contactors are heavier and more robust, so too are the solenoid coils. A stronger electromagnetic field must be developed to pull the armature down against the spring pressure, bringing the contact sets together. Since these coils function by using induction, there is an inrush current associated with the action. More current flows when the coil is first energized and is working to pull the armature down. Once the assembly is drawn tightly together, the current falls to a lower level and remains consistent until it is deenergized.

As mentioned previously, larger contactors, especially three-pole models, often have replaceable/interchangeable coils. Coil voltages from 24 VAC to 240 VAC are readily available for them. 24 VAC coils are the most widespread, since they are used in the majority of residential and commercial equipment. Coils are often marked as dual voltage, such as 208–230 VAC or 208–240 VAC. These coils are designed to reliably operate across the stated range of voltages. Like relay coils, contactor coils have pickup and dropout voltage ratings.

Switched Voltage

Definite-purpose contactors are rated to switch up to 600 VAC unless otherwise noted. That covers the vast majority of residential through industrial systems. Their construction is sufficient to effectively isolate one leg of power from the other. Most definite-purpose contactors have a current rating that is the same for voltages from 120 VAC to 600 VAC.

Current Capacity

Selecting the correct current capacity is very important to the reliability and longevity of the contactor. It is acceptable to oversize from an electrical point of view, but doing so needlessly increases the cost to the customer.

Like relays, contactors have separate current ratings for inductive and resistive loads. A three-pole contactor label is shown in **FIGURE 9-5**. The **full-load amps (FLA)** current rating shown, 25 FLA, refers to the maximum continuous current for inductive loads. Note that the current rating for resistive loads (designated as RES, for resistive, on the label) is 40 amps, since there is no inrush of current involved in those loads.

Also, note that the horsepower (hp) rating for a contactor increases as the applied voltage increases. This is because a given hp is produced with less current at a higher voltage. Therefore, the current through the contacts of a 15 hp three-phase motor or compressor operating at 480 VAC is not substantially different from the current supplying a 10 hp three-phase motor operating at 240 VAC.

The **locked rotor amps (LRA)** column shows the peak value of current that the contactor can handle as the motor starts, or if a serious failure occurs that causes the motor to seize. Note that in Figure 9-5 the LRA capacity is reduced as the voltage applied increases. Remember that the contacts can handle this level of current for only a few moments before sustaining damage. Passing this amount of current for too long will likely weld the contacts together permanently due to the heat produced.

Before selecting a replacement contactor, determine the continuous FLA of the load. The most significant loads will normally include the compressor and the condenser fan motor in residential and commercial equipment, but the wiring diagram must be consulted to find out all the loads served through its contacts.

Installing Contactors

Definite-purpose contactors used in most HVACR applications were designed for simple mounting. The universal baseplates (**FIGURE 9-6**) and the holes for the mounting screws are designed for flexibility, such that changing the brand of a contactor doesn't usually result in having to drill new holes. NEMA and IEC contactors have standard mounting hole patterns. IEC contactors that snap into DIN rails are also available.

It is essential that the wires be properly reconnected when replacing a contactor. Accidentally placing a control-wiring lead on a line-voltage terminal, or vice versa, can cause serious electrical damage to other components. Also, be sure to correctly connect the wires back to the load side of the contactor when replacing a single-pole contactor. Keeping the wrong portion of the circuit continuously energized with one leg of power can lead to functional problems as well as safety issues. For three-phase contactors, both the incoming power (L1, L2, and L3) and load (T1, T2, and T3) wiring are reconnected to the contactor exactly as found. Mismatching the leads—swapping the T1 and T2 leads, for instance—will result in the motor operating, but it will rotate in the opposite direction. Swapping motor wiring from one pole to another

▶ TECHNICIAN TIP

Contactor-Coil Voltage

Coil voltages from 24 VAC to 240 VAC are readily available for larger two-pole and three-pole contactors. A technician can carry several different sizes of contactors, along with coils of several different voltages, and be well-prepared for commercial and industrial contactor repairs. Smaller single- and two-pole contactors, such as those found in residential equipment, are readily available with 24 VAC coils, but other coil voltages may not be as easy to acquire. Small contactors with coils of a higher voltage are rarely applied in residential or commercial HVACR equipment, and therefore, local supply houses are less likely to stock them. A special order may be required. Supply and demand is always a factor in the inventory of an HVACR parts wholesaler.

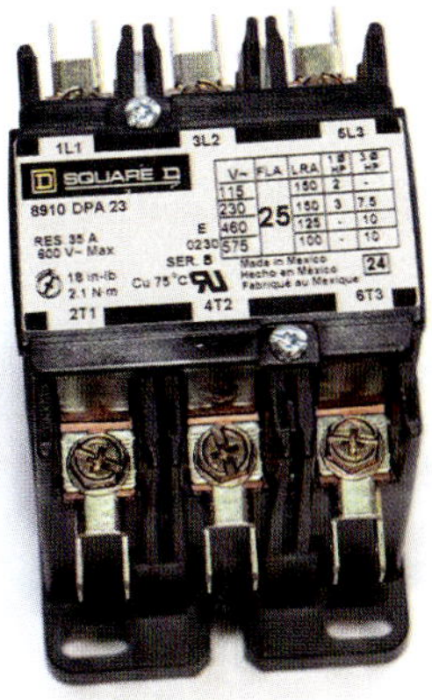

FIGURE 9-5 Label of a three-phase contactor.

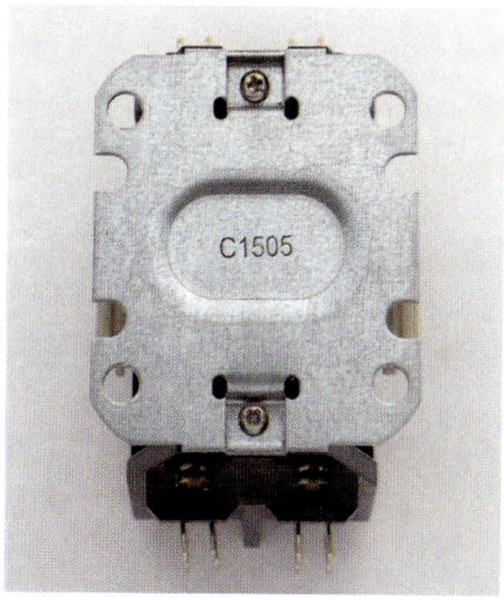

FIGURE 9-6 Baseplate of a definite-purpose contactor.

FIGURE 9-7 The tawny crazy ant.

© Michael Bentley/ourBIGbackyard

on three-phase motors is how the direction of rotation is changed, which will be discussed more in Chapter 14.

To select and install a replacement contactor, follow the steps in **SKILL DRILL 9-1**.

▶ CORE CONNECTION

Ants!

HVACR technicians will, in some locations, encounter a contactor covered in ants, both dead and alive. In some locations, this occurs regularly. Ants, especially certain species, are drawn to the electromagnetic field created by the coil. Although it does sometimes occur with regular relays, the field is not nearly as strong and attractive as that of a contactor coil.

In Texas alone, the tawny crazy ant (*Nylanderia fulva*) shown in **FIGURE 9-7**—also known as the Raspberry crazy ant after an exterminator who noted its spread—is responsible for over $100 million of electrical equipment damage each year. But other ant species can also create a problem, and a swarm of ladybugs has also been known to descend upon a contactor. Once the ants make their way to the contactor, they often get shocked, causing them to release pheromones that represent a cry for help. As more ants come in to defend the injured ant, they are fried between the contacts, interfering with power transfer. Eventually, something shorts out or the contacts burn, shutting down the equipment. In response to the problem, one or more manufacturers have developed special contactors that prevent the ants from accessing the contacts. Tawny crazy ants, which were first discovered in South America, have been spreading throughout the southern US since their discovery in Texas in 2002. They now appear to be slowly moving north, as well as east and west.

SKILL DRILL 9-1 Selecting and Installing a Replacement Contactor

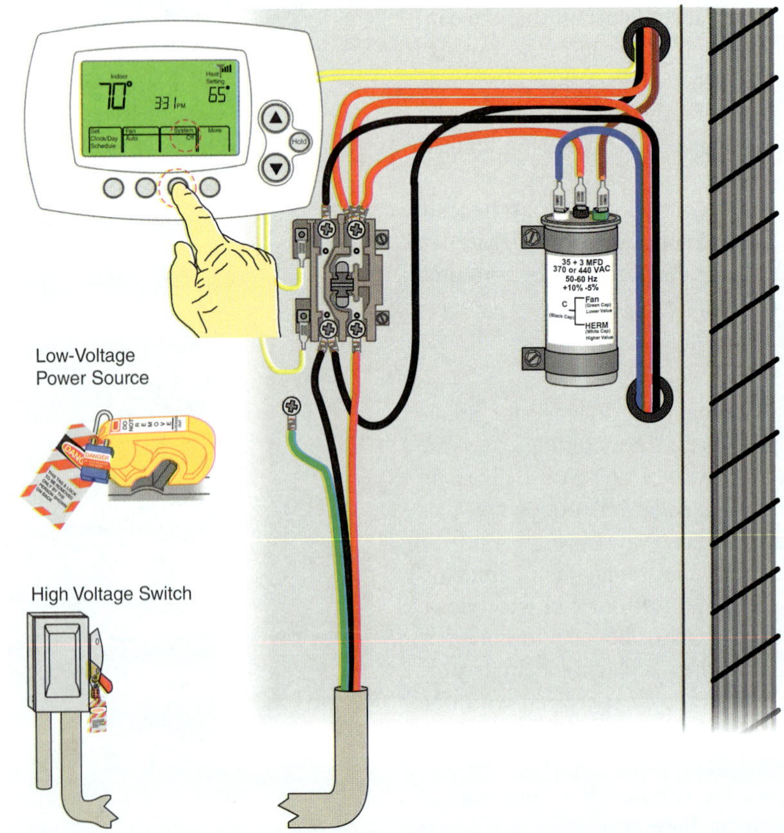

1. Ensure that all controls, including the thermostat, are set to the Off position. Deenergize, lock out, and tag the equipment being serviced. Be sure that the process deenergizes not only the line voltage feeding the equipment but also the low-voltage control power source that may be located in a separate piece of equipment.

SAFETY TIP

Working with electrically energized circuits is hazardous. Avoid contact with any surface that could potentially be energized. When working inside energized enclosures, maintain a stable position and, whenever possible, avoid placing both hands inside the enclosure. All work on energized equipment by unqualified individuals should be supervised by a qualified instructor or technician.

Continued

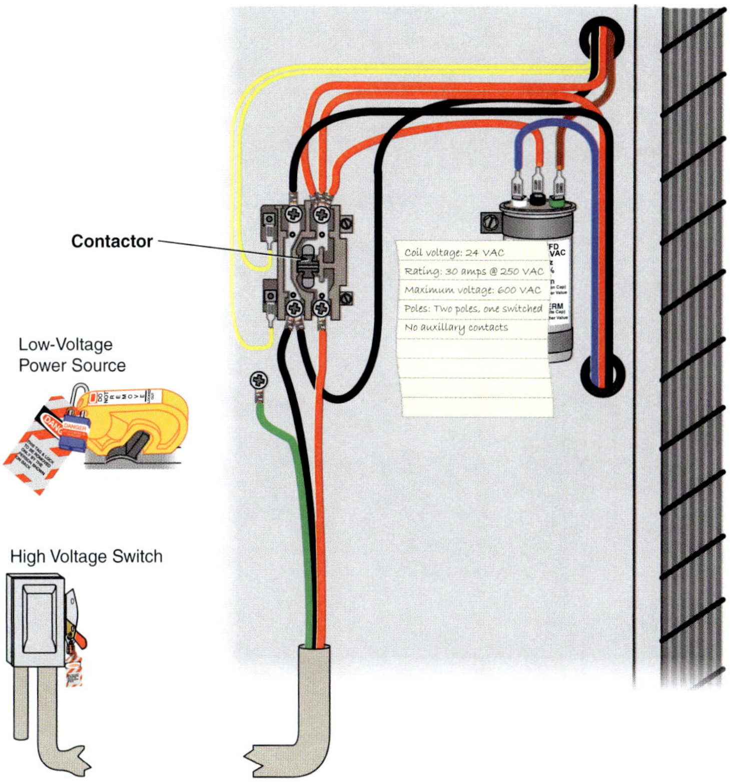

2. Locate and examine the contactor to be replaced. Record the relevant information from its label: the coil voltage; the inductive current rating at the voltage to be switched; and the maximum permissible voltage if shown. Also, visually determine how many poles are needed and whether any auxiliary contacts have been mounted to the contactor. If there are auxiliary contacts, examine them to determine whether they are normally closed or normally open and whether multiple throws are needed. A replacement can also be selected based on a part number shown on the contactor, from either the equipment or contactor manufacturer. However, the auxiliary-contact sets would not likely be included in that part number; they typically have separate part numbers since they are accessories.

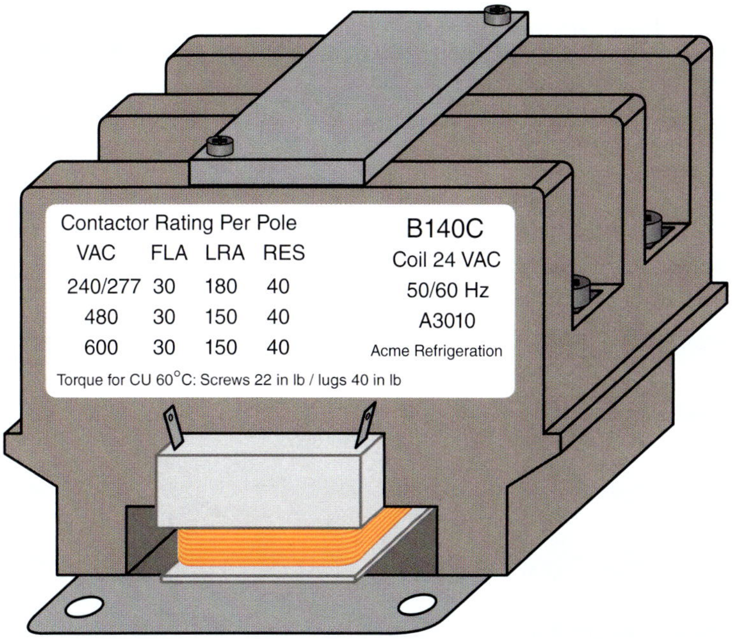

3. Select a proper replacement contactor based on the recorded information. Remove it from the box and examine the characteristics and label to ensure that it is compatible.

Continued

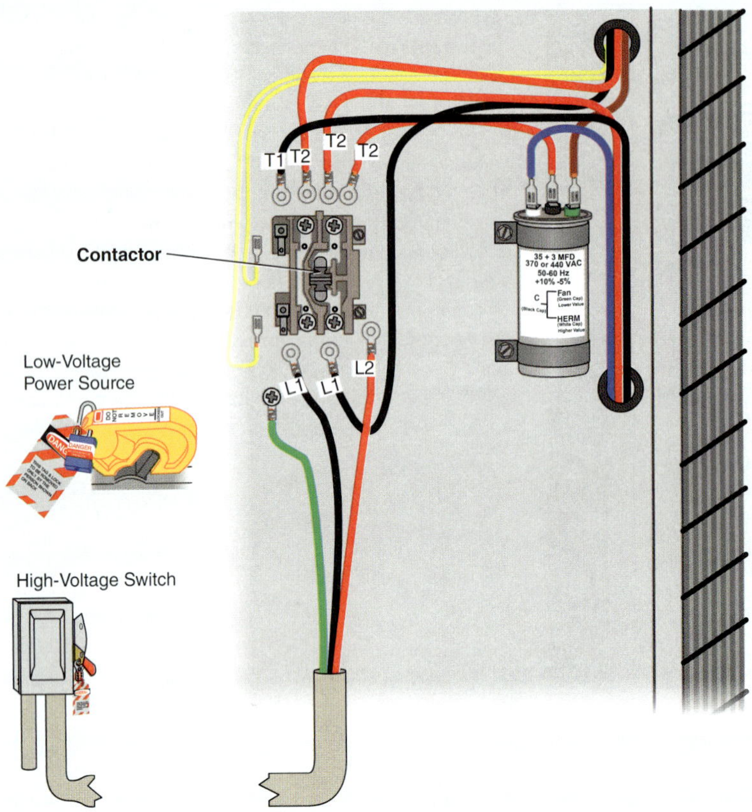

4. Examine the wiring and unit wiring diagram to ensure that all wiring routed to the contactor is understood. Then mark or label the wires while removing them from the contactor terminals to ensure they are properly reconnected. This is especially true for three-phase contactors, since switching the positions of two power leads will reverse the rotation of a motor.

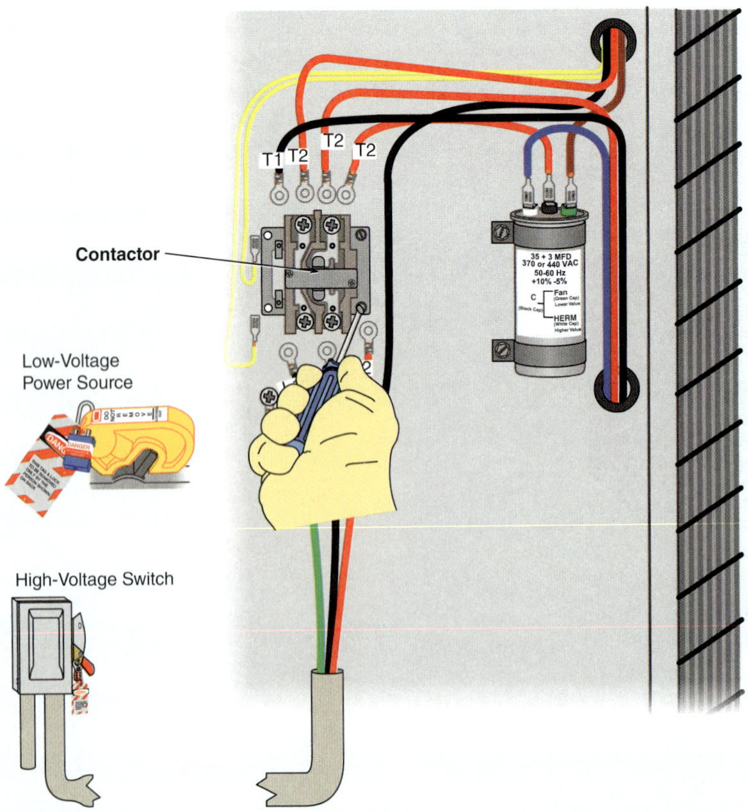

5. Remove the old contactor and mount the new contactor in a secure manner. A different style of contactor may necessitate drilling one or more new mounting holes. Always take great care when drilling in the control cabinet, to ensure that the drill does not damage other wiring, control components, or refrigerant tubing.

Continued

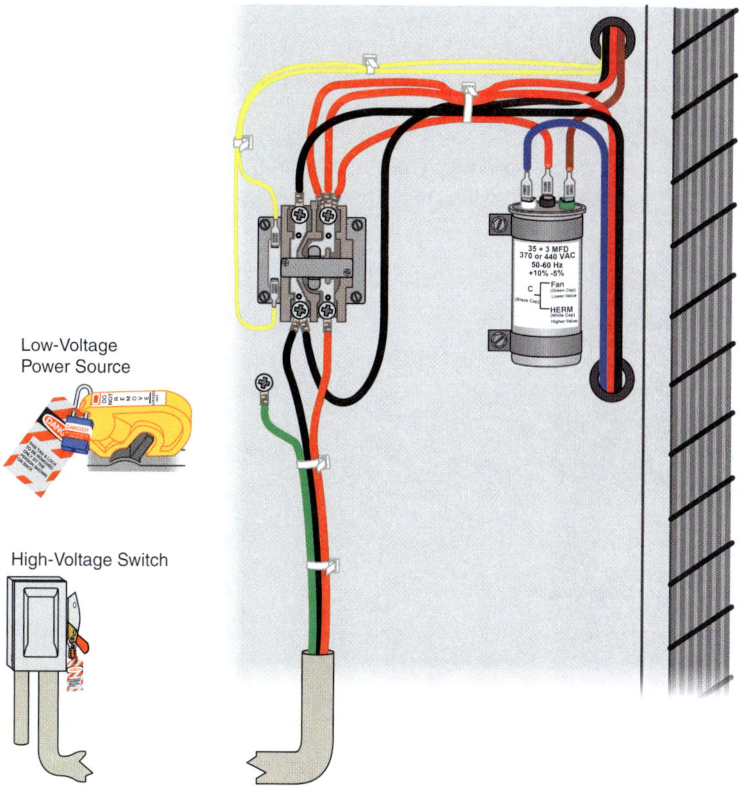

6. Connect the wiring to the appropriate terminals. Inspect the condition of the termination hardware before making the connection, to ensure a good electrical connection exists. If it is in questionable condition, cut off and terminate the wire with a new connector. Recheck all work upon completion, and ensure that the wires are properly connected and that no wire or termination hardware is in danger of making contact with another conductor or the cabinet. Use wire ties as necessary to neatly control wires.

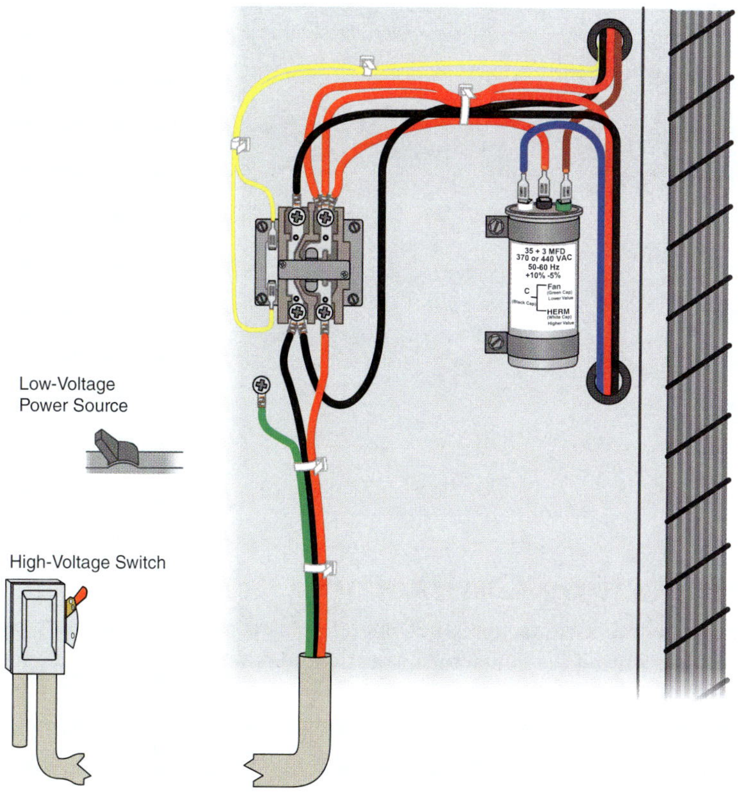

7. Ensure that all controls, including the thermostat, remain set to the Off position. Remove the locks and tags from the power source(s). Restore power to the system.

Continued

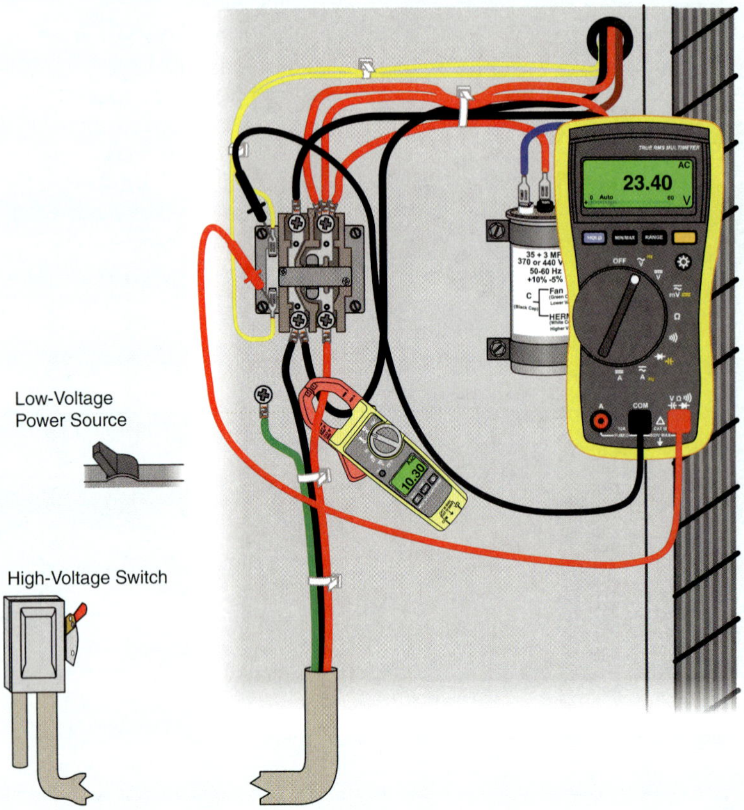

8. Start the equipment and ensure that the contactor participates in the sequence of operation as expected. Use a clamp-on ammeter to monitor the current passing through each set of contacts to confirm that it is within the capabilities of the contactor. Use a voltmeter to check the voltage applied to the contactor coil to confirm that it is steady and sufficient to reliably hold the contactor armature.

9. Once testing is complete, carefully reinstall control panel covers and equipment access doors, and then return the system to service.

K09003 Explain how to troubleshoot a contactor.

S09002 Troubleshoot a contactor.

▶ Troubleshooting Contactors

If there is reason to believe a contactor is not operating as it should, begin with visual and tactile inspections. During the contactor inspection, look for any obvious damage or burning, broken or loose wires, unusual noises, or burning odors. A contactor coil would be suspected when the motor or compressor won't attempt to start and when preliminary checks indicate that control power is available but no safety devices have opened the control circuit. These preliminary checks include verifying the thermostat, or a similar device controlling the contactor is set and its contacts are closed.

Contactors fail in the same manner as relays. The solenoid coil can become open or shorted. The contactor armature can also bind, due to being fouled with dirt or grease (**FIGURE 9-8**). The contacts can become excessively pitted or even welded shut. Since the contacts of most contactors are accessible, the condition of the contacts can be checked visually. Each time the contacts open or close, an electric arc occurs. Over time, the arc damages the contact surface. Refer to **FIGURE 9-9**. Contacts that carry a great deal of current must have faces that meet smoothly and have plenty of contact area. A perfect match of the faces on definite-purpose contactors is rare. As a result, when the contacts come together, only a small portion of the contacts touch first and the current is very high at that moment. As damage continues, the problem gets worse and more burning occurs, leaving high and low spots on the contacts. Once the contacts begin to stick together or meet poorly—or the material has melted in spots, causing what appears to be weld splatter—the contactor (or contact sets) must be replaced. Note that contact damage can create significant problems for the load as well, due to voltage drop and the increased current that results. Voltage drop across a contactor can be measured during testing under load.

While observing appropriate electrical safety precautions, check the freedom of movement of the contactor. Push the armature down toward the base to check freedom of motion. No resistance or jamming should be detected through the full range of motion. There will be some minor resistance offered by the springs as the contactor is depressed. When released, the armature should freely move up and the contact faces should fully separate. If the armature doesn't move freely, replace the contactor. Larger contactors and NEMA-standard contactors can be disassembled to determine what the problem is and possibly be repaired.

FIGURE 9-8 Dirt and debris can prevent the contactor from moving freely.

Electrical Tests

Electrically troubleshooting a contactor is very similar to electrically troubleshooting a relay. Check for electrical power to the coil with the contactor in place and connected to the system. If there is no response from the contactor with voltage applied to the coil, it's likely the coil is open. Turn off the power, disconnect the leads from the coil terminals, and use an ohmmeter to check for continuity through the coil. If there is no continuity through the coil, it is open and the contactor (or coil) must be replaced. If there is continuity through the coil, check for a ground path between a coil terminal and the baseplate. This check should indicate an open circuit. If there is continuity here, the coil is shorted to ground. A short between the coil and ground also requires replacement of the contactor (or coil).

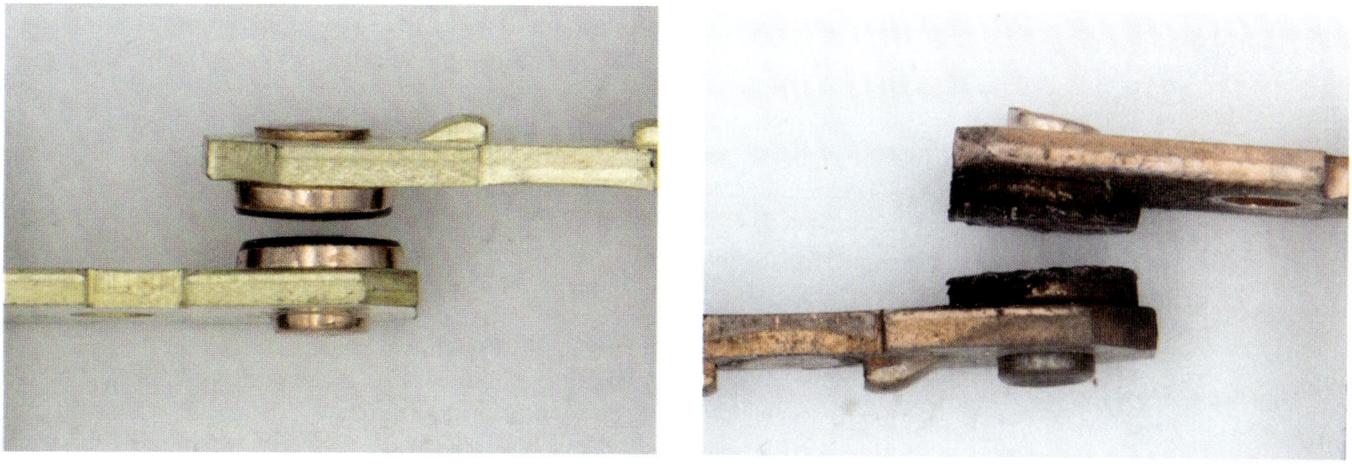

FIGURE 9-9 Example of new and worn contacts.

Short circuits to ground generally cause the control-circuit fuse to blow and/or damage the transformer, due to instantaneous high current.

In rare cases, the coil winding is internally shorted from turn to turn. In other words, a short circuit has eliminated some of the coil windings. Without those windings, the coil may not be able to generate a strong enough magnetic field to draw the armature down. This can be determined only by knowing the precise specifications for the coil resistance and then reading the resistance with an ohmmeter to compare. This is not generally necessary. If the coil has the correct voltage applied to it and does not pull the armature down, it is defective regardless.

The contacts can be tested electrically, but there is limited value to doing so unless the contacts are under load. Since they can be seen, unlike most relays, the fact that they come together and properly separate can be seen when manually operating the contactor. If the contacts touch at all, they will show continuity with the power off, but they may still be in poor condition. Of course, if a high resistance is measured across a contact, there is certainly cause for concern. But a visual inspection should also reveal their poor condition. A better test of contact condition, beyond a visual inspection, is checking the voltage drop across each contact set while under load.

A three-phase contactor will be used here as an example of testing for voltage drop. Voltage measurements are taken with the contactor under load, as follows: L1 to T1; L2 to T2; and L3 to T3. With a two-pole contactor, only two readings are available (L1 to T1; L2 to T2); only one for a single-pole contactor. Since the contacts are closed, the reading should be in the millivolt range. Then take a current reading on each leg of the contactor as well. With this information, the power loss across the contacts can be determined. For instance, assume that the reading of L1 to T1 is 30 millivolts and that the current is 20 amps. Multiply the two values to determine watts:

$$E \times I = P$$

$$0.03 \text{ volts} \times 20 \text{ amps} = 0.6 \text{ watts}$$

This is a reasonable loss and no cause for concern. A second example would be a drop of 750 millivolts at a current of 30 amps:

$$0.75 \text{ volts} \times 30 \text{ amps} = 22.5 \text{ watts}$$

This is a significant loss. It will cause more heat at the contactor, leading the contacts to degrade more rapidly. Note that a small voltage drop across a contact may be insignificant at very low currents, but as the current increases with the same amount of voltage drop, the power loss becomes progressively worse. As a general rule, technicians feel that a voltage reading higher than 500 millivolts (0.5 volts) across a set of loaded contacts, regardless of current, indicates that it is time to replace the contactor. A power loss of more than 20 watts should also be cause for concern. Before condemning the contactor, though, the next course of action is to try cleaning the contact surfaces.

FIGURE 9-10 Electrical contact cleaner.

Contact cleaning can be done by using any one of a variety of products, such as the one shown in **FIGURE 9-10**. Use the spray to wet the contacts and then wipe away any dissolved residue. More aggressive cleaning can be done with a pencil eraser. The contacts are coated with a thin layer of metal, such as cadmium or silver, to improve their conductivity and durability. This layer is very thin on definite-purpose contactors and will be removed quickly if sandpaper or similar abrasives are used. If the contacts of a definite-purpose contactor need this level of cleaning, it's time to replace the contactor. High-current, NEMA-standard contactors can perhaps be cleaned more aggressively because the protective layer is thicker and because the contact sets can be replaced as well. If contact cleaning does not reduce the voltage drop, replace the contactor or contact set. Replace all of the contact sets on a contactor together, rather than just one pole.

To troubleshoot a contactor, follow the steps in **SKILL DRILL 9-2**.

SKILL DRILL 9-2 Troubleshooting a Contactor

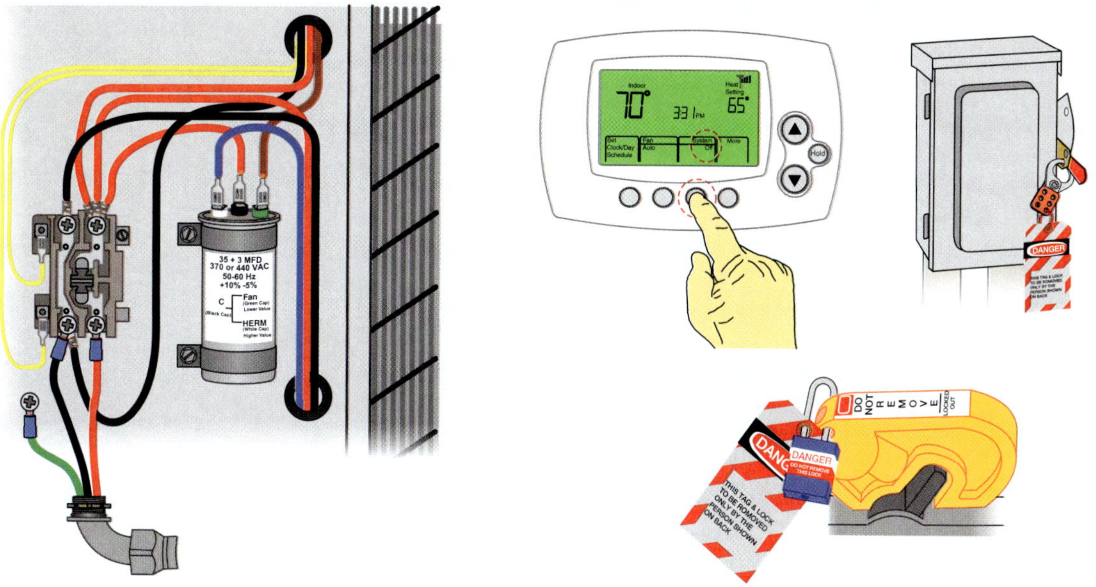

1. Ensure that all controls, including the thermostat, are set to the Off position. Deenergize, lock out, and tag the equipment to be serviced. Be sure that the process deenergizes not only the line voltage feeding the equipment but also the low-voltage control power source that may be located in a separate piece of equipment.

SAFETY TIP

Working with electrically energized circuits is hazardous. Avoid contact with any surface that could potentially be energized. When working inside energized enclosures, maintain a stable position and, whenever possible, avoid placing both hands inside the enclosure. All work on energized equipment by unqualified individuals should be supervised by a qualified instructor or technician.

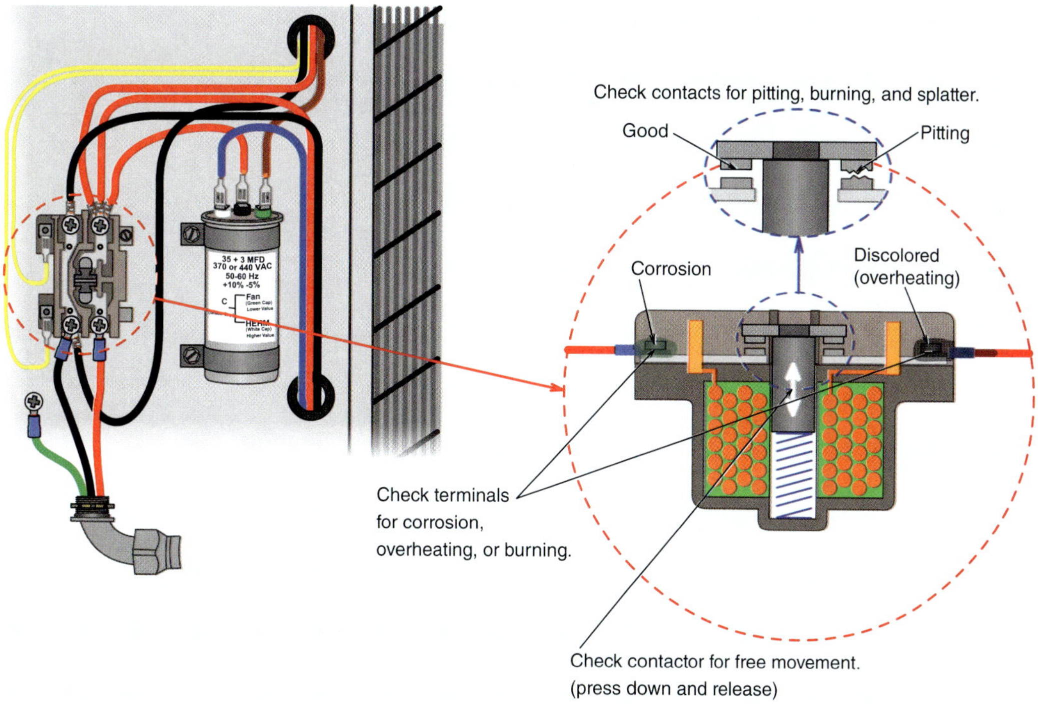

2. Visually inspect the contactor carefully. Look for evidence of overheating, corroded or burned terminations, and other signs of damage. Examine the contacts carefully for excessive pitting, burning, and splatter. Move the contactor up and down to ensure it has full freedom of movement.

Continued

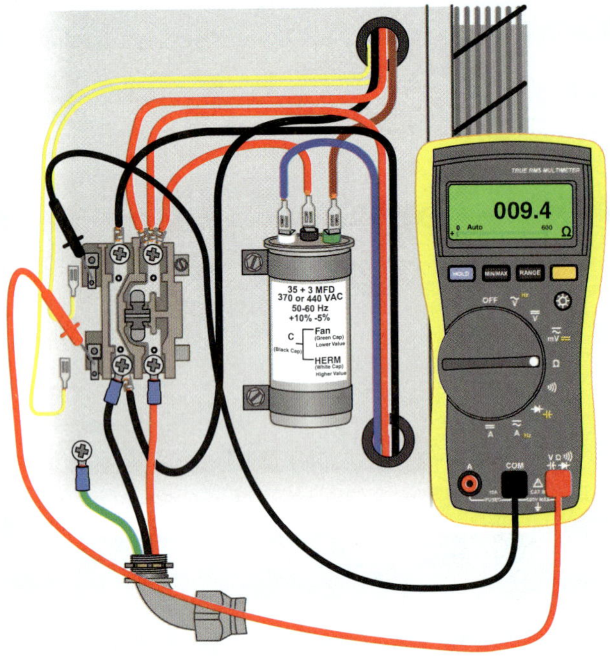

3. If the contactor shows no significant signs of damage or wear, begin testing by checking the contactor coil. Remove the wires from the coil, and use an ohmmeter to check for continuity through the coil. If there is no continuity, the coil is open and must be replaced. Continuity with a small amount of resistance should be detected.

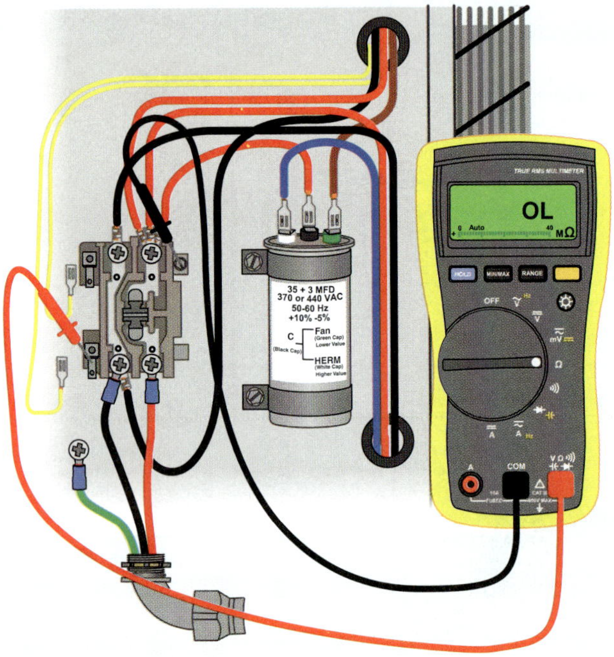

4. Next, check the coil for a grounded condition. With one meter probe on a coil terminal and the other touching the unpainted base of the contactor, check for continuity. No continuity should be noted; continuity indicates a grounded coil.

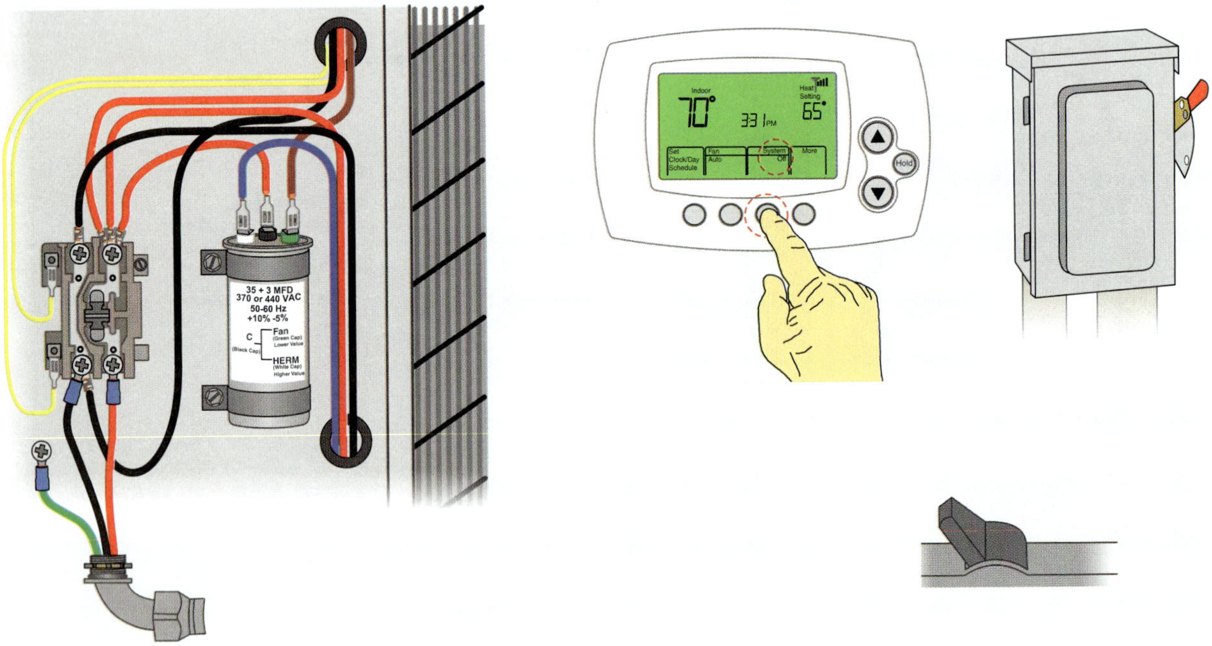

5. Reconnect the wires to the contactor coil. Ensure that the thermostat or other primary control is still in the Off position, and restore power to the unit.

Continued

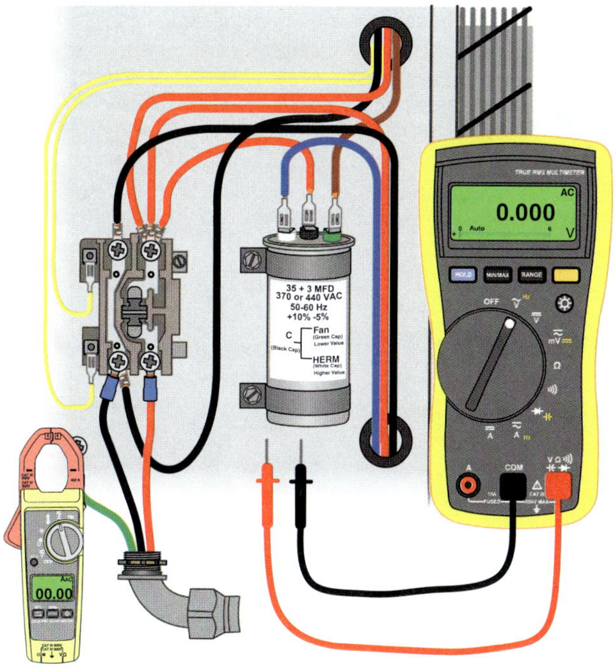

6. Prepare the meter to read voltage, and prepare a clamp-on ammeter to read current on one leg of power on the load side of the contactor. Be sure the meter probes can safely contact each terminal of the contactor; practice if necessary with power off. Once prepared, start the system.

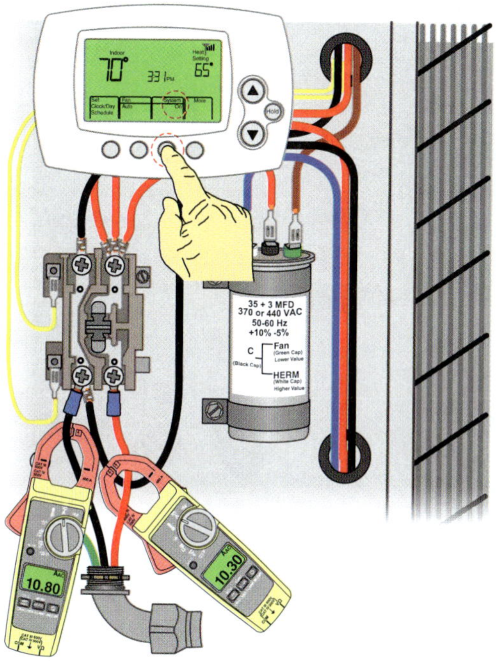

7. Check the reading on the ammeter, and ensure that it is as expected for the load and within the current rating of the contactor. Check and document the current on each leg (T1, T2, and T3, if the contactor is three-phase).

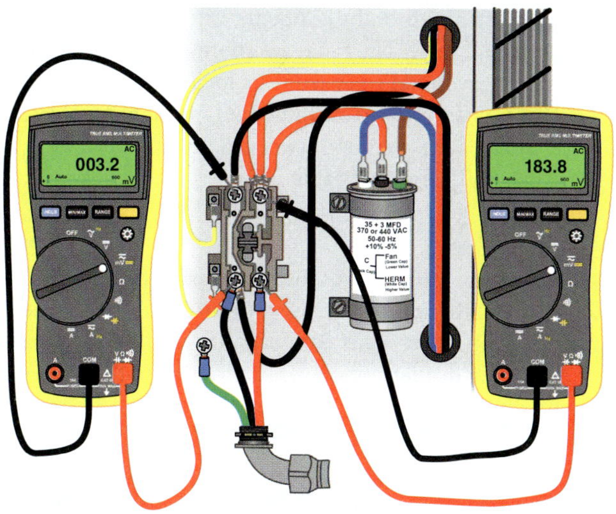

8. Use the voltmeter to check the voltage across each set of contacts (L1 to T1, L2 to T2, and L3 to T3, if present). This requires two measurements for single-phase systems and three measurements for three-phase systems. The reading should be in the millivolt range. A reading exceeding 300 millivolts (0.3 volts) is cause for immediate concern. Record the voltage reading(s).

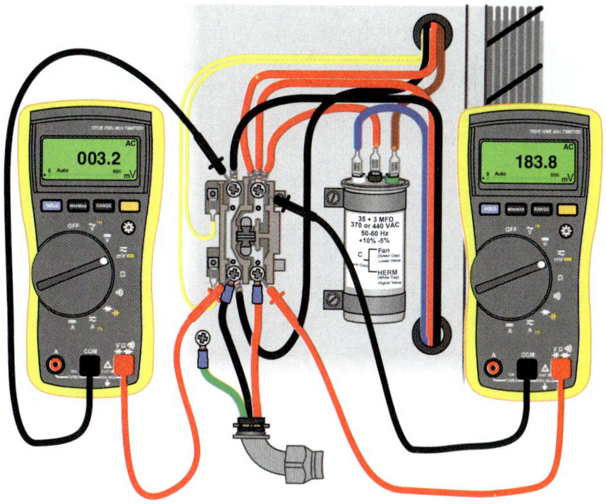

Watts = Amps x Volts

Power Loss L1–T1
Watts = 10.8A x 0.003V
Watts = 0.032 Watts

Power Loss L2–T2
Watts = 10.3A x 0.183V
Watts = 1.885 Watts

9. With the voltage and current readings complete, calculate the power loss across each set of contacts. Remember that a power loss of more than 20 watts is cause for concern.

Continued

10. If the readings and calculations indicate the contactor is in good condition, carefully reinstall the equipment access doors and return the system to service.

▶ Motor Starters

K09004 Identify and describe the operation of motor starters.

Contactors alone can be used for starting and stopping smaller inductive and resistive loads. However, to provide protection for larger and more expensive motors, a more precise method of protection is needed. Fuses cannot provide that level of protection. This protection is provided by an **overload relay** that monitors compressor current. This combination of contactor and overload relay assembly results in what is known as a motor starter.

Motor-Starter Structure

The physical arrangement of motor starters is fairly predictable, dictated by national and/or international standards. A typical motor starter consists of a contactor, often one that can be used independently if desired, mated to an overload relay assembly. Line voltage comes to the top of the contactor. It may be abbreviated L1, L2, and L3. The load terminals of the contactor are directly connected to the line terminals of the overload relay. They may be abbreviated T1, T2, and T3. Wiring to the motor leaves the overload relay from its load terminals at the bottom of the assembly. **FIGURE 9-11** shows two examples of typical motor starters. In NEMA-standard starters, most of these components can be replaced in the field. IEC-standard starters are generally less repairable.

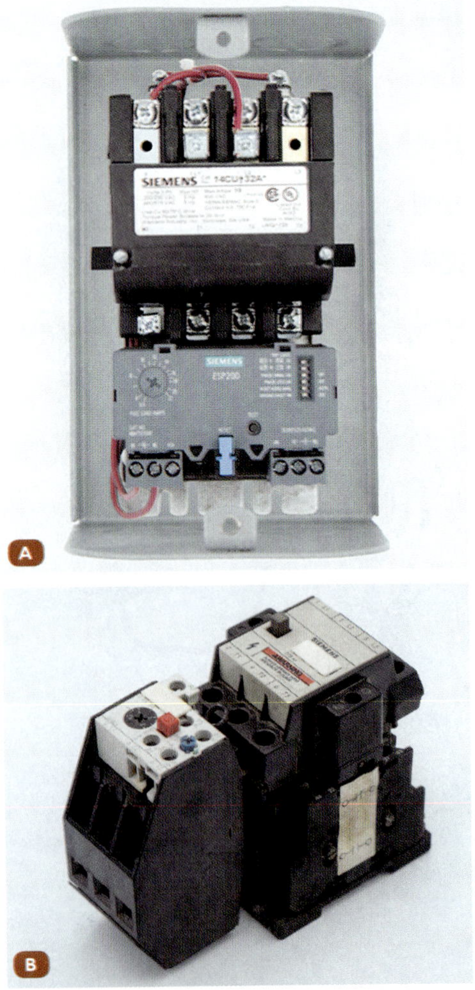

FIGURE 9-11 **A.** NEMA Starter **B.** IEC Starter.

Overload Relays

The addition of an overload relay separates a motor starter from a basic contactor. The overload relay protects the motor against current loads that exceed the continuous-current rating of the motor. This condition causes the motor's windings to overheat and the winding insulation to eventually fail. On the other hand, a grounded motor winding creates instantaneous, extremely high currents that can seriously damage anything in the circuit path. Circuit breakers and fuses are uniquely designed to handle these major events quickly and protect the circuit. Overload relays cannot respond as quickly as needed. However, it is common (and appropriate) to see a circuit fused at 30 amps serving a motor that has a full-load current rating of only 23 amps. If the motor operates for a significant amount of time above 23 amps, it will eventually fail, and the fuse will never sense that there is a problem until the failure actually occurs. This is where the protection of an overload relay enters the picture.

The current of each phase or leg of the motor starter passes through the overload relay. When an overload condition is sensed, the overload-relay contact, which is in series with the contactor-coil circuit, opens and stops the motor.

Overload relays are classified based on the speed at which they respond to an overload condition. Some time is allowed by all classes for a temporary problem to clear. They are classified as follows:

- *Class 10*—10-second response time
- *Class 20*—20-second response time
- *Class 30*—30-second response time.

There are two primary types of overload relay. Thermal overload relays rely on heat, while solid-state overload relays are more sophisticated and flexible in their application.

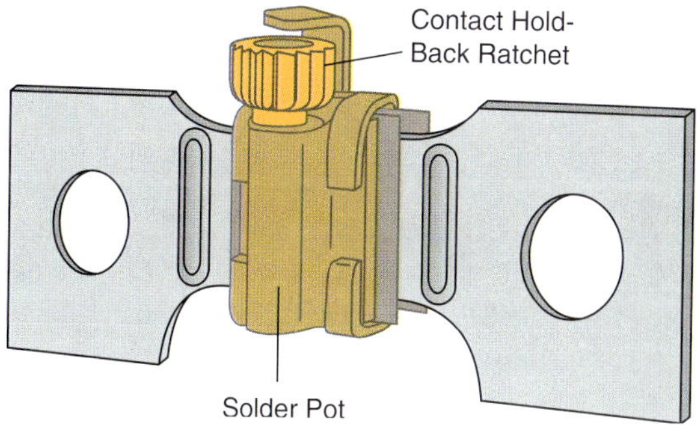

FIGURE 9-12 Melting-alloy thermal element.

Thermal Overload Relays

Thermal overload relays use one of two kinds of elements. The first is called a melting-alloy element (**FIGURE 9-12**). The element is fitted with a solder pot containing a solder-like material that melts when the temperature becomes too high. The molten metal cannot hold the ratcheting linkage in place, and the mechanism opens the overload-relay contact. The heating portion of the element is designed with precision to produce a specific amount of heat at a given current. Melting-alloy elements are selected based on the motor operating characteristics and the environment of the starter. Once the starter trips and the motor shuts down, the solder cools and solidifies, and the starter can be reset after a short time.

The other type of element is called a bimetal element (**FIGURE 9-13**). Like bimetal elements discussed elsewhere, they bend as they are heated. In this case, the bimetal element is an integral part of the overload relay. The heating elements that warm the bimetal, examples of which are shown in **FIGURE 9-14**, are replaceable and are also selected to match the motor and the starter environment. These too are designed to produce a specific amount of heat at a given current. One advantage they have over melting-alloy elements is that the bimetal element can usually be adjusted up or down ±15%, providing some flexibility in their application. Due to the nature of melting-alloy elements, they are not adjustable. To change the current setting, a different set of melting-alloy elements must be installed. A range of bimetal heating elements or melting-alloy elements are available for each size of NEMA-standard starter.

Solid-State Overload Relays

Solid-state overload relays do not require the careful selection of thermal elements. There are several factors that affect the choice of thermal elements for a starter/motor combination that do not affect solid-state overloads. Solid-state overload relays may also have some additional features, such as a setting to change the standard manual-reset function to an automatic-reset setting.

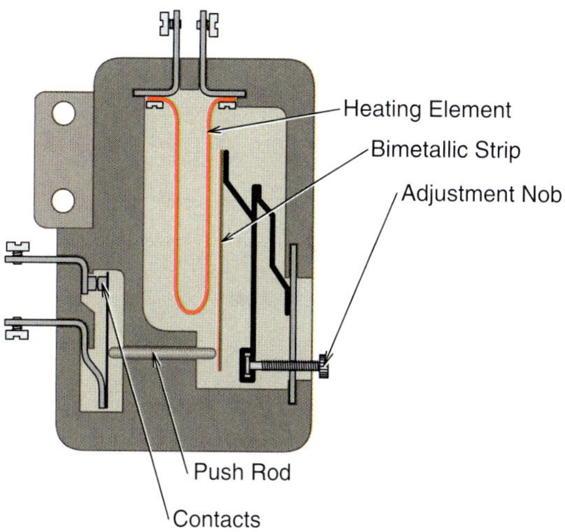

FIGURE 9-13 Bimetal thermal element of an overload relay.

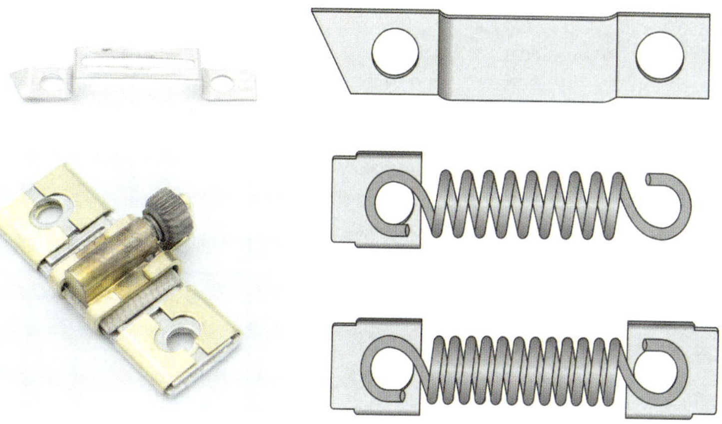

FIGURE 9-14 Heating elements for bimetal thermal elements.

▶ CORE CONNECTION

Voltage Imbalance

Voltage imbalance in three-phase systems describes the situation where the phase-to-phase voltage of a circuit is not equal. More specifically, it describes the situation when there is more than a 2% difference. Anything less than that is generally considered to be acceptably balanced, even though it's not perfect. A voltage imbalance can occur for a variety of reasons. It can result from the configuration of a utility transformer, for example. Another common cause is found in facilities where the loads are not balanced across the three phases. If, for example, 40 single-phase motors are powered by one phase, 10 motors by another, but no motor loads are placed on the third phase in a facility, a major load imbalance is created. The placement of single-phase loads on a three-phase system must be considered carefully. In a new facility, everything is usually well engineered electrically. As time passes and equipment comes and goes, a voltage imbalance can progressively develop.

Since voltage imbalance significantly affects inductive loads like motors, the topic will be covered in greater detail in Chapter 14. But give some thought now as to how the necessary calculations for voltage imbalance in a three-phase system might be made, and then determine whether that was the correct approach when the topic is later presented.

▶ TECHNICIAN TIP

Selecting Motor Starters

The world of motors and starters is more complex than it may initially appear. This text can introduce only the most basic elements of motors and motor starters. In some applications, motor starters must be able to control multi-speed motors automatically and reverse motor direction in others. As technicians continue on in their career, they will learn about variable-speed motor applications and the advanced starters and electronic drives used to control speed with great precision. Do not attempt to specify a motor starter for an application at this stage of training. If a starter needs troubleshooting and/or requires replacement or repair, consult qualified electricians and other resources, such as the equipment manufacturer, to ensure that a compatible replacement is provided.

One of the more important advantages of solid-state relays is that they provide much better protection when a phase of power is lost. When a three-phase motor loses one phase of its power source, the current instantly rises in the other two phases. Thermal elements may respond in time to save the motor but not necessarily, especially if the thermal elements are somewhat oversized. Solid-state overload relays typically respond to the loss of a phase in three seconds or less. It is worth noting that there are other power-monitoring devices beyond the overload relay that can very quickly provide protection for the loss of a phase or a **voltage imbalance**. Voltage imbalance causes the current through the motor windings to differ, potentially harming the motor. Motors and equipment operating in critical environments require the maximum protection offered by both external power-monitoring devices and overload relays.

As mentioned previously, thermal elements are selected based on motor characteristics and the environment of the starter. For example, the motor starter may be in an enclosure, and the enclosure itself may be in a warmer location than the motor. This occurs frequently in HVACR applications, since many large blower motors are in the cool airstream of the system. The starter for the motor may be located on the wall of a hot equipment room. This alters the selection of a melting-alloy or bimetal thermal element to compensate for the environmental difference.

Solid-state overload relays cover a range of current values by design. The selection of an appropriate relay only requires that the motor's current falls within the relay's specified range. Final adjustments in the current settings can be made on-site based on the protected motor's current. Note that technicians should seek qualified help when they select motor-starter components.

Motor-Starter Control Circuit

The overload relay provides a normally closed contact that is wired into the circuit of the contactor solenoid coil. If there is an overload condition sensed, the contact will open to shut down the motor.

Motor starters, technically the contactor itself, can be directly controlled through a low-voltage control circuit with a compatible low-voltage coil in the contactor. Although this is convenient in some cases, large contactors demand a significant amount of current to operate. As a result, many contactors in motor starters have line-voltage coils.

FIGURE 9-15 shows a simple low-voltage, motor-starter control circuit. The bold lines represent line voltage, whereas the thinner lines represent the low-voltage circuit. Various

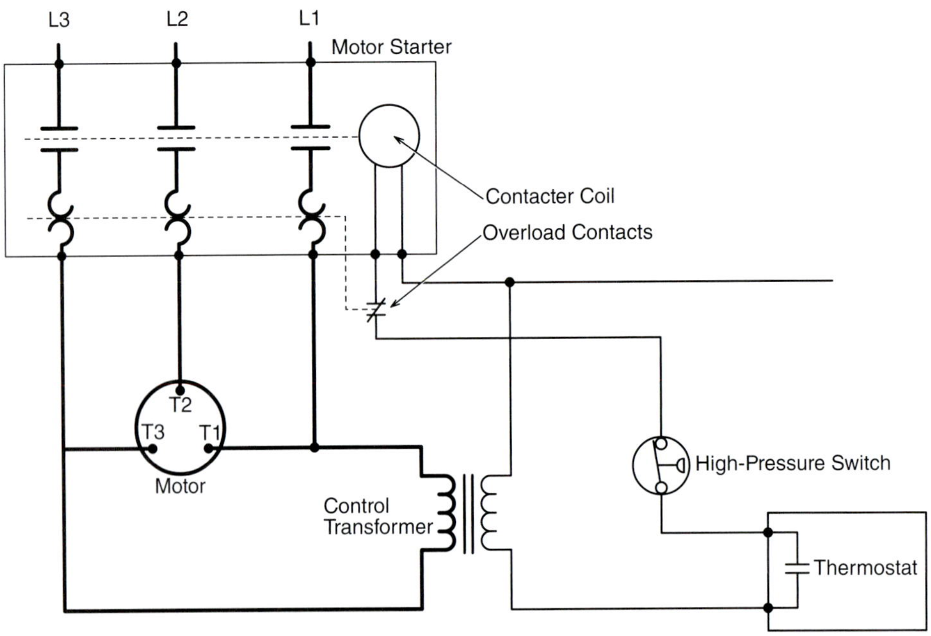

FIGURE 9-15 Motor starter with a low-voltage control circuit.

FIGURE 9-16 Start-Stop and HOA controls.

safety devices, such as pressure switches and other interlocks with the motor-control circuit, may be found in the control circuit. The pressure switch shown in the figure is just an example of what can be placed in the circuit. Any number of different safety and control devices are used in applications. The control switch could be a thermostat, pressure switch, or any one of many possible options as well. For instance, a valve in the piping may need to open fully before the motor can start. A normally open proving switch for the valve's position would then be placed in the starter control circuit.

Controlling a starter with line voltage supports other practical control options. For example, starters can be equipped with a Start-Stop or Hand-Off-Auto (HOA) control switch (**FIGURE 9-16**). Those with a Start-Stop switch, which can also be installed away from the starter, are considered manual starters since they require human control. With an HOA switch, the user can manually start the motor, or turn control of it over to an external control circuit. That circuit may be at line voltage as well, or it may be a low-voltage circuit interlocked through a relay (**FIGURE 9-17**).

▶ CORE CONNECTION

Full-Voltage and Reduced-Voltage Motor Starters

In a standard motor starter, when the contactor closes, the full line voltage from the supply circuit is applied to the motor. This results in a very large inrush current until the motor gets up to speed. Starting a large motor can place an unacceptably large current load and/or voltage drop on the power source. Utilities charge some users more based on their peak demand for current. To eliminate this problem, motor starters have been developed to lower the inrush of starting current. While the technical details of each type of starter are beyond the scope of this course, do be aware of some fundamental characteristics of reduced-voltage starters:

- Primary-resistor starters insert resistors into the motor supply circuit that reduce starting voltage. They are automatically switched out as motor speed increases.
- Reactor starters perform the same service, except that inductors are used to drop the AC (alternative current) supplied to the motor.
- Transformers can reduce the motor supply voltage to a lower value while it is starting and then be switched out of the circuit when the motor is at full speed.
- Solid-state electronic starters (also called soft-start starters) are the most sophisticated types, relying on semiconductor devices to control the AC waveform and frequency of the line current supplied to the motor while the motor is starting.

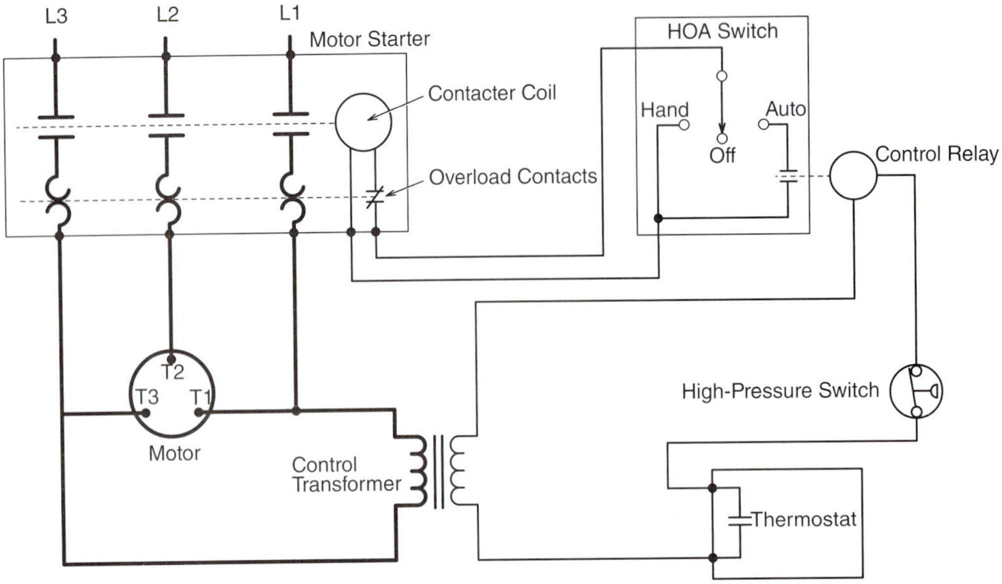

FIGURE 9-17 Motor starter with an HOA switch interlocked with an external low-voltage control circuit.

FIGURE 9-18 An add-on auxiliary contact for a motor starter.

Auxiliary Contacts

Motor starters may include one or more auxiliary contacts, just like contactors. These may be built into the starter with their own terminals, or the auxiliary-contact housing may be mounted on the side of the starter (**FIGURE 9-18**). An interlocking finger from the main contactor armature protrudes through the side of the starter frame and positions the contact mechanically.

▶ Troubleshooting a Motor Starter

K09005 Explain how to troubleshoot a motor starter.

The approach to testing a motor starter is very similar to testing a contactor, since a contactor is a primary component of the starter. Refer to that section of this chapter if necessary. The main difference is the testing of the overload relay. This process can vary somewhat from starter to starter, especially with solid-state overload relays. In general, however, basic testing of the overload relay includes checking the relay control-circuit contacts for their status (open or closed). For thermal overload relays, testing might also include the melting-alloy elements or the heating elements for bimetal switches.

It is important to point out that during field practice, it is difficult to separate starter troubleshooting from motor troubleshooting. A technician is often confronted with the following problem: a motor that has stopped because the overload relay has tripped, opening the control circuit to the contactor. In this and similar situations, it is best to begin by assuming the starter has done its job and the problem is external to it. For instance, a loss of power from one phase of a three-phase system is likely to cause a starter to trip as the current rises through the remaining two phases. Once power is restored, the starter is manually reset, the motor starts, and all is well. Information about power issues can usually be obtained from the client who might be aware that there was a momentary loss of power. However, it is always wise to check the motor windings with a meter for continuity or a grounded condition before it is restarted. Motor testing and troubleshooting is covered in Chapter 14.

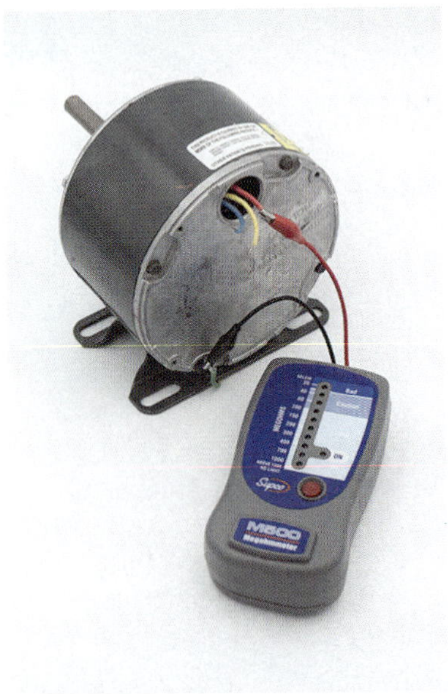

FIGURE 9-19 Testing motor windings with a megohmmeter.

▶ TROUBLESHOOTING TIP

Megohmmeter Testing

A megohmmeter is designed to check very high resistance values in motors and other devices, including motor starters. These instruments generate a very high voltage that is applied to a motor winding or motor-starter assembly to detect the slightest weakness in the insulation system. Although a standard multimeter may indicate a motor winding is not grounded, the voltage from a common meter cannot provide the voltage needed for a serious test. Applying *thousands* of volts makes a big difference. High-resistance grounds can be detected in this way. Analyzing a trend of megohmmeter readings taken over time makes the electrical death of a motor reasonably predictable. These instruments are often used for testing motors and motor starters as part of an industrial maintenance program (**FIGURE 9-19**). However, motor-starter failures due to high-resistance grounds are rare unless they have been subjected to extreme currents and other abuse.

The starter load circuits can be tested from the line-voltage terminals of the contactor to the load terminals of the overload relay with the power off. There should be a very low and balanced resistance through each circuit. If there are unusual variations, it is most likely due to damaged contacts in the contactor. The contacts can be accessed for inspection and cleaned or replaced as necessary. The heating elements of the overload relay can also be tested to rule them out as the source of an unusually high resistance, confirming that the problem lies at the contacts. Thermal elements in starters can overheat and open, but it is a rare occurrence. Some starters develop problems in the reset-button mechanism, and because of this, the control-circuit contacts will not close. If this occurs, the overload-relay assembly must be replaced.

Before resetting a starter and restarting a motor, always be prepared to monitor current by placing a clamp-on ammeter on one of the load wires. The expected inrush of current should be noticeable, followed by a drop in current until it stabilizes. Once it stabilizes, the current should be equal to or less than the rated full-load current of the motor. If the start-up appears to be normal after the first few seconds, quickly move the ammeter from one leg to another, looking for an unusual current reading that could indicate a problem. If the current remains extremely high and the motor isn't rotating, immediately deenergize the starter. If the starter has been inspected and tested while deenergized, the problem is likely in the motor.

Damaged or failed contactors, contacts, solenoid coils, and overload relays can be replaced in NEMA-standard starters. Many IEC starters permit replacement of the contactor or overload modules only. In some cases, the entire IEC starter must be replaced. HVACR technicians often consider IEC starters expendable items because of their lower replacement cost. Although IEC-standard starters are not built to the same standards of durability (by design) as NEMA-standard starters, there is no reason to avoid them in the HVACR industry. Going forward, expect to see their use increase in the United States.

Due to the complexity of solid-state overload relays, the appropriate service literature should always be consulted when they require testing and troubleshooting. Indeed, technicians should refer to the manufacturer's documentation to troubleshoot all motor starters.

▶ Wrap-Up

Summary

HVACR contactors are heavy-duty, remotely operated switches similar to relays. They are designed to start and stop larger electrical loads associated with HVACR equipment and systems. An electromagnetic solenoid, when energized, closes or opens the contacts. A contactor control circuit will often include various switches, relay contacts, thermostats, pressure switches, and similar devices that can start or stop the connected load as needed.

A motor starter provides the same service as a contactor but is also uniquely designed to protect motors. A motor starter is composed of a contactor and an overload relay that monitors motor current. The overload relay protects against excessive motor current.

Both contactors and motor starters are designed and built to different standards that affect their application and flexibility. The HVACR technician needs to be familiar with the various types to effectively operate, troubleshoot, and repair HVACR systems that rely on these devices.

Key Terms

auxiliary contacts Electrical contacts that are connected to and operate at the same time as the main contacts of a contactor or motor starter but provide switching for circuits other than the main power supply to the load.

definite-purpose contactor A magnetic contactor designed and built for use primarily in HVACR equipment, where frequent cycling of high-current loads is common.

full-load amps (FLA) The current carried by a motor or other inductive load when operating at its maximum designed load.

locked rotor amps (LRA) The maximum current that is drawn by a given motor that is seized or so heavily loaded so that its rotation significantly slows or stops. The LRA may also be exhibited monetarily during motor start-up.

overload relay A special type of relay that monitors current passing through and opens a control circuit when the current exceeds a predetermined value. Overload relays provide overload protection for large inductive loads, such as motors.

voltage imbalance The largest difference between the average voltage of the phases and the voltage of any individual phase, divided by the average voltage. Voltage imbalance, also known as phase imbalance, is generally expressed as a percentage.

Review Questions

1. Which of the following statements about contactors is true?
   a. Contactors are more like circuit breakers than any other electrical component.
   b. Two or three control relays working together can replace a contactor.
   c. Definite-purpose contactors were developed to satisfy the demands of HVACR equipment manufacturers.
   d. IEC-standard contactors are the most durable type available for all applications.
2. Which of the following characteristics sets a contactor apart from a relay?
   a. Contactors are often repairable in the field.
   b. A contactor does not have a magnetic coil.
   c. Contactors have no fewer than three poles.
   d. Contactors are always enclosed, alone, in a dedicated enclosure.
3. In the comparison of a contactor built to NEMA standards to one built to IEC standards, which of the following statements would be true?
   a. IEC-standard contactors are more expensive.
   b. NEMA-standard contactors are not considered as reliable.
   c. NEMA-standard contactors are used mainly in new equipment.
   d. IEC-standard contactors are designated for more specific applications.
4. What is the main concern associated with the use of single-pole contactors?
5. Contactors often have current ratings for both inductive and resistive loads listed on their label. Which of the two load types is likely to have the higher current rating on the label?
6. Swapping two of the three motor leads at the contactor or starter terminals for a three-phase motor results in ______ ______________________.
7. What voltage are definite-purpose contactors rated for, unless otherwise labeled?
   a. 250 VAC
   b. 400 VAC
   c. 600 VAC
   d. 1,000 VAC
8. Voltage drop through one or more circuit paths of a motor starter causes ______________.
9. Which of the following is a correct statement about contactor and starter troubleshooting?
   a. There should not be continuity from a contactor-coil terminal to ground.
   b. The resistance measured through a set of good contacts may be as high as 10–15 ohms.
   c. The voltage drop measured from L1 to T1 under load should be at least 3–5 volts.
   d. A single-pole contactor is allowed to have three times the voltage drop as one pole of a three-phase contactor.
10. Which of the following electrical formulas is used to determine the power loss through a set of contacts?
    a. $R \times I = E$
    b. $P \div E = I$
    c. $E^2 \div P = R$
    d. $E \times I = P$
11. When checking the power loss through a set of contacts, losses greater than ______ are cause for concern.
    a. 2 watts
    b. 5 watts
    c. 10 watts
    d. 25 watts
12. How do the contacts of a definite-purpose contactor differ from those in a NEMA-standard contactor, in terms of durability?
13. Which of the following is *not* a common category of overload relay?
    a. Bimetal element
    b. Reactive element
    c. Melting-alloy element
    d. Solid state
14. Which device would likely respond more quickly to a grounded motor winding—a motor starter or a fuse?
15. Class 10 motor starters are designed to respond to an overload condition within ______ seconds.
16. Solid-state overload relays usually respond to the loss of a phase of power in ______ seconds or less.
17. Control of a motor starter can be manual or turned over to an external control circuit using a(n) __________ switch.
18. When observing the ammeter during a motor start-up, if the current remains extremely high and the motor isn't rotating, what should be done?

19. The instrument used to test the insulation condition of motor windings and similar electrical components is called a(n) ______.
  a. oscilloscope
  b. frequency modulator
  c. Bourdon meter
  d. megohmmeter

20. A reduced-voltage motor starter is designed to ____________________.
  a. reduce the inrush current associated with motor start-up
  b. increase the full-load amps (FLA) capacity of the motor
  c. convert a single-speed motor to multi-speed
  d. provide a method of braking a motor

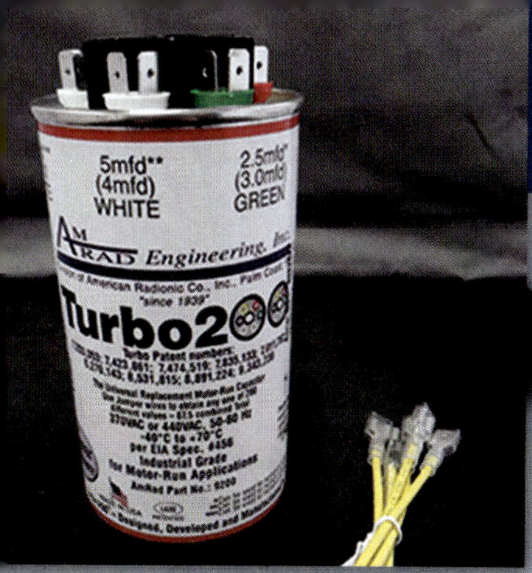

CHAPTER 10
Capacitors

Knowledge Objectives

After reading this chapter, you will be able to:

- **K10001** Describe the construction and function of capacitors.
- **K10002** Describe the types of capacitors used in HVACR systems and their operation.
- **K10003** Explain how to troubleshoot motor start and run capacitors.
- **K10004** Explain how to select and replace motor start and run capacitors.

Skill Objectives

Skills required to meet the objectives of this chapter:

- **S10001** Test a motor start and run capacitor.
- **S10002** Wire a start relay and start capacitor into an existing compressor circuit.

▶ Introduction

FIGURE 10-1 Examples of capacitors used in electronic circuits.

Capacitors are simple yet important electrical devices that are designed to store electrical energy in the form of an electric field. Some capacitors are quite small and are used in electronic circuits (**FIGURE 10-1**), whereas others are quite large and are used in power-distribution substations, where they help eliminate fluctuations and disturbances in the power supply (**FIGURE 10-2**).

Many packaged electronic components and printed-circuit boards are used in HVACR systems. But these devices are rarely repaired in the field. Service technicians will troubleshoot printed-circuit boards and determine whether they are functioning as expected, but verifying exactly what has failed at the board level is left to shops where the necessary tools and equipment are readily available for bench testing. Field service technicians will remove and replace a complete printed-circuit board or control as a unit. Troubleshooting capacitors at this level is not necessary for field technicians.

The capacitors that technicians must be thoroughly familiar with are those related to the operation of single-phase motors (**FIGURE 10-3**). This chapter will explain how capacitors work and their function in motor power circuits. In addition, it will explain how to troubleshoot capacitors and choose a suitable replacement when necessary.

▶ Capacitor Construction and Function

K10001 Describe the construction and function of capacitors.

Capacitors are constructed in a variety of ways. Most consist of at least two plates that serve as electrical conductors. The two conductive plates are separated by a non-conductive **dielectric** material. Some common dielectric materials used in capacitors include glass, porcelain, paper film, ceramics, and various polymers. Although a dielectric material is non-conductive, it is capable of storing an electrostatic charge. Storing a charge is precisely what the capacitor is designed to do. A simple diagram of a capacitor is shown in **FIGURE 10-4**. Note that the capacitor is connected to a simple power source with positive and negative poles.

FIGURE 10-2 A high-voltage capacitor bank serving a power substation.

Since the dielectric material will not allow the passage of electrical energy, opposing charges develop on each side of it. A positive charge accumulates on one plate, and a negative charge accumulates on the other. This leads to a difference in electrical potential between the two plates, creating an **electrostatic charge**. The amount of energy that a capacitor can store is referred to as its **capacitance**.

In some ways, a capacitor is similar to a battery. Both devices store electrical energy, but a battery releases its energy slowly

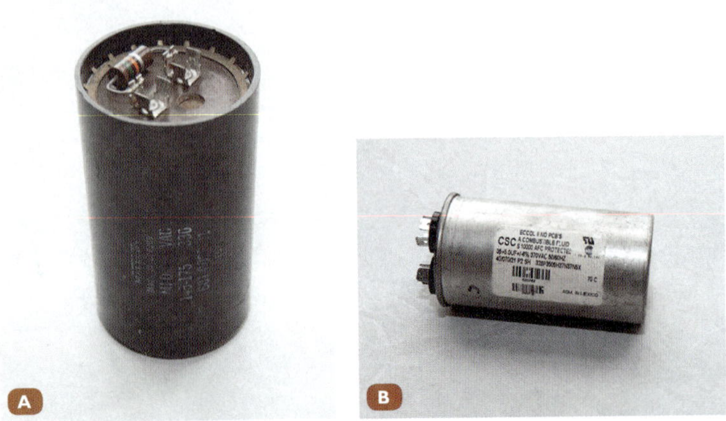

FIGURE 10-3 **A.** Motor Start Capacitor. **B.** Motor Run Capacitor

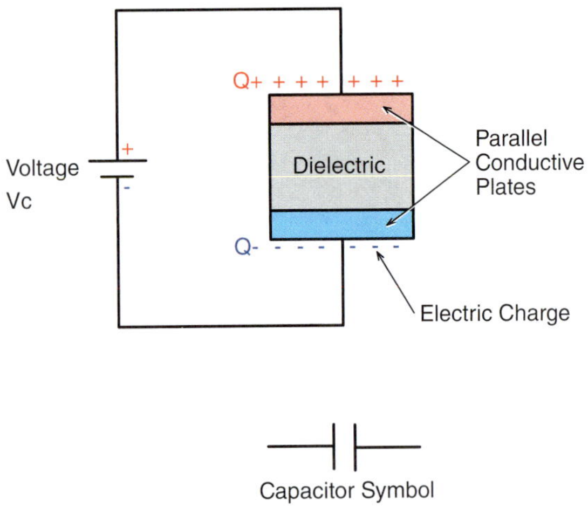

FIGURE 10-4 Capacitor construction.

over a long period of time, whereas a capacitor is often designed to release the stored energy very quickly. For example, in a common HVACR application where a capacitor is used to help start a single-phase motor, electrical energy from a power source is first applied to the capacitor to charge it. When the motor is energized and begins to start, the capacitor instantly releases the stored electrical energy to increase the starting torque of the motor. Once the motor has reached approximately 75% of its normal operating speed, the capacitor is no longer needed, and then a switch or relay is used to remove the capacitor from the circuit.

Capacitor Ratings

The primary characteristics used to rate capacitors are the amount of electrical energy they can store (capacitance) and the voltage at which they are designed to operate. For a parallel-plate capacitor, such as the one shown in Figure 10-4, the capacitance value depends on the area of the conductive plates, the distance between the plates, and the dielectric constant of the dielectric material. **TABLE 10-1** shows the dielectric constants of several common dielectric materials.

Capacitance is expressed using a standard unit called a **farad** (abbreviated F), which is named after the nineteenth-century English physicist Michael Faraday. A capacitor with a capacitance of 1 farad can store 1 **coulomb** of charge on its plates when a potential of one volt is applied across its terminals. The following is the equation showing the relationship between capacitance (C), the charge on the plates (Q), and voltage (V):

$$C = Q \div V$$

A single farad is actually a very large amount of capacitance. For that reason, most capacitors are rated in fractions of a farad. Common capacitance ratings include the following:

- *millifarad*—one-thousandth of a farad, typically expressed as mF
- *microfarad*—one-millionth of a farad, typically expressed as μF or MFD
- *nanofarad*—one-billionth of a farad, typically expressed as nF
- *picofarad*—one-trillionth of a farad, typically expressed as pF.

Microfarads

Capacitors used in HVACR motor applications are rated in microfarads (μF). For instance, the capacitance rating of the start capacitor shown in **FIGURE 10-5** is 540–648 μF. A capacitance rating represented as a range, rather than as a single value, is common for motor start capacitors.

FIGURE 10-5 Capacitance rating on a capacitor label.

TABLE 10-1 Dielectric Constants of Common Dielectric Materials

Dielectric Material	Dielectric Constant
Air	1.0006
Aluminum Oxide	8.5
Barium Titanate (Class 2)	3,000–8,000
Impregnated paper	2.0–6.0
Mica	6.8
Paraelectric Ceramics (Class 1)	5–90
Polyester PEN	3.0
Polyester PET	3.3
Polypropylene PP	2.2
Polypropylene Sulfide PPS	3
Tantalum Oxide	27.7
Vacuum	1

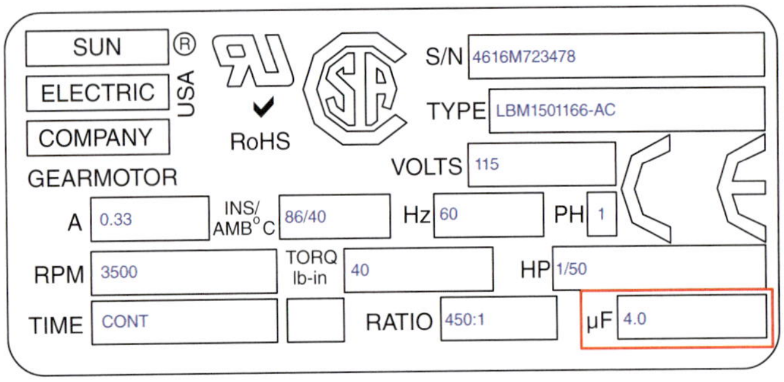

FIGURE 10-6 Motor nameplate.

The microfarad rating of a capacitor must always match the requirements of the motor with which it is being used. The capacitor requirements are sometimes shown on the motor's nameplate (**FIGURE 10-6**). Unfortunately, that is not always the case.

Using a capacitor with a lower-than-required microfarad rating can reduce the starting torque, the rpm, and/or the overall efficiency of the motor. Using a capacitor with a higher-than-required microfarad rating has the same effect. In either case, using a capacitor with the wrong microfarad rating can lead to excessive heat in the motor's windings and decrease the lifespan of both the motor and the capacitor.

Voltage

Capacitor ratings are also based on voltage. The voltage rating of a capacitor is the maximum, or peak, voltage that can be supplied to the capacitor without damaging or destroying it. Keep in mind that this voltage rating is not the same as the system operating voltage, nor does it refer to the voltage of the stored charge. It is simply the maximum voltage that the capacitor can safely tolerate without risking a breakdown of the dielectric material inside. There are moments during a capacitor's operation when the voltage to which it is exposed is higher than the system operating voltage. This is the reason for the rated voltage to be significantly higher than the operating voltage. For example, many of the run capacitors used with alternating current (AC) motors in HVACR applications are rated at 370–440 VAC at a frequency of 50 to 60 Hertz (Hz). **FIGURE 10-7** shows this voltage range on the label of a run capacitor for a compressor motor. The voltage may be reported as a range or as a specific voltage. Start capacitors need to match only the supplied voltage since they are in the circuit for less than a second.

FIGURE 10-7 Voltage rating on capacitor label.

Unlike the microfarad rating of a capacitor, which should always match the requirements of the motor, the voltage rating of a capacitor can be increased. In other words, a 440 VAC capacitor can be used in place of a 370 VAC capacitor. In this case, the 440 VAC rating would simply provide greater assurance that the capacitor could handle the potential voltage without affecting its performance. The 440 VAC capacitor might actually last longer, but using a capacitor with a higher voltage rating could cost a little more, and it may be larger than the mounting location can easily accommodate.

Although it is acceptable to use a capacitor with the same or higher voltage rating than what is recommended, it is important to note that the voltage rating of a capacitor should never be decreased. If the applied voltage exceeds the capacitor's voltage rating, the dielectric material in the capacitor can be damaged.

▶ TECHNICIAN TIP

Positive Temperature Coefficient (PTC) Thermistors

PTC thermistors are temperature-sensitive resistors that can provide an economical way to assist with starting a motor. Single-phase motors used in HVACR applications usually have a separate start winding. If a capacitor is used, the stored electrical energy is quickly discharged to temporarily increase the starting torque of the motor. Once the motor starts, the capacitor is electrically removed from the circuit. However, a PTC thermistor can be used to achieve similar results in starting. The resistance of a PTC thermistor increases significantly as temperature rises. This heating effect comes into play when a PTC thermistor is connected in series with the motor's start-up winding (**FIGURE 10-8**). When voltage is applied to the circuit, the resistance of the PTC thermistor is low and current flows through the start-up winding of the motor. As the current flow continues, the PTC thermistor begins to heat up. The heat causes its resistance to increase significantly—ultimately to a point where the current flow to the start-up winding stops completely. When compared to conventional start-capacitor arrangements, PTC thermistors can take up less space, cost less, and reduce the number of wires needed to accommodate motor start-up. HVACR equipment manufacturers often offer them as optional accessories for new equipment, or they can be purchased at an HVACR parts distributor as an aftermarket device.

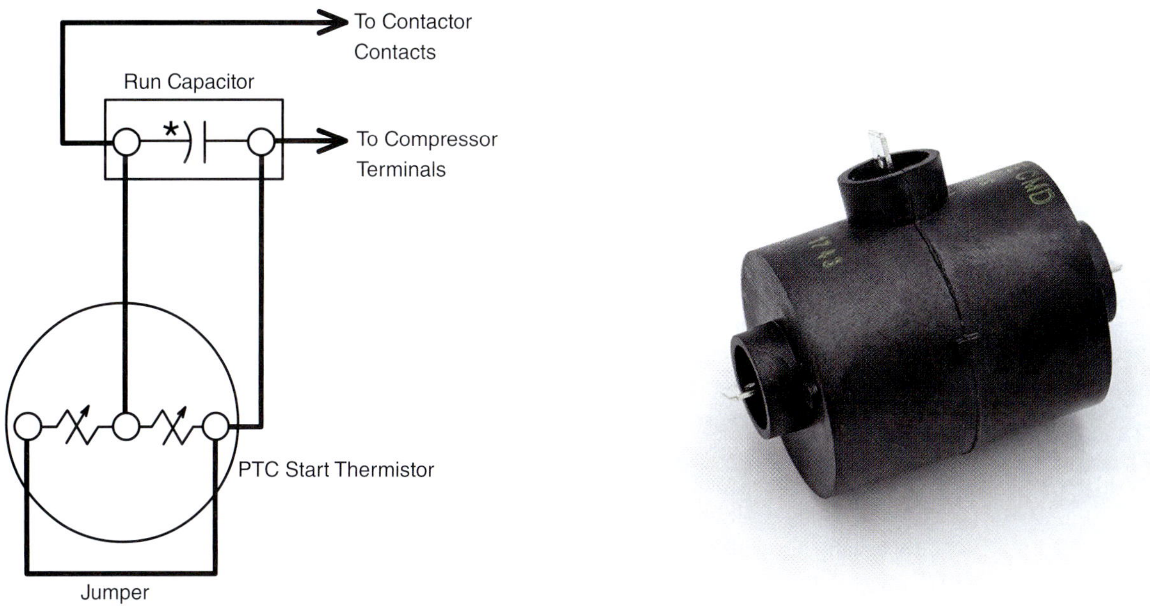

FIGURE 10-8 PTC start thermistor.

▶ Start and Run Capacitors

K10002 Describe the types of capacitors used in HVACR systems and their operation.

HVACR technicians work with two basic types of capacitors: start capacitors and run capacitors. Both are related to the operation of single-phase motors. As their names suggest, start capacitors assist in motor starting only, while run capacitors are in operation throughout a motor's operating cycle. Start capacitors have higher capacitance values and are used briefly to increase the torque of a motor during start-up. Run capacitors have lower capacitance values and are used continuously to help maximize a motor's performance and efficiency.

Motor Start Capacitors

Motor start capacitors are designed for intermittent and brief use—typically no more than a few seconds. As mentioned previously, they have relatively high capacitance values that are often listed as a range rather than as a specific number.

FIGURE 10-9 shows a circuit diagram for a typical single-phase electric motor that has both a start capacitor and a run capacitor. This type of motor is commonly referred to as a capacitor-start, capacitor-run (CSCR) motor. Note that there are other types of motors that will be described in chapters dedicated to motors. In this case, the two capacitors are connected in parallel. There is also a switch in the circuit that is used to connect and disconnect the start capacitor from the circuit. Smaller motors generally use a **centrifugal switch** for this purpose. When the motor is started, the centrifugal switch or relay is closed and power flows through both capacitors and the windings in the motor. The start capacitor creates a 90-degree phase shift between voltage and current. This phase shift creates the optimal starting torque. Once the motor reaches about 75% of its normal operating speed, the centrifugal switch or relay opens to disconnect the starting capacitor from the circuit. The motor continues to run with only the run capacitor remaining in the circuit.

Due to the higher capacitance of start capacitors, they can deliver a significant electric shock. Start capacitors are often equipped with a bleed resistor mounted across the two terminals (**FIGURE 10-10**). For safety reasons, this external 10,000–20,000 (10–20K) ohm resistor bleeds off any residual charge whenever power to the capacitor is disabled. When the capacitor is energized and charging, the resistor is an insignificant electrical load.

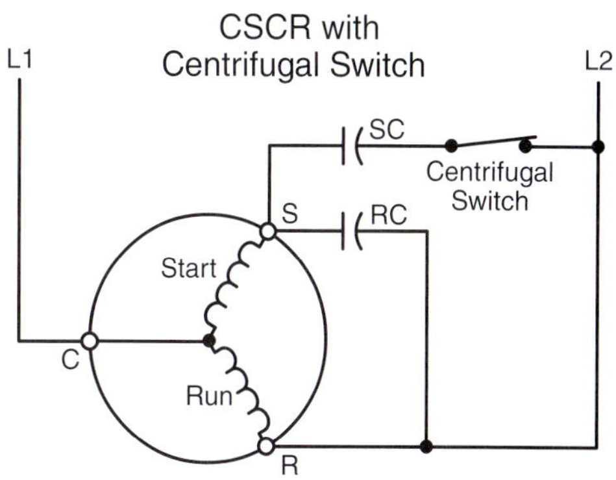

FIGURE 10-9 Circuit diagram of CSCR motor with a centrifugal switch.

FIGURE 10-10 Bleed resistor on a start capacitor.

SAFETY TIP

Capacitor Handling

Never touch the terminals of a capacitor until any residual electric charge has been safely bled off. A capacitor can deliver a significant electric shock. The process of bleeding a capacitor will be discussed in this chapter, along with capacitor testing procedures.

Start Relays

While some motor circuits use a centrifugal switch to connect and disconnect the start capacitor, others use a start relay. A common application for a start relay is on a single-phase compressor used in HVACR systems. Many single-phase fan motors do not require start capacitors, because they have less resistance to overcome when starting.

A relay is a type of switch that is operated electrically. Inside is an operating coil and one or more sets of contacts. When energized, the coil creates a magnetic field that acts on the contacts, drawing them closed or open, depending upon the relay design. Start relays usually have only one set of contacts.

The precise operation of a start relay depends on the type. The two basic types used in HVACR motor applications are the **current relay** and the **potential relay** (**FIGURE 10-11**). The latter is sometimes called a voltage relay. The two types function differently but accomplish the same basic function as a centrifugal switch: connect and disconnect start capacitors from the motor electrical circuit.

As the circuit diagram in **FIGURE 10-12** shows, the coil of a current relay is connected in series with the run winding of the motor, whereas the normally open (NO) contacts in the relay are connected in series with the start capacitor and start winding. When power is applied, the start winding of the motor, due to the open contacts of the relay, does not experience any current flow. At that instant, the motor will not turn, but it will draw a great deal of current through its run winding. The current relay's low-resistance coil is energized by the high inrush of current, closing the NO contacts to energize the start winding. The motor begins to turn and accelerates quickly. Once the motor is rotating, the current drops sharply. As it approaches full speed, the current through the run winding is no longer sufficient to keep the relay coil energized and the contacts open. With the start-relay contacts open and the start capacitor removed from the circuit, the motor settles into normal operation. This sequence occurs very quickly, over the course of one to three seconds.

FIGURE 10-13 shows a circuit diagram of a motor that uses a potential relay. Potential relays are used with motors that require a higher starting torque. This type of relay is connected in parallel with the motor's start winding and operates on **counter-electromotive force**

▶ TECHNICIAN TIP

Current Relay Application

Compressor motors can be challenging to start. One of the factors working against them is the pressures that they develop when in operation. A significant amount of pressure difference develops between the low-pressure and high-pressure sides of the compressor when the compressor runs, by design. If there is still a significant pressure difference present when the compressor tries to start again, it may be unable to do so. Thus, current relays are most often applied on systems that use simple refrigerant-metering devices, like a capillary tube. When the compressor stops running, refrigerant pressures equalize faster with these metering devices. This makes it easier for the compressor to start soon after it has been cycled off. When more starting torque is needed to overcome some expected pressure difference, potential relays work best.

FIGURE 10-11 A. Current start relay. **B.** Potential start relay.

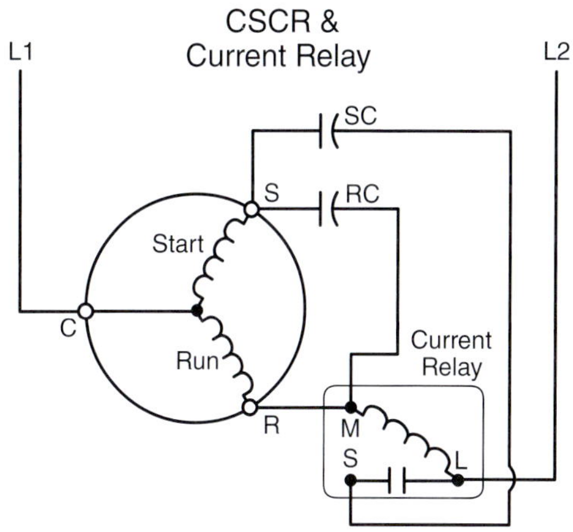

FIGURE 10-12 Circuit diagram of CSCR motor with current relay.

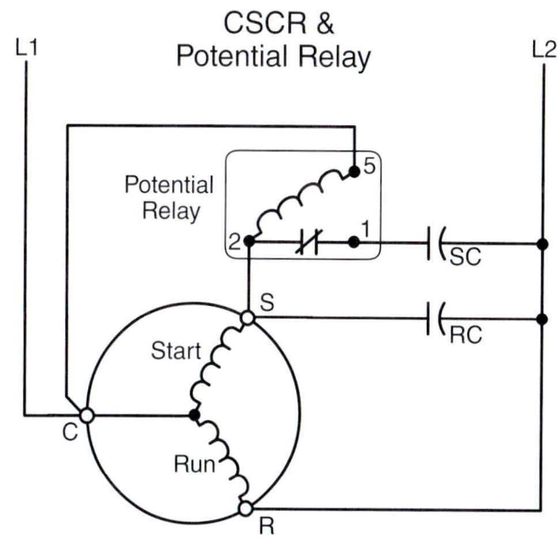

FIGURE 10-13 Circuit diagram of CSCR motor with potential relay.

(CEMF) generated in the winding. You may also see this referred to as back EMF (electromotive force). The contacts of a potential relay are normally closed (NC)—unlike those of a current relay, which are NO.

When power is applied, the motor's start and run windings are both immediately energized because the relay contacts are closed. This applies the start capacitor's charge to the motor's start winding from the onset. As motor speed increases, the motor's rotor turning through the magnetic field of the windings causes CEMF to be generated. As the motor reaches 75–80% of full speed, the CEMF generated is sufficient to energize the relay coil, opening the relay's NC contacts. Once the contacts open, the start capacitor is removed from the starting circuit, and the motor continues to run. As long as the motor's speed is sufficient to generate the needed CEMF, the relay contacts remain open. Once power is cycled off and the motor begins to slow down, the relay contacts close again, prepared for the next start cycle. As is the case with current relays, the start sequence occurs in just one to three seconds.

Motor Run Capacitors

Run capacitors are designed for continuous use. Their functions are to improve the operating efficiency of the motor and to provide some minimal starting torque. Compared to start capacitors, run capacitors have lower capacitance values—typically in the 3–70 μF range in HVACR applications. Whereas start capacitors are electrically removed from the circuit once a motor is running, run capacitors remain in the circuit to maximize the motor's performance and efficiency.

Looking back at the circuit diagrams in Figure 10-12 and Figure 10-13, tracing the circuits shows that the run capacitor is unaffected by the operation of the start relay; its source of power does not flow through the relay contacts. In cases when a single-phase motor does not require extra starting torque, there is no start relay or start capacitor; only a run capacitor is needed. Some motors need no capacitor at all, but they are also limited in their performance and application as a result.

▶ TECHNICIAN TIP

Multi-Range Capacitors

A special type of capacitor often found among an HVACR technician's supplies is called a multi-range capacitor (**FIGURE 10-14**). Multi-range capacitors, sometimes called universal run capacitors, enable technicians to select various capacitance values from a single capacitor. This feature is accomplished either by setting switches on the capacitor or by using jumpers across several capacitor terminals. For instance, the capacitor pictured in Figure 10-14 has a capacitance range of 2.5 to 67.5 μF for 370 VAC or 440 VAC applications. By attaching jumpers across the multiple terminals on top of the capacitor, a technician can configure it to provide a variety of capacitance values. The example of a multi-range capacitor in Figure 10-14 shows how to set the device up in order to produce 27.5 μF for a compressor motor. In most cases, technicians use this capacitor as a temporary tool to get a system back in operation. Once a standard capacitor is acquired and installed, the universal capacitor is removed and placed back on the truck. This is primarily due to its cost and its physical size, which may make it difficult to mount in the available space.

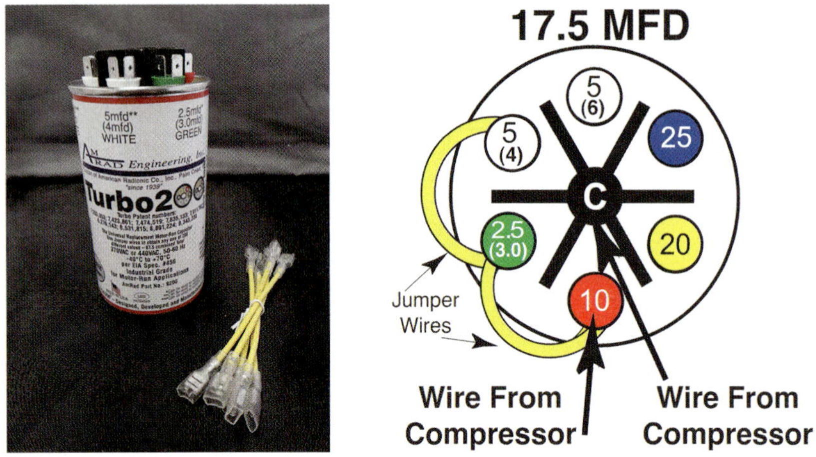

FIGURE 10-14 Multi-range run capacitor.

FIGURE 10-15 Dual run capacitor terminals.

FIGURE 10-16 Dual run capacitor label.

Dual Run Capacitors

Dual run capacitors are very popular in HVACR applications and were likely developed with the HVACR industry in mind. A dual run capacitor has three terminals and can serve two motors simultaneously. For example, a residential condensing unit operates on single-phase power, and both the compressor motor and the condenser fan motor generally require a run capacitor. Instead of using two separate capacitors, a dual capacitor with the correct ratings can serve both motors.

FIGURE 10-15 shows the top of a typical dual run capacitor. The terminal designated C is the common terminal where incoming power is applied. The terminal marked HERM is for the compressor motor, which will require the greatest capacitance of the two motors. The HERM label stands for **hermetic compressor**, although not every compressor it is used with will necessarily be hermetic. The terminal marked FAN is for the condenser fan motor.

Since there are two outputs involved, the capacitor must be marked to show the capacitance available from each side. **FIGURE 10-16** provides an example of a dual run capacitor label. Note that there are two values shown—35 μF and 5 μF. The larger value will always be associated with the HERM terminal. When connecting the wiring, it is very important to ensure that the correct terminals are chosen.

▶ Troubleshooting Capacitors

K10003 Explain how to troubleshoot motor start and run capacitors.

S10001 Test a motor start and run capacitor.

Capacitors are reliable, but they do fail. There are several reasons why capacitors fail, including normal wear, electrical anomalies like power surges and lightning strikes, and overheating. A motor with deficiencies of its own can overstress the capacitor. If an improperly sized capacitor is used, such as one with a lower voltage rating than required, premature failure is likely. Since start capacitors are designed only for intermittent service, a start relay that fails and causes the capacitor to remain in the circuit longer than necessary usually leads to a failure.

Capacitors are more prone to failure during periods of high ambient temperatures, long operating cycles and/or frequent motor starts. When ambient temperatures rise and the system is operating under heavy load, the additional heat can cause changes in the dielectric material inside the capacitor. This affects the capacitance value of the capacitor and may degrade the dielectric material. Eventually, the change in capacitance can affect the operation of the motor.

HVACR systems that cycle on and off frequently are more likely to experience problems with start capacitors. The constant charging and discharging required under these conditions takes a toll on the capacitor over time.

Troubleshooting a capacitor involves several steps, including monitoring the motor for abnormal sounds or performance during start-up and operation, visually inspecting the capacitor for physical damage, and testing it with a capacitor tester or multimeter.

When monitoring a motor with a defective start and/or run capacitor, technicians might detect several abnormalities. If the motor is energized and makes a humming sound but does not actually rotate, a failed start capacitor or start relay may be the problem. A failed or failing run capacitor might be the problem if the motor starts but tends to run at a higher-than-nameplate current and overheats.

Visual inspection of a capacitor can also reveal problems. A common sign of a failed capacitor is a swollen outer casing (**FIGURE 10-17**). In the most severe failures, the casing may rupture. The capacitor might also emit a pungent smoke or leak oil from inside. Most failures are less dramatic, visually. If the casing is swollen, regardless of the results of testing, it should be replaced.

FIGURE 10-17 Failed capacitor.

▶ TROUBLESHOOTING TIP

Capacitor Testing

Capacitor testers can be purchased as stand-alone devices. Many digital multimeters are manufactured with this feature built in. Regardless of which instrument is used, always read and follow the manufacturer's instructions for the specific type and brand of test instrument to be used. Failure to follow the proper instructions could compromise safety and cause damage to the instrument.

Testing Capacitors

By design, capacitors store a significant electric charge. Before handling a capacitor, it is necessary to place a 1K Ω (minimum) resistor (**FIGURE 10-18**) across the capacitor terminals for a few seconds to discharge the stored electrical energy. There are few tools on the market that are reliable and built for this purpose. Most technicians, however, construct a simple discharge tool by using a resistor and some wire. A fully charged start capacitor can deliver a major electric shock. Although some technicians are known to simply place a screwdriver across the terminals, this is not recommended. The resulting arc can damage the screwdriver and/or the user. Even if the shock does not cause significant injury, secondary injuries that result from electric shock—such as falling backward, causing head injury—are common. For safety's sake, do not assume that a capacitor with permanent bleed resistor has been discharged; these are best discharged as well, using an appropriate tool, before they are handled.

One way to test a capacitor is to measure the capacitance across the terminals by using a suitable digital multimeter (DMM), such as the one shown in **FIGURE 10-19**. Not all DMMs,

FIGURE 10-18 Capacitor discharge tool.

FIGURE 10-19 Typical digital multimeter (DMM).

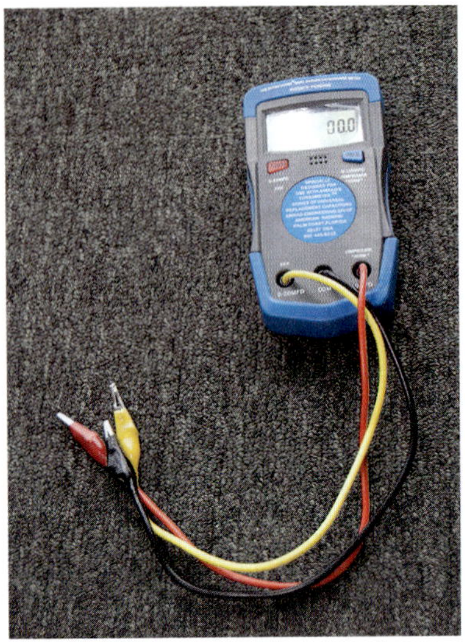

FIGURE 10-20 Capacitor tester.

▶ TECHNICIAN TIP

Keeping Those Wires Straight

A good way to document which wires connect to which terminals on a part is to take a photo of the wiring prior to disconnecting anything. Later, after testing is complete or the part has been replaced, simply refer to the photo to ensure proper reconnection of the wiring.

however, have a capacitance-measurement feature available. Alternatively, a stand-alone capacitance tester can be used (**FIGURE 10-20**).

Capacitance readings should fall within the margin of error listed on the capacitor label (**FIGURE 10-21**). Typically, the measured microfarad value for a start capacitor should be within ±20% of the value shown on the capacitor label. For a run capacitor, the measured value should generally be within ±10% of the value shown on the capacitor. However, check the label to see whether the manufacturer has established other values. If the measured value is near the range limits or outside of them, it is best to replace the capacitor.

Using a Digital Multimeter to Test a Capacitor

When testing a capacitor, there are three characteristics that must be checked. The capacitor should be checked to see whether it is open (no continuity) and to see whether it is shorted internally. If both tests are satisfactory, the remaining test measures the capacitance. If the capacitor is open or shorted, it is defective regardless of capacitance.

The first step is to shut off the power supply to the equipment and verify that the equipment is electrically isolated. Then use a capacitor discharge tool to bleed off any stored electrical energy. This should be done even before any wiring is removed. Note the color coding of the connected wiring and then disconnect the wires from the capacitor terminals. For accurate readings, the capacitor must be completely removed from the circuit. If a permanent bleed resistor is attached, it must be disconnected. Set the DMM for a continuity/resistance test.

Because there is no direct circuit between the terminals due to the dielectric material, capacitors behave in a unique manner when tested for continuity with a DMM. Since the DMM uses its internal battery as a power source for measuring continuity/resistance, the battery places a very small charge on the capacitor. Once it has absorbed as much charge as the battery can deliver, the power flow stops. As a result, when the meter is first connected to the capacitor terminals, there will be a slight jump toward continuity or zero ohms, followed immediately by the resistance reading falling back toward infinity (open). This jump in resistance occurs quickly and is easier to see on analog meters with a moving needle. It is easy to miss altogether on a DMM. If the test is repeated in the same manner, there will be no response, since the capacitor now has a charge. In order to repeat the test, either reverse the polarity of the meter probes on the terminals or fully discharge the capacitor and try again. If continuity is displayed on the meter continuously, rather than a momentary sign of continuity, the capacitor is shorted internally.

To check for shorts, connect one meter probe to a capacitor terminal and the other to the outer case. This works only if the case is metal. Many start capacitors have an outer casing constructed of hard plastic, which is non-conductive. Run capacitors generally have metal cases. If continuity is detected, the capacitor is shorted internally and therefore defective. If there is no continuity, check for continuity between the other terminal and the case to complete the test.

Assuming the capacitor has passed these tests, set the DMM to the appropriate position to make a capacitance test. When the meter probes are touched to the capacitor terminals, the capacitance should be visible on the display. Ensure that the value is one of the following:

- within the range of accuracy shown on the label
- within ±20% of the capacitance rating shown on the label for a start capacitor
- within ±10% of the capacitance rating shown on the label for a run capacitor.

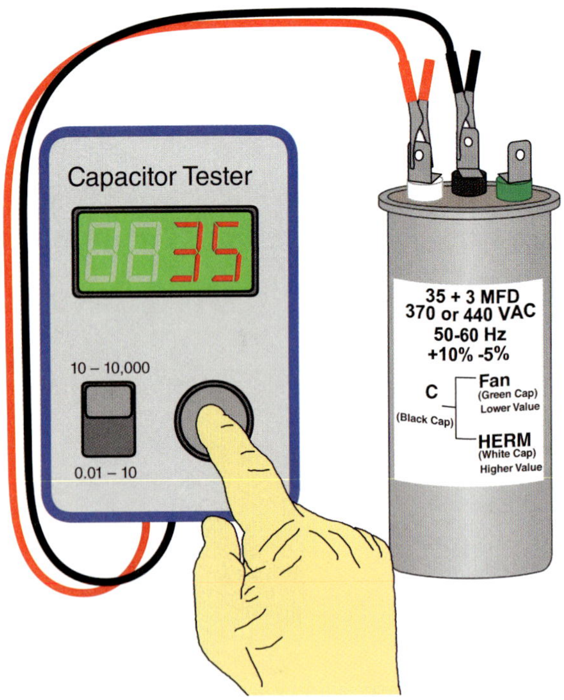

FIGURE 10-21 Capacitor tester showing capacitance value.

To test a capacitor, follow the steps in **SKILL DRILL 10-1**.

SKILL DRILL 10-1 Testing a Capacitor

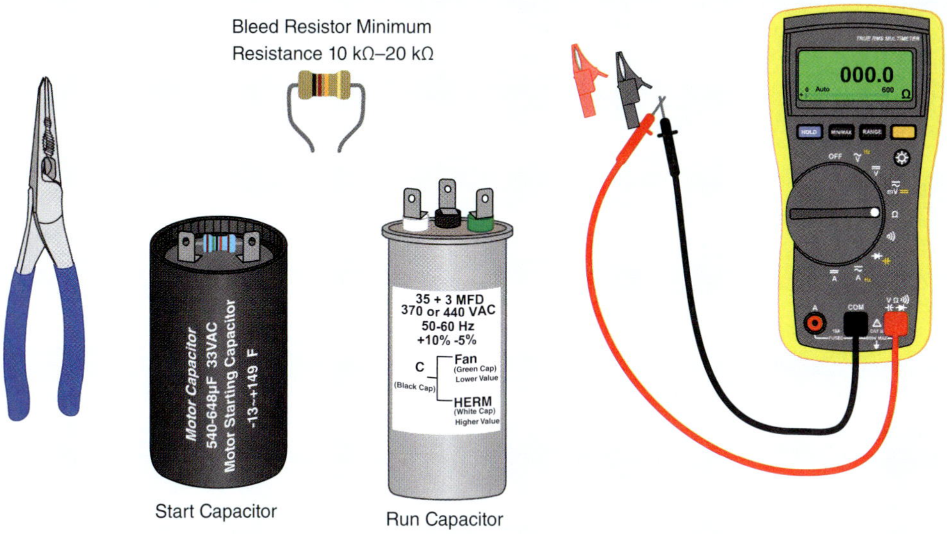

1. Test a start and a run capacitor for a single-phase motor found in a typical air conditioning unit. Select an appropriate DMM for the job—one which has the capability of measuring capacitance. Alternatively, an analog multimeter can be used for the continuity/resistance tests since its needle movement is easy to visually detect and thus works better than a digital display. Visually inspect the meter for any signs of damage. Test it to verify that it is functioning properly.

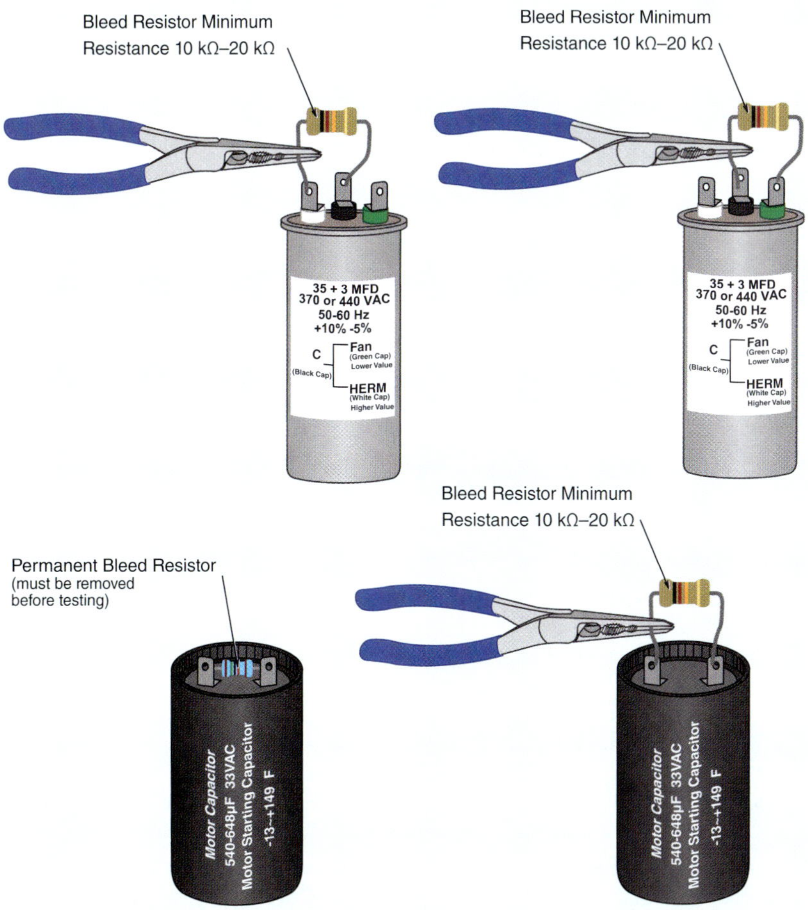

2. Use the bleed-resistor tool to bleed any remaining electrical charge from both capacitors by spanning their terminals with the tool. If the start capacitor has a permanent bleed resistor installed, it must be removed for testing purposes.

Continued

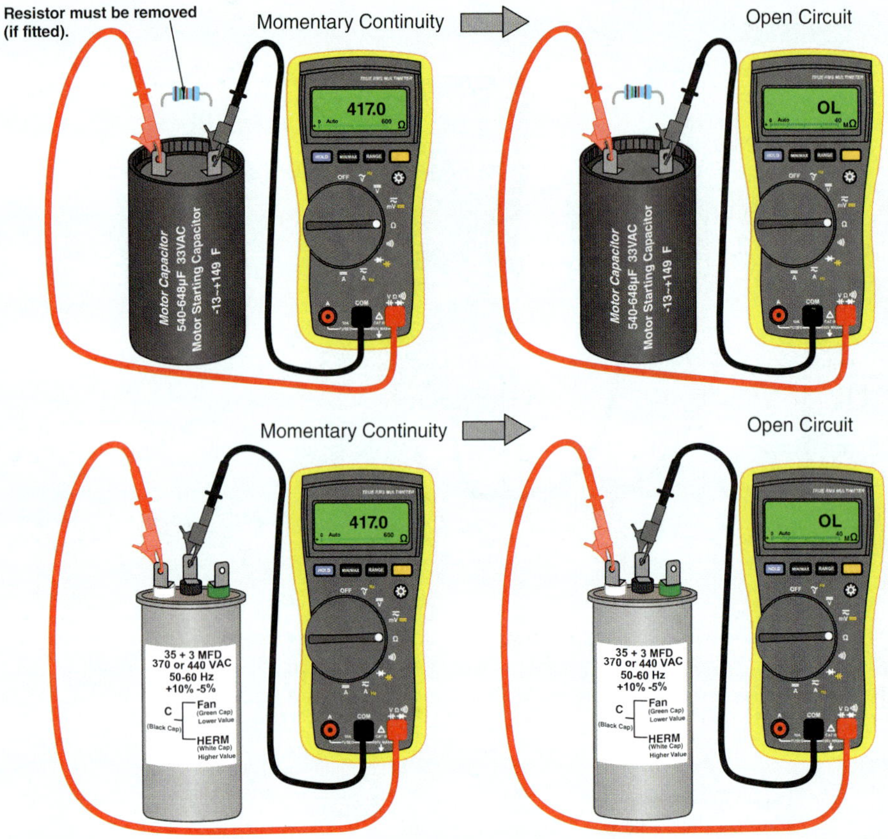

On dual run capacitors, repeat this test across the black and green terminals.

3. Set the DMM to measure continuity/resistance. For the test to be valid, however, the bleed resistor must be disconnected from the start capacitor. While visually monitoring the display, touch the meter probes to the start-capacitor terminals.

Remember that any indication of continuity will be very quick, before the meter returns to an infinity (open) reading. In the event that a test needs to be repeated, bleed off the capacitor completely or reverse the polarity of the test by placing the meter probes on the opposite terminals of the capacitor. If there is continuity and a low-resistance value continuously displayed, the capacitor is shorted internally. If there is no momentary indication of continuity, followed by the meter showing an open circuit, then the capacitor is likely open. However, this is easily confirmed through the capacitance test as this task progresses. Repeat the test as described above on the run capacitor, which is less likely to have a permanent bleed resistor installed.

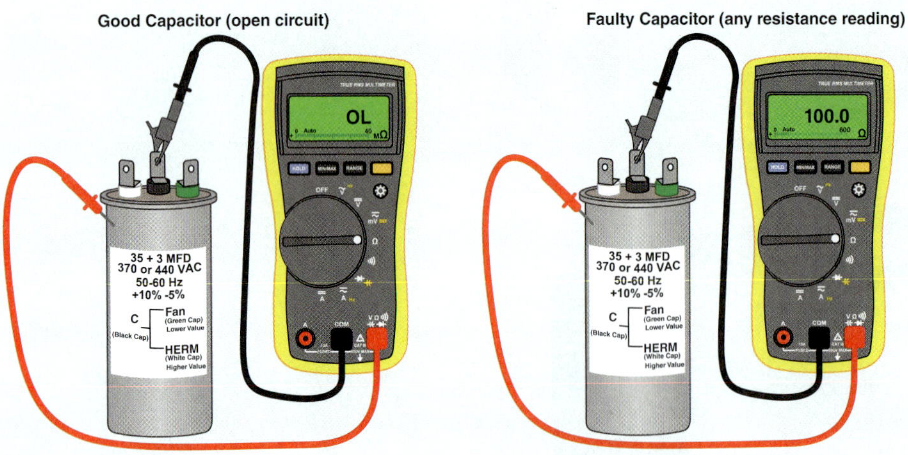

4. Now test the run capacitor to see whether it is shorted to the case. Remember that few start capacitors have a conductive case; this test is of no value if the case is non-conductive. With the meter still set to test continuity/resistance, touch one meter probe to a run capacitor terminal and the other to the outer case. Any reading of resistance other than infinity indicates that the capacitor is internally shorted and defective.

Continued

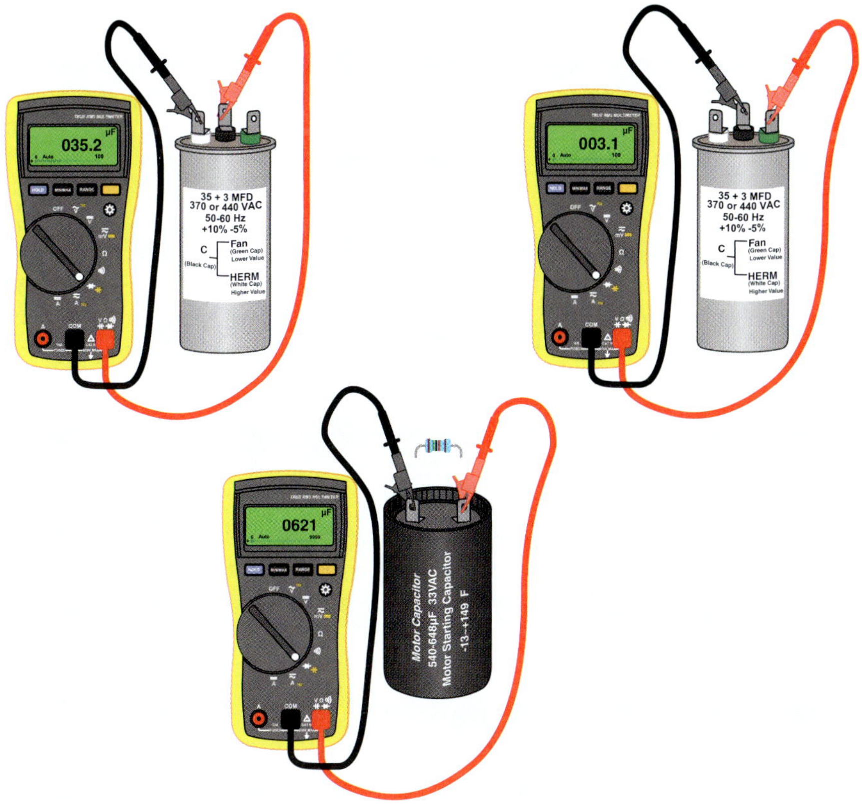

5. Test the capacitance of both capacitors and compare the values to the label rating. Set the DMM to measure capacitance and ensure that the two probes are connected to the correct meter jacks. Attach or touch the two meter probes to the start-capacitor terminals first, wait a few seconds, and then read the capacitance value being displayed. The reading should fall within the margin of error listed on the capacitor label, at least within ±20% of the value. If the capacitance reading falls outside of the allowable margin of error, the capacitor is defective. If there is no capacitance value displayed, the capacitor is open internally. If the earlier test for continuity gave the impression that the capacitor is open and yet a valid capacitance value is displayed, then it is not open after all. Repeat the procedure on the run capacitor. Remember that the standard margin of error for run capacitors, unless otherwise displayed on the label, is ±10%.

▶ Selecting and Installing Capacitors

K10004 Explain how to select and replace motor start and run capacitors.

S10002 Wire a start relay and start capacitor into an existing compressor circuit.

When testing reveals that a capacitor is defective, the technician needs to make sure that a suitable replacement is obtained and installed in the correct manner. In addition, when a motor is replaced, it is common practice to replace any associated capacitors as well. In some cases, the new motor may require a different size(s). But even when it does not, a new capacitor is still commonly installed.

As a general rule, a replacement capacitor with ratings that precisely match the original capacitor is chosen. The details about capacitance and voltage ratings can be found on the old capacitor label. However, it is important to remember that the capacitance rating must match the needs of the motor. Don't assume that an existing capacitor is the correct part; it is best to check the motor data plate to confirm this information. For some motors and compressors, the information may be available only from the manufacturer's documentation. It is not provided on the label or data plate of every motor or compressor encountered.

Never use a capacitor with a capacitance rating that is lower or higher than the motor requires. The same holds true for a capacitor with a lower voltage rating than is required. A capacitor with a higher voltage rating can be used since that rating simply identifies the peak voltage that the capacitor can withstand.

For start capacitors, a bleed resistor is recommended for HVACR applications. If a new capacitor is provided without a resistor, obtain one from a supply house: most supply houses sell resistors with factory-soldered female blade terminals. It is much easier to add than to solder an appropriate resistor to the terminals.

When installing a replacement capacitor, it is critical to make the proper wiring connections. The color coding of the wiring, if any, should be noted prior to removing the old one. This way, the same wiring connections can be made to the replacement capacitor. However, a technician needs to understand where each wire leads and why it is connected as it is. This is the professional way to eliminate uncertainty. Through continued training, technicians will learn to use a wiring diagram to guide them on every unit.

Two things to consider when replacing a capacitor are shape and size. Capacitors are sometimes mounted in a confined area, and they do come in different shapes and sizes. The majority of them are either oval or round. A readily available replacement with the same electrical characteristics may be a different shape than the original capacitor. As a result, the mounting-clamp style may have to be changed. When mounting a capacitor, avoid penetrating its case with a screw. Capacitors are not position-sensitive, unlike current relays that often must be mounted in an upright position to operate correctly.

▶ TECHNICIAN TIP

Identifying Compressor Terminals

You may notice that few compressor motors have their three electrical terminals identified. One terminal represents common, or C, which is the terminal that is common to both the run and start windings. The other two terminals are start (S) and run (R), and they represent the opposite ends of those two windings (**FIGURE 10-22**). An ohmmeter can help identify the terminals easily enough. Use the ohmmeter to make and record a resistance reading between each pair of the three terminals.

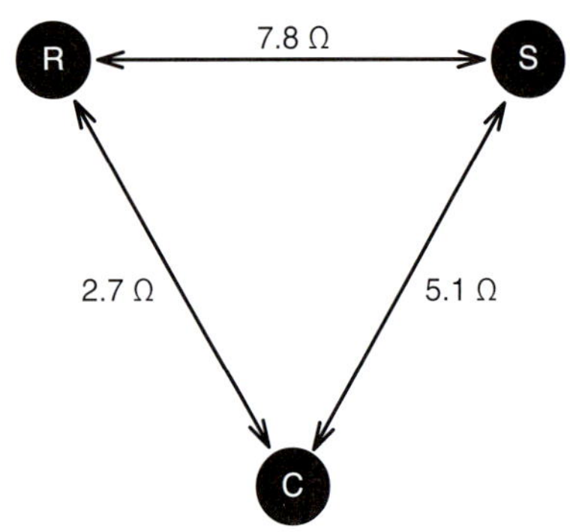

FIGURE 10-22 Identifying single-phase compressor terminals.

One reading should be very close to the sum of the other two. This represents the reading of both the start and run windings; therefore, the "odd man out" is the common terminal. The pair with the lowest reading represents the run winding, and the other, higher value represents the start winding. The figure here represents an example of what measurements and recordings might look like and the respective terminals.

To install a start relay and a start capacitor into an existing compressor circuit, follow the steps in **SKILL DRILL 10-2**.

SKILL DRILL 10-2 Installing a Start Relay and a Start Capacitor into an Existing Compressor Circuit

This exercise provides an opportunity to wire a start relay and start capacitor into an existing air conditioning or refrigeration compressor circuit.

SAFETY TIP

Working with electrically energized circuits is hazardous. Avoid contact with any surface that could potentially be energized. When working inside energized enclosures, maintain a stable position and, whenever possible, avoid placing both hands inside the enclosure. All work on energized equipment by unqualified individuals should be supervised by a qualified instructor or technician.

Continued

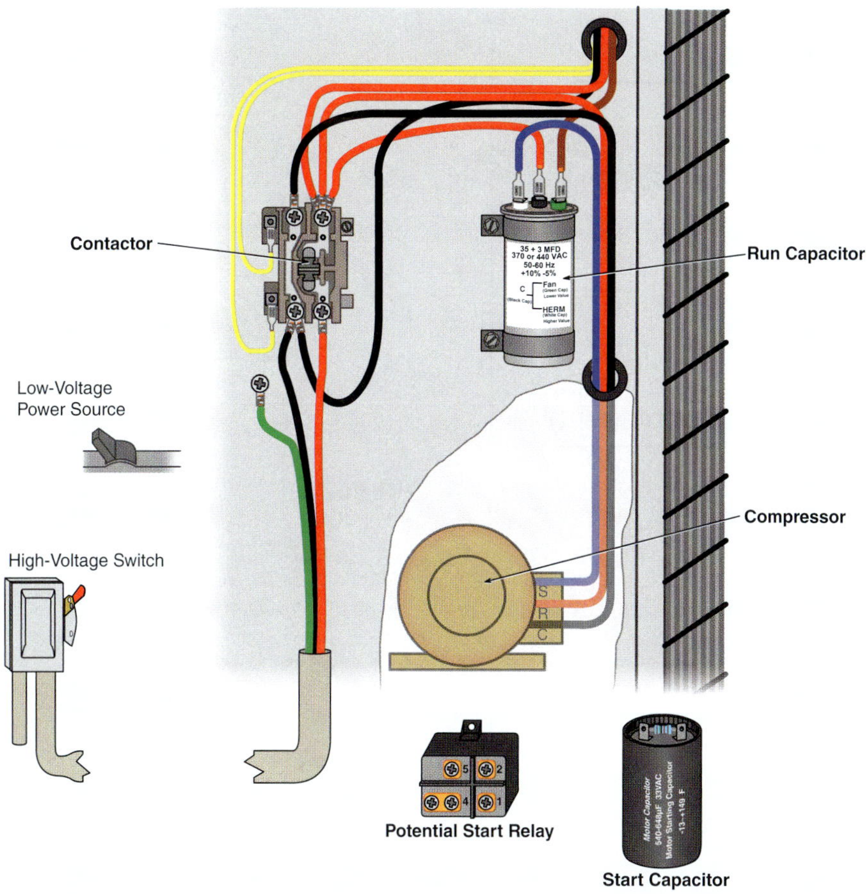

1. Assemble the necessary tools for connecting the components into the motor circuit. You will need an air conditioning unit with a single-phase compressor and an existing run capacitor in its circuit. In addition, the appropriate potential start relay and start capacitor for the compressor in use is required.

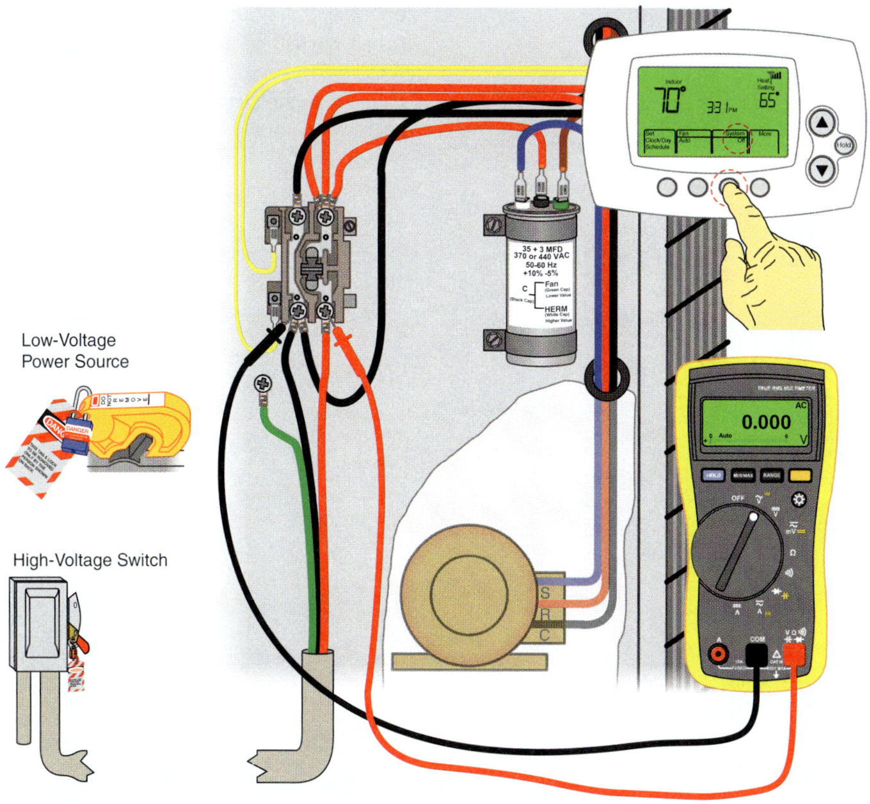

2. Ensure that both the low- and high-voltage power sources for the unit have been disabled. Lock out and tag the equipment disconnect(s) as required. Verify that the power is off, by using a multimeter.

Continued

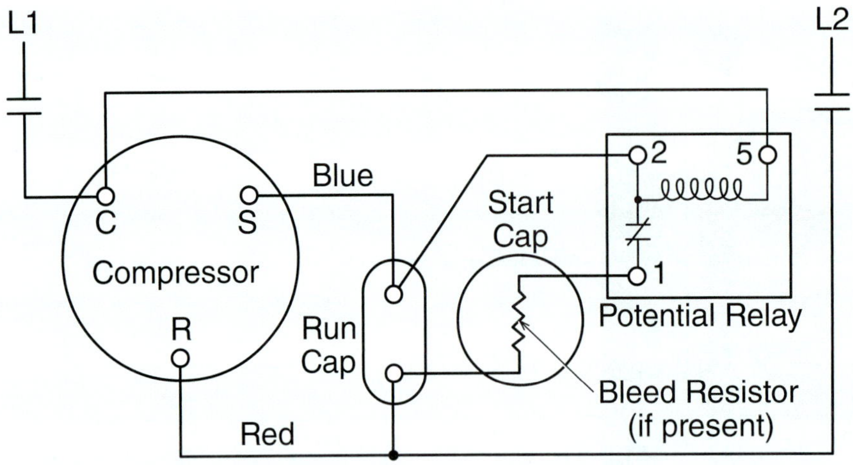

3. Refer to the circuit diagram shown here for guidance, or refer to the wiring diagram available for the specific condensing unit and compressor used for this task. Note that the wire colors may differ from those in the wiring diagram shown here. Attach each circuit wire needed to the proper terminal on the compressor, start relay, and capacitors.

Note that some new wiring and terminations are required to complete the task. Ensure that the AWG wire size used matches the wire sizes presently used in the unit or are of a larger size. The compressor terminals themselves need not be disturbed. Remember that they typically are a specific type of terminal that is not always readily available.

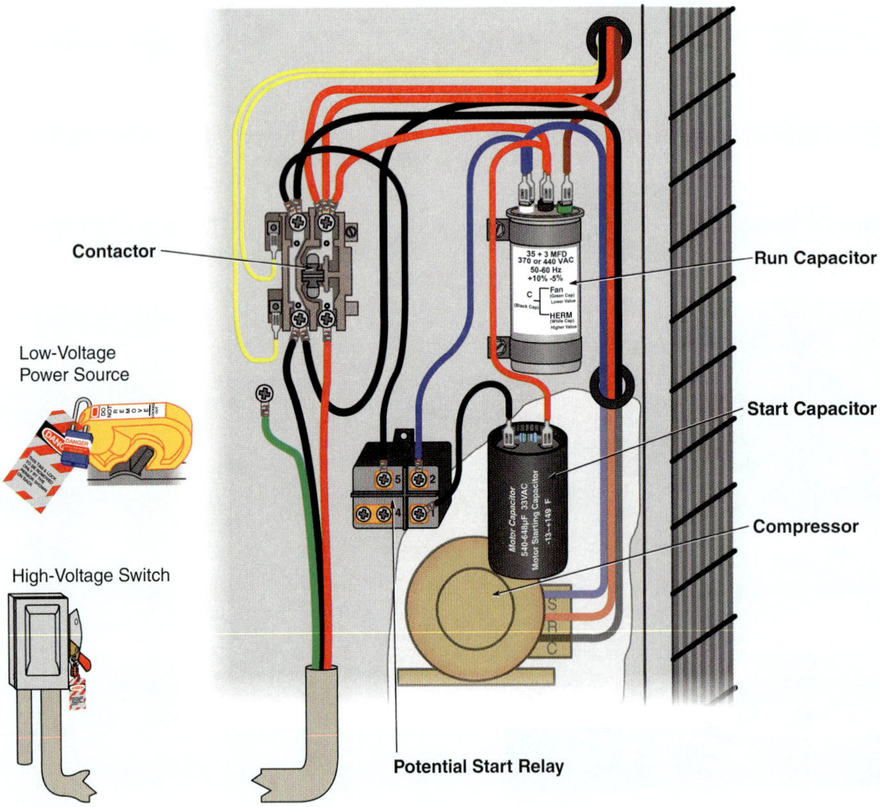

4. Recheck the completed work to ensure accuracy, and have the instructor also check the wiring before the circuit is energized. Make sure that no terminals or conductors are touching other metal components or nearby terminals. Also inspect the work area to ensure that no tools, materials, or scraps remain inside the unit cabinet or on top of it.

Continued

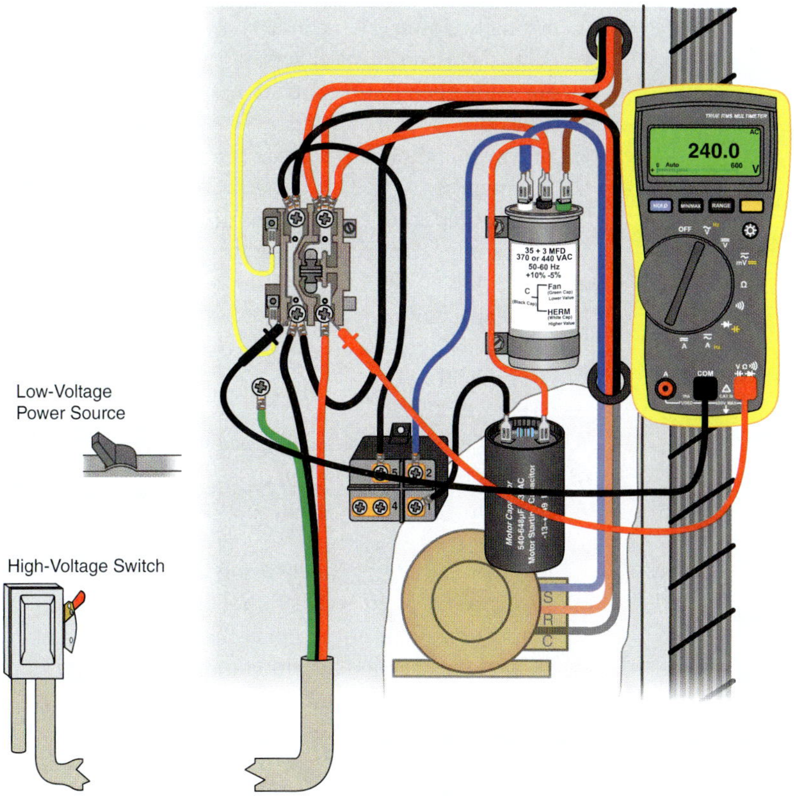

5. Remove any lockout/tag-out devices and restore power to the system. Verify that line voltage is now present inside the unit, by using a multimeter. With the assistance of an instructor, the unit can be restarted and tested for proper operation by using the compressor current being monitored in the process.

▶ Wrap-Up

Summary

Although capacitors have many uses, those related to HVACR equipment apply primarily to the operation of single-phase motors. Because they are common to so many HVACR systems, it is essential for technicians to have a thorough understanding of them and their related components. It is also beneficial to understand how a failed capacitor affects motor operation. A technician's responsibilities include using test equipment such as multimeters and capacitance testers, selecting proper replacement capacitors, and safely removing and replacing the components in motor electrical circuits.

Safety is always the first priority when working on or around any type of electrical equipment. All service technicians share a responsibility to work safely for their own protection and for the protection of their coworkers. Since capacitors can store an electrical charge even after a circuit has been deenergized, technicians must properly discharge capacitors before removing and handling them.

Technicians should always abide by the following basic rule when replacing capacitors: use replacement capacitors that have the characteristics required by the motor. The failed capacitor is usually the correct type, but mistakes do occur. It is best to refer to the motor manufacturer's data to confirm that the correct replacement capacitor is used. Whenever in doubt, access and acquire the proper information from the motor or equipment manufacturer.

Key Terms

capacitance A measure, in farads, of the electric charge that a capacitor can store.

capacitor An electrical device designed to store electrical energy in the form of an electric field.

centrifugal switch An electric switch that opens or closes due to the application of centrifugal force generated by a spinning motor or shaft.

coulomb An SI (International System) unit of electrical charge, equal to the power conveyed in one second by a current of 1 ampere. A capacitor with a capacitance of 1 farad can store 1 coulomb of charge on its plates when a potential of one volt is applied across its terminals.

counter-electromotive force (CEMF) An electromotive force, commonly called back voltage or back EMF, that is created in a motor and that opposes the electromotive force being applied by the circuit.

current relay A relay with a low-resistance coil and a set of normally open (NO) contacts that responds to the passage of current through its coil.

dielectric A non-conductive material positioned between the two conductive plates of a capacitor that store an electrostatic charge.

electrostatic charge An electric charge that develops on the surface of an insulated body, such as the metal plates inside a capacitor.

farad A standard unit of electrical capacitance.

hermetic compressor A compressor that is factory sealed, making its internal components inaccessible without destroying the external shell.

potential relay A relay with a coil and a set of normally closed (NC) contacts that responds to an increase of voltage applied to the coil.

Review Questions

1. The two conductive plates in a capacitor are separated by a non-conductive _____ material.
2. The amount of energy that a capacitor can store is referred to as its _____.
   a. resistance
   b. conductance
   c. capacitance
   d. back voltage
3. Most capacitors used in HVACR applications are rated in _____.
   a. millifarads
   b. microfarads
   c. nanofarads
   d. picofarads
4. What does the voltage rating of a capacitor indicate?
5. The ___________ of a capacitor must always match the requirements of the motor with which it is being used.
6. Unless otherwise specified on a run capacitor's data plate, its measured capacitance should be within _____.
   a. ±5% of the labeled value
   b. ±10% of the labeled value
   c. ±15% of the labeled value
   d. ±20% of the labeled value
7. The most common voltage range for capacitors used with single-phase motors in HVACR applications is _____.
   a. 115–220 VAC
   b. 370–440 VAC
   c. 480–600 VAC
   d. 600–1,000 VAC
8. A type of temperature-sensitive resistor that is sometimes used in place of a start capacitor to assist with motor starting is a(n) ____________.
9. Which motor capacitor type is designed with a high capacitance range and is used only briefly in a motor circuit?
10. Another name for a potential relay is a _____.
    a. current relay
    b. grounding relay
    c. transformer relay
    d. voltage relay
11. Inside a current relay is a coil and a set of _____.
    a. normally closed (NC) contacts
    b. parallel closed (PC) contacts
    c. normally open (NO) contacts
    d. series closed (SC) contacts
12. What type of motor start relay uses a set of normally closed (NC) contacts?
13. The type of capacitor that remains in a motor circuit during the operating cycle, maximizing its operating efficiency, is a(n) _____.
    a. start capacitor
    b. run capacitor
    c. NC capacitor
    d. PTC capacitor
14. A dual run capacitor has two capacitance values, the higher of which is used for the _____.
15. A common sign that a capacitor is failing, or has failed, is _____.
    a. a swollen or bulging outer casing
    b. the display of a red FAIL symbol on the casing
    c. a loss of magnetic attraction at the terminals
    d. the presence of condensation on the outer casing
16. Prior to disconnecting and testing a capacitor, it is necessary to first _____.
    a. use a condenser or battery to fully charge the capacitor
    b. temporarily connect all wiring to an alternate capacitor
    c. discharge the capacitor by using a suitable bleed-resistor tool
    d. permanently attach a resistor across its terminals
17. The common range of resistance for bleed resistors permanently installed on start capacitors is _____.
    a. 10–20 Ω
    b. 100–200 Ω
    c. 1,000–2,000 (1K–2K) Ω
    d. 10,000–20,000 (10K–20K) Ω
18. After a capacitor has been tested for continuity/resistance with a multimeter and a test needs to be repeated, what two options are there to receive a valid result?
19. When testing reveals that a run capacitor has failed, the best thing to do is replace the capacitor with a(n) _____.
    a. start capacitor that has less capacitance
    b. run capacitor with a lower voltage rating
    c. exact duplicate of the failed one
    d. type that matches the specified needs of the motor
20. Before removing and replacing any capacitor, a service technician should always _____.
    a. remove any lockout/tag-out devices
    b. make note of all the wiring connections
    c. ensure that the capacitor is charged
    d. attach a coulomb filter to the capacitor

CHAPTER 11

Solid-State Devices

Knowledge Objectives

After reading this chapter, you will be able to:

- **K11001** Describe solid-state components.
- **K11002** Describe the design and purpose of a printed circuit board.
- **K11003** Identify and describe specific components of a printed circuit board.
- **K11004** Describe the operation of printed circuit boards commonly found in HVACR equipment.
- **K11005** Describe basic troubleshooting and replacement procedures for HVACR printed circuit boards.

Skill Objectives

Skills required to meet the objectives of this chapter:

- **S11001** Troubleshoot a solid-state control device.
- **S11002** Replace a solid-state control device.

▶ Introduction

The control circuits for many HVACR systems and subsystems manufactured today rely on **solid-state components**. These components provide a level of control and logic-based operation that is not available from conventional electromechanical components. Examples of the uses of solid-state components in HVACR systems can be found in defrost controls, variable-speed motors, variable refrigerant flow (VRF) systems, electronically commutated motors (ECMs), and integrated building-management systems, to name just a few.

Many solid-state devices are mounted together on open printed-circuit (PC) boards or are encased in sealed modules where the devices themselves are not visible to the technician. HVACR technicians rarely need to troubleshoot at the individual (discrete) electronic component level. They are far more likely to isolate a control-circuit problem to a PC board or electronic module and replace it with a new one. Board-level repairs in the field are discouraged by manufacturers for a variety of reasons. However, it is helpful to understand the various solid-state components, the functions they perform, and how to troubleshoot systems containing these components.

▶ What Is a Solid-State Component?

K11001 Describe solid-state components.

The term "solid-state" refers to electronic components that are made from **semiconductor** material. Silicon, germanium, and selenium are among the many materials used as semiconductors.

Most substances can be classified as conductors or insulators. Semiconductors, on the other hand, are neither. They have a neutral electrical charge. They are solid materials in which current flow is controlled by varying the voltage applied to the semiconductor material. Semiconductors are made positive or negative by a process called **doping**. By adding certain substances to the semiconductor material, it can be changed from a neutral charge to a positive charge, or P-type semiconductor. Negative, or N-type, semiconductors are created in the same way, but different doping substances are used to produce positively and negatively charged materials.

Integrated Circuits and Microprocessors

Integrated circuit (IC) chips are mass-produced sealed units that can contain thousands of microminiature components, such as **transistors**, capacitors, **diodes**, and resistors. IC chips are designed to perform specific functions. They can serve as counters, computer memory chips, and **amplifiers**, as well as other functions. **FIGURE 11-1** shows examples of IC chips.

A **microprocessor** (**FIGURE 11-2**) is a type of user-programmable IC designed to act as an data processor. Microprocessors serve as the central processing unit (CPU) for

FIGURE 11-1 Examples of IC chips.

FIGURE 11-2 A microprocessor chip on a printed circuit (PC) board.

computers. These so-called chips can be programmed to perform specific tasks, such as mathematical calculations. In HVACR systems, microprocessors collect information and determine, based on their programming, how to manipulate controlled devices. Microprocessors used in HVACR systems help improve functionality and efficiency by analyzing information and adjusting system operation to effectively adapt to changing environmental conditions. For example, in a system that uses some volume of outdoor air during cooling, heating, and ventilation modes, sensors send temperature and humidity information to a microprocessor. It then determines when it is suitable to use outdoor air and what the mixture ratio to indoor air should be. The microprocessor, in turn, sends signals to a damper actuator to assume a position that results in the correct amount of air.

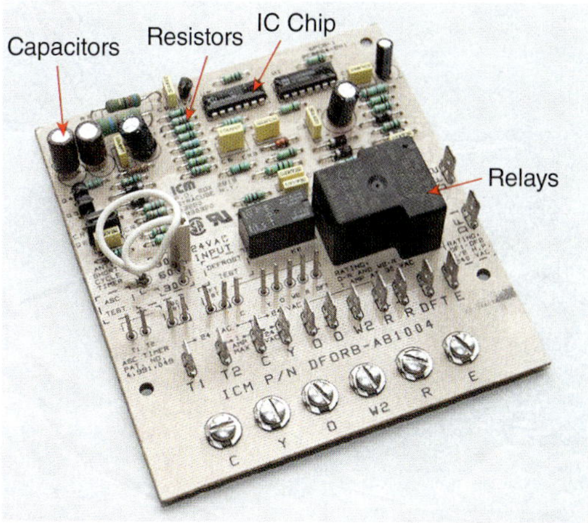

FIGURE 11-3 An HVAC defrost-control board.

▶ Printed Circuit Boards

PC boards used in modern HVACR systems take advantage of the versatility of solid-state devices, ICs, and microprocessors and provide users with maximum comfort and efficiency. As seen in **FIGURE 11-3**, a PC board contains components such as relays, IC chips, capacitors, fuses, and resistors. As mentioned previously, a PC board might be packaged in a sealed module, like the furnace ignition control shown in **FIGURE 11-4**. Regardless of whether the control board components are visible or encapsulated, the PC board or module is replaced as a unit if it fails testing.

K11002 Describe the design and purpose of a printed circuit board.

A typical PC board has components mounted on one side and printed circuits on the other side, the latter of which is referred to as the foil side (**FIGURE 11-5**). During the board manufacturing process, the conductive pathways are automatically printed on the foil side of the board by a machine. The terminals of components to be mounted on the top of the board are inserted into the holes in the board and soldered to the copper foil with great precision by automated equipment.

Some PC boards are electrically connected to a system by the use of an edge connector (**FIGURE 11-6**), which is inserted into a female connector that has been attached to the equipment. Both the male edge connector and female connector are shown in Figure 11-6. Other PC boards use screw terminals to make the wiring connections (**FIGURE 11-7**), and yet other use blade terminals.

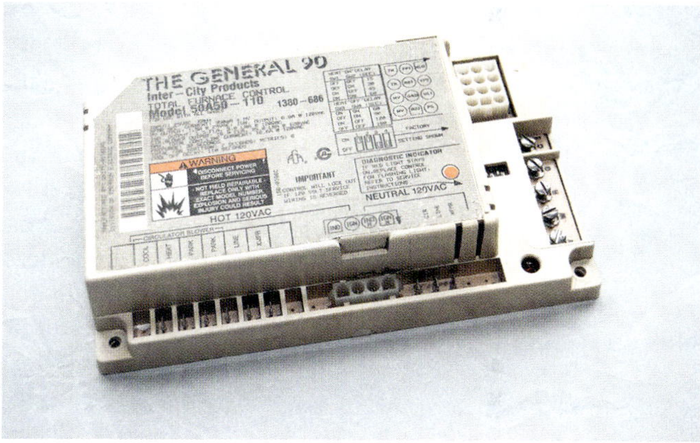

FIGURE 11-4 A furnace ignition-control module.

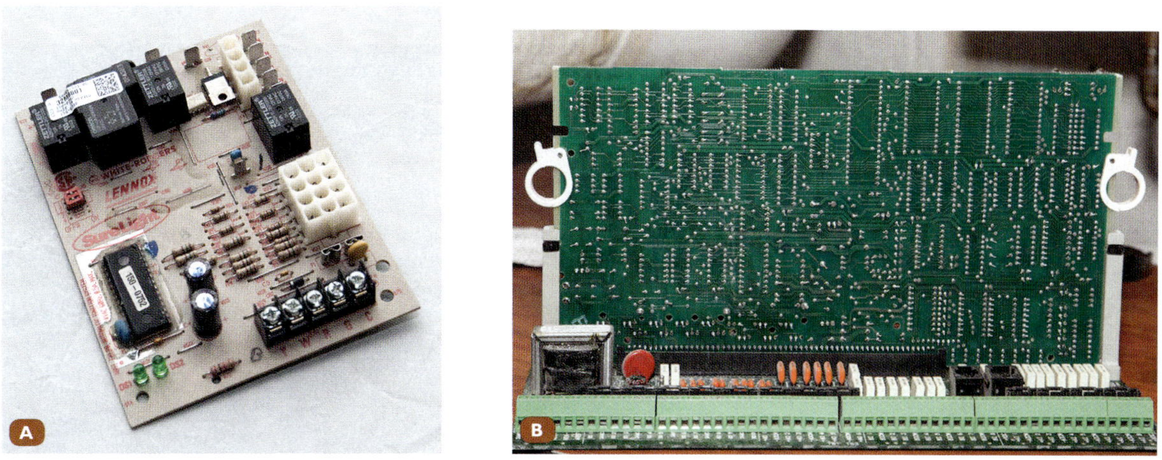

FIGURE 11-5 The component side and foil side of a PC board. **A.** Component side. **B.** Foil side.

FIGURE 11-6 Male and female edge connectors. **A.** Male edge connector. **B.** Female edge connector examples.

▶ Solid-State Circuit Board Components

K11003 Identify and describe specific components of a printed circuit board.

PC boards are designed to perform specific functions in one or more types of equipment. Each PC board must contain the specific components necessary to perform the needed functions.

FIGURE 11-7 Screw terminals on a PC board.

Fusing

Solid-state components such as relays, IC chips, and microprocessors are very delicate and can be quickly damaged by excessive current. There are also solid-state fuses, as shown in **FIGURE 11-8**. Compared to conventional fuses, these possess a super-fast interrupting capability. These fuses are sometimes included on PC boards to protect sensitive components from current surges. Since they are soldered to the board, they are not considered field-replaceable.

In addition to solid-state fuses, technicians might also find conventional fuses used on PC boards. **FIGURE 11-9** shows an example of what is known as a blade fuse. Glass cartridge fuses might also be found on PC boards. Both require an appropriate fuse holder to be mounted on the board.

Blade and cartridge fuses that are not soldered to the board are field-replaceable units. If a board is receiving power but is not responding in any way, carefully remove and test the fuse with an ohmmeter. If there is no continuity through it (open), be sure to replace it with a fuse that has matching size, current, and voltage characteristics.

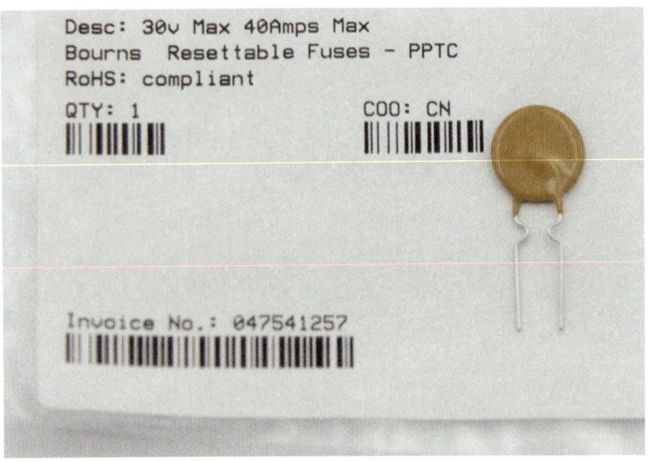

FIGURE 11-8 Solid-state fuse.

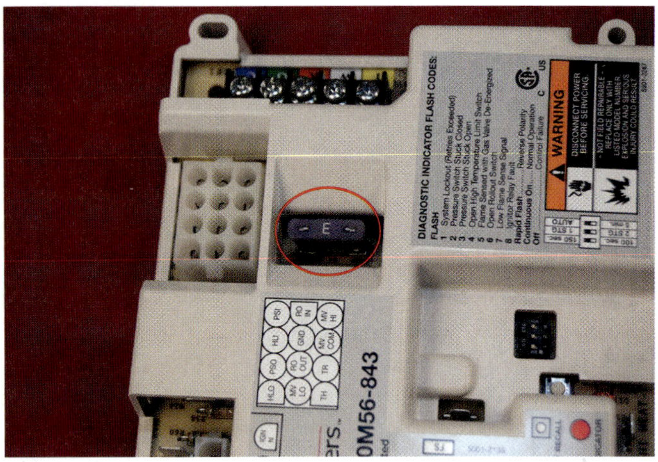

FIGURE 11-9 Example of conventional blade fuse on a PC board.

If the fuse opens again when power is applied, there is obviously a problem. The problem may lie in the board itself. However, these fuses are often protecting an alternating current (AC) load circuit that passes through a board-mounted relay. If that load is being energized when power is applied to the board itself, then the load or the interconnecting wiring are more likely sources of the problem, rather than the board.

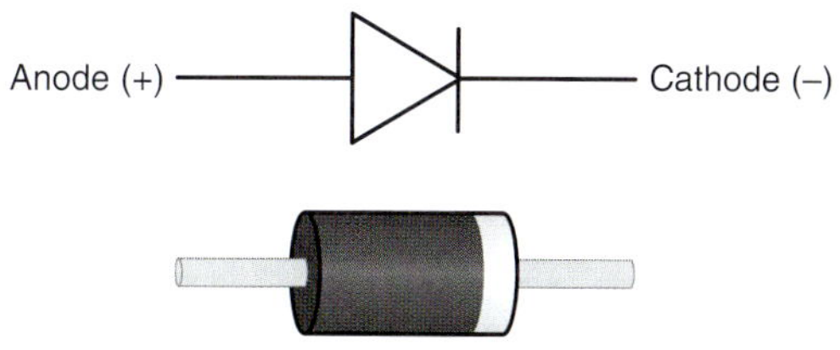

FIGURE 11-10 Diode and Diode symbol.

Diodes

Conventional controls such as relays and contactors require AC power, but solid-state circuits commonly require low-level direct current (DC) power, such as 5 VDC. To provide a DC power source, **rectifier** circuits are incorporated into the board. A rectifier converts AC to DC by using diodes.

A diode is the most basic of solid-state devices (**FIGURE 11-10**). It consists of one piece of P-type material, called the anode, and one piece of N-type material, called the cathode. The surface where they are joined is referred to as a PN junction. Because of this construction, current will flow in only one direction through the device. If an AC voltage is applied to the anode of the diode, it will conduct only on the positive half-cycle of the sine wave, producing a fluctuating DC voltage with gaps that represent the period of the negative half-cycle of the sine wave. A diode with a positive voltage applied to its anode is said to be forward biased. With a negative voltage applied to the anode, the diode is reverse biased. A rectifier with a single diode is known as a half-wave rectifier. Note that when positive or negative voltage is referred to here, it relates to the polarity on the other side of the device.

As **FIGURE 11-11** shows, the DC voltage produced by a single diode in a half-wave rectifier is of poor quality because it skips every other half-cycle of the AC voltage and has a great deal of residual AC **ripple**. Using two diodes in a full-wave rectifier circuit produces a cleaner DC voltage because one of the diodes conducts on each half-cycle of the AC waveform. Because the AC transformer is center tapped, a positive voltage is applied to the anode of one of the diodes on each half-cycle. This arrangement produces an uninterrupted DC voltage. To produce the smooth DC voltage of the quality needed for solid-state circuits, rectifiers commonly use four diodes in what is known as a bridge rectifier arrangement (**FIGURE 11-12**). A capacitor is added to the circuit to filter out the

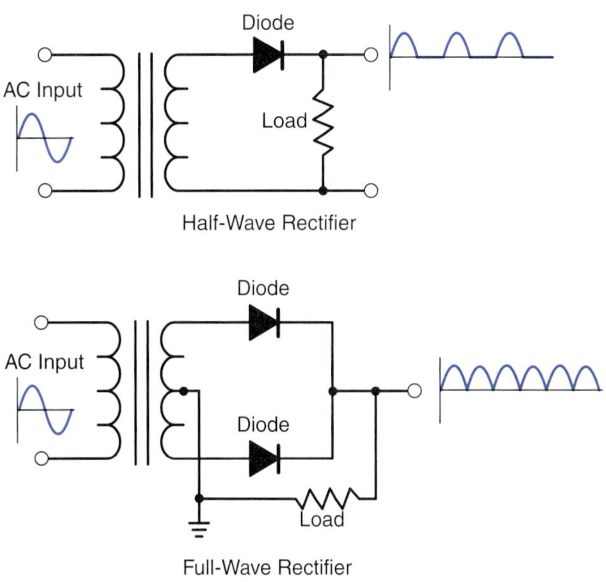

FIGURE 11-11 Half-wave and full-wave rectifiers.

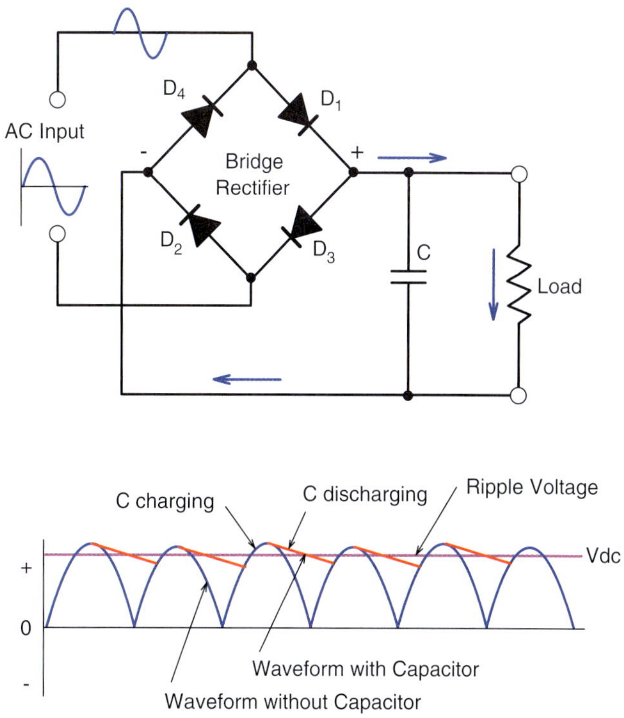

FIGURE 11-12 Bridge rectifier arrangement.

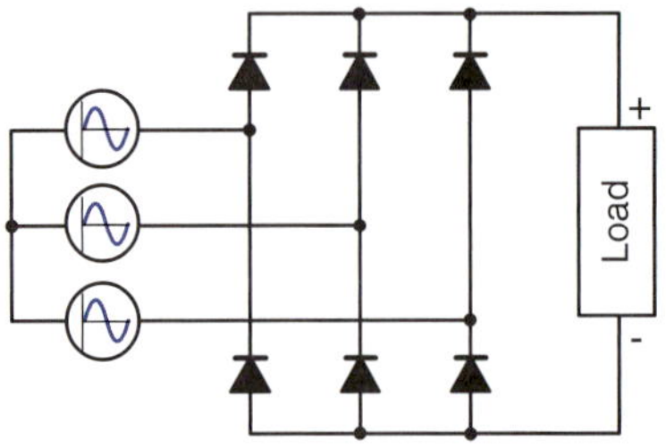

FIGURE 11-13 Three-phase rectifier circuit.

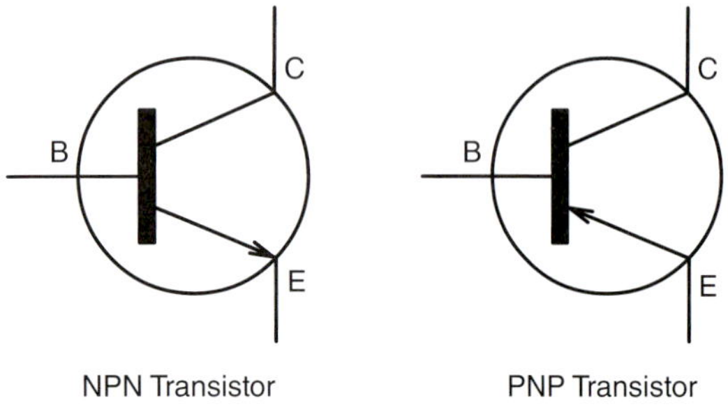

FIGURE 11-14 Transistor symbols.

ripple, producing a much smoother and more exact DC voltage. A rectifier containing six diodes is used to rectify three-phase voltages, as shown in **FIGURE 11-13**.

Transistors

A transistor consists of three layers of semiconductor material and has two PN junctions. Transistors have two basic configurations. An NPN transistor is made from two layers of N-type material with one layer of P-type material sandwiched between them. A PNP transistor has the opposite construction. The semiconductor layer in the center of the transistor is called the base, identified by B in **FIGURE 11-14**. The other layers are the collector (C) and emitter (E). Notice the direction of the arrows in the figure. These arrows indicate the direction of current flow. In the PNP type of transistor, for example, the current flows from the emitter to the base and then to the collector. The voltage applied to the base determines how much current will flow through the transistor.

A transistor can function as an amplifier or switch. At one time, it was common to find individual transistors in electronic circuits. Today, they are part of the microminiature world; they are embedded in ICs and microprocessors. There may be thousands of them in a single IC chip.

Diacs and Triacs

A **diac** is a type of diode that conducts current only when a voltage level, known as the **breakover voltage**, is exceeded. It is used as a solid-state switching device. The term "diac" is an abbreviation for "diode alternating current." **FIGURE 11-16** shows the schematic symbol for a diac. Once a diac has begun to conduct, it will continue to do so until the voltage falls below a certain level, known as the **holdover voltage**.

A **triac** (an abbreviation of "triode alternating current"), the symbol for which is also shown in Figure 11-16, is basically a diac with a gate electrode that is used to initiate current

▶ CORE CONNECTION

Vacuum Tubes

During the early days of TV in the 1950s and 1960s, plug-in vacuum tubes (**FIGURE 11-15**) were used to serve as amplifiers, rectifiers, frequency generators, and switching devices. If there was a problem with the TV, owners could turn off the TV, pull the vacuum tubes out, and take them to the local drugstore or hardware store, which had vacuum-tube testers and replacement tubes. Vacuum tubes were later replaced by discrete transistors and subsequently by microminiature solid-state circuits mounted on PC boards.

FIGURE 11-15 Amplifier vacuum tubes.

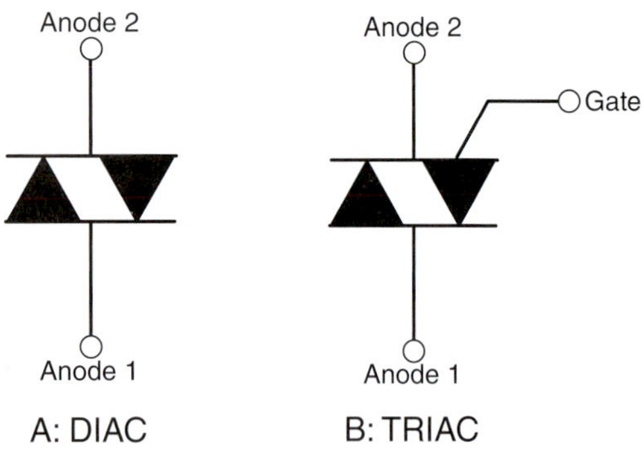

FIGURE 11-16 Diac and triac symbols.

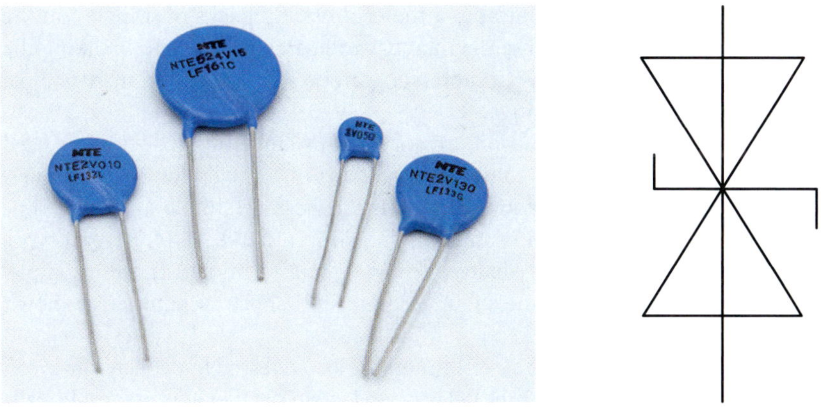

FIGURE 11-17 A selection of metal oxide varistors (MOVs).

flow. Once gated on, the triac will continue to conduct until the current falls below a certain level. Unlike the diac, the triac does not require a breakover voltage to be reached before it will conduct. Triacs are used in speed controls for variable-speed fan motors.

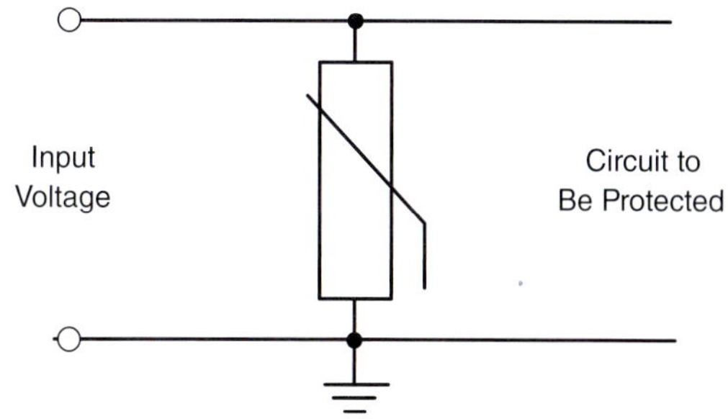

FIGURE 11-18 A protective varistor circuit.

Varistors

The term "**varistor**" is short for variable resistor. Varistors (**FIGURE 11-17**) are solid-state variable resistors in which the resistance level varies in relation to the applied voltage. Varistors are used to protect solid-state circuitry from **electrostatic discharge (ESD)**, a topic that will be covered later in this chapter. ESD is like a miniature lightning strike. It most commonly occurs when a person who has built up an electrical charge touches the terminals of an IC chip or PC board. The voltage it produces can cause solid-state components to fail. The varistor reacts instantly to counteract such a discharge. When such a discharge occurs, the resistance of the varistor is instantly reduced, causing the surge voltage to be directed to ground instead of to the sensitive semiconductor circuit (**FIGURE 11-18**).

Thermistors are also variable-resistance devices, but their resistance value varies in relation to temperature rather than voltage. Thermistors are widely used in HVACR systems to sense temperature. Although some are mounted on PC boards, they are more often mounted where a specific temperature needs to be known, such as on the tube of a condenser coil. The resistance is translated into temperature values by microprocessors.

▶ Operation of Printed-Circuit Boards

K11004 Describe the operation of printed circuit boards commonly found in HVACR equipment.

Typical PC boards found in HVACR equipment include but are certainly not limited to compressor short-cycle protection, heat-pump defrost-control boards, and furnace ignition-control modules. They contain devices and circuits needed to monitor system operation and perform specific tasks. Some examples of how these boards and modules interact with system components to perform their control functions follow.

Compressor Anti-Short-Cycle Timer Operation

Hermetic and semi-hermetic compressors are driven by electric motors that operate in unique conditions, buried inside an enclosure that includes oil and refrigerant. These motors require significant starting torque, even under ideal conditions. If there is an attempt to start the compressor soon after it has shut down, the opposing refrigerant pressures (high on one side of the valves and low on the other side) will work against the motor, making it difficult to start and possibly causing winding damage. This can occur when there is a momentary power outage or perhaps when there is a thermostat-setting battle going on in a home. A user may move the thermostat setting up and down several times in a minute, during the course of such a battle. In the days when thermostats with mercury

FIGURE 11-19 Compressor anti-short-cycle timer.

bulbs were common, a roofer dropping packs of shingles on the roof could cause the mercury to jiggle, making and breaking the circuit rapidly. Compressors can be short-cycled in many simple and obscure ways.

To prevent compressor short-cycling, timing devices are used to ensure the compressor remains off, usually for 3–10 minutes, once it shuts down. Most of the devices are simple solid-state timing relays. An example is shown in **FIGURE 11-19**. Several types are available, usually with two to four terminals. In this example, we will examine a four-terminal anti-short-cycle timer, as shown in the figure.

Anti-short-cycle timers are also referred to as delay-on-break relays. Starting the delay period when the thermostat initially calls for cooling (delay-on-make) would mean that the system cannot start for several minutes. Instead, the delay period needs to begin when the thermostat satisfies, the control circuit opens, and the compressor stops—hence the term "delay-on-break." Then it should not interfere with normal system operation unless something unusual occurs (thermostat jiggle or momentary power outage).

> **▶ TECHNICIAN TIP**
>
> **Duplicating Short-Cycle Protection**
>
> Today's solid-state thermostats are often equipped with a setting to prevent compressor short-cycling. The setting is usually found in the installer menus, and it can be enabled or disabled. Some thermostats allow the length of the delay period to be adjusted as well. Before considering or recommending the installation of an anti-short-cycle timer, check to see whether the thermostat already has this feature. Using both devices for protection may not always create a problem, since their timing periods will usually coincide. But it is unnecessary to use both in the vast majority of installations.

A schematic for timer wiring is shown in **FIGURE 11-20**. Although there are four terminals, the two center terminals are one and the same electrically. The Common wire from the control transformer connects to one of the blade terminals. The second Common terminal is used to transfer power from Common to one terminal of the compressor contactor coil, as shown. The second leg of power from the control transformer, referred to as R on most systems, passes through the thermostat, leaves on the Y terminal, and is then connected to Terminal 3 of the timer. Terminal 1 is used to transfer power from Y to the other terminal of the contactor coil, providing a close circuit at the end of the timing period. When both the timer and thermostat circuits are closed, the coil has 24 VAC applied to it.

Note that this is just one example of how such a timer is wired. There are a number of variations; for each device, follow the diagrams provided by the manufacturer.

FIGURE 11-21 shows the internal schematic of the timer. Assume that the timer has been sitting without power to Terminal 3 for more than five minutes. When it is initially

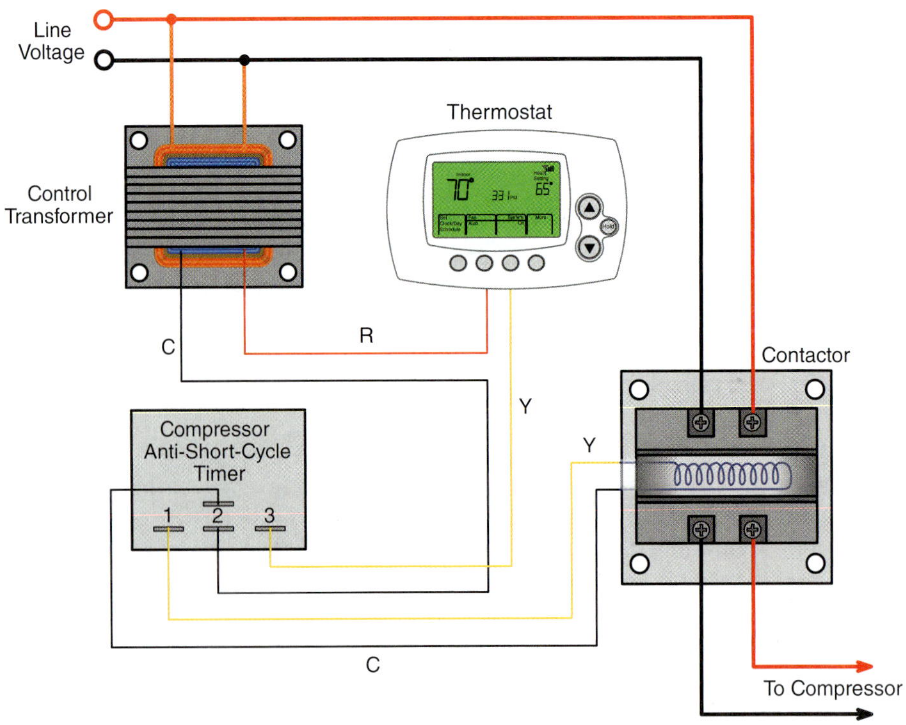

FIGURE 11-20 Wiring for a four-terminal anti-short-cycle timer.

energized at Terminal 3 by the thermostat, the logic chip allows the internal contact to close, energizing the compressor contactor coil. Once power is lost to Terminal 3 (thermostat opens), the chip commands the switch contacts to open. They will not reclose again until the time-delay period has elapsed. After that, it is ready for the next cycle.

Furnace Ignition-Control Module Operation

A gas furnace goes through a specific sequence of events necessary to ensure safe operation. The solid-state components of an ignition-control module (**FIGURE 11-23**) control and monitor the sequence and make sure each critical event occurs before the next one is allowed. An example connection diagram is shown in **FIGURE 11-24**. The following is a typical sequence of operation for this type of solid-state control. Although this information likely precedes a thorough presentation on the operation of furnaces, it will provide an introduction to ignition systems and act as an example of how solid-state devices are applied to furnace control.

1. The board is activated when the thermostat calls for heat through the TH terminal. On receiving the call for heat, the control module initiates a pre-purge period that allows any unburned gas or combustion byproducts in the heat exchanger to dissipate.
2. Following the pre-purge period, the control energizes the hot-surface ignitor (HSI) for a warm-up period. This module allows a warm-up period of either 17 or 45 seconds.
3. Following the warm-up period, the gas valve is opened for while the HSI remains energized. If the gas does not ignite during a relatively trial-for-ignition period, the gas valve is deenergized. At this time, the control will follow optional programming: it will either lock out the ignition system until the control power is cycled off and back on or restart the sequence and try again, up to two attempts. If the gas does ignite, burner operation will continue.
4. In the diagram shown here, the HSI also contains a flame sensor. The sensor informs the module through electrical current production (or the lack of it) if the gas has ignited and a flame is present. On other systems, the terminal FP (flame-proof) is used for a stand-alone flame sensor. It is essential to safety that the presence of a flame to consume the fuel gas is proven before the gas valve can remain open.
5. Limit switches in the thermostat control circuit, or the thermostat itself, can interrupt the operation of the module and, in turn, the burner at any time.

There are a number of possible variations in the sequence of operation for various ignition controls. Some furnace controls also manage and control the operation of the induced-draft fan, and safety controls or limit switches may be connected directly to the module instead of being wired in series with the temperature controls. This particular control requires having external control over the induced-draft fan.

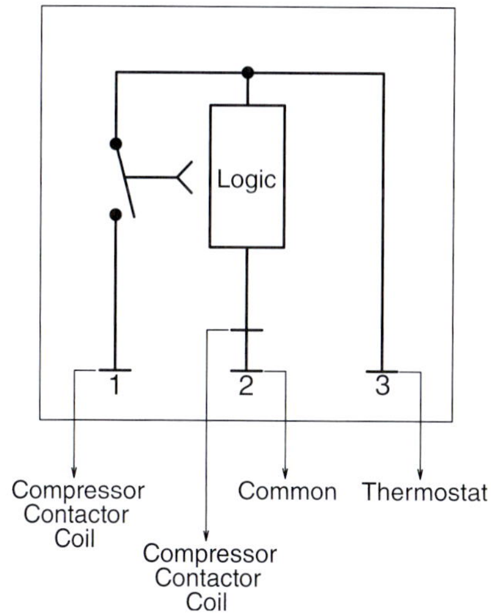

FIGURE 11-21 Example of an internal design of a four-wire anti-short-cycle timer.

> ▶ **TECHNICIAN TIP**
>
> **Anti-Short-Cycle Timers**
>
> Anti-short-cycle timers are available with fixed or adjustable timing. Most fixed timers have a period of three minutes, whereas adjustable models allow shorter and longer periods. Some systems may require the longer delay to allow time for the high and low pressures on each side of the compressor to equalize. Note that there are also two-wire versions of these devices, which offer a simpler installation. A two-wire adjustable model is shown in **FIGURE 11-22**. Timers that begin the delay period when the thermostat initially calls for cooling (delay-on-make) are also available and have the same basic appearance. These might be used, for example, to delay the start of a second or third compressor to reduce the inrush current to the unit.

FIGURE 11-22 Adjustable two-wire anti-short-cycle timer.

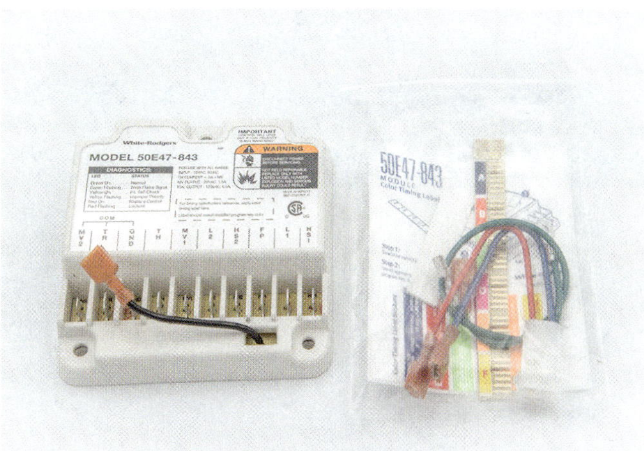

FIGURE 11-23 A furnace ignition-control module.

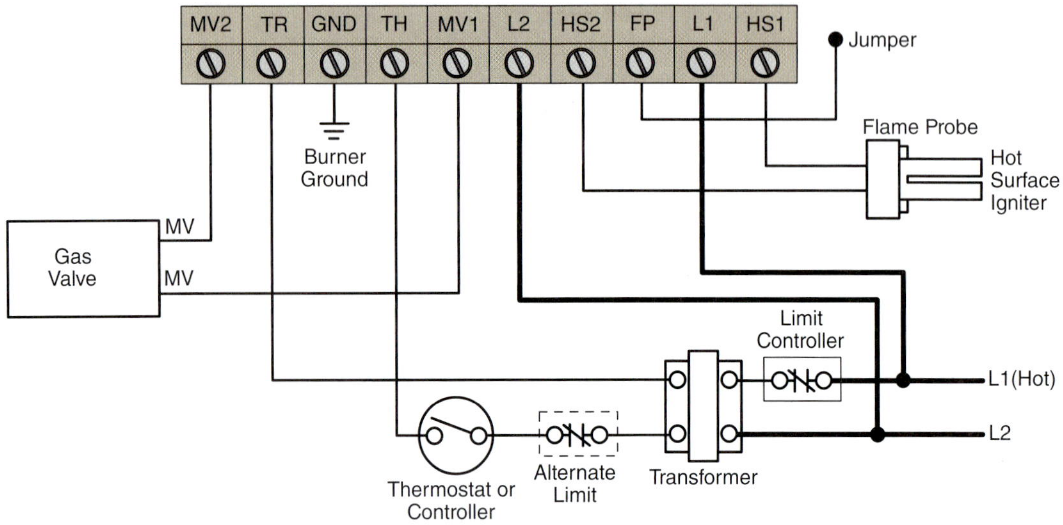

FIGURE 11-24 Connection diagram for an ignition-control module.

Since the focus of this chapter is on solid-state devices and PC boards, it is important to recognize several important factors. One is that the described sequence of operation above can largely be reproduced without the use of a module loaded with solid-state devices. An array of electromechanical devices could be employed to do much the same thing. However, the devices needed, and the wiring involved, take up a great deal of space, cost significantly more, and have a far greater potential for failure. Solid-state controls enable complex control operations to be done economically, reliably, and without requiring as much space. Another factor is that solid-state devices, like ignition-control modules, can provide information about the status of the control and inform a technician about what type of problem has been encountered.

Troubleshooting a problem in a unit containing a PC board often begins with evaluating the operation of light-emitting diodes (LEDs) on the board. Although this feature will be discussed further in the next section, consider how this type of information could be effectively conveyed without using solid-state devices. Keep in mind that the LEDs are also solid-state devices, functioning with far less power and heat production while also being able to illuminate and darken faster than standard incandescent lamps.

Note in Figure 11-23 the lettered and color-coded items to the left of the control. These are referred to as program keys. One of the six keys must be inserted into a mating socket provided on the module. Rather than requiring a technician to interface with the module by using a tablet or laptop computer or by providing switches or buttons on the module, the keys are used to set the desired parameters for the furnace. **FIGURE 11-25** shows the programming offered by each key when inserted.

Program Key Timing Specifications Quick Reference

Program Key (Color)	Timing and Retry			
	Trial for Ignition	Retries	Prepurge	Interpurge Ignitor (Warmup)
A (blue)	4 Sec.	0	30 Sec.	45 Sec.
B (red)	4 Sec.	2	30 Sec.	45 Sec.
C (green)	7 Sec.	0	30 Sec.	45 Sec.
D (violet)	7 Sec.	2	30 Sec.	45 Sec.
E (orange)	4 Sec.	2	30 Sec.	17 Sec.
F (yellow)	7 Sec.	2	30 Sec.	17 Sec.

Note: Program keys are lettered and color coded.

FIGURE 11-25 Ignition-control module program keys.

These advantages and the flexibility they offer illustrate the degree to which solid-state devices have changed how HVACR systems are applied and controlled. The development and application of solid-state devices has revolutionized the HVACR industry on many fronts.

▶ Solid-State Component Troubleshooting and Replacement

Although a PC board may contain many components, the HVACR technician is not expected to isolate and replace failed components on the board itself. Instead, the entire board or sealed module is replaced. Indeed, many manufacturers place statements such as the following on the documentation for PC boards and similar parts: "No user serviceable components—replace the control as a unit only." One reason is liability, especially when it comes to ignition-control devices and similar controls. If a technician elects to make a board-level repair in the field and a serious incident happens as a result, the manufacturer has eliminated or limited any liability through the added statement. In general, it is a poor decision to undertake a board-level repair in the field.

K11005 Describe basic troubleshooting and replacement procedures for HVACR printed circuit boards.

S11001 Troubleshoot a solid-state control device.

S11002 Replace a solid-state control device.

Electrostatic Discharge

It was mentioned earlier in the chapter that semiconductor components such as IC chips and microprocessors can be damaged by ESD. A common source of ESD is static charges that build up in the human body and are then released when a grounded object is touched. ESD charges are most likely to build up in low-humidity environments and are produced by walking across a carpet or vinyl floor, handling plastic bags, or handling polystyrene foam. This is especially true in winter, when the air tends to be dry. Humid air tends to suppress static buildup.

Thousands of volts can be produced in this way. Even a minor ESD release can damage sensitive electronic components. For that reason, special precautions must be taken when handling a PC board. First, avoid touching the pins of any IC chip or microprocessor. Touching the pins of an edge connector or the terminals on a terminal strip can also cause damage to components that are directly connected to these points. If it is necessary to handle a PC board, a grounding wrist strap (**FIGURE 11-26**) should be worn. At a minimum, touch a grounded surface to discharge any buildup of static electricity before touching a PC board, each and every time. New PC boards or modules are often stored in metallized (conductive) bags made specifically for their protection (**FIGURE 11-27**).

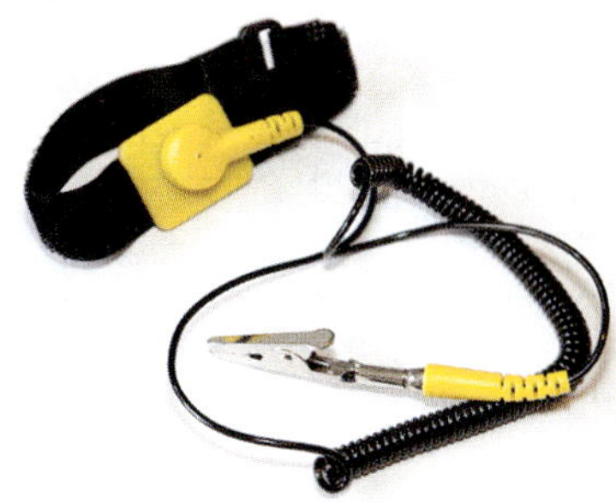

FIGURE 11-26 Grounding wrist strap.

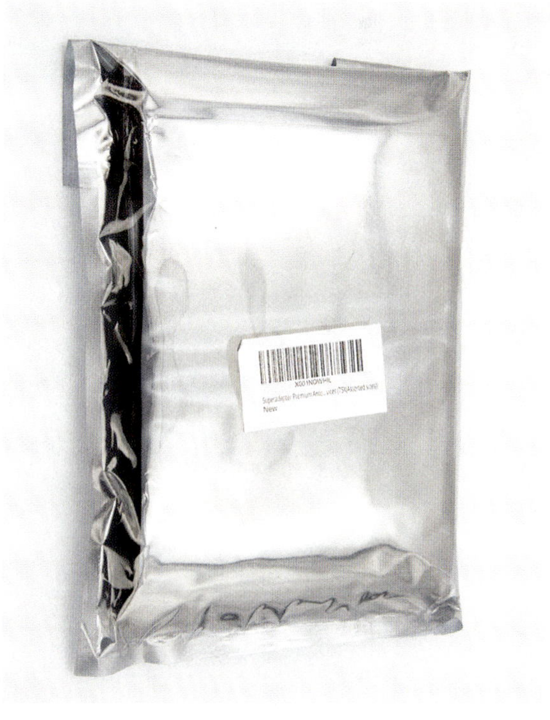

FIGURE 11-27 PC-board storage bag.

Troubleshooting Solid-State Controls

The goal of troubleshooting these devices is to determine whether the complete board assembly is functioning as designed and configured. There are many cases of PC boards being replaced that are not defective. Sometimes a technician is unable to identify another problem, and the board represents something of a mystery. Components that are misunderstood often become the focal point when no other source of a problem can be readily identified. That mysterious device must be the problem, right?

Keep in mind that control boards and modules are designed to interact with other control devices. PC boards and modules typically have both inputs and outputs. Inputs provide the board with information, such as temperature and pressure values, or perhaps a signal indicating that furnace burner operation has been requested. Outputs represent the response of the board to the inputs, signaling other components or powering them to action. Furnace and/or ignition-control modules, for example, are linked to components such as the room thermostat, induced-draft fan, indoor blower, draft-proving pressure switch, gas valve, igniter, flame sensor, and limit switches. The thermostat, draft-proving pressure switch, flame sensor, and limit switches would be considered inputs. The board does not control these devices; they simply provide information to be acted upon. Devices like the gas valve and igniter are controlled by the board, so these connections to the board are considered outputs.

Any of the input devices, or the devices controlled by the outputs, can fail. Therefore, it is important to determine the real source of a problem during the troubleshooting process. The best approach is to make an effort to eliminate the board from the list of possibilities, rather than focus on proving it to be the problem. The difference in thinking is subtle but generally more productive.

Many HVACR-related PC boards and solid-state modules contain a self-diagnostic feature that will identify the nature of the problem, if not the likely source. Many PC boards contain one or more LED lights that flash a coded signal to indicate the cause of a failure. **FIGURE 11-28** shows the diagnostic chart related to the White-Rodgers 50E47 ignition-control module discussed in the previous section. A label with the diagnostic information is often attached to the unit or the device itself. Some commercial systems have more complex diagnostic displays that display a numeric or alphanumeric diagnostic code that can be interpreted using a provided label or the installation and operation manual (IOM) for the unit or component.

Even if the diagnostic feature of a PC board indicates an internal failure, it is best to verify that the board has failed before replacing it. The way to do that is to confirm that the board is receiving the correct inputs and providing the correct outputs. Of course, if the PC board's own source of power (rectifier to provide DC voltage) has failed and it will not show any indication of life with its AC power source confirmed, there is little more that can be done. If the board has replaceable fuses, remove and test them. Replace any defective fuses and retest the board.

To troubleshoot any solid-state control device, it is crucial to understand the operating sequence of the device and connected components and know precisely what to expect. Technicians will need to know how the board should respond to specific inputs and how the output responses can be measured. With the correct knowledge, testing the inputs and outputs of a PC board will lead to an accurate diagnosis and replacement of the correct part(s) the first time.

▶ TROUBLESHOOTING TIP

Testing PC Boards

When testing a PC board, it is best to disconnect as many of the output connections as possible during testing. Outputs go somewhere, often to loads or load controls, and the connected devices may interfere with testing. If one or more output recipients is defective, it may be affecting board operation. Test with inputs connected and outputs disconnected when it does not prevent the board from doing what it should do. Then output switch closures, voltages, and currents can be freely tested with a multimeter.

LED	Condition
Green Solid On	Normal
Green Rapid Flashing	Weak flame signal
Red Rapid Flash	Control in lockout Flame sensed when there should be none
Red 1 Flash	Control in lockout Ignition retries exceeded
Red 2 Flash	Control in lockout Ignition recycles exceeded
Yellow Solid On	Internal self check
Yellow Rapid Flashing	Improper Polarity
OFF	Internal Failure
Red Solid On	Gas valve miswired or Internal error detected

FIGURE 11-28 Diagnostic chart for an ignition-control module.

Many PC boards have troubleshooting charts available, allowing a technician to evaluate board operation in an organized manner. **FIGURE 11-29** shows the troubleshooting flow chart for the White-Rodgers 50E47 ignition-control board. Flow charts like this provide an excellent troubleshooting guide for controls with complex functions.

To troubleshoot a solid-state control device, follow the steps in **SKILL DRILL 11-1**.

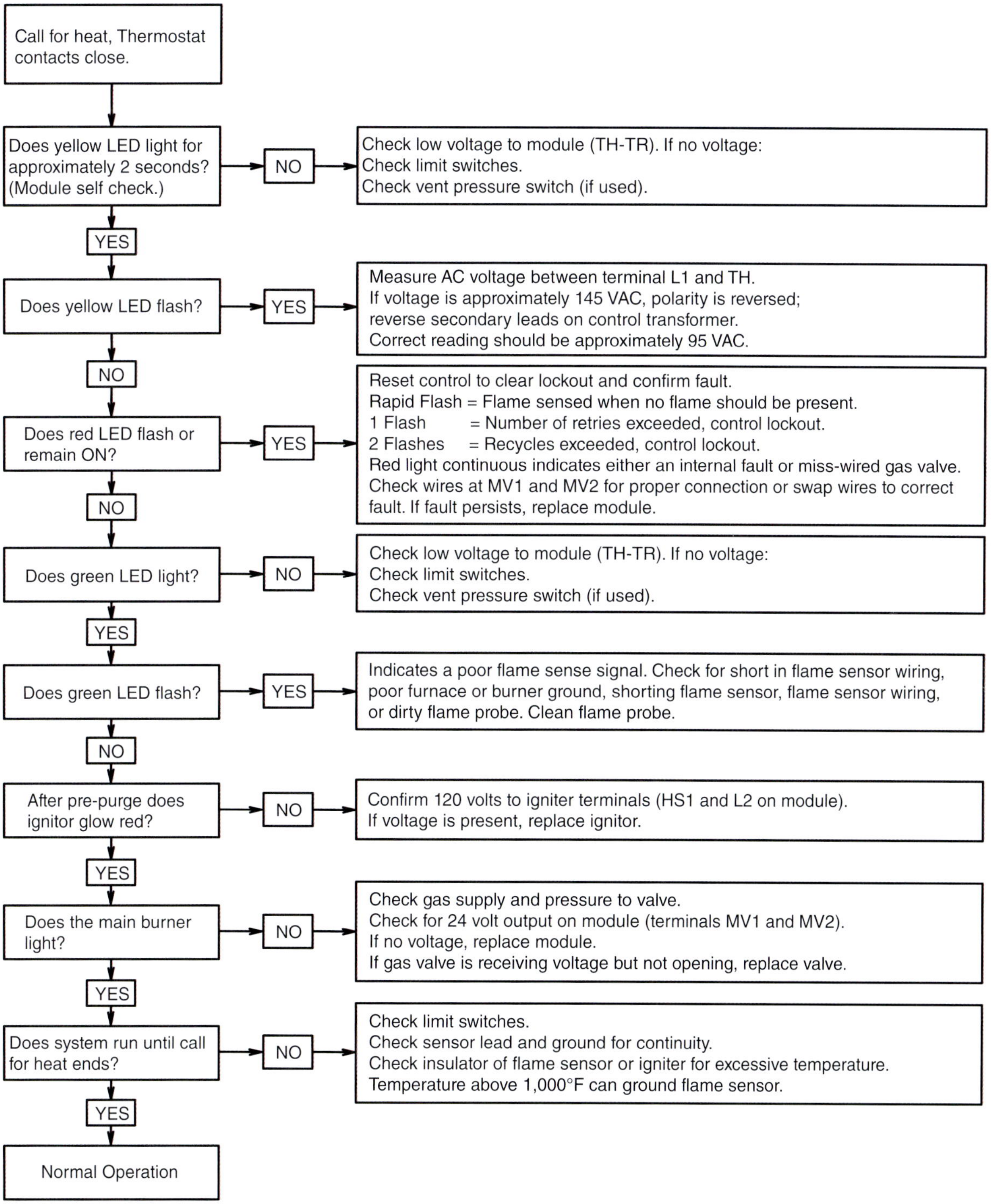

FIGURE 11-29 Ignition-control module troubleshooting flow chart.

SKILL DRILL 11-1 Troubleshooting a Solid-State Control Device

This exercise is based on the four-wire compressor anti-short-cycle timer discussed in this chapter, installed in a residential condensing unit or heat pump. The timer can be fixed or adjustable. The testing process should be similar for all three- and four-wire delay-on-break timers. If the system is equipped with a two-wire timer or other configuration, an instructor will assist in adjusting the troubleshooting process.

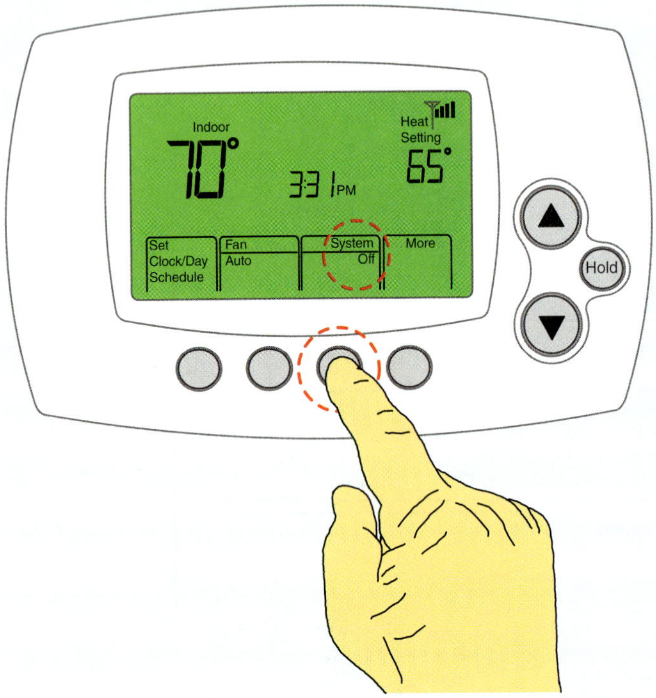

1. Ensure that the system thermostat is set to Off or that the cooling set point is well above room temperature. *Note:* this exercise can be done with the line-voltage power source also disabled unless the low-voltage transformer is in the unit being serviced. This is usually the case with packaged systems. If the transformer is located outside of the serviced equipment (common for split systems), deenergize, lock out, and tag the line-voltage power source to the unit to maximize safety. Leave the power source to the control transformer active.

SAFETY TIP

Working with electrically energized circuits is hazardous. Avoid contact with any surface that could potentially be energized. When working inside energized enclosures, maintain a stable position and, whenever possible, avoid placing both hands inside the enclosure. All work on energized equipment by unqualified individuals should be supervised by a qualified instructor or technician.

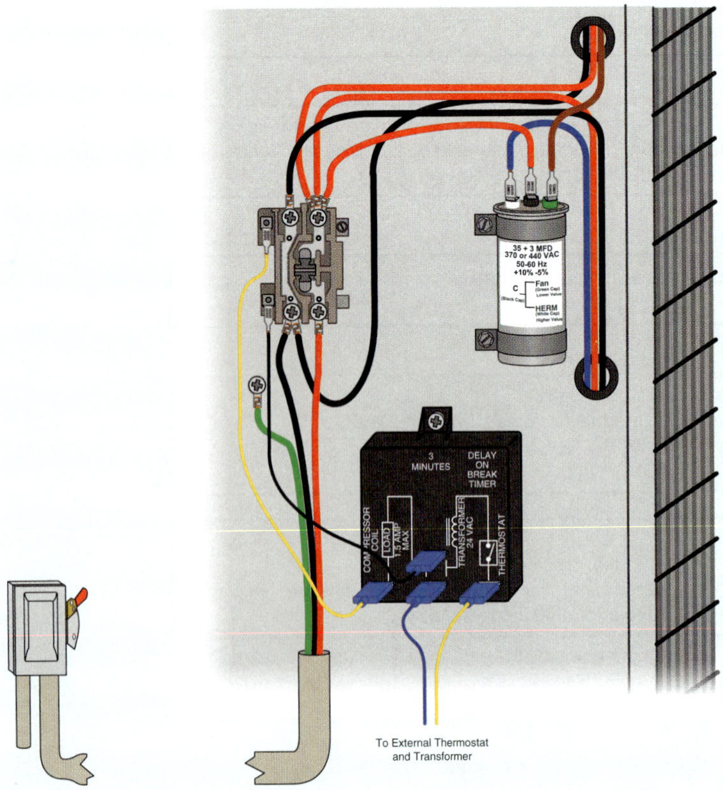

2. Remove the condensing unit or heat-pump access door and control panel cover. Locate the timer, which is typically in the same compartment as the compressor contactor.

Examine the wiring and ensure that it matches expectations. Terminal 1 should be wired to one terminal of the compressor contactor coil. Terminal 3 should be wired to the Y (cooling) connection from the thermostat-wire bundle or indoor-unit terminal board. One terminal of Terminal 2 should be connected to the Common wire from the transformer. The other terminal of Terminal 2 should be connected to the remaining terminal of the compressor contactor.

Continued

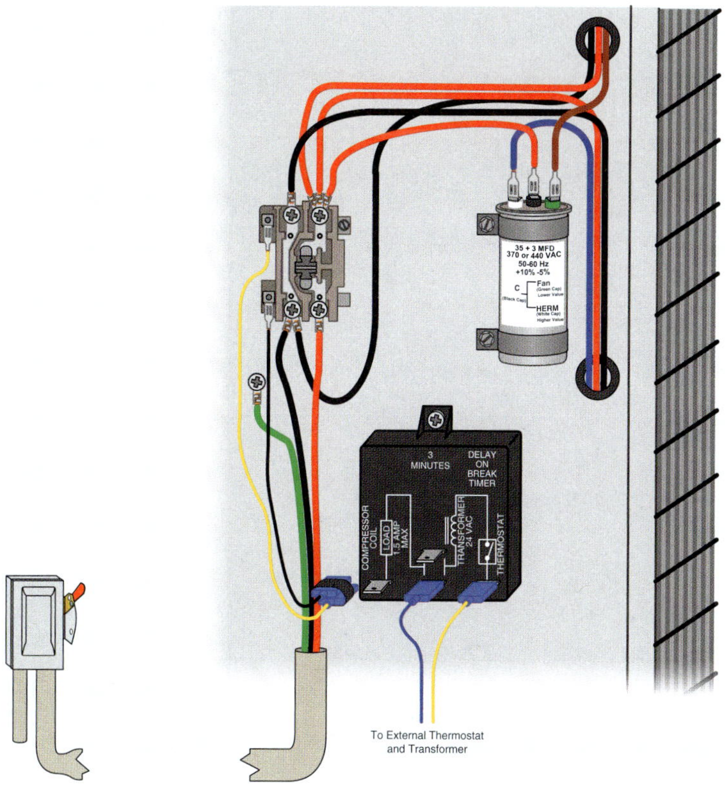

3. The timer can be tested without starting the compressor for added safety; starting and stopping the compressor for this test is unnecessary. Carefully disconnect the two wires routed to the compressor contactor coil from Terminals 1 and 2. Apply electrical tape to the terminal hardware, and secure the two wires away from the timer. The Common wire from the low-voltage power source on Terminal 2 must remain connected.

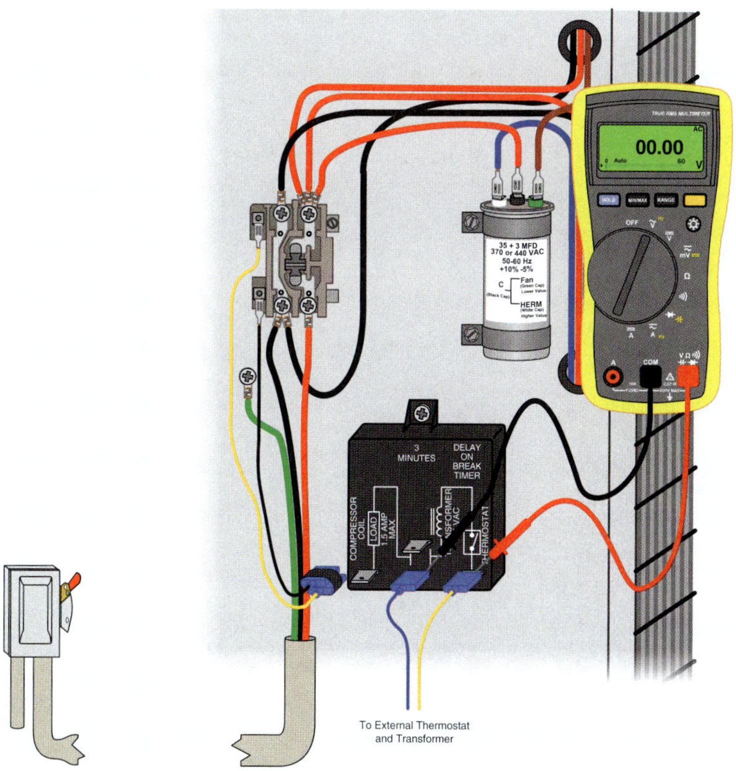

4. Use a voltmeter to check for the presence of 24 VAC at Terminals 2 and 3 of the timer; control power should not be present, since the thermostat is not calling for cooling, but remember that the transformer remains powered, providing power to Terminal 2.

Continued

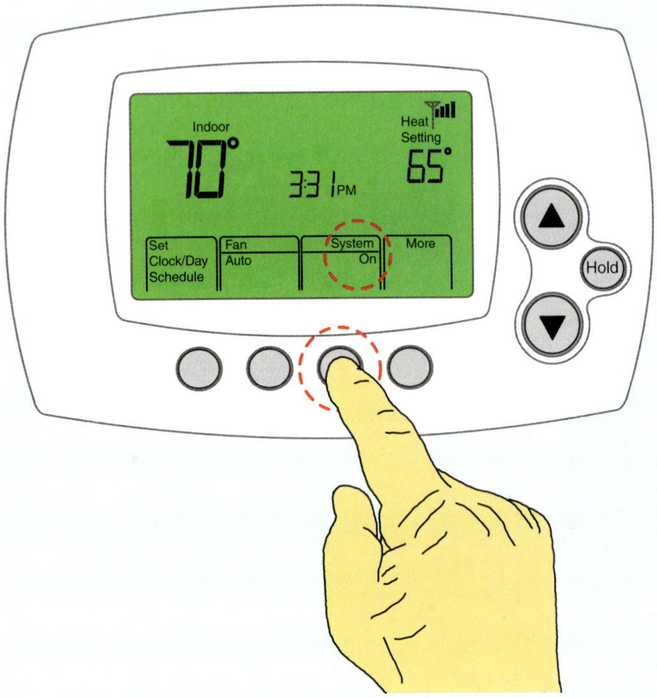

5. Ensure that all wiring and tools remain clear of energized wires and controls, then set the thermostat to Cool or Auto, and lower the set point well below room temperature.

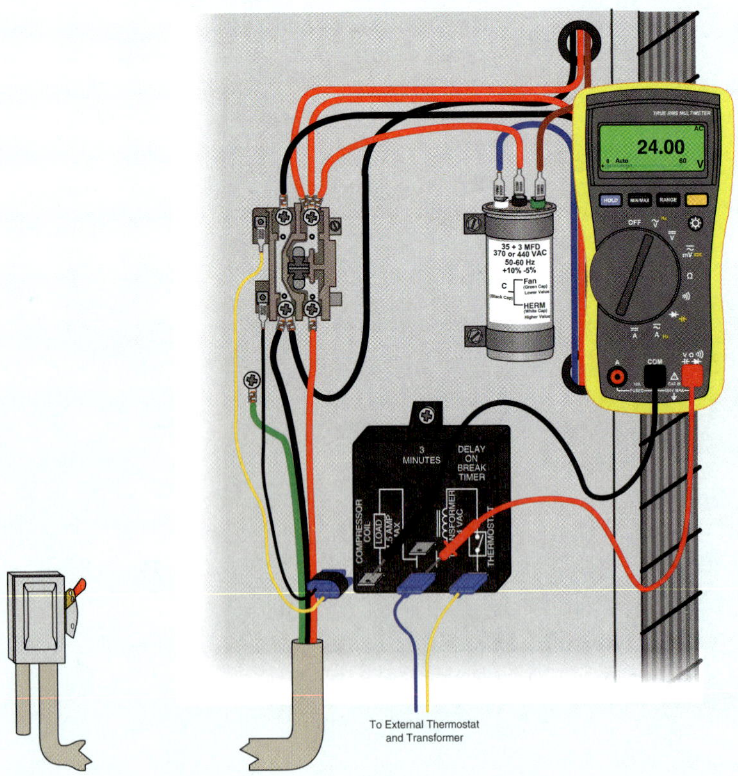

6. Return to the unit and test for the presence of 24 VAC again at Terminals 2 and 3. The 24 VAC control voltage should now be present. If so, transfer the meter probes to measure the voltage between Terminals 1 and 2. If the timer switch has closed, 24 VAC control power will be present there as well.

Continued

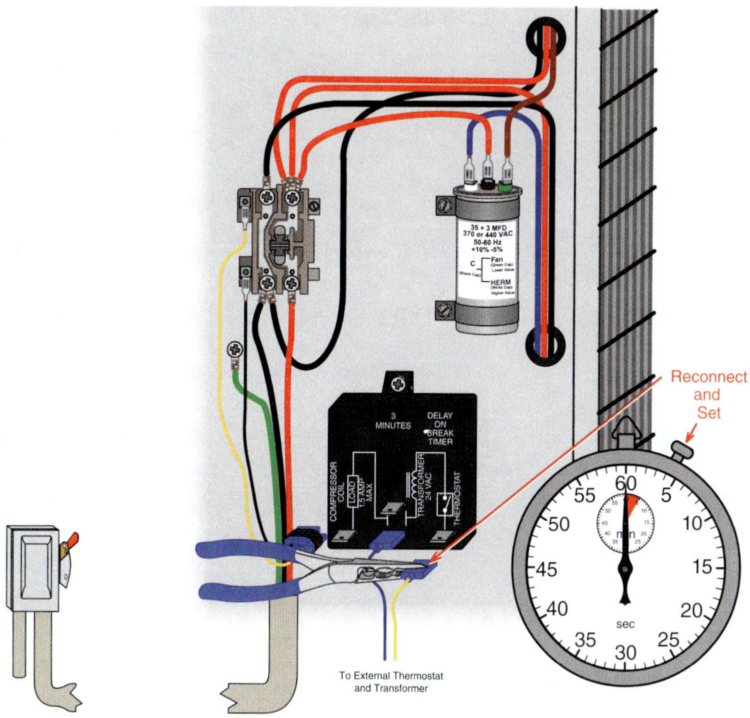

7. To test the timer function, control power to Terminal 3 must now be broken. Prepare a watch or smartphone to time the delay period. The process can begin by returning to the thermostat, raising the set point well above room temperature for 15–20 seconds, and then again lowering the set point well below room temperature. This resets the timer logic. The short waiting period gives electronic thermostats time to respond to the change; some have a slight delay before responding. Alternatively, the wire connected to Terminal 3 of the timer can be carefully removed for a few seconds with needle-nosed pliers, and then reconnected.

The moment that the thermostat responds to the change and calls for cooling or the moment that the wire is reconnected to Terminal 3 of the timer, begin timing the delay period with a suitable watch or smartphone. The delay period should be the fixed value shown on the timer or the set point of the timer dial if it is adjustable.

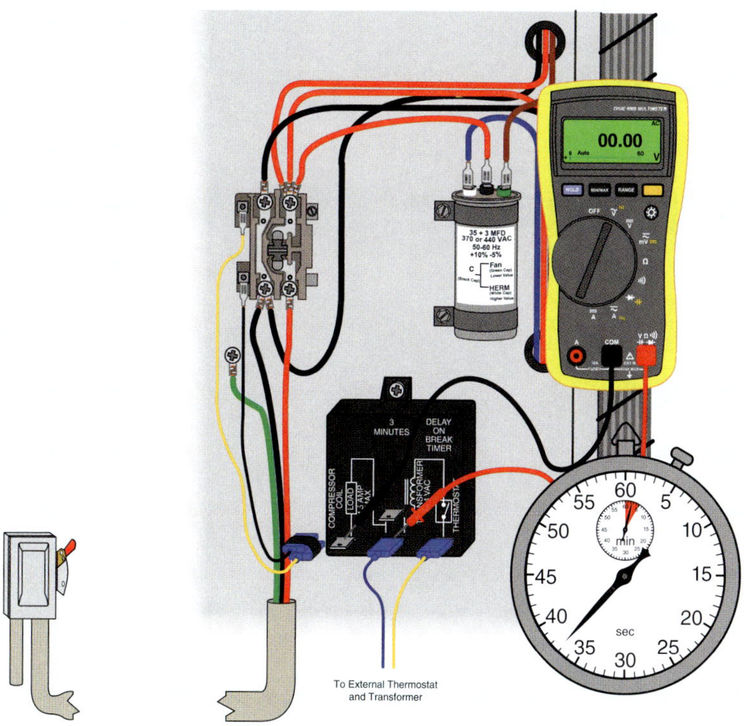

8. Immediately use the voltmeter to check for power across Terminals 1 and 2 of the timer. This time, control power should not be present, even though the thermostat is calling for cooling. This shows that the timer logic opened the contacts as expected but has initiated the time delay before reclosing them.

Continued

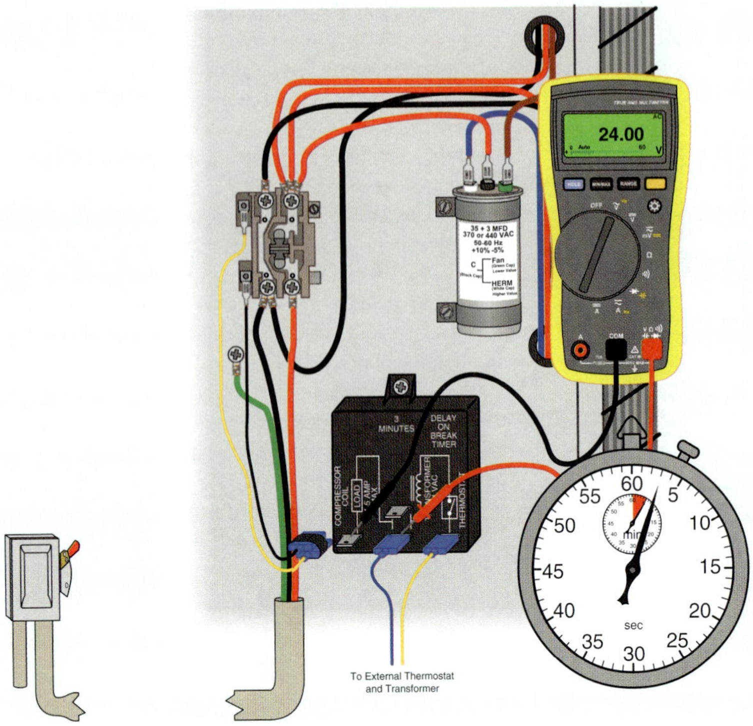

9. About 30 seconds before the delay period elapses, touch or connect the meter probes to Terminals 1 and 2 of the timer; 24 VAC power should still not be present until the time elapses. Continue monitoring for 24 VAC power as the delay period continues. Once the period has elapsed, the meter should display 24 VAC, indicating that the contacts have closed. If they do not reclose at the end of the timing period with control power applied to Terminals 2 and 3 throughout the countdown, the timer is defective. Always be sure, however, to allow an appropriate amount of time for the action to occur; laboratory accuracy in the timing should not be expected.

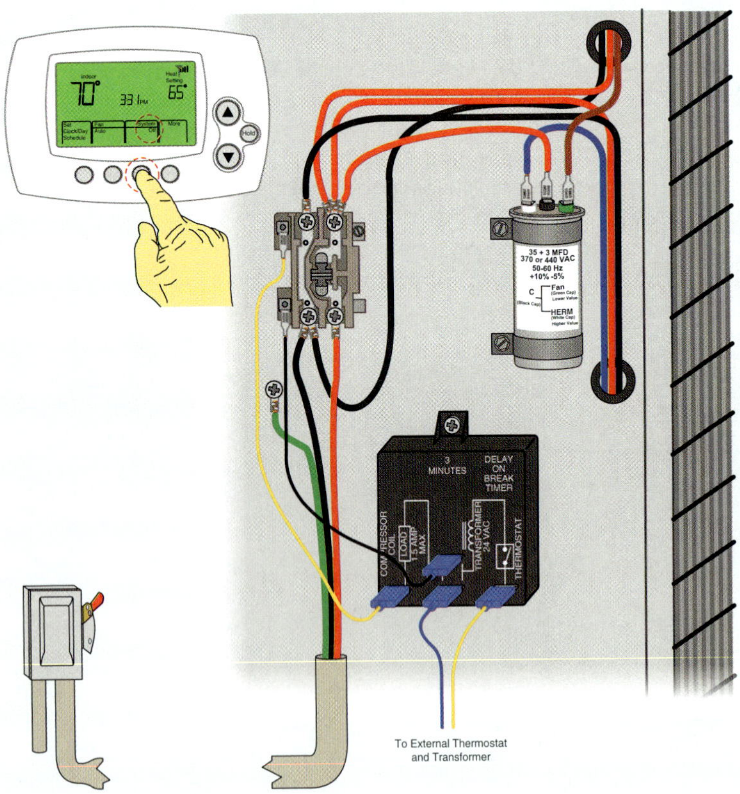

10. Testing is now complete. First, set the thermostat to Off and/or adjust the set point well above room temperature. Then reconnect the wiring from Terminals 1 and 2 of the timer to the compressor contactor coil.

Continued

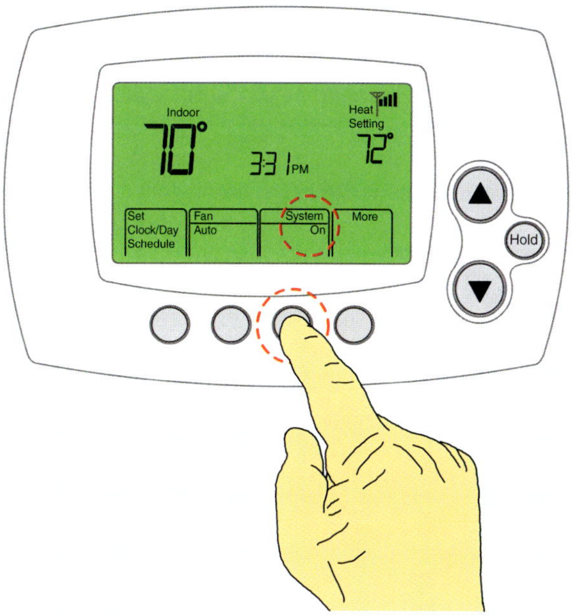

11. Remove the locks and restore line-voltage power to the unit if it was deenergized. Reinstall the access doors. Return to the thermostat, turn the system on, and adjust the set point below room temperature. After the timing period has elapsed (remember that it began when the thermostat was turned Off), the system should start and operate normally. Adjust the thermostat to the desired temperature to complete the task.

Replacing a Solid-State Control Device

As shown in **FIGURE 11-30**, PC boards are usually mounted on standoffs. The standoffs may be made of non-conductive material, such as plastic, or of conductive metal. Metal standoffs are used when the goal is to ground the board through the mounts. The standoffs keep the foil side of the board from contacting conductive surfaces and creating a short circuit. There are holes in the corners of the board that are used to attach the board to the standoffs with screws. As described in the ESD section above, it is vital to discharge any static electricity through grounding before touching component leads and connection terminals.

FIGURE 11-30 PC-board mounting standoffs.

Simple devices without exposed electrical paths, especially those that are enclosed—such as an anti-short-cycle timer—do not generally require special handling. They may be mounted with only one or two screws. If a board or solid-state device is enclosed, standoffs for mounting are not necessary.

The first step in replacement, after discharging any static through grounding, is to examine the new control and ensure that it is the correct part. In some cases, the component may have been updated or replaced by one that is visually different and even has different terminal markings. This is especially true for complex boards, such as defrost and ignition controls. Ensure the board is a proper match, and review the installation instructions carefully before proceeding further. If a new component has been provided to replace an older generation, technicians must understand how to connect it to the system. Documentation of this should be provided by the manufacturer.

The next step is to disconnect all wiring from the input and output terminals of the board and mark each one so it can be correctly attached to the new board. If the PC board has an edge connector or a ribbon cable connected, disconnect it while the board is still firmly attached to the unit. Otherwise, it will be difficult to figure out how to delicately hold the board while trying to remove what might be a stubborn connector. Note that some ribbon cables and edge connectors are equipped with locks to securely hold the mating halves together (**FIGURE 11-31**). Examine the connection closely to ensure that any locking features have been released.

Before removing the screws holding the device in place, inspect the position of all control DIP (dual inline package) switches and jumpers that are present. DIP switches are tiny switches that usually have two positions, but possibly more. Usually provided in groups, their position establishes board behavior, increasing the flexibility of the control. Small jumpers, usually soldered to the board on one end and connected to a pin terminal on the other, provide the same basic function. For example, a defrost control board for a heat

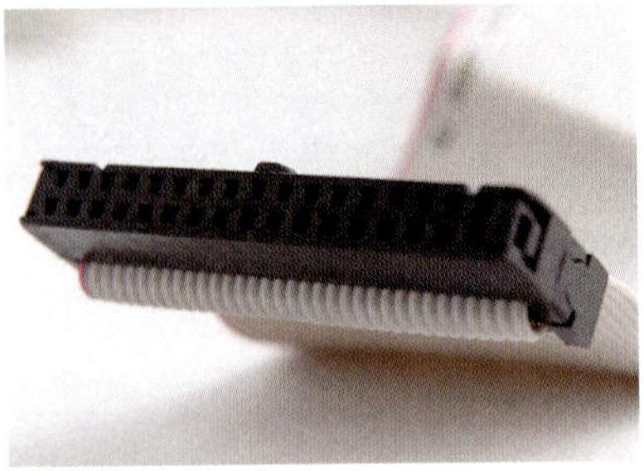

FIGURE 11-31 Ribbon cable connection with locks.

pump often has a jumper and several possible pins from which to choose to complete the connection. Each pin represents the time between defrost cycles, such as 30, 60, and 90 minutes. In climates that rarely fall below freezing, 90 minutes might be a reasonable choice. In cold, snowy climates, the shorter 30-minute cycle would be appropriate. DIP switches and jumpers allow field technicians to make program choices without interfacing directly with a control's programming. Before removing the board, document the position of switches and jumpers so that the configuration can be duplicated on the new board. This can also be done by simply configuring the new board to match while inspecting the existing control in place.

Once all wiring terminations are disconnected and the board is properly configured, remove the screws that secure the board to the equipment. Then grasp the board by its edges and gently lift it off the standoffs. Note that some standoffs may be attached to the board or to the mounting surface or, in other cases, may not be attached at all, allowing them to freely slide off the mounting screw. If they are free to move, ensure that they are not lost, since they will be needed to mount the new board. Remember to remain grounded throughout the entire process.

Do not over-tighten the mounting hardware when installing the new board; doing so may damage the board. Once it is secure, connect the wires and wiring harnesses to the board. Inspect the work area for leftover hardwire and tools, restore power to the system and check for the proper AC voltage at the board terminals. Then test system operation to make sure the new board is working correctly.

To replace a solid-state control device, follow the steps in **SKILL DRILL 11-2**.

SKILL DRILL 11-2 Replacing a Solid-State Control Device

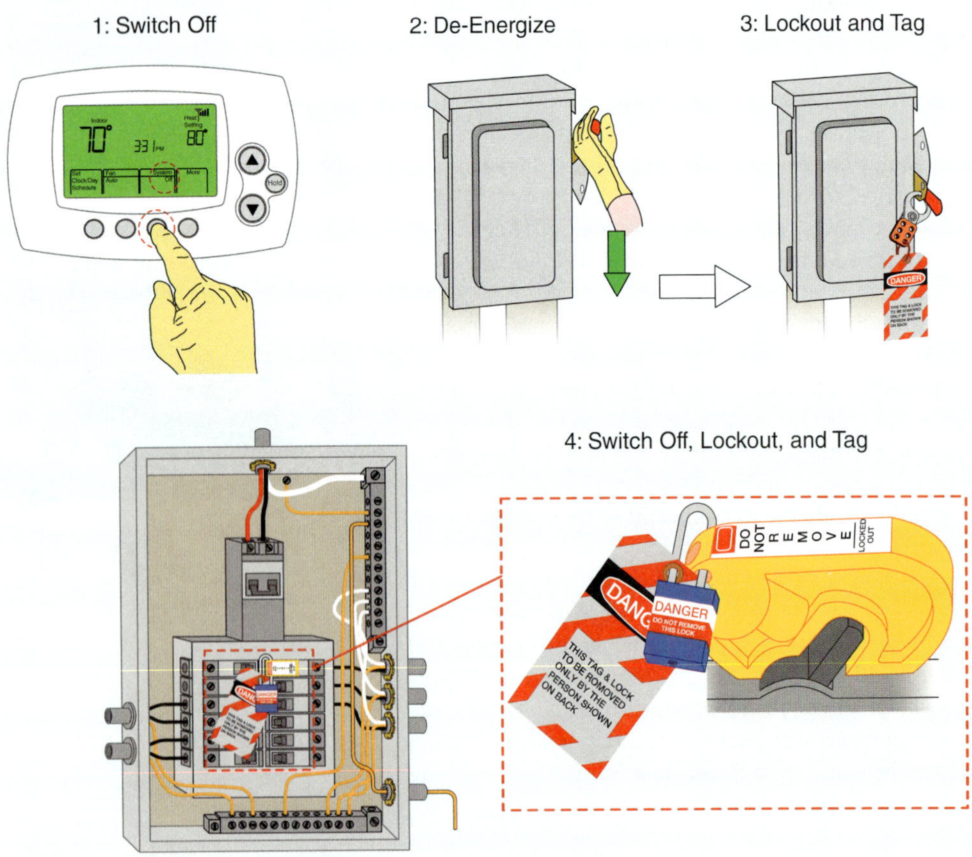

1. Ensure that all system controls are set to the Off position. Deenergize, lock out, and tag all related energy sources. Be sure that the process deenergizes not only the line voltage feeding the unit but also any low-voltage control power source that may be in a separate piece of equipment. Verify that everything is deenergized, by using a voltmeter.

SAFETY TIP

Working with electrically energized circuits is hazardous. Avoid contact with any surface that could potentially be energized. When working inside energized enclosures, maintain a stable position and, whenever possible, avoid placing both hands inside the enclosure. All work on energized equipment by unqualified individuals should be supervised by a qualified instructor or technician.

Continued

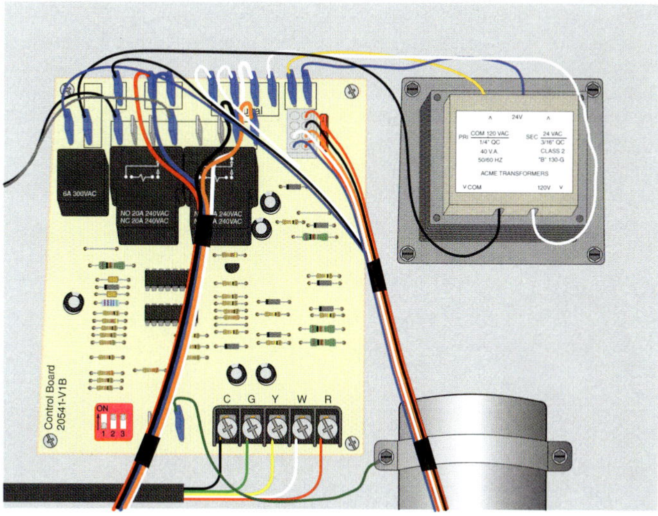

2. Remove the necessary access doors, and locate the solid-state control to be replaced.

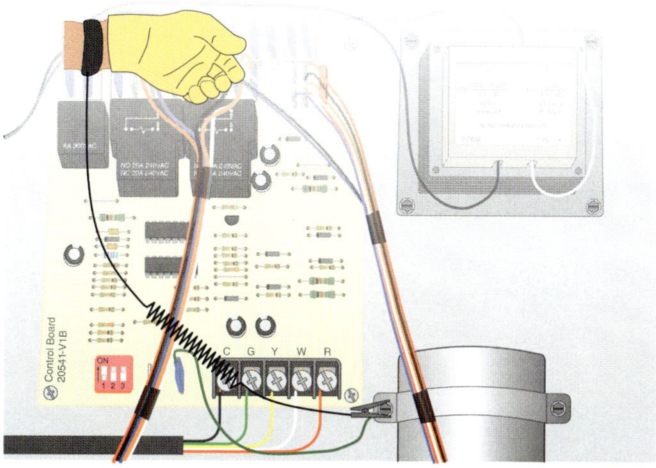

3. Don a grounding wrist strap before proceeding further, to avoid any ESD.

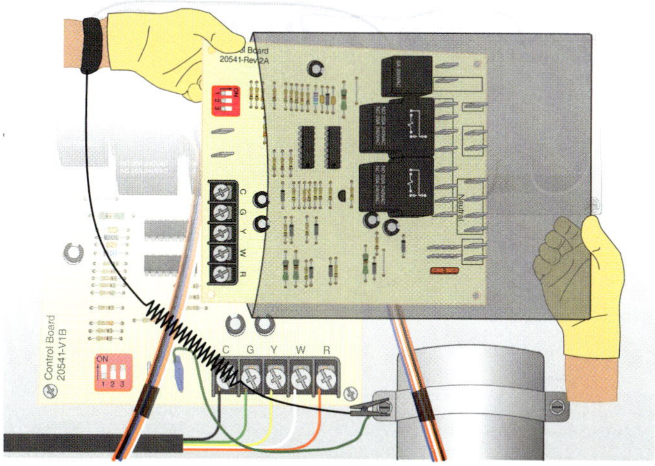

4. Remove the new device from its packaging and compare it to the device to be replaced. Ensure that it is either the same device, or that instructions are included for its use in replacing a device from a previous generation.

Continued

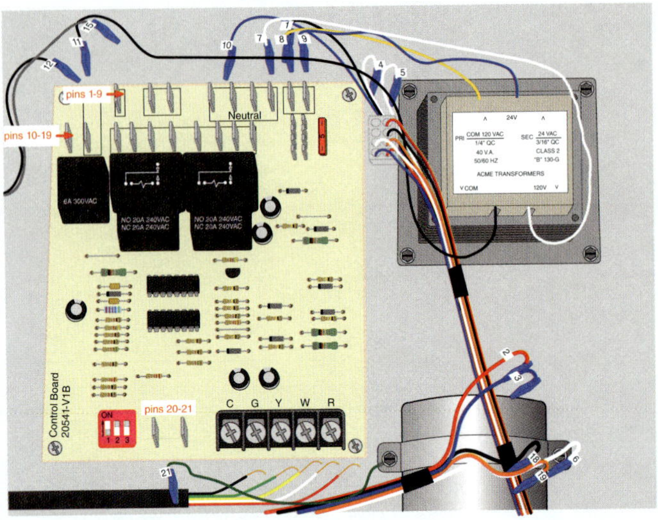

5. Mark or tag, then disconnect all external wiring from the control. Check for locks on ribbon cables, edge connectors, and harnesses, disengage any locks as necessary, then disconnect the cables from the device as well.

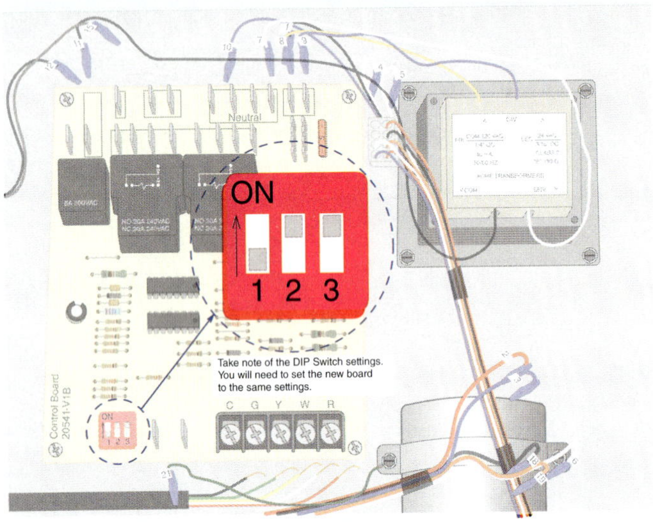

6. Examine any DIP switches and jumpers on the board, and set the DIP switches and jumpers on the new board to the same positions.

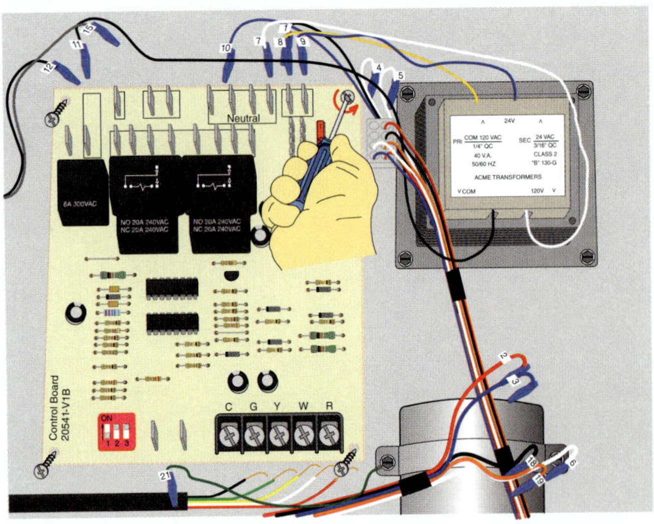

7. Remove the screw(s) that secure the device to the equipment. Be sure to capture the mounting hardware, such as loose standoffs holding the board away from the panel surface. Grasp the device carefully by its edges and remove it.

Continued

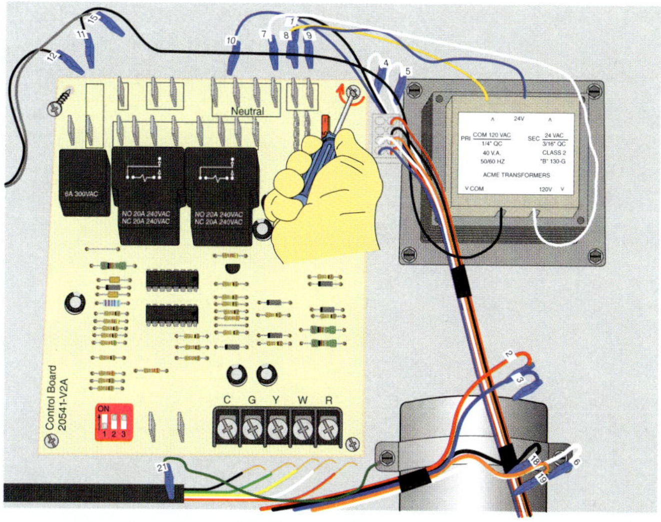

8. Place the new device into position, on the standoffs as required, and secure it with the original hardware.

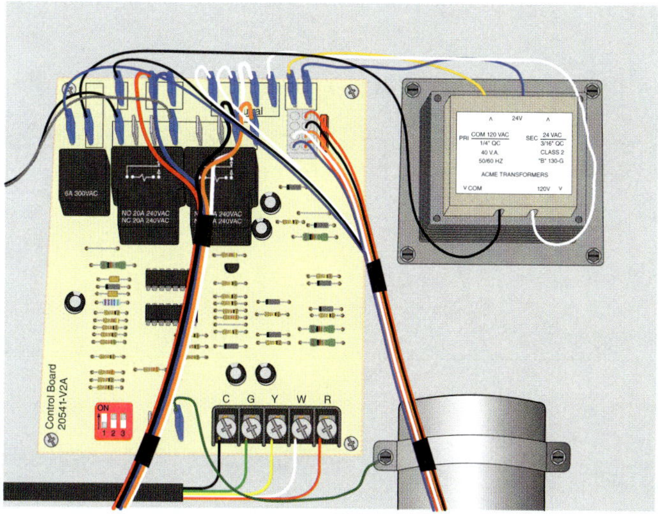

9. Reconnect all the wiring, as marked, to the new device.

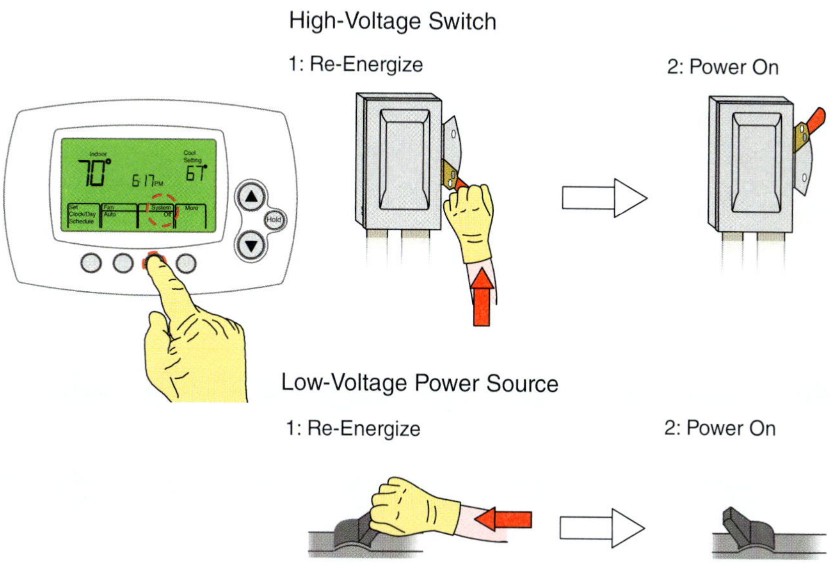

10. Ensure that all controls, including the thermostat, remain set to the Off position. Remove the locks and tags from the power source(s) and restore power to the system.

Continued

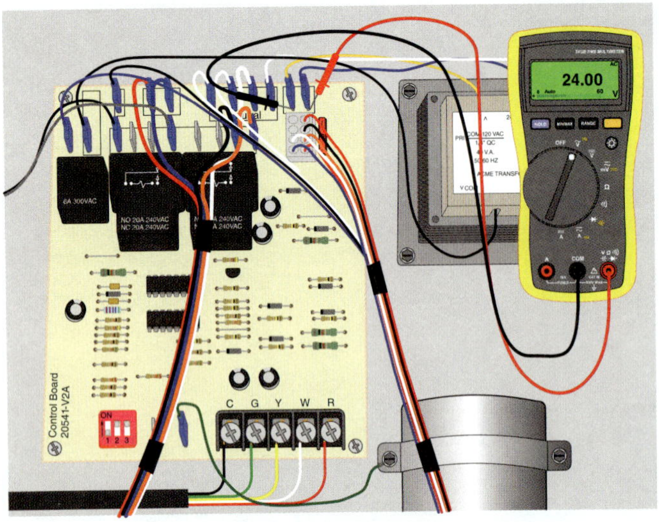

11. Use a voltmeter to check for the proper voltage at the unit and at the component if it is constantly powered. Some control devices are energized only in the midst of an operating sequence.

12. Complete the installation by testing the device in accordance with the equipment sequence of operation or the device's own installation instructions. Once testing is complete, reinstall the access doors and return the unit to service.

▶ Wrap-Up

Summary

In today's world, the management and control of HVACR systems often relies on solid-state devices and microprocessors. Components such as IC and microprocessor chips enable modern systems to collect information from various system inputs, evaluate the information according to their programming, and produce outputs based on that evaluation. For example, in an HVAC system that mixes outdoor air with return air, the temperature and humidity of the two airstreams are continuously sensed and reported to a controller. The appropriate position of dampers in the duct system is then determined, and outputs are sent to initiate their movement to a new position as necessary.

PC boards have a component side and a foil side. The connection pathways between components are printed on the foil side with copper foil, and the component leads are inserted through openings in the board and soldered to the copper foil. In addition to IC chips and microprocessor chips, a board may contain fuses, relays, resistors, capacitors, and a rectifier.

The solid-state components found on PC boards are very sensitive to electrostatic discharge (ESD), so it is important to avoid touching component leads and connection terminals. Anyone handling PC boards or modules containing semiconductor devices should wear a grounding wrist strap. Take care when handling PC boards to avoid damage, and be sure to use standoffs as needed to isolate the foil side of the board from other conductive components.

Many PC boards and control modules used in HVACR systems have diagnostic aids that will help to determine the nature of a problem. If a failed PC board or control module is indicated by the diagnostics, it is a good idea to verify it by checking to see whether it is receiving the correct inputs and producing the required outputs. Replacing a solid-state device that has not actually failed, an error that occurs all too often, can result in an embarrassing customer callback.

Key Terms

amplifiers Electronic devices used to boost the amplitude of a voltage.

breakover voltage The positive voltage at which a diode switches into the conductive state.

diac A semiconductor diode that conducts electricity only after its breakover voltage has been reached. The term "diac" is an abbreviation of diode for alternating current.

diodes Semiconductor devices containing N-type and P-type semiconductor material that allows current to pass only when a positive voltage is applied to the anode.

doping The intentional introduction of impurities to a pure semiconductor to change its electrical properties.

electrostatic discharge (ESD) The release of accumulated static electricity that occurs when two objects touch. If not discharged to ground before handling solid-state components, ESD can damage or destroy electronic components.

holdover voltage The voltage at which a diac ceases to continue conducting current. A diac will conduct current when the breakover voltage is reached, and cease to do so once the voltage falls below the holdover voltage.

integrated circuit (IC) A semiconductor chip containing a variety of microminiature electronic circuits.

microprocessor A user-programmable electronic device containing one or more integrated circuits (ICs).

rectifier An electronic circuit containing diodes that is used to convert AC voltage to DC voltage. Rectifiers are needed to obtain the DC power required for electronic control circuits.

ripple The small, undesirable periodic variation in a DC output that has been derived from an AC power source.

semiconductor A material that is electrically neutral and serves as the building block for solid-state components. Its neutral status can be changed to positive (P-type) or negative (N-type) by doping.

solid-state components Electronic components made of semiconductor material that has been doped to make it positively or negatively charged.

thermistors Semiconductor devices in which the resistance varies in proportion to the temperature. Thermistors are used extensively in HVACR systems to sense temperature.

transistors A semiconductor device capable of controlling a proportionally large amount of current with a small amount of voltage. Transistors are used in solid-state circuits as amplifiers and switches.

triac A diac with a gate circuit that must be triggered for the triac to conduct current. The term "triac" is short for tr*iode for alternating current.*

varistor A solid-state protective device in which the resistance varies in relation to the voltage applied to it.

Review Questions

1. Which of the following is a correct statement about printed-circuit (PC) boards?
   - a. It is often necessary to isolate a fault to a specific component on a PC board.
   - b. Technicians rarely, if ever, need to replace an electronic component on PC boards.
   - c. Components are generally mounted on one side of the PC board and wiring is applied to the other side.
   - d. All PC boards use edge connectors for system connections.
2. The process of adding substances to semiconductor material to make it positively or negatively charged is called ______.
   - a. breaking over
   - b. doping
   - c. dumping
   - d. charging
3. The integrated circuit device that performs computing functions is the ______.
   - a. triac
   - b. diac
   - c. diode
   - d. microprocessor
4. On a PC board, the copper pathways are printed on the ________ side.
5. The solid-state component that can serve as a counter or provide computer memory is a(n) ______.
   - a. capacitor
   - b. varistor
   - c. semiconductor
   - d. IC chip

6. Compared to conventional fuses, solid-state fuses possess _____.
   a. super-fast interrupting capability
   b. the ability to handle extreme current
   c. the ability to handle extreme voltages
   d. their own microprocessor for control
7. In a diode, the negatively charged terminal is called the __________.
8. With a positive AC voltage applied to the anode of a diode, it will conduct only on the positive half-cycle of the sine wave and is referred to as being ____________.
   a. forward-biased
   b. reverse-biased
9. A common use of diodes is to _____.
   a. amplify AC voltage
   b. store electrical energy
   c. rectify AC voltage to DC
   d. act as a switch
10. A triac differs from a diac in that it _____.
    a. has a gate electrode
    b. only conducts on one half-cycle
    c. is used primarily as an amplifier
    d. is not a solid-state device
11. Because the output of a rectifier contains ripple, it must be ____________.
    a. amplified
    b. filtered
12. A proper timing relay used for compressor short-cycle protection is also known as a ________________relay.
    a. delay-on-break
    b. delay-on-make
13. An LED is _____.
    a. a special type of incandescent light
    b. a miniature microprocessor
    c. a diode that emits light
    d. an adjustable delay-on-make relay
14. Explain why an anti-short-cycle timer is used for compressors.
15. Which of the following is considered an input for an ignition-control module, as opposed to an output?
    a. Hot-surface igniter (HSI).
    b. Gas valve.
    c. Induced-draft fan relay.
    d. Flame sensor.
16. Which of the following should be done when replacing a PC board?
    a. Touch the board on the bottom only.
    b. Remove the board from its mount before disconnecting the wiring.
    c. Ground yourself using a grounding wrist strap.
    d. Ensure that a dedicated power source is used to power the board.
17. If you are troubleshooting a furnace ignition-control board, the most likely first step in the operating sequence is _____.
    a. a control lockout if there is no gas pressure
    b. a purge cycle to clear out the heat exchanger
    c. starting of the indoor blower
    d. the opening of the gas valve
18. ESD is more likely to be a problem when the ambient air is ____________.
    a. wet
    b. dry
19. A common method used by an HVACR solid-state control, such as an ignition-control module, to provide diagnostic information is _____.
    a. flashing and/or colored lights
    b. a printed report
    c. a data key that records operating history
    d. an audible alarm
20. Regarding inputs and outputs, what is the best approach for troubleshooting a solid-state control like a PC board?
    a. Disconnect both inputs and outputs from the board.
    b. Disconnect as many outputs as possible without affecting normal board operation.
    c. Jumper inputs to outputs to simulate board operation and evaluate the results.
    d. Leave any output that represents a load connected to the board.

CHAPTER 12

Comfort-System Thermostats

Knowledge Objectives

After reading this chapter, you will be able to:

- **K12001** Identify and describe bimetallic comfort-system thermostats.
- **K12002** Identify and describe digital comfort-system thermostats.
- **K12003** Identify common thermostat mounting requirements and wiring terminal designations.
- **K12004** Describe common digital-thermostat features and settings available to users.
- **K12005** Describe common digital-thermostat settings available to installers.
- **K12006** Explain how to troubleshoot a thermostat.

Skill Objectives

Skills required to meet the objectives of this chapter:

- **S12001** Wire a thermostat to control an electric-cooling/gas-heating system.
- **S12002** Wire a thermostat to control a heat-pump system with supplemental electric heat.
- **S12003** Configure a digital thermostat.
- **S12004** Troubleshoot a thermostat.

▶ Introduction

Comfort-system thermostats are very important components to the systems they control, providing the interface between the user and the equipment. But another important consideration is that, to the average user, the thermostat *is* the system. In some cases, the rest of the equipment is completely out of sight and possibly inaccessible to the user. Thermostats probably experience far more human abuse, verbal as well as physical, than the rest of the system.

In technical terms, a thermostat is simply a switch that responds to changes in temperature. To get the job done, it must provide the user with a way to set a given temperature, and then it must respond to that setting in some fashion. All comfort-system thermostats have that simple characteristic in common. But today's thermostats have taken control to a completely different level, where all types of operating requirements and preferences are entered and used to control system behavior.

In spite of their capabilities, today's digital thermostats, so named because they make use of solid-state components, are not that complicated. However, one of the challenges in the industry is that there are several generations of thermostats still in use, and several generations of people operating them. What is second nature to one user may seem like astrophysics to another. A technician should have a clear understanding of all comfort-system thermostats in order to help the user. Since many see the thermostat as the system itself, technicians will instantly appear to be an expert when they can explain clearly and accurately how it works and how the user can make best use of it.

Before going further, it is important to note that it would take a much larger publication to cover the world of temperature controls in HVACR. There are many applications and many variations in the approach to temperature control yet to be learned. This chapter will focus on comfort-system thermostats that are familiar residents on the walls of homes and many businesses. However, many of the principles and concepts learned here will also apply to other temperature-control devices and applications.

▶ Bimetallic Thermostats

K12001 Identify and describe bimetallic comfort-system thermostats.

There are many types of comfort-system thermostats on the market. Bimetallic thermostats have been used for many years to control HVACR comfort systems, and many remain in service today. Understanding these simple thermostats will not only contribute to service knowledge but will also inspire an appreciation for the features that digital thermostats now offer.

The bimetallic thermostat (**FIGURE 12-1**) is the most basic thermostat design still in common use, with many versions also referred to as snap-action thermostats. A wound bimetallic element responds to temperature by unwinding or winding tighter. This is the result of the two dissimilar metal sheets that are bonded together, each of which expands and contracts at different rates. In **FIGURE 12-2**, note the magnet placed at the end of the coil. Without the magnet to positively bring the two contacts together and close the circuit reliably, the contacts would come together very slowly. The result is that the first contact may be electrically unreliable and intermittently open and close the circuit, causing arcing across the contacts in the process. The magnet helps close and reopen the circuit with authority. This minimizes arcing and ensures that good electrical contact is made. Not all bimetallic thermostats have magnets, however. Some use spring-loaded mechanisms or other approaches for firm contact closure. Any bimetallic design that incorporates a way to close the contacts with added energy can be considered a snap-action model.

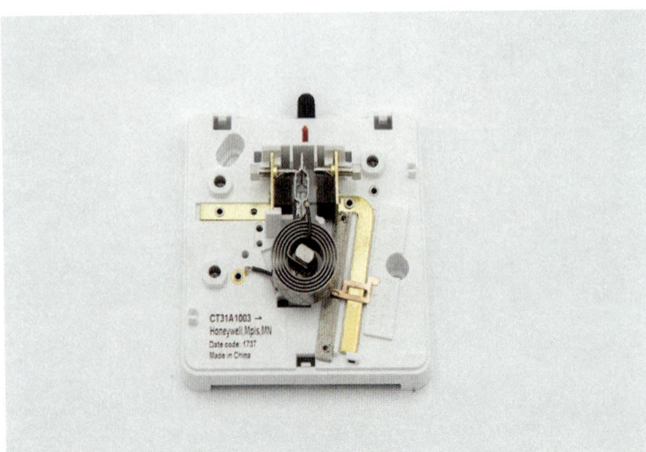

FIGURE 12-1 Bimetallic thermostat with cover removed.

Simple bimetallic thermostats are used in simple applications where precision and elaborate control features are unnecessary. Combination heating and cooling models are available, as well as heating-only and cooling-only models (**FIGURE 12-3**). They are certainly capable of controlling a simple residential system, but they do not offer the level of sophistication preferred by today's user.

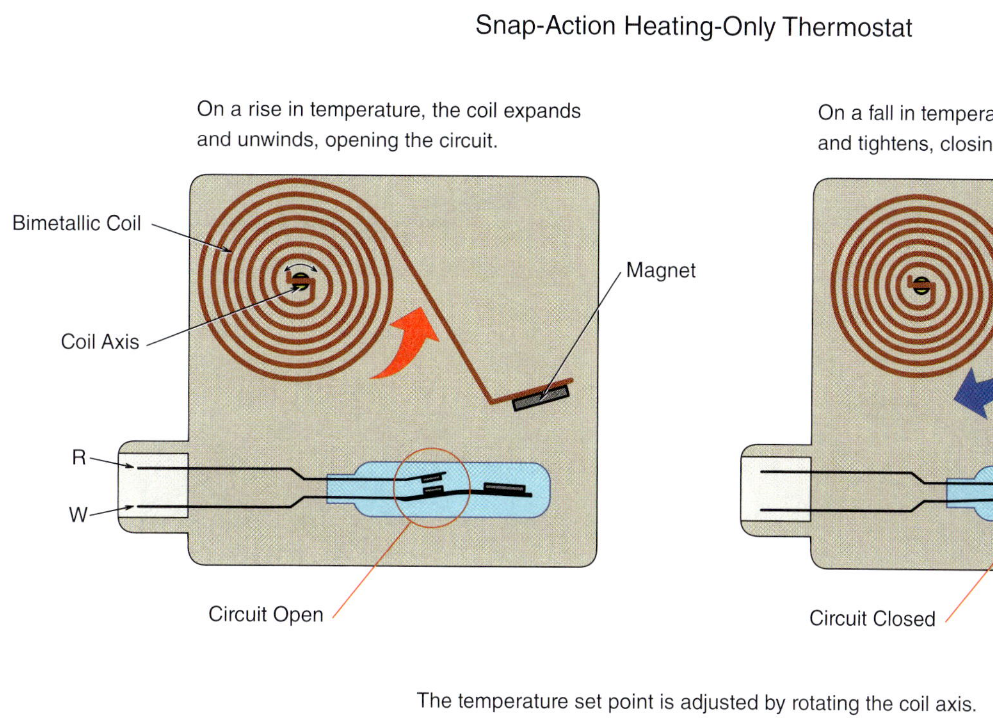

FIGURE 12-2 Bimetallic thermostat operation.

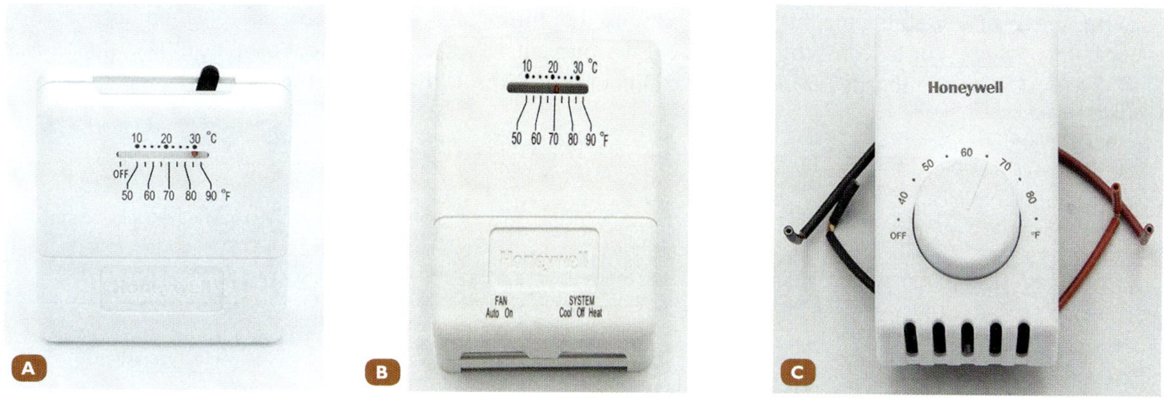

FIGURE 12-3 Typical bimetallic wall thermostats. **A.** Heating-only thermostat. **B.** Heating/cooling thermostat with fan switch. **C.** Line-voltage electric heat or ventilator thermostat.

There are also line-voltage versions used to directly switch the load current of electric baseboard heaters, unit heaters, or ventilation fans. Today, there are digital models that can serve the same purpose with the contact structure to manage the load current. Bimetallic thermostats are durable and the least expensive option for these applications, though they are not as precise.

Mercury-Bulb Thermostats

A mercury-bulb thermostat (**FIGURE 12-4**) is a variation of a bimetallic thermostat. One difference is that no snap action is needed to positively close a set of contacts. A mercury-filled bulb (**FIGURE 12-5**) is mounted directly to the bimetallic element. The mercury inside serves as the switch. As it rolls from one end to the other, the conductive mercury completes a circuit. Another difference is that a single mercury bulb can have switch contacts placed at each end, with one set used to control heating and one set used to control cooling.

Sealing the contact assembly in glass keeps the contacts clean and unexposed to contaminants. Technicians will likely encounter some very old mercury-bulb thermostats in homes and will note that the bulbs are still well sealed and functional.

FIGURE 12-4 The classic T87F thermostat from Honeywell.

Thomas Northcut/Photodisc/Getty images

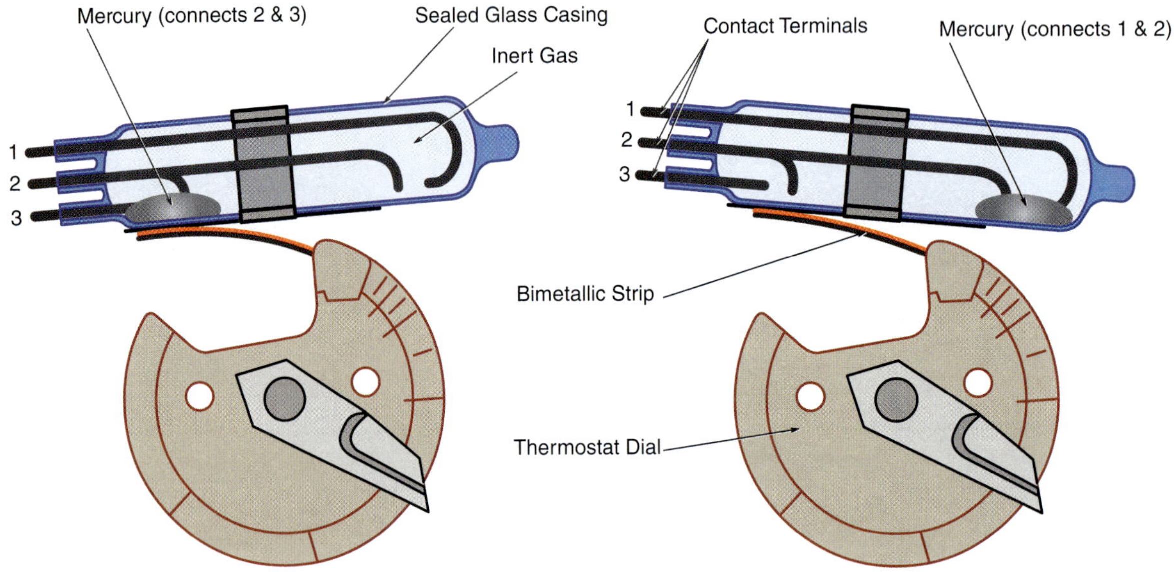

FIGURE 12-5 Mercury-bulb switching action.

▶ CORE CONNECTION

Mercury in Thermostats

According to the Thermostat Recycling Corporation (TRC), the average amount of mercury used in a thermostat is roughly 4 grams (0.14 oz) (**FIGURE 12-6**). Based on that small amount, it is hard to believe how much mercury actually went into thermostats. In 2001, approximately 14.6 tons (13.2 metric tonnes) of mercury were installed in thermostats. Through regulation and education, that volume dropped to 3.9 tons (3.5 metric tonnes) by 2007. The volume has now become insignificant. The TRC was established through the cooperative efforts of Honeywell, White-Rodgers (now an Emerson subsidiary), and General Electric. The organization focuses on the collection of mercury-bulb thermostats for separation and recycling. For more information on the TRC program, visit www.thermostat-recycle.org.

FIGURE 12-6 Liquid mercury.

Mercury is a unique metallic element that is highly toxic and causes a wide variety of serious health concerns. For that reason, it has already been banned from use in many controls and thermometer applications, and many states also prohibit the sale and installation of new mercury-bulb thermostats. Regardless of location, any removed thermostats containing mercury must be disposed of through a wholesaler that accepts them for recycling or by contacting the EPA (Environmental Protection Agency) or a local hazardous waste authority for instructions. Do not simply throw them in the trash.

SAFETY TIP

Mercury should never be handled with the bare hands. Mercury-bulb thermostats should be handled with care to ensure the bulb is not broken. Properly dispose of every mercury-bulb thermostat that is removed from service, by taking them to an authorized recycling facility or collection point.

Heat Anticipation

One of the problems with basic bimetallic designs is that they do not respond as quickly as needed. Once the system is running in a heating or cooling mode, conditioned air must flow around the bimetallic element and transfer heat to or from it before it begins to change shape. This relatively slow process results in systems running too long and overshooting the set point. To enable these thermostats to respond more quickly in the heating mode, a **heat anticipator** (**FIGURE 12-7**) is added. A heat anticipator is a variable resistor that produces a small amount of heat near the bimetallic element. When the thermostat heating circuit closes, the heat anticipator is energized and begins to produce heat. Heat from the anticipator works along with heat added to the surrounding air to speed up the response of the element.

The heat anticipator is usually adjustable, which means the amount of heat it produces can be changed. The initial setting is typically made based on the current in the active heating control circuit. This value is measured with an ammeter, and then the position of the anticipator is set to that value. Since this may not produce the ideal **cycle rate** for the user, the anticipator resistance can be adjusted up or down to change it. In Figure 12-7, notice the word "Longer" on the dial and the directional arrow.

Increasing the setting decreases the heat production of the resistor, and the element thus unwinds more slowly. This lengthens the heating cycle. Decreasing the setting increases the heat output of the resistor, adding heat and making the element unwind faster. This shortens the heating cycle. This adjustability also helps the thermostat function better with different types of heating systems. A warm-air heating system, for example, will warm the air in a room faster than a baseboard hydronic system will, and the airflow keeps air moving past the thermostat. On the other hand, if a baseboard hydronic system is allowed to operate too long using a higher anticipator setting, the area may become uncomfortably warm from the residual heat in the baseboard units.

In summary, setting the heat anticipator may begin with setting it to the current of the heating control circuit, but changes commonly match the type of heating system and structure and provide occupant comfort. Note two important things about the anticipator setting:

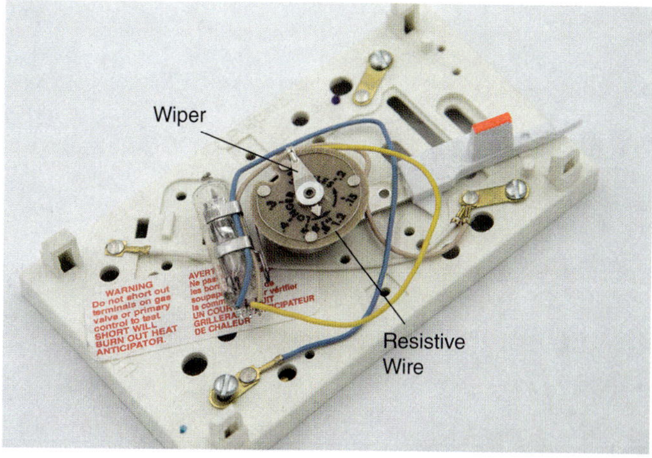

FIGURE 12-7 Bimetallic thermostat heat anticipator.

- Longer cycles improve the operating efficiency of most HVACR systems, especially fossil-fuel furnaces and cooling systems.
- Long cycles may result in unacceptable comfort for some users, but **short-cycling** reduces efficiency and increases wear and tear on the equipment.

There are several heat anticipator styles. Another common type is shown on the classic Honeywell T87F thermostat in **FIGURE 12-8**. What they have in common is a wire filament with a precise resistance per unit length. When the heat anticipator is adjusted, the length of the wire placed in the control circuit is changed. The setting lever, through its contact with the wire, actually completes the anticipator circuit.

FIGURE 12-9 shows the circuits through a common thermostat. In this example, the thermostat heating circuit is closed, so the heating system is running. The heat anticipator is shown using the drawing symbol for a variable resistor, and it is energized, providing heat.

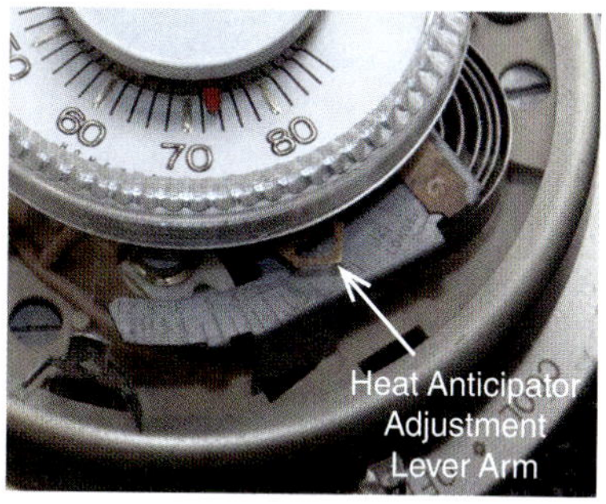

FIGURE 12-8 Honeywell T87F thermostat heat anticipator.

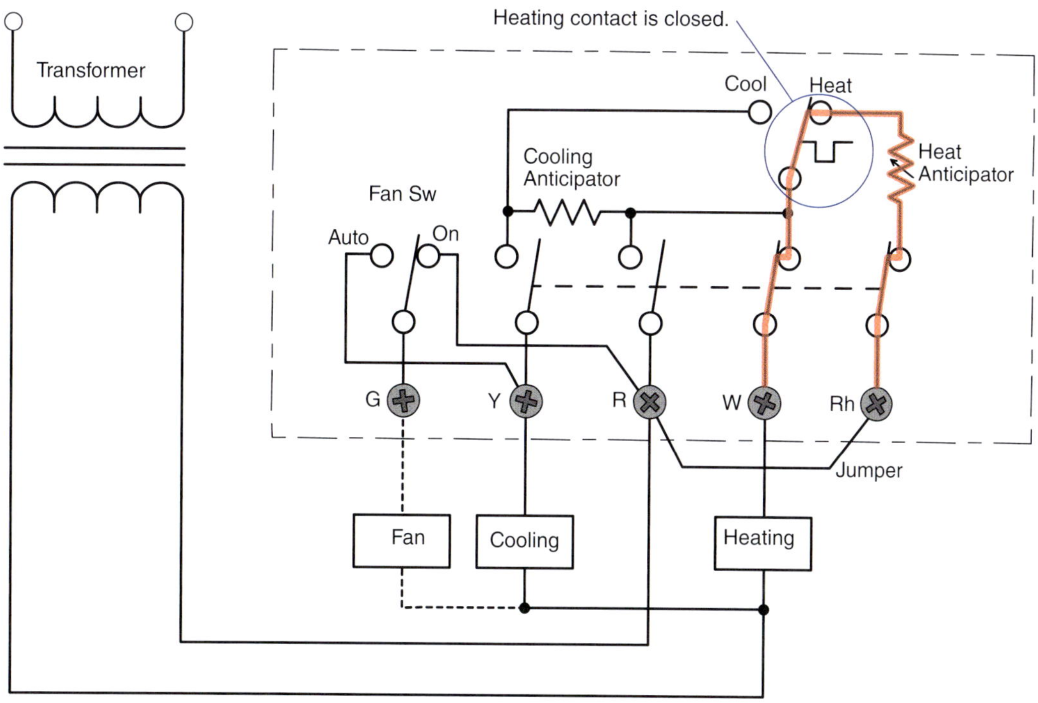

FIGURE 12-9 Heat anticipator circuit, with heating circuit closed.

FIGURE 12-10 Fixed cooling anticipator.

A cooling anticipator is also provided (**FIGURE 12-10**). However, it is rarely given much thought, since it isn't adjustable. It is a simple fixed resistor, and it is not in series with the contacts of the thermostat. Instead, the added warmth is needed when the cooling circuit is open (during the off cycle), so it is wired in parallel and energized whenever control power is available to it.

The cooling anticipator circuit is shown in **FIGURE 12-11**. The cooling circuit in this example remains open (room temperature is below the set point), but the thermostat mode switch is in the Cool position. But since the resistor is wired in parallel with the cooling contacts, it is powered and provides warmth while the cooling system is off.

Multistage Thermostats

Since heat-pump systems typically require more than one stage of heat, they prompted the wider development of the **multistage thermostat** (**FIGURE 12-12**). Prior to that, there were multistage models available for larger commercial systems that had more than one stage of heating and/or cooling. But the entrance of the heat pump significantly increased their numbers and the addition of other switching features.

Multistage thermostats have two bulbs mounted to a single bimetallic coil, and they usually have more than one coil assembly, as shown in the figure. One coil is used for heating, and the other for cooling. The bulbs are mounted at slightly different angles so that the mercury rolls to close the contacts of one stage of heating or cooling first. The difference in bulb mounting angles is chosen so that the closure of the second-stage control circuit lags two to three degrees behind the other. In heating for example, the first-stage circuit closes as the temperature falls to, or just below, the set point. If the first stage handles the demand and the temperature rises, the circuit will open and the system will shut down. If it doesn't and the temperature continues to fall, the coil winds tighter and the mercury in the second bulb will roll over to close the second-stage heating circuit. The

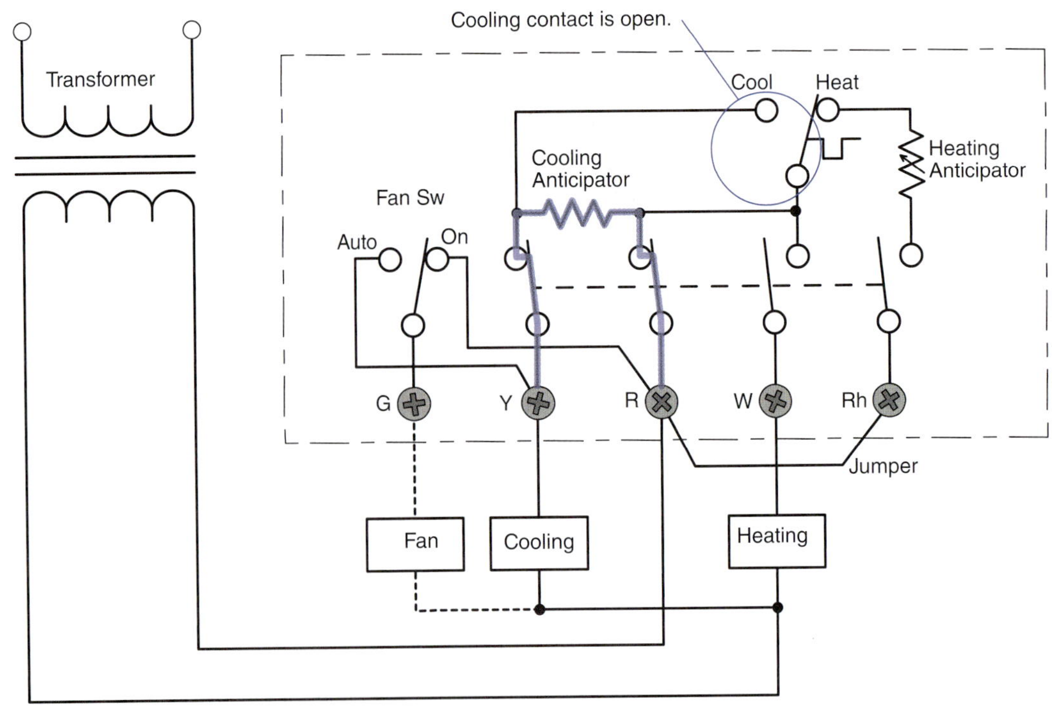

FIGURE 12-11 Cooling anticipator circuit, with cooling circuit open.

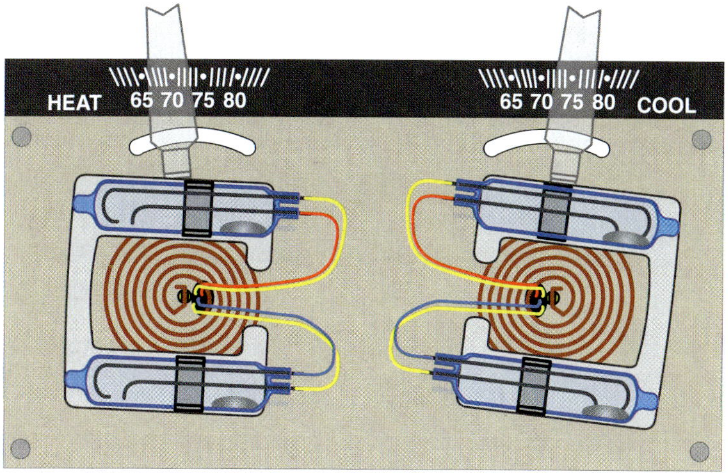

FIGURE 12-12 A mercury-bulb multistage thermostat.

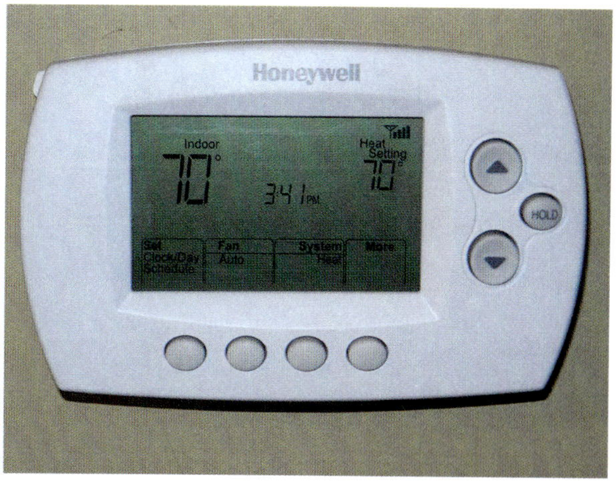

FIGURE 12-13 Typical digital thermostat.

second stage of heating or cooling will be called upon at the same time as the first if the user makes a set point adjustment of three degrees or more at once.

▶ Digital Thermostats

K12002 Identify and describe digital comfort-system thermostats.

Digital thermostats have nearly taken over the market for new installations and replacements. Their cost has fallen to a competitive level, especially given their additional flexibility and features. Even simple systems can benefit from the flexibility and reliability of a digital thermostat.

There are several names used for electronic thermostats: electronic, solid-state, programmable, smart—all of which apply to at least some degree. In fact, some thermostats could be called by all of these names with accuracy. When these terms are used, they generally refer to thermostats that operate by using solid-state components and provide a digital display.

FIGURE 12-13 shows an example of a digital thermostat. The digital display provides a completely different look than that of a standard thermostat, so they are not hard to identify. There are a wide range of models and features from which to choose.

Most digital thermostats are designed to interface with a standard 24 VAC control system. To that end, not everything about them is necessarily solid state. Most have miniature relays attached to the PC board that control 24 VAC circuits through the thermostat. The electronics then control the relay. Some also have traditional switches that are opened and closed manually.

Sensing Temperature

Instead of a bimetallic coil, digital thermostats use one or more **thermistors** (FIGURE 12-14) to sense temperature. A thermistor is a special resistor that has a highly predictable and accurate resistance determined by the surrounding temperature. The thermostat's microprocessor translates the resistance to a temperature value for display and uses that value to determine room temperature.

FIGURE 12-14 Thermistor on a digital thermostat.

Heat Anticipation

Heat anticipation by adding heat near the thermistor is not necessary. Instead, digital thermostats are furnished with a setting to adjust the cycle rate. In most cases, the number entered reflects the number of times the system will be allowed to cycle per hour. The microprocessor examines trends in the operating cycles and

FIGURE 12-15 Outdoor thermistor.

FIGURE 12-16 A thermostat being programmed.

> **▶ TECHNICIAN TIP**
>
> **Thermistors**
>
> Some digital thermostats can accept inputs from several thermistors positioned around a space or home. The microprocessor then averages the values to have a better idea of what the conditions are like throughout the area. Some also offer a thermistor accessory (**FIGURE 12-15**) that is placed outdoors, enabling the thermostat to display the outdoor temperature and determine the trend of outdoor temperature changes. This information is then used to predict the heating or cooling load with some precision. Outdoor thermistors are also used to limit the operation of heat-pump supplemental heat, and to determine when to switch from one fuel to the other in a dual-fuel system.

may adjust length of the operating cycles to compensate for the trend. If the operating trend has been an increasing demand for heat, each heating operating cycle will be slightly longer, or shorter if the demand has been decreasing. This is just one example of how solid-state components have made the thermostat a far more intelligent and flexible temperature-control device.

Programmability

As a general rule, a programmable thermostat is one that can accept a schedule of temperature-setting changes during a single day and, often, for each day of the week. Most digital thermostats have a variety of settings and configurations that can be changed via the touchscreen and/or keypad, but these features are not what are generally thought of when the term "programmable" is used.

FIGURE 12-16 shows a digital-thermostat display in the programming mode. This is referred to as five-day programming. Some thermostats like this one, allow for a series of set point changes that are fixed for five days of the week, and then a separate set of set points can be programmed for the remaining two days of the week. This is referred to as five/two programming.

For example, during the week, a homeowner may rise at 7 a.m. and all family members have left by 8:30 a.m. The family begins returning at 4 p.m., and then everyone is in bed by 10 p.m. On the weekends, some family members are home all day, and the times for rising and sleeping may be different as well. A seven-day programmable thermostat can be set as shown in **TABLE 12-1** to save energy but preserve comfort when desired.

Programmable thermostats typically offer a Hold setting that can be used to set a specific temperature and hold it for a specific period of time. This setting ignores the programmed schedule. Of course, regardless of the programmed schedule and the active set point, users can adjust the temperature settings. The thermostat will respond to the change until the next scheduled event arrives and then return to its scheduled settings.

TABLE 12-1 Example thermostat programming schedule.

PERIOD	Time	Heating Set Point	Cooling Set Point
Wake	7:00 a.m.	74°F (23°C)	78°F (25.6°C)
Leave	8:30 a.m.	65°F (18°C)	82°F (27.8°C)
Return	4:00 p.m.	74°F (23°C)	78°F (25.6°C)
Sleep	10:00 p.m.	68°F (20°C)	73°F (22.7°C)
Weekend Wake	8:30 a.m.	74°F (23°C)	78°F (25.6°C)
Weekend Sleep	11:00 p.m.	68°F (20°C)	73°F (22.7°C)

Another common feature of programmable thermostats is replaceable batteries (**FIGURE 12-17**). Batteries are provided to allow the thermostat to keep its programming in the event of a power failure. There are usually two to four standard AA or AAA batteries. Some newer thermostats use coin-cell batteries, like the style found in watches. New thermostats may or may not ship with batteries included.

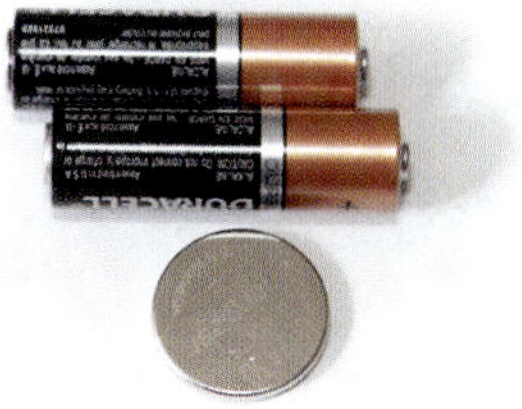

FIGURE 12-17 Thermostat batteries.

Other Digital-Thermostat Features

Digital thermostats have come a long way since their early introduction and are now capable of joining home networks, communicating through portable devices, and even communicating by voice.

Touchscreen Operation

Interacting with computers, tablets, and smartphones through the use of touch-enabled screens is very common today. This is also a common feature found in digital thermostats. Some thermostats are operated exclusively through a touchscreen, while others combine touchscreens with traditional switches and buttons.

Wi-Fi Capabilities

Although the concept of an "Internet of Things" is not new, the fact that it now exists and continues to evolve is new. Today, all sorts of appliances can be connected to home and business networks. To the average user, the heating/cooling system is an appliance, and it can now be a part of the Internet of Things through a wireless connection to the local network. Thermostats with Wi-Fi capabilities are becoming more common, providing owners with great flexibility. Apps for smartphones, tablets, and computers (**FIGURE 12-18**) are readily available and generally free with the purchase of the thermostat. Users can have the thermostat set on Hold while on vacation and change it on the way home by connecting to their thermostat via the Internet, at which point they can raise or lower the temperature set point before they return home. While they are gone on that vacation, they can also check the temperature at home and see if everything is working as it should. Note that an Internet connection also allows thermostats to synchronize with a central time source.

FIGURE 12-18 Apps can communicate with a capable thermostat, which is becoming increasingly popular.

Voice Control

More devices are now able to listen and respond to voice commands. Gaming systems were some of the early proponents of voice-controlled functions in the home. Some of today's digital thermostats can also be controlled by using a wake-up command, such as "Hi, thermostat," and then speaking the correct words to change settings or modes of operation.

▶ Mounting and Wiring Comfort-System Thermostats

K12003 Identify common thermostat mounting requirements and wiring terminal designations.

S12001 Wire a thermostat to control an electric-cooling/gas-heating system.

S12002 Wire a thermostat to control a heat-pump system with supplemental electric heat.

Regardless of the type of thermostat, mounting and making the wiring connections is much the same. Digital thermostats with lots of features and in control of sophisticated systems may have more wires to deal with, but the process doesn't change.

Article 725 of the *National Electric Code* (*NEC*) lists requirements for the installation of thermostat wiring. Thermostat wiring is considered to be Class 2 wiring (low voltage), and therefore, it does not have to meet the same requirements for higher-voltage installations. Here are several important points to remember:

- Thermostat wiring must often pass through ceiling or floor areas that serve as a **plenum** for the HVAC system. To pass through these areas, and through any type of ductwork

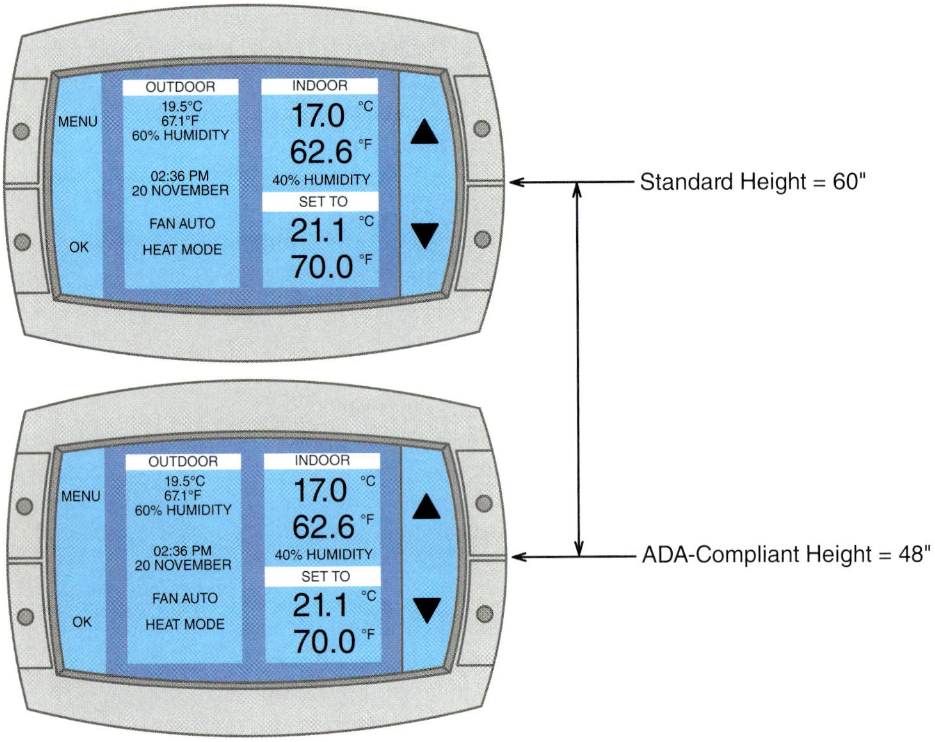

FIGURE 12-19 Thermostat mounting height.

that provides air for comfort purposes, the wire must meet the requirements for a CL2P or CL3P classification. The "P" stands for plenum. These materials are tested to ensure they do not substantially support combustion and they generate limited smoke when burned. Review this topic in the *NEC* when in doubt about the type of wire required for a given situation.

- Class 2 wiring cannot be placed inside **conduit**, junction boxes, or any other kind of **raceway** with Class 1 wiring (consider this to mean line-voltage power wiring). This is due to the thermostat wire not having the same level of electrical insulation as line-voltage conductors.
- It is often convenient for thermostat wiring to be routed near conduit or other types of raceway material for power wiring. Note that *NEC Article 725.143* specifies that Class 2 wiring cannot be strapped or taped to any type of conduit or raceway for support.
- The *NEC* lists many conditions and building areas where certain types of thermostat wire cannot be installed because they're unsafe. This is especially true in commercial or multi-family residential construction. Consult the *NEC* and/or the construction specifications before selecting the wire.
- The *NEC* does not specify a mounting height for wall thermostats. However, the Americans with Disabilities Act (ADA) specifies that thermostats must be mounted no higher than 48" (122 cm). Any structure that must comply with ADA specifications must use this mounting height. Otherwise, the normal mounting height is 60" (152 cm) (**FIGURE 12-19**).

Installing a Thermostat

Most thermostats consist of two parts: the thermostat assembly and a subbase. The subbase may also be referred to as a wall plate. The subbase provides the mounting holes for screws, a path for the thermostat wiring to enter through the back, and the thermostat-wiring terminals.

Thermostat Location

Keep these tips in mind when considering where to place a thermostat:

- Choose a centrally located area. The thermostat is best located near the return-air grille for the system. This provides good access to the average temperature and sufficient air movement.
- Do not place the thermostat where a supply-air register can blow warm or cold air onto it. This will make the system cycle too quickly.
- Avoid locations that expose the thermostat to direct sunlight, as well as locations near a heat source, such as a lamp.
- Do not install a thermostat on an outside wall or near outside doorways where cold or hot air coming and going will affect it. Also avoid a location that is exposed to intense sunlight.
- Avoid walls that can shake or vibrate. This is especially a concern for bimetallic and mercury-bulb thermostats.

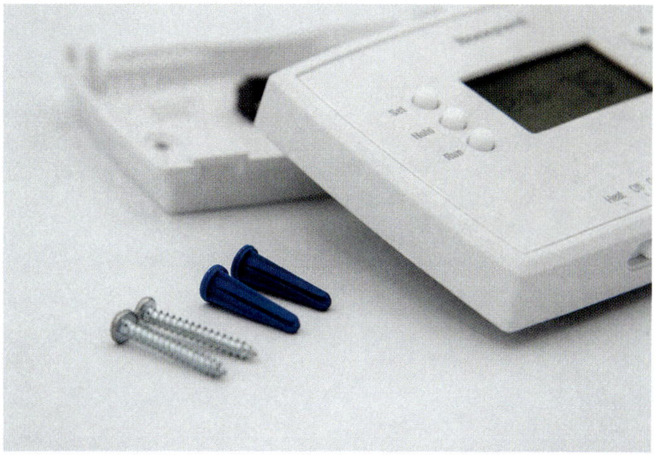

FIGURE 12-20 Thermostat hardware kit provided with a new thermostat.

Before the walls were enclosed on each side, thermostat wiring should have been routed (roughed-in) to the mounting location and through the wall, passing through a small hole at the correct mounting height. It is best to keep this hole small so that it does not interfere with the wall-plate mounting screws and so as to minimize drafts through the wall.

Sheetrock is a popular wall covering, and one that doesn't hold a screw very well. It is best to use some type of wall anchor, and most new thermostats come with some type of anchor included with the mounting screws (**FIGURE 12-20**). The thermostat wire may be routed through the wall next to a wall stud. If so, this might allow a mounting screw to be put into the stud, wood or metal, for a more secure mount.

Mounting the Subbase

To begin the installation, ensure that system control power has been deenergized. Use a voltmeter to confirm it. Since the control power originates from a line-voltage piece of equipment, lock out and tag its power source. Place the subbase on the wall where the wire protrudes, threading the wire bundle through the opening in the subbase. Refer to **FIGURE 12-21**. Most mounting holes are oblong so that they provide leveling adjustment after mounting. Align the subbase on the wall and use a level across the two points identified for this purpose (**FIGURE 12-22**). Once level, mark the center of the two mounting holes.

▶ FOCUS ON CUSTOMER SERVICE

A Level Thermostat

Mercury-bulb thermostats must be mounted level to operate accurately. The position of the bulbs is factory calibrated to a level position. Although snap-action thermostats and digital thermostats do not need to be level to operate accurately, a thermostat should always be level on the wall, regardless of what kind it is. Nobody likes to look at a crooked thermostat, any more than they want to see a crooked picture on the wall. Clients are likely to wonder about the quality of the work that they can't see if the work that they can see is unsatisfactory.

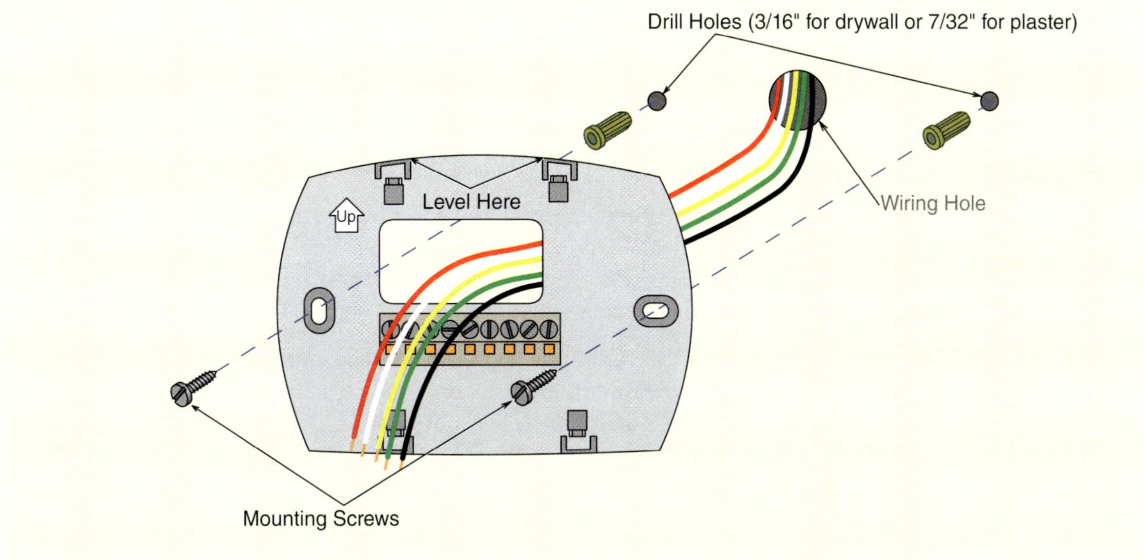

FIGURE 12-21 Mounting a subbase.

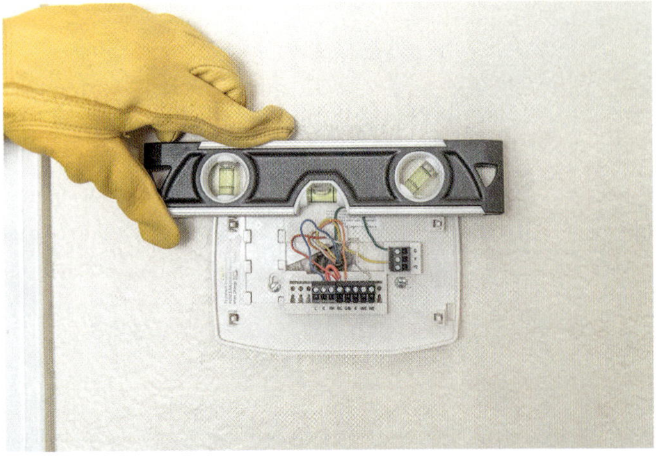

FIGURE 12-22 Leveling the subbase.

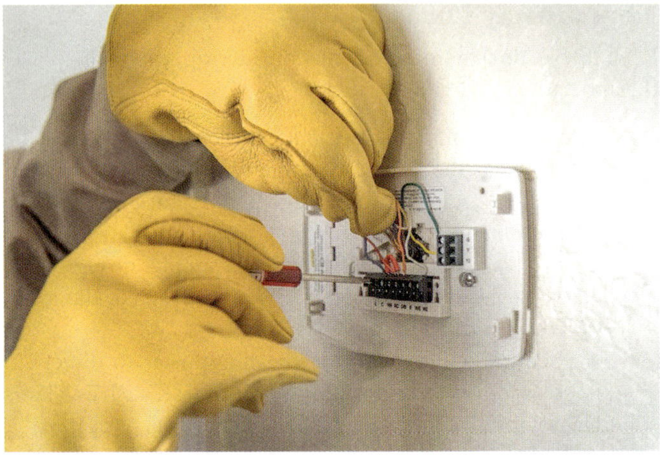

FIGURE 12-23 Terminating the thermostat wires.

Install the wall anchors at the marked locations. Place the subbase on the wall again, with the thermostat wire threaded through the opening, and insert the screws. Snug the screws, use the level once again to set the subbase in a level position, and then tighten the screws.

The next step is to terminate the individual thermostat wires. Strip the wire only as necessary; the insulation near the terminations needs to stay intact and undamaged so that shorts between the wires cannot occur. Examine the connector carefully to determine how much needs to be stripped. Neatly arrange the wires in order of the terminal markings before making any connections. Insert the wires in their proper terminals and tighten the terminal screws (**FIGURE 12-23**).

If batteries are needed, install them into the battery holder on the subbase or on the back of the thermostat. Note that as soon as the batteries are inserted, the display will usually power up. Therefore, do not to install the batteries until just before the system is ready for start-up, to keep the batteries from being drained. Then, following the manufacturer's instructions carefully, screw, snap, or push the thermostat into position onto the subbase. Once the rest of the system is prepared for start-up, control power can be restored.

> **▶ TECHNICIAN TIP**
>
> In the process of attaching the thermostat to its subbase, in many cases a technician must also align a series of small pins with a matching electrical connector on the subbase. These pins connect the incoming wires to the thermostat's PC board. Ensure that thermostat is properly aligned with its subbase, and do not force it into position. Permanent damage to the pins can result.

Thermostat-Wiring Terminals

The HVACR industry has a reasonably standardized method of identifying the role of 24-VAC thermostat wires. The term "reasonably" is used because there are certainly variations used by some manufacturers. Years ago, there were only a few terminals to manage. Today, there are many more possibilities since the thermostats as well as the systems have become more sophisticated. When variations are encountered, consult the manufacturer's information and determine what the role of the wire is, then match it with the appropriate terminal on the opposite end. When in doubt, consult the manufacturer.

In a perfect world, there would be a thermostat wire color in every bundle that perfectly matches the terminal marking. Of course, it isn't a perfect world. The most common terminal designations and wire colors for thermostats are as follows (**FIGURE 12-24**):

- C—(Common) power from the common side of the transformer to power thermostat functions; black wire.
- R—the other leg of power from the transformer, which will be switched to all the other terminals; red wire.
- Y—cooling operation; yellow wire.
- W—heating operation; white wire.
- G—indoor blower operation; green wire.

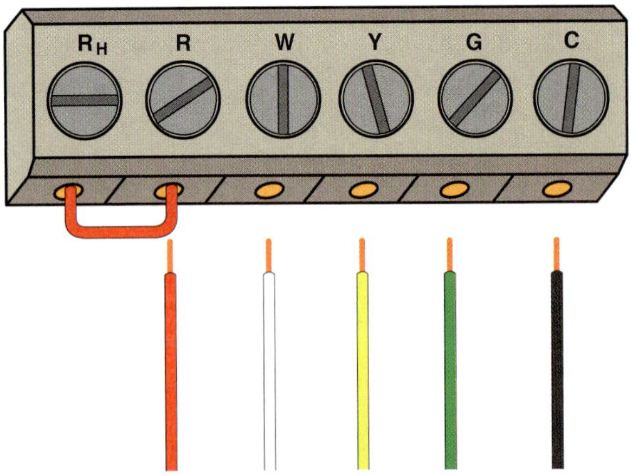

FIGURE 12-24 Standard thermostat-wiring terminals.

Although the "C" for the Common wire doesn't match the black color, it is generally the color in the bundle used for this purpose. The wire color of the other listed terminals corresponds to the respective terminal. The problem comes when terminals like Aux/E and X enter the picture. With so many terminals possible today, it is rare that there will be a matching color in the bundle for all the terminals. The result is that thermostat wiring is not about matching letters and colors without any thought; it is about matching each wire to its corresponding terminal on the other end that allows it to control its function. Keep that in mind during every installation and service call.

TABLE 12-2 shows standard terminal designations for thermostats. It is important to note that some of these designations can change from manufacturer to manufacturer, and from generation to generation of equipment. The system and thermostat manufacturers' instructions should always be consulted when there is any doubt about the use of a terminal.

It is important to understand the relationship of R to most of the other terminals. **FIGURE 12-25** shows a simplified diagram of a cooling circuit using a digital thermostat. C is used to power the thermostat electronics, along with R; many non-digital thermostats do not require C, because there is nothing internally to power. When cooling is needed, the thermostat connects R to Y and to G. The wires connected to these terminals take that leg of power back to one side of the compressor contactor coil (Y) and to the indoor fan relay (G). C is routed to one side of all common control components already and is not switched. Thus, when the R circuit is completed through the thermostat to Y and on to the contactor coil, then it has 24 VAC power and the coil is energized.

TABLE 12-2 Common Thermostat Terminal Designations

Terminal Designation	Common Color	Description
C	Black (sometimes Blue)	Common leg of power from the control transformer; typically provided to power the thermostat itself.
R or R_C	Red	The other leg of power from the control transformer; this leg is switched by the thermostat and routed to other terminals to energize system control components.
R_H	Red	For systems that have more than one source of control power, this terminal is used to power heating control. In systems with only one control power source, R/R_C is physically connected to R_H with a jumper wire.
G	Green	Indoor blower control
Y or Y_1	Yellow	Cooling control, or the first stage of cooling if there is more than one. For heat pumps, Y energizes the compressor in both the heating and cooling modes. The reversing valve position determines the operating mode of the outdoor unit (see O and B).
Y_2, Y_3, etc.	N/A	Cooling control for the remaining stages of cooling.
W or W_1	White	Heating control, or the first stage of heating if there is more than one.
W_2, W_3, etc.	N/A	Heating control for the remaining stages of heating.
E	N/A	Related to the use of emergency heat (backup heat) for heat pump systems. When emergency heat is selected, an LED or an icon on a digital display indicates the choice. First stage of heat (the heat pump) is locked out; in some systems, emergency heat control may also be switched to the W/W_1 terminal to accurately maintain the thermostat set point, eliminating the common 2–3-degree differential in the second stage of heat.
Aux	N/A	Usage may vary, but often the same as E, or may be used in a related way for backup heat operation in a heat pump.
O	Orange	Energized whenever the thermostat is switched to the cooling mode. Can be used to energize a damper or other system component, but most often used to energize the reversing valve of a heat pump that defaults to the heating mode.
B	Blue or Brown	Energized whenever the thermostat is switched to the heating mode. Can be used to energize a damper or other system component, but most often used to energize the reversing valve of a heat pump that defaults to the cooling mode.
S_1, S_2, etc.	N/A	Varies; often used to connect external thermistors to the thermostat.
L	N/A	Varies; often used to illuminate an LED or display and icon when energized by the outdoor unit. Also used to communicate the use of Emergency Heat to zone panels.

▶ TECHNICIAN TIP

Stealing Power

Some digital thermostats do not require a Common wire to function. Instead power is stolen from the R terminal and through the other connected wires, many of which lead back to Common through a relay coil. The thermostat can steal a tiny amount of power through the wiring for itself, but stealing an excessive amount would cause equipment to operate with no request to do so. This technology was developed for digital thermostats in the early days, since many homes and businesses did not have a sufficient number of wires installed to connect a Common wire to the new thermostat. All manufacturers wanted to market simple installation features to homeowners as well as installers. In some cases, the approach can create problems. There have been reports of systems running when they should not as the battery voltage becomes too low. It is always best to power a digital thermostat by providing a Common wire, whenever possible.

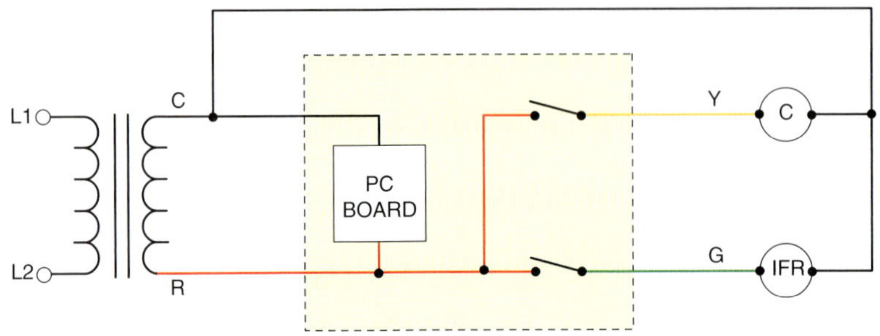

FIGURE 12-25 Simplified diagram of transformer control power.

Note that in the heating mode, R is not usually connected to the G terminal on a call for heat. Gas and oil furnaces start the blower from the furnace, not the thermostat. Heat-pump thermostats, however, do start the indoor blower when started in the heating mode.

Common should never be switched in an HVAC control circuit. It is referred to as Common because it provides a common leg of power to control-circuit components without being switched. R is switched many different ways and places in order to control the devices. When learning about control circuits, it is vital to remember the following: R is the switched leg of control power, and its path to a given device is often what needs troubleshooting when things don't work as they should.

To wire a thermostat to control an electric-cooling/gas-heating system, follow the steps in **SKILL DRILL 12-1**.

To practice wiring a thermostat for a heat-pump system with one stage of supplemental electric heat, follow the steps in **SKILL DRILL 12-2**.

SKILL DRILL 12-1 Wiring a Thermostat to Control an Electric-Cooling/Gas-Heating System

For this task, an appropriate heating/cooling thermostat is required, as well as a functional heating/cooling system so that testing can be done upon completion of the wiring.

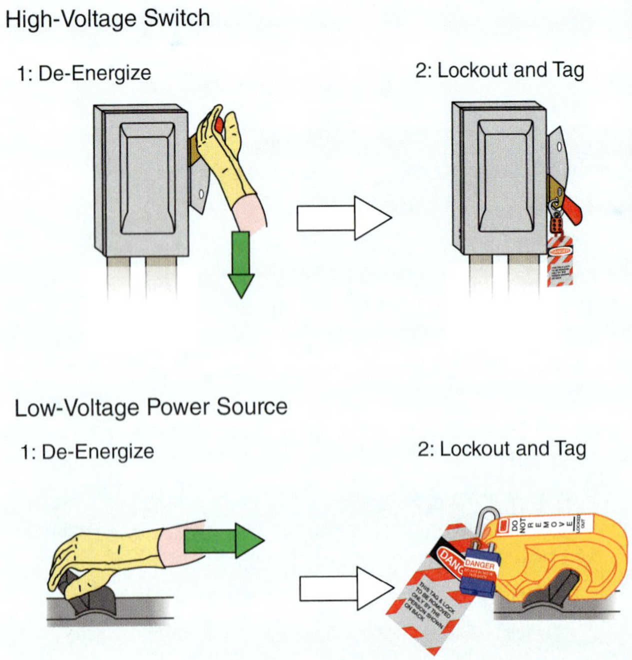

SAFETY TIP

Working with electrically energized circuits is hazardous. Avoid contact with any surface that could potentially be energized. When working inside energized enclosures, maintain a stable position and, whenever possible, avoid placing both hands inside the enclosure. All work on energized equipment by unqualified individuals should be supervised by a qualified instructor or technician.

1. Deenergize, lock out, and tag the equipment being serviced. Be sure that the process deenergizes not only the line voltage feeding the equipment but also the low-voltage control-power source that may be located in a separate piece of equipment.

Continued

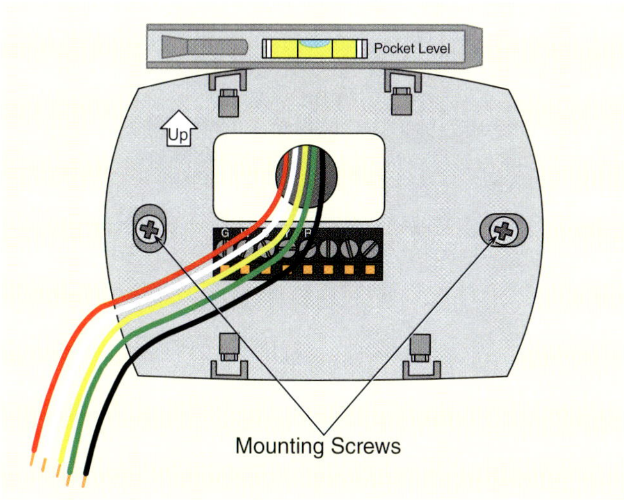

2. Remove the thermostat from its packaging and locate the mounting hardware. Route the thermostat wire bundle through the opening in the back of the subbase and mount the subbase to the wall per the manufacturer's instructions. Be sure to level the subbase on the wall before marking the position of the mounting screws.

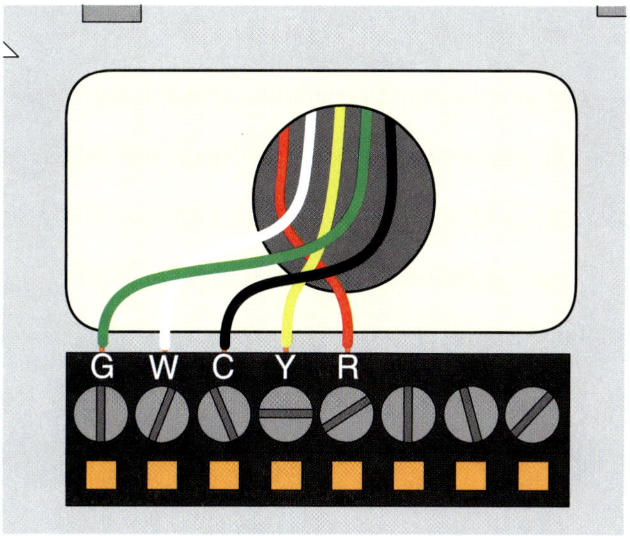

3. Examine the equipment wiring diagram and the thermostat control wiring connections at the equipment terminal board. Identify which wire color in the thermostat bundle is connected to R, Y, W, and G at the equipment, and record this information as necessary. If the thermostat requires a Common wire for power, examine this terminal as well. Strip an appropriate amount of insulation from each wire and connect it to the proper terminal of the subbase. Maintain the same color code as was used at the equipment terminal board. Make sure any exposed wires cannot make contact with each other or another terminal. If the thermostat wire bundle is lacking one of the primary colors, a color substitution will be required.

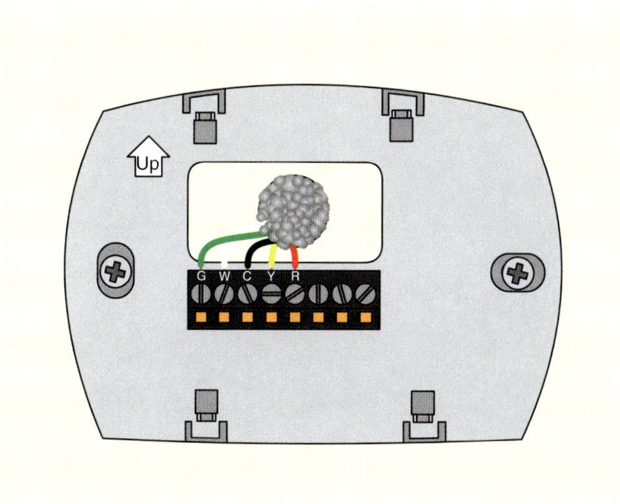

4. Neatly and carefully tuck any excess wire back into the wall; do not kink the wires too sharply. Protruding wire will interfere with mounting the thermostat. Seal the opening tightly with a ball of fiberglass insulation, soft putty, or caulk.

Continued

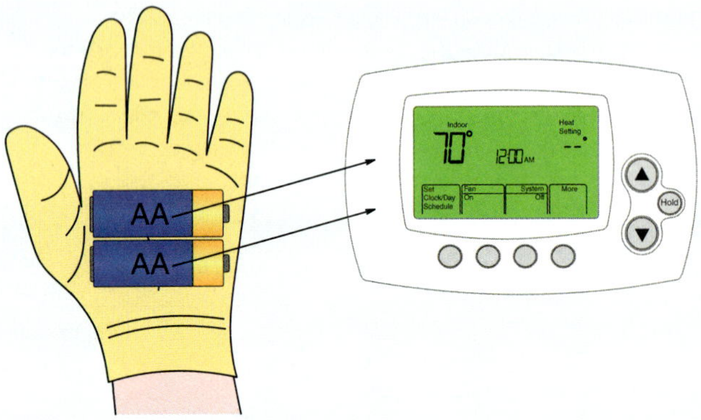

5. Following the manufacturer's instructions, mount the thermostat to its subbase. If the thermostat is digital and requires batteries, install them before mounting the thermostat. Note that as soon as the batteries are installed, the thermostat display becomes active. Check the thermostat settings and ensure that the mode is set to Off. Also ensure that the fan switch is set to Auto.

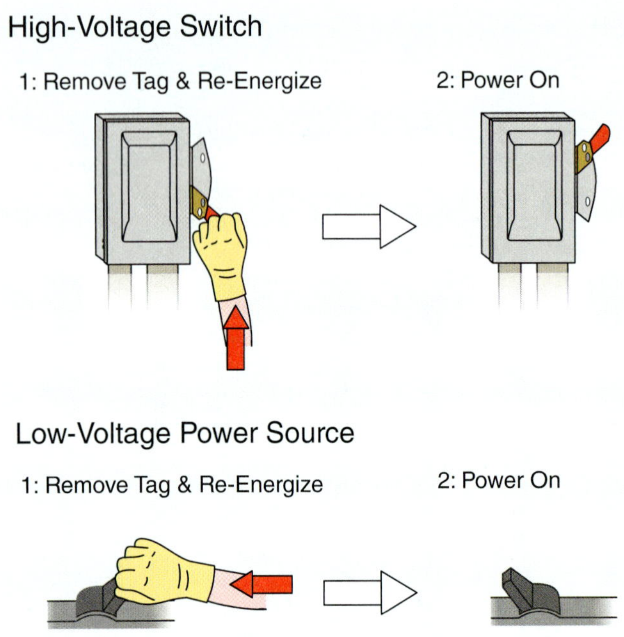

6. Remove the locks and tags from the power source(s). Restore power to the system and verify with a meter that both line voltage and control power are available.

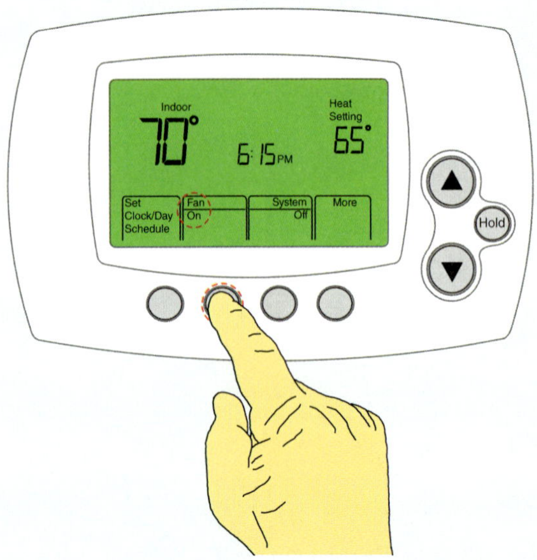

7. Begin with a simple test of indoor blower control. Set the thermostat to Fan On for continuous air circulation. Verify that the system indoor blower started.

Continued

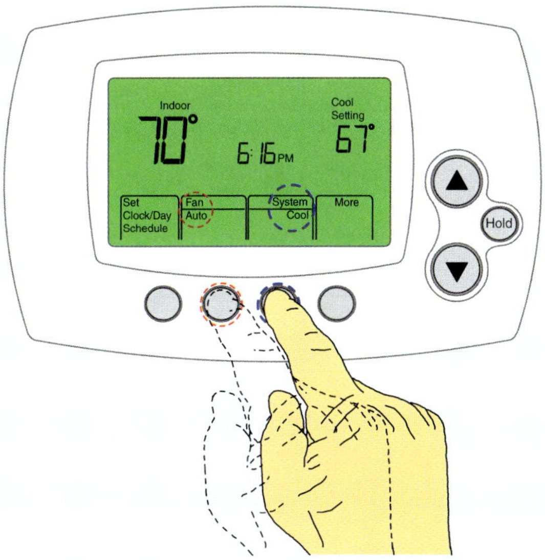

8. Once fan operation has been verified, return the fan switch to Auto. Next, change the mode to Cool and lower the set point until it is several degrees below room temperature. Verify that both the indoor blower and the condensing unit have started. If the condensing unit does not start immediately, it may be locked due to short-cycle protection. If so, allow the remaining time to elapse.

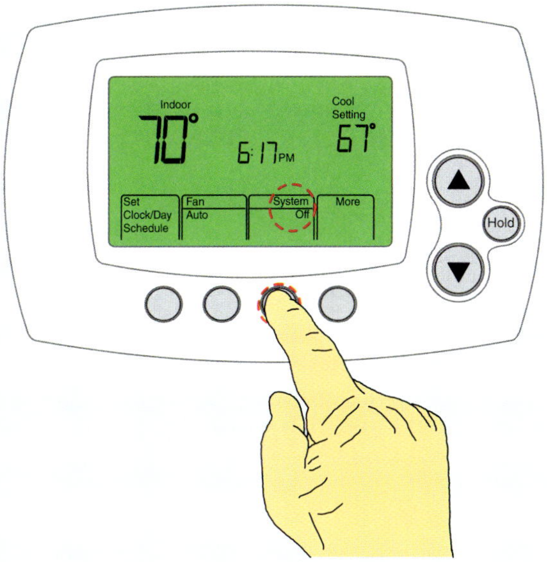

9. Once the operation of the cooling equipment has been verified, return the mode switch to Off. Allow time for the indoor blower to shut down, because it may be controlled by an off-delay timer. Ensure all cooling system components shut down as expected.

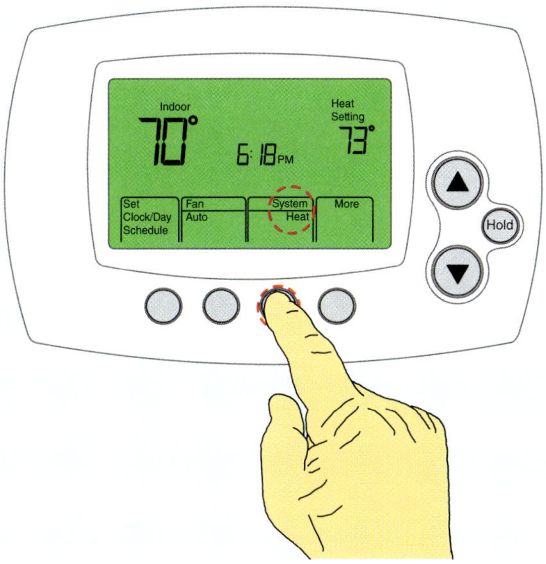

10. Next, check the operation of the heating equipment. Remember that the indoor blower of fossil-fuel heating system will be started by the heating unit controls, not the thermostat. Set the mode to Heat and raise the set point until it is several degrees above room temperature. Verify that the furnace has initiated a heating cycle and that the indoor blower starts soon after the burner has started operating.

Continued

11. Set the mode switch to Off and verify that the heating equipment and indoor blower shut down in the expected sequence. If the heating system responds as expected, the task is complete. Clean the work area, reinstall any equipment access doors, and return the system to service.

SKILL DRILL 12-2 Wiring a Thermostat to Control a Heat-Pump System with Supplemental Electric Heat

For this task, an appropriate heating pump thermostat is required, as is a functional heat-pump system so that testing can be done upon completion of the wiring.

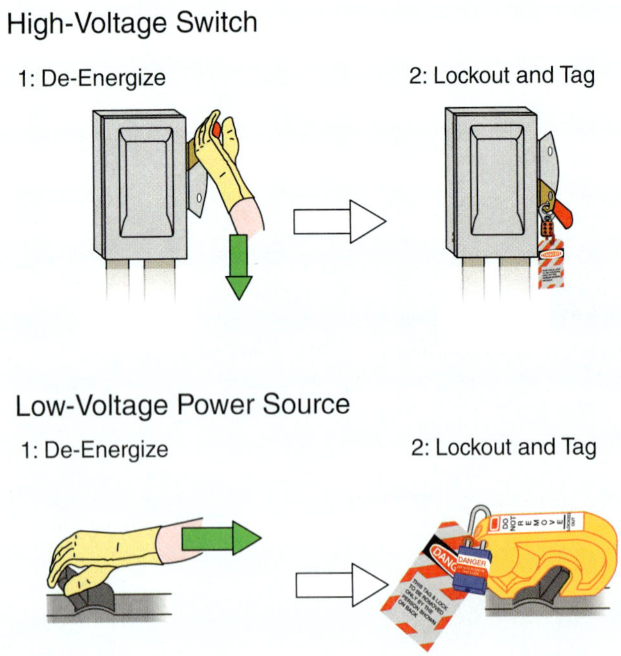

SAFETY TIP

Working with electrically energized circuits is hazardous. Avoid contact with any surface that could potentially be energized. When working inside energized enclosures, maintain a stable position and, whenever possible, avoid placing both hands inside the enclosure. All work on energized equipment by unqualified individuals should be supervised by a qualified instructor or technician.

1. Ensure that all controls are set to the Off position. Deenergize, lock out, and tag the equipment being serviced. Be sure that the process deenergizes not only the line voltage feeding the equipment but also the low-voltage control-power source that may be located in a separate piece of equipment.

Continued

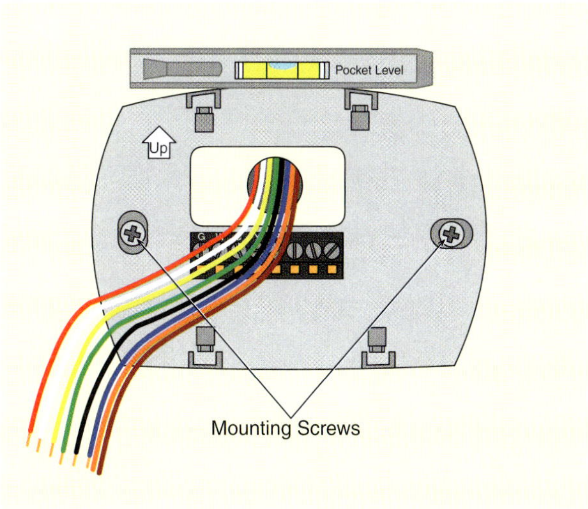

2. Remove the thermostat from its packaging and locate the mounting hardware. Route the thermostat wire bundle through the opening in the back of the subbase and mount the subbase to the wall per the manufacturer's instructions. Be sure to level the subbase on the wall before marking the position of the mounting screws.

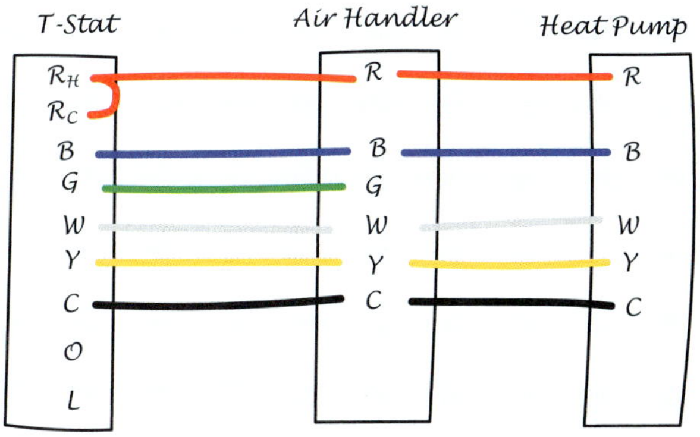

3. Examine the unit wiring diagram and the thermostat control wiring connections at the unit terminal board. Heat-pump control wiring can differ from system to system, primarily in the terminal designations for reversing-valve control. The R, C, Y, W, G, O, and B terminals are reasonably standardized among products. Heat pumps that energize the reversing valve for cooling use the O terminal. Those that energize the reversing valve for the heating mode use the B terminal. Determine what terminals are needed for the system from the unit wiring diagram, then compare them to the thermostat's wiring information. Make a simple sketch of the control wiring connections and the colors to be used, based on the colors available in a thermostat wire bundle. In the example sketch shown here, B is being used instead of (O is not needed), so the reversing valve energizes in the heating mode. Depending on the equipment and thermostat, terminal designations may not match precisely. Be sure to note any permanent jumpers that must be installed at the thermostat, such as between R and R_H or R_C.

4. Return to the thermostat. Strip an appropriate amount of insulation from each wire and connect it to the proper terminal of the subbase, maintaining the sketched color code. Make sure any exposed conductors cannot make contact with each other or another terminal. Any unused wires in the bundle should be left unstripped and folded back into the wall opening.

Continued

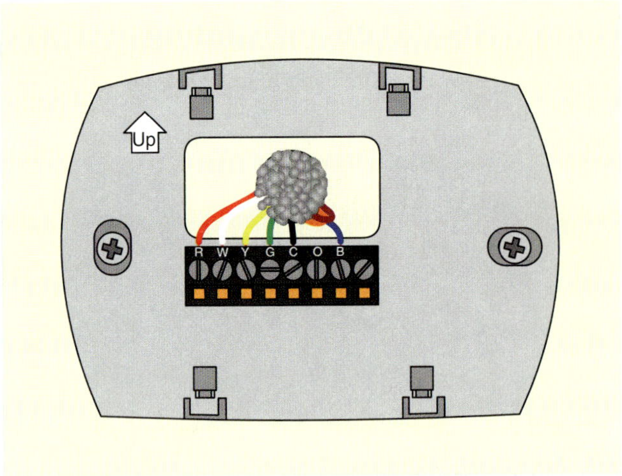

5. Neatly and carefully tuck any excess wire back into the wall. Seal the opening to prevent air from interfering with thermostat operation.

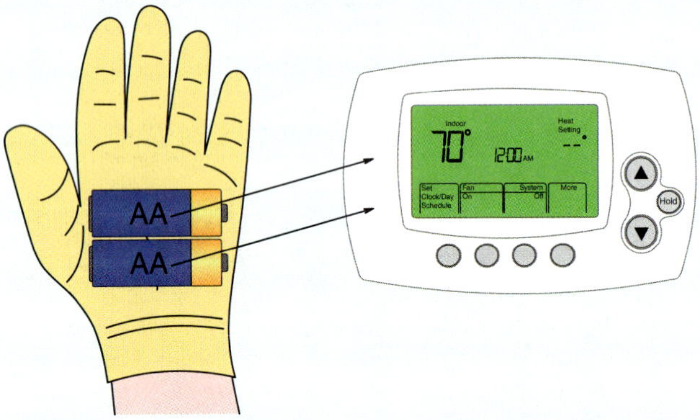

6. If the thermostat is digital and requires batteries, install them before mounting the thermostat. Following the manufacturer's instructions, carefully mount the thermostat to its wall plate. Note that as soon as the batteries are installed, the thermostat display will likely become active. Check the thermostat settings and ensure that the mode is set to Off. Also ensure that the fan switch is set to Auto.

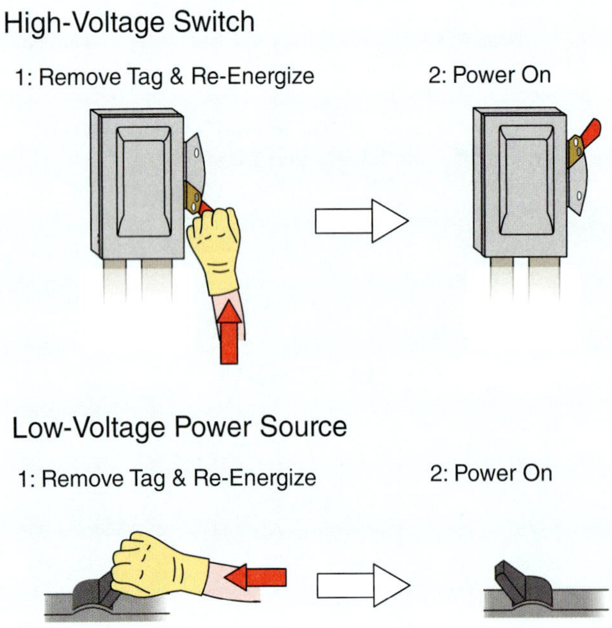

7. Remove the locks and tags from the power source(s). Restore power to the system, and verify with a meter that both line voltage and control power are available.

Continued

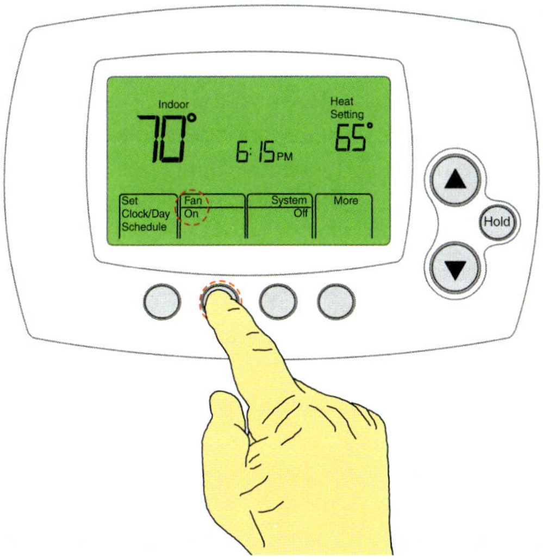

8. Begin with a simple test of indoor blower control. Set the Fan switch to On for continuous air circulation. Verify that the indoor blower started.

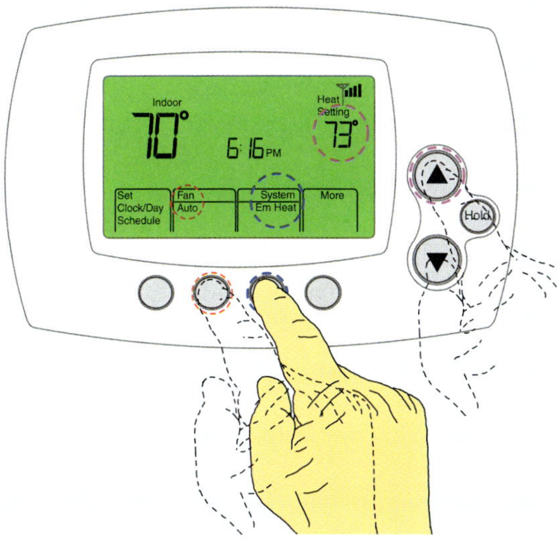

9. Once fan operation has been verified, return the Fan switch to Auto. Continue testing each operating mode in the following sequence:

- Cooling—Indoor blower starts with the heat pump compressor in the cooling mode.
- Heating—Indoor blower starts with the heat pump compressor in the heating mode. Check this mode first, with the set point no more than two degrees above the room temperature.
- Supplemental heat (not directly controlled by a thermostat switch)—Heat pump and electric heat operate together, which requires the thermostat setting to be at least three degrees above room temperature. Note that operation of the supplemental heat can be affected by the existence of an outdoor thermostat or thermistor that disables it above a certain temperature, or by the configuration of a digital thermostat.
- Emergency heat (if equipped)—Heat pump does *not* run on a call for heat; only the indoor blower and supplemental electric heat operate on a call for heat.

10. If the thermostat and system function as expected, set the mode switch to Off and verify that the equipment and indoor blower shut down in the expected sequence. Clean the work area, reinstall any equipment access doors, and return the system to service.

K12004 Describe common digital-thermostat features and settings available to users.

▶ Digital-Thermostat User Features and Settings

Although this section focuses on the user features of digital thermostats, many of these user features are common to all thermostats. Users need an intuitive thermostat interface. Most manufacturers recognize this and have designed their product accordingly.

FIGURE 12-26 shows an example of a manufacturer's guidance in understanding the thermostat display. Of course, there are as many display layouts as there are thermostats, but most have the same information in common. The example shows a thermostat that combines touchscreen control with a few traditional buttons.

The common user features are presented in the sections that follow. Refer to Figure 12-26 as necessary.

▶ TECHNICIAN TIP

Use the Manual!

It cannot be emphasized enough that technicians must have and use the correct literature for a digital thermostat. There are many models, and although many look identical, they are often differences in their setup and features. After gaining experience with schedule programming on several different models, others will become easier to navigate without a manual. But when completing the installation setup, the correct manual is essential. It is OK to use that manual in front of the homeowner and to encourage them to use it as well when they wish to make program changes. Use it as a training aid for users when appropriate.

Displays

Digital thermostats display the room temperature, and often the outdoor temperature with an accessory outdoor temperature sensor (**FIGURE 12-27**). The humidity can also be displayed if the thermostat has a humidity sensor. They display current set points and provide an indication when the heating or cooling equipment is being called upon to run. Note that the thermostat has no way of knowing if the equipment is actually running. It knows only that the equipment has been called upon.

In the heating mode, heat-pump systems will generally show whether the heat pump is running, auxiliary/supplemental heat is running with it, or the system is running in the emergency heat mode.

The time of day, day of the week, and sometimes the date are shown on many displays, unless it is a non-programmable model. There may also be an icon or the letters "DST" displayed to indicate the thermostat has switched to daylight savings time.

A filter reminder can often be programmed, prompting the user to replace the system air filters. Since the thermostat has no idea of the actual condition of the air filter, this is simply a timer that will display the reminder after some number of days or operating hours have elapsed.

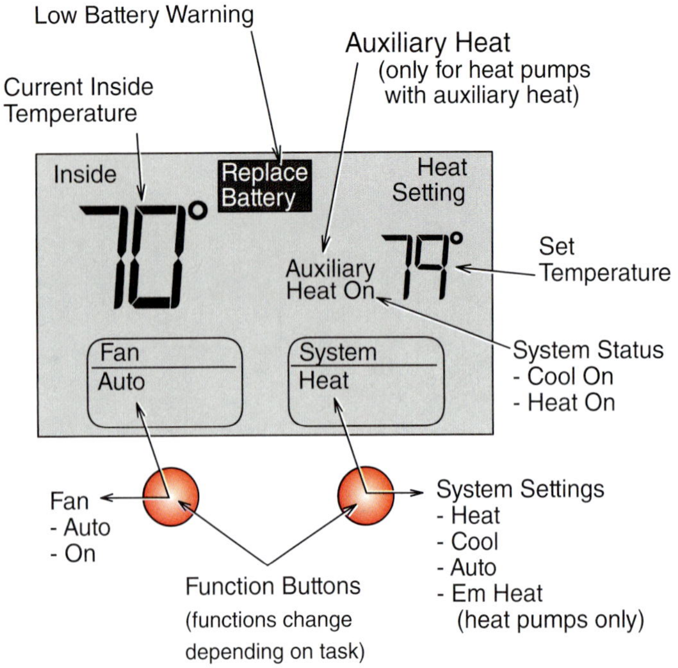

FIGURE 12-26 Digital-thermostat display guidance.

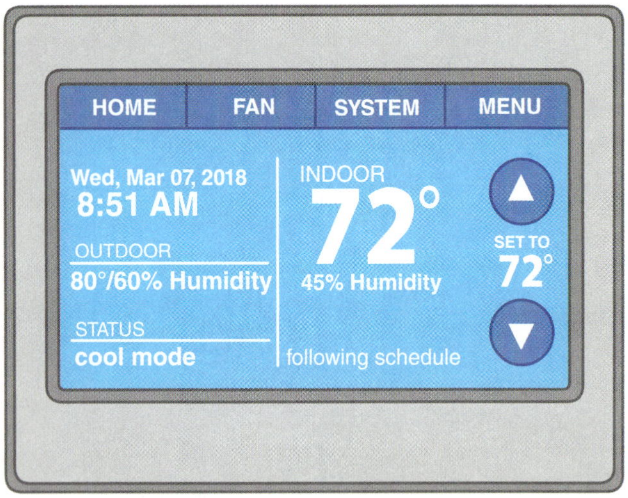

FIGURE 12-27 Example of a thermostat display.

When the batteries fall below a preset voltage, a message or icon is displayed to bring attention to it. Wi-Fi-enabled thermostats will show the status of the network connection and its strength, much like a smartphone or tablet.

Mode Selection

The user can typically choose between the following modes, using a touchscreen or traditional switches:

- Cooling—Only the cooling system will operate, regardless of how low the temperature drops.
- Heating—Only the heating system will operate, regardless of how high the temperature rises.
- Auto—The thermostat will use both modes to maintain the temperature set point. There is usually a 3°F (1.7°C) range between the heating and cooling set points that the user cannot eliminate. This 3°F (1.7°C) range is called the **deadband**.
- Emergency Heat (*heat pump only*)—This mode locks out a heat pump and operates only the supplemental heat source (commonly electric heat) to maintain the heating set point.
- Fan On or Auto—This mode allows the user to operate the fan continuously for air circulation or allows it to cycle on and off along with cooling and heating control.

Temperature Settings

Refer to **FIGURE 12-28**. Temperature settings are usually made by using arrows on a touchscreen or by using traditional buttons while the value shows on the display. Changes in temperature can usually be made during programmed periods. The same method is used to set the temperature during schedule programming. When the set point is adjusted during a scheduled period, it is temporary. When the time arrives for the next scheduled event, the setting reverts to the programmed set point.

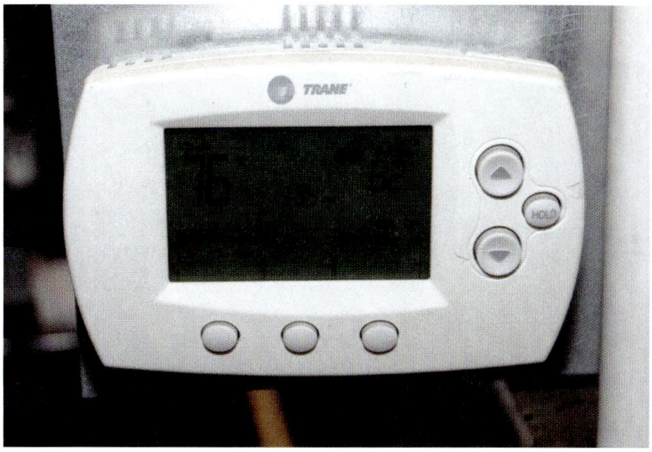

FIGURE 12-28 Programmable thermostat display.

Date and Time

The date and time settings are accessed through a dedicated button or touchscreen icon. Wi-Fi-enabled models can take time and date information through the Internet connection. It's important that these settings not be ignored, since the accuracy of all programming depends on the displayed time and day of week. The date allows the thermostat to track the change from daylight savings time to standard time and to know the day of the week from year to year.

Schedules

Programmable thermostats usually offer a Hold function (**FIGURE 12-29**). It serves no purpose on non-programmable models. The Hold button allows a set point to be chosen and held until the user releases the hold. All schedules can be ignored when the Hold button is used. Most allow a hold to be set for a specific number of hours or days, and then the thermostat automatically returns to the preprogrammed schedule at the end of the hold period.

Programming operating schedules is often the most challenging part for users. Manufacturers provide complete information with step-by-step instructions, but many users either misplace the manual or fail to use it. An example of a manufacturer's programming guidance is shown in **FIGURE 12-30**.

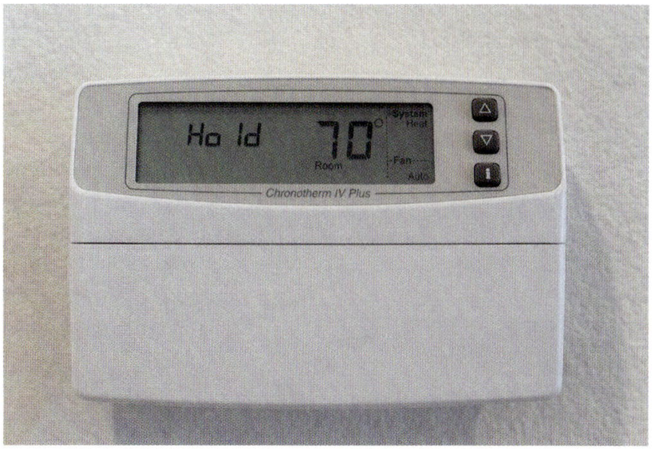

FIGURE 12-29 A thermostat operating in Hold mode.

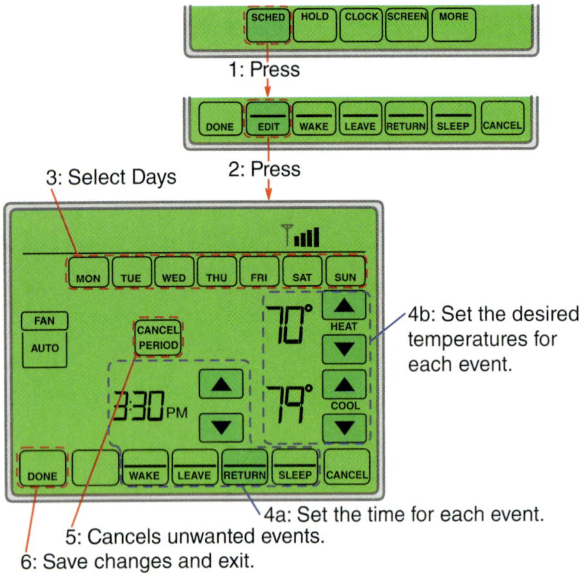

FIGURE 12-30 Programming guidance for a specific digital thermostat.

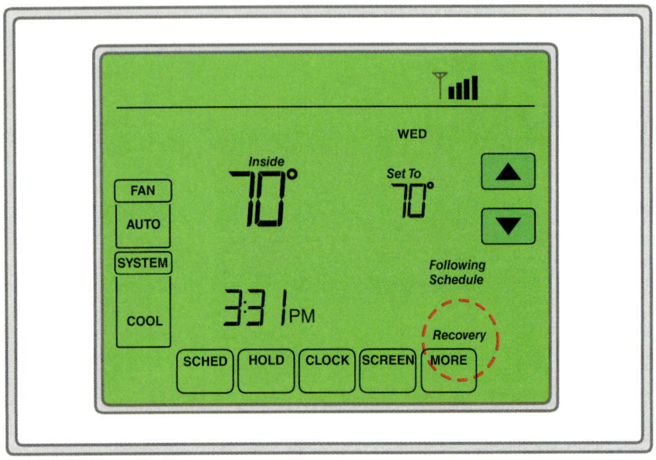

FIGURE 12-31 Announcing that a recovery process is underway.

Programming requires the selection of the day(s) of the week affected, and the period of the day. Wake, Leave, Return, and Sleep are common labels for the periods. Once the period is selected, the time the period begins and the temperature settings are made. Encourage users to plan their times and heat/cool settings on paper first, and then they will better prepared to focus on the entries at the thermostat.

Intelligent Recovery Features

Intelligent recovery refers to the ability of a thermostat to make decisions regarding when to start a system to achieve a programmed set point. This has a major effect on how the thermostat responds to user programming. Manufacturers have a variety of trade names for this feature.

When an intelligent recovery process is active, a future programmed set point is evaluated long before the time arrives. For example, assume a Sleep set point for heating is 68°F (20°C). The user also programs a WAKE set point of 74°F (23°C) for 7 a.m. With an intelligent recovery process, the thermostat starts the system before 7 a.m., with the goal of reaching the new set point by that time. To determine when to start the heat, the thermostat evaluates the length and number of heating cycles and examines system performance in previous mornings. By evaluating this information, it can make an educated guess about the starting time. If it achieves the new set point too early or too late, that will generally effect when it starts the next time. In other words, it learns from past performance.

Note that not every model uses this approach. Some may simply operate by assuming that it requires 5 minutes to raise the temperature 1°F (0.55°C) and therefore would start the system 30 minutes early (5 minutes × 6°F, or 3.3°C, increase). No evaluation of past performance or logic is involved here.

It is important that users understand intelligent recovery. When a digital thermostat activates this feature and starts a system before the scheduled time, it will generally display the word "Recovery" or an icon to indicate it is working on achieving the next set point by the preset time (**FIGURE 12-31**).

▶ Digital-Thermostat Installer Setup

K12005 Describe common digital-thermostat settings available to installers.

S12003 Configure a digital thermostat.

Today's digital thermostats are incredibly flexible. With the previous generation of thermostats, a specific thermostat model needed to be selected to suit the installation. Many digital thermostats are capable of handling anything from a simple gas furnace without cooling to a heat pump with multiple stages of heating and cooling.

To inform a thermostat about the system it serves and outline the behavior wanted from it, an installer setup process is provided. These menus are not generally accessed by users, although the means to enter them is not difficult to find out if an effort is made. Entry into the menus usually requires a specific sequence of entries. An example would be holding two separate buttons at the same time for five seconds. Since a user can easily introduce operating problems through the installer menus, manufacturers at least make menu access challenging enough so that it isn't accidental. **FIGURE 12-32** shows an example of a process to gain access to an installer menu.

Menus identify the functions and settings in a variety of ways. Some use a four-digit function code, while others may use a single digit. Some may use a **clear language display**, where the choices are made clear, but that is rare. Most screens aren't large enough to accommodate the needed text, and encoding the functions also helps to deter someone from making changes without the correct information.

Note that even thermostats of the same model may have slight differences in these menus. For instance, do not assume that function code 0120 is the year setting on every thermostat that appears to be the same. When working within an installer's menu, a technician must be armed with the correct information. Fumbling through the settings and guessing is not productive.

The list of installer menu functions can be quite long and can differ from product to product. Here we will review some common functions that can be configured. **FIGURE 12-33** shows part of an installer's function menu. The following provides an explanation of various settings:

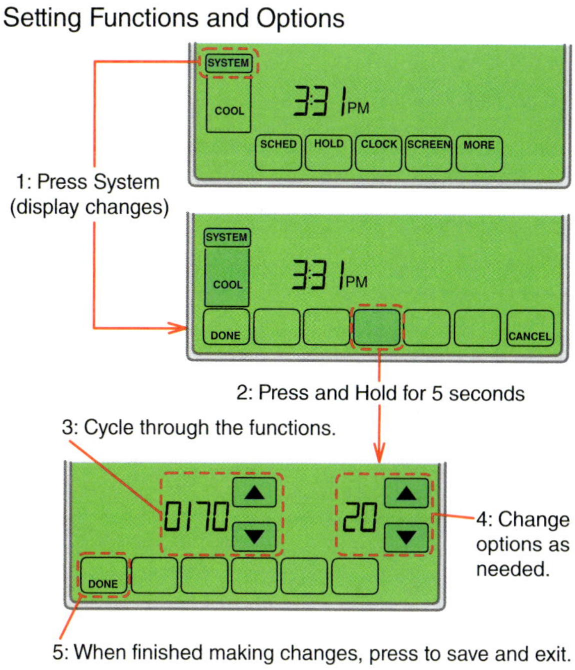

FIGURE 12-32 Gaining entry into an installer menu.

- *Function 1*—This where the type of system in use is entered. Each choice establishes a slightly different behavior. For example, if SS1 is selected, the thermostat knows that there are no extra stages of heating or cooling, and no Emergency Heat function is made available. This selection also disables other choices on the menu that are related exclusively to heat pumps.
- *Function 4*—This setting determines whether the user can program schedules. If so, the installer chooses the type of programming available. If it is set to non-programmable, then all schedule-related menu items will be disabled.
- *Function 7*—In this example, Energy Management Recovery refers to the thermostat's ability to start recovering to a new set point early. When controlling a heat pump, some thermostats will try to recover early enough to avoid using supplemental heat during the process, saving energy. The feature can be turned off, and the thermostat will function like a traditional time clock.
- *Functions 8–10*—This is where the cycle rate is set, which is the digital thermostat's answer to heating and cooling anticipation.
- *Function 12*—This function forces the compressor to remain off for several minutes before it is restarted. If short-cycle protection is provided by the equipment, this feature can be programmed off.

Setting Functions and Options

You can change options for a number of system functions.
Available functions depend on the type of system you have.

This thermostat is pre-set for a single-stage heating/cooling system.
Setting function 0170 for a heat pump will adjust the default settings.

1 Press SYSTEM
You'll see several blank buttons on the bottom of the display.

2 Press and hold the center blank button until the screen changes (approximately 5 seconds).

3 Cycle through the functions, press ▲▼ next to the four digit number on the left.

4 As needed, change options for any function by pressing ▲▼ next to the number on the right.

5 When you have made all changes, press DONE to save and exit.

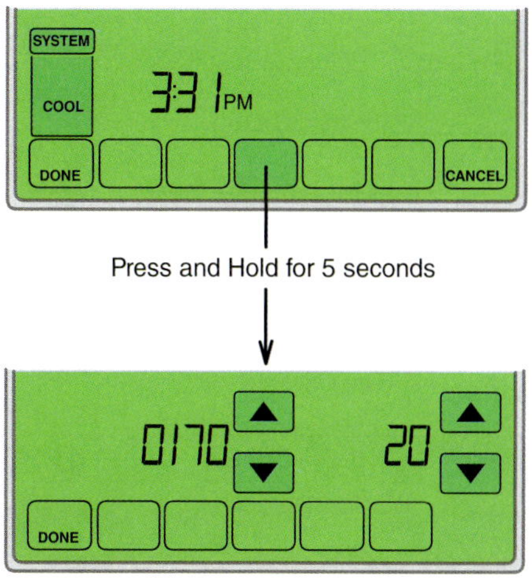

FIGURE 12-33 Partial list of settings available on an installer's menu.

- *Function 14*—This allows the installer to adjust the room temperature display if it is inaccurate when checked against reliable temperature-measurement equipment.
- *Functions 20 and 21*—This a rather complex feature. Long air conditioning operating cycles do a better job of lowering the indoor humidity. In the cooling season, humans feel equally comfortable at a higher temperature when the humidity is low. This feature takes advantage of this fact by changing the set point incrementally as a long cycle continues. Later, when the operating cycle length becomes shorter, indicating the load has reduced, it will gradually return to normal programming.

Refer to **FIGURE 12-34** for a description of the functions discussed below.

- *Function 22*—In this case, compressor optimization provides a time delay for indoor blower operation, both on and off. For example, a heat pump may run in the heating

CONFIGURATION MENU						
Screen Reference Number	SS1 MS2	HP1 HP2	Press Key	Displayed Factory (default)	Press ◁ or ▷ to select from listed options	Comment
1	●	●	△	MS 2	HP 1, HP 2, SS 1	Selects Multi-Stage (MS2, No Heat Pump), Heat Pump 1 (HP1, 1 compressor), Heat Pump 2 (HP2, 2 compressor or 2 speed compressor), or Single Stage.
2	●	●	△	(gas)	ELE	GAS setting: furnace controls blower. ELE setting: thermostat controls blower.
3		●	△	Ob (0)	b	Selects Reversing Valve (This item is only to appear if HP1 or HP2 is selected above.)
4	●	●	△	Days, (7) P	5 or 0	Programs per week. (7 days, 5+1+1 days or non-programmable)
5	●	●	△	(4) PS	2	Programs per day. 4 = Morning, Day, Evening, Night 2 = Day, Night Not available if 4 is 0
6	●		△	Cool-Off-Heat-Auto	Cool-Off-Heat, Heat Off, Heat, Cool-Off, Auto Off	System switch configuration in non heat pump mode.
		●	△	Cool-Off-Heat-Em-Auto	Cool-Off-Heat-Em, Off-Em-Auto	System switch configuration, heat pump mode.
7	●	●	△	(On) E	OFF	Selects Energy Management Recovery. Not available if 4 is 0
8	●	●	△	(FA) Heat, Cr	SL	Selects Adjustable Anticipation, cycle rate, Heat
9	●	●	△	(FA) Cool, Cr	SL	Selects Adjustable Anticipation, cycle rate, Cool
10		●	△	(FA) Cr/AU, Em	SL	Selects Adjustable Anticipation, cycle rate auxiliary, (This item is only to appear if HP1 or HP2 is selected above).
11	●	●	△	(OFF) SC	On	Select stage cycle completion On or Off.
12	●	●	△	(OFF) CL	On	Selects Compressor Lockout.
13	●	●	△	(OFF) dL	On	Selects Continuous Display backlight.
14	●	●	△	0 (Temperature)	5, LO to 5, HI	Selects Adjustable Ambient Temperature Display [range -5 (LO) to +5 (HI)].
15	●	●	△	°F	°C	Selects °F/°C Display (temperature units in Fahrenheit or Celsius).
16	●	●	△	(On) b	OFF	Selects audible Beeper On/Off.
17	●	●	△	(On) dS	OFF	Selects Daylight Saving Time calculation.
18	●	●	△	(On) Heat, AS	OFF	Selects Automatic Schedule for comfort temperature Programming, heat mode. Not available if 4 is 0
19	●	●	△	(On) Cool, AS	OFF	Selects Automatic Schedule for comfort temperature Programming, cool mode. Not available if 4 is 0
20	●	●	△	(OFF) CS	On	Selects Cool Savings Feature On of Off.
21	●	●	△	CS Cool Savings (3)	1-2-3-4-5-6	Selects amount of Cool Savings adjustment.

FIGURE 12-34 Additional settings available on an installer's menu.

mode for a few seconds to warm the indoor coil first, before the blower starts. The fan can also be delayed off to take advantage of hot or cold refrigerant remaining in the coil when the compressor stops.

- *Functions 26 and 27*—Some digital thermostats offer keypad lockout to keep others from being able to make unauthorized changes. This feature can be popular in both residential and commercial applications. In this case, a three-digit combination is set to unlock the keypad. Both full and partial lockout options are available. With a partial lockout, a user might be able to temporarily change the current set point, but not the programming.
- *Functions 33 through 35*—**Dual-fuel systems** use something other than electric heat for supplemental/emergency heat, such as a gas furnace, which we will use as an example here. The heat pump and the gas furnace cannot be allowed to operate at the same time. These settings allow the installer to configure it as a dual-fuel system, then determine when the system changes from one heat source to the other.
- *Functions 38 through 41*—These choices determine whether an air filter or ultraviolet (UV) lamp replacement reminder is used and how many hours of operation are allowed before the reminder displays. The hours selected are based on actual blower operating hours.

Note that there are several pages of notes in the installer's manual that are related to the functions. Again, to properly set up a thermostat, it is necessary to have the correct literature in hand and to carefully consider every choice. Although many different thermostats may share features and functions, the terminology may differ. Incorrect choices can cause obvious as well as very subtle problems. A great deal of testing and troubleshooting can be wasted, only to find out that an incorrect installer setting is the source of a problem. Consider each and every choice and how it will affect both the system and the user.

Wi-Fi Setup

Connecting a device to a home or business network has now become second nature for most people. People often connect their smartphone or tablet to a local network when they visit friends or businesses that provide access. In many cases, users are directly involved with thermostat Wi-Fi setup since an account must be established and passwords are needed.

The first step in the process is connecting the thermostat to the local network. A Wi-Fi menu must be accessed in the thermostat configuration menus. Most thermostats then display a list of wireless networks that are transmitting within range, and the user selects the appropriate network. Assuming that the network is secured, as most are and should be, a password will be required to connect. A touch-sensitive keyboard is often provided on the screen to enter the password.

Next, the thermostat must be registered, and an account established, online at the appropriate website. Users should generally handle this step because they will communicate with the thermostat in the future and because the process is often a part of the app installation. Like many other devices, the thermostat has unique identifying information so that it can be found and identified on a network. Most thermostats have a MAC (media access control) address and a MAC CRC (cyclic redundancy check) number. Both numbers are entered during registration, providing the path for communication. Once the app has been installed and the appropriate information has been entered, it is simply a matter of the user becoming familiar with the app and navigating the choices.

One thing to remember about Wi-Fi-enabled thermostats is that they may disconnect themselves from the network and shut down external communications when the batteries become weak, to save power.

To practice how to configure a digital thermostat, follow the steps in **SKILL DRILL 12-3**.

SKILL DRILL 12-3 Practicing How to Configure a Digital Thermostat

This activity requires an installed and powered digital thermostat with an accessible installer's configuration menu. While the most elaborate thermostat available is not necessary, a versatile programmable model is preferred.

SAFETY TIP

Working with electrically energized circuits is hazardous. Avoid contact with any surface that could potentially be energized. When working inside energized enclosures, maintain a stable position and, whenever possible, avoid placing both hands inside the enclosure. All work on energized equipment by unqualified individuals should be supervised by a qualified instructor or technician.

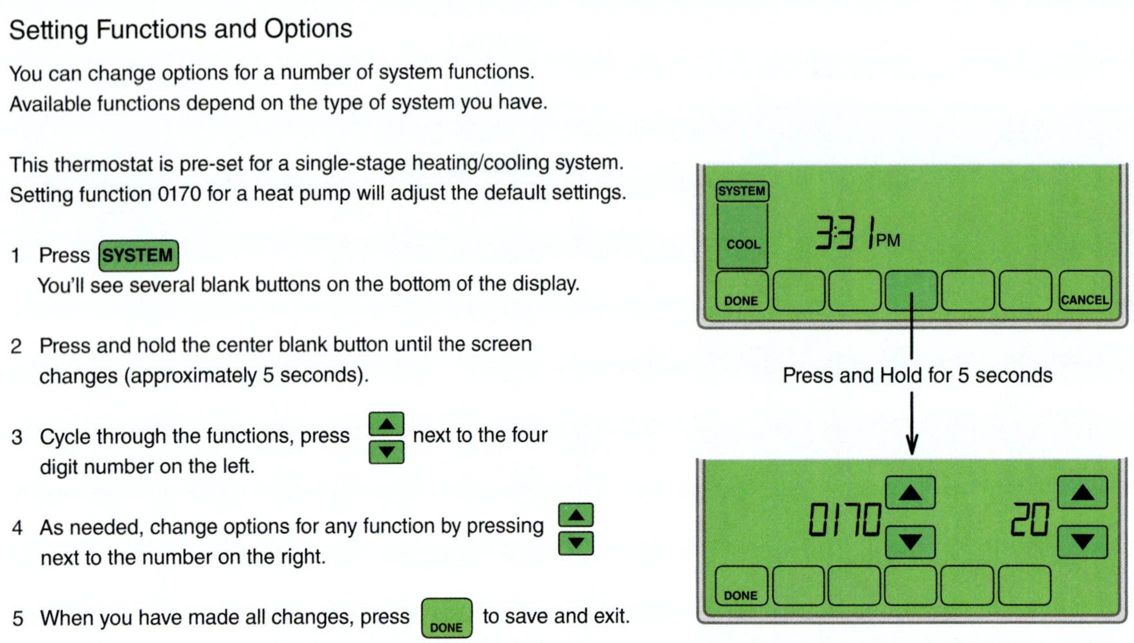

Setting Functions and Options

You can change options for a number of system functions. Available functions depend on the type of system you have.

This thermostat is pre-set for a single-stage heating/cooling system. Setting function 0170 for a heat pump will adjust the default settings.

1 Press SYSTEM
You'll see several blank buttons on the bottom of the display.

2 Press and hold the center blank button until the screen changes (approximately 5 seconds).

3 Cycle through the functions, press ▲▼ next to the four digit number on the left.

4 As needed, change options for any function by pressing ▲▼ next to the number on the right.

5 When you have made all changes, press DONE to save and exit.

1. Acquire the necessary installation instructions for the thermostat to be configured. Review the menu of items to configure. The menu shown here is simply an example and should not be followed for this drill. Also, before beginning, become familiar with the details of the connected system.

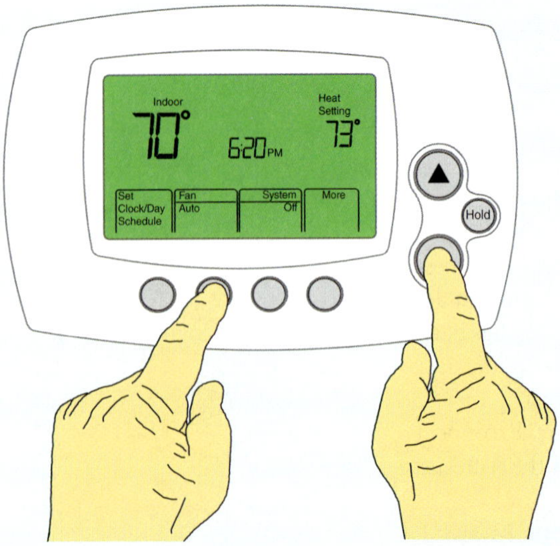

2. Using the manufacturer's instructions, access the installer's configuration menu. This usually requires a unique combination of key presses.

Continued

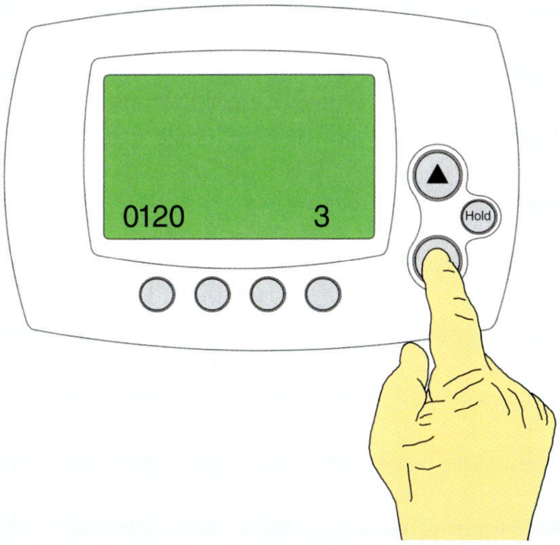

3. Once the menu is open, check each configuration setting and pencil in the final value entered on the menu. Follow the manufacturer's instructions for changing a value. Be sure to check any footnotes that may be related to the entries for additional information and guidance. Note that the menu may disappear by default if it is open for too long without any activity.

PERIOD	Time	Heating Set Point	Cooling Set Point
Wake	6:00 a.m.	72°F (22°C)	75°F (24°C)
Leave	7:30 a.m.	67°F (19°C)	80°F (27°C)
Return	5:30 p.m.	72°F (22°C)	75°F (24°C)
Sleep	10:30 p.m.	68°F (20°C)	72°F (22°C)

4. Once the configuration is complete, return to the main display. Following the manufacturer's instructions, set the current day and/or date and the current time. Then, access the programming menu to enter a schedule. Enter the schedule shown in the table for practice.

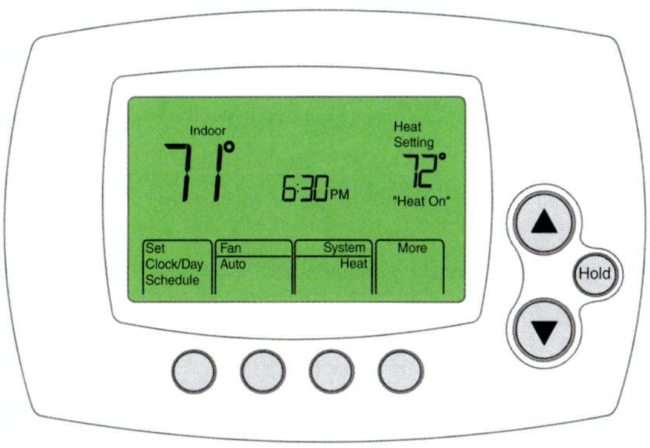

5. Once the entries are complete, return to the main display. Ensure that the thermostat indicates the correct heating and cooling set points for the period as entered in the schedule, and it begins to respond accordingly.

▶ Troubleshooting a Thermostat

Troubleshooting a thermostat is not difficult if the technician understands how it works and knows what to expect from it. Most of the time, a technician is looking for a reason why something *is not* happening, as opposed to why something *is* happening. Troubleshooting digital thermostats requires the same approach as troubleshooting other solid-state controls—testing for specific outputs when the correct inputs are present.

K12006 Explain how to troubleshoot a thermostat.

S12004 Troubleshoot a thermostat.

For digital thermostats that appear to be functional, it is always a good idea to start by reviewing the selections in the installer's menu, for two reasons. First, a selection that isn't correct and is possibly the source of the problem might be found. Second, the present configuration provides a basis for the expected responses. For example, if compressor short-cycle protection is enabled, a cooling signal from the thermostat Y terminal shouldn't be expected until the time delay has elapsed between cycles.

Troubleshooting Charts

FIGURE 12-35 shows an example of a troubleshooting chart for a digital thermostat. Review the troubleshooting information after checking the thermostat's configuration. Although most of the guidance is very general, it is a good idea to investigate simple problems first. Since batteries are used, for example, the thermostat display does not usually go blank when power to the thermostat is lost. Instead, it will usually display an appropriate message when power is lost and maintain an active display.

Once any relevant guidance on the troubleshooting chart has been considered and simple possibilities have been eliminated, more detailed troubleshooting begins. Manufacturers rarely provide this type of troubleshooting guidance. At this point, technicians must rely on their skills and knowledge of system operation.

Electrical Testing

Recall that a Common (C) wire is not usually provided to non-digital thermostats, unless there are lights to illuminate. Digital thermostats often do have a Common wire. But all have an R wire, which is the leg of control power switched thorough the thermostat and

CONFIGURATION MENU						
Screen Reference Number	SS1 MS2	HP1 HP2	Press Key	Displayed Factory (default)	Press ◁ or ▷+ to select from listed options	Comment
22	●	●	△+	(Off) CO	On	Select Compressor Optimization (not available on earlier models)
23	●	●	△+	(OFF) CA	On	Selects Comfort Alert Feature On or Off. (not available on earlier models)
24	●	●	△+	(99) Heat, HL	62-98	TEMPERATURE LIMIT, HEAT (max. heat set point).
25	●	●	△+	(45) Cool, LL	46-82	TEMPERATURE LIMIT, COOL (min. cool set point).
26	●	●	△+	OFF, 🔒 Keypad Lockout	L (total), P (partial), Temperature Limit (limited temperature range)	Selects Keypad Lockout.
27	●	●	△+	000	001-999	Selects Keypad Lockout Combination (active only if keypad Lockout is selected).
28	●	●	△+	(On) Heat, FS	OFF	Fast second stage of heat (not available if SS1 is selected above).
29	●	●	△+	(On) Cool, FS	OFF	Fast second stage of cool (not available if SS1 or HP1 is selected above).
30	●	●	△+	Remote (OFF)	On	Remote temperature sensor, enable/disable.
31	●	●	△+	Remote, In	Outdoor Remote	Remote temperature sensor (Indoor/Outdoor).
32	●	●	△+	(On) LS	OFF	Local temp. Sensor enable/disable (only when Indoor Remote is selected On).
33	●	●	△+	(OFF) dF	On	Selects Dual Fuel feature On or OFF (this item appears if HP1 or HP2 is selected above).
34	●	●	△+	(35) dF	-5 - 50	Selects Dual Fuel setpoint (°F), dF selected On with outdoor sensor available.
35	●	●	△+	(05) dF	0 - 09	Selects Dual Fuel setpoint (°F), dF selected On with no outdoor sensor.
36	●	●	△+	(60) Cd	0-99	Selects compressor off delay in seconds, dF selected On
37	●	●	△+	(80) AO	-5 - 79	Selects Auxiliary Heat cut out temperature. This item appears if HP1 or HP2 is selected and outdoor sensor is installed and enabled.
38	●	●	△+	(OFF) Change UV Lamp	On	Selects Change UV Lamp feature.
39	●	●	△+	350 Days	25-1975	Change UV Lamp duration days.
40	●	●	△+	OFF Change Filter	On	Selects Change Filter feature.
41	●	●	△+	200 Hrs	25-1975	Change Filter duration hours.

FIGURE 12-35 Example of a thermostat troubleshooting chart.

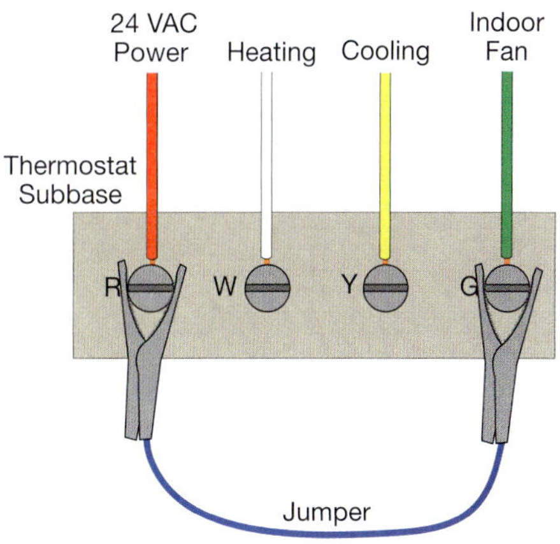

FIGURE 12-36 Checking indoor blower control-circuit operation.

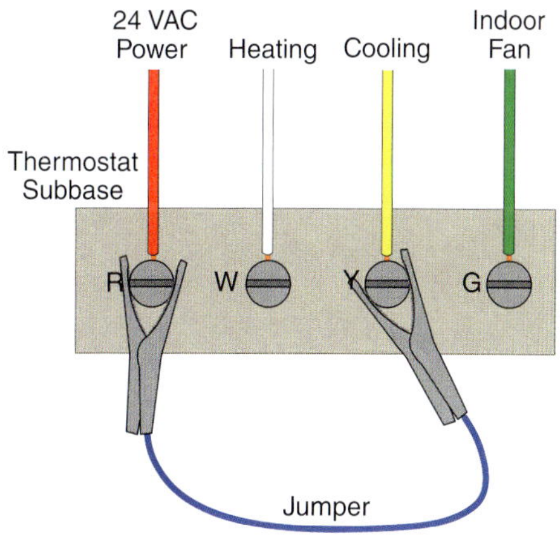

FIGURE 12-37 Checking cooling control-circuit operation.

distributed to the other terminals. A few manufacturers will not label it as R, but it will be a leg of power from the control transformer.

When a system will not respond to thermostat inputs, the quickest way to narrow down the search is to begin at the thermostat subbase. Using a jumper wire, quickly determine whether the equipment and other controls work properly by bypassing the thermostat. Alternatively, the tests that follow can be conducted at the terminal board of the equipment where the transformer is. The advantage of working from the thermostat subbase is that the interconnecting thermostat wiring is also being tested in the process.

Before beginning, remove the thermostat from its subbase. Refer to **FIGURE 12-36**. When a jumper wire is used, power must be on for the system to respond. Remember to practice proper electrical safety. Although a 24 VAC circuit is virtually harmless, do not allow a short circuit to occur between the C and R terminals, since the control transformer can be damaged or destroyed instantly unless it is fuse protected.

Begin by testing the indoor fan operation. Attach a jumper wire to the R terminal and touch the other end to the G terminal. This test does exactly what the thermostat does when it calls for indoor blower operation: it closes a 24 VAC circuit between R and G. If the indoor blower starts, it and the controls are functional and the interconnecting wiring is intact.

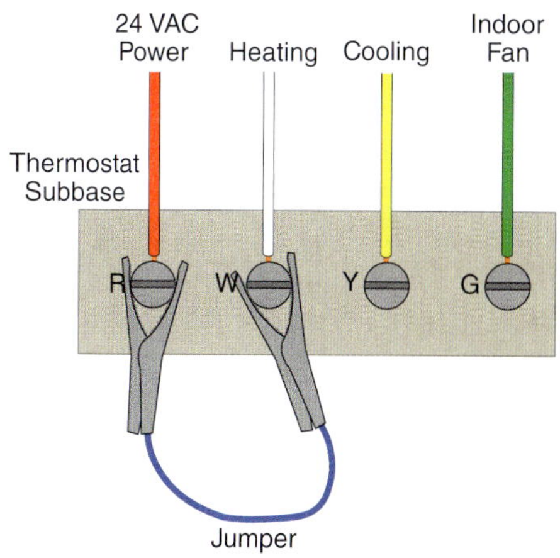

FIGURE 12-38 Checking heating control-circuit operation.

Continue testing by maintaining the R terminal connection and touching the other end to the Y terminal (**FIGURE 12-37**). It is possible that a short-cycle protection timer in the unit will delay the start of the compressor; be aware of this before determining that something isn't working. Otherwise, the condensing unit should start when this jumper connection is made. Note that this test must be brief. It should only be long enough to determine whether the control system is working. The condensing unit should not be operated for any substantial amount of time without the indoor blower running. A few seconds is all it takes to determine whether the circuit is functional.

> **▶ TROUBLESHOOTING TIP**
>
> While using jumper wires for troubleshooting with conventional 24-VAC control systems is common and acceptable, systems that use advanced communication protocols and/or DC signals should not be tested in this manner.

Test the heating system next by maintaining the R terminal connection and connecting the other end to the W terminal (**FIGURE 12-38**). This should initiate heating operation for a non-heat-pump system. If the furnace starts an ignition cycle, that circuit is intact as well.

These quick tests help narrow down the location of a problem. If the equipment does not respond to one or more of the above tests, there is a problem in the interconnecting wiring or equipment. If the system operates as it should when using the jumper wire, reinstall the thermostat and prepare for additional tests using a meter.

With the control circuit energized, set the thermostat to call for cooling, heating, or indoor blower operation, depending on the complaint. For this example, assume the problem is that the indoor blower will not run, but it does run without a problem when using the jumper wire. Set the thermostat fan switch to On. At the unit terminal board, use a voltmeter to check for 24 VAC between R and G. If the thermostat is set for continuous fan operation and is operating properly, the meter should show only the background voltage between R and G. Testing the voltage between C and G should yield 24 VAC. If the indoor blower is running, the circuit is obviously working. If the blower did not start and 24 VAC

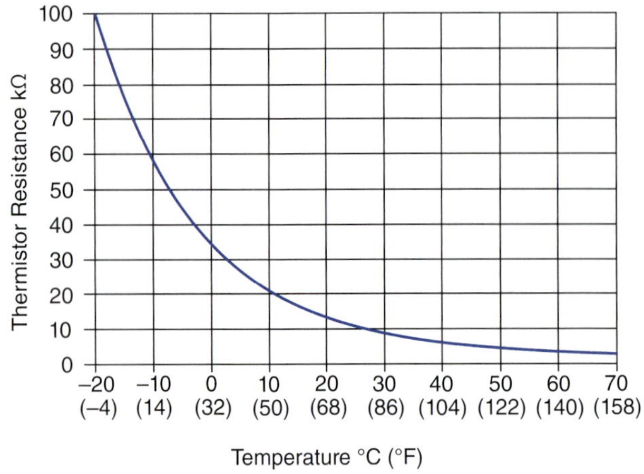

FIGURE 12-39 Example of a temperature/resistance chart for a thermistor.

> ▶ **TECHNICIAN TIP**
>
> **Practice Testing and Troubleshooting Skills**
>
> One of the best ways to prepare for troubleshooting is to make voltage tests on, and closely observe, systems that are working just fine. By taking voltage measurements at various control-circuit locations, technicians learn what to expect when a system is operating as it should and become familiar with meter responses to open and closed circuits.

is displayed between R and G, it is an indication that the circuit through the thermostat is still open. This identifies the thermostat as the source of the problem.

All of the primary functions of the thermostat—one or more stages of cooling, one or more stages of heating, and indoor blower operation—can be tested using the process above. The processes above can also be reversed—voltmeter testing can be done first if desired. But if a problem is detected, the interconnecting thermostat wiring will still need to be eliminated as the problem source before condemning the thermostat. This is why testing with a jumper wire is often completed first.

Terminals O and B are normally used to energize the reversing valve of a heat pump. If the O connection is used, the reversing valve is energized in the cooling mode. If the B connection is used, the reversing valve is energized in the heating mode. To test the thermostat's control over these circuits, use the same two methods: a jumper wire to see whether the circuit functions and a meter to see whether the thermostat is closing the switch between R and O, or R and B, as it should.

Digital thermostats may have thermistors connected to them to provide an external temperature input. Thermistors are easily tested if they are suspected of failure. There are two types of thermistors. A positive temperature coefficient (PTC) thermistor increases its resistance as the temperature increases. A negative temperature coefficient (NTC) thermistor decreases its resistance as temperature increases. PTC thermistors are more common for this application.

A thermistor can be checked with an ohmmeter to determine whether it is open; an open circuit indicates failure in either type. If resistance is present, measure the temperature at the thermistor's location and compare it to a temperature/resistance chart for the thermistor in use. **FIGURE 12-39** provides an example of a chart. If the thermistor is accurate, its resistance should match the chart at a given temperature. There are various thermistor values and scales, so be sure to use the correct chart.

To troubleshoot a thermostat, follow the steps in **SKILL DRILL 12-4**.

SKILL DRILL 12-4 Troubleshooting a Thermostat

This activity requires an installed and powered thermostat. The drill is based on a standard single-stage electric-cooling/gas-heating split system.

> **SAFETY TIP**
>
> Working with electrically energized circuits is hazardous. Avoid contact with any surface that could potentially be energized. When working inside energized enclosures, maintain a stable position and, whenever possible, avoid placing both hands inside the enclosure. All work on energized equipment by unqualified individuals should be supervised by a qualified instructor or technician.

Continued

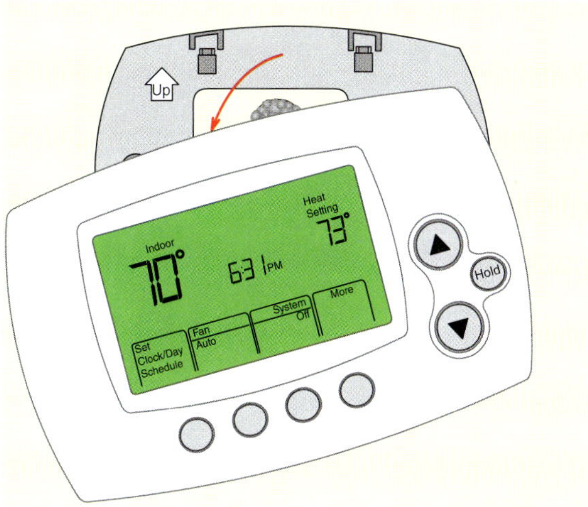

1. Ensure that thermostat mode switch is set to Off, and the fan switch is set to Auto. The system can remain powered, but once the thermostat terminals are exposed, avoid short-circuiting the terminals. Follow the manufacturer's instructions to remove the thermostat from its subbase.

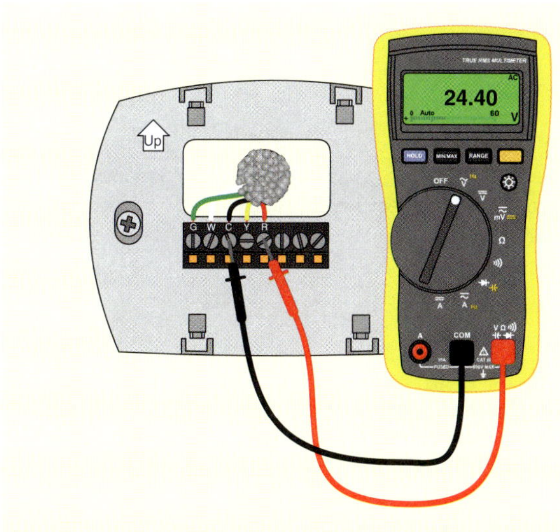

2. If desired, power to the thermostat can be confirmed first. If there is both a C wire and an R wire connected, use a voltmeter to check for 24 VAC between the two terminals.

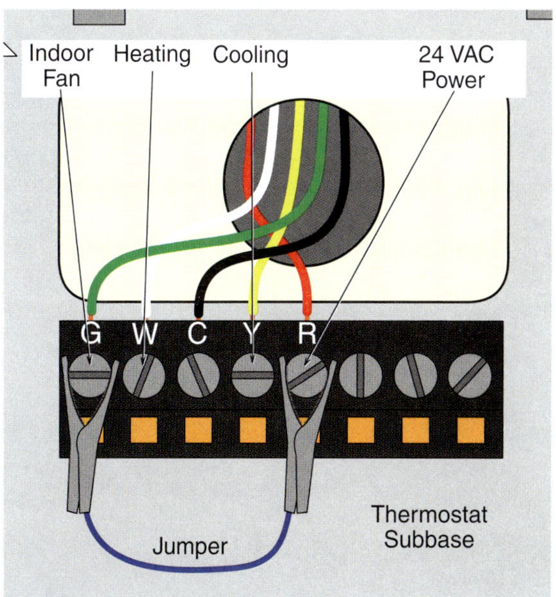

3. Carefully attach one end of a jumper wire to the R terminal. Then touch the other end to the G terminal. This should cause the indoor blower to start.

Continued

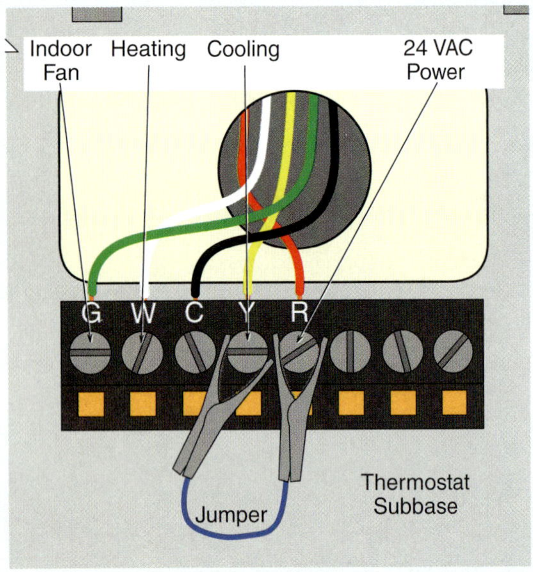

4. With one end of the jumper still attached to R, touch the opposite end to the Y terminal. Make the connection smoothly and firmly to avoid energizing and deenergizing the compressor rapidly. Unless the unit has short-cycle protection that is actively timing, the condensing unit should start. As soon as it is confirmed, remove the jumper. A helper may be required to stand by the outdoor unit unless it can be seen or heard from the thermostat location.

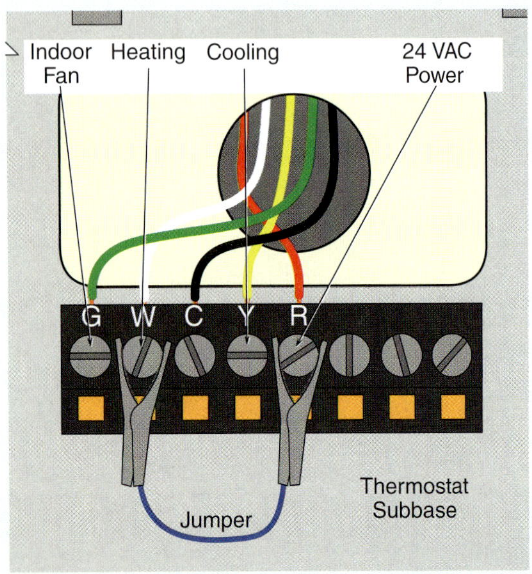

5. With one end of the jumper still attached to R, touch the opposite end to the W terminal. This should start the furnace ignition sequence. Once the sequence has been confirmed to be proceeding, remove the jumper wire. If the blower, cooling, and heating circuits operated normally, the equipment and interconnecting thermostat wire is OK.

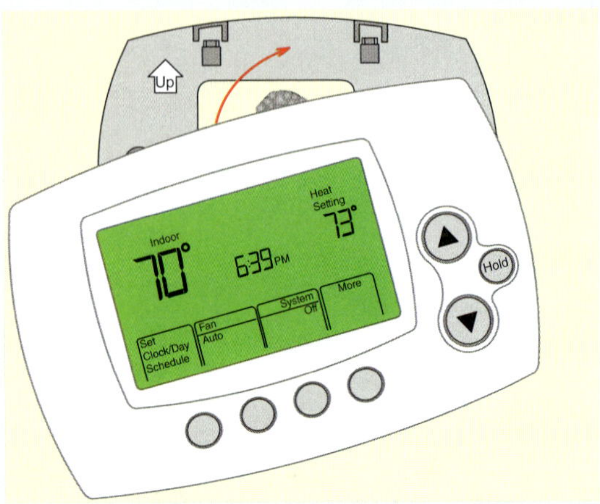

6. Carefully reinstall the thermostat on the subbase.

Continued

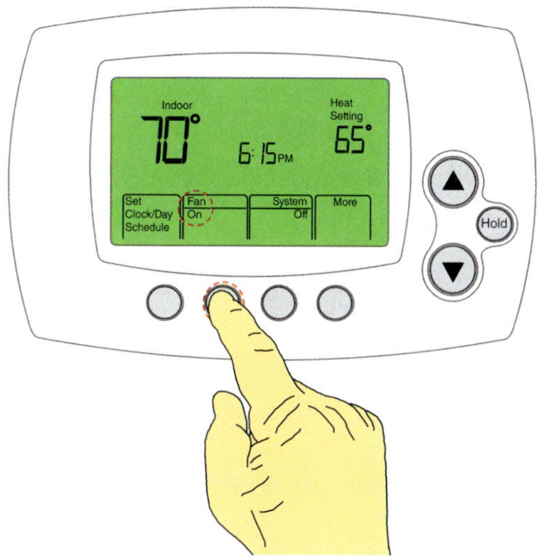

7. At the furnace, remove the blower door to access the thermostat wiring terminal board. Since this door typically has an interlock switch, it will need to be temporarily bypassed by connecting the wires passing through it. Note that this is line-voltage power; appropriate electrical safety precautions must be taken before opening the junction box. Alternatively, if it can be done reliably, apply a piece of tape over the rocker or button switch to hold it down. *Never* leave a door switch defeated or bypassed after a troubleshooting task!

8. Prepare a voltmeter, and ensure there is clear, safe access to the thermostat wiring connections to make voltage measurements. Return to the thermostat and set the fan switch to the On position.

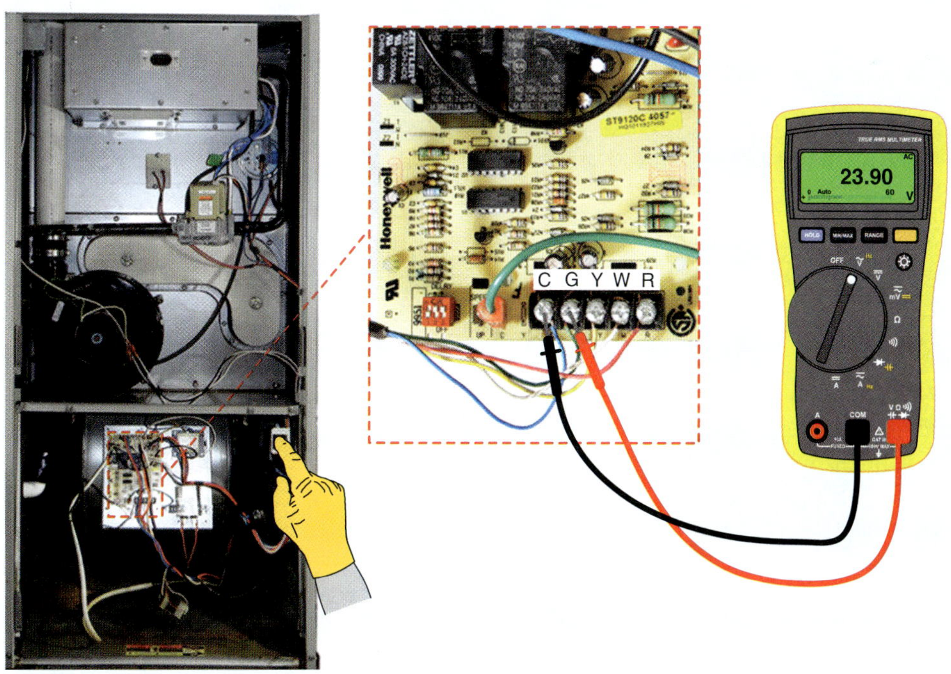

9. Return to the furnace and check for power between the C and G terminal. If the blower is running, obviously the circuit is energized, and the meter should display 24 VAC between C and G. If the thermostat did not close the circuit and the blower is not running, the meter will display 24 VAC between R and G. Since the interconnecting wiring and equipment have been confirmed to be good, this would indicate a failed thermostat.

Continued

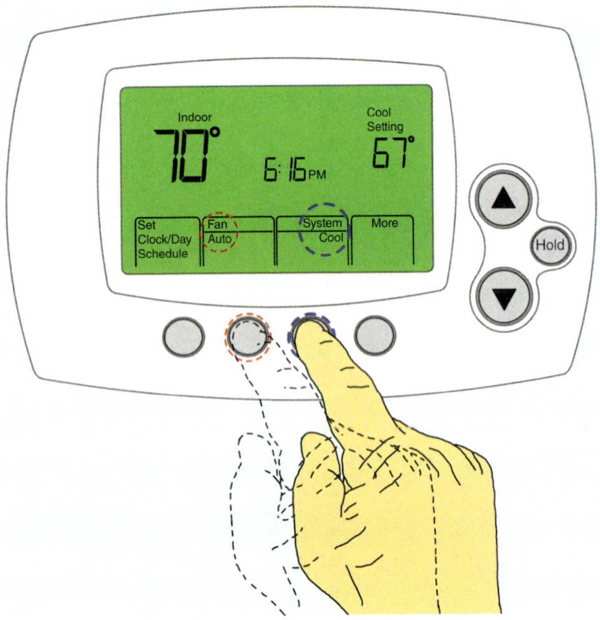

10. Return to the thermostat and move the fan switch from On to Auto. Then set the mode to cooling and lower the set point until it is several degrees below room temperature.

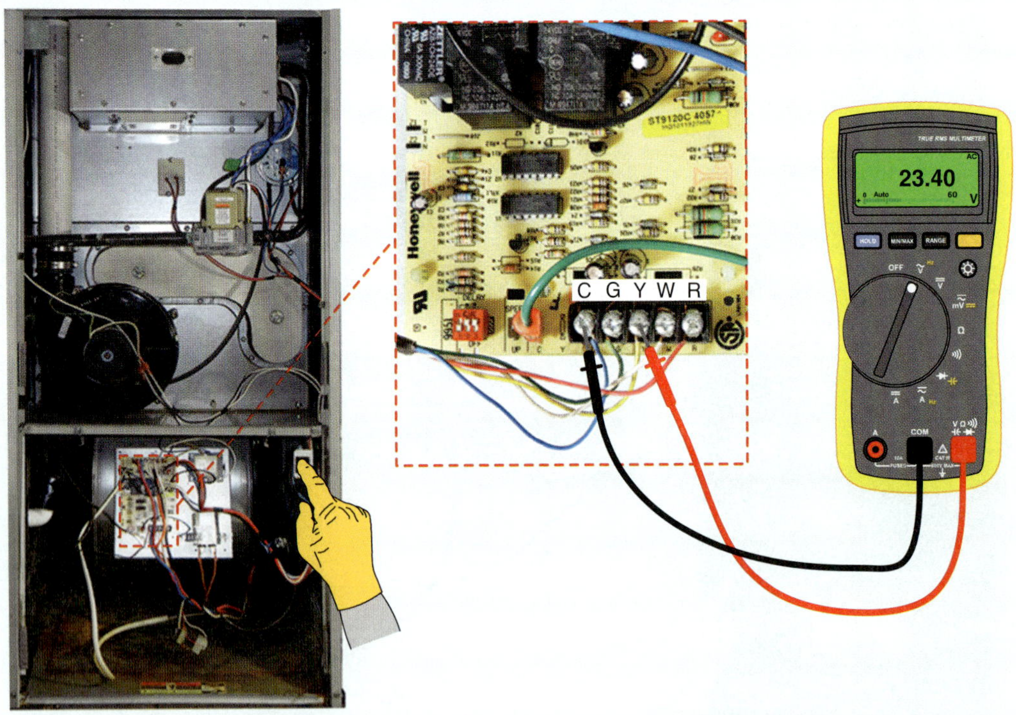

11. Return to the furnace and check for power between C and Y, then C and G. The indoor blower and condensing unit should both be running (unless thermostat short-cycle protection is active), and roughly 24 VAC should be detected. Remember that the control transformer is carrying some additional load now, so the voltage may be less than measured in the previous test. If the blower and/or condensing unit is not running and only background voltage is displayed, the thermostat has not closed one or both circuits.

Continued

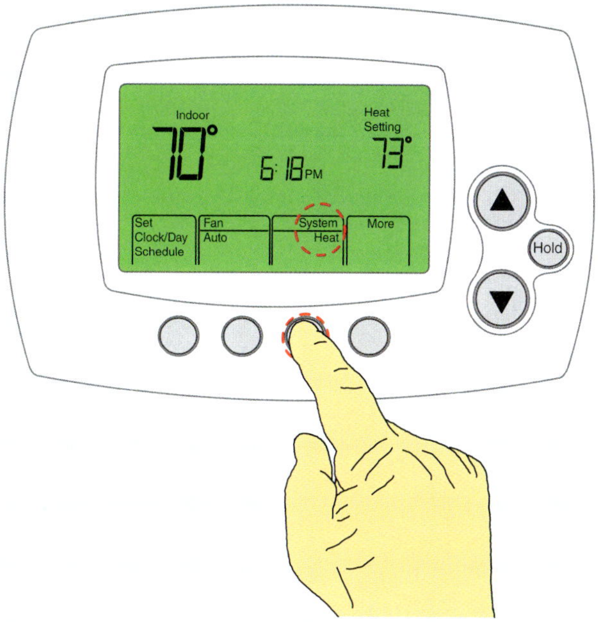

12. Return to the thermostat. Change the mode switch from Cool to Heat, and raise the set point until it is several degrees above room temperature.

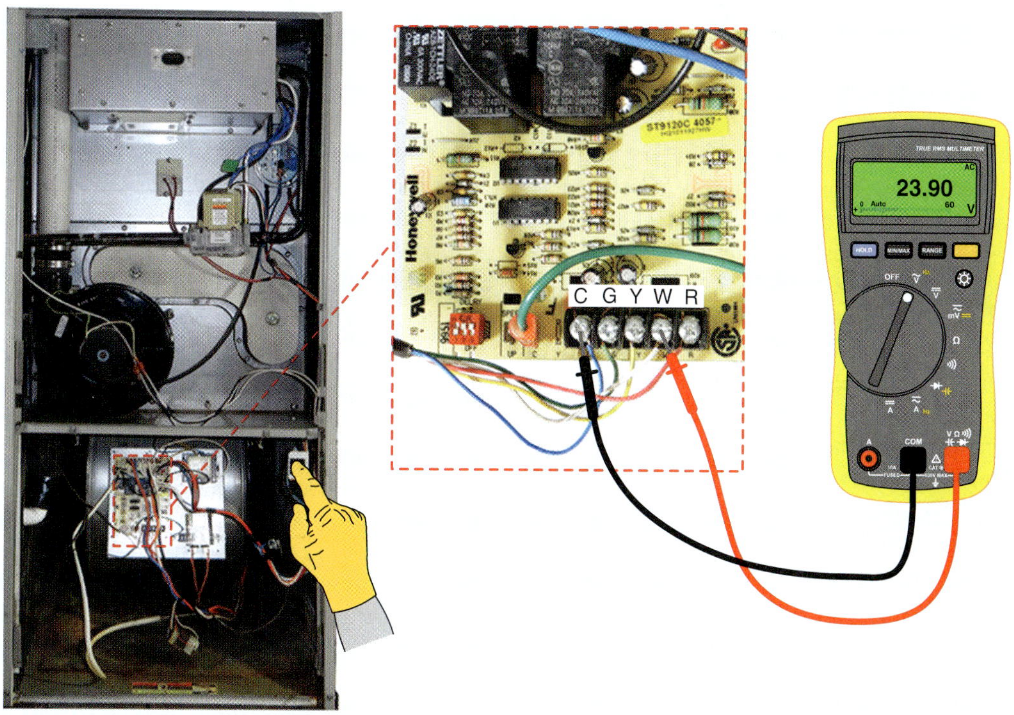

13. Return to the furnace and check for power between C and W. The furnace should have started its ignition sequence, and roughly 24 VAC should be detected. If the furnace has not responded and only background voltage is displayed, the thermostat has not closed the circuit.

Continued

14. If the equipment has responded to the thermostat settings as expected, the thermostat is functional. Other circuits through a thermostat, such as O and B for the reversing valve of a heat pump, can be tested in the same manner. If preliminary testing with a jumper wire at the wall plate went well but those same circuits are not being energized with the thermostat in place, then the thermostat is the likely source of the problem. Rewire or remove the tape applied to the blower door switch, reinstall the access door, and then return the system to service.

▶ Wrap-Up

Summary

Comfort-system thermostats represent the HVAC system to many users, especially when the system is out of sight. In any case, they provide the means for users to interface with and control the system.

There are a variety of thermostats that have been developed over the years, but all provide the same basic functions. Today's digital thermostats, however, can do far more than their predecessors. They can provide operating-schedule programming and be configured in a variety of ways to suit the needs of different systems. In addition, some digital thermostats can be connected to a Wi-Fi network and communicate with the user via the network or Internet. Their capabilities and versatility are the reason that they are beginning to dominate the market. However, many non-digital thermostats and those without Wi-Fi capabilities will remain in service.

Troubleshooting a thermostat requires a logical yet simple sequence of tests made to determine whether it is the source of a problem. All troubleshooting should begin with a review of the thermostat settings and its configuration. The manufacturer's troubleshooting guide should always be followed to eliminate simple possibilities. This should be followed by a review of the system and its characteristics to ensure compatibility and to know what to expect during functional testing. Any electrical testing should always follow these preliminary steps.

Key Terms

clear language display A display that provides clearly understood language to guide the user, as opposed to codes or other indications that must be interpreted.

conduit Piping or tubing, usually steel, aluminum, or plastic, which is used to route and protect electrical wiring.

cycle rate The number of heating or cooling cycles that occur per hour. Cycle-rate limitations can be established in most digital thermostats since they have no resistor for heating and cooling anticipation.

deadband A neutral zone between heating and cooling set points where no action occurs. For an auto-changeover mode setting, a deadband is needed to ensure the system does not rapidly alternate back and forth between modes to maintain a set point.

dual-fuel systems Systems that use more than one heating fuel. An example is a heat pump paired with a gas furnace. Note that a heat-pump system with supplemental electric heat is not dual-fuel, since both operate on electricity.

heat anticipator A resistor in the thermostat or subbase that generates a small amount of heat to warm the thermostat during a heat operating cycle.

multistage thermostat A thermostat that controls more than one stage of heating and/or cooling.

plenum An air-filled space that in HVAC systems is commonly used as a path for air. For example, the area above a suspended ceiling is often used as a path for return air to the system, in which case it is referred to as a return-air plenum.

raceway A track or channel used to route electrical wiring. Conduit is typically considered to be a type of raceway.

thermistors Resistors whose resistance varies predictably with the surrounding temperature.

Review Questions

1. The __________ thermostat is the most basic comfort-system thermostat design.
2. When replacing a thermostat that contains mercury, it _____.
   a. must be sent to an authorized collection point as hazardous waste
   b. can be disposed of with common trash
   c. must be reported to the EPA
   d. should be burned
3. A furnace heat anticipator is energized _____.
   a. when the cooling system is on
   b. when the cooling system is off
   c. when the heating system is on
   d. when the heating system is off
4. How do cooling anticipators differ in construction from heat anticipators?
5. When there are multiple stages of heating and/or cooling, the second stage lags behind the first stage by _____.
   a. 7°F–10°F (3.8°C–5.5°C)
   b. 5°F–7°F (2.8°C–3.8°C)
   c. 3°F–5°F (1.7°C–2.8°C)
   d. 2°F–3°F (1.1°C–1.7°C)
6. Thermistors are a type of _____.
7. The term "programmable thermostat" describes one that _____.
   a. is digital
   b. can be connected to a wireless network
   c. can accept a schedule of set point changes
   d. has an installer's configuration menu
8. Class 2 thermostat wiring is covered by the *National Electric Code* under Article _____.
9. How far above the floor are thermostats mounted in ADA-compliant locations?
   a. 66 inches (168 cm)
   b. 60 inches (152 cm)
   c. 54 inches (137 cm)
   d. 48 inches (122 cm)
10. Which type of thermostat must be level to function properly?
    a. mercury-bulb
    b. bimetallic
    c. snap action
    d. digital
11. Why should you wait to install digital-thermostat batteries until it is time for start-up?
12. The color of the Common wire to a thermostat is typically _________.
13. For which type of system does the thermostat start the indoor blower on a call for heat?
    a. gas furnace
    b. oil furnace
    c. heat pump
14. The O and B terminals of a thermostat typically control what component?
15. Which of the following is a true statement about the Hold feature of a thermostat?
    a. The thermostat ignores scheduled events and maintains a continuous temperature for the length of the hold.
    b. A Hold feature can only be set for the cooling mode.
    c. When the next scheduled event is encountered, the thermostat leaves the Hold mode.
    d. The heating or cooling system runs continuously when the Hold feature is used.
16. During the cooling season, when the humidity is low, humans feel equally comfortable at a _____ temperature.
    a. higher
    b. lower
17. In a dual-fuel system, __________.
    a. the furnace and heat pump operate together to produce maximum heat
    b. the furnace and heat pump cannot operate at the same time
    c. cooling is not available
    d. a fossil fuel is used to power the cooling system
18. Assuming the display is functional, what is the first step you should take before starting to troubleshoot a digital thermostat?
19. A thermostat is set to the Heat mode, and the set point is several degrees above room temperature. If the thermostat does *not* close the R–W circuit, a meter reading voltage between R and W at the furnace terminal board should display _____.
20. At a furnace terminal board, 0 volts or background voltage only is read between the C and B terminals. This means that the thermostat R–B circuit is ________.
    a. open
    b. closed

CHAPTER 13
Pressure Switches

Knowledge Objectives

After reading this chapter, you will be able to:

- **K13001** Identify different types of pressure switches and explain how they function.
- **K13002** Describe the operation and maintenance of high- and low-pressure switches.
- **K13003** Describe the operation and maintenance of differential-pressure switches.
- **K13004** Describe the operation and maintenance of fan-cycling pressure switches.
- **K13005** Describe the operation and maintenance of air-pressure switches.

Skill Objectives

Skills required to meet the objectives of this chapter:

- **S13001** Troubleshoot a high-pressure safety switch.
- **S13002** Replace a high-pressure safety switch.
- **S13003** Troubleshoot an adjustable low-pressure cycling switch.
- **S13004** Replace an adjustable low-pressure cycling switch.

▶ Introduction

HVACR systems rely on a number of pressure switches to both control and protect key components during operation. The majority of the pressure switches that HVACR technicians work with monitor refrigerant pressures. A high-pressure safety switch (**FIGURE 13-1**), for example, shuts down the compressor if the refrigerant discharge pressure is excessive. However, there are other applications for pressure switches as well. Since pressure-switch operation relies on various approaches to sensing pressure, the concept of pressure and how it is measured will first be reviewed.

Pressure and Pressure Units

Recall that pressure is the force that a fluid exerts per unit area. According to a principle of physics called **Pascal's law**, a pressure change in an enclosed fluid is transmitted equally throughout the fluid. When applied to HVACR systems, this means that we can remotely monitor and respond to a system pressure by connecting a pressure-sensing instrument to a point of interest in the system.

Pressure can be measured in a wide variety of units. One pressure that we are all exposed to consistently is atmospheric pressure. The average pressure at sea level is one standard atmosphere (1 atm). In the United States, pressure is commonly measured in pounds per square inch (psi). One atmosphere is equal to 14.7 psi (101 kPa). Very small pressures may be measured using a **manometer**, calibrated in inches of a water column (in. WC).

The international (metric) unit of pressure is the pascal (Pa). Since the pascal is a small unit compared to psi, the kilopascal (kPa) is often used as a conversion unit in HVACR applications instead of the pascal. One atmosphere is 101,300 Pa, or 101.3 kPa, the same as 14.7 psi. Another metric unit of pressure sometimes encountered in HVACR is the bar, which has no special abbreviation. One bar is slightly less than one atmosphere of pressure; it is equal to exactly 100,000 Pa (100 kPa).

The sea-level atmospheric pressure and its equivalent value in other units are examples of **absolute pressure**. Absolute-pressure system abbreviations are shown by adding the suffix "a" on the pressure abbreviation—for example, 250 psia (1,723 kPa). However, we are usually interested in a system pressure as measured on a gauge. If a refrigerant circuit's absolute pressure is the same as the atmospheric pressure, 0 psi is indicated on a connected gauge, because a gauge's zero pressure *is* atmospheric pressure. In other words, 0 psi on a gauge = 14.7 psia. System pressure measured with a gauge is called **gauge pressure**, and gauge-pressure unit abbreviations must include the suffix "g" on the end. For example, a compressor may have a discharge pressure of 195 psig (1,446 kPa). To convert this to absolute-pressure units, add the atmospheric pressure of 14.7 psia for a total of 209.7 psia (1,344 kPa). See **TABLE 13-1** for a brief summary of pressure units and conversion information.

FIGURE 13-1 A high-pressure safety switch.

TABLE 13-1 Atmospheric Pressures and Pressure Units

Pressure Unit	Abbreviation	1 ATM Equals
Atmosphere	atm	1 atm
Pounds per square inch	psi	14.7 psi
Inches water column	in. WC	406.8 in. WC
KiloPascals	kPa	101.3 kPa
Bar	bar	1.01325 bar

▶ Types of Pressure Switches

Pressure switches operate by using one of a variety of mechanical sensors to detect pressure. These are mainly **bellows**, **diaphragms**, and **Bourdon tubes**, as well as a few other devices. The switches themselves may perform different functions, depending on where and how the pressure switch is connected to a system.

K13001 Identify different types of pressure switches and explain how they function.

Pressure-Sensing Elements

The pressure-sensing element of a pressure switch is the component that responds to system pressure to actuate a switch. All such elements change shape as pressure inside or against them changes. A bellows is an enclosed pleated chamber made of a thin, flexible material (**FIGURE 13-2**). The bellows lengthens or shortens as the pressure inside changes relative to the pressure outside.

A diaphragm is a thin, flexible disk or sheet that may be flat or ribbed for added flexibility (**FIGURE 13-3**). The diaphragm divides two sides of a chamber, one side of which is connected to the source of pressure. It flexes away from the higher pressure exerted on one side compared to the other.

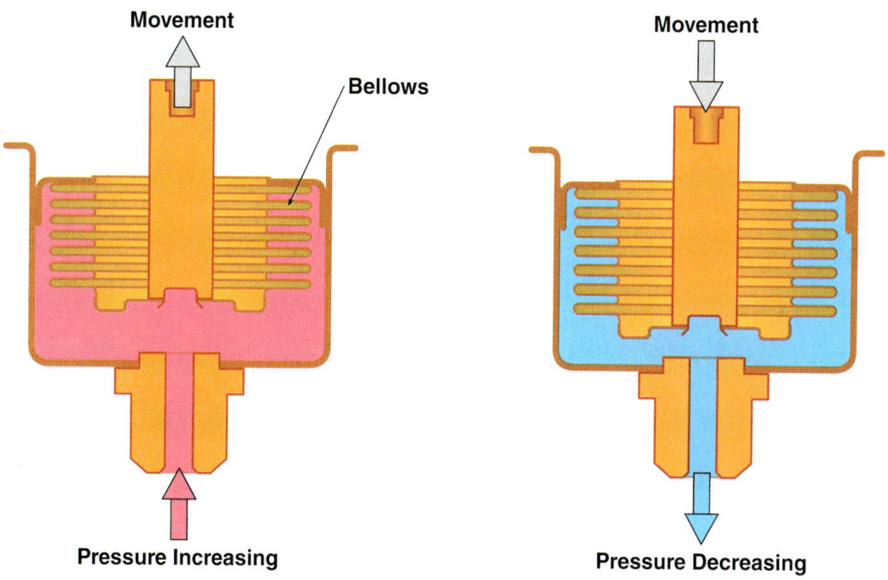

FIGURE 13-2 Pressure-sensing bellows used in pressure switches.

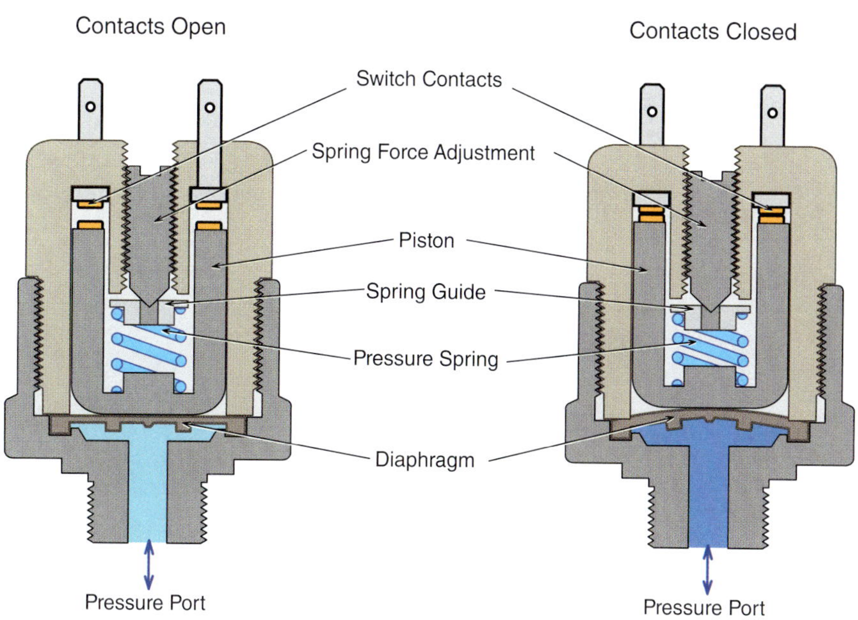

FIGURE 13-3 A diaphragm-operated switch.

A Bourdon tube is a flexible C-shaped or **helical** tube connected at one end to the fluid being monitored. The tube is sealed at the other end. As fluid pressure inside the tube increases, the Bourdon tube tends to straighten out (**FIGURE 13-4**), or unwind in the case of a helical tube.

As part of a pressure switch, each type of sensing element is connected to electrical contacts, either directly or indirectly through a mechanical linkage. The movement of the sensing element opens and closes the switch contacts.

Every pressure-sensing element is technically a differential-pressure sensor. For high- or low-pressure switches, the monitored system pressure is compared to atmospheric, or **ambient**, pressure. The difference between system and ambient pressures determines how much the sensor changes shape. True differential-pressure switches, however, monitor two pressures within a system. One side of the sensing element is connected to the higher system pressure, and the other side is connected to the lower system pressure. These are usually diaphragm- or bellows-type switches.

In most pressure-switch designs, the pressure-sensing element presses against an adjustable spring or has built-in spring-like characteristics that resist a change in shape. The spring force acting on the pressure sensor is important because it determines how the pressure sensor responds to system pressure. As the pressure increases, the sensing element moves against the spring force. With further increases in pressure, the sensing element further compresses the spring, and thus the spring's resistance increases. This design permits the position of the pressure-sensing element to be proportional to pressure. Manually changing the spring compression with a setscrew is one way of adjusting a pressure switch's set point. In other adjustable switches, the set point can be changed by changing the distance between the electrical contacts. In switches with fixed set points, they are either set at the factory or mass-produced for a given standard setting.

Electrical Switch Operation

Pressure-switch contacts also have normal electrical positions that are defined when no pressure is applied to the switch sensor. The spring force mentioned earlier tends to position the switch contacts to this default position in the absence of a pressure input. As discussed in Chapter 8, pressure switches have normally open (NO) and normally closed (NC)

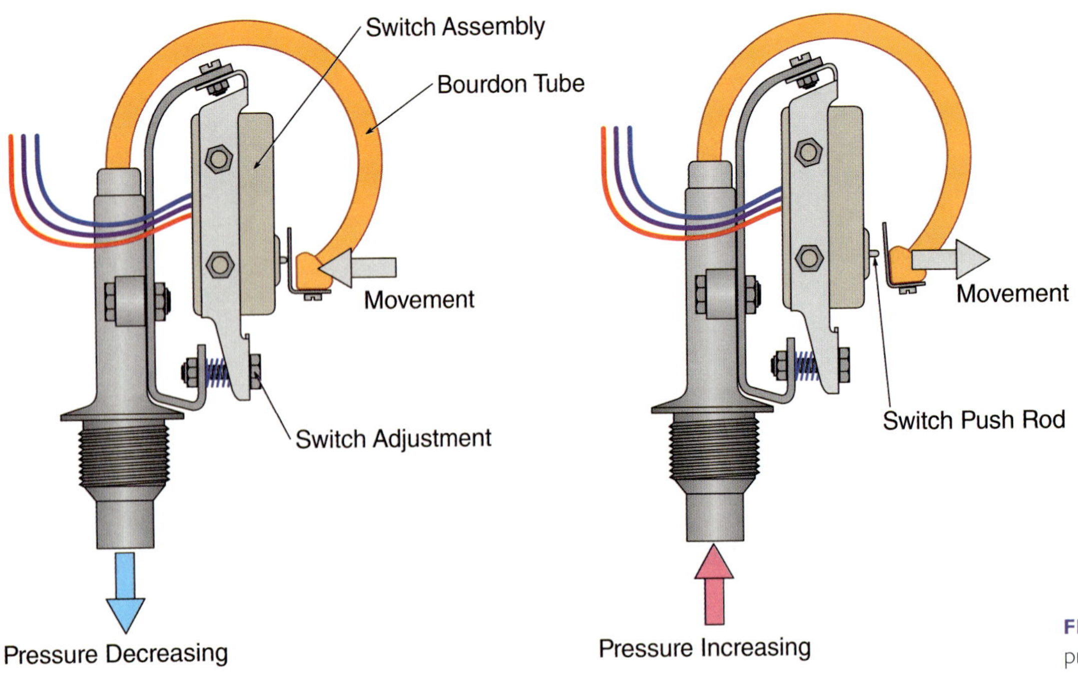

FIGURE 13-4 Bourdon-tube pressure switch.

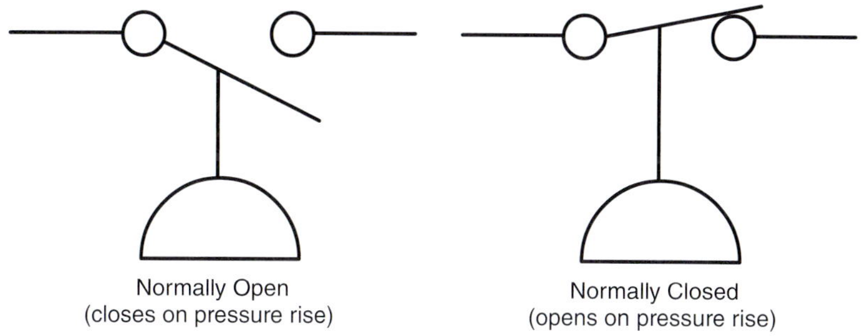

FIGURE 13-5 Schematic symbols for NO and NC pressure switches.

positions, like other switches and relays. Electrical schematic symbols for NO and NC pressure switches are shown in **FIGURE 13-5**.

The symbols shown in the figure are also drawn in such a way that indicates their intended function. Note the switch arm on the NO switch is below the circuit line. Since the arm must rise to close the switch, this pressure switch closes on a rise in pressure. On the NC switch to the right, the arm is above the circuit line. If the arm rises, the switch opens. This switch, therefore, opens on a rise in pressure.

The electrical contacts of pressure switches must make and break cleanly for reliable operation, so they are constructed to snap into position when activated. These snap switches may be self-contained microswitches installed inside the pressure-switch housing and actuated by a plunger attached to the pressure-sensing device. Other switches might be constructed from discrete, spring-like components contained within the switch housing connected by a linkage to the pressure sensor.

▶ High- and Low-Pressure Switches

K13002 Describe the operation and maintenance of high- and low-pressure switches.

S13001 Troubleshoot a high-pressure safety switch.

S13002 Replace a high-pressure safety switch.

S13003 Troubleshoot an adjustable low-pressure cycling switch.

S13004 Replace an adjustable low-pressure cycling switch.

Pressure switches commonly perform several important functions in HVACR systems. For example, they can act as safety switches, shutting down a system when refrigerant pressure is too high or too low. They can also be used to control a refrigeration system instead of a thermostat.

High-Pressure Switches

High-pressure safety switches are commonly installed in the discharge piping of refrigerant compressors (**FIGURE 13-6**). The refrigerant at this location of a typical air conditioning or refrigerant plant is a hot, high-pressure gas. If the pressure gets too high, the compressor could overheat due to the load. It is also very important to prevent the system from reaching a pressure that could lead to the circuit rupturing at a component or piping joint. Common causes of high discharge pressure include the following:

- condenser coils fouled
- condenser fan not running
- restricted air flow through the condenser due to damaged fins
- **noncondensable gases** in the refrigerant.

When the discharge pressure rises to the high-pressure switch's set point, it opens its NC contacts. The switch contacts are in series with the compressor control circuit. Once the contacts are open, the compressor shuts down. Some high-pressure safety switches include a built-in manual-reset button, as shown in Figure 13-6. To restart the system, the switch must be reset by hand. This prevents the system from continually cycling on and off as the switch cycles repeatedly. Other high-pressure switches reset automatically once the pressure falls to a safe level.

FIGURE 13-6 A typical nonadjustable high-pressure safety switch.

Systems with safety switches that automatically reset use another method of stopping the unit until a technician intervenes. A lockout relay in the compressor control circuit that opens when the control circuit opens is often used in this case. The lockout relay is then reset by turning the control power to the unit off and then back on again. The system will then restart while a technician monitors its operation to determine what went wrong. It is important to note that not all systems are equipped with high-pressure safety switches.

Low-Pressure Switches

Low-pressure switches are installed to protect a system from abnormally low pressures. In a typical refrigeration circuit, the lowest pressure occurs in the return piping at the compressor inlet. A low suction pressure often results in abnormally low suction-line temperatures. In air conditioning applications where the coil temperature should remain well above freezing, ice formation on the suction line is generally an indication of a problem (**FIGURE 13-7**). Ice can also build on the evaporator coil, as shown in **FIGURE 13-8**. Significant ice formation often results from a complete loss of indoor airflow. Note that in low-temperature refrigeration applications, frost on the suction line as well as on the evaporator coil is normal.

Thus, an important purpose for a low-pressure switch in the compressor suction line is to protect the system from a loss of refrigerant charge due to leaks, a persistent problem in the HVACR industry. Most refrigerant compressors depend on the cooling effect of the returning refrigerant vapor to cool themselves. As a result, not only does a low refrigerant charge cause functional problems in the system, but the compressor can overheat as well. The following are some of the other potential causes of low refrigerant pressure in the low-pressure side of the system:

- a clogged or defective metering device, such as a thermostatic expansion valve (TXV) or **orifice**
- loss of cooling load on the evaporator coil, perhaps due to a blocked return-air duct or a failed indoor blower motor
- undersized piping that causes excess pressure drop.

Low-pressure safety switches typically reset automatically as the pressure rises; manual-reset versions are relatively rare. Most low-pressure switches used in a safety role for air conditioning systems are fixed (nonadjustable), **encapsulated** switches in most cases.

Like high-pressure switches, low-pressure switches can be used in ways that go beyond their role as a safety device. In small refrigeration systems, for example, such as those serving a reach-in cooler, a low-pressure switch can be the system control in lieu of a thermostat. A low-pressure control for this purpose is shown in **FIGURE 13-9**. As shown on the scale, a

▶ TECHNICIAN TIP

Loss-of-Charge Switch

While some systems have a low-pressure safety switch, others, especially heat pumps, may have what is known as a loss-of-charge (LOC) switch. Both are designed to shut the system down if the pressure is too low, but in different ways and at different times. A low-pressure safety switch is located on the low-pressure side of the system, whereas an LOC switch may be located on the high-pressure side of the system. The LOC is also set to a significantly lower pressure. It is designed to keep the system off before start-up occurs if the refrigerant pressure is too low, but it is unlikely to open while the system is in operation, since it is on the high-pressure side of the system. Low-pressure safety switches, located on the low-pressure side, are more likely to intervene while the system is running. Note that the fixed pressure setting of LOC switches is generally lower than that of a low-pressure safety switch.

FIGURE 13-7 Suction-line ice accumulation.

FIGURE 13-8 Ice on an evaporator (indoor) coil.

FIGURE 13-9 Refrigeration low-pressure cycling control.

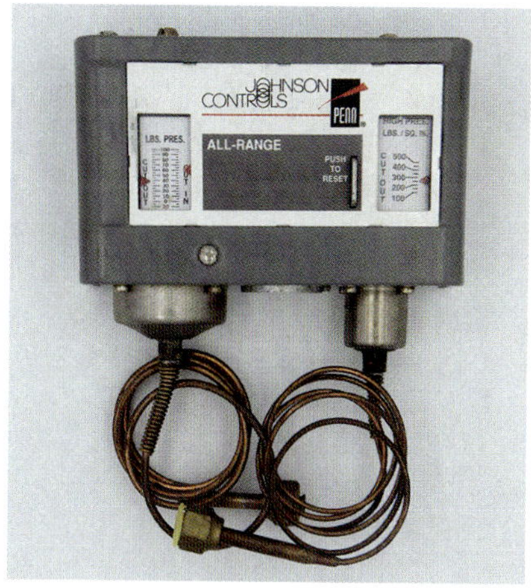

FIGURE 13-10 Combination high-pressure safety switch and low-pressure cycling control.

cut-in pressure is set to start the compressor, and then a differential is set that determines the cut-out setting. In systems that use these controls, the suction pressure reflects the temperature in the box. As it warms, the pressure rises and the compressor starts. Once it reduces the pressure by the amount of the differential setting, the compressor cycles off. These controls may also be combined with a high-pressure safety switch, as shown in **FIGURE 13-10**. Both the pressure switches just described and pictured are rarely applied in residential and commercial comfort-cooling equipment, but they are very common in refrigeration systems.

High- and Low-Pressure Switch Troubleshooting and Replacement

High- and low-pressure switches of all types are relatively reliable. When an HVACR compressor stops unexpectedly, assume that the safety switches are doing their job and that some type of problem exists. Only after verifying that everything is working properly attention be turned to the pressure switches.

There are two common failure modes for pressure switches. First, the microswitch or electrical contacts in the pressure switch can fail mechanically, or they might be overloaded and become scorched or even welded together. Second, pressure switches can fail to sense pressure accurately due to binding or sticking of the linkage to the electrical contacts.

An internal leak that equalizes the pressures across the sensing element, such as a crack or pinhole leak in a bellows or diaphragm, is also a source of failure. However, a leak here will also deplete the system of refrigerant eventually. Few leaks are traced to pressure-switch internal assemblies, but many are traced to leaking pressure-switch system connections.

Pressure switches stuck in their operating position due to a mechanical or electrical problem act like they aren't even in the system. They provide no protection or control.

Troubleshooting Pressure Switches

Any time a pressure switch is suspected of failing to operate as designed, the technician should rule out obvious problems first. Perform a careful inspection of the mechanical and electrical connections for signs of external damage, breaks, electrical shorts, wear, leakage, or tampering. High-pressure safety switches may or may not be equipped with a reset button. If a reset button is present, attempt to reset the switch. Then check the switch contacts to see if they closed as expected. When an inspection and operational test does not reveal any external causes for excessive pressure to develop, then the switch must be tested.

If a system is intact and operable, most pressure-switch testing can be done in place. This is especially true of switches connected to the low-pressure side of the system. Using an accurate gauge manifold to monitor pressure and a voltmeter attached across the switch contacts, a technician can monitor the pressure and know exactly when the contacts open or close, For this type of testing, it is best to connect the gauge as close to the pressure connection of the switch as possible so that they are reading pressure at the same point. A tee fitting at the connection point is ideal for this. However, the low-pressure setting of a safety switch cannot be reached if the system is operating normally without manipulating system valves. Closing or partially closing the valve controlling the flow of liquid from the condenser will cause liquid refrigerant to back up in the condenser and reduce the suction pressure. By controlling the position of the valve manually, the suction pressure for testing can be reduced. Note, however, that the system should not be allowed to run for long periods this way, because compressor cooling may suffer. An LOC switch, however, can be tested effectively only by removing refrigerant from the system or relevant section of the refrigerant circuit.

Most high-pressure safety switches used in comfort cooling are self-contained and nonadjustable. Their set point can be tested on an active system, but it does require forcing the high-side pressure to rise to the set point of the switch. It is essential that the pressure not be allowed to continue rising well beyond the set point if the switch does not open when it should. To compensate for calibration errors and the expected accuracy of the switch, it can be allowed to rise approximately 15 psig (103 kPa) beyond the set point. This value (±15 psig) is a common tolerance for the accuracy of a snap-action, nonadjustable pressure safety switch. If the switch does not open within ±15 psig of its set point, shut the system down and replace the switch. However, it is important to note that high-pressure safety switches do not reset the position of their contacts when the pressure has fallen only 15 psig. It is common for high-pressure safety switches that automatically reset to return to their NC position after the pressure has dropped 100–150 psig (700–1,050 kPa). However, the reset value differs some based on the high-pressure set point and what constitutes high pressure for a given refrigerant.

▶ TECHNICIAN TIP

Pressure Safety-Switch Selection

Not so many years ago, three to four refrigerants dominated the HVACR industry and its many applications. Some of those systems have been converted to operate on today's newer refrigerants. There are many other refrigerants on the market today, used in all types of new equipment. When replacing a pressure safety switch, make sure to replace it with one that has an appropriate set point for the refrigerant in use. For example, installing a high-pressure safety switch set for R-22 refrigerant will cause nuisance shutdowns on an R-410A system. Always test the replacement switch to ensure it functions as intended.

To test the set point, connect an accurate gauge manifold to the system. Attach the probes of a voltmeter to the pressure-switch electrical leads and check for voltage. Remember that reading the control-circuit voltage on the leads indicates that it is open; if the reading is 0 volts, the switch is closed. Start the compressor and observe the discharge pressure; the switch should remain closed during normal operation.

If the compressor shuts down well below the switch set point and the meter display changes to indicate the system's control voltage, then it is clear that the switch opened, and it should be replaced. If not, and the system is again operating normally, then an increase in the pressure must be initiated in order to reach its set point and test the switch.

There are two primary ways to cause the pressure to rise. One is to block an air-cooled condenser with cardboard or plastic sheets, starving the airflow. Another is to disconnect the condenser fan motor from the circuit. Either should cause the pressure to rise. However, it is vital to be in control of the system and be able to shut it down quickly. Allow the system to run as the pressure increases to the set point of the switch. If the switch has not opened within ±15 psig (103 kPa) of the set point, shut the system down immediately and replace the unreliable switch.

FIGURE 13-11 Nitrogen cylinder and regulator.

In some applications, especially in a very cold environment, it may be difficult to get the system pressure to rise as high as needed. The alternative in these situations, and other occasions when a pressure switch cannot be safely tested in place, is to remove the switch and test its operation by using pressure from a nitrogen tank (**FIGURE 13-11**). Use a regulator and a gauge manifold to control the nitrogen pressure output. The pressure switch can be connected to a hose or port of the manifold, with the nitrogen source connected to the charging hose. Nitrogen pressure is then applied to the manifold until the switch set point is reached. An ohmmeter is needed, attached to the pressure-switch leads, to monitor switch activity. Alternatively, the system refrigerant can be recovered and a controllable nitrogen source can be connected directly to the system.

To practice how to troubleshoot a high-pressure safety switch, follow the steps in **SKILL DRILL 13-1**.

SKILL DRILL 13-1 Practicing How to Troubleshoot a High-Pressure Safety Switch

SAFETY TIP

Working with electrically energized circuits is hazardous. Avoid contact with any surface that could potentially be energized. When working inside energized enclosures, maintain a stable position and avoid placing both hands inside the enclosure whenever possible. All work within energized equipment by unqualified individuals should be supervised by a qualified instructor or technician.

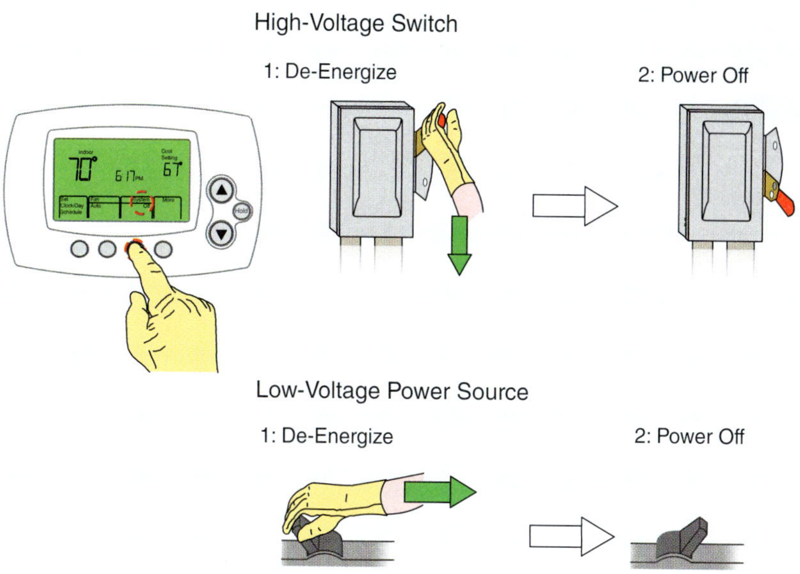

1. Ensure that all controls, including the thermostat, are set to the Off position. Deenergize the equipment to be serviced using the disconnect switch, including the low-voltage control power source. However, do not lock and tag, since the unit must operate for the test. Use a voltmeter to verify the equipment is deenergized. Prepare some appropriate material, such as plastic sheets or cardboard, to block the airflow through the condenser when needed. Plastic sheets work better because they will lay flat against the coil and are more effective at blocking airflow.

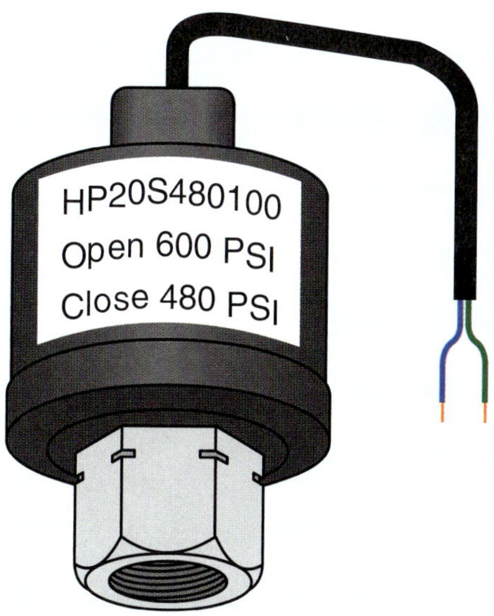

2. If the switch set point is clearly listed on the switch labeling, then the required pressure can be found. However, switch labeling can be difficult to decipher and misleading. If the pressure is not clearly shown, consult the equipment manufacturer's documentation to determine the set point.

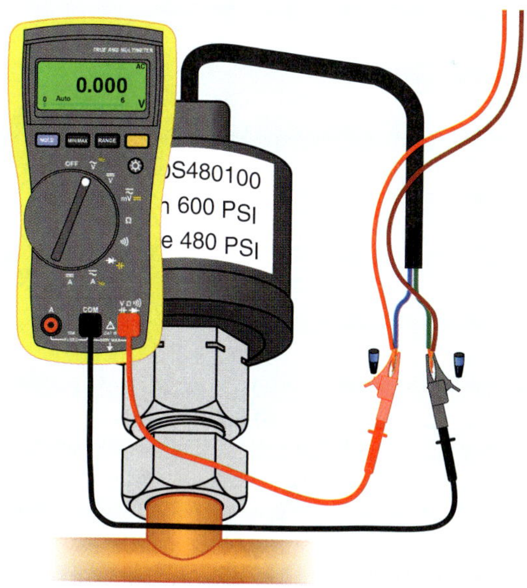

3. Locate the electrical leads from the high-pressure safety switch and determine how and where they are connected. This may include the use of wire nuts or a terminal board. Leaving the leads connected to the control circuit, attach the voltmeter leads to the same point. Ensure that the meter leads and control wiring are secure and will not be drawn into the condenser fan when it starts.

Continued

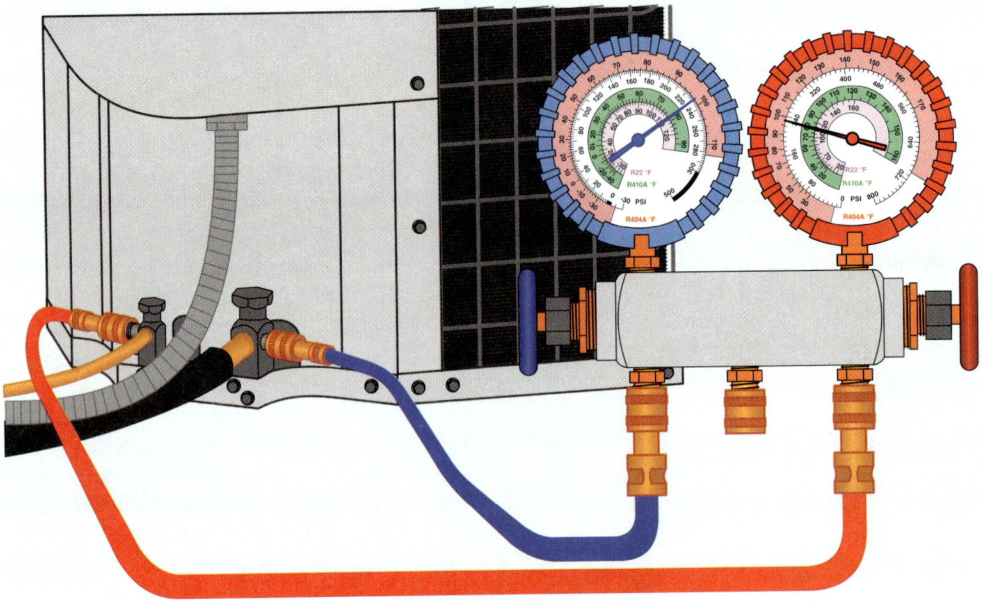

4. Connect both the high-side and low-side hoses of a gauge manifold to the service valves or other refrigerant-line ports.

SAFETY TIP

Working with refrigerants under pressure is hazardous. To avoid injury when making refrigerant-gauge manifold connections to equipment, wear safety glasses and appropriate gloves for protection.

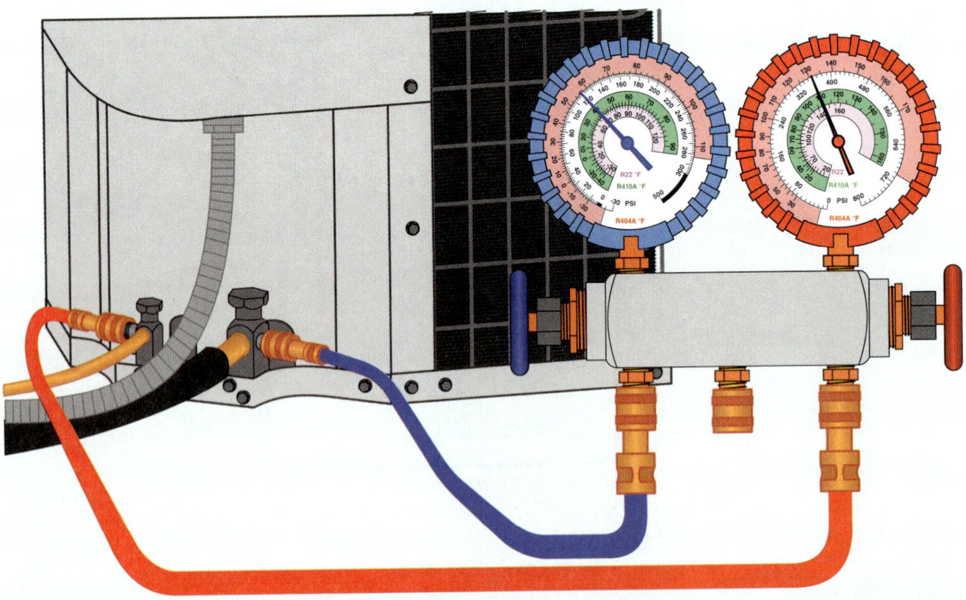

5. Restore control and line-voltage power. Set the thermostat to start the system, and allow the refrigerant pressures to stabilize.

Continued

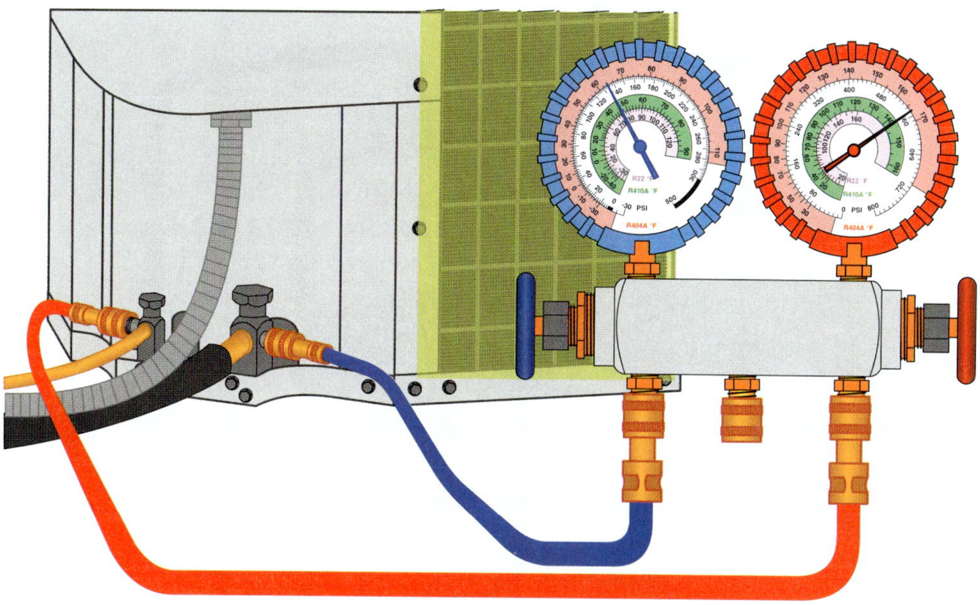

6. Carefully wrap the plastic sheet or cardboard material around the condenser to increase the high-side pressure; do not bend or otherwise damage the coil fins in the process. Monitor both the gauges and the voltmeter as the pressure rises to the switch set point. At this point, the voltmeter should show no voltage, since the contacts of the switch are still closed.

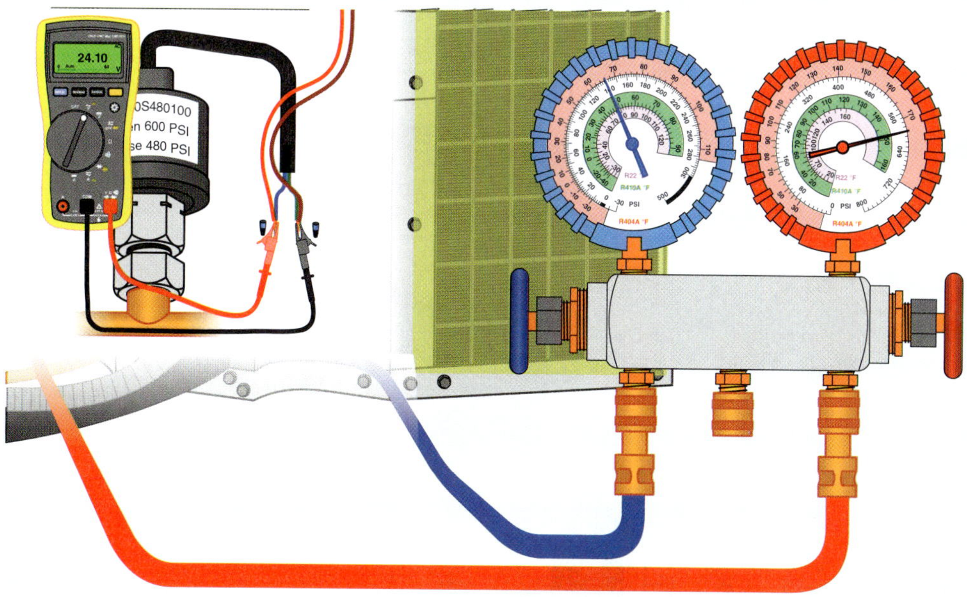

7. The switch should open at or very near its set point. If the high-side pressure rises to as much as 15 psig (103 kPa) above the high-pressure switch setting, but the switch does not open, shut the system down immediately using the nearby disconnect switch. At this point, the pressure switch should be considered unreliable It should also be considered unreliable if it opens more than 15 psig *below* the set point. If the switch does open in the correct pressure range, the voltmeter will display the control-circuit voltage and the system should shut down on its own. Thus, the switch has passed the test. Shut the system down immediately using the nearby disconnect switch.

Continued

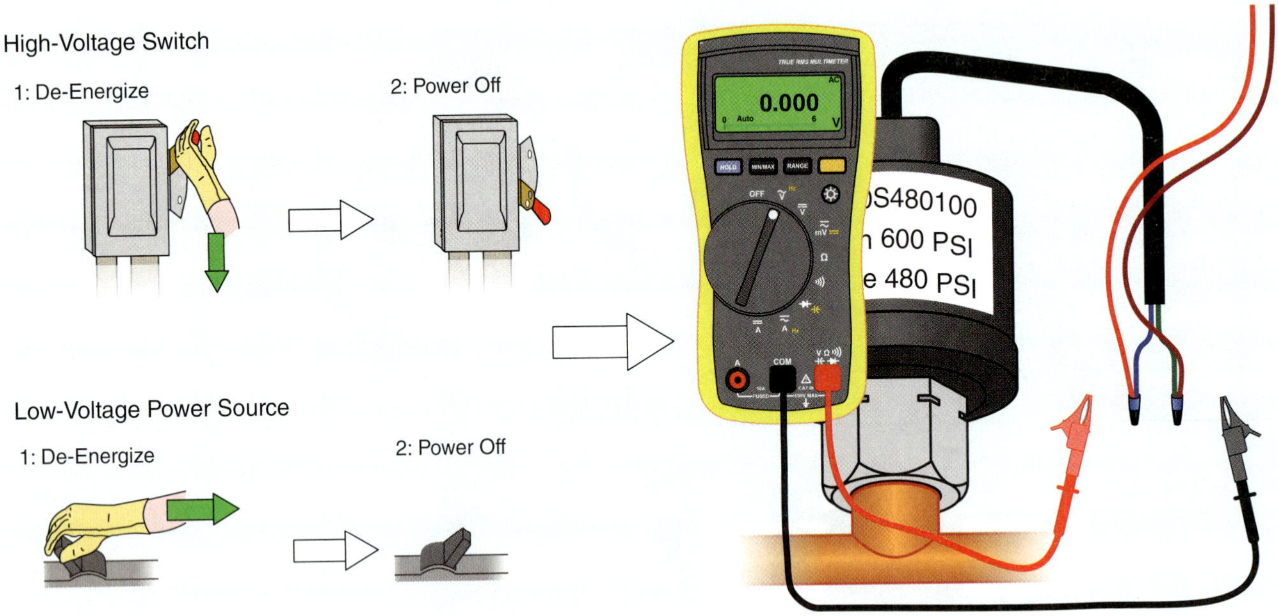

8. Ensure that all controls, including the thermostat, are set to the Off position. Deenergize the equipment at the disconnect switch if it is not already off. Disconnect the voltmeter and reconnect the pressure-switch leads.

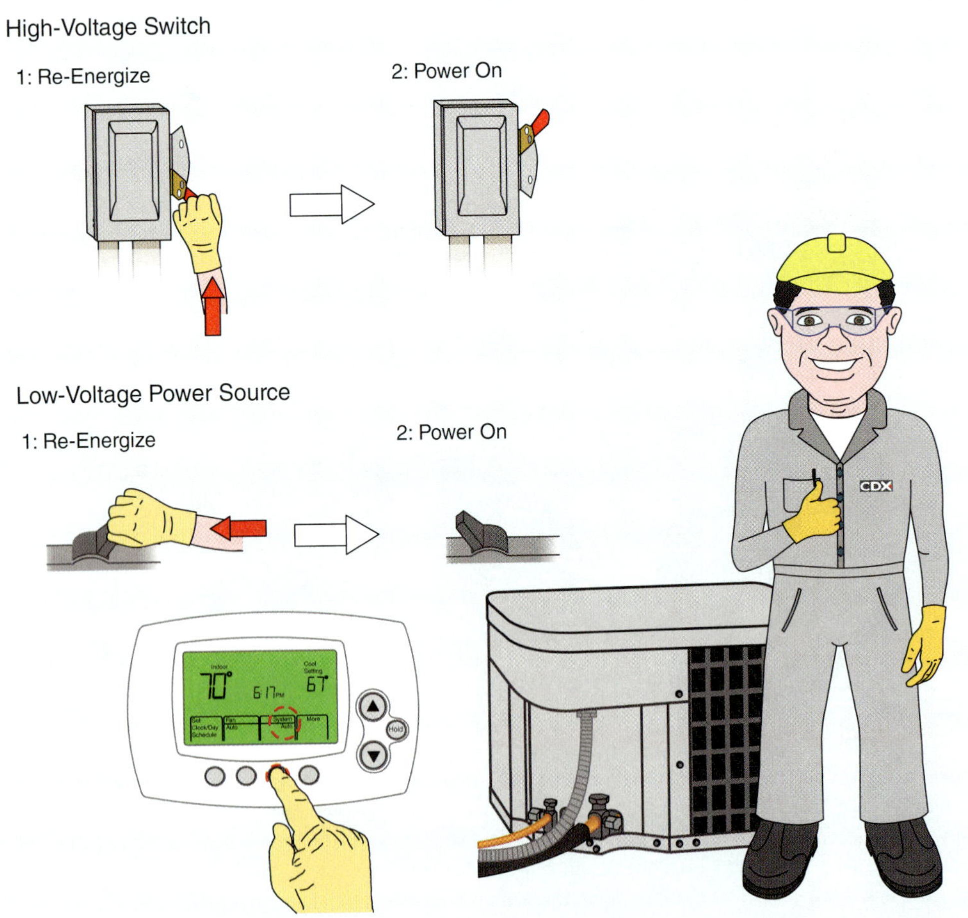

9. Disconnect the gauge manifold from the unit. Remove any materials used to block the condenser airflow. Carefully reinstall control-panel covers and equipment access doors, and then return the system to service by restoring power. Start and operate the unit to ensure that the high-pressure switch has reclosed and that the system is operating normally.

Replacing a Pressure Safety Switch

A pressure safety switch is relatively easy to replace in most cases. The primary factor in the effort required is related to its connection to the system.

Nonadjustable pressure safety switches are attached to the refrigerant circuit in one of two ways. The preferred method, as far as replacement is concerned, is a male flare connection fitted with a valve core. With a pressure switch that has a valve-core depressor inside (**FIGURE 13-12**), it can be removed by simply unscrewing it from the fitting. The valve core will seat and close off the port as it is unscrewed. This eliminates the need to recover refrigerant to accomplish the task. The other method used by manufacturers is to braze the switch onto a short section of copper tubing, which is also brazed onto the refrigerant line. This latter method is popular with equipment manufacturers because it represents a cost savings, but this method does not provide any means of convenient shutoff from the refrigerant circuit. When pressure safety switches must be replaced, many technicians choose to install a flared-tee fitting on the tubing and install a switch with a ¼" (6.35 mm) female flare fitting (or a metric equivalent). The tee provides an extra point of connection for testing purposes. Both the port connected to the pressure switch and the port for the gauge connection should be equipped with valve cores. The parts and fittings shown in **FIGURE 13-13** are one option in making the transition to a flared connection that allows a new pressure switch to be removed easily in the future.

However, it is important to note that high-pressure safety switches are generally installed without a valve core at the factory, regardless of which way it is installed. This is because the valve core could fail to open properly and provide an accurate pressure to the switch—an important safety device. Without testing, this problem may exist and remain undiscovered indefinitely, jeopardizing the safety of technicians and the equipment. Never assume that a flared high-pressure switch has a valve core. Where no valve core exists, removing the flared high-pressure fitting allows refrigerant to be released uncontrollably.

Use a pair of flare-nut or open-end wrenches to remove pressure switches mounted to a flared connection. Be sure to use two wrenches on the fitting to apply counter-torque. Small fittings can be deformed when using an incorrect technique or by over-tightening. Refer to the common torque values for flare fittings shown in **TABLE 13-2**. Leak testing must be done after the installation of a new switch.

To practice how to replace a high-pressure safety switch, follow the steps in **SKILL DRILL 13-2**.

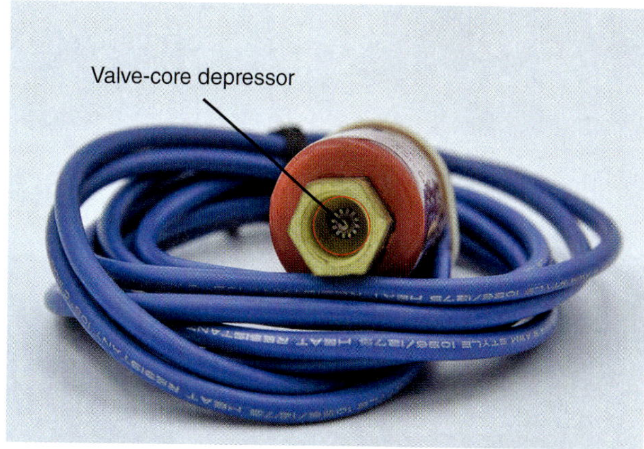

FIGURE 13-12 The valve-core depressor inside a pressure switch.

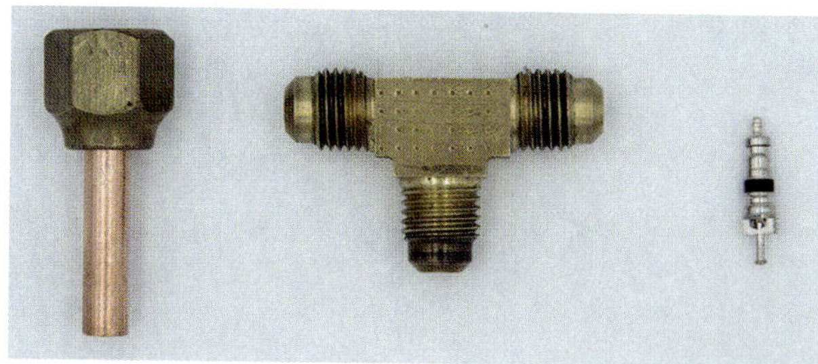

FIGURE 13-13 Fittings used to change from a brazed connection to a flared connection.

TABLE 13-2 Flare Nut Torque Specifications

Brass Flare Tubing Size (in.)	Recommended Seating Torque in Foot-Pounds (Newton-meters)
¼ (6.35 mm)	8–10 (11–13)
⅜ (9.5 mm)	15–18 (20–24)
½ (12.7 mm)	28–32 (38–43)
⅝ (15.8 mm)	38–42 (52–57)
¾ (19.05 mm)	50–55 (68–75)

▶ FOCUS ON CUSTOMER SERVICE

Finishing Up

One difference between a professional and someone just holding down a job is the way work is completed. The tendency in a world where it seems that only the bottom line matters is getting the work done quickly and moving on to the next job. However, taking the time to complete the task at hand properly will earn the customers' gratitude and loyalty. Neatly secure wiring harnesses with tie-wraps. Secure **capillary tubes** so that they don't vibrate. Keep track of the cabinet fasteners, and properly reinstall the access doors so that they don't rattle. Finally, leave every jobsite and piece of equipment cleaner than it was before arrival: not as clean, but cleaner. The few minutes it takes to do these things can help avoid dissatisfaction and callbacks, but they also help to ensure the customer will call again when service is needed. A customer may not know that a beautiful brazed joint was made inside the unit, but they can certainly see how the work area was left, and they *are* looking.

SKILL DRILL 13-2 Practicing How to Replace a High-Pressure Safety Switch

This task is prepared under the assumption that the pressure switch is threaded for flare-joint connection, but it is unknown if a valve core is present in the joint.

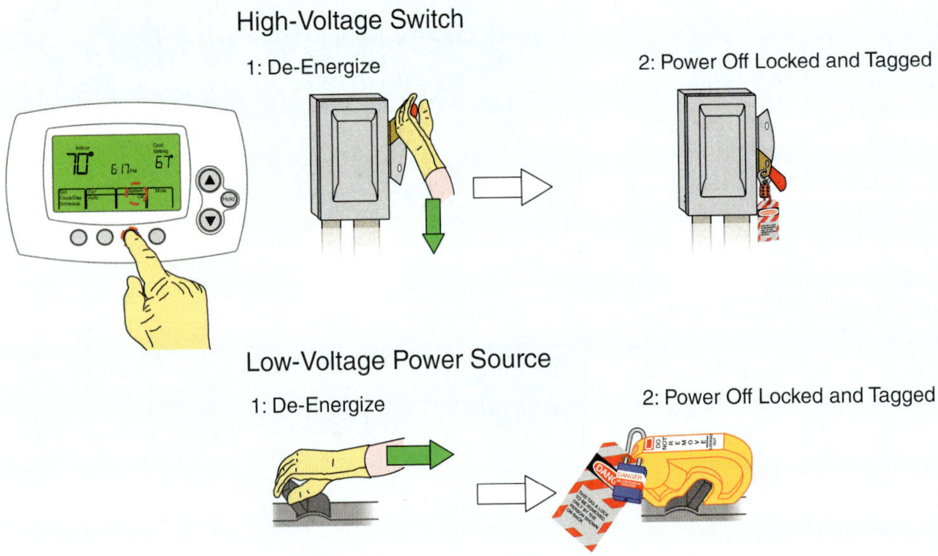

1. Ensure that all controls, including the thermostat, are set to the Off position. Deenergize, lock out, and tag the equipment to be serviced. Be sure that the process deenergizes not only the line voltage feeding the equipment, but also the low-voltage control power source that may be located in a separate piece of equipment.

SAFETY TIP

Working with electrically energized circuits is hazardous. Avoid contact with any surface that could potentially be energized. When working inside energized enclosures, maintain a stable position and, whenever possible, avoid placing both hands inside the enclosure. All work on energized equipment by unqualified individuals should be supervised by a qualified instructor or technician.

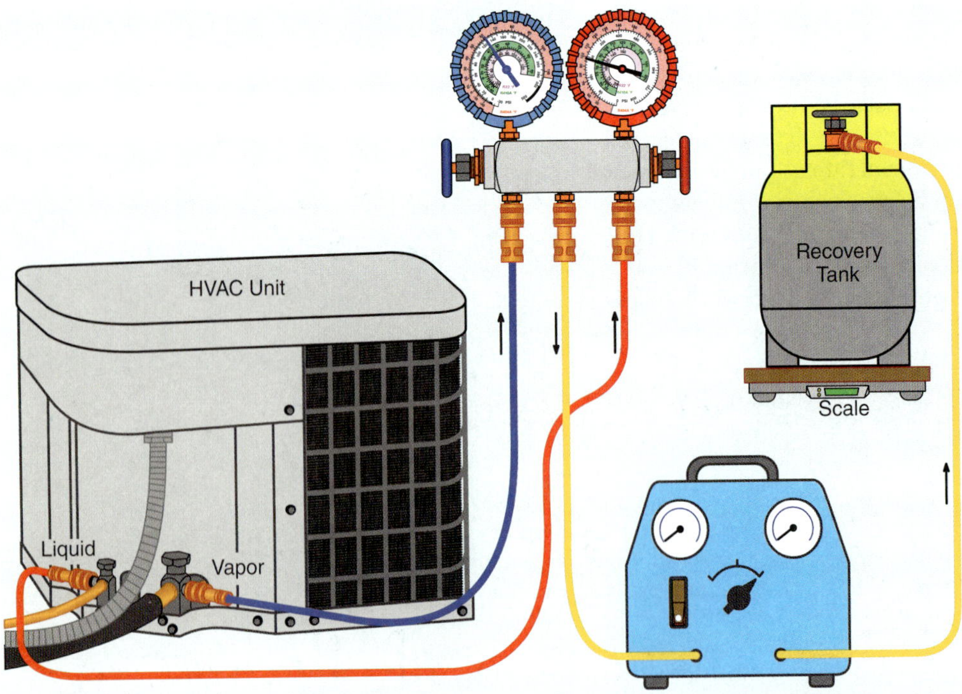

2. The system refrigerant must be recovered when the presence of a valve core and valve-core depressor in the switch fitting is unlikely or cannot be confirmed. Recover the system refrigerant as required. A very slight positive pressure (2–3 psig, or 14–21 kPa) remaining in the system after the recovery is ideal. If the refrigerant pressure goes to 0 psig (0 kPa) due to an installation delay, a 500-micron vacuum will be required.

SAFETY TIP

Working with refrigerants under pressure is hazardous. To avoid injury when making refrigerant-gauge manifold connections to equipment, wear safety glasses and appropriate gloves for protection.

Continued

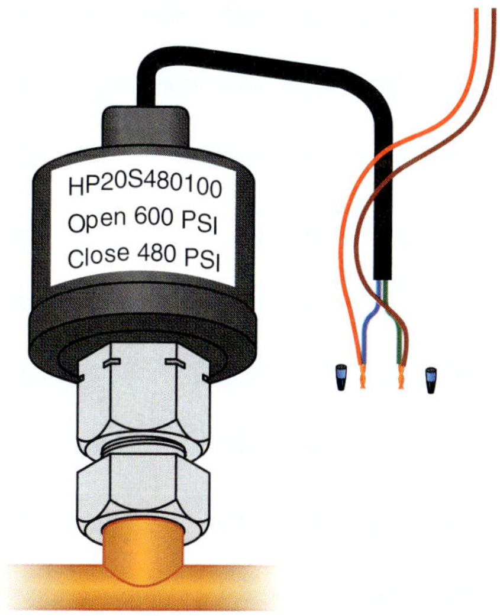

3. Disconnect the pressure-switch electrical leads from the control circuit. If the leads are embedded in a harness, carefully cut the harness ties and remove the wires from the harness.

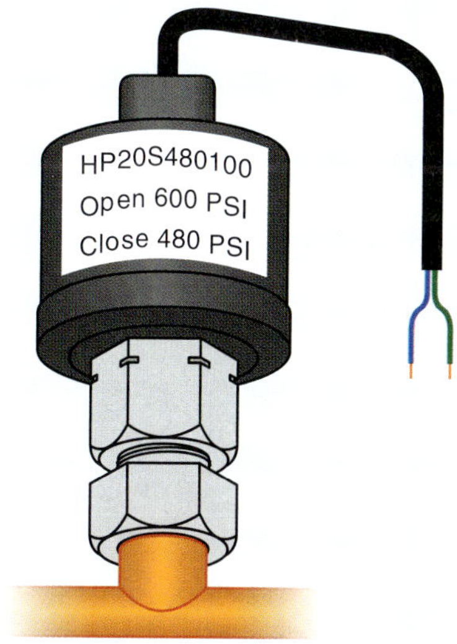

4. Remove the new pressure switch from its packaging. Confirm that it is the correct type and has the correct high-pressure set point. Also confirm that the leads provided are long enough to reach the existing points of connection. To minimize the loss of refrigerant as the two switches are exchanged, the new switch must be threaded onto the fitting very quickly once the old switch is removed.

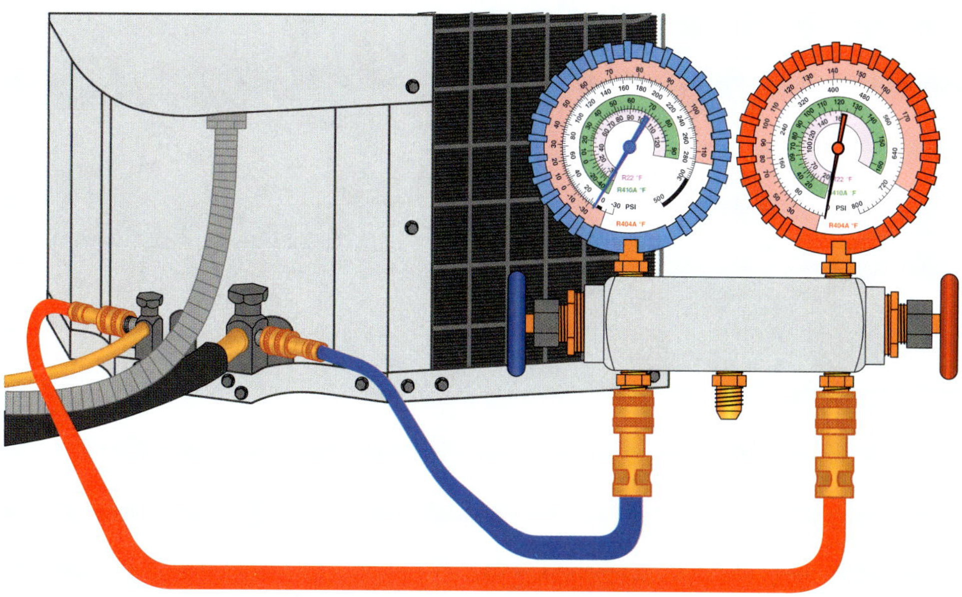

5. Check the refrigerant gauges once more before proceeding and ensure the pressure is still appropriate; 2–3 psig (14–21 kPa) is ideal. With a slight positive pressure remaining in the unit, the old switch can be removed quickly while a minimal amount of refrigerant escapes; the escaping gas will prevent air from entering the refrigerant circuit.

Continued

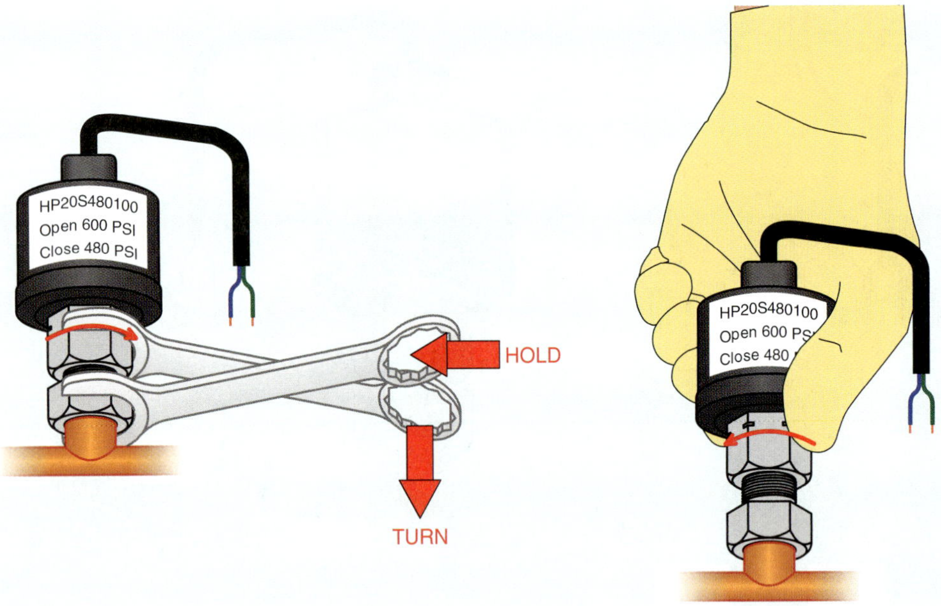

6. With the new switch within easy reach, loosen the pressure-switch flare nut with two appropriate wrenches. Once it is loose, quickly remove the switch and thread the new switch onto the fitting while a small amount of refrigerant vapor continues to escape. Tighten the joint to the proper torque of 8–10 foot-pounds (11–13 Newton-meters) for a fitting of this size. Always use a backup wrench to tighten as well as loosen the fitting. As long as a small amount of vapor is noticeably escaping, air is not entering the refrigerant circuit.

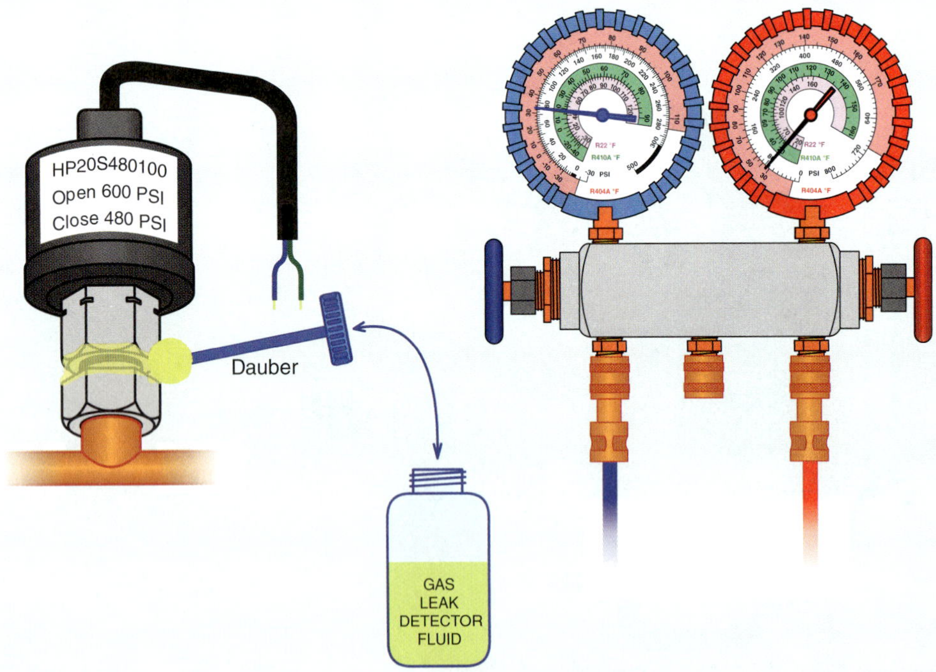

7. To perform a leak test on the connection, return enough refrigerant to the system to raise the pressure to 50–75 psig (345–520 kPa). If a positive pressure was present during the switch exchange, no air should have entered the system and thus an evacuation is not necessary. Perform a leak test on the joint with a leak-detection solution and examine the joint for bubbles or foam. Alternatively, an electronic leak detector appropriate for the refrigerant in use can be used.

Continued

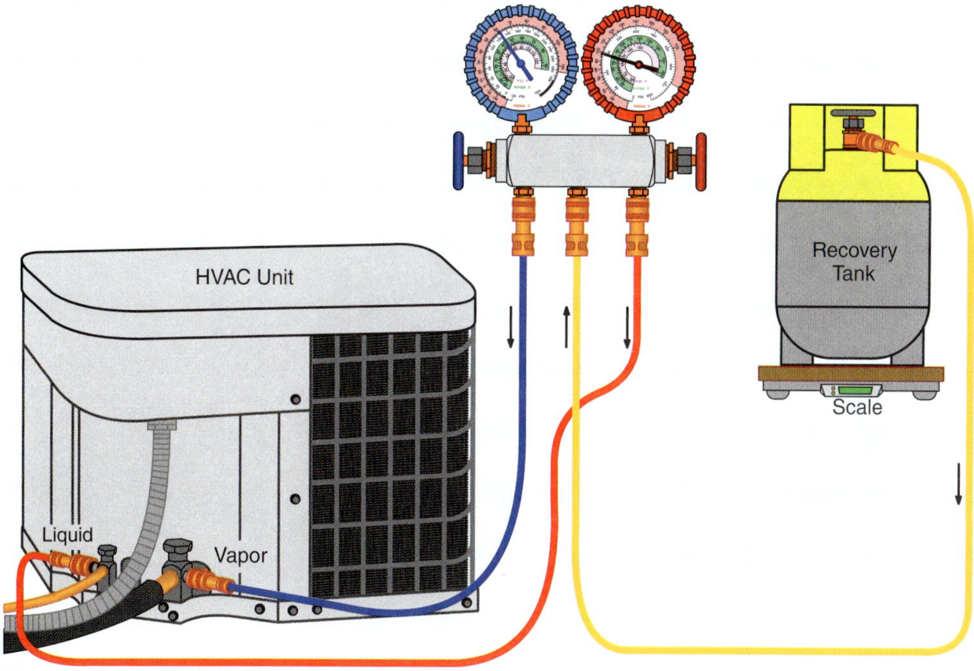

8. If there are no leaks, return any remaining refrigerant into the system. The final charge can be adjusted with the equipment in operation.

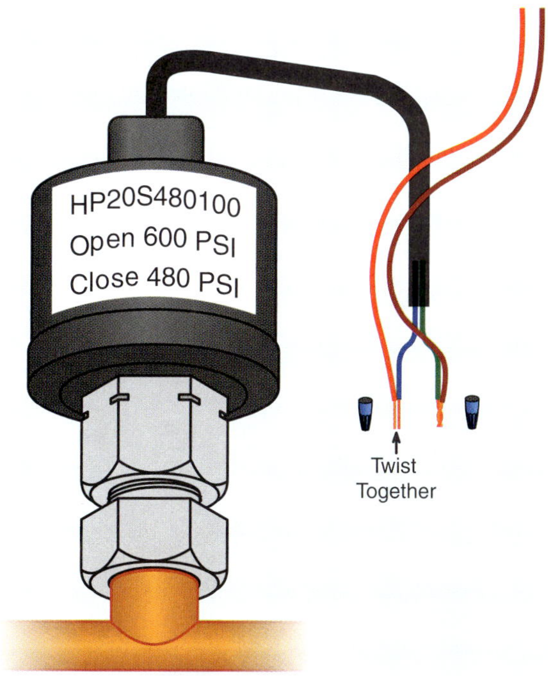

9. Route the pressure-switch wiring back to the point of connection and wire the switch into the control circuit.

Continued

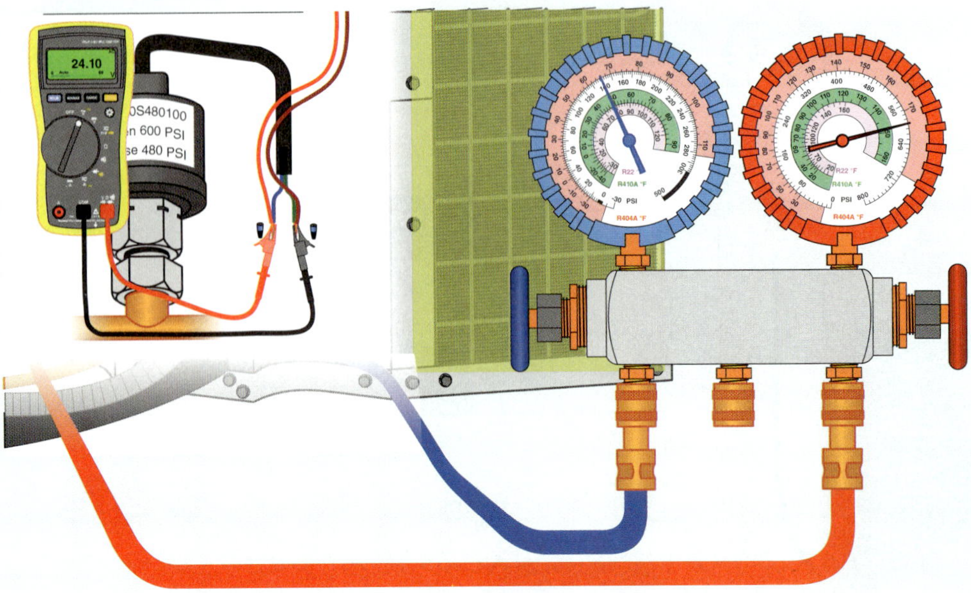

10. Restore power and start the system. Ensure that the refrigerant charge is appropriate. Complete the job by testing the switch as described in the previous exercise.

Troubleshooting an Adjustable Pressure Switch

Adjustable pressure switches used in the refrigerant circuit are different in design than the encapsulated, nonadjustable switches. To accommodate the adjustment mechanism and more robust electrical contacts, the assembly must be much larger. As a result, they are in enclosures mounted away from the pressure source and connected to the source by a copper capillary tube (**FIGURE 13-14**). Wiring is routed to the screw terminals of the control, as shown in **FIGURE 13-15**; there are no leads provided. Some switches combine a high-pressure safety switch with a low-pressure cycling control in the same enclosure, which means that they have two capillary tubes. Some of today's switches are connected using special flexible hoses instead of copper capillary tubes. They are less fragile and more resistant to vibration than the copper capillaries. In combination models, the high-pressure safety switch is tested in the same manner as other safety switches. The only difference is that the set point is adjustable. As a result, this section will focus on testing a low-pressure cycling control.

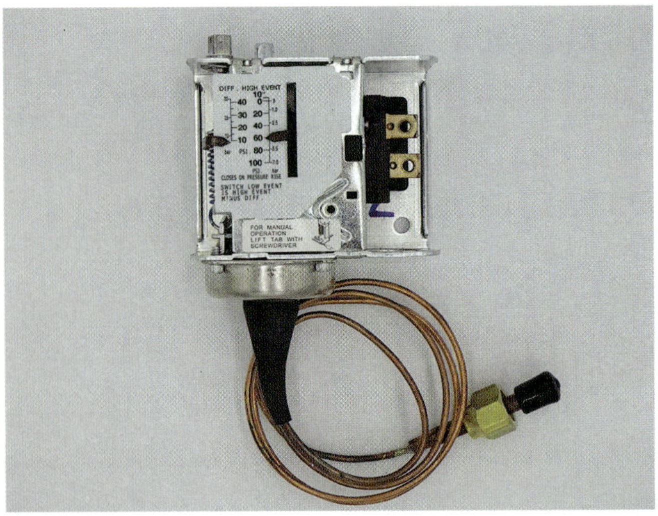

FIGURE 13-14 An adjustable pressure switch.

Low-pressure cycling controls are connected to the low-pressure side of the system to monitor refrigerant pressure and cycle the compressor. There are two settings to be made, as shown in **FIGURE 13-16**. The first setting is the cut-in pressure. When the pressure rises to the set cut-in pressure, the switch contacts close and start the compressor. The other setting is the differential. As noted on the pressure scale, the cut-out pressure is the cut-in pressure minus the differential. For example, with a cut-in setting of 35 psig (241 kPa) and a differential of 12 psig (83 kPa), the switch will open (reach the cut-out pressure) when the pressure falls to 23 psig (158 kPa). Note the switches in many controls are single-pole, double-throw (SPDT) instead of single-pole, single-throw (SPST). That provides a little more flexibility for the user, allowing the control to be used in other applications.

There are several types of failure associated with low-pressure cycling controls:

- crimped or blocked capillary tube
- capillary-tube leaks
- a leak in the pressure-sensing element
- breakage or excessive wear of the switch components or moving parts.

FIGURE 13-15 Switch terminals of an adjustable pressure switch.

Capillary tubes can easily be crimped or otherwise damaged. If the tube passage is partially or fully closed, the switch will not respond properly to the system pressure. A partially blocked tube may cause the control to react too slowly, since the pressure will not rise and fall in unison with the system.

A leak in the capillary tube releases system refrigerant. Regardless of whether the switch functions properly, it must be replaced. Capillary tubes cannot be replaced or effectively repaired. However, the new switch can be purchased with male flare connections at the switch instead of capillaries. Hoses designed for this purpose can then be installed instead of capillary tubes. They are far more durable. A leak in the switch assembly itself is also cause for replacement.

Testing a low-pressure cycling control is relatively simple. It involves operating the system with an attached gauge manifold while monitoring the activity of the switch. A low-pressure cycling control is used in refrigeration applications in lieu of a thermostat. Consequently, setting the switch to start and stop the system as desired, without short cycling the compressor yet maintaining the desired box temperature, is the most important result. If the compressor is starting before the box temperature has risen to the desired temperature, the cut-in setting must be increased. If it shuts down too quickly as the pressure falls rapidly after starting, the differential setting must be increased.

As a low-pressure cycling control is tested, remember that the low-pressure reading of the system when the compressor is not running (once stabilized) indirectly reflects the box temperature. The concept is to use a pressure switch to cycle the compressor on and off at specific

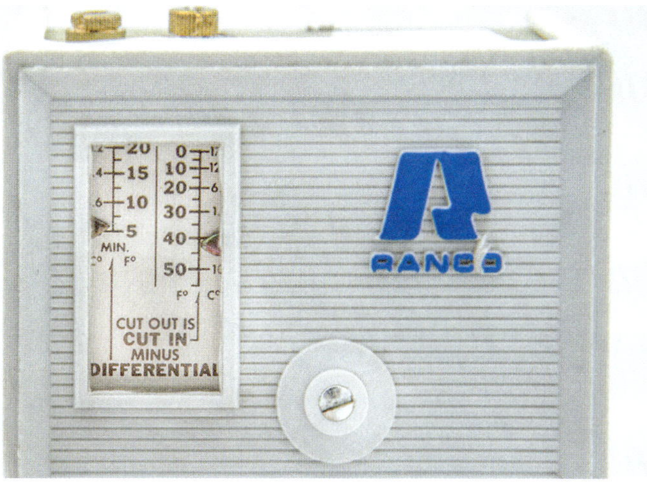

FIGURE 13-16 Setting scales of a low-pressure cycling control.

pressures, but the ultimate goal is to have the system start and stop at temperatures that are appropriate for the box and the products it protects. Keep in mind that setting up a low-pressure cycling control requires patience and some trial and error to ensure that it functions as it should.

To practice how to troubleshoot an adjustable low-pressure cycling switch, follow the steps in **SKILL DRILL 13-3**.

SKILL DRILL 13-3 Practicing How to Troubleshoot an Adjustable Low-Pressure Cycling Switch

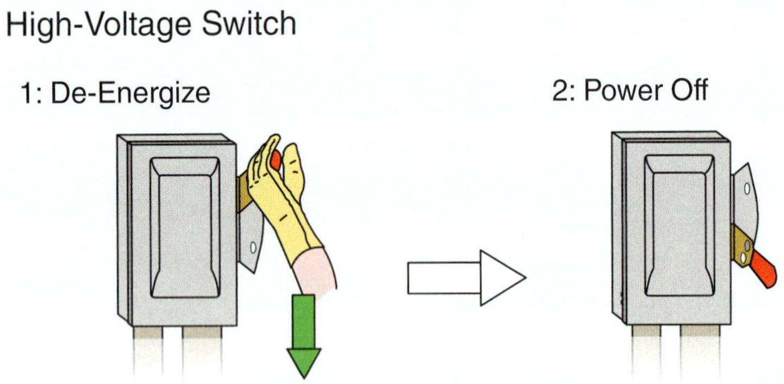

1. Ensure that all controls, including the thermostat (if any), are set to the Off position. Deenergize the equipment to be serviced using the disconnect switch, but do not lock and tag, since the unit must operate for the test.

SAFETY TIP

Working with electrically energized circuits is hazardous. Avoid contact with any surface that could potentially be energized. When working inside energized enclosures, maintain a stable position and, whenever possible, avoid placing both hands inside the enclosure. All work on energized equipment by unqualified individuals should be supervised by a qualified instructor or technician.

Common refrigeration system controls are not powered by low-voltage sources. Line voltage is applied to all controls and loads. Maintain the appropriate clearance from all energized surfaces at all times.

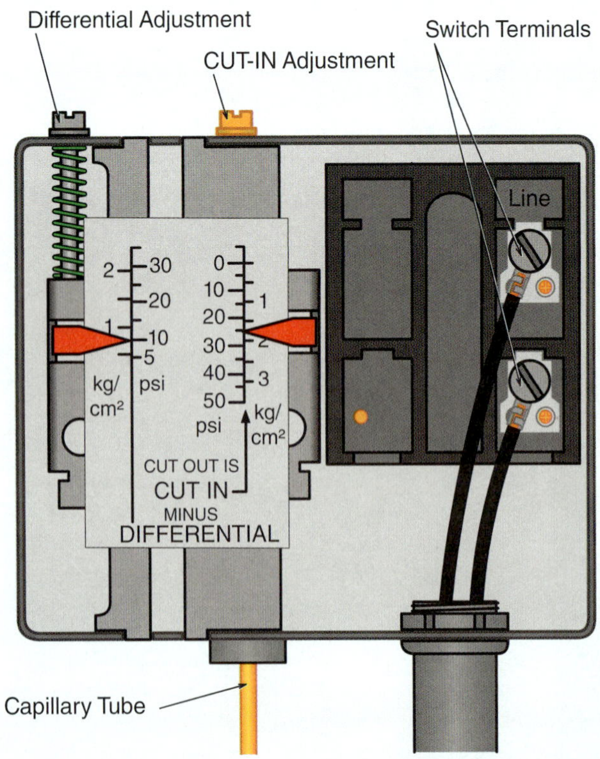

2. Remove the cover of the pressure cycling control and inspect the mechanical and electrical connections to the pressure switch, internal and external, for damage, signs of leaks (such as oily stains), and loose connections. Ensure that the wires are connected to the correct terminals of the switch (NO for standard cycling-control applications).

Continued

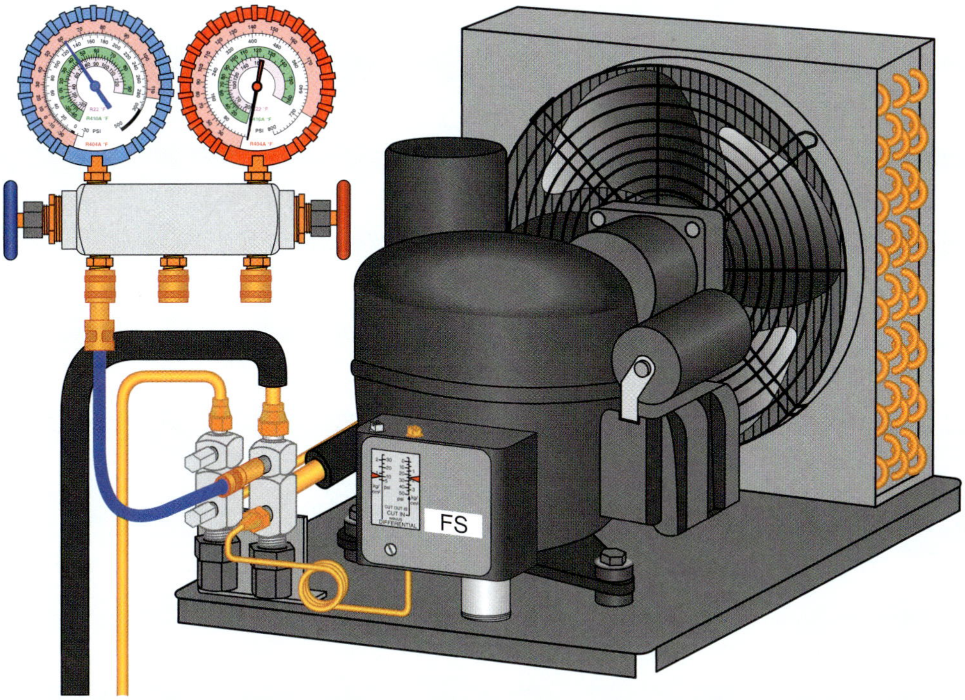

3. Connect an accurate gauge manifold to the system. Connect the low-pressure gauge to the same location as the low-pressure control if at all possible. Most cycling controls are connected to a service port on the back-seating service valve, where it can easily be isolated. A tee can be placed on this port to provide a gauge connection if necessary. Back-seat the valve, remove the cap, connect the low-pressure gauge of the manifold, and then reopen the valve two turns. This will apply refrigerant pressure to both the gauge and the switch.

SAFETY TIP

Working with refrigerants under pressure is hazardous. To avoid injury when making refrigerant-gauge manifold connections to equipment, wear safety glasses and appropriate gloves for protection.

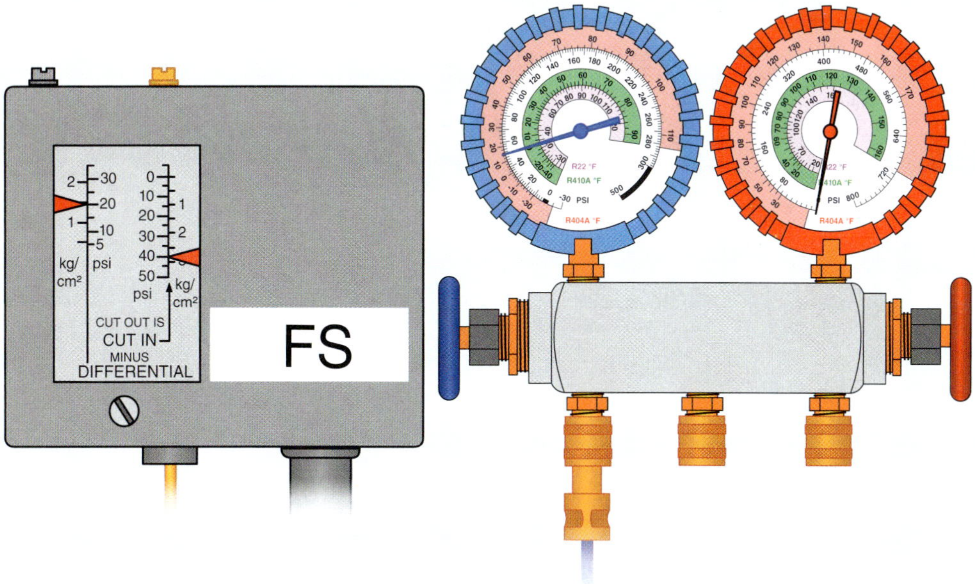

4. Compare the present reading on the low-pressure gauge to the cut-in setting of the switch. If the gauge pressure is obviously higher than the cut-in setting, the contacts should be closed. If the pressure is lower than, or very near, the cut-in setting, the contacts are likely open. To determine the position of the contacts with certainty, an ohmmeter can be used with the power off to check for continuity through the contacts. This step provides an idea of what to expect once power has been restored; the compressor and condenser fan will either start or remain idle. Reinstall the pressure switch cover before proceeding further.

Continued

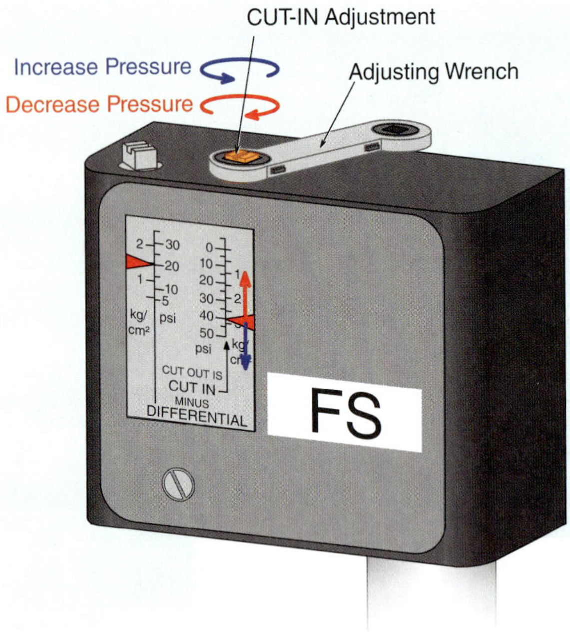

5. Restore power to the system. If the pressure is above the cut-in setting, the compressor should start immediately. Once the system starts, the operating pressures must be allowed to stabilize while observing the gauge. Once power has been restored, if there is any doubt about the position of the pressure switch contacts during the task, a voltmeter can be used to check the contacts. As always, the voltmeter will read the line-voltage applied if the contacts are open, but will read 0 volts or background voltage if the contacts are closed.

If the contacts are closed and the compressor did not start, there is another problem beyond the pressure switch. If the contacts are open when they should be closed, based on the setting and the system pressure, the switch is inaccurate or possibly defective. Note that the scales of these switches are short and compact; do not expect laboratory precision. If the box is at a high enough temperature (measured with a thermometer) for the system to be operating, attempt to slowly adjust (lower) the cut-in setting of the switch to start the system. If the pressure switch is significantly inaccurate or never closes, it is defective and must be replaced.

If the switch is closed but the box temperature is already cold enough, the pressure cut-in setting can be raised to start the compressor at a higher box temperature.

If the system pressure is below the cut-in setting (the box is cold), preventing the switch from closing, the pressure must be encouraged to rise for testing. This can be done by imposing a load on the refrigerated space; briefly opening the door of a reach-in, for example, will cause it to warm up more quickly. Monitor both the gauge pressure and the switch as the pressure rises, to determine whether the compressor starts at the appropriate pressure. If it does not start and the pressure continues to rise beyond the cut-in setting, the switch is defective or poorly calibrated.

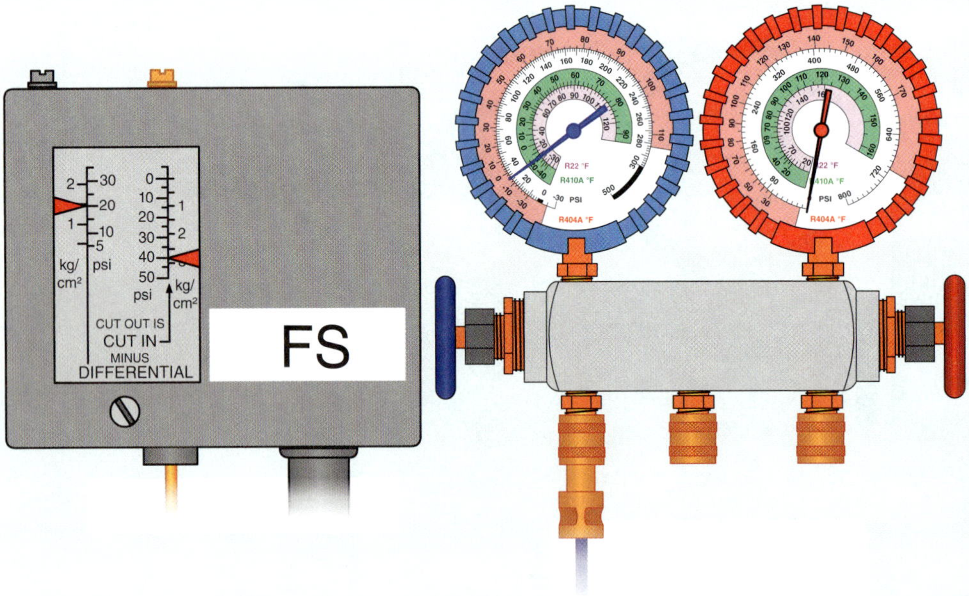

6. Now the focus moves to the differential setting, which determines the cut-out setting. The cut-out setting equals the cut-in setting minus the differential. Some switches do have a labeled cut-out setting, but most have a differential setting instead. For a cooler operating on R-134a, for example, the desired cutout pressure might be around 20 psig (138 kPa). Monitor the pressure gauge and pressure switch while the compressor runs. The low-side pressure decreases as the box gets cooler.

Continued

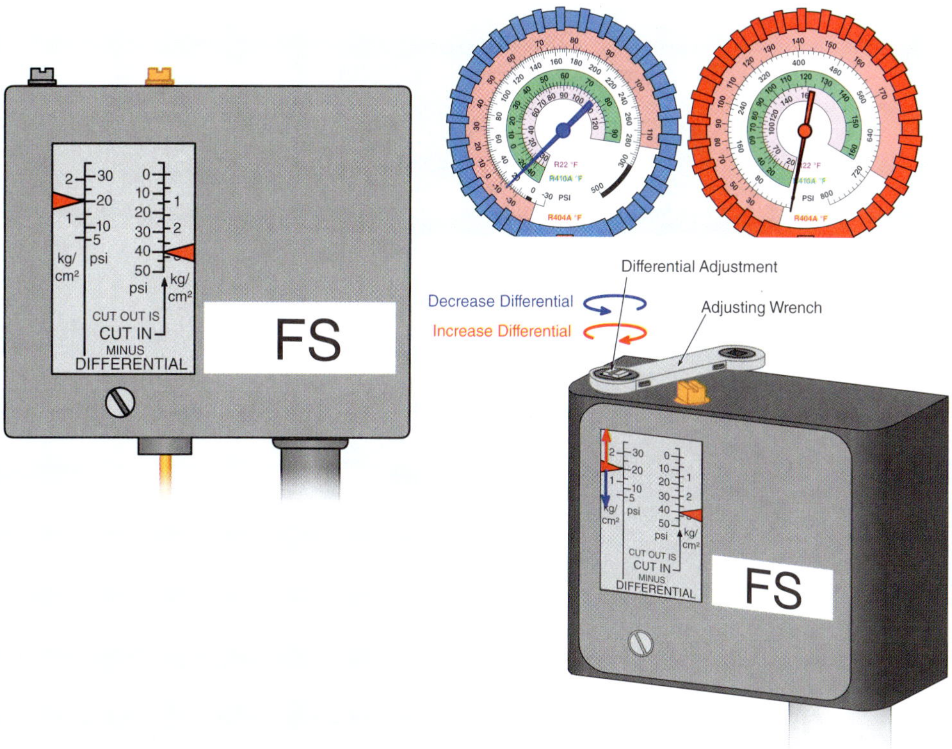

7. Once the pressure has fallen to the cut-out point (cut-in minus differential), the switch contacts should open and the compressor should shut down. If it does not and the pressure continues to fall while the compressor runs, the switch may be defective or inaccurate. Attempt to make a minor adjustment in the differential setting to determine whether the switch responds favorably. Lowering the differential would be the appropriate action if the box is becoming colder than desired. If it does not respond, the switch is unreliable and should be replaced. If it does respond as it should, allow the compressor to shut down. The low-pressure reading should immediately rise slightly, and then begin a slower increase as the box warms.

8. If any adjustments were made in the switch settings, allow the complete cycle to repeat itself. This ensures that the new settings are cycling the compressor as desired. Getting the settings just right often takes patience and several adjustments. It is best to monitor system operation all the way through a complete cycle without interfering (by artificially warming up the box, for example) whenever possible. If a series of adjustments cannot produce reliable operation, the pressure switch may need to be replaced.

Continued

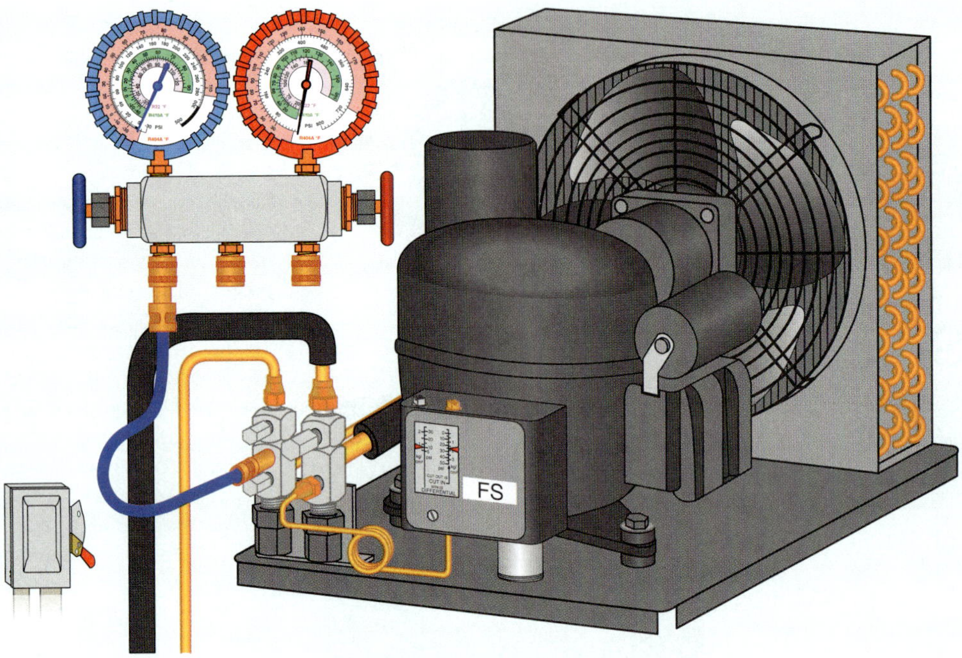

9. If the switch is operating reliably, the compressor will shut down at the cut-out set point. Once it does, deenergize the system. Back-seat the refrigerant service valve, and disconnect the gauge manifold from the system. After reinstalling the cap on the port, reopen the service valve one-half to one turn to provide an active pressure source to the cycling control for normal operation.

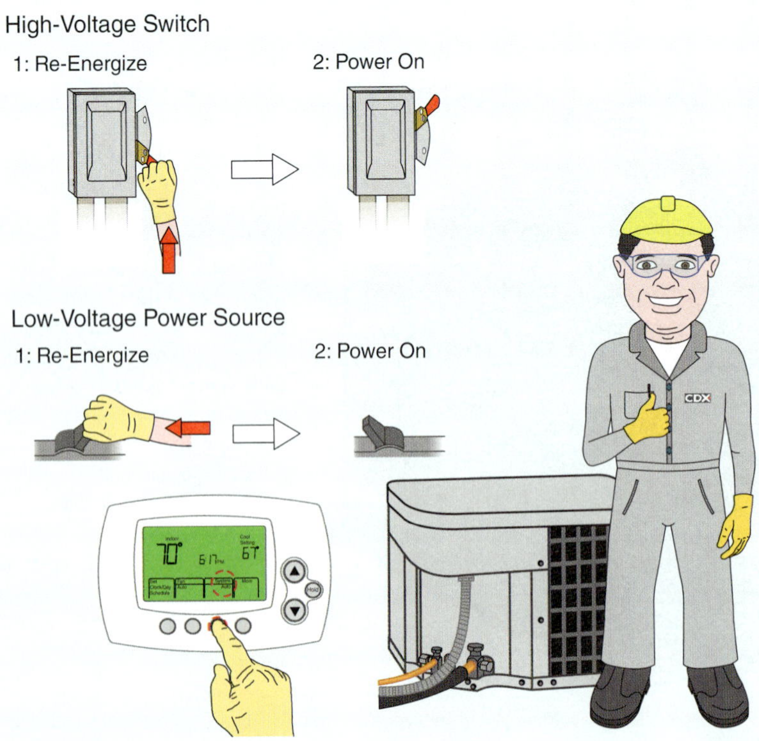

10. Restore power to the system and return the refrigeration unit to service. Make sure that any refrigerant port caps are reinstalled and that the cover for the pressure switch is secure.

Replacing an Adjustable Pressure Switch

As has been discussed already, adjustable pressure switches are usually mounted away from the source of pressure. They are connected to the system by capillary tubes or hoses, and the control-circuit wiring is routed to the switch terminals. Replacement, therefore, requires that the switch be disconnected from the pressure source; the wiring disconnected from the switch terminals; and the switch assembly removed from its mount.

Pressure-switch mounting plates are relatively universal. It most cases, the new switch can be mounted to the same mounting plate as the failed switch. A new mounting plate is shipped along with most pressure switches and can be used if needed.

Since the capillaries of these switches are generally connected to the compressor/condensing unit service valves, the switch can easily be isolated from the refrigerant circuit without a loss of refrigerant.

▶ TECHNICIAN TIP

The Handy-Dandy, All-Purpose Service Ratchet

All HVACR technicians should own a service ratchet **FIGURE 13-17**. This tool is very versatile. It is used to operate service valves of all sizes, since they have a square stem, without damage. Using other types of wrenches on square service-valve stems will eventually damage the stems. Most pressure switches are also adjusted with a square adjusting stem that protrudes from the top of the control enclosure. The service ratchet is also perfectly designed for adjusting them.

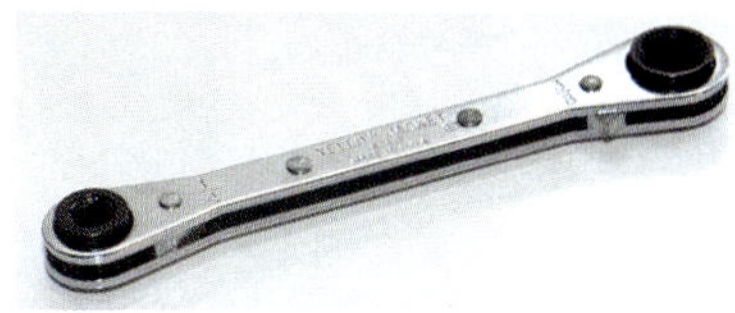

FIGURE 13-17 HVACR service ratchet.

To practice how to replace an adjustable low-pressure switch, follow the steps in **SKILL DRILL 13-4**.

SKILL DRILL 13-4 Practicing How to Replace an Adjustable Low-Pressure Cycling Switch

SAFETY TIP

Working with electrically energized circuits is hazardous. Avoid contact with any surface that could potentially be energized. When working inside energized enclosures, maintain a stable position and, whenever possible, avoid placing both hands inside the enclosure. All work on energized equipment by unqualified individuals should be supervised by a qualified instructor or technician.

Common refrigeration system controls are not powered by low-voltage sources. Line voltage is applied to all controls and loads. Maintain the appropriate clearance from all energized surfaces at all times.

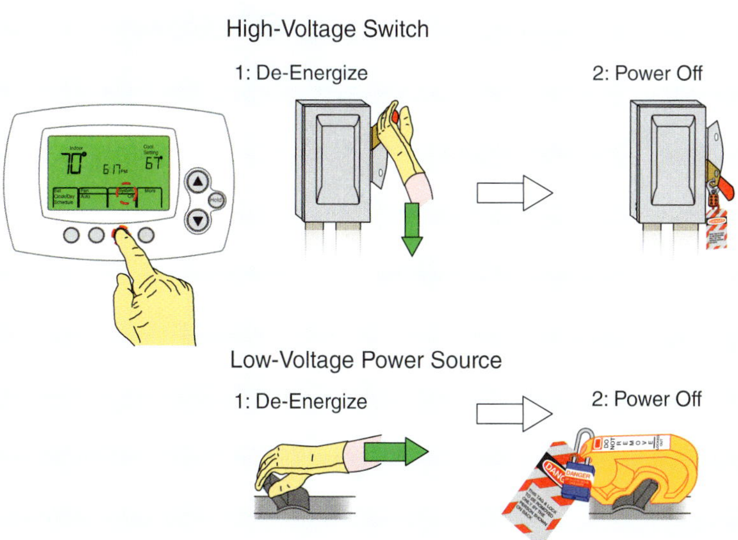

1. Ensure that all controls are set to the Off position. Deenergize, lock out, and tag the equipment to be serviced.

Continued

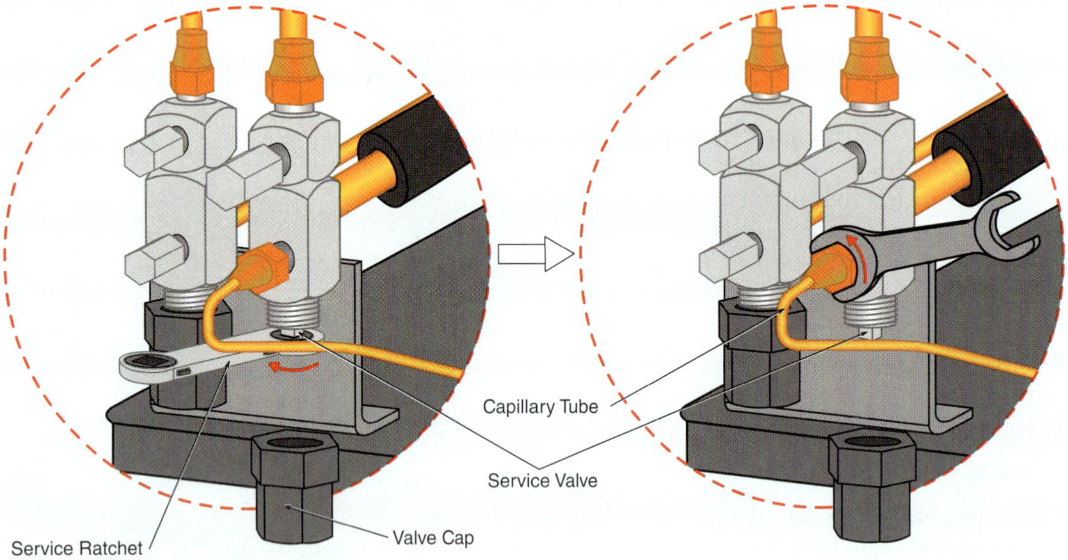

2. Isolate the pressure-switch from the refrigerant circuit by fully back-seating the compressor suction service valve. Vent any remaining refrigerant pressure by slowly loosening the capillary-tube flare fitting, then remove the connection completely.

SAFETY TIP

Working with refrigerants under pressure is hazardous. To avoid injury when making refrigerant-gauge manifold connections to equipment, wear safety glasses and appropriate gloves for protection.

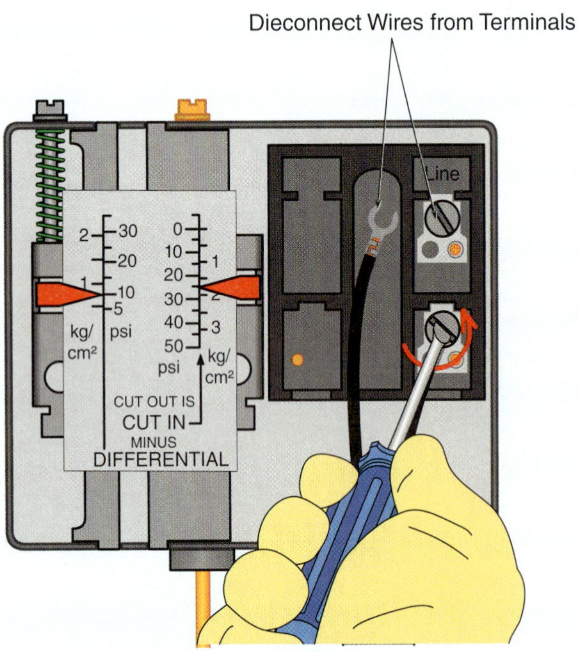

3. Remove the cover and disconnect the wiring from the pressure switch.

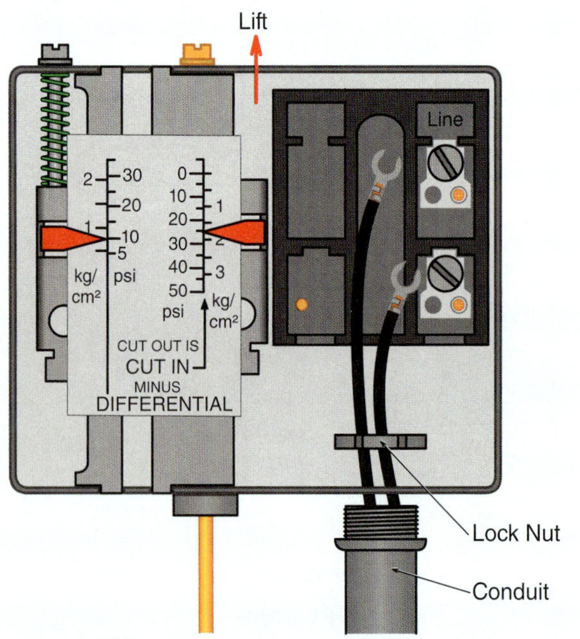

4. Remove the locknut mounting the pressure-switch assembly to the conduit, or the fasteners holding it to its mounting plate, and remove the pressure switch from the unit.

Continued

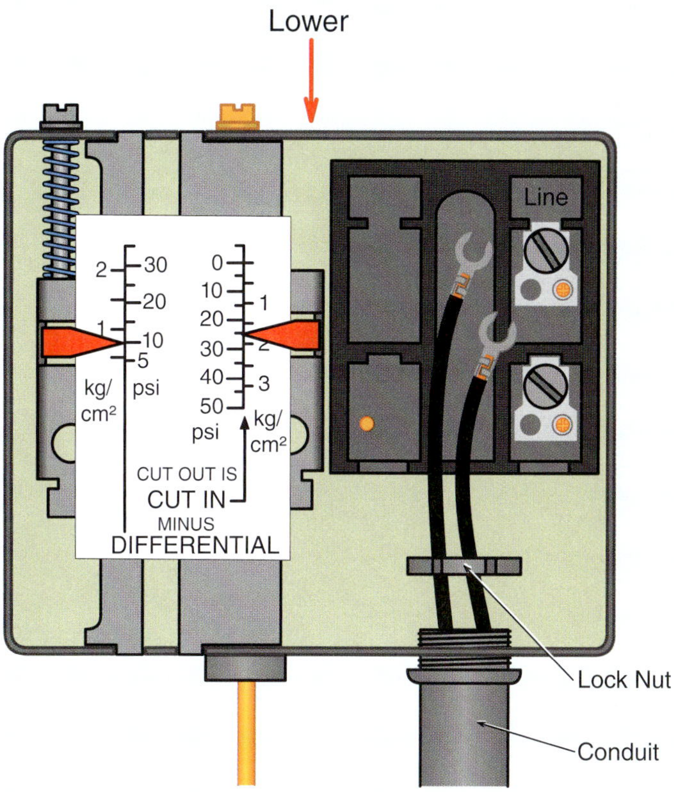

5. Remove the new switch from the packaging. Inspect the capillary tube for damage and kinks. Install the pressure switch on its mounting plate by using the fasteners removed in Step 4, or use the new fasteners that came with the switch.

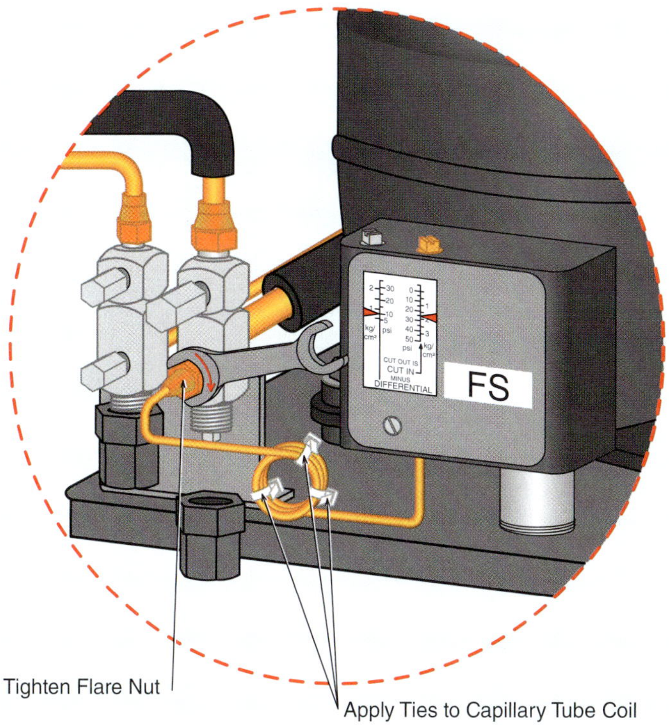

6. Carefully uncoil the capillary tube while avoiding tight bends and kinks. The excess capillary should remain neatly rolled up in a circle that is about 3" (7–8 cm) in diameter. This helps dampen vibration. The coils should remain slightly separated so that they don't rub against each other. Alternatively, the coils can be banded tightly together with several wire ties, to prevent them from vibrating against each other. Connect the flare nut to the same port as the previous switch. Remember to observe proper torque values of 8–10 foot-pounds (11–13 Newton-meters) of torque.

Continued

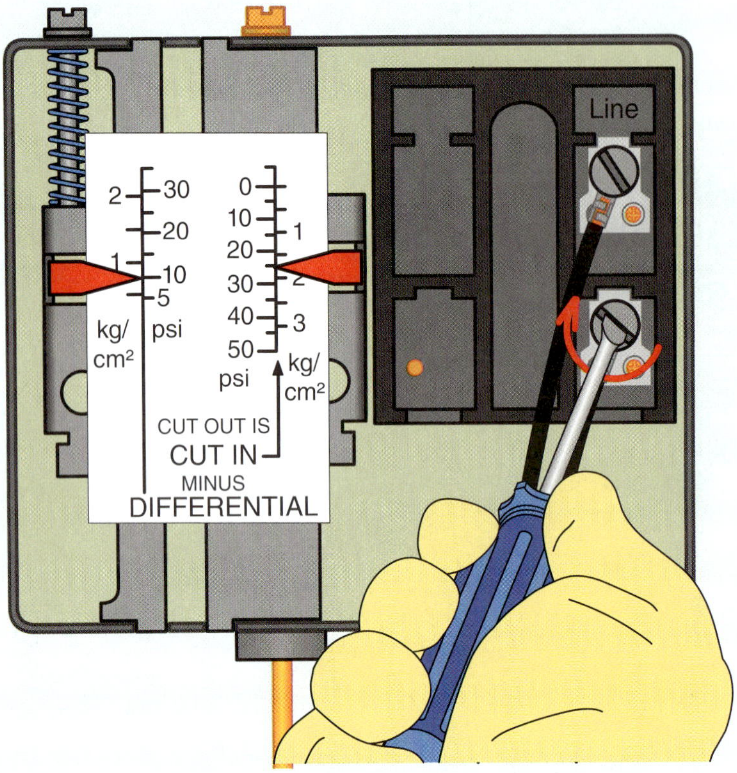

7. Terminate the wiring at the appropriate switch terminals and reinstall the switch cover.

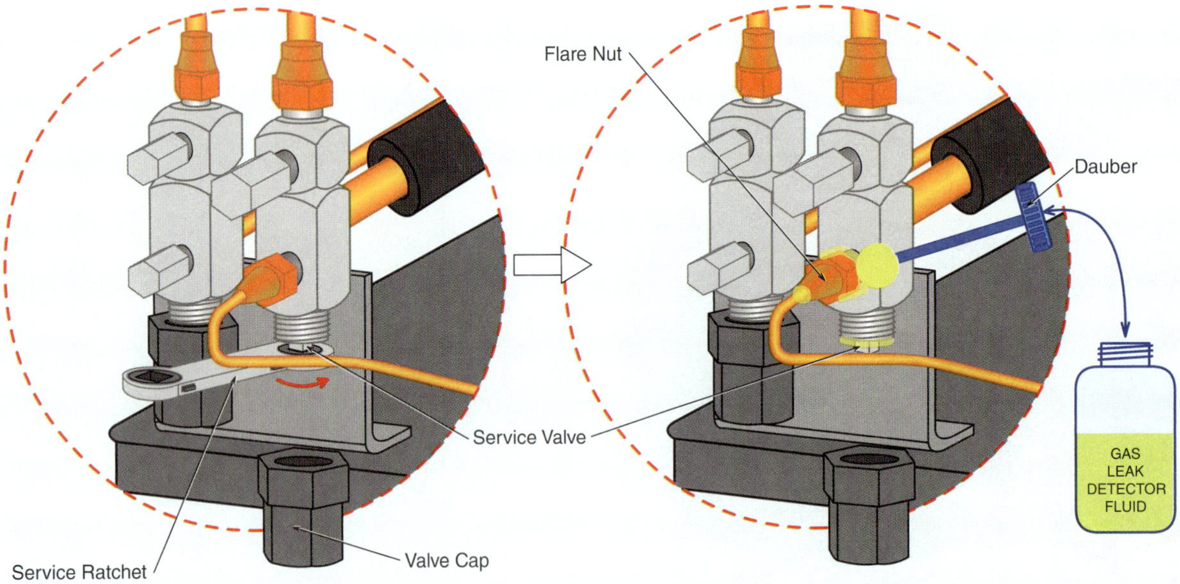

8. Crack open the suction service valve (clockwise) approximately one full turn to pressurize the switch capillary. The system must operate with the service valve cracked open from its back-seated position to function properly when there is a pressure switch attached to the service-valve port. Apply a leak-detection solution to the flare joint(s) and packing nuts of the service valve to ensure there are no leaks.

Continued

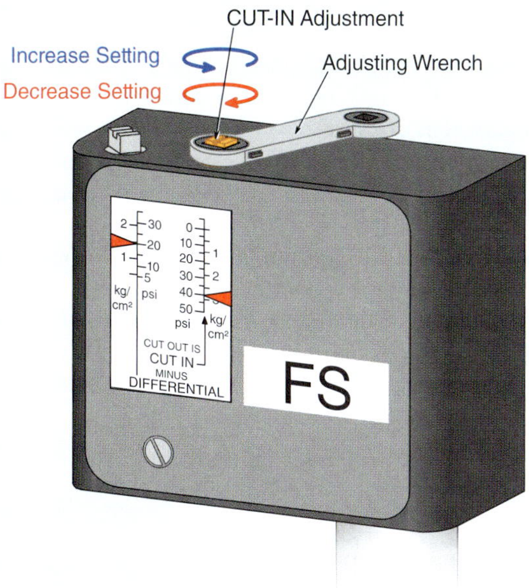

9. Set the switch as close as possible to the desired cut-in and differential settings.

10. Restore power to the system. Test and adjust the switch as outlined in the previous drill.

▶ Differential-Pressure Switches

K13003 Describe the operation and maintenance of differential-pressure switches.

As defined earlier, differential-pressure switches detect the difference in pressure between two separate pressure sources. In other words, it compares the pressure of one source to another. The switch then opens or closes a set of contacts based on the difference between their pressures.

In HVACR systems, a differential-pressure switch may be used to monitor pump operation—for example, closing its contacts if the difference between the suction and discharge pressure of the pump is too small, indicating the pump isn't doing what it should. The switch closure might set off an alarm or other means of notification that something is wrong, stop the pump and start another one, or start a second pump to assist the first one. Differential-pressure switches may also be used to monitor the air pressure in a duct and compare it to atmospheric pressure outside the duct (**FIGURE 13-18**).

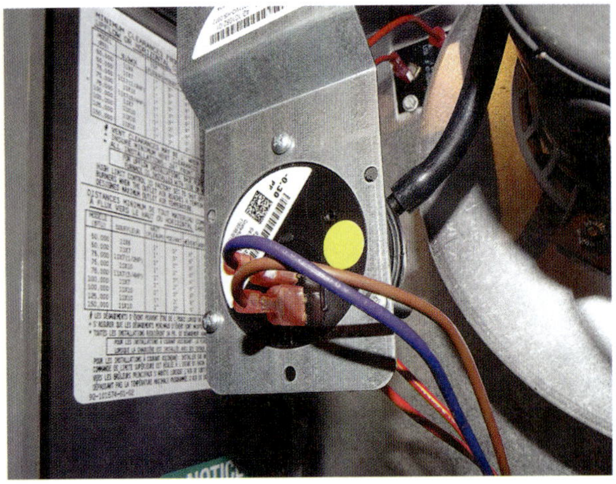

FIGURE 13-18 Differential-pressure switch used to monitor pressure in a furnace heat exchanger.

This brings up an important point about differential-pressure switches. Although they are designed to compare the pressure from two sources, they are not always physically connected to two sources. They are often used to compare the pressure from one source to atmospheric pressure. In this case, there is no need to connect anything to one of the pressure ports, as long as that port is open to the surrounding atmosphere. The pressure switch in Figure 13-18 is an example of this. The black tube is connected to the heat exchanger, and the opposite side of the internal diaphragm is open to the atmosphere. Differential-pressure switches can be applied in many ways to HVACR systems.

Differential-Pressure Switch Operation

One of the common differential-pressure switch applications in the HVACR trade is the oil-pressure safety switch. This section will focus on this specific type, since its operation is somewhat unique.

Oil-pressure safety switches are used on many semi-hermetic and open-drive compressors, especially those used in refrigeration. A few large hermetic compressors (10 hp and larger) may also be equipped with one, but very few hermetic compressors have the necessary ports to make the connections. Since they are hermetic, a failure of its lubrication system results in replacing the compressor. Hermetic compressors cannot be repaired in the field.

As expected for a differential-pressure switch, an oil-pressure safety switch has a high-pressure port and a low-pressure port (**FIGURE 13-19**). It is important to understand that these names do not refer to the high- and low-pressure sides of the system. In this case, the terms "high" and "low" refer only to the relationship between the two pressure sources. The highest pressure monitored is connected to the high-pressure port of the switch.

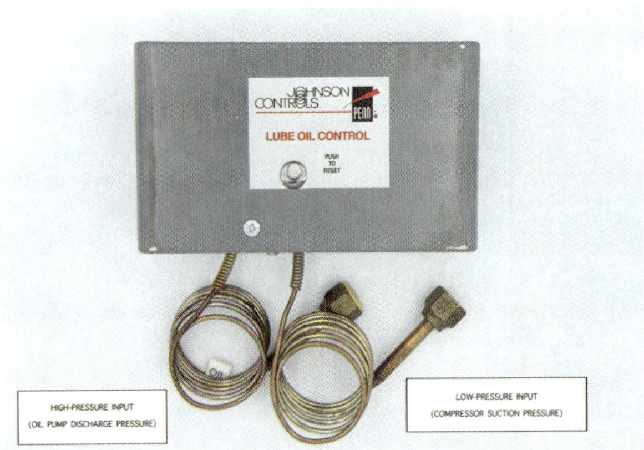

FIGURE 13-19 Oil-pressure safety switch on a semi-hermetic compressor.

In a typical semi-hermetic compressor, the crankcase serves as the oil reservoir. The crankcase is also in the path of the incoming refrigerant vapors, which travel through the crankcase on the way to the compression process. The point here is that the crankcase of the compressor, where the oil collects, is exposed to the low-side, or suction, pressure of the system. The oil pump (**FIGURE 13-20**) picks up oil from the crankcase and distributes it to the moving parts and areas of high friction throughout the compressor. The pump's job is to increase the oil pressure enough to make the oil move effectively into and through these areas. A great deal of pressure is not necessary for most units. The crankshaft of the compressor drives the oil pump.

Since the oil in the crankcase is exposed to the refrigerant vapor on the low-pressure side of the system, the pressure exerted at the oil-pump outlet is the sum of the refrigerant suction pressure and the pressure added by the work of the pump. This is referred to as the total or gross oil pressure.

Net oil pressure is, however, the value that counts. Net oil pressure is the difference between the pump discharge pressure and the refrigerant suction pressure, measured in the crankcase. For proper operation, the oil pump must produce a sufficient amount of net oil pressure to move the oil through the compressor, regardless of the suction pressure. For example, the minimum net oil pressure for some compressors is 12 psig (83 kPa). To achieve that, the pressure at the discharge of the oil pump must remain at least 12 psig above the refrigerant suction pressure (in the crankcase) at all times. The oil-pressure safety switch measures the difference between the pressure in the crankcase and gross oil pressure to determine net oil pressure and respond accordingly.

FIGURE 13-20 Location of the oil pump on a semi-hermetic compressor.

There are occasions when the net oil pressure falls below the required minimum pressure setting of the safety switch. One occasion is compressor start-up. At start-up, there is no net oil pressure until the compressor gets moving, the oil pump is turning at full speed, and the pressures begin to stabilize. To compensate for this, oil-pressure safety switches have a built-in time-delay circuit that allows the NC contacts in the control circuit to remain closed long enough for the oil pressure to build.

The net oil pressure may also fall below normal when there is a great deal of foam in the oil, often due to liquid refrigerant entering and/or migrating to the crankcase. A rapid reduction of pressure in the crankcase causes liquid refrigerant to vaporize rapidly (boil), resulting in foam. Since the foam is mostly vapor, the pump has difficulty producing an acceptable pressure differential. The time-delay feature of the safety switch is also active in this scenario, allowing some time for a temporary problem (90 to 120 seconds) to resolve itself. If the pressure remains too low after the time has elapsed, then the switch contacts open, breaking the compressor control circuit. Oil-pressure safety switches typically have a manual-reset switch to ensure any problem with oil pressure is investigated before a restart is allowed.

Oil-Pressure Safety-Switch Troubleshooting and Maintenance

To test or troubleshoot an oil-pressure safety switch, it is important to understand its sequence of operation. Refer to **FIGURE 13-21**. Note that this schematic is typical of an electromechanical device with bellows-type pressure sensing. There are electronic versions of oil-pressure safety switches that provide the same function but operate differently.

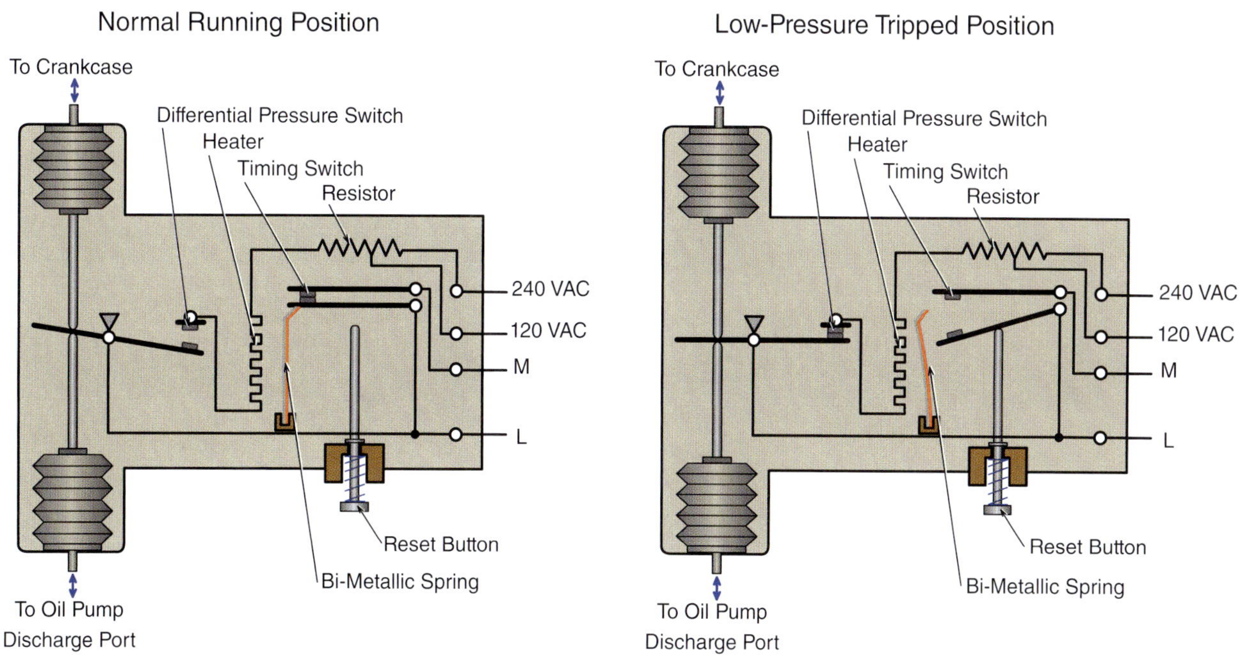

FIGURE 13-21 Typical oil-pressure safety-switch schematic.

▶ TECHNICIAN TIP

Electronic Oil-Pressure Safety Switches

Although many electromechanical oil-pressure safety switches remain in service and are still installed today, electronic devices are also popular (**FIGURE 13-22**). While the function of the switch remains unchanged, the switch operates very differently. No capillary tubes or hoses are required, for example. A pressure sensor is installed directly into the oil-pump housing. The housing is equipped with passages that have access to both the inlet and outlet pressures of the pump. A cable connects the sensor to the switch assembly. Manufacturers have also introduced even more sophisticated controls that provide not only lubrication protection but also integral overload protection for the compressor. Always consult the manufacturer's documentation when troubleshooting any control device that is not entirely familiar.

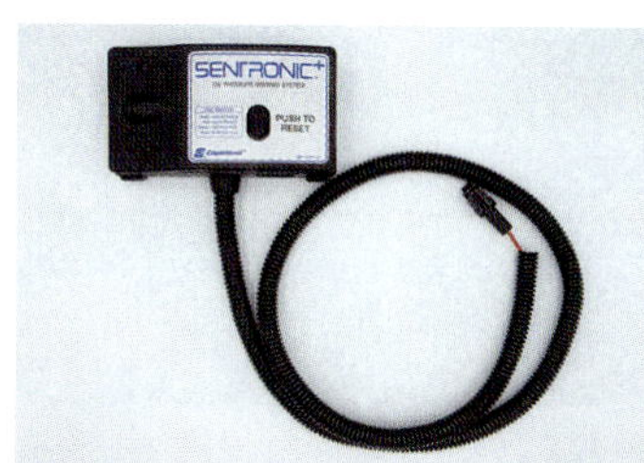

FIGURE 13-22 Electronic oil-pressure safety switch.

On the right side of the schematic, notice that the control is designed to accept a power source of 230 or 115 VAC. Although 24 VAC switches are also available, they are rare because 24 VAC control circuits are not often used in refrigeration applications. A power source is needed to power the time-delay thermal element. One leg of power to the switch is always available, while the other leg is cycled on and off by the compressor contactor or motor starter.

The differential-pressure switch inside the control is NC, but it is not in the compressor control circuit. The timing-switch contacts are in control of the compressor, and they are NC as well. This allows the compressor to start.

When the compressor is started, the NC contacts of the differential-pressure switch remain closed while oil pressure begins to build. This energizes the heater in the time-delay circuit with power now available. Most oil-pressure safety switches are designed for a time delay of 90 seconds to two minutes. However, some models provide only 45 seconds of delay, as required by some compressor manufacturers. The compressor has this time to establish an acceptable oil pressure.

Once the pressure builds beyond the minimum, the differential-pressure contacts open. This deenergizes the heater in the time-delay circuit, and the timing contacts remain closed as a result. All is well.

If the compressor does not develop sufficient oil pressure to open the differential-pressure switch contacts, the warmth of the heater eventually bends the bimetal strip and the timing contacts open. This breaks the compressor control circuit, shutting it down. As mentioned earlier, a manual reset is required to reclose the timing contacts after the switch heater has cooled off.

If an oil-pressure safety switch is suspected of being faulty, it must be tested. However, a faulty oil-pressure switch should not be the first thing to consider. A technician must first assume that oil pressure is lacking at some part of the operating cycle and instead investigate operational problems first, such as foaming oil or a low oil charge.

Although they are reliable devices, oil-pressure safety switches are subject to both mechanical and electrical wear and tear. They may also experience a lot of vibration, depending on their mounting location. One possible mode of failure is related to the heating element of the timing relay. If it fails, the timing-switch contacts will never open and the compressor could potentially operate without developing any oil pressure, but not for very long. However, this is not a common failure.

Like other types of switches, the first thing to determine is the deenergized position of the switch contacts. Referring to Figure 13-22, there should be continuity between the L and M terminals, which are part of the contactor/starter control circuit, with little or no resistance. There should also be continuity between the line-voltage terminal used (230 or 115 VAC terminal) and the L terminal. This part of the internal circuit will have resistance since there is a resistive heating element in it.

If the deenergized testing of the oil-pressure switch is satisfactory, an operational test is needed. A flare tee should be installed at the oil-pump discharge so that both a gauge and the oil-pressure switch can be connected at the same time (**FIGURE 13-23**). Connect an oil-pressure gauge to this point, and then connect the gauge manifold as it would normally be connected, to monitor suction and discharge pressure. For this purpose, the low-pressure gauge of the manifold is best connected to the same point as the oil-pressure switch connection to the crankcase.

FIGURE 13-23 Flare tee for service access at the oil-pump outlet.

The switch can be tested without running the compressor to determine whether it will open its timing contacts at the appropriate time. As already mentioned, the contactor or motor starter energizes the oil-pressure safety control. With the system deenergized, disconnect the compressor power leads from the contactor and move them aside. The terminals of the wires can be taped to ensure they do not accidentally contact anything during the test. With the system energized and started, the contactor will close, but the compressor will not operate. The oil-pressure safety switch will be energized and working through its time-delay period. Once the time elapses, the NC timing-switch contacts should open, deenergizing the contactor solenoid coil. This test confirms that the switch will open the control circuit when there is no oil pressure. The next test ensures that it remains closed when an appropriate amount of oil pressure does exist.

After reconnecting any control wiring and removing the jumper, the system is then started, and the suction pressure and the oil-pump discharge pressure are monitored. You can easily subtract one from the other to mentally track the net oil pressure. Once the net oil pressure reaches the switch setting, the differential-pressure switch should open, deenergizing the heater. The compressor should continue running. With this test complete, the pressure switch should be considered functional and not the source of any operational problems.

It is important to note that compressor oil pumps are extremely durable gear pumps and very reliable. It is highly unlikely that the pump will fail to produce the appropriate oil pressure due to an internal failure, but it is possible.

Oil-Pressure Safety-Switch Replacement

Removal and replacement of an oil-pressure safety switch requires the same general steps as the replacement of a high- or low-pressure switch. The main difference is the fact that there are two pressure-sensing connections to the system. The pressure in the compressor crankcase must be reduced to just above 0 psig to disconnect the switch without allowing refrigerant to escape or air to enter the system.

Before disconnecting the switch capillaries or hoses, it is often best to disconnect the wiring and remove the old switch from its mounting plate. The defective switch can be laid aside and the new switch mounted to the plate. The wiring terminations can also be completed if desired. With the new switch in place, the old capillaries or hoses can be quickly disconnected and quickly reconnected to the new switch. It is important to ensure that the capillaries are connected to the proper port. The capillaries might be labeled "Low" and "Oil," instead of "High" and "Low." The Oil connection is connected to the outlet of the oil pump, whereas the Low connection is made to the low-pressure side (crankcase) of the system. If they are reversed, the switch will not sense a positive pressure differential and will soon shut the system down on each start attempt.

Some oil-pressure switches have a fixed setting because they are designed to operate on a specific brand or model of compressor. Others have adjustable settings. It is important to acquire the correct information for the compressor in service from the compressor or equipment manufacturer to ensure that the correct settings are made.

▶ Fan-Cycling Pressure Switches

K13004 Describe the operation and maintenance of fan-cycling pressure switches.

Air conditioning and refrigeration systems must be designed with condensers that can reject enough heat in the hottest environment for which they are rated to operate. In environments that are cooler than the maximum design temperature, the condensers may be so effective that the reduced pressure can create operational problems. Some systems must continue running to cool loads inside a building, even though it is below freezing outside.

There are several ways to control the capacity of the condenser and keep the high-side pressure as high as needed to properly feed refrigerant to the evaporator. The simplest method is condenser fan cycling. A pressure switch can be used to monitor the high-side pressure and cycle the condenser fan motor(s) off when the pressure falls to a predetermined point (cut-out). The pressure switch later restarts the fan(s) as the pressure rises to another predetermined point (cut-in). This method, however, can result in the system high-side pressure modulating up and down through a range of pressures too quickly, especially if there is only one condenser fan. This leads to instability in the refrigerant pressure applied to the metering device. As a result, this method of pressure control requires the use of an expansion valve as opposed to a fixed method of refrigerant flow control to accommodate the variations in pressure.

FIGURE 13-24 A commercial refrigeration unit with multiple condenser fans.

A fan-cycling approach works much better when there is more than one condenser fan motor (**FIGURE 13-24**). Unless the unit must operate in extreme cold, one condenser fan may remain running at all times while the remaining fans are cycled based on pressure. This typically results in smoother pressure control. Although it is not the ideal method of control, it does offer simplicity and works well for many refrigeration systems. It can also be used on comfort-cooling equipment that must operate in cool weather.

It is important to note that stopping and starting an electric motor repeatedly can shorten its life. To minimize motor stress, the pressure settings should reflect the widest range of pressures possible that still allows the refrigerant circuit to operate as it should. However, a range that is too big may create some significant problems with refrigerant flow control and performance. Again, although it is important to understand their operation, there are better methods to control head pressure under varying conditions.

▶ TECHNICIAN TIP

Head-Pressure Control

This section refers to methods that are preferred over fan-cycling to control system head pressure under varying environmental conditions. Fan-cycling controls do work well on systems with three or more condenser fans. Large air-cooled chillers, for example, may have a dozen condenser fans or more. But for smaller systems with one or two condenser fans, modulating speed controls work much better. With a modulating speed control, the speed of the condenser fan can be controlled to a specific pressure or liquid-refrigerant temperature, providing much smoother control. These controls, however, are significantly more expensive than simple cycling controls. Electronically commutated motors (ECMs) are also being used in condenser fan applications now, providing another method to control motor speed. ECMs are presented in Chapter 15.

Setting a Fan-Cycling Pressure Switch

Fan cycling is achieved with an adjustable pressure switch (**FIGURE 13-25**). Some of them are similar to low-pressure cycling controls, with a set point for cut-in and a set point for the differential to determine the cut-out pressure. Many switches designed for this purpose have a range of 100–470 psig (690–2,750 kPa) and are compatible with a wide variety of refrigerants. They are connected to the high-pressure side of the system, typically on the liquid refrigerant line leaving the condenser. The settings are based on the type of refrigerant and the needs of the system. As mentioned previously, it is important to minimize cycling the fans more often than necessary.

FIGURE 13-25 A condenser fan-cycling switch.

Troubleshooting a Fan-Cycling Pressure Switch

Fan-cycling pressure switches do not perform a protective or safety function. The failure of a fan-cycling switch, however, may hamper system performance in several ways. If the switch contacts fail to close and start a condenser fan, the high-pressure safety switch will likely shut the system down. If the fan-cycling switch closes but does not reopen, the high-side pressure might fall too low, creating abnormally low pressures and possibly preventing enough liquid from entering the evaporator to maintain the desired indoor conditions. Note that even when the switch is set to the best settings and operating properly, the ambient temperature will have a significant effect on the time between fan cycles. In other words, the fan(s) will cycle more or less often depending on the ambient temperature.

Testing fan-cycling switches is relatively simple. These switches use NO contacts that close when the cut-in setting is reached, and then reopen when the pressure falls and the cut-out setting (or cut-in minus differential) is reached. Testing their operation is a matter of ensuring that the switch contacts close and open at the appropriate settings.

Determining the settings for the control is probably the most challenging. As mentioned previously, short-cycling motors can significantly reduce their life. Consequently, providing the widest differential setting that still allows the system to effectively fulfill its mission is best.

Although pressure set points are what is being set, the settings are more about condensing temperature than pressure. One approach is to consider that 95°F (35°C) is a common design ambient temperature for equipment. An old rule-of-thumb is to add 30°F (17°C) to the ambient design to determine a "normal" condensing temperature. This results in a 125°F (52°C) condensing temperature. For a system operating on R-134a refrigerant, 125°F equals a condensing (saturation) pressure of 185 psig (1,276 kPa). This would be a reasonable starting point for the cut-in setting on R-134a. Using the same temperatures with R-404A, the cut-in pressure would be 333 psig (2,296 kPa). Although these are very different settings, they both reflect 125°F condensing temperatures for the two refrigerants.

A reasonable starting point for the cut-out pressure is a pressure equal to the 95°F design ambient temperature discussed earlier. For R-134a, this pressure would 114 psig (786 kPa). If the setting is made by using a differential, the differential setting would be 185–114 psig—a difference of 71 psig (490 kPa). With these values as a starting point, based on the refrigerant in use, the system should be operated and observed in conditions that will allow the switch to cycle the fan while other vital system factors, such as subcooling and superheat, are monitored to evaluate performance.

In some weather conditions, the set points just discussed may result in the fan cycling too quickly. Again, this is one of the disadvantages of this form of control. The system must operate properly in a wide range of ambient temperatures, and therefore a technician cannot focus solely on the operation of the switch at a single temperature or set of operating conditions. If the fan starts too often across a range of ambient temperatures, widening the range to the extent the system continues to perform well will help. When system performance starts to deteriorate due to a wide range of pressures being applied to the expansion valve, the range of operating pressures may need to be narrowed. Another approach is to raise the range by raising both the cut-in and cut-out settings. To maintain a wide range, the cut-in temperature can possibly be increased more than the cut-out. In summary, it takes some time to ensure the settings work for the system. No single pair of settings will work for every system and environment. A balance between the number of motor cycles and efficient system operation must be found for each system and its operating environment.

One thing to remember is that if a significant problem in performance is noted, it will likely be at the lower (cut-out) pressure setting. If the high-side pressure is allowed to fall too low, the expansion valve may not be able to open wide enough to admit sufficient refrigerant into the evaporator and cool effectively. The superheat will rise as a result. If the superheat begins to rise significantly on each cycle, impeding cooling, the cut-out set point must be raised to a point where the expansion valve can still function properly with the pressure applied. The valve can typically manage the superheat effectively at or near the cut-in point.

Replacing a Fan-Cycling Pressure Switch

In most cases, the fan-cycling pressure switch will be connected to the system at a tee located at the compressor high-side service valve. In some cases, it may be connected to a port installed in the liquid line for this purpose. The same style of mounting brackets used with other adjustable pressure switches are used, as well as mounting the switch on a short section of conduit.

Removing the fan-cycling pressure switch requires removing the refrigerant pressure at the point of connection, disconnecting the capillary, disconnecting the wiring, and removing the switch from its mounting bracket. As is the case with other switches, the entire mounting bracket can also be removed and replaced if needed or desired.

Installing the replacement switch is just the reverse of the removal steps. Once the switch has been fully installed and refrigerant pressure restored to the port, the system can be started and the final adjustments made during testing.

Switches may be added or replaced at a time when the ambient temperature is high, making it difficult to fully test the switch. In these cases, it is wise to plan for a return visit when the temperature is more favorable for testing.

▶ Differential Air-Pressure Switches

K13005 Describe the operation and maintenance of air-pressure switches.

The differential air-pressure switch is widely used in the HVACR industry as an effective control to prove the movement of air. More accurately, the control is a practical application to monitor the flow of air or other gas and ensure that a sufficient gas volume is flowing. This section will present three unique applications of the differential air-pressure switch.

In the refrigeration application, a differential air-pressure switch might sense the pressure on either side of a refrigerant coil. As the coil becomes restricted with frost, the differential air pressure increases. Once the difference increases to the control set point, one or more electrical contacts either open or close. This approach might be used to initiate defrost cycles in large systems.

In the HVACR industry, carefully monitoring the condition of air filters is essential ensure the continued and correct airflow through the duct system. This is especially true in large commercial systems, hospitals, and similar applications. Thus, another application of the differential-pressure switch is to monitor the loading of the air filters. By monitoring the pressure on both sides of the filters, the amount of contamination can be monitored. Once the differential pressure increases to the control set point, it triggers a notification.

An application of the differential-pressure switch in a fossil-fuel heating system is different from the two previous examples. Specifically, on a forced-draft furnace, it is important that by-products of combustion continue out of the combustion chamber. To measure the required flow, a differential air-pressure switch is used to measure the pressure difference between the inlet and outlet of the induced-draft blower, or between atmospheric pressure and the heat exchanger.

The induced-draft blower overcomes the resistance to air flow through the heat exchanger. If the proper pressure difference exists, the combustion process continues. If the pressure differential decreases below the control set point, it is an indication that an insufficient amount of air is flowing through the heat exchanger. A set of contacts then open, stopping the combustion process.

Air-pressure switches will be encountered in various applications throughout a technician's career. Understanding their function in the system, and the parameters for their settings, will allow them to be tested accurately and efficiently.

▶ Wrap-Up

Summary

Pressure switches serve many purposes in HVACR systems. They protect refrigeration systems from damaging conditions that can occur during operation, such as high and low refrigerant pressures and low oil pressure. Other pressure-switch applications include proving the flow of air in ducts and condenser fan control.

The service technician must be familiar with the wide variety of pressure-switch applications and configurations, how they sense pressure, and the specific troubleshooting and replacement techniques for each type. Since safety is always a priority, the technician must also understand the potential hazards of working near or around energized electrical equipment and pressurized refrigerant circuits.

Remember that part of every pressure-switch service task includes ensuring that repairs conclude with pressure-switch capillaries and hoses properly secured to prevent vibration and provide leak-free connections to the refrigerant circuit. The task also includes securing wiring with tie-wraps and properly reinstalling access doors and covers with all the required fasteners.

Key Terms

absolute pressure The actual pressure exerted by a fluid with reference to a vacuum (zero pressure). Atmospheric pressure + gauge pressure = absolute pressure.

ambient Relating to the immediate surroundings or environment.

bellows In pressure-measuring instruments, a thin, enclosed, flexible, pleated cylinder that lengthens and shortens with changes of internal pressure.

Bourdon tubes C-shaped tubes closed on one end and attached to a source of pressure. Increased internal pressure causes the tube to straighten. The motion can be linked to an indicating or control mechanism.

capillary tubes Small-diameter flexible tubing, usually made of copper, used for providing pressure inputs to instrumentation; named for the very tiny blood-carrying capillaries in humans and other animals.

diaphragm In pressure-measuring instruments, a thin, flexible disk that separates two sides of a chamber. Differences of pressure across the disk cause it to deflect in the direction of the lower pressure.

encapsulated Relative to switches and electrical devices, an encapsulated device is enclosed or contained in a sealed housing.

gauge pressure The pressure of a system measured in reference to atmospheric pressure. Gauge pressure = total pressure - atmospheric pressure.

helical In the shape of a helix—a three-dimensional spiral wrapped around a cylinder.

manometer A U-shaped tube containing a fluid such as water or mercury where one end is open to atmosphere and the other end is connected to a pressure source. Alternatively, both ends can be connected to a pressure source, for a differential reading. The pressure is measured by the difference in the displaced liquid height on the two sides of the U-shaped tube. Modern digital manometers, without using fluid, can supply and measure small air pressures for testing purposes.

net oil pressure The difference between the crankcase pressure of a compressor and the measured oil pressure at the outlet of its pump.

noncondensable gases Gases that, under all operating pressures and temperatures of a given system, remain in the gaseous state. For systems using condensable fluids such as refrigerants, the presence of noncondensable gases such as air can significantly interfere with their operation.

orifice An opening or passage. In the HVACR field, it commonly refers to a fixed refrigerant metering device that provides the pressure drop for refrigerant entering an evaporator coil.

Pascal's law A principle of physics stating that, for an enclosed liquid system, a pressure change at one point in the system is transmitted equally throughout the system.

Review Questions

1. You can measure very small pressures, using inches of water column (in. WC) as a unit of measure, using a __________.
2. The pressure normally indicated by a pressure instrument attached to an HVACR system is __________.
   a. relative pressure
   b. suction pressure
   c. gauge pressure
   d. absolute pressure
3. A pressure-sensing element made of a thin, flexible disk or sheet is called a __________.
   a. Bourdon tube
   b. diaphragm
   c. bellows
   d. helix
4. For proper and reliable make-and-break switch operation, pressure switches are usually made with __________.
   a. snap-action switches
   b. double switches
   c. NO switches
   d. NC switches
5. Which of the following is *not* likely to be the cause of high compressor discharge pressure?
   a. Condenser fan not running.
   b. Blockage in the condenser piping.
   c. Noncondensable gases in the refrigerant.
   d. Low refrigerant charge.
6. One of the visible symptoms of a low refrigerant charge in an operating air conditioning system is __________.
7. When testing a pressure switch, the gauge or gauge manifold used to read pressure should be ______.
   a. equipped with a relief valve
   b. connected at least three pipe diameters away from the switch connection
   c. connected as close as possible to the point of connection for the switch
   d. calibrated for the current atmospheric pressure before using it
8. After the pressure switch contacts have opened due to excessive pressure, it is common for auto-reset high-pressure safety switches to return to their NC position after the pressure has dropped __________.
   a. 15 psi (103 kPa)
   b. 25 psig (172 kPa)
   c. 100–150 psig (700–1,050 kPa)
   d. 200–250 psig (1,380–1,720 kPa)
9. Manufacturers typically install high-pressure safety switches on flare fittings that __________.
   a. do not have a valve core
   b. cannot leak
   c. have a valve core
   d. have left-hand threads
10. For most adjustable low-pressure cycling controls, the cut-out pressure is the cut-in pressure __________.
    a. plus 50 psi (345 kPa)
    b. minus 50 psi (345 kPa)
    c. plus the differential setting
    d. minus the differential setting
11. On a system using a low-pressure cycling control in place of a thermostat, if the compressor starts before the box temperature has risen to the desired temperature, __________.
    a. the cut-in setting must be decreased
    b. the cut-in setting must be increased
    c. the differential setting must be increased
    d. the differential setting must be decreased
12. Pressure-switch mounting plates are generally __________.
    a. unique to every switch
    b. universal
13. When troubleshooting a low-pressure cycling switch in a system, which of the following would be a sign that the switch is faulty?
    a. The NO switch closes when the pressure rises to the cut-in set point.
    b. The NO switch opens when the pressure equals cut-in minus the differential.
    c. The NO switch contacts are closed with no pressure applied.
    d. The NO switch contacts are open with no pressure applied.
14. In reference to oil pressure, the net oil pressure represents __________.
    a. the sum of crankcase pressure and oil-pump output pressure
    b. the difference between the crankcase pressure and oil-pump output pressure
15. What feature of a compressor oil-pressure safety switch permits the compressor to start?
16. Which of the following is likely to cause a low oil-pressure condition to be sensed by an oil-pressure safety switch?
    a. Foaming of the compressor oil.
    b. A high discharge pressure.
    c. A crimped capillary connected to the compressor crankcase.
    d. High ambient temperatures.
17. The differential-pressure switch inside an oil-pressure safety switch controls power to the __________.
    a. compressor control circuit
    b. thermal element of the time delay
    c. timing switch
    d. low-pressure cycling control

18. When deenergized with no pressure applied, what is the position of the differential-pressure switch and the timing-switch contacts of an oil-pressure safety switch?
   a. differential-pressure switch open; timing switch open
   b. differential-pressure switch closed; timing switch open
   c. differential-pressure switch open; timing switch closed
   d. differential-pressure switch closed; timing switch closed
19. The intent of a fan-cycling pressure switch is to __________.
   a. control the speed of an indoor blower motor
   b. turn off condenser fans before the high-side pressure gets too high
   c. control the head pressure to ensure that it does not fall too low
   d. alternate condenser fan motors to balance wear
20. A reasonable starting point for the cut-in setting of most fan-cycling pressure switches is a refrigerant pressure equal to the condensing temperature of the refrigerant at __________.
   a. 75°F (24°C)
   b. 95°F (35°C)
   c. 125°F (52°C)
   d. 150°F (66°C)

CHAPTER 14

Motor Basics

Knowledge Objectives

After reading this chapter, you will be able to:

- **K14001** Describe the basic operating theory and components of an AC motor.
- **K14002** Identify different types of single-phase motors and their common applications.
- **K14003** Describe the construction and operation of three-phase motors.
- **K14004** Describe basic motor troubleshooting procedures for single- and three-phase motors.

Skill Objectives

Skills required to meet the objectives of this chapter:

- **S14001** Troubleshoot a single-phase motor and related starting components.

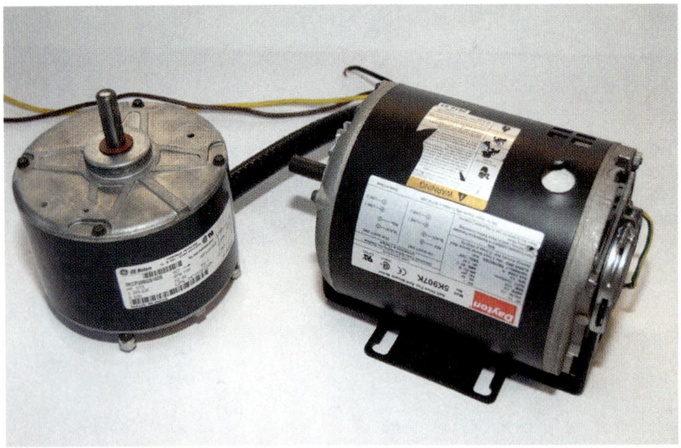

FIGURE 14-1 Examples of electric motors.

▶ Introduction

Alternating current (AC) induction motors (FIGURE 14-1) are used extensively in HVACR systems to drive a variety of loads, including compressors, fans, and pumps. While some single-phase motors develop only a small fraction of one **horsepower (hp)**, larger motors can develop far more power. However, due to the excessive current requirements and a lack of **torque**, single-phase motors are rarely used in HVACR applications beyond 2–3 hp (1.5–2.2 kW). Three-phase motors produce a lot more torque than single-phase motors and draw less current in the process. For that reason, they are used in many HVACR applications when more than 1.5 hp (1.1 kW) is required.

▶ Motor Theory and Components

K14001 Describe the basic operating theory and components of an AC motor.

Motors are devices that convert electrical energy into mechanical energy, which explains why they have so many applications. In a motor, electricity is first converted to electromagnetic energy, which is in turn converted to mechanical energy through a rotating shaft.

The rotation of the shaft is a response to the attraction and repulsion occurring between magnetized components. In an AC motor, those components are referred to as the **stator** and the **rotor** (FIGURE 14-2). The stator is fixed, whereas the rotor can rotate freely, as the name implies. The motor shaft is an extension of the rotor. The load, such as a fan blade, is mechanically attached to the rotating shaft.

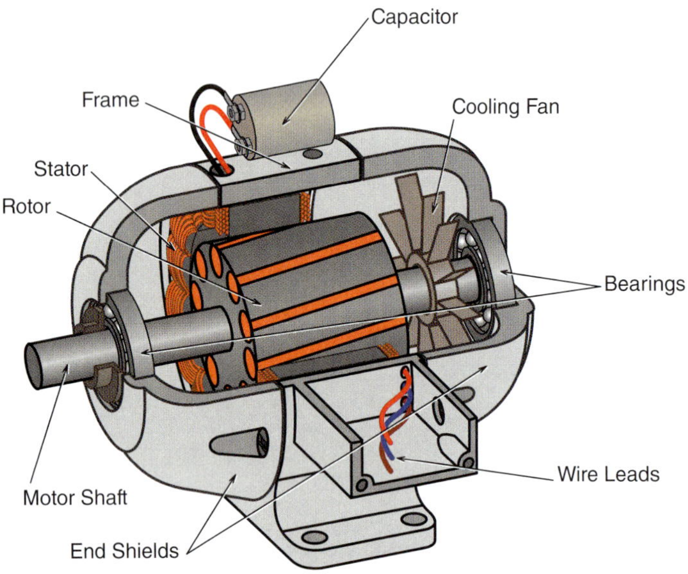

FIGURE 14-2 Basic motor construction.

The principles of magnetism and how those principles apply to the operation of a motor are reviewed in this section. A permanent magnet emanates a magnetic field made up of magnetic lines of force (flux), as shown in FIGURE 14-3. Think about what happens when two magnetized objects are placed in contact with each other. If the north and south poles of the two objects are placed in proximity, they will attract each other. However, when like poles are placed in proximity (N to N or S to S, for example), they repel each other. If the magnets are strong, it may be challenging to even make them touch. Any **ferrous metal** material influenced by magnetic flux lines will also become magnetized.

When a voltage is applied to a coil of wire wrapped around an iron core, the core becomes magnetized. This was demonstrated in the principles of transformers presented in Chapter 7. Thus, when an AC voltage is applied to the poles of an electromagnet such as the stator poles of a motor, magnetic lines of force produced by the stator will induce an electric current in the rotor. The magnetic field developed by the rotor opposes that of the stator. This is the action that causes the rotor to spin. As the AC voltage applied to the stator windings fluctuates through its positive and negative half cycles, the north and south magnetic poles of the stator windings will alternate. This creates a rotating magnetic field that is constantly attracting one part of the rotor while repelling another part.

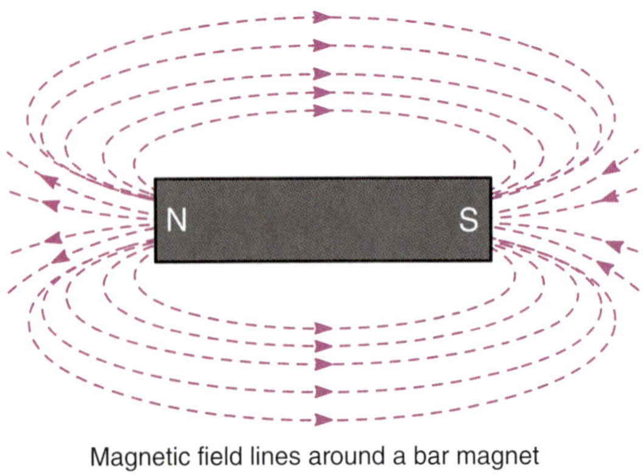

FIGURE 14-3 Magnetic lines of force.

Motor Components

The stator and rotor (FIGURE 14-4) are the main components of an induction motor. The stator (FIGURE 14-5) is made from layers of steel laminated together. Hundreds of feet of thin copper wire are wound around these

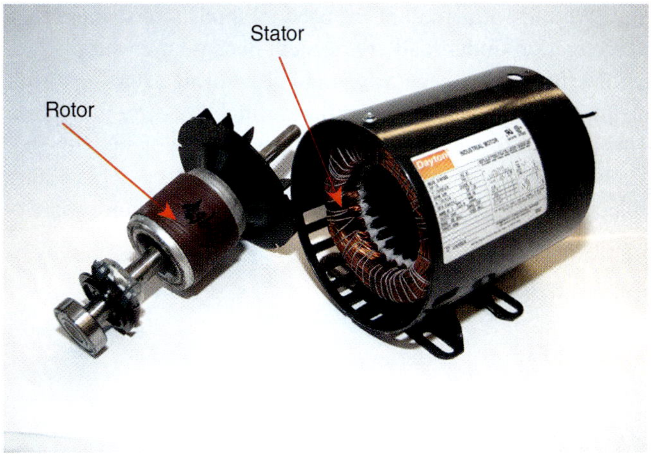

FIGURE 14-4 Stator and rotor of a motor.

FIGURE 14-5 The stator of a motor.

laminations, half in a counterclockwise direction and half in a clockwise direction. In an AC motor, a single-phase or three-phase voltage is applied to the stator windings, turning them into electromagnets that generate a magnetic field. This magnetic field passes through the rotor, inducing a voltage in it.

The frame of the motor is the case in which the stator, rotor, and bearings are housed. There are different sizes and styles of frames, based on the size and application of the motor. Some have mounting feet, while others are designed for mounting with a band around the motor frame.

Motor bearings may be of either sleeve or ball construction. Sleeve bearings are used in situations where the load is fairly light and noise transmission is a concern. (Ball bearings are generally noisier than sleeve bearings.) Noise needs to be considered when the motor is close to an occupied area. Ball bearings are used on larger motors, which are typically installed in areas away from occupied spaces. Sleeve bearings maintain their needed lubrication through a wicking system, where oil from a reservoir is wicked through the sleeve material to the shaft. They may be permanently lubricated or have an oil reservoir that needs to be periodically replenished. Ball-bearing motors may also be permanently lubricated or require periodic greasing. Ball bearings generally require grease as opposed to a liquid lubricant.

▶ TECHNICIAN TIP

Ball Bearings

Most motors used in applications where the speed is modulated are equipped with ball bearings. Sleeve bearings generally do not maintain proper lubrication at low rotational speeds and tend to fail prematurely in these applications. Note that this applies to speed-controlled motors and does not apply to standard multi-speed motors, since their lowest selectable speed is typically sufficient to support proper sleeve-bearing lubrication.

Rotors

The rotor of a motor may be one of two types: squirrel-cage rotor or wound rotor. The squirrel-cage rotor (**FIGURE 14-6**) is the most common. It is made by placing aluminum or steel bars into slots made in the iron core of the rotor. The end rings secure the bars and short them together electrically. The following are some of the advantages of using squirrel-cage rotors:

- They have no brushes or slip rings and thus have low maintenance requirements.
- They are less expensive because they lack slip rings and have less conductor material than wound-rotor motors. The absence of brushes and slip rings also eliminates sparking, which makes squirrel-cage motors more suitable for use in environments containing flammable dust or vapors.
- They are able to provide variable torque at a fairly constant speed, so they adapt well to load variations.

The primary disadvantage of squirrel-cage motors is their low starting torque and high starting current. The starting currents of these motors is often six to eight times that of the full-load current. Despite those disadvantages, squirrel-cage motors are widely used to drive HVACR blowers, pumps, and compressors because of their relatively low cost and low maintenance requirements.

In a wound-rotor motor, the rotor windings are connected to external resistances through **slip rings**. The speed and torque characteristic of the motor can be adjusted by

FIGURE 14-6 Squirrel-cage rotor.

varying the external resistance. Wound-rotor motors are used in applications where high starting torque and smooth acceleration under load are needed. Because the wound-rotor motor has more turns of wire than the squirrel-cage type, it can provide high starting torque with low inrush current. The low starting current makes them suitable for use in applications where the power systems will not tolerate the high-inrush currents of squirrel-cage motors. Wound-rotor motors have been used in a variety of variable-speed applications. However, motors with variable-speed drives in control do not necessarily require some of the advantages of wound rotors.

▶ CORE CONNECTION

DC Motors

DC motors operate on the principle of a rotating magnetic field, just like AC motors. However, since DC voltage has a constant amplitude, the DC voltage applied to the stator cannot develop that rotating magnetic field. It has to be done mechanically. The commutator (**FIGURE 14-7**) is the answer. It is a split ring attached to the rotor (called an armature on a DC motor). The DC voltage is applied to the commutator through carbon brushes that are spring-loaded to remain in constant contact with the commutator. As the motor rotates, the split commutator causes the DC voltage to reverse direction every 180 degrees, simulating an AC voltage. This voltage applied to the rotor creates the rotating magnetic field action between the rotor and stator.

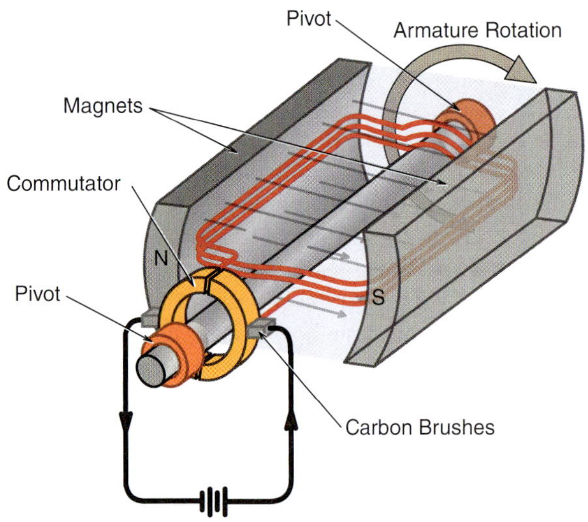

FIGURE 14-7 DC motor principles.

▶ Single-Phase Motors

K14002 Identify different types of single-phase motors and their common applications.

There are several types of single-phase induction motors, and they all have one thing in common. They need some means of creating a push–pull effect on the rotor to overcome the starting torque and make the rotor turn. Once it starts, it will continue to run until the voltage is removed from the stator. The starting impetus is obtained by temporarily creating a second phase of the applied voltage so that the voltages applied to the stator poles are momentarily out of phase. All single-phase induction motors use some means of creating a second phase of the voltage applied to the stator.

Shaded-Pole Motor

In a shaded-pole motor, the phase shift needed to start the motor is created by using grooved stator windings (**FIGURE 14-8**). A small segment of each stator winding is separated from the main winding by the groove. A metal shading band is placed around that small segment.

This band acts as a coil, slightly shifting the phase of the shaded segment of the winding. This type of motor offers a very low starting torque, so the small amount of phase shift is enough to start turning the rotor. When the pole is shaded to one side or the other, it will turn in only one direction. The motor will rotate in the direction of the shading coil. If the pole is shaded in the center, it can be made to rotate in either direction. These motors operate on AC, which creates a positive and negative voltage 60 times each second (60 hertz). When the positive and negative peaks of this voltage drop toward zero voltage, the shaded pole creates a concentrated magnetic field that moves the rotor toward the shading pole or shading slot. Even if the motor is locked up, the direction of rotation can be identified by finding the shading coil. The motor turns in the direction of the shading coil, so in Figure 14-8 the rotor will move in the clockwise direction, often expressed as CW.

Shaded-pole motors are used in fractional horsepower applications. In HVACR systems, they may be used to drive small propeller fans and blower wheels.

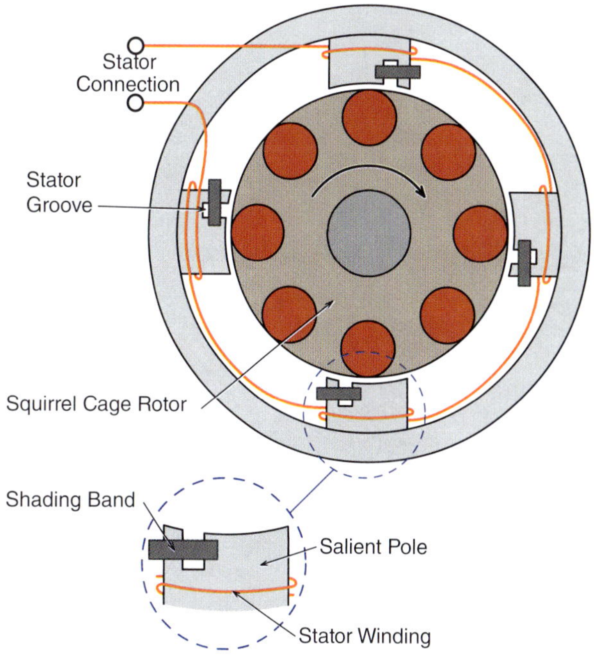

FIGURE 14-8 Shaded-pole motor arrangement.

Permanent Split Capacitor Motor

The permanent split capacitor (PSC) motor may sometimes be referred to as a capacitor-run motor. A run capacitor is placed in series with the start winding (**FIGURE 14-9**). The main purpose of the run capacitor is to reduce running amps even though it will offer a minimal amount of starting torque to help the motor get started. It remains in the circuit to improve operating efficiency throughout each run cycle; no relay or other device is needed to switch it out of the circuit after the motor starts. The characteristics of a PSC motor allow it to operate at a relatively constant speed even when the load tends to vary. They are quite efficient and one of the most common types found in HVACR systems, powering fans, blowers, and compressors.

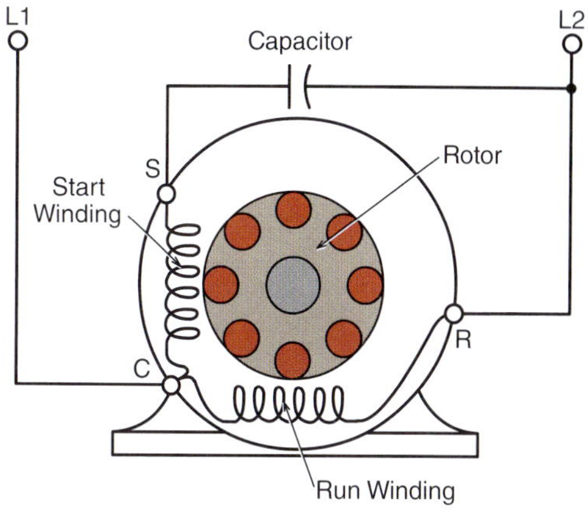

FIGURE 14-9 PSC motor arrangement.

Split-Phase Motors

Like the PSC motor, the split-phase motor has both a start winding and a run winding (**FIGURE 14-10**) that are 90 degrees out of phase with each other. The electrical characteristics of the start winding are different from those of the run winding. The start winding often has more turns of smaller wire than the run winding, but there are many variations. The phase difference between the two windings creates the force that starts the rotor turning. The higher **inductance** of the run winding causes the phase shift, and the magnetic field of the start winding develops first.

The arrangement of the start winding will cause it to generate excessive heat, however, so it must be switched out of the circuit by the **centrifugal switch** as soon as the motor is up to about 75% of its running speed. The centrifugal switch is located on the rotor (**FIGURE 14-11**). As the rotor spins, centrifugal force acts to open the switch contacts, switching the start winding out of the circuit. They have better starting torque than PSC motors and may be found on belt-driven blowers.

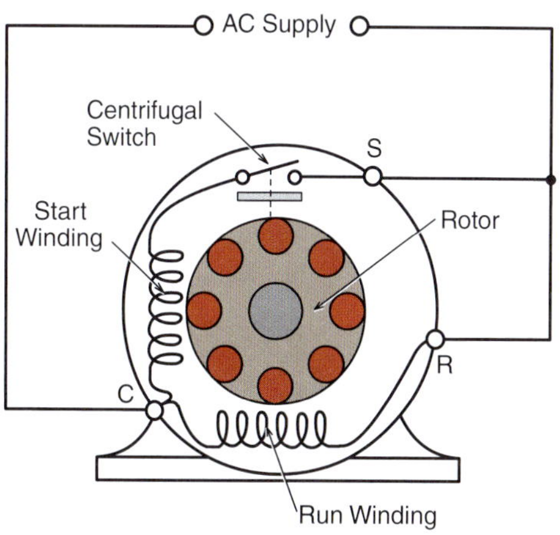

FIGURE 14-10 Split-phase motor arrangement.

Capacitor-Start Motors

The capacity of capacitor-start motors ranges from fractional horsepower to as much as 10 hp. Such motors rely on an electrolytic capacitor in series with the start winding to provide the 90-degree phase shift needed for the motor to start (**FIGURE 14-12**). The start capacitor and start winding are not designed to operate continuously. Therefore, they must be switched out of the circuit when the motor has reached 75–80% of its running speed. This is done either with a centrifugal switch, as previously described, or by using a current or potential relay. The combination of the electrolytic capacitor and start winding enable a starting torque that is about four times that of the running torque. This extra torque enables the capacitor-start motor to overcome the high starting torque of some loads.

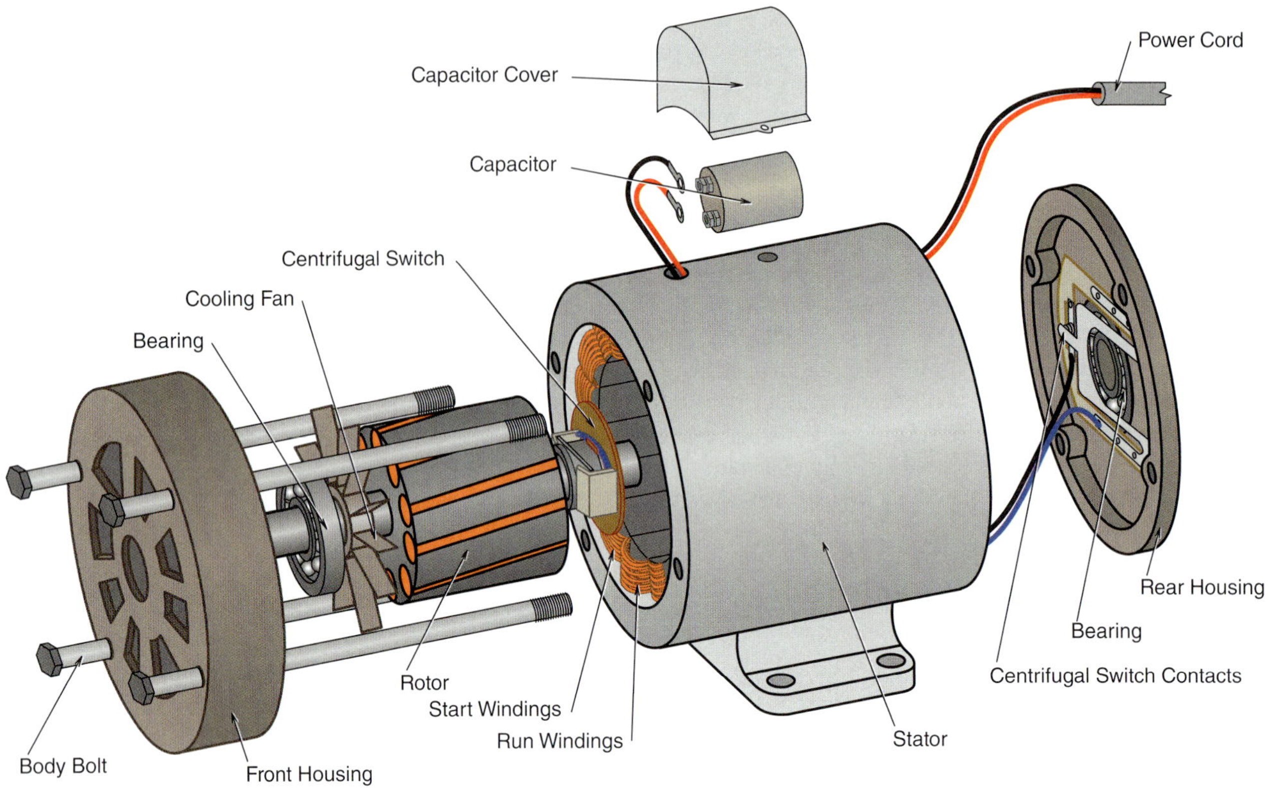

FIGURE 14-11 Parts of a split-phase motor.

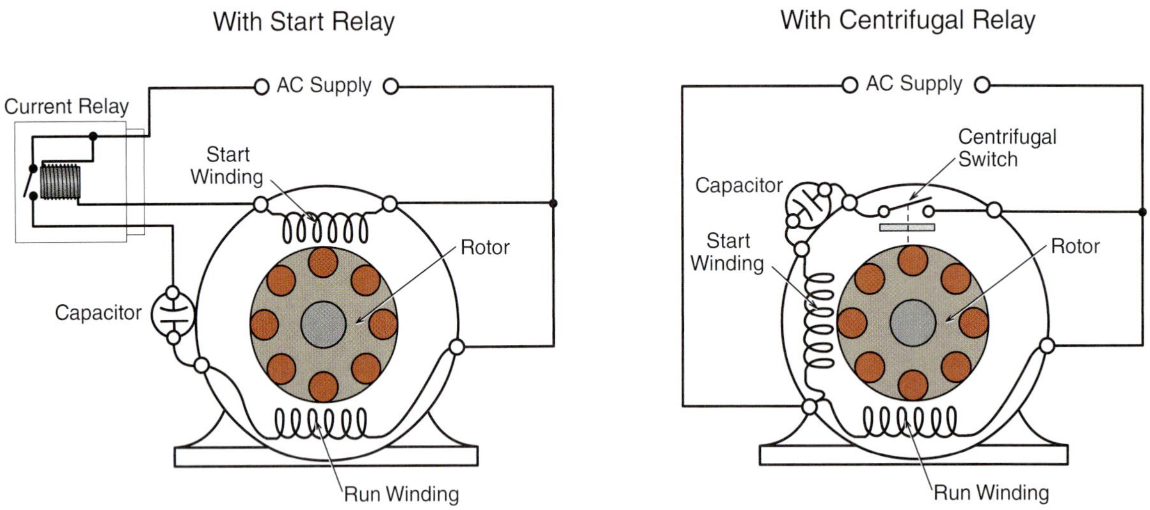

FIGURE 14-12 Arrangement of a capacitor-start motor.

Start-Relay Circuits

Both potential and current relays are used in motor-start circuits. A potential-relay coil is energized by the reverse (back) EMF developed by the stator once the motor starts to rotate. The back-EMF voltage increases to higher than the input voltage as the motor comes up to speed and soon reaches the point where the voltage is high enough to energize the relay. The voltage required to energize the coil is called the pickup voltage. For example, in a 230 VAC motor, the back-EMF developed may be in the range of 400 VAC. This will open the normally closed relay contacts in series with the start capacitor and start winding. The name *potential relay* stems from the fact that the level of voltage applied determines when they are energized.

A current relay, on the other hand, is responsive to elevated current. It uses the high-inrush current drawn by a motor when it is starting to energize the coil of the relay and allow the start capacitor to assist. The coil of this relay is wired in series with the run winding. Unlike the potential-relay contacts, the current relay contacts are normally open. The contacts are wired in series with the starting circuit. As the motor attempts to start, the inrush current reaches a high level at which the contacts close, closing the starting circuit. Once the motor starts and the operating current falls toward a normal level, the contacts reopen to remove the capacitor from the circuit.

Another device used in starting motors is the positive temperature coefficient (PTC) thermistor. In a PTC relay circuit, the PTC thermistor is in series with the starting components. As current flows through the thermistor, it heats up rapidly, causing its resistance to grow rapidly as well. The resistance increases to a point where current becomes very low, since there are other paths for current to flow against less resistance. This effectively removes the start winding from the circuit. The PTC thermistor remains in the circuit, however, so it will continue to produce heat. Once the motor has stopped, the thermistor needs time to cool down before restarting. Its resistance must be low when a start is attempted.

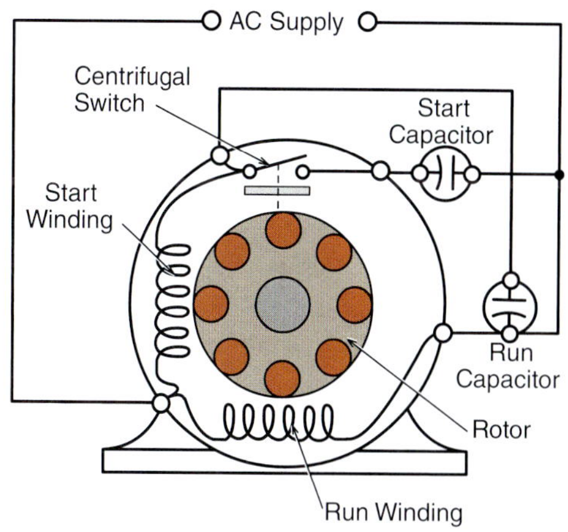

FIGURE 14-13 Arrangement of a CSCR motor.

Capacitor-Start Capacitor-Run Motor

A capacitor-start capacitor-run (CSCR) motor is a variation of the capacitor-start motor, with the addition of a run capacitor in series with the run winding (**FIGURE 14-13**). The run capacitor remains in the circuit while the motor is running. The addition of the run capacitor enables smooth, quiet operation. Like the capacitor-start motor it has high starting torque, though not quite as high as that of the capacitor-start motor. It is commonly used to drive single-phase hermetic compressors.

▶ CORE CONNECTION

Horsepower

The traditional interpretation of horsepower is that 1 hp equals the energy required for a horse to raise an object weighing 33,000 pounds a distance of 1 foot in 1 minute. However, knowing the horsepower rating of a given motor will not help determine the effect that the motor might have on an electrical circuit. For that, the horsepower rating must be converted to a value that will work in one of the common electrical formulas. Fortunately, it has been determined that 1 hp is equal to 746 watts. Therefore, to figure out how much current the motor would draw, simply use a variation of the power formula. If the applied voltage and the horsepower are known, for example, then use I = P ÷ E. Assuming a 240 VAC supply voltage and a 3 hp motor:

$$I = P \div E$$

$$I = (746 \times 3) \div 240$$

$$I = 2{,}238 \div 240$$

$$I = 9.325 \text{ A}$$

Multi-speed Motors

Many motors, especially blower motors used in residential and light-commercial equipment, are multi-speed PSC motors (**FIGURE 14-14**). These motors have several different speeds that are used as needed to suit the installation. For example, a high-speed lead may be used in the cooling mode for maximum airflow, whereas a medium speed might be used in the heating mode to minimize drafts and elevate the air delivery temperature. These two leads would then be connected to the indoor blower relay and to the furnace control circuit for the fan, respectively. Any unused leads must be secured and protected with a wire nut secured to each lead to prevent arcing when the motor is energized. While some motors may have only two or three speeds available, some have five or six.

FIGURE 14-14 A multi-speed blower motor.

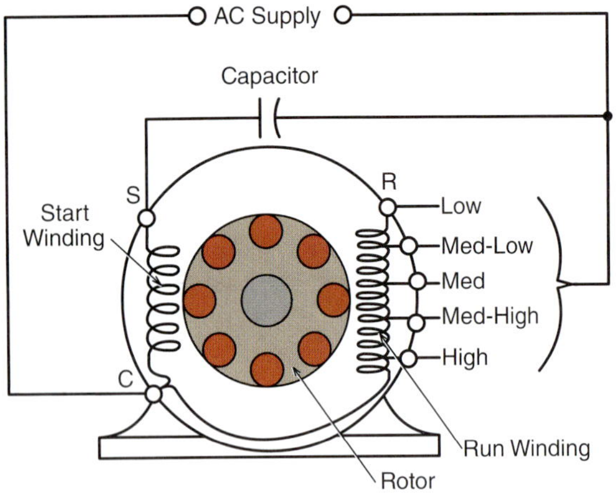

FIGURE 14-15 Winding arrangement of a multi-speed motor.

K14003 Describe the construction and operation of three-phase motors.

Multi-speed motors have a run winding and start winding, like most other motors. The run winding is tapped in different locations to produce the different speeds (**FIGURE 14-15**). The colored leads from the motor are connected to these tap points. The data plate of the motor identifies which leads provide the various speeds.

▶ Three-Phase Motors

Three-phase motors, such as the one shown in **FIGURE 14-16**, are used where it is necessary to overcome the starting torque and consistent operating challenge of larger loads. Three-phase motors are widely used in applications to drive large compressors, pumps, and fans that require more power than a single-phase motor can practically provide. Three-phase motors are also more efficient and reliable than single-phase motors.

Recall from the discussion of single-phase motors that a second phase of the applied voltage is developed though a separate start winding, start capacitor, or shaded pole. This is not necessary in a three-phase motor, because the applied voltages are already 120 degrees out of phase with each other. In addition, the stator windings are physically 120 degrees apart. This phase difference and physical separation of stator windings combine to provide the rotating magnetic field needed to start the motor and keep it turning. This is illustrated in **FIGURE 14-17**.

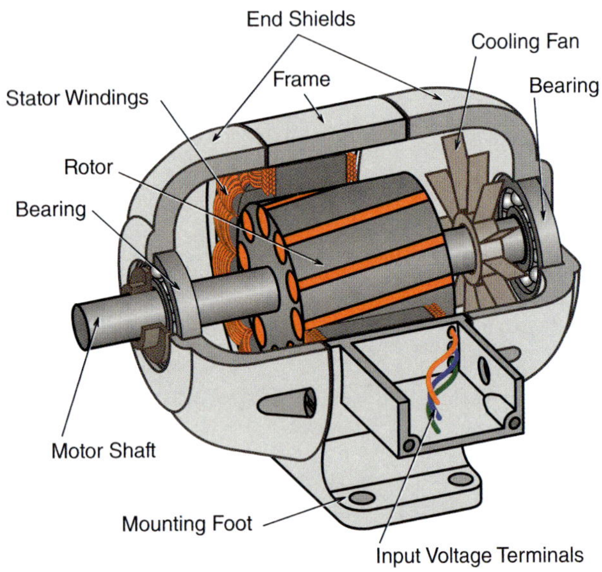

FIGURE 14-16 Cutaway view of a three-phase motor.

Phase A is applied to the north and south poles of the stator winding at 12 o'clock and 6 o'clock. Since the rotor rotates counterclockwise, Phase B is applied to the stator winding at 10 o'clock and 4 o'clock, followed by Phase C applied at 8 o'clock and 2 o'clock. As in single-phase motors, the electromagnetic field developed in the stator windings induces currents in the rotor windings. These currents produce their own magnetic field that attempts to line up with the stator field. However, since the stator field is constantly rotating, the rotor field winds up chasing it.

The number of stator poles determines the rotational speed of the motor, measured in revolutions per minute (rpm). Each phase may have anywhere from two to four poles, so the entire stator may have as few as six poles and as many as 12. The rpm of a motor is calculated as follows:

$$\text{rpm} = (120 \times \text{frequency}) \div \#\text{ of poles}$$

Therefore, a 60-Hz motor with four poles would have an rpm of 1,800 as determined here:

$$\text{rpm} = (120 \times \text{frequency}) \div \#\text{ of poles}$$
$$\text{rpm} = (120 \times 60\text{ Hz}) \div 4\text{ poles}$$
$$\text{rpm} = 7{,}200 \div 4$$
$$\text{rpm} = 1{,}800$$

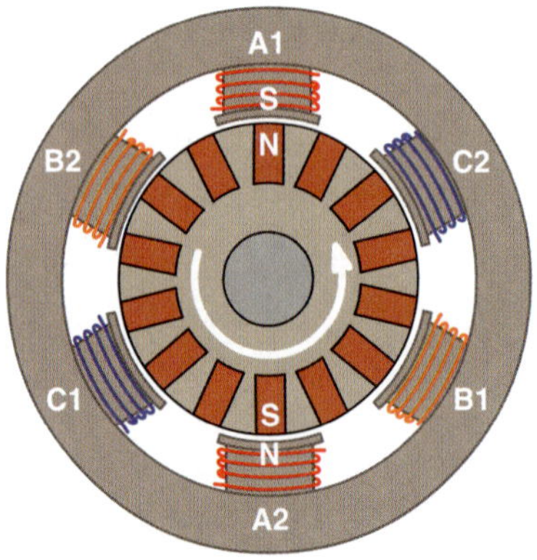

FIGURE 14-17 Stator arrangement of a three-phase motor.

As these calculations show, the greater the number of poles, the slower the motor rotates.

The direction of rotation for a three-phase motor is easily changed by reversing two of the three stator connections. It is therefore important to ensure that the connections are made properly. If they are accidently reversed, the motor will turn in the wrong direction. This can cause a serious problem with fans and pumps, for example, because they would not perform as intended. On the other hand, this characteristic makes it easy to change the rotation on new equipment installations to suit the phasing of the power supply. This feature is also used occasionally to try to dislodge a mechanically stuck three-phase compressor. Bumping the compressor motor in the opposite direction can sometimes free the compressor. The leads are then returned to their previous phase. Note that some compressors will pump in either direction but others need to rotate in a specific direction.

The stator of a three-phase motor may be wound in a wye (star) or delta configuration (**FIGURE 14-18**). From a service and maintenance point of view, the two types are not dramatically different. The chosen approach to winding design is generally left to engineers that base their choice on the available voltage and load type.

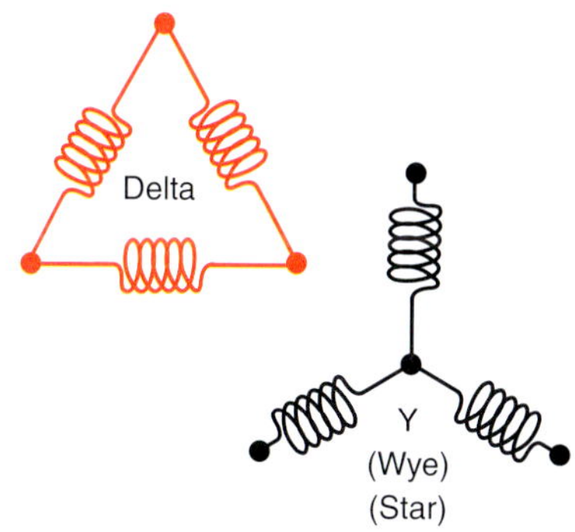

FIGURE 14-18 Three-phase motor stator configurations.

K14004 Describe basic motor troubleshooting procedures for single- and three-phase motors.

S14001 Troubleshoot a single-phase motor and related starting components.

▶ AC Motor Troubleshooting

Motors are expensive, so the decision to condemn a motor should be made only after verifying that it has failed. It is also a good idea to determine, as much as possible, what caused the motor to fail. This approach encourages consideration of external conditions that could have caused the failure and may also cause the replacement motor to fail. It is also important to determine the nature of the failure—i.e. grounded, open, internally shorted, or mechanically seized. With this, the customer can be given a definitive reason for replacing the motor. It's better to be able to reassure a customer by saying something like, "I tested the motor and determined that the windings are grounded, so the motor needs to be replaced," instead of "The motor is bad." Reporting the specific nature of the failure illustrates troubleshooting knowledge and conveys confidence to the client as well as employers.

Before testing motor windings with a multimeter or other instrument, remember to take the following safety precautions:

- Disconnect, lock out, and tag out the power supply to the motor at the disconnect or breaker panel.
- Discharge any start and run capacitors.

When testing an energized system is necessary, make sure safety guards are in place and keep hands and clothing away from rotating components. Take every precaution to protect against unseen electrical hazards and against the hazard of rotating shafts, fans, and blower wheels.

▶ CORE CONNECTION

Consistent Speed

An induction motor will slip under load, so its rotation speed is not quite the same as that of the rotating magnetic field. If rotation speed is critical, a **synchronous motor** is used. A synchronous motor is one that operates at a constant speed, irrespective of load variations. To accomplish this, a DC excitation voltage is applied to a winding on the rotor.

▶ CORE CONNECTION

Motor Alignment

One source of motor failures is misalignment between the motor shaft and the shaft of the load that the motor is driving. This is not really a problem with blower wheels and fan blades that are installed directly onto the motor shaft. Pumps and open-drive compressors are another story.

In these cases, a coupling connects two separate shafts. Flexible coupling styles are sometimes used because they allow for slight alignment errors. However, if the misalignment is too great, it can result in excessive wear or damage to the bearings or either shaft. Precise alignment is preferable.

Motor alignment is done by using dial indicators or laser alignment equipment designed for the purpose (**FIGURE 14-19**). This is an advanced skill to be learned as a technician's career progresses.

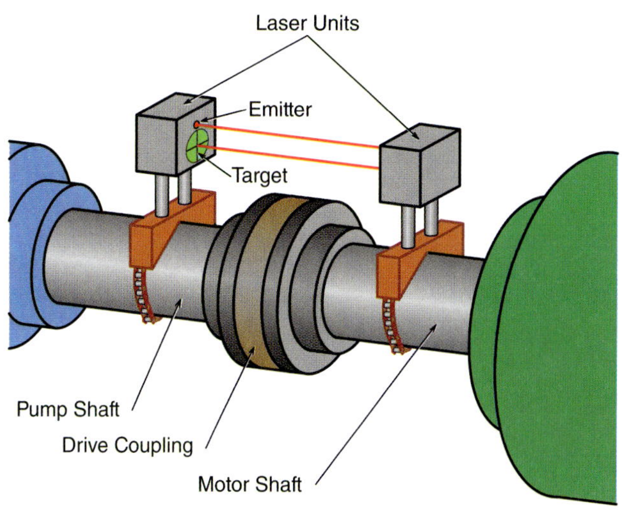

FIGURE 14-19 Pump/motor shaft alignment.

Motor Inspection

A number of visual and physical checks can be performed that may help in ascertaining the cause of failure:

- Verify that the wiring to the motor is correctly connected by comparing it to wiring diagrams.
- Look for loose or broken mounting structures and hardware.
- Check the motor enclosure for darkened paint that would indicate overheating.
- Look to see if dirt and debris has entered through openings in the motor enclosure. A buildup of dust or dirt could short stator windings when they absorb airborne moisture.
- Try spinning the motor shaft. If it does not turn freely, a bearing may have failed or some other mechanical component has seized. Look for side-to-side movement that would also indicate bearing failure. A small amount of end play in the shaft (movement of the shaft in and out of the assembly; axial movement) is acceptable in many motors, but it cannot be excessive. Although the allowable end play depends on the motor and manufacturer, anything beyond ¼" (6 mm) should generally be cause for concern. Sleeve-bearing motors tend to have more end play than ball-bearing motors.
- If the motor has a run and/or start capacitor, check for evidence of leakage or a bulging case. Also check the microfarad rating of the capacitors and their actual capacitance. Run capacitors should be within ±10% of the rating. Start capacitors generally have a wider range of acceptance, but a value more than 15% lower than the rating is cause for replacement. Replacing a failed capacitor may be all that is needed to make the motor operational again.

Energized Testing

The first step in troubleshooting a motor that isn't running is to check for power at the motor leads. Use a multimeter to verify that the motor is receiving the correct input voltage. This requires creating the conditions under which the motor would normally run. For example, troubleshooting an indoor blower motor may need to begin by ensuring that the thermostat is calling for cooling or heating or is set to operate the fan continuously. A voltmeter connected to the motor leads or their source (such as relay or contactor terminals) should read the appropriate voltage. If there is no voltage, the troubleshooting process moves on to other areas.

Motors may be rated for a single voltage, such as 230 VAC, or for a range of voltages. Many motors are rated to operate at 208–230 VAC to increase their flexibility across the range of voltages commonly encountered. The input voltage range for motors is as follows:

- For both single-phase and three-phase single-voltage motors, the voltage must be within ±10% of the nameplate value. For instance, if the motor is rated at 230 volts, the input voltage must be within the range of 207–253 VAC (±23 volts, representing 10% of the nameplate rating).
- For both single-phase and three-phase dual-voltage motors, the input voltage must be within ±10% of the lowest voltage and ±10% of the highest voltage listed. For example, if a motor is rated for 208–230 VAC operation, then the minimum voltage is 187 VAC and the maximum voltage is 253 VAC.

When a motor that has failed is energized, it will respond (or not) based on the type of failure. For example, if the motor is grounded, a fuse or circuit breaker will likely open very quickly. If a fuse was blown or a circuit breaker was tripped upon arrival, a grounded condition would explain why.

If the windings of the motor are open, it will do nothing at all when power is applied. This is because no magnetic fields are being generated. If the windings are shorted turn-to-turn, it may try to operate, but the magnetic fields will be out of balance or weak, and the motor will likely struggle to run, draw an unusual amount of current, and/or overheat quickly. Any time a motor is being energized for testing purposes, a clamp-on ammeter should be applied so that the current can be monitored.

Finally, if the problem is mechanical, such as a seized bearing, the magnetic fields will not be strong enough to turn the shaft at all, resulting in locked rotor amps (LRA) being drawn along with a humming sound.

When any of these responses occur, it's time to test the motor windings or recheck the motor shaft to determine whether it is free to move. Electrical problems in motors can sometimes make it appear as if the shaft is mechanically seized when it is not.

Deenergized Testing of Motor Windings

Motor windings can fail in three primary ways:

1. open—there is a break in the winding path
2. shorted turn to turn—one or more individual wires have made electrical contact somewhere amid the winding
3. grounded, or shorted to ground—wires in the winding have made electrical contact with a path to ground.

The ohmmeter function of a multimeter can be used to check a motor for these conditions. However, a standard meter test may not accurately diagnose a shorted winding.

Testing for Open Windings

Resistance testing for an open winding involves checking for continuity between the common winding and all other motor leads. Start by setting the meter to read the lowest resistance range (R×1). If the winding is intact, the resistance reading should be in the single-digit range. If the winding is open, a resistance reading will not be displayed (infinity; no continuity). Some motors may have internal thermal protection that will open a switch in the windings if the motor overheats. The motor must cool enough for the protective device to reclose its contacts before testing. If the thermal protector never closes, the motor must still be replaced. Since it has the same effect as an open winding, the circuit is open and the motor is not going to run.

For a multi-speed motor, the resistance between common and the lead for the highest speed will show the lowest resistance of the speed-related leads. The highest common-to-run winding resistance will be found on the lead for the lowest speed. Note that differences in the readings will be rather small since all the readings will be in the single-digit range.

Windings that are shorted in the midst of their turns are more difficult to diagnose. For a multi-speed motor, there may be unusual differences in the resistance values from winding to winding. Unusual readings on a motor that fails to run as expected, or draws a higher current than it should, generally indicates an internal problem that requires motor replacement, even though it is difficult to confirm a shorted winding is the cause.

▶ TROUBLESHOOTING TIP

Identifying Single-Phase Motor Windings

Motor terminals may not be clearly marked. But the terminals can be identified by using some simple resistance measurements. Start by sketching a quick diagram of the terminals on paper and numbering the terminals as 1, 2, and 3. Measure the resistance from 1 to 2, from 1 to 3, and from 2 to 3, then record them on the sketch (**FIGURE 14-20**). Analyze the results as follows:

- The highest resistance is found between the run and start terminals. The third terminal can now be considered the common terminal.
- The highest resistance recorded between the common terminal and the other two identifies the start winding. The one with the lower resistance identifies the run terminal.
- In many cases, the terminal at the top on the compressor plug is the common winding, or C (short for "common"). But this is not guaranteed.

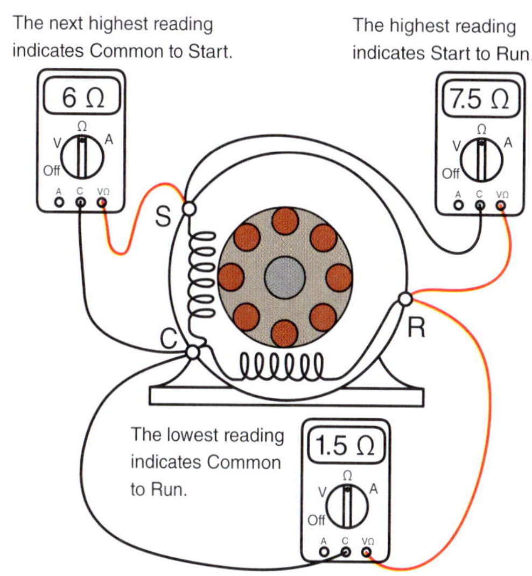

FIGURE 14-20 Motor winding identification.

Testing for Grounded Windings

A winding that is grounded will cause a fault current that will trip the circuit breaker or blow a fuse quickly when the motor is energized. To test for a grounded winding, connect one lead of the ohmmeter to the common winding and the other to the motor housing. The housing connection must be made on a surface that offers good electrical conductivity, such as an unpainted surface. If necessary, scar the paint on the motor housing a bit to reveal shiny metal that will provide a good electrical path. Most motors, though, have an unpainted surface somewhere that can be accessed with the meter probe. If the motor has continuity through all its windings to common, the test for a ground from only the common winding is valid.

In this case, the ohmmeter should be set to its highest setting of R × 10,000 or a similar value. The reading should be open (infinity). In rare cases, a very high resistance may be measured. Any value that is less than 1,000 Ω per volt indicates a grounded motor. For instance, if the motor is designed to operate at 120 VAC, any reading below 120 kΩ (1,000 Ω × 120 VAC = 120,000 Ω or 120 kΩ) indicates a ground. Seeing any level of continuity to ground in functional motors, however, is very rare with a standard ohmmeter.

Larger, more valuable, or critical motors may be tested with a **megohmmeter**, which applies a very high voltage to the winding. A megohmmeter can reveal ground paths in the winding that a common ohmmeter will not identify.

SAFETY TIP

Working around compressor terminals that have pressure behind them can be dangerous. The terminals could blow out and cause a serious injury if the assembly is damaged in some way. To minimize the hazard, measure voltage and resistance on the connected wiring, away from the terminals. Do not place your face in the direct path of the terminal assembly, in case it fails unexpectedly. It is rare, but it can happen. On the other hand, if testing the compressor motor windings through the wiring leads indicates a fault, the terminals and wiring must be inspected to ensure they are not the real problem.

Testing Three-Phase Motors

Testing the windings of a three-phase motor is generally quicker and easier than testing a single-phase motor. All three windings should have the same, or very close to the same, resistance readings. Significant variations may indicate that a winding is shorted turn-to-turn, or that the motor was improperly wound. This condition will also reveal itself through the amount of current flowing through the affected winding.

A three-phase motor is tested for grounds in the same manner as a single-phase motor. One lead of the ohmmeter is connected to a terminal, and the other is connected to a metal surface of the motor frame. Again, this point on the frame must offer good electrical contact, so a point without paint and rust must be used. If necessary, clean the paint from a small area to expose bright metal.

Grounded windings in a three-phase motor will show the same response as expected for a single-phase motor when the motor is energized; fuses and/or circuit breakers will react quickly to a grounded condition. However, if there is an open condition in one winding only, it will behave as if the motor is single-phasing. The motor will hum, the current will be very high, and it will not likely rotate at all. The motor starter (if equipped) will likely trip soon, perhaps before a fuse or circuit breaker. A single-phase motor with an open winding will do nothing when energized, unless it also has a grounded winding.

▶ TROUBLESHOOTING TIP

Single-Phasing of a Three-Phase Motor

An estimated 15% of three-phase motors fail because of single-phasing, which is a condition that occurs when the input voltage to one of the three windings is lost. This might occur when one fuse alone in a three-phase service opens. When this happens, the current through the other two windings increases sharply and immediately. If the motor-starter overload devices or other protective controls do not react swiftly, the increased current will damage the two active windings. The problem with the power may not have been taken care of, and a number of tripped motor starters may result before a technician arrives on-site. When it is obvious that a number of three-phase loads have been affected, ensure that proper power has been restored before resetting and observing the operation of the equipment. If only one motor starter in a plant was affected, the problem likely lies within the system itself or its own power supply.

Voltage and Current Imbalance

Voltage imbalance in three-phase systems describes the situation where the phase-to-phase voltages of the circuit are not equal. A 2% imbalance is enough to cause significant problems and is considered the limit. Anything less than that is generally considered acceptable, even though it is not ideal.

Current imbalance occurs when the current flowing through each of three windings of a motor is not equal. The measured current through each winding should be equal, or very close to it. The allowable current imbalance is 10%.

Current imbalance is most often caused by voltage imbalance. However, problems in the motor windings can create current imbalance when no voltage imbalance exists. It is important to note that a small percentage of voltage imbalance causes a significantly larger percentage of current imbalance. The current imbalance can be six to ten times the amount of voltage imbalance. In other words, a voltage imbalance below 2% (acceptable) can still create an unacceptable level of current imbalance at the motor.

Voltage imbalance and current imbalance are calculated in the same way. To calculate voltage imbalance, measure the voltage between the three legs of power that serve an operating motor. Then average the three values. The voltage with the largest deviation from the average is then identified and divided by the average. To change the result to a percentage, multiply that result by 100.

The formula used to calculate voltage imbalance is as follows:

$$\text{voltage imbalance} = \frac{\text{max voltage deviation}}{\text{average voltage}} \times 100$$

For example, assume the following voltage measurements are taken on a system: 208, 213, and 206 volts. First, determine the average voltage:

$$208 + 213 + 206 = 627$$
$$627 \div 3 = 209 \text{ average voltage}$$

Now determine which of the three values is farthest from the average of 209. In this case, it is the 213-volt reading, since it deviates by 4 volts. Divide the deviation of 4 volts by the average:

$$\text{4-volt deviation} \div 209 \text{ average voltage} = 0.019$$
$$0.019 \times 100 = 1.9\% \text{ voltage imbalance}$$

In this example, the imbalance is within tolerance but is higher than desired. It could result in reduced motor life and should be reported to the client for evaluation by the utility or an electrical contractor. It may also result in a current imbalance that will be outside the 10% tolerance.

As noted earlier, current imbalance is calculated the same way. Here is the current imbalance formula:

$$\text{current imbalance} = \frac{\text{max current deviation}}{\text{average current}} \times 100$$

For example, assume that the three current readings on an operating three-phase motor are 12 amps, 14 amps, and 11 amps. First, determine the average current as follows:

$$12 + 14 + 11 = 37$$
$$37 \div 3 = 12.33 \text{ average current}$$

The largest deviation from the average current is found in the 14-amp reading, with a deviation of 1.67 amps. Now divide the deviation by the average current and multiply the result by 100 to determine the percentage:

$$1.67 \div 12.33 = 0.135$$
$$0.135 \times 100 = 13.5\%$$

Since this exceeds the tolerance of 10%, it is unacceptable. If there is a significant voltage imbalance also present in this scenario, it is the likely cause. If not, then the motor windings must be considered the likely cause. Testing the motor windings with a megohmmeter can assist in confirming that the resistance of each winding differs enough to cause the problem. If so, the motor must be rewound or replaced. Note that loose connections, long and/or undersized conductor runs, and pitted contactor points can also cause high or imbalanced current readings.

Motor-Starting Circuits

Every single-phase motor, other than the shaded-pole motor, can have external components that assist in motor starting. These components include start and run capacitors, start relays, and centrifugal switches. In many cases, a motor is still functional but cannot start or run properly due to the failure of one or more starting-circuit components. For example, a potential start relay that fails to open when it is supposed to can cause a start winding to overheat and fail. Failed start and run capacitors also create problems for the motor, but they will not necessarily damage it before the problem is identified and resolved. Note that capacitors and their testing procedures will be covered in detail in Chapter 10.

Start Relays

Both current and potential (voltage) relays are used on HVACR motors. A potential start relay, which is by far the most common, has normally closed contacts that allow voltage to be applied to the start winding when the motor is energized (**FIGURE 14-21**). As the motor comes up to speed, back-EMF voltage builds in the winding until there is enough to activate the potential-relay coil and open the contacts. Remember that the pickup voltage is significantly higher than the line voltage feeding the motor.

A current relay has normally open contacts. The coil of the relay is energized by the high current experienced when the motor is first energized. The contacts then close, enabling the capacitor to assist in starting. Once the motor approaches full speed and the current has dropped significantly, the relay contacts will reopen.

If the contacts of either type of relay fail to change position when they should, the motor will likely overheat and potentially fail, especially after a number of start attempts. This condition can be caused by either damaged contacts, an open relay coil that prevents it from energizing, or even a mechanical failure inside the relay.

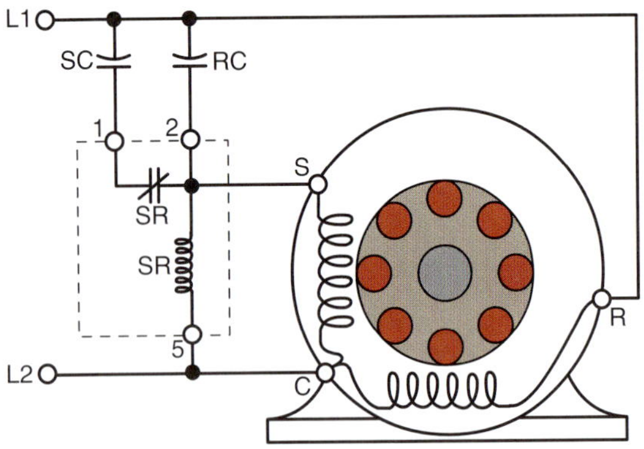

FIGURE 14-21 Potential relay in a motor-starting circuit.

The ohmmeter function of a multimeter can be used to check these relays when they are deenergized. With power turned off and tagged out, disconnect the relay from the circuit and test for continuity through the contacts. In the case of a potential relay, the resistance should read 0 Ω or very close to that, because the contacts are normally closed. If the meter shows an open circuit or any significant resistance (100–200 Ω, for example), it indicates that the contacts are either stuck in the open position or damaged. The contacts of a current relay should yield the opposite result because they are normally open; there should be no continuity through the contacts when it is deenergized.

The coil can also be tested using an ohmmeter. The potential-relay coil has very high resistance, perhaps as much as 6 kΩ. Use an ohmmeter that can measure this level of resistance. If the coil of a current relay is intact, a single-digit resistance will be displayed, perhaps less than 1 Ω.

Since potential relays represent the majority of start relays used in HVACR systems, the remainder of the discussion will focus on them alone.

Energized Potential-Relay Testing

As mentioned previously, potential starting relays have normally closed contacts. That means the start capacitor is in the starting circuit when the motor is initially energized. To develop enough back-EMF to energize the contacts, the motor needs to reach 75–80% of full speed. Once enough voltage is generated to open it, it will remain energized with the contacts open as long as the motor continues running.

To test relay operation, place a clamp-on ammeter on the wire connecting the start capacitor to the start-relay contacts. Monitor the current in the circuit as the motor is energized. A current reading indicates the capacitor is in the circuit and power is flowing. As the motor nears full speed and the relay coil is energized, the current reading should drop to zero. If it does not, the contacts of the relay are remaining closed. As such a test is conducted, remember that it happens very quickly.

To practice how to troubleshoot a single-phase motor and related starting components, follow the step in **SKILL DRILL 14-1**.

SAFETY TIP

When a potential relay is replaced, an exact replacement is required. Potential relays are designed to energize within a specific voltage range that is significantly higher than the line voltage. If a relay has the incorrect pickup voltage (the voltage required to energize the relay), it can damage the motor and/or capacitor. Consult the manufacturer of the motor if there is any doubt about the proper replacement.

SKILL DRILL 14-1 Practicing How to Troubleshoot a Single-Phase Motor and Related Starting Components

A single-phase motor equipped with a potential relay and start capacitor is required for this exercise.

SAFETY TIP

Working with electrically energized circuits is hazardous. Avoid contact with any surface that could potentially be energized. When working inside energized enclosures, maintain a stable position and, whenever possible, avoid placing both hands inside the enclosure. All work on energized equipment by unqualified individuals should be supervised by a qualified instructor or technician.

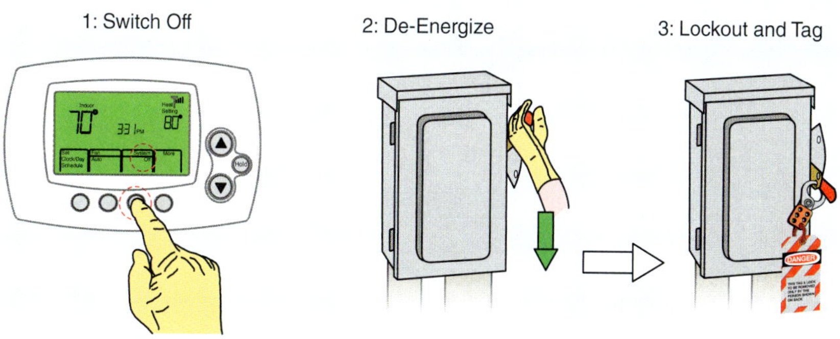

1. Ensure that all controls, including the thermostat, are set to the Off position. Deenergize, lock out, and tag the equipment being serviced. Be sure that the process deenergizes not only the line voltage feeding the equipment but also any low-voltage control power source that may be located in a separate piece of equipment.

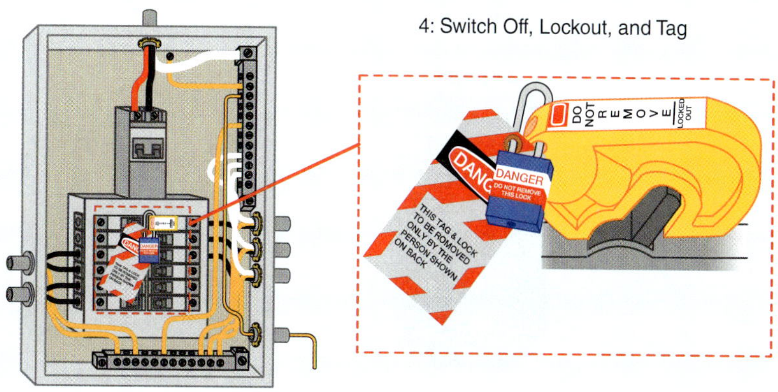

Continued

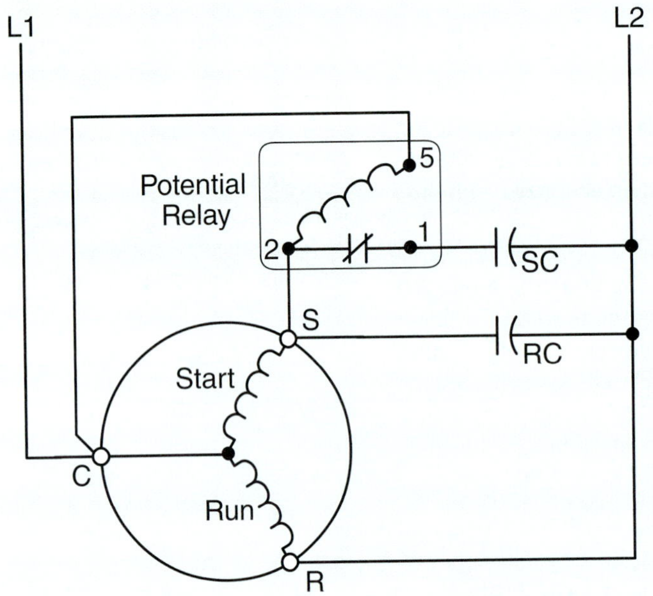

2. Refer to the wiring diagrams for the unit to identify and locate the compressor or motor, start relay, start capacitor, and run capacitor (if present). Study the arrangement of the wiring to confirm that it agrees with the wiring diagram and identify each wire serving the motor.

3. Bleed any remaining electrical charge from all capacitors by spanning their terminals with an appropriate tool or resistor.

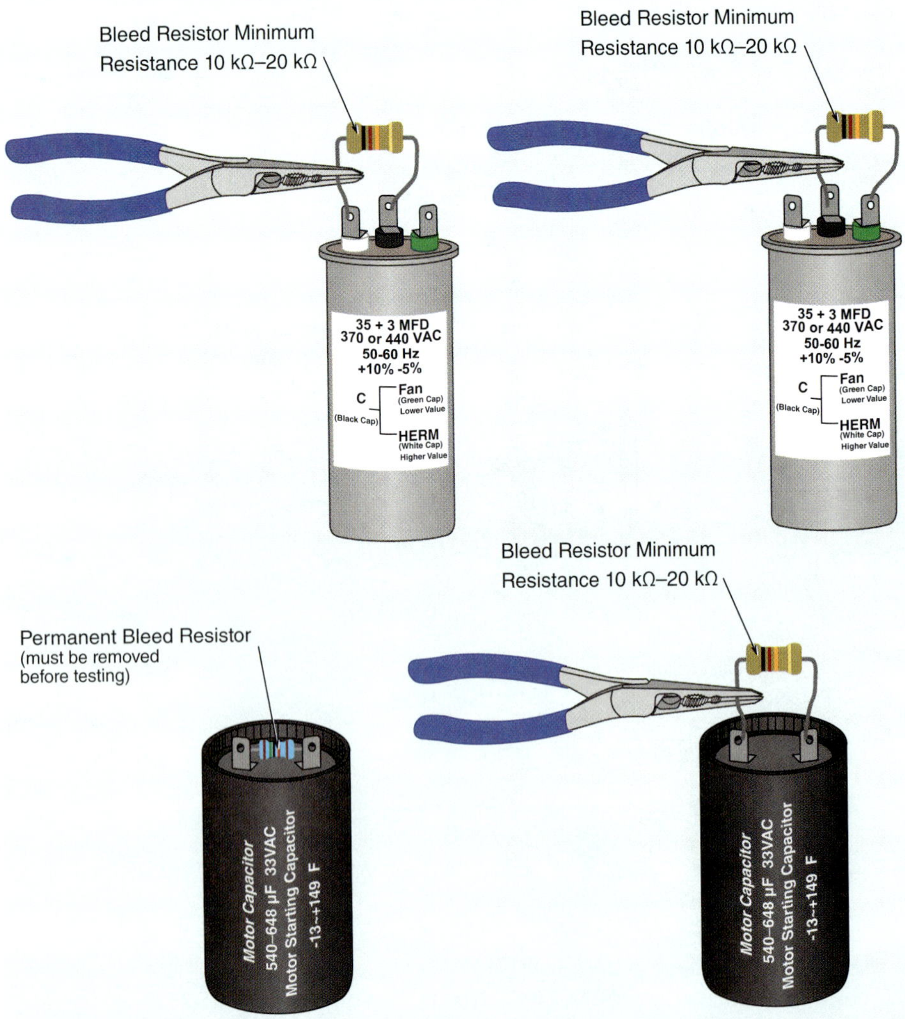

Continued

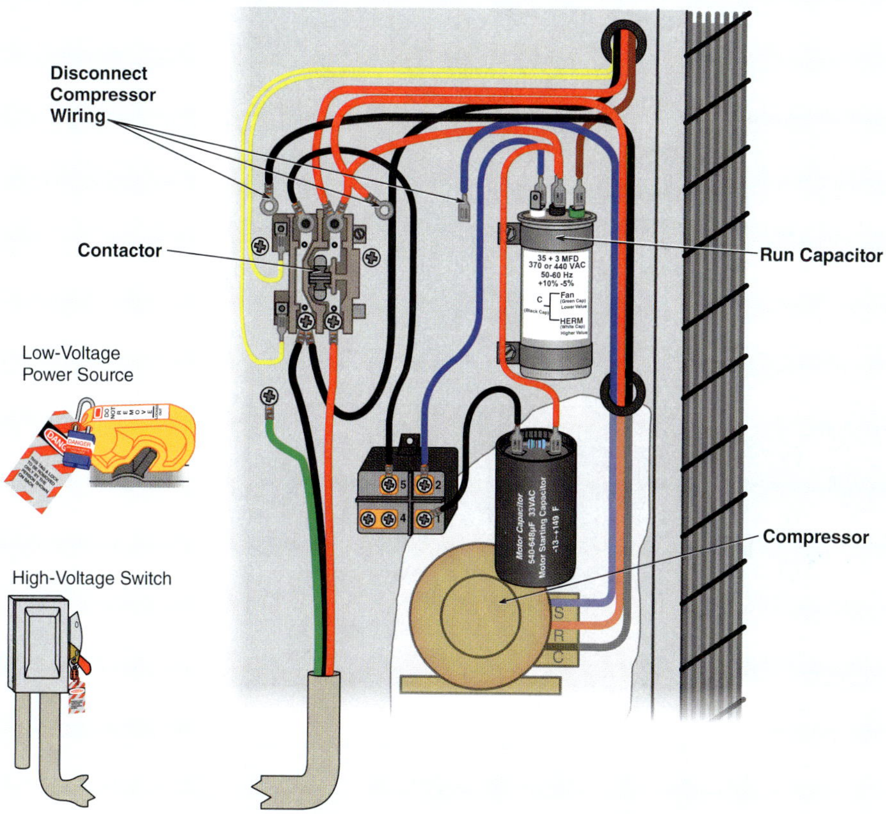

4. Identify the wiring connected to the Common, Start, and Run motor terminals using the wiring diagram. Tag the wires for reconnection, and then disconnect the three leads from the contactor, start relay, and capacitor. The leads will be used for testing to avoid working directly at the terminals. However, it is important to note that the wiring or its terminals could be flawed, although such a flaw is rare.

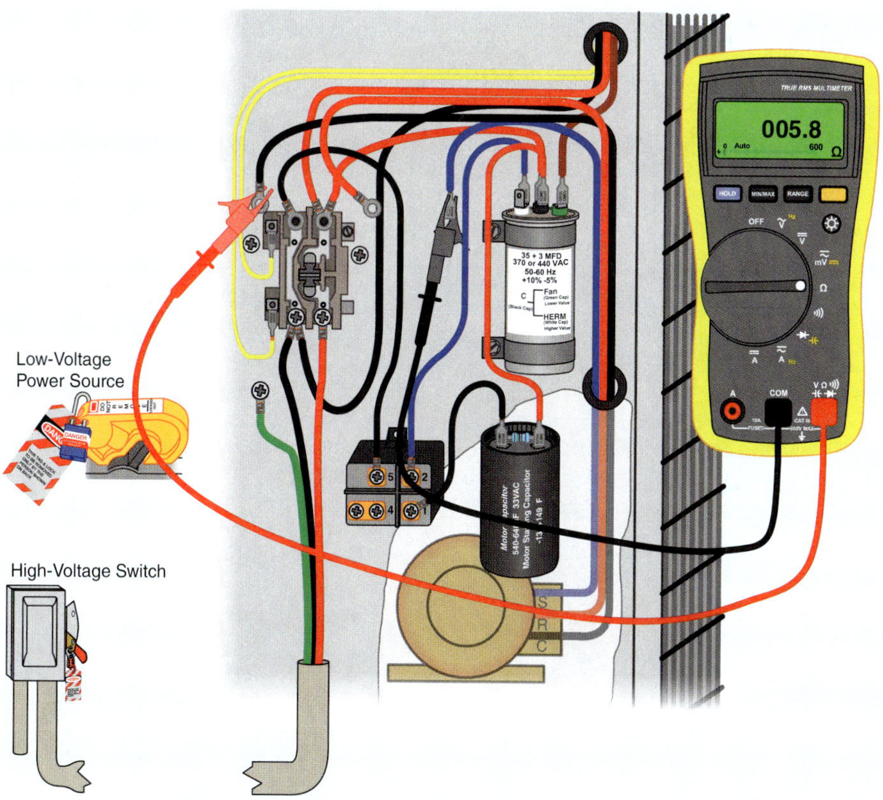

5. Set the ohmmeter to the lowest resistance range (R × 1) if it is not auto-ranging. Measure the resistance of the start winding by placing the meter probes on the Common and Start leads to the compressor. If the winding is intact, the resistance reading should be in the single-digit range. If the winding is open, the resistance reading will be infinity.

Continued

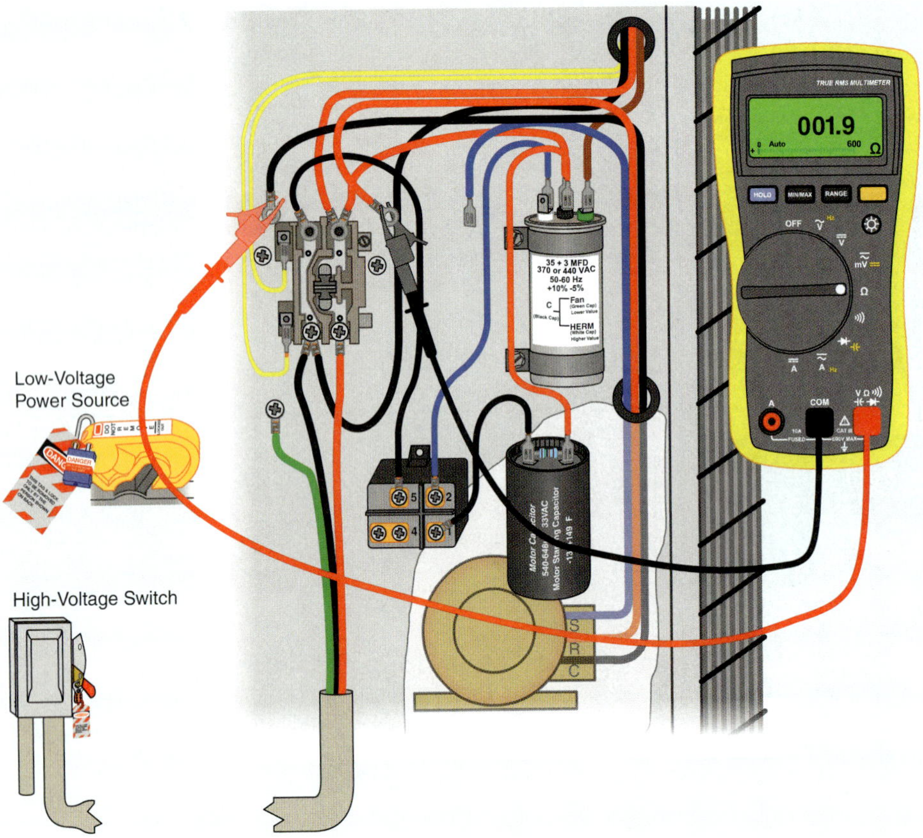

6. Next, measure the resistance of the run winding by placing the meter probes on the Common and Run leads to the compressor. If the winding is intact, the resistance reading should be in the single-digit range. If the winding is open, the resistance reading will be infinity.

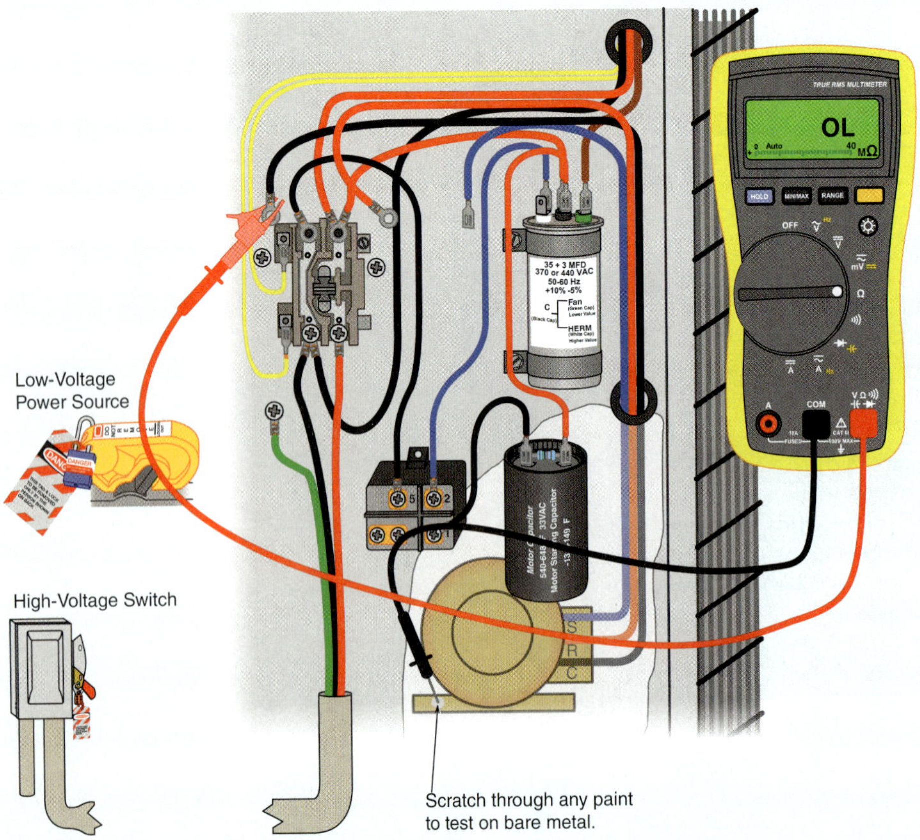

7. After it is determined that the start and run windings are intact, any one of the three terminals can be used to check for grounded windings. Most technicians, however, use the Common terminal. Set the ohmmeter to its highest range. Connect one meter probe to any terminal and the other to a clean, unpainted metal portion of the compressor or motor outer shell. If no such point can be found, create one by filing or scraping the paint from a small area on the compressor base. The windings are electrically isolated from the motor case, so no evidence of continuity (infinity) should be shown on the display of a common ohmmeter. It is a good idea to move from terminal to terminal and to different locations on the compressor or motor shell to ensure there are no indications of an electrical path to ground. Remember that ohmmeters that can measure very low resistance values may indicate continuity with a very high resistance value. The resistance value displayed must be at least 1,000 Ω for each volt of the motor's design operating voltage.

Continued

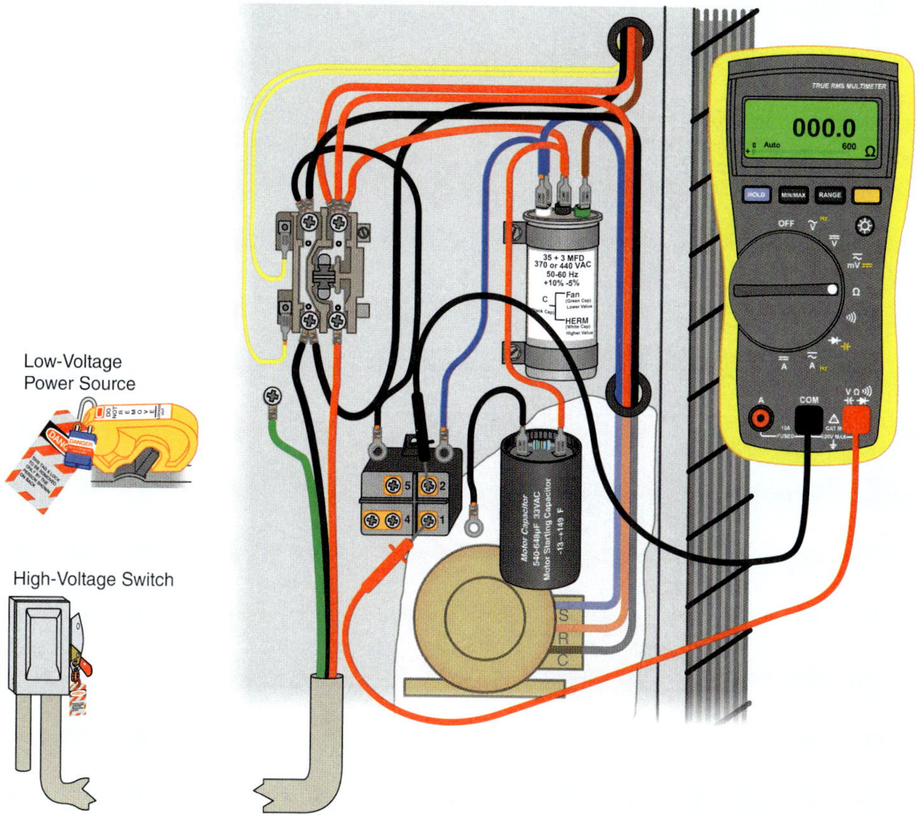

8. Before proceeding further, carefully reconnect the three leads for the compressor to the appropriate terminals. Make sure each connection is sound.Now focus on the potential start relay. Locate the terminals of the relay that represent the contacts and remove the wires. Use the ohmmeter to ensure that the contacts are closed while the system is deenergized. If any significant resistance is noted, such as 100 Ω, the contacts are likely damaged, so the relay should be replaced.

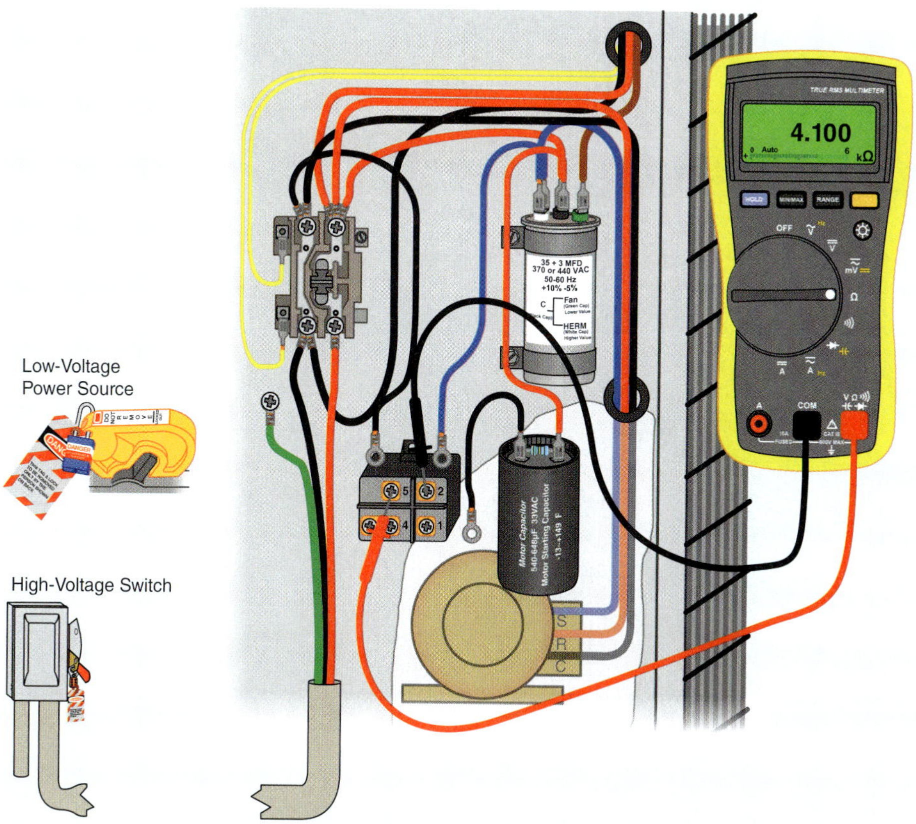

9. Next, use the ohmmeter to test the coil of the start relay for continuity. In the example here, the coil is connected to Terminals 2 and 5 of the relay. Note that the coil will likely have a very high resistance (thousands of ohms); setting the ohmmeter range to R × 100 is an appropriate choice.

Continued

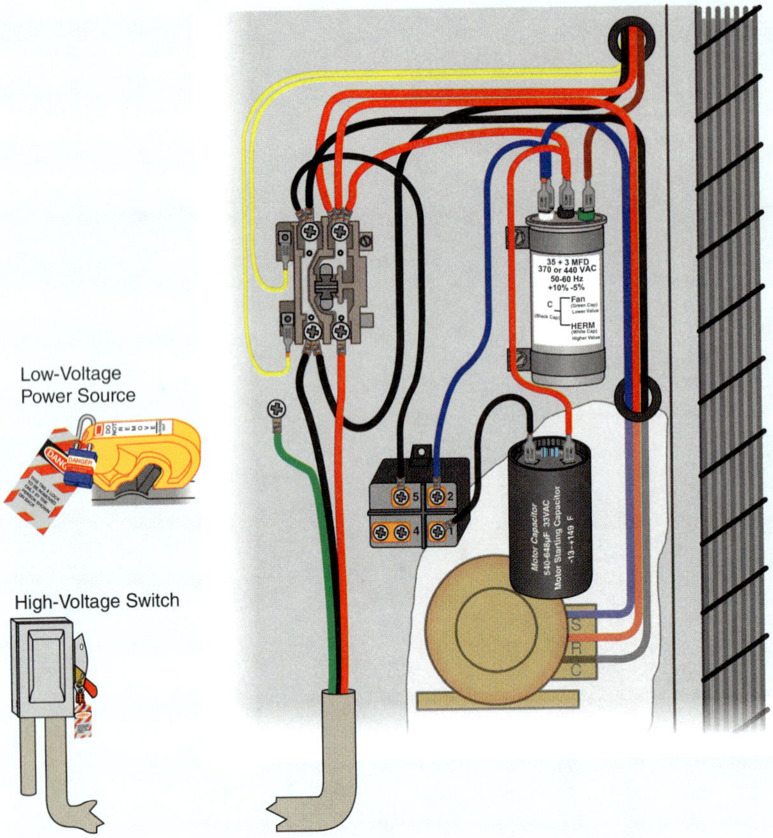

10. Carefully reconnect the wiring to the start relay. Ensure that each connection is sound.

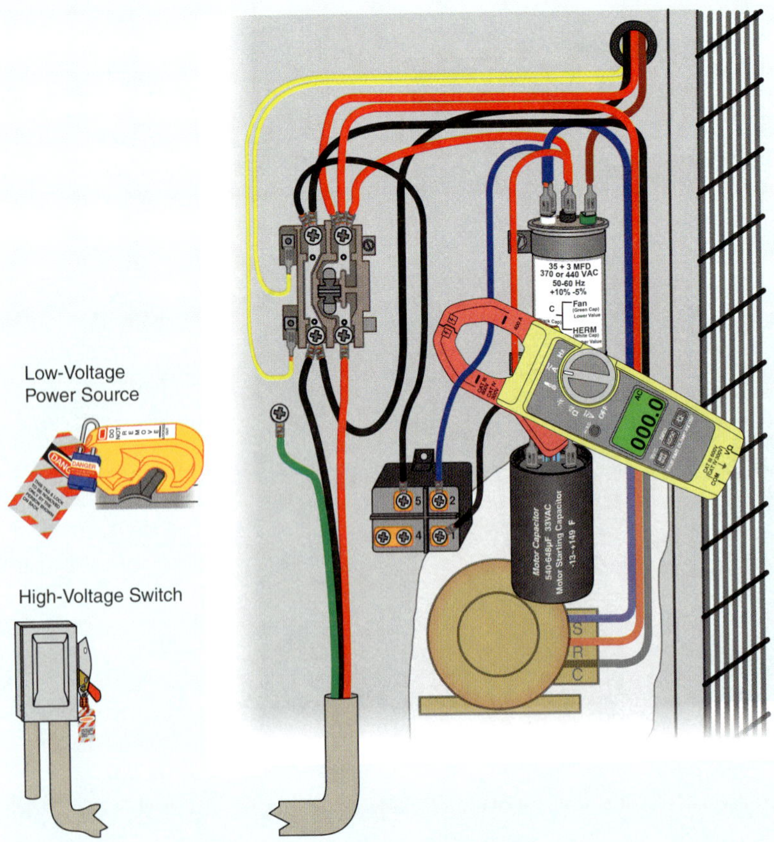

11. Place a clamp-on ammeter on the wire connecting the potential-relay contacts to the start capacitor. If the feature is available, set the ammeter to capture the highest current measured. Ensure the display can be viewed clearly as the compressor starts.

Continued

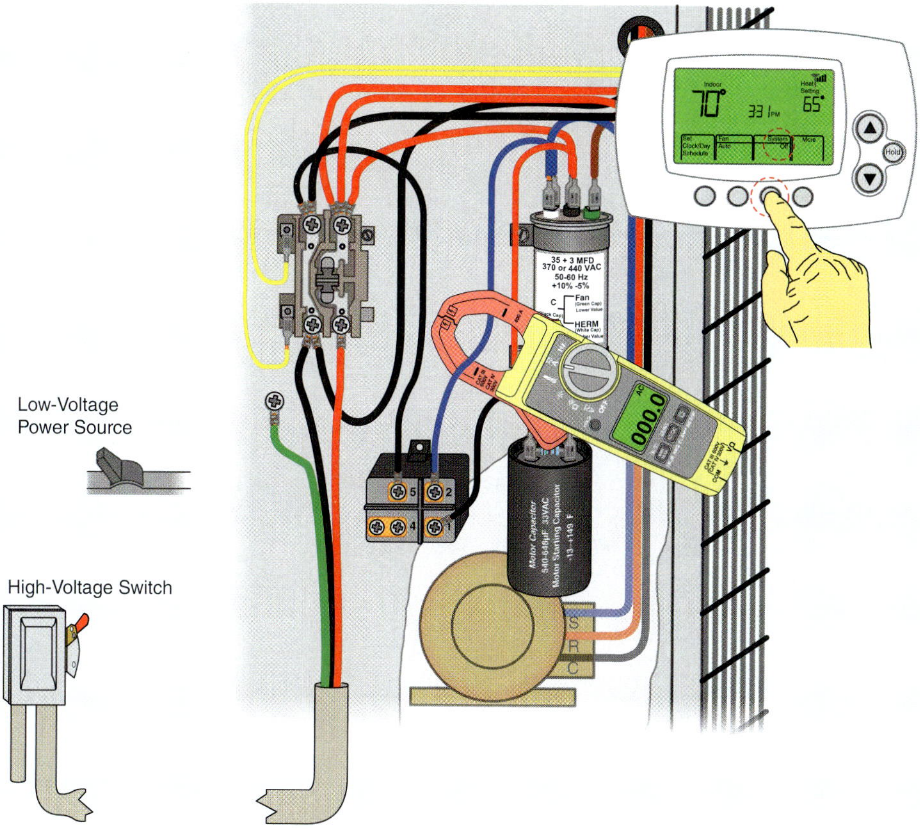

12. Ensure that all controls, including the thermostat, remain set to the Off position. Remove the locks and tags from the power source(s). Restore power to the system.

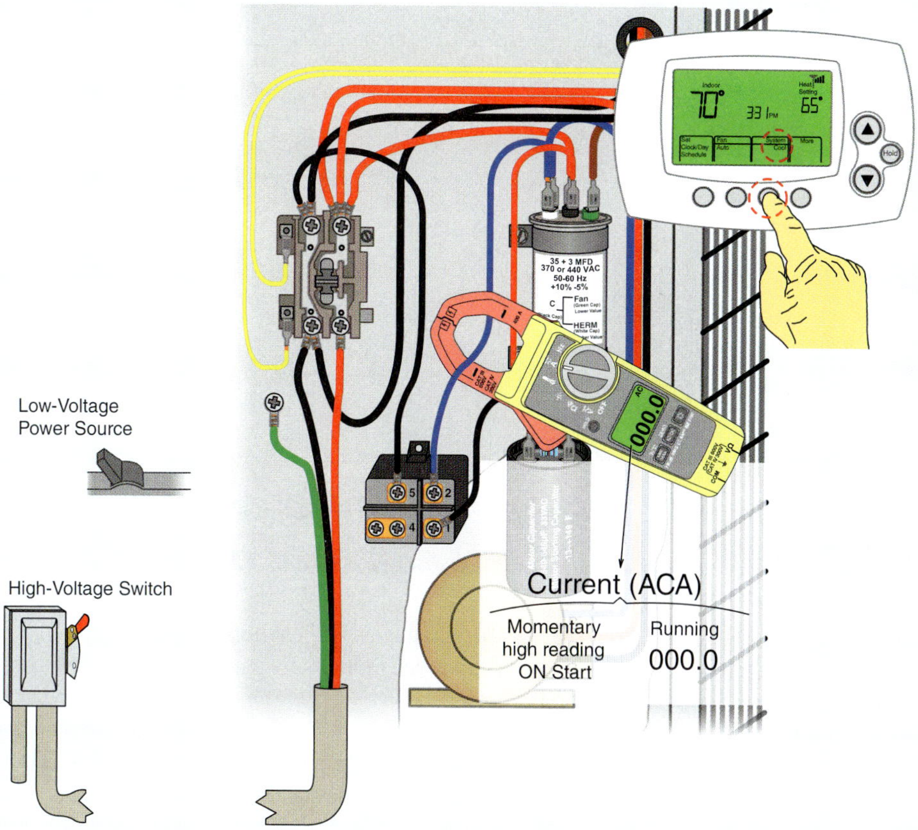

13. The ammeter should display current flow as soon as the compressor is energized, but for a very short time. Once the motor reaches roughly 75% of full speed, the relay contacts should open and the current flow will fall to zero.

Start the system and observe the ammeter display. If the start is normal, the current flow will be observed briefly, followed quickly by a reduction to 0 amps. This confirms that the contacts of the start relay have opened and removed the start capacitor from the circuit. If the current does not fall to 0 amps during the start sequence, quickly disconnect power to shut the motor down before damage occurs.

Continued

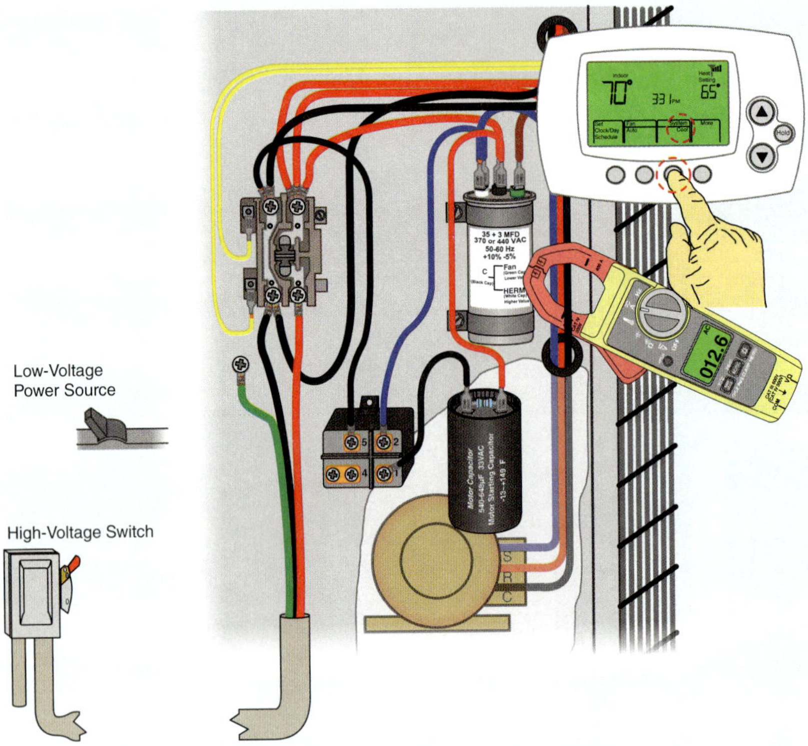

14. Relocate the ammeter to the Common compressor lead, and measure the operating current to ensure it is within the rated load amperage stated on the compressor, motor, or unit nameplate.

15. Once the testing is complete, carefully reinstall equipment covers and access doors, and then return the system to service.

▶ Wrap-Up

Summary

AC motors are used to drive compressors, pumps, and fans in HVACR systems. Single-phase motors are common in residential and light-commercial systems due to the limited current draw of systems in this size range and the lack of availability of three-phase power.

There are several types of single-phase motors. Low-torque, low-horsepower motors, such as the shaded-pole motor, are used to drive small fans. Higher-horsepower motors, such as capacitor-start and PSC models, are used to drive larger blowers and compressors.

Three-phase motors are used in larger commercial and industrial environments, where such power is available. Three-phase motors have the high starting torque needed to drive larger compressors, pumps, and blowers. Also, their operating efficiency is higher than that of a single-phase motor.

When any motor fails to operate as it should, it must be tested to determine the cause of the failure. In the case of single-phase motors with starting circuits to increase torque, the failure may be due to overheating of the start winding due to a failed capacitor or start relay. When a motor with a starting circuit is replaced, it is important to test the start components when the new motor is placed in service.

Key Terms

centrifugal switch A switch that is influenced by centrifugal force as a motor nears full rotational speed. Centrifugal switches are generally normally closed, opening when a sufficient centrifugal force is applied.

current imbalance An imbalance in the current consumption of the three windings in a three-phase load. A current imbalance up to 10% is typically allowed before motor damage is likely to result.

ferrous metal Any metal containing iron, such as steel or cast iron. Aluminum is an example of a nonferrous metal.

inductance A property of an electric circuit by which an electromotive force is induced in it through a variation of current, either in the circuit itself or in a nearby circuit.

horsepower (hp) A unit of power measurement that equals the power of 745.7 watts; it is also stated as 550 foot-pounds (ft-lb) per second.

megohmmeter A special type of ohmmeter used to measure very high resistance values (megohms, or MΩ), such as those associated with insulators. In HVACR applications, meggers, as they are often called, are used to evaluate the condition of motor-winding insulation.

rotor The rotating assembly of an electric motor, to which the motor shaft is usually attached.

slip The difference between the rotational speed of the rotor and the synchronous speed of the magnetic field.

slip rings Circular devices that transfer electrical energy between a rotating object and a stationary object.

stator The stationary portion of the motor that contains the motor windings that, when energized, generate electromagnetic fields.

torque A force applied to an object that causes it to rotate; it is commonly used to describe the force that must be overcome to start the rotor of a motor turning.

voltage imbalance An imbalance in the voltage supplied on each of the three legs serving three-phase electrical loads. A voltage imbalance up to 2% is typically allowed. However, a lesser voltage imbalance can cause a current imbalance that is beyond an acceptable range.

Review Questions

1. The magnetized component of a motor that does not rotate is called the __________.
2. In a DC motor, the component that converts the constant DC voltage to a pulsating DC voltage is the __________.
3. AC motors used in speed-controlled applications function more reliably when the motor is equipped with which type of bearings?
4. The type of rotor that has magnetic bars inserted into its iron core is the __________ rotor.
5. Which of these single-phase motor types does not have a separate start winding?
   a. PSC motors
   b. Capacitor-start motors
   c. Shaded-pole motors
   d. Capacitor-start, capacitor-run motors
6. A three-phase motor does not require a start winding because _____.
   a. it has a DC excitation voltage applied to the stator
   b. it has a start capacitor
   c. its windings are 120 degrees apart
   d. it uses a squirrel-cage rotor
7. A start-assist device for single-phase motors that does not require a potential or current relay to switch it in or out of the circuit is called a(n) _____.
   a. permanent split capacitor (PSC)
   b. PTC thermistor
   c. catalytic capacitor
   d. EMF thermistor
8. A motor rated to operate at 120 VAC should be provided with a minimum input voltage of _____.
   a. 117.5 VAC
   b. 114 VAC
   c. 110 VAC
   d. 108 VAC
9. A motor rated to operate at 440–480 VAC should be provided with a maximum operating voltage of _____.
   a. 550 VAC
   b. 528 VAC
   c. 490 VAC
   d. 480 VAC
10. A 60 Hz motor with six poles would have a rotation speed of _____ rpm.
11. When energized, a single-phase motor may behave as if it has a seized shaft (locked rotor), even when the bearings and shaft/rotor is fine, due to __________.
12. The start winding of a split-phase motor has _____.
    a. few turns of large wire
    b. many turns of large wire
    c. few turns of fine wire
    d. many turns of fine wire
13. When identifying the windings of a single-phase motor by using resistance measurements, the highest resistance measured will be between which two windings?
14. When identifying the windings of a single-phase motor by using resistance measurements, the lowest resistance will be found in which individual winding?
15. A potential start relay has __________ contacts.
    a. normally open
    b. normally closed

16. Why *doesn't* a potential start relay energize as soon as the motor is energized, since the same voltage is applied to its terminals at that time?
17. The maximum voltage imbalance allowed for reliable motor operation is ______.
    a. 1%
    b. 2%
    c. 5%
    d. 10%
18. Calculate the percentage of voltage imbalance present in a three-phase power supply when the voltage measured on the three legs of power are 458 volts, 462 volts, and 467 volts.
19. The maximum current imbalance allowed for reliable three-phase motor operation is ______.
    a. 1%
    b. 2%
    c. 5%
    d. 10%
20. Calculate the percentage of current imbalance present in a three-phase motor when the current measured on the three windings are 37 amps, 39 amps, and 43 amps.

CHAPTER 15

Electronically Commutated Motors

Knowledge Objectives

After reading this chapter, you will be able to:

- **K15001** Compare ECMs with conventional AC and DC motors.
- **K15002** Explain the operation of an ECM.
- **K15003** Describe the common wiring connections used with ECMs.
- **K15004** Explain how to troubleshoot ECMs.
- **K15005** Describe how to remove and replace an ECM.

Skill Objectives

Skills required to meet the objectives of this chapter:

- **S15001** Troubleshoot an ECM.
- **S15002** Replace an ECM.

▶ Introduction

Conventional AC (alternating current) motors have been the workhorses of the HVACR industry for many years. However, the demand for greater energy efficiency and application flexibility has caused manufacturers to seek alternatives for HVACR applications. The motors used in HVACR systems account for 90% of their energy consumption. The electronically commutated motor (ECM) has become an important part of the drive toward greater efficiency and has added the option of versatile variable-speed operation to small systems.

ECMs are three-phase DC (direct current) motors used primarily in residential and light-commercial systems to drive blowers and fans. Premium systems that offer higher efficiency and dehumidification options are where they are most often found. ECM power consumption is significantly less than that of a comparable AC induction motor. As time goes on, they will likely become the standard offering at all equipment efficiency levels.

There are three basic types of ECMs, all shown in **FIGURE 15-1**:

- A **constant-airflow ECM**, also known as a constant-cfm or constant-volume ECM, is used for indoor blowers in the most efficient comfort cooling and heating systems. They provide exceptional flexibility in their programming and are able to **modulate** their speed to accommodate changes in the **external static pressure (ESP)** of the air-distribution system. In most cases, the data required to perform as desired is contained within the attached module.
- A **constant-torque ECM** provides a preprogrammed **torque** value to the motor shaft as directed by an external control. They are also used in indoor blower applications, as well as some other HVACR applications.
- A **constant-speed ECM** runs at a preprogrammed speed, with one or two speeds usually available. They are not generally used in indoor blower applications, since they offer no real advantage t here. They are more commonly used in axial or propeller fan applications.

> ▶ **CORE CONNECTION**
>
> **Brushless DC Motors**
>
> The motor at the core of an ECM is also referred to as a **brushless** DC motor. Keep in mind, however, that they do require an AC power source. Another term that may be encountered is permanent magnet synchronous motor, or PMSM.This is the same as a brushless DC motor, but it may use a different method of commutating the voltage. Finally, a VFD, or variable-frequency drive, is something altogether different. A VFD is an electronic controller used to obtain precise control and better part-load operation by modulating the speed of an AC induction motor. VFDs are most commonly used with three-phase motors.

Some ECMs, such as the X13 type, look similar to comparable standard motors. The electronics for these motors are integrated into the motor case. Other ECMs, such as the ECM 2.0, 2.3, 2.5, and 3.0 constant-airflow motors, have a separate, removable electronic control module attached to the motor.

A major feature of ECMs is that the electronics in the motor control module can be programmed by the equipment manufacturer to provide equipment-specific performance under a wide variety of operating conditions.

FIGURE 15-1 Examples of ECMs.

▶ Electronically Commutated Motor vs. Standard Alternating Current and Direct Current Motors

An ECM is based on a combination of AC and DC motor construction. Although it is powered by AC voltage, the voltage used to drive the motor is actually a three-phase DC voltage.

K15001 Compare ECMs with conventional AC and DC motors.

Alternating Current Induction Motors

Chapter 14 described the operation of single-phase AC motors, such as the permanent split capacitor (PSC), shaded-pole, and capacitor-start types, as well as three-phase motors. These motors rotate because of a revolving magnetic field created by fluctuations in the alternating current applied to the stator windings of the motor (**FIGURE 15-2**).

FIGURE 15-2 Rotor and stator of an AC motor.

Single-phase motors often need a means of producing additional starting torque. The PSC and capacitor-start motors, for example, use a separate start winding that creates a phase shift. Three-phase motors naturally offer good starting torque because their stator windings are inherently out of phase with each other.

PSC motors, commonly used to drive indoor blowers, have the disadvantage of being unable to directly compensate for reduced airflow that results from variations in ESP. Variations result from zone-damper position changes, dirty air filters, and similar conditions. PSC motors do have multiple speeds to select from, but they are unable to change speed as conditions change. Manual speed selection through a wiring change is generally required. Therefore, for a standard PSC blower motor, the volume of airflow changes as the ESP changes. Specifically, as ESP increases, the air volume is reduced.

Direct Current Motors

DC motors are used in applications that require a high starting torque. DC voltage has a constant amplitude, so the DC voltage applied to the motor cannot develop a revolving magnetic field. In DC motors, it must be done mechanically. The **commutator** (**FIGURE 15-3**) is the answer. It is a segmented ring attached to the **armature** of the DC motor. The DC voltage is applied to the commutator through carbon **brushes** that are spring-loaded to remain in constant contact with the commutator as they wear. As the motor rotates, the split commutator causes the DC voltage to reverse direction. This **excitation voltage** being applied to the wound rotor simulates the action of a revolving magnetic field between an AC motor rotor and stator.

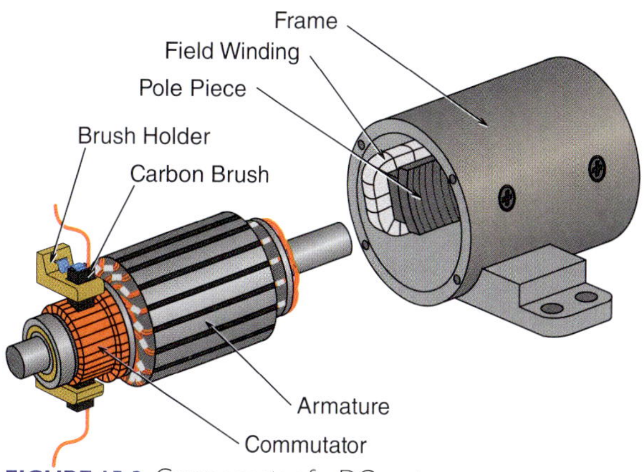

FIGURE 15-3 Components of a DC motor.

FIGURE 15-4 shows a simplified schematic diagram of a DC motor. The physical arrangement of a DC motor is the opposite of the AC motor. In an AC motor, the stator consists of magnets wound with wire and the rotor is a magnet with no windings. In the DC motor, the stator is made of magnets and the rotor is wound with wire.

A common problem with standard DC motors is that the brushes are relatively soft and wear out, though they can be replaced. Another issue with DC motors is that they require a substantial DC power supply, since the power supplied by utilities is AC.

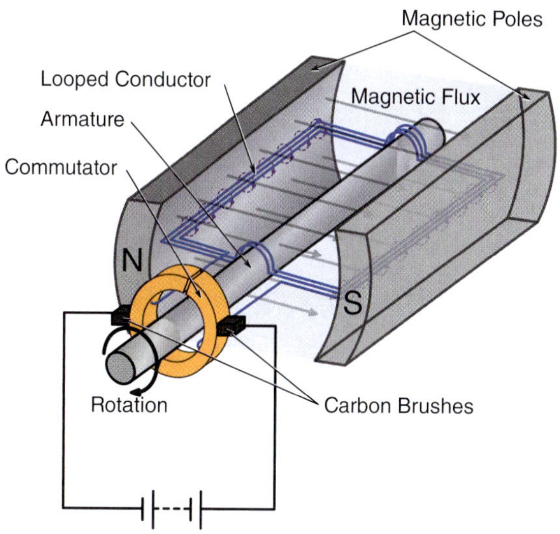

FIGURE 15-4 Simplified schematic diagram of a DC motor.

Electronically Commutated Motors

An ECM (**FIGURE 15-5**) is powered by a three-phase DC voltage that is obtained by rectifying the single-phase AC supply voltage. One important advantage they have over standard DC motors is that ECMs have a brushless design. They use electronics to commutate the DC voltage,

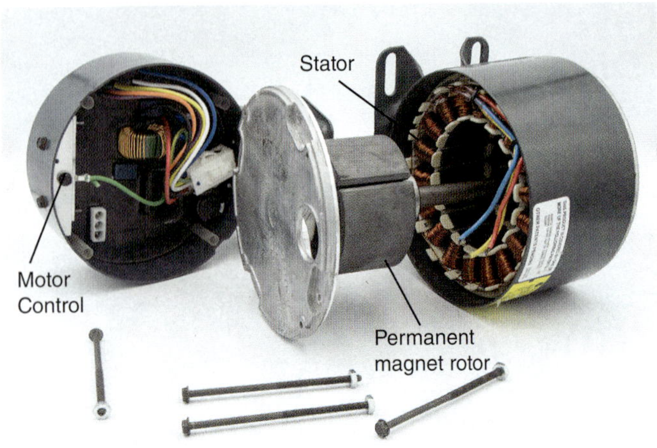

FIGURE 15-5 Components of an ECM.

FIGURE 15-6 An ECM control module.

converting it into a three-phase voltage. The electronics in the control section of the ECM (FIGURE 15-6) convert the AC input to a pulsed DC square wave, possessing a frequency that represents the desired motor speed or torque. Unlike the standard DC motor, the excitation voltage is applied to the stator windings. The electronic control module separates the DC voltage into three distinct segments, then applies them sequentially to the three sets of stator windings, so that the motor essentially acts as a three-phase motor. This is the source of the term "electronically commutated." Whereas conventional DC motors use a split commutator and brushes attached to the rotor to create an alternating DC voltage, ECMs use solid-state electronics to develop a three-phase DC voltage.

As noted earlier, ECMs are designed for constant-torque, constant-airflow, or constant-speed operation. Each of the three types is designed to provide specific operating characteristics. They also represent three different levels of cost. As one might expect, the most versatile ECM, the constant-airflow type, is the most expensive.

▶ Electronically Commutated Motor Operation

K15002 Explain the operation of an ECM.

An ECM is notable for its self-contained electronic controls, which perform both power and control functions for the ECM. The power section of the control module contains a rectifier, capacitors, and filters that convert the AC supply voltage to a DC voltage that is nearly ripple-free. Microprocessor circuits then convert the DC voltage to a three-phase DC voltage that is applied in sequence to the stator windings of the motor, thereby creating a rotating magnetic field. The stator is divided into three segments, similar to the stator of a three-phase AC motor (FIGURE 15-7).

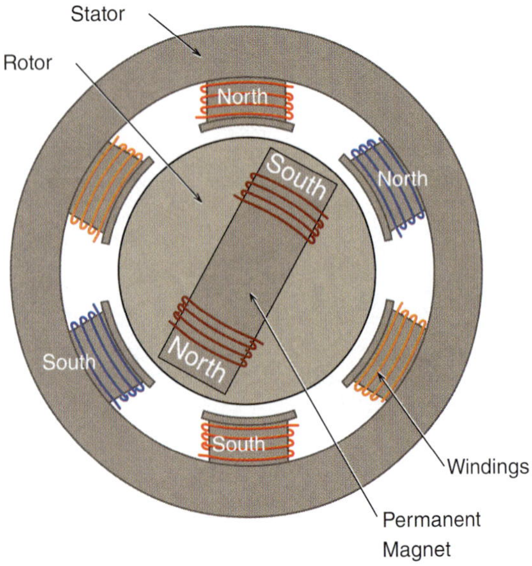

FIGURE 15-7 ECM stator arrangement.

The AC supply voltage is always present at the power connector of the ECM. It is not switched through a traditional blower relay. However, the motor will not run unless it receives a control voltage or signal from the system.

As shown in FIGURE 15-8, a call for heating, cooling, or continuous-fan operation from the thermostat initiates blower operation. The control board in the unit reacts to the call by sending out a command that reflects the fan performance needed. This signal is received by the ECM control module and is used by the microprocessor circuits to **modulate** the voltage applied to the stator of the motor. Most ECMs use 24 VAC control signals, although some use 18 VDC.

The programming of ECMs must be considered based on the characteristics of the HVACR equipment. The programming of the ECM allows designers to select the options that the ECM will be able to execute. For example,

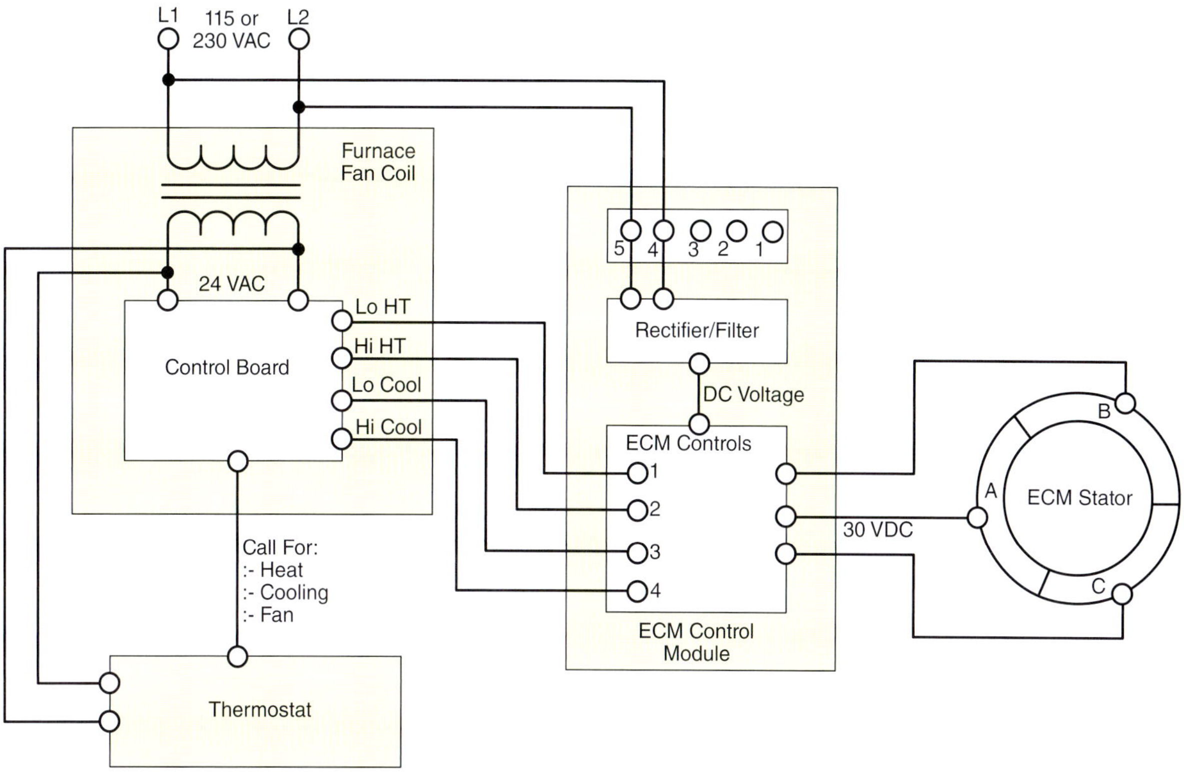

FIGURE 15-8 Simplified schematic diagram of an ECM application.

- a high-efficiency furnace may have several stages of heat, each of which requires a different airflow value
- a furnace or fan coil unit may be designed to function with several condensing units within a range of capacities, such as 2- through 4-ton units.

Constant-Airflow Electronically Commutated Motors

The constant-airflow ECM is the most versatile of the ECMs. Models such as the ECM 2.3, 2.5, and 3.0 can deliver a consistent air volume over a wide range of ESPs. If, for example, a dirty filter or change in damper position results in a higher static pressure, the microprocessor in the ECM control module will compensate by ramping up the torque, thus increasing the blower speed, to ensure that the required amount of air is delivered.

These ECMs have a separable control module attached to the end bell of the motor. They are very quiet and very efficient. On start-up, they begin rotating softly and ramp up to the required speed smoothly. They offer versatility in blower operation to support various modes of operation, to the benefit of the system and the user. A disadvantage is their cost, although this can be recovered over time through reduced operating costs.

A wide variety of airflow options are available, determined by the programming and the selections made at the system control board by using dual in-line package (DIP) switches or pin jumpers. Systems that require different air volumes for the heating and cooling modes are good candidates, and this describes most systems that combine fossil-fuel furnaces with electric cooling systems. Systems that employ a dehumidification mode, which often requires a reduced air volume, are also good candidates.

DIP switches or pin jumpers usually provide the means to select characteristics unique to the system. **FIGURE 15-9** shows a popular PC board that employs jumpers to make selections. Each pin-jumper position on the board connects to a specified set of parameters that determine the motor's operating characteristics. For example, when the capacity of the heat pump or air conditioner is established by a jumper position, a range of airflow values required to support the tonnage is selected. In essence, the unit PC board tells the ECM control module which airflow chart in its memory to follow. Roughly 400 cfm per ton of cooling is a long-standing rule of thumb.

> ▶ TECHNICIAN TIP
>
> **Stall Detection**
>
> A number of things can interfere with a blower or axial fan. This is especially true of condenser fan motors. Debris can fall through the grille and jam the blade, or the mount can break and allow the motor to fall to one side, jamming the blade against the **volute**. Ice buildup on heat pumps can also prevent the fan from turning. Of course, the motor itself can seize due to a failed bearing. To protect an ECM under these conditions, many are programmed with stall-detection programming. If the module is sending power to the motor but the motor isn't turning, power to the motor will be disabled. In most cases, it will try again, but the number of retries may be limited.

FIGURE 15-9 Example of a PC board with jumpers to select configuration characteristics.

▶ TECHNICIAN TIP

Dehumidification

Air conditioning and refrigeration systems remove moisture from the air as a natural part of the cooling process. The evaporator coil of a system used for comfort cooling is usually well below the dew point temperature of the air flowing through it. Consequently, moisture in the air collects in droplets on the surface of the coil, since the cooled air can no longer hold the water vapor in suspension. In some areas of the country, comfort can be improved significantly by reducing the humidity of the room, but what can be done to dehumidify a room when it isn't experiencing a significant heat load? In other words, no significant cooling is needed, but the humidity needs to be reduced.

An ECM, with the ability modulate its speed, offers a productive answer. By operating in the cooling mode but at a reduced airflow setting, the air remains in contact with the coil surface for a longer time. The longer the air is in contact, the cooler it becomes, forcing it to release more excess moisture. The result is that the coil's latent cooling capacity is increased. Another major factor in dehumidification is the thickness of the evaporator coil. Thicker coils also allow the air to spend more time in contact with the coil surface, increasing moisture removal.

The system manufacturer's service and installation instructions provide guidance in selecting the various options. Motors are preprogrammed to satisfy the needs of the equipment manufacturer and are then fine-tuned at the point of installation with the use of switches and jumpers on the unit PC board.

A constant-airflow ECM responds to changes in total external static pressure (TESP). As TESP increases, more resistance to airflow has been added and the motor must respond with more speed and torque to move the same amount of air. With a common PSC motor, the air volume is reduced when TESP increases and the motor has no means to compensate. Note that the current of a PSC blower motor drops as TESP increases, while the current draw increases for an ECM. This is because the PSC motor's operating current is proportional to the load, but the PSC motor sees load as the volume of air moving through the fan, not resistance to airflow. Consequently, the PSC motor current falls along with the volume of airflow when resistance increases.

This is not true of ECMs, since they add torque to compensate and meet the airflow requirement. However, the power factor associated with their DC design still allows them to use less AC current than a typical PSC motor under the same conditions. Note that the noise level also rises as the motor works harder, but this is not usually a problem in a properly designed system.

Manufacturers experiment with motors in specific system combinations. The system is operated against various levels of resistance, and the torque of the motor is adjusted at each test point. This creates a chart of torque requirements to move a given amount of air at various resistance values. Motor speed is also tracked, providing a second chart of data. Once a motor is programmed with the air volumes to be moved, along with the data from the charts, it is ready for operation. The unit will signal the motor to operate in the cooling (or other) mode, and it will start at a charted torque and speed. If the ESP is different than what it expected, the torque and speed will not precisely match the motor's expectations for the desired airflow. Torque is then adjusted up or down, causing a speed change. When the torque and speed values match chart data for the programmed air volume, the motor settles in and runs at these values. No direct data regarding ESP is gathered or shared with the motor during operation. ECMs can be loaded with data for a wide range of air volumes at the factory, making them extremely versatile.

FIGURE 15-10 helps to illustrate how the ECM responds as TESP changes. The red line represents the tested profile of the motor when delivering 800 cfm of air. This is the nominal airflow requirement for a 2-ton cooling system. On command to deliver 800 cfm, the motor starts and operates at Point A, roughly 390 rpm, and 28% of torque is required

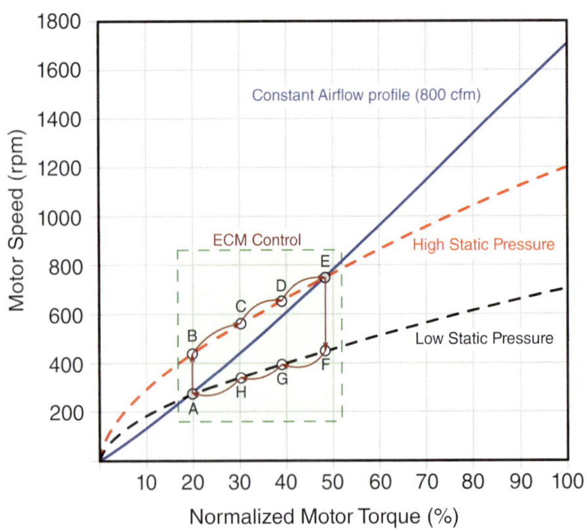

FIGURE 15-10 Example speed/torque chart for a constant-airflow ECM.

for a low TESP. In the example, the TESP increases while the motor is operating (Point B). The result is that the motor speed increases at the same torque; the increase in speed for an ECM is caused by the reduction of resistance and the resulting reduction of air volume, but no change in torque. The motor's programming recognizes that this speed and torque combination do not exist on the plotted red line. As a result, torque is increased and speed is monitored until the combination of the two again lands on the red line (Points C through E). A reduction in TESP while the motor operates (Point F) causes the opposite reaction: torque is reduced until the torque and speed relationship find their way back to the red line (Points G through A).

One note of caution is related to the motor being operated against high static pressures or near the top of its capacity for long periods of time. Pushing these motors to their maximum for extended periods of time has a tendency to decrease ECM-module life. The point is that they should not be considered a solution to poorly designed air-distribution systems.

Constant-Torque Electronically Commutated Motors

A constant-torque ECM, such as the X13, can deliver consistent torque to the motor shaft under changing conditions. The original equipment manufacturer (OEM) determines the level of torque required for various air volumes, then programs the speed taps of the ECM to produce the airflow required at a design static pressure. ESPs above and below the design point do result in fluctuations in the air volume. But the motors are significantly more efficient and effective than PSC motors. The programming may be done as a percentage of maximum torque or entered as a specific torque value for each combination of ESP and air volume. Constant-torque ECMs help increase the **seasonal energy-efficiency ratio (SEER)** of equipment and thus reduce operating cost. In addition, users of these motors may experience improved humidity control.

Constant-torque ECMs are commonly used for driving indoor blowers for residential and light-commercial furnaces, heat pumps, and cooling-only systems. They generally have five speed taps that are activated by 24 VAC signals from the host unit control board. Although the taps may be labeled from High to Low or simply numbered, the labels do not represent speed values directly; they represent torque settings. However, adding more torque does increase speed. The motor is programmed to respond to control inputs at the speed taps by operating at a preprogrammed torque value for a specific mode of operation, such as Stage 1 heating, Stage 1 cooling, or continuous fan operation.

Note that an extreme amount of resistance and a high torque selection could cause the motor to ramp up and over-speed. For that reason, each motor is programmed with a maximum speed limit. Regardless of the demand for torque, the motor will not exceed its programmed limit.

> ▶ TROUBLESHOOTING TIP
>
> **A State of Confusion?**
>
> When an ECM starts, it may briefly rotate in the wrong direction before reversing to rotate in the proper direction. It may also move back and forth several times. There is no cause for concern in this case—unless that's *all* it does. The motor is simply orienting itself to determine the correct direction of rotation.
>
> Also, manually rotating the shaft of an unpowered ECM might give the impression that the bearings are defective. The permanent magnets in the motor makes the effort to rotate the rotor uneven.

Constant-Speed Electronically Commutated Motors

Constant-speed ECMs possess the construction characteristics of other ECMs but are less complex and perform much like a multispeed AC motor. Speed taps provide selection, and when one tap is energized, the motor runs at a predetermined consistent speed. Speed choices vary according to the application and system characteristics and may be programmed by the motor manufacturer or the manufacturer of the related system. However, they may also be designed for a single speed.

Constant-speed ECMs are most often used to drive axial-blade (propeller type) condenser fans or refrigeration-coil evaporator fans. Models have also been developed to replace the compact unit-bearing motors used on small refrigeration evaporators found in reach-ins, beverage dispensers, and similar units (**FIGURE 15-11**). They are typically single- or two-speed motors with speed values programmed to match the application. Because of the limited number of speed taps, these motors can be controlled directly through thermostat signals like their AC induction-motor counterparts.

FIGURE 15-11 Constant-speed ECM for small refrigeration evaporator fans.

▶ TECHNICIAN TIP

ECM Retrofits

Retrofitting motors that drive axial fans with ECMs is growing in popularity. Constant-airflow ECMs are unnecessary for these applications. Constant-speed ECMs are the better and less expensive choice. Several motor manufacturers offer ECMs specifically designed for the field replacement of standard motors. Replacing existing induction motors with ECMs to power axial evaporator fans on refrigeration systems can result in improved motor life and operating efficiency, which reduces the operating cost. Typical payback for a retrofit is reported to be less than one year when tax credits and/or rebates further offset the cost. Retrofits in ducted comfort systems may not be as effective, because a high ESP may reduce the operating efficiency and therefore the return on investment.

K15003 Describe the common wiring connections used with ECMs.

▶ Electronically Commutated Motor Connections

Each type of motor has its own unique electrical connections. At the same time, they also have some similarities in that terminals are provided for power inputs and communication or speed inputs.

Constant-Airflow Electronically Commutated Motor Wiring Connections

Constant-airflow ECMs have two separate and unique sockets for external wiring connections (**FIGURE 15-12**). One is a five-pin connector that receives the 120 or 240 VAC line voltage. The other connector receives control signals from the equipment in which it is located. ECM 2.3 and 2.5 constant-airflow models have a 16-pin signal connector, whereas the ECM 3.0 model constant-airflow ECM uses a four-pin connector for the same purpose.

All three versions use a wire jumper to configure the motor for 120 VAC operation in lieu of 240 VAC (**FIGURE 15-13**). The jumper wire serves to reconfigure the internal rectifier for the applied voltage. The jumper is left off for 240 VAC operation. Note that the wiring connectors often have locks to keep the connection from separating.

The 4- or 16-pin connections are used for communication. There are three primary communication approaches used with these motors:

1. *Thermostat mode*: The equipment controls send inputs to the motor's electronics based on the thermostat inputs it receives. The voltage received could indicate that the system wants to run in its first stage of cooling or perhaps its second stage of heating. Preprogrammed data within the motor's electronics are then consulted, and the motor is instructed to run accordingly.
2. *PWM mode*: **Pulse-width modulation (PWM)** signals are also developed by the equipment controls and transferred to the motor electronics as a DC voltage. The digital signal is pulsed (off and on), meaning that the voltage is there momentarily and then it's not, repeatedly. The amount of time it is pulsed on is generally an indication of need. For example, if the pulse is present to the motor electronics 70% of the time, the motor responds by operating at a point that represents 70% of its maximum capability. Like the thermostat mode, the necessary data for each individual installation is stored in the motor electronics.

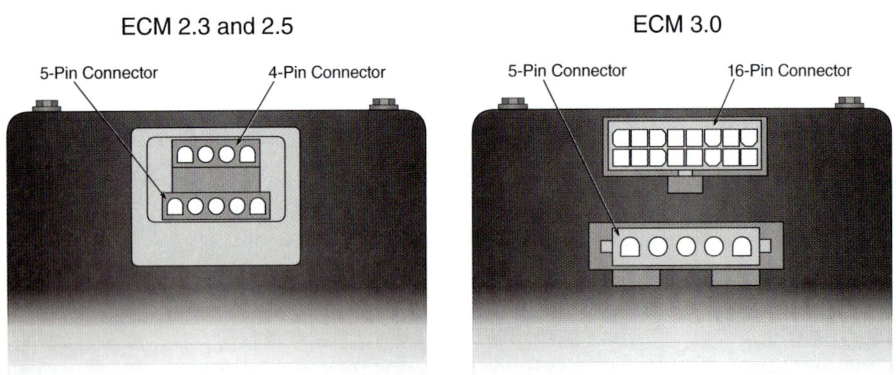

FIGURE 15-12 Connection sockets on constant-airflow ECMs.

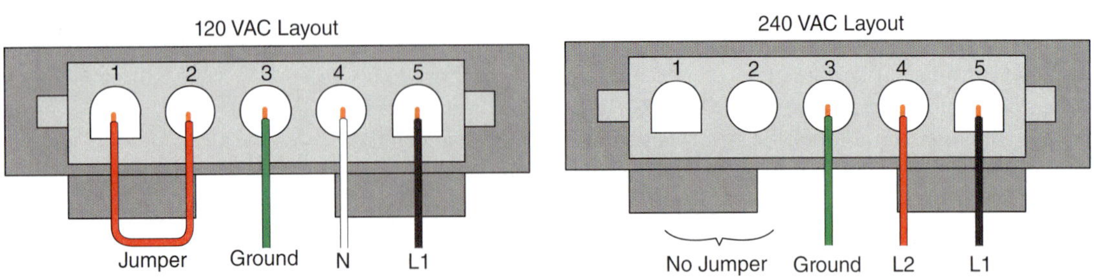

FIGURE 15-13 Power connections for a 120- and 240-volt constant-airflow ECM.

	Communication Connector Pin Allocation	
	Thermostat Mode	PWM Mode
1	Common	Common
2	W1 (first stage heat)	not connected
3	Common	Common
4	Delay/Profile select	not connected
5	Cool select	not connected
6	Y1 (first stage A/C or heat pump)	not connected
7	Adjust select	not connected
8	Output -	Output -
9	O (reversing valve)	not connected
10	BK (dehumidification)	PWM signal
11	Heat select	not connected
12	R (24ACV)	not connected
13	W2 (second stage heat)	not connected
14	Y2 (second stage A/C or heat pump)	not connected
15	G (continuous fan)	Start/Stop Signal
16	Output +	Output +

FIGURE 15-14 Thermostat mode and PWM mode communication examples.

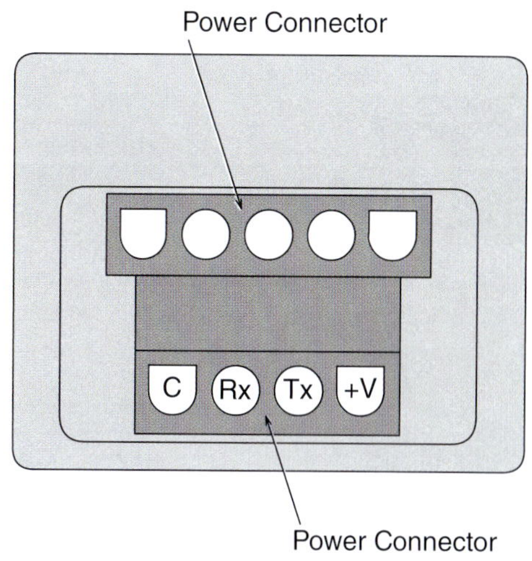

FIGURE 15-15 DSI Mode communication connection.

3. *DSI mode*: DSI stands for digital serial interface. Serial communication has been in use with computers for many years. Binary code, a unique series of 1s and 0s, is used to communicate information. This is the simplest and least versatile communication approach. In this case, operating details are not stored within the motor electronics; all instructions come from the equipment controls.

FIGURE 15-14 shows an example of communication wiring used for 16-pin motors controlled through the thermostat mode and the PWM mode. Note that the thermostat mode makes use of many of the 16 pins, whereas the PWM mode does not.

FIGURE 15-15 shows the configuration of the four-pin connector style, which is primarily used for DSI communication. However, there are a few that use this connector style with PWM communication. The terminals are typically used as follows:

- *C*—common or ground
- *Rx*—the data input from the equipment controls to the motor electronics
- *Tx*—the data output to send information back to the equipment controls
- *+V*—the voltage path for data output.

Constant-Torque Electronically Commutated Motor Wiring Connections

Constant-torque ECMs have two rows of terminals provided in a single opening in the shell. X13 constant-torque motors use three pins in the upper row for the power source (**FIGURE 15-16**). Those pins are designated L, G, and N—representing line, ground, and neutral. The terminals are used as shown in the figure to accommodate voltages of 120–460 VAC.

Note that the C terminal is not used for line-voltage power wiring. The common lead from the 24 VAC control transformer is connected here. Also note in Figure 15-16 that the three terminals designated L, G, and N appear smaller than C and all the terminals in the bottom row. They are smaller for good reason. Line voltage is applied to the three smaller

▶ TECHNICIAN TIP

They Look the Same, But ...

It is important to remember that each constant-airflow ECM has been uniquely programmed to serve the needs of the specific HVACR equipment model. For example, an ECM programmed for a Rheem gas furnace is unlikely to have the same programming in it as the motor for a comparably sized Trane furnace. When looking at the two motors, there may be no visible difference other than labeling. However, the two will not perform the same way, even if the method of communication is compatible. For this reason, a failed ECM must be replaced with one that is appropriate for that specific unit, including its programming. A growing service offered by wholesalers is to maintain an inventory of unprogrammed constant-airflow ECMs and then provide unit-specific programming at the point of sale. To provide this service, wholesalers must access the equipment manufacturer's programs for specific equipment models and have the equipment to do the job. However, note that manufacturers release programming information only to their network of authorized distributors. In other words, an equipment wholesaler aligned with one brand cannot assist with the programming of a motor for another brand. Perhaps this will change as ECMs become increasingly common.

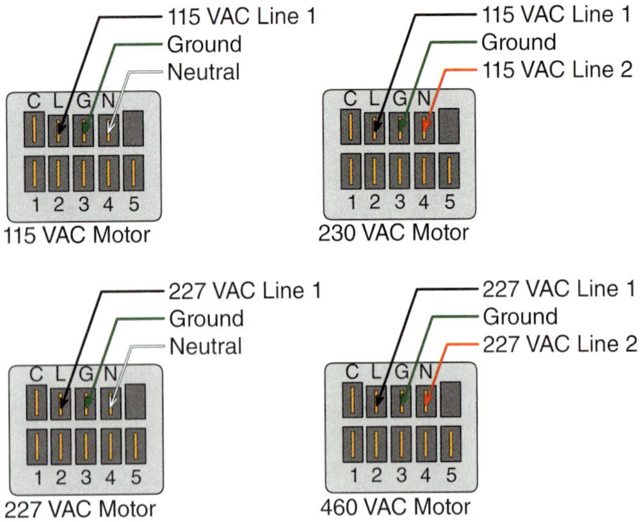

FIGURE 15-16 Power wiring for an X13 constant-torque ECM.

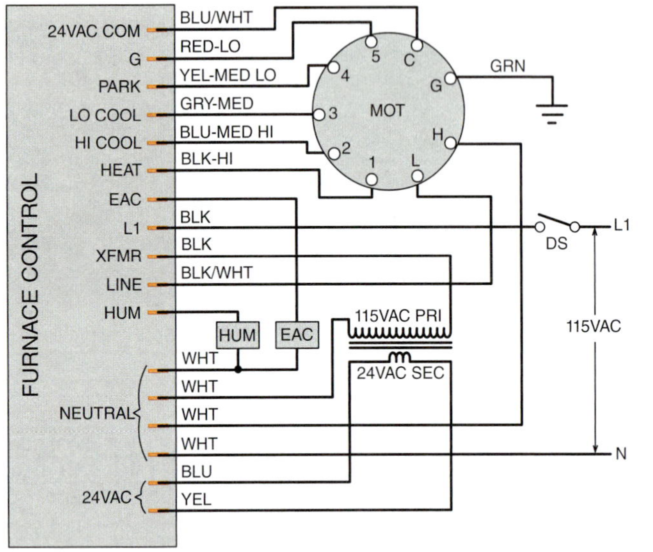

FIGURE 15-17 Example of system wiring for an X13 constant-torque ECM.

terminals, and only the much lower control-circuit voltage is applied to the remaining terminals. The difference in pin size helps prevent wiring errors.

The wiring harnesses from the HVACR unit are designed so that they can fit into the plug on the ECM only if they are properly oriented. It is impossible to insert them upside down. Once inserted, the connector must be firmly pushed in to ensure it latches in place.

FIGURE 15-17 shows a diagram of the system connections for an X13 ECM. Note how the five pins on the lower row are used to select a specific torque value programmed into the motor.

Inside the motor assembly, ECMs that have a removable control module are equipped with a wiring harness with molded plugs to interface with the motor (**FIGURE 15-18**). To separate the two components for replacement of one or the other, this harness must be disconnected.

Constant-Speed Electronically Commutated Motor Wiring Connections

Constant-speed ECMs are primarily used to drive axial fans such as condenser fans and refrigeration-system axial evaporator fans. Their single- or two-speed design is similar to that of AC motors that have served in these roles for years. However, like other ECMs, the motor is continuously provided with line voltage, as long as the system is energized. The command to operate comes from a 24 VAC input or through PWM, which is similar to the PWM operation of a constant-airflow ECM.

FIGURE 15-19 shows the motor and electronic module separated. The black cable electrically connects the two pieces. It also allows the module for this type of motor to be

FIGURE 15-18 Motor-to-electronic module wiring connection.

FIGURE 15-19 Constant-speed ECM with motor and module separated.

installed away from the motor. System connections are made using the preinstalled leads from the electronic module. These motors do not use wiring harness connections or plugs.

FIGURE 15-20 shows the three wiring arrangements for constant-speed motors. For two-speed operation using 24 VAC control, one tap or the other is energized with 24 volts. The motor will operate at the speed corresponding to the energized tap. For PWM control of two speeds, the system is usually programmed with just two PWM pulse patterns, each corresponding to one of the two speeds.

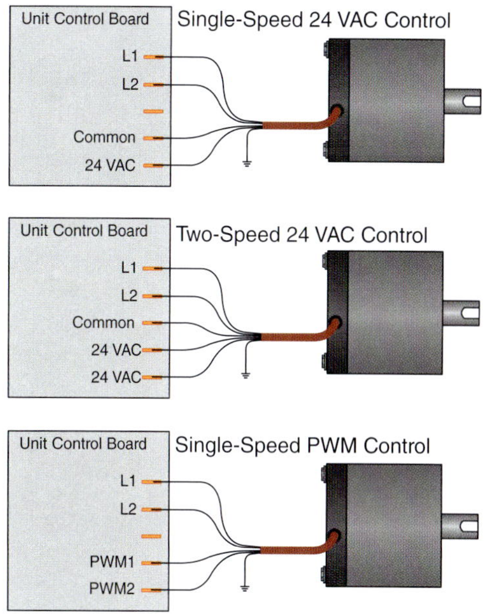

FIGURE 15-20 Constant-speed ECM wiring configurations.

▶ Troubleshooting Electronically Commutated Motors

Before condemning an ECM, it is essential to verify the failure. A series of tests similar to those used in troubleshooting induction motors can be used for this purpose. These tests are used to accomplish the following:

- verify that the motor is configured for and is receiving the correct line voltage
- determine whether the host equipment is supplying the required communication signals
- determine whether the motor windings are grounded, shorted, or open.

One of the most important factors in successfully troubleshooting ECM motors is the acquisition of the right documentation. As has been shown already, there are a number of unique wiring configurations and methods of communication. While this publication is being developed, manufacturers are undoubtedly working on something new. As mentioned previously, most of these motors are programmed with data as specified by the equipment manufacturer. Although the troubleshooting process for each major type of ECM is similar, don't expect to make accurate determinations without the proper literature in hand. Flying blind, thinking that they are all the same, is not the way to go. Acquire the needed troubleshooting information. The equipment manufacturer is typically the best source because it knows specifically how the motor was tested and programmed. ECM manufacturers also offer literature for this purpose, as do the manufacturers of special test instruments designed for ECM troubleshooting.

K15004 Explain how to troubleshoot ECMs.

S15001 Troubleshoot an ECM.

Testing for Power

Unlike standard motors that receive power only when they are operating, ECMs receive power any time the system is energized. Troubleshooting an unresponsive motor usually begins with testing for the proper line voltage at the motor connector. Since power is always there, this test does not require that the system fan be commanded on by the thermostat or other control.

Start by determining whether the ECM is intended for 120–240 VAC operation. It is powered by the same line voltage provided to the host equipment. Use the OEM installation instructions to determine how the motor-module supply voltage connection should be configured for the provided voltage. As previously shown, this can vary from one motor type to another. For example, constant-airflow ECMs use a jumper wire between Terminal 1 and Terminal 2 for 120 VAC, but no jumper for 240 VAC. Constant-torque ECMs use a plug that fits into an otherwise unused opening on the motor connector to configure the voltage. Constant-airflow ECMs always use wiring harness connections, but constant-torque ECM connections may be made through a wiring harness or by connecting individual wires to the terminals. One possibility for a problem is that a jumper was left in place when it should have been removed, or vice versa. This would likely damage the motor, but it is not a given. Even if this condition is found, motor testing should proceed. Of course, this error is rarely found in equipment that has been operating properly.

To check for the correct input power, first turn off power to the unit, then disconnect the five-pin connector or power leads.

Restore the system power and use a multimeter to check for the proper voltage at the correct wiring harness pins (**FIGURE 15-21**). Do not jam the probes into connectors or behind pins; this can damage the connector or pin and create a new set of problems. Work

SAFETY TIP

Never connect or disconnect the harness or leads with power applied, because it could result in an arc that damages the motor electronics. Always deenergize the system first and use a meter to confirm the power is off. Once the connector is removed, wait five minutes before touching the motor connectors with a meter or opening/removing the motor module. This allows time for the internal capacitors to discharge.

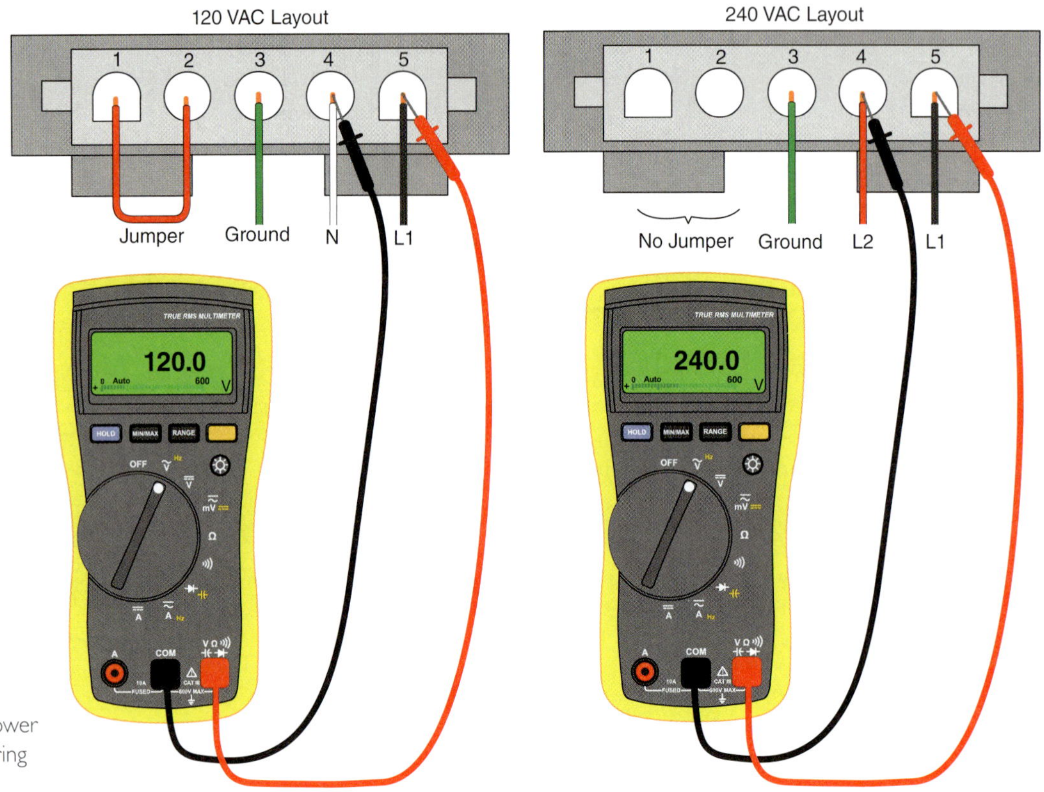

FIGURE 15-21 Testing for power at a constant-airflow ECM wiring harness.

in a way that keeps hands free, so that both hands can be used to manipulate the probes, and simply touch each meter probe to the pins. As a general rule, the voltage should be within ±10%.

For constant-torque ECMs, if there is no harness connection (individual wires connected to the terminals), the power can be tested at the motor-module terminals (**FIGURE 15-22**). Otherwise, the harness is disconnected and tested at the harness pins.

If the line voltage is missing, the problem is not with the motor but with the power source for the motor. If the voltage is good, then the next step is to test the electronics module and communication with the host equipment.

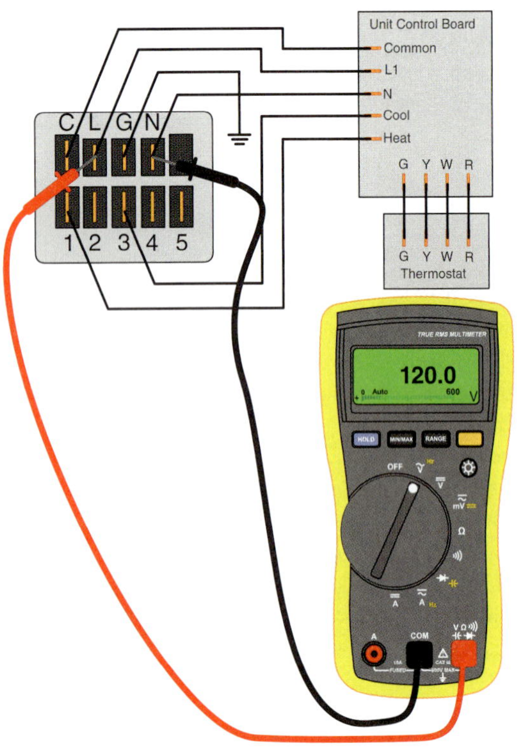

FIGURE 15-22 Testing a constant-torque ECM for line-voltage power.

Testing Control Communications

Since each type of ECM has its own unique features, there are some differences in how the modules are tested. Following this section, testing of the motor itself will be covered. Motor testing for all types is relatively simple and more consistent in the approach than module testing.

Constant-Airflow Electronically Commutated Motor Communication Testing

Before proceeding, note that some equipment manufacturers provide diagnostic tools built into their PC boards to assist in ECM testing. Again, always check the equipment manufacturer's documentation first and follow the guidance provided. The procedure discussed here assumes there is no built-in diagnostic tool provided and that there is no indication of a problem with the host equipment.

For constant-airflow motors, there are special diagnostic tools available for testing. These test sets are simple to use, versatile in that they can test most of the constant-airflow motor types, and reasonably priced.

FIGURE 15-23 shows one popular type. The TECINspect diagnostic tool is designed to test ECM 2.0, 2.3, and 3.0 motors, as well as some other models. It comes with a 16-pin connector that fits the 2.0 and 2.3 modules and now includes a four-pin adapter for testing ECM 3.0 modules. Note that it was formally called the TECMate PRO, which is now out of production but widely distributed among active technicians. The Universal Zebra tester (not pictured here) appears to have more testing features, but regardless of which tool is chosen, ensure that it is fully compatible with the motor and module being tested.

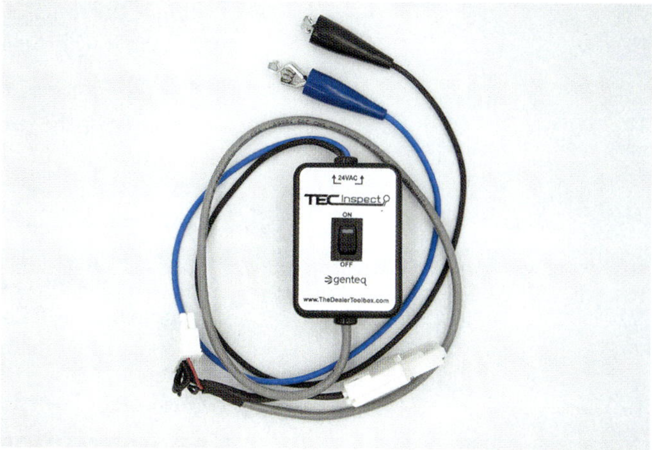

FIGURE 15-23 ECM test sets.

If a diagnostic device like this is not available, a voltmeter can be used to determine whether appropriate control voltages are being applied to the motor control module. Both AC and DC voltage will need to be tested. To do so, refer to the specific OEM testing instructions to determine which pins will receive a control voltage and what operating conditions initiate the signal. Disconnect the 16- or 4-pin connector and perform the tests at the pins of the wiring harness, not at the equipment PC board. Testing at the harness also tests the interconnecting wiring and connections to the board. This type of testing is often referred to as pin-out testing.

Although the information may be available for detailed testing of each pin for the proper voltage at the proper time, it can be a tedious task and often requires the technician to follow a complex wiring diagram. At the same time, it can be a great learning experience, as long as such training isn't done at a client's expense. Since pin-out testing procedures can vary widely from system to system, troubleshooting here will revolve around the use of the TECINspect diagnostic tool.

Before connecting the tool, verify that the proper voltage is being provided to the motor as described earlier. After reconnecting the power connector but before restoring power to the unit, connect the TECINspect, as shown in **FIGURE 15-24**. The power connector to the motor remains connected throughout the test, because it is needed to power the motor. The switch on the tool remains in the Off position until power has been restored.

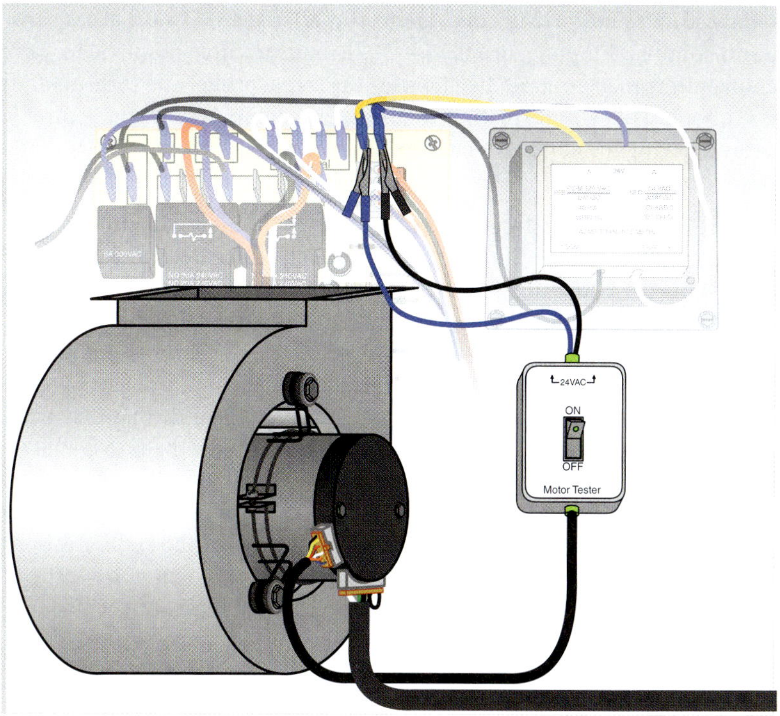

FIGURE 15-24 The TECINspect tool connects to the communication connector of the motor and 24 VAC power from the unit.

Once the tester is plugged into the motor, provided with a 24 VAC power source, and turned on, it will run a preprogrammed test. The results are interpreted as follows:

- If the motor operates at any speed at all, the motor and module are not the likely source of the problem. From this point, testing and inspection should focus on the unit wiring harnesses and control devices, including the PC board communicating with the motor.
- If the motor fails to run at all, it is likely that the ECM control module and/or motor has failed. To determine whether one or both components have failed, continue with motor testing as described in an upcoming section. In most cases, just one of the components has failed. If the motor tests are good, then it is likely that the module can be replaced while the existing motor remains.

Constant-Torque Electronically Commutated Motor Communication Testing

Since the constant-torque ECM communicates through speed taps, a diagnostic tool is not necessary. When the speed taps are energized, the motor is commanded to run at the preprogrammed torque value. Do remember, however, that the speed taps can be energized by either 24 VAC or 15–33 VDC. The vast majority operate using 24 VAC, but before beginning, it is essential to know how they are powered. In the procedure discussed here, we will assume that the voltage is 24 VAC.

There are several important points to remember about the speed taps:

- Not all speed taps are used in every application. For instance, a manufacturer may choose to use only three of the five available taps. This information is usually shown on the unit wiring diagram or on the motor labeling. Energizing an unprogrammed speed tap gets no response from the motor.
- Some systems may energize more than one speed tap at the same time, for a variety of reasons. This doesn't negatively affect the motor; it will simply respond to the speed tap that is requesting the highest torque.
- Some motors may be programmed with a delay period when commanded off. Many manufacturers, for example, operate the indoor blower for a brief period after the condensing unit or heat pump has been shut down by the thermostat. The reason is to extract all the stored cooling capacity possible from the boiling refrigerant still in the evaporator coil. If it is part of the motor programming, it can't be eliminated or adjusted. The motor will continue to run after the thermostat has satisfied and the run command is gone, usually for 30 seconds to three minutes. However, some manufacturers may program two taps for the same torque, one with an off delay and one without. This is information that only the equipment manufacturer can provide, and having the correct documentation is crucial if changing this characteristic is desired.

Again, testing begins by testing the power supply to the motor. If power is not available, that must be corrected first. Additional testing may not be needed once power is restored.

Begin testing the communication between the motor and system by deenergizing the system, waiting five minutes, and then disconnecting the wiring or wiring harnesses from the motor. If individual wires are connected instead of harnessed with connectors, the line-voltage power wiring can remain connected to the motor. All wiring related to the 24 VAC (or 15–33 VDC) inputs must be disconnected. Remember that they will be energized during testing, so keep them controlled and do not allow them to contact anything else conductive as testing progresses.

With the communication wiring disconnected, system power can be restored. Determine which of the wires or terminals will be energized when the system is operating in one mode or another, and then identify the 24 VAC Common that was attached to the C terminal of the motor. All 24 VAC readings will be taken between this wire and the other speed-tap wires.

Operate the system thermostat and call for heating or cooling, one operating mode and one stage of heating or cooling (if more than one stage exists) at a time. Check for

24 VAC power between Common and the speed-tap wire that should be energized (**FIGURE 15-25**). In this figure, only two of the five speed taps are active, so there are only two modes of operation to check. Also note that the motor terminals are shown as the point of measurement in the figure, but the measurements would be taken at the pins of the wiring harness if one is present.

If 24 VAC power is present on the proper wires or terminals for each mode of operation, the equipment controls and interconnecting wiring are not the problem. The module and/or the motor is likely defective. Further testing is required to determine whether one or both components are at fault.

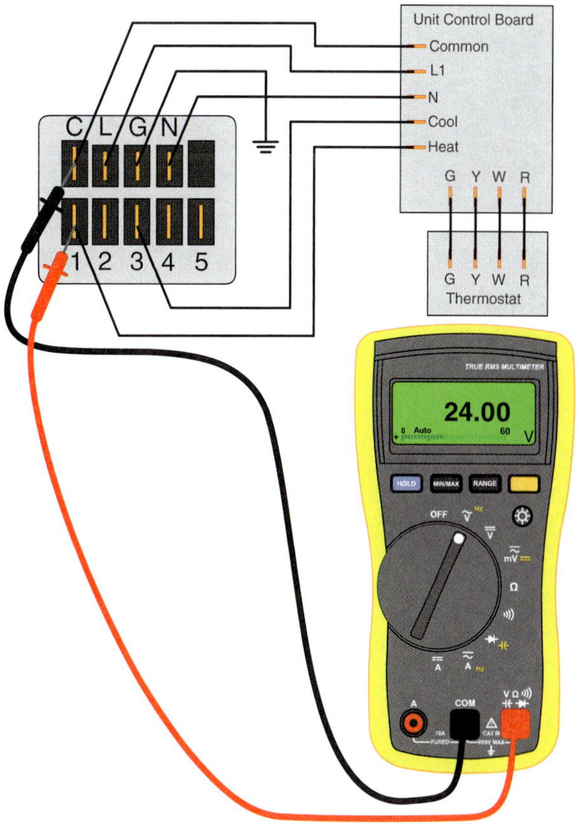

FIGURE 15-25 Testing for communication signals to a constant-torque ECM.

Constant-Speed Electronically Commutated Motor Communication Testing

For constant-speed ECMs, the command to run at a given speed is provided in the same manner as constant-torque ECMs. Most are commanded to run by a 24 VAC signal, although a few are commanded to run by 0–10 VDC signals, and possibly using the PWM approach. These are rare, however. This is yet another situation when it is essential to have the correct documentation from the manufacturer before diving in.

Constant-speed ECMs are generally limited to just two speeds and many have only one speed. Since these motors are most often found driving an axial fan, a condenser fan motor rather than a blower motor is probably being tested. The motor may also be mounted on a refrigeration evaporator coil.

Since these motors do not typically have wiring harness connections, testing is done where the individual leads connect to the host equipment. All ECM wiring can usually remain connected.

With the power supply to the motor already confirmed, operate the related system to command the motor to run. If the motor is not running, check for the appropriate voltage between the Common terminal and the speed-tap terminal (**FIGURE 15-26**). If 24 VAC is present but the motor isn't running, the module and/or motor have likely failed. Additional testing is needed to determine whether one or both components have failed.

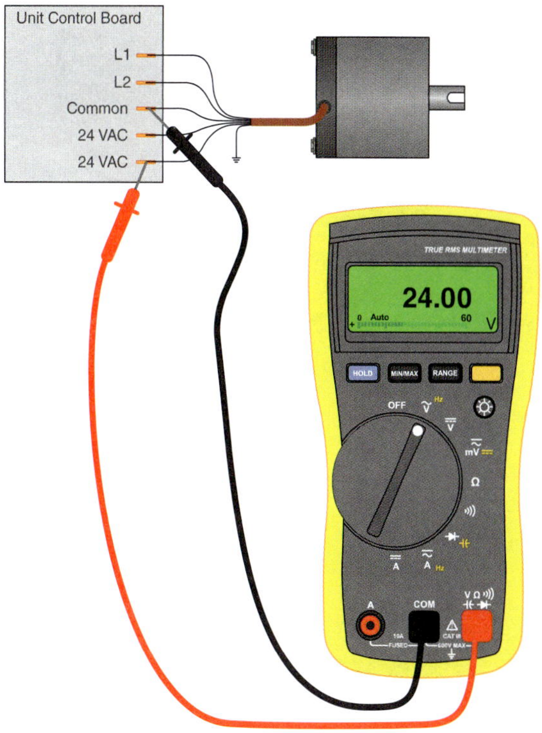

FIGURE 15-26 Testing for a communication signal to a constant-speed ECM.

Electronically Commutated Motor Winding Tests

Each of the preceding sections related to communications testing ends with mentioning the need for additional testing to determine the condition of the motor itself. When a motor with power does not respond to communication signals, it does not necessarily mean that the module is defective. If the motor itself has failed, it makes no difference what the module commands it to do. Sometimes only one component has failed, and sometimes both have failed. Indeed, one may have failed due to the failure of the other.

Direct testing of the motor requires that the motor and module be separated as shown in **FIGURE 15-27**. The system must be deenergized and the wiring harnesses disconnected first. The module is mounted to the motor with a pair of long, slender bolts. Wait five minutes after the motor has been deenergized, then remove the bolts and cradle the module to prevent it from falling. Then disconnect the three-pin connector that electrically connects the motor to its module.

With the module separated from the motor, the motor windings can be tested. Refer to **FIGURE 15-28** for testing the integrity of the motor windings phase to phase.

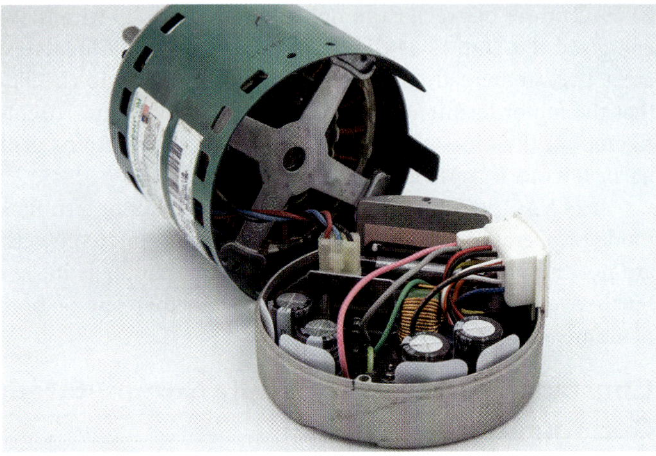

FIGURE 15-27 Separating the module from the motor.

- There are three pins. Consider them to be Pins 1, 2, and 3. Check for continuity between 1 and 2, 1 and 3, and finally 2 and 3.
- The readings must be less than 20 Ω and within 10% of each other. In Figure 15-28, the average of the three readings is 5.3 Ω. All readings must be within 0.53 Ω (10%) of this average, and they are.
- If any of the readings indicates an open circuit, the motor is defective and must be replaced. Readings outside of the 10% tolerance from the average generally indicate a winding is shorted turn to turn.

Next, test the windings to determine whether they are grounded. Refer to **FIGURE 15-29**.

- Place one probe of the ohmmeter on a clean, paint-free metal part of the motor frame.

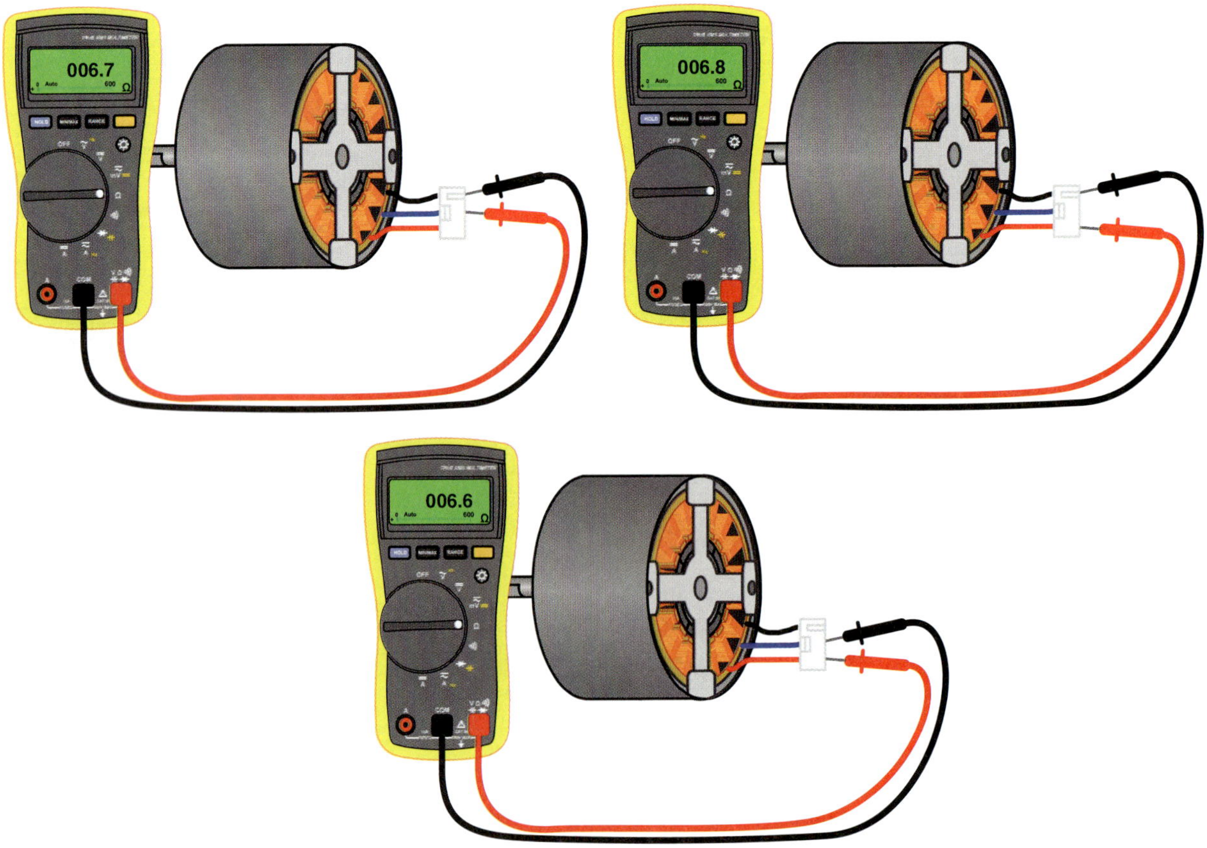

FIGURE 15-28 Testing the motor windings for continuity.

- Touch the other meter probe to each pin, one at a time. There should be no indication of a closed circuit to ground. If there is a reading but it is above 100 kΩ, the motor is OK. Any reading below that value indicates a grounded winding. However, a high reading is relatively rare. Once a motor winding becomes grounded, the resistance is usually much lower than 100 kΩ and there is no question that it is grounded.

Of course, the motor can also suffer a failed bearing or a similar mechanical breakdown. Before reassembling the motor and module, grasp the shaft and turn it by hand. If it is seized and won't turn, or if it takes quite a bit of power to move it at all, replace the motor. Note, though, that the shaft of this type of motor does not generally spin freely like an induction motor. As the shaft is manually turned, resistance will be felt as it passes through the strong magnetic fields of the permanent magnets. In between these areas, the shaft turns easily. This is normal for an ECM.

To practice how to troubleshoot an ECM, follow the steps in **SKILL DRILL 15-1**.

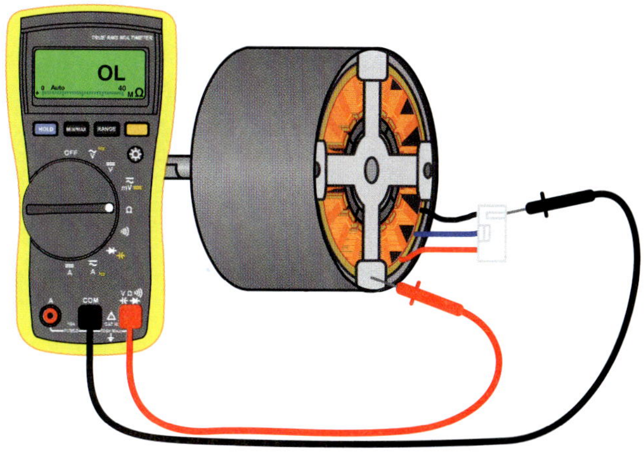

FIGURE 15-29 Testing for grounded windings.

SKILL DRILL 15-1 Practicing How to Troubleshoot an ECM

This drill is based on a constant-airflow ECM tested with the TECInspect or TECMate PRO diagnostic tool. Remember that this particular tool is not designed for testing of ECM models 1.0 or 2.5. For other diagnostic tools, refer directly to the tool's documentation for testing procedures. If direct pin testing is to be done, refer directly to the equipment manufacturer's documentation for the procedure.

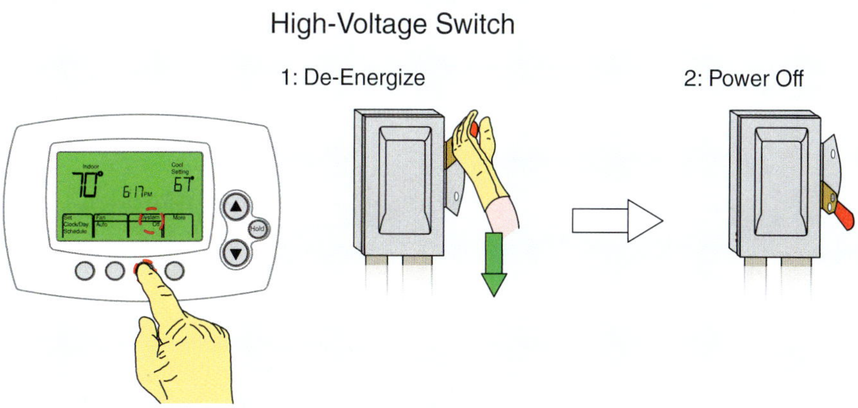

1. Turn the system off at the thermostat, deenergize the unit, and us a voltmeter to verify that power has been disabled. Note the power supply voltage serving the equipment, which would be the same for the ECM.

SAFETY TIP

Working with electrically energized circuits is hazardous. Avoid contact with any surface that could potentially be energized. When working inside energized enclosures, maintain a stable position and, whenever possible, avoid placing both hands inside the enclosure. All work on energized equipment by unqualified individuals should be supervised by a qualified instructor or technician.

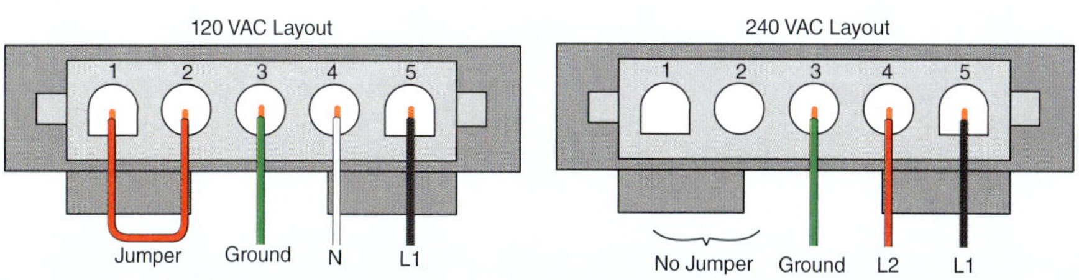

2. Examine the power wiring connections to the motor and verify that the motor is correctly configured for the applied voltage, i.e. a jumper is present between Terminals 1 and 2 for 120 VAC or not present for 240 VAC.

Continued

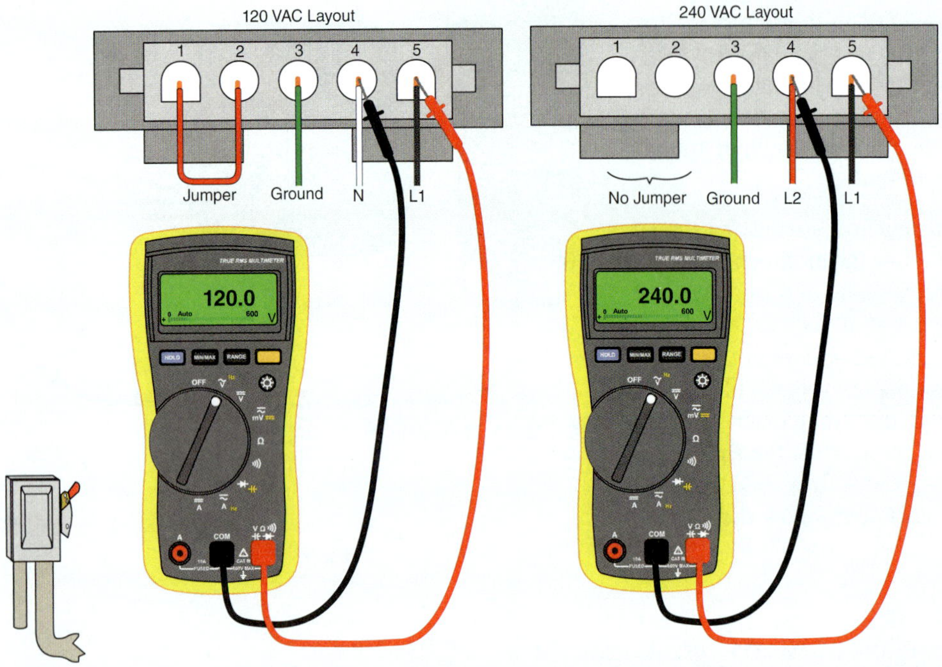

3. Allow several minutes to pass before disconnecting the wiring harnesses at the motor, providing enough time for the internal capacitors to discharge. Disconnect the two harnesses from the motor, ensure the harness pins cannot touch anything conductive, and then restore power to the unit.

Use a voltmeter to check for the proper voltage at Pins 4 and 5 of the harness connector. If the expected voltage is present and within 10% of the nameplate power-supply requirement, the power source to the motor has been verified.

SAFETY TIP

The pins of the power wiring harness are now energized, as is the rest of the unit; proceed with caution.

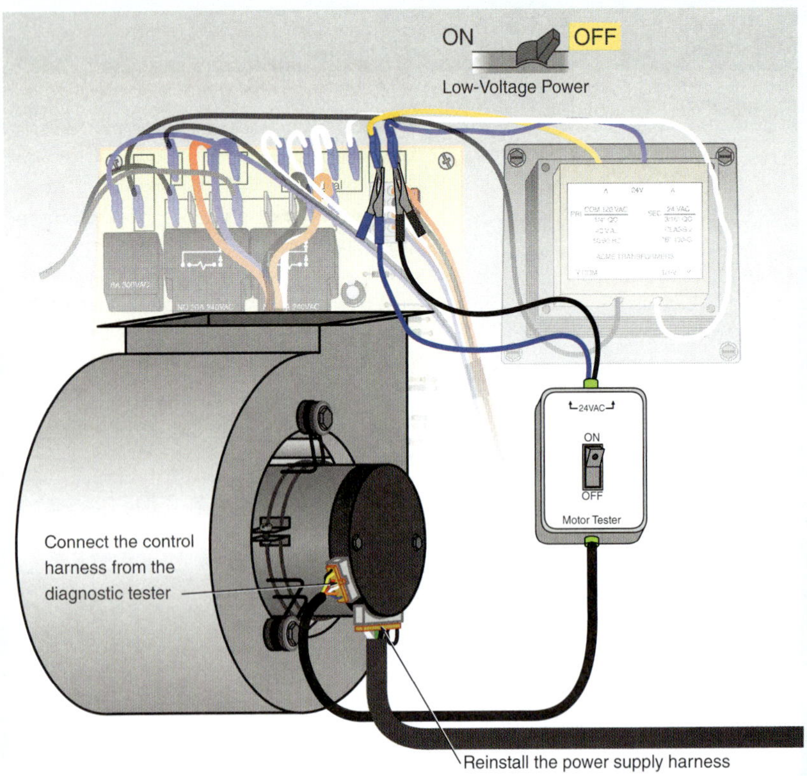

4. Once the motor's power supply has been verified, deenergize the system again. Reconnect only the power wiring harness to the motor so that the motor can run during the next step of testing. Connect the diagnostic tool to the motor by using either the 4- or the 16-pin connector, as required by the motor configuration. Ensure that the diagnostic-tool power switch is in the Off position. The alligator clips of the test tool are then attached to a 24 VAC power source, such as the terminals of the transformer.

Continued

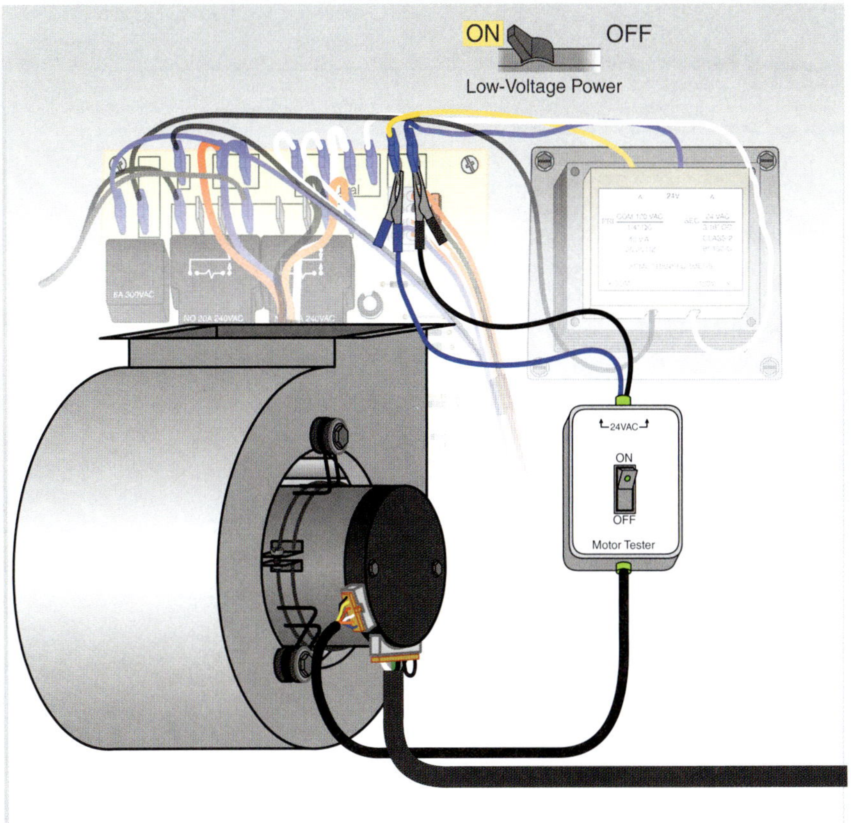

5. Restore power to the system and make sure the indicator light on the tool is on to indicate the presence of 24 VAC. The light should be On even though the power switch is still set to Off.

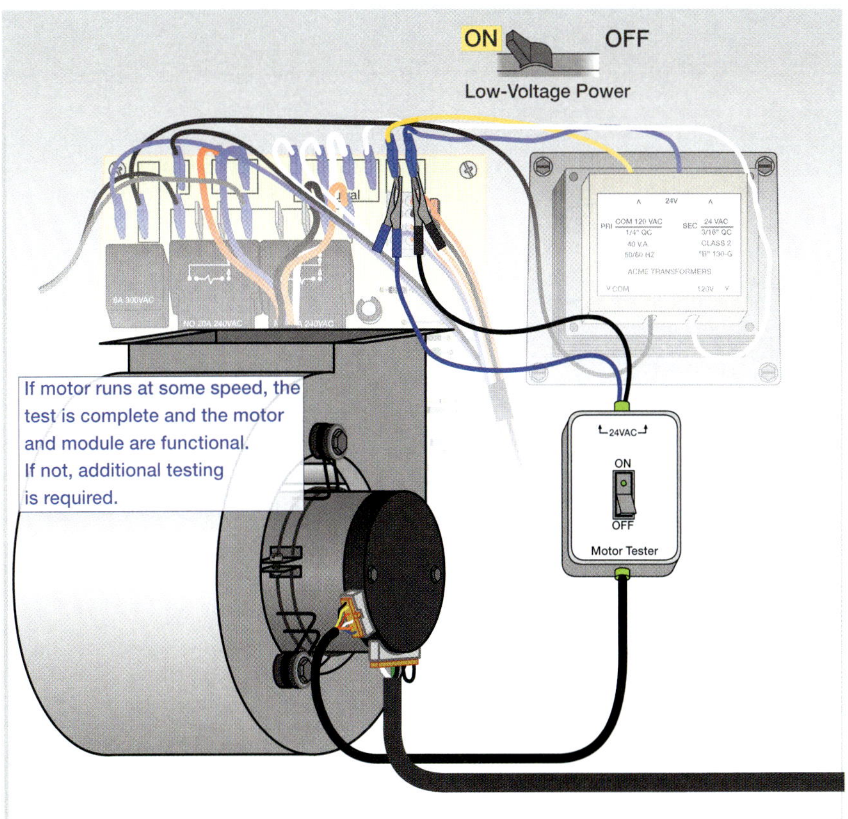

6. Ensure that the blower motor is clear to operate without obstructions. Slide the diagnostic switch to the On position. If both the motor and module are functional, the motor should start and turn at some reasonable speed. If so, the troubleshooting of the ECM is complete. Any problem with motor operation that the system still experiences is likely in the equipment controls. If the motor does not run, then the motor and/or module have failed. In this drill, we will assume that the motor did *not* run as expected.

Continued

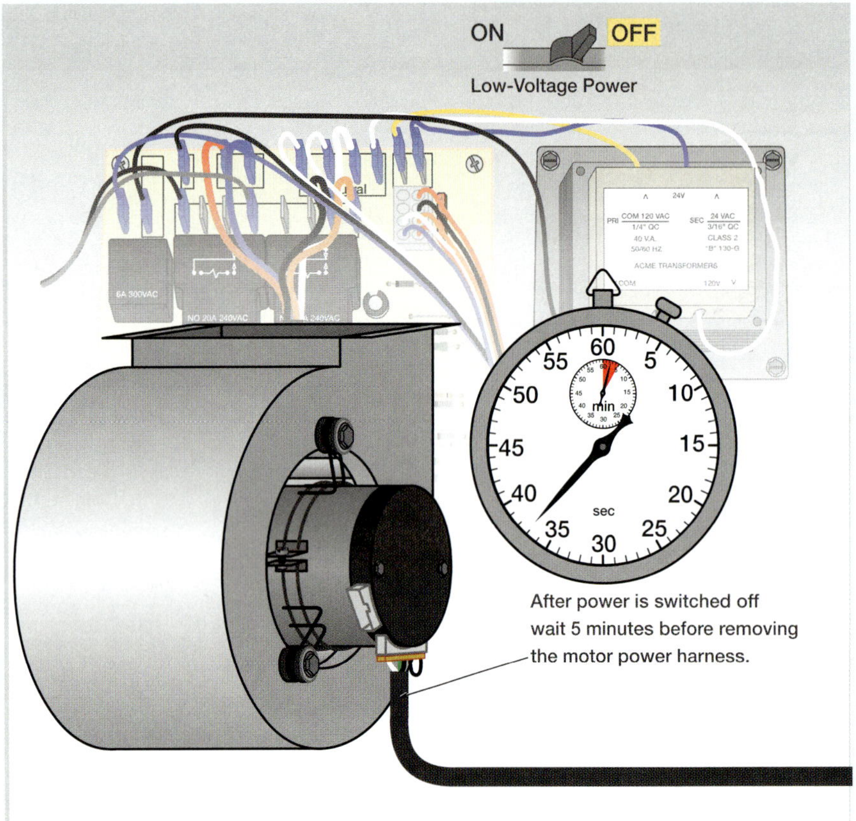

7. Deenergize the system and verify that power has been disabled with a voltmeter. Disconnect the diagnostic tool completely. Wait five minutes and then disconnect the power wiring harness from the motor.

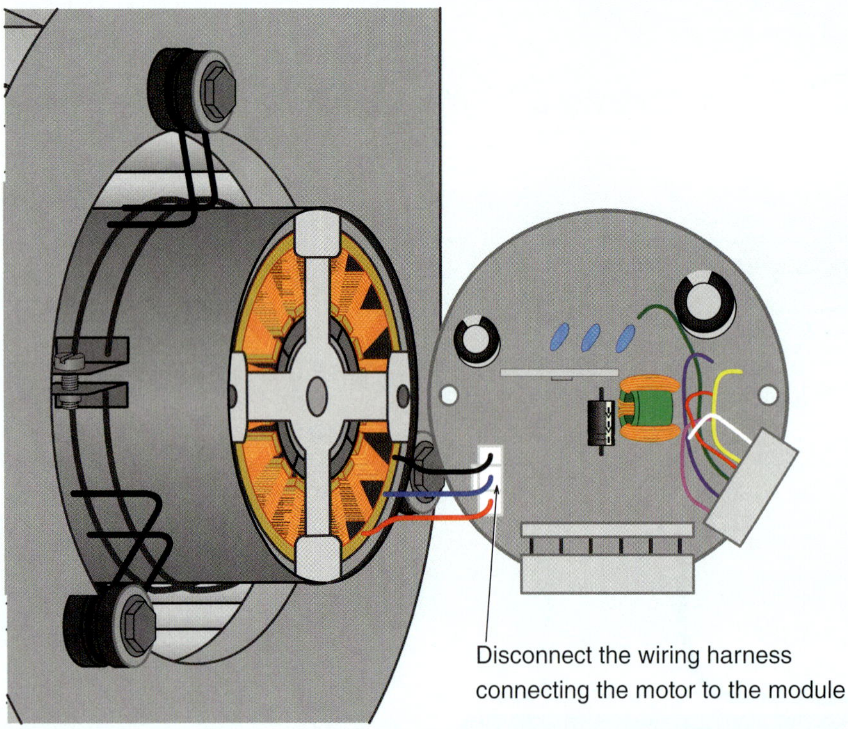

8. Remove the bolts that secure the module to the motor. Cradle the module carefully as it is separated from the motor, so that it doesn't drop and stress the interconnecting wiring. Disconnect the three-pin wiring harness to fully separate the two components.

Continued

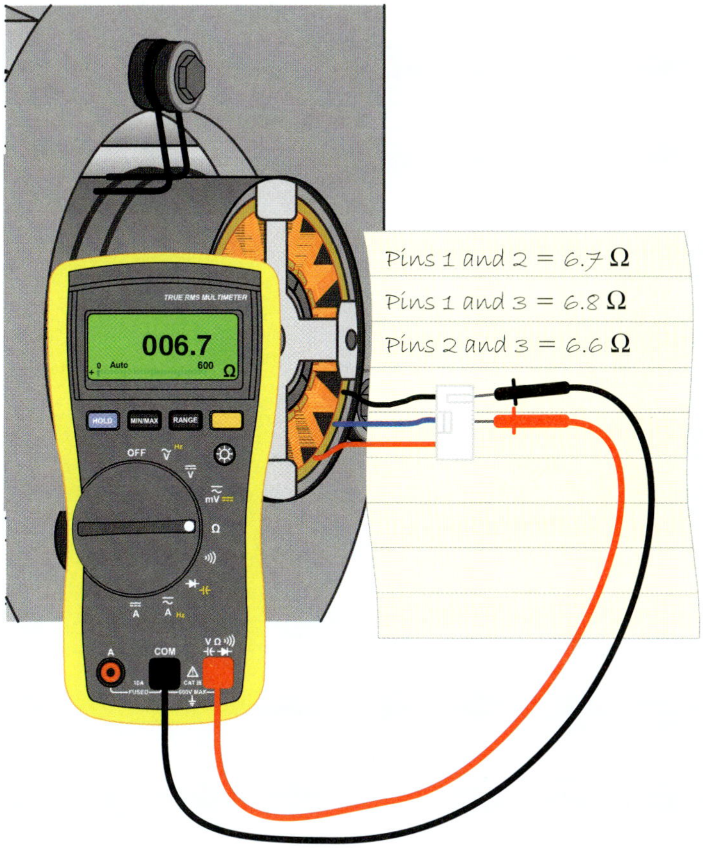

9. Prepare a multimeter to check the resistance of the motor windings. Touch the probes sequentially to Pins 1 and 2, 1 and 3, and then 2 and 3 of the motor's three-pin connector. Each measured resistance should be less than 20 Ω and within 10% of each other. An infinite or open reading indicates an open winding. Values that are significantly lower or higher than the other two often indicate a turn-to-turn short in the winding. Either condition is cause for motor replacement.

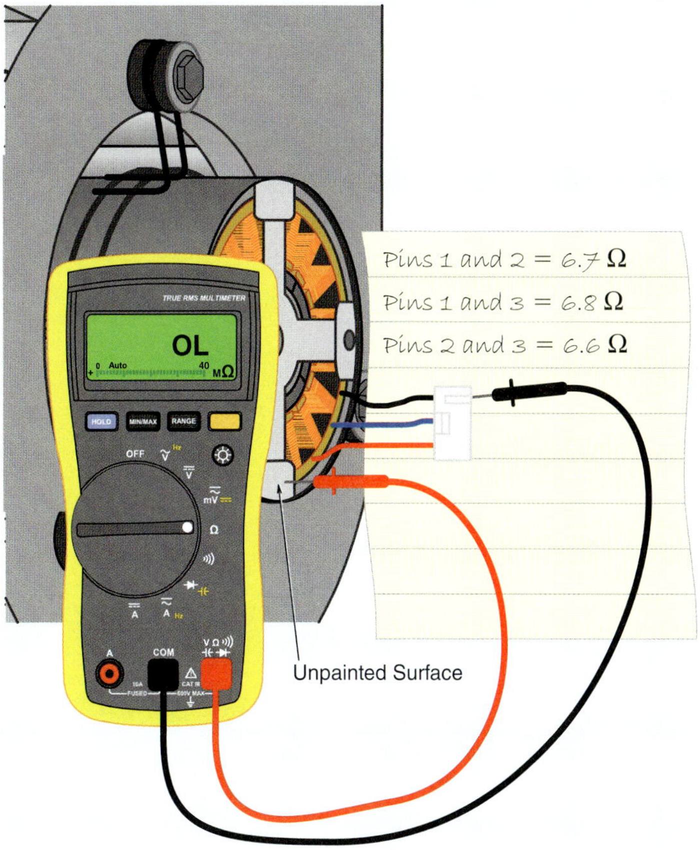

10. The windings are now tested for a grounded condition, using the ohmmeter. Touch one meter probe to a pin and the other to a clean, unpainted, conductive spot on the motor casing. Observe the meter display. Ideally, the display has no reaction (OL), indicating an open circuit. If there is a resistance value displayed, it must be greater than 100 kΩ; anything less is cause for rejection. Repeat the test for all three windings, to ensure accuracy.

Continued

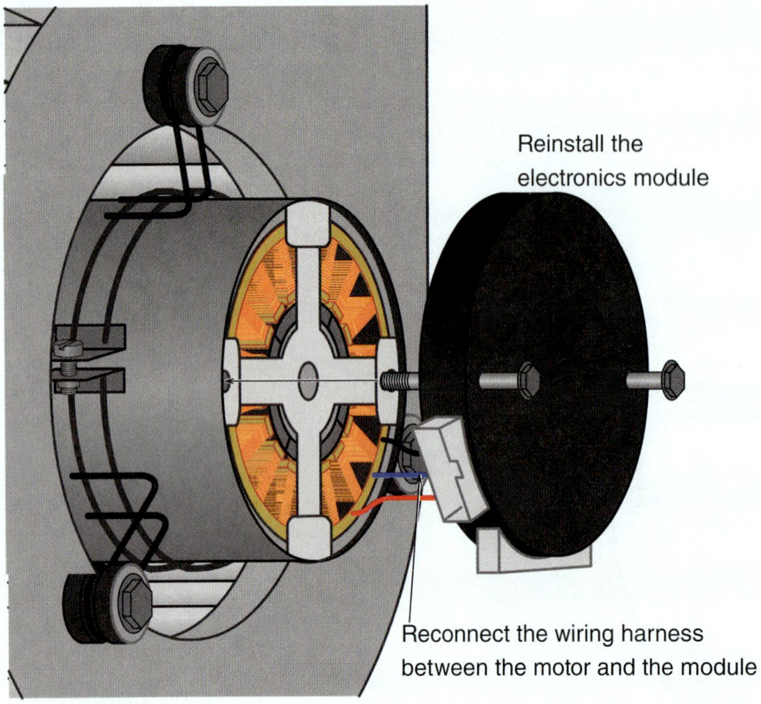

11. If the motor and module have passed all the tests, reassemble the motor and module. Begin by connecting the harness between the motor and module. Position the module at the end of the motor and insert the mounting screws. Tighten them, but do not overtighten.

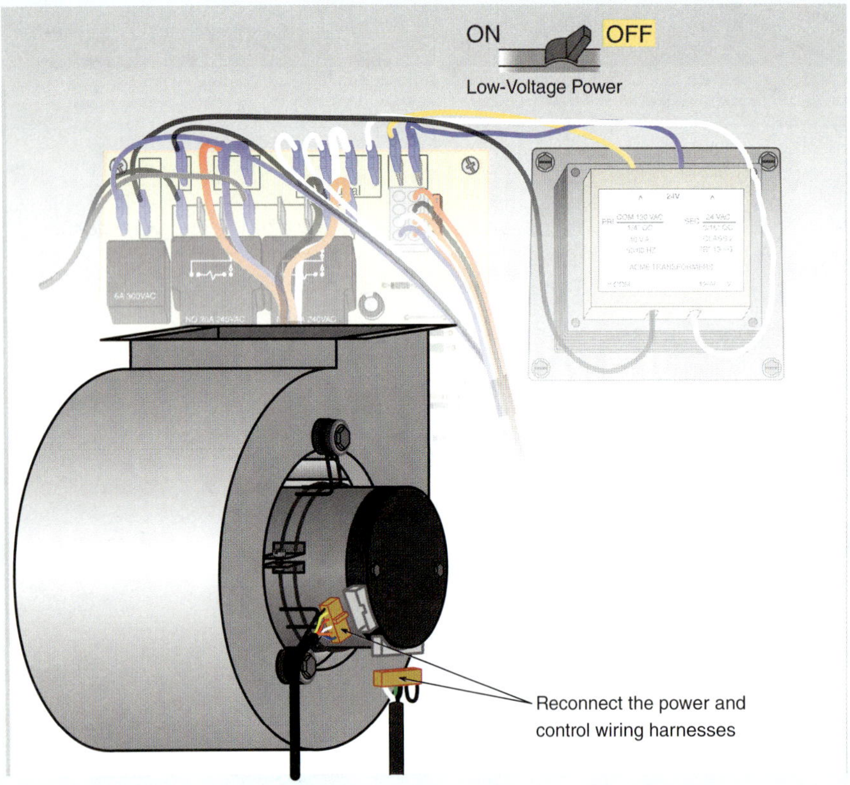

12. Inspect each pin on both remaining wiring harnesses for bent or otherwise damaged pins. If there is no apparent damage, carefully reconnect the wiring harnesses to the motor. Do not force the connectors, because they must be properly aligned and oriented before they will mate and latch. If there is unusual resistance, stop and reinspect the pins and connectors, ensuring that they are properly oriented and aren't damaged in some way.

Continued

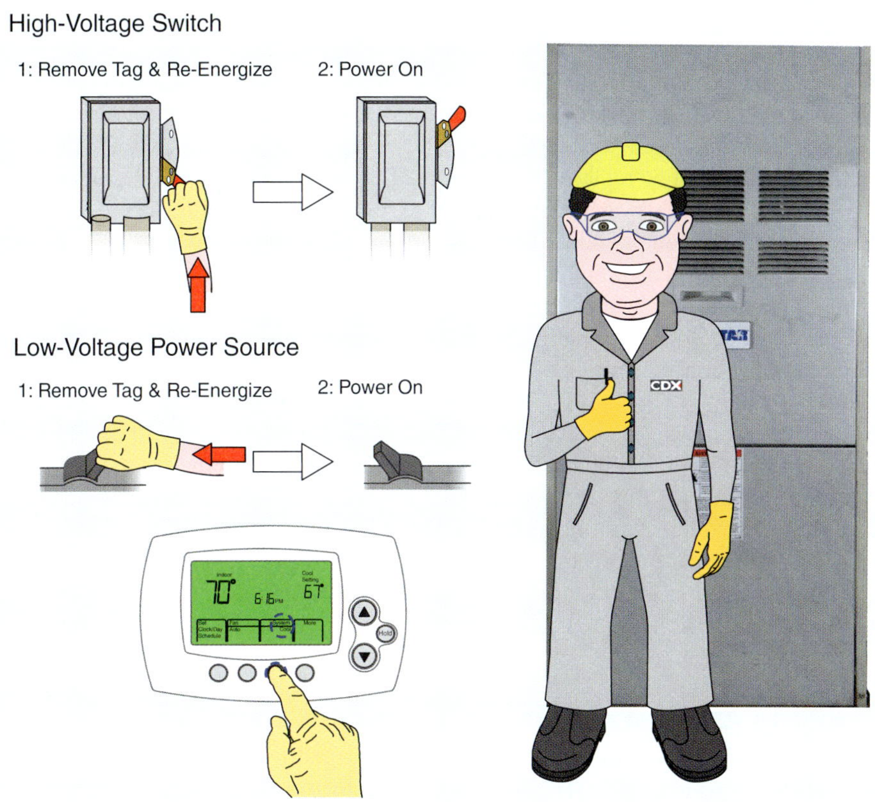

13. Make sure all tools and instruments have been removed from the work area and that the blower wheel is clear of obstructions. Then restore power. Use the thermostat to start and test the system, to ensure that the motor functions as expected.

▶ FOCUS ON CUSTOMER SERVICE

Don't Experiment Using Client Money

Constant-airflow ECMs are the costliest of the three types of ECMs. The cost of a replacement motor may even exceed the cost of the system compressor. Since the module can usually be separated from the motor, either component can be replaced. Practice testing ECMs and ensure that an accurate diagnosis can be made. If something unexpected occurs during testing, communicate with the manufacturer before proceeding. Complex parts are often replaced needlessly by technicians who do not fully understand the technology and are unable to identify the real problem. Don't experiment with the client's money. The HVACR career field needs professionals that respect the client–technician relationship and work hard to understand and keep pace with new technology. If this advice is taken seriously, a technician's professional services will be in constant demand.

▶ Electronically Commutated Motor Replacement

K15005 Describe how to remove and replace an ECM.

S15002 Replace an ECM.

Before discussing replacement procedures, it might be helpful to analyze and summarize the results of the troubleshooting process.

- If the motor and module are separable, it is likely that either component can be replaced independently.
- If the ECM does not respond during communication testing but the motor windings show no opens, shorts, or grounds, replace only the module.
- If the ECM does not respond to communication testing and the motor windings are open, shorted, or grounded, replace only the motor. However, there is still a chance that the module is also defective. An accurate diagnosis of the module is difficult to make without a functional motor. In this case, it is probably best to have both components on hand for replacement. Replace the obviously defective motor first and test the operation again. This will help determine whether the module also needs to be replaced.

ECMs are programmed by the OEM to match the needs of the host equipment. Therefore, when it is necessary to replace a constant-airflow or constant-torque ECM, it must be replaced with an ECM that contains the proper programming. If only the motor needs to be replaced, the concern is reduced, since one motor can operate with many physically compatible modules. The module usually contains the programming (except in the case of DSI communication).

If the module must be replaced, it is essential that it contain the proper programming for the system. In this case, it is best to return to an authorized wholesaler of the equipment brand to ensure compatibility. To enable equipment distributors to stock fewer, more versatile components, many of them now have the capability of programming a somewhat generic module to operate with a specific equipment model.

Repeated failures in a system might be an indication of excessive static pressure. As mentioned earlier, constant-airflow ECMs are not a fix for air-distribution system flaws or poor design. If the motor is constantly forced to operate near the top of its torque curve, this can reduce its life significantly. Since AC induction blower motors respond to excessive static pressure by providing less airflow and operating at a lower current, they are not affected in the same way. Measure system static pressures following a failure to determine whether excessive static pressure is a factor. It should be noted that constant-speed ECMs, typically used to drive axial fans, are not negatively affected by excessive static pressure.

In the next section, the discussion of motor removal and installation will focus on constant-airflow and constant-torque motors used to drive indoor blowers in furnaces, fan coil units, and packaged units.

Removing an Electronically Commutated Motor

In many cases, the entire blower assembly must be removed from the unit to execute a motor replacement (**FIGURE 15-30**). This is most often true for fossil-fuel furnaces, but not necessarily for fan coil units, since access from the side is generally available. The module can usually be replaced without removing the blower assembly or the motor, assuming there is sufficient access to remove the long screws and separate the two components. However, pulling the complete motor out requires additional clearance.

Before beginning, set all controls to the Off position, deenergize the system, and apply locks and tags. Then disconnect the wiring harnesses or leads from the motor. Check the blower assembly visually to ensure that there are no other wires or tubing secured to the housing by the manufacturer for vibration control.

Next, remove the screws that are holding the blower assembly in place in the unit cabinet (**FIGURE 15-31**). The hardware is generally sheet metal screws. Instead of the assembly being suspended from a handful of screws, the flanges of the blower assembly generally slide into a track and are secured with just two or four screws to lock it in place. Once all the screws have been removed, slide the assembly out of the unit.

FIGURE 15-30 Removing a blower assembly.

FIGURE 15-31 Blower assembly attachment to the unit.

The bore of the blower wheel hub is designed to fit the motor shaft snugly. It is secured to the blower housing with a collar and setscrew (**FIGURE 15-32**). The setscrew may have a square head or require an Allen wrench (hex key). Some blower assemblies are built with a small hole in the housing, allowing a technician to insert a long Allen wrench through the housing wall and into the setscrew. The hole and the setscrew are aligned by rotating the wheel to the proper position. This allows the blower wheel to be separated from the motor without removing the entire assembly from the unit. Loosen this setscrew before loosening the hardware holding the motor assembly in place.

Flip the blower housing over and remove the bolts that secure the belly-band mount to the blower housing (**FIGURE 15-33**). With the blower wheel already loose on the motor shaft, pull up on the motor assembly to remove it from the assembly.

Before removing the belly-band mount, note the position of the motor in the band. Doing this enables placing the band in the correct position on the new motor. If the motors have different lengths, some additional adjustment in the position might be necessary. Then loosen the band clamping screw (**FIGURE 15-34**). Note that there are various styles of belly-band mounts; not all styles look exactly like the one shown here, but all serve the same purpose.

FIGURE 15-32 Blower wheel attachment to the motor shaft.

FIGURE 15-33 Remove the three or four bolts holding the belly-band mount to the blower housing.

Installing the New Motor

It is important to orient the motor so that the interconnecting wiring will form a drip loop at the bottom of the motor (**FIGURE 15-35**). Envision the blower housing in its mounting position in the unit. Rotate the motor in the belly band as necessary so that a drip loop can be formed once the assembly is reinstalled into the housing. Ensure that the wiring harnesses will reach the motor connectors as the motor is positioned in the bracket and blower housing.

To install the new motor, first attach the belly-band mount on the motor, positioning it as it was on the old motor. Do not cover any motor-ventilation openings with the mounting band. Position the motor assembly in the blower housing with the motor shaft inserted through the bore of the blower wheel hub.

FIGURE 15-34 Belly-band mount for a blower motor.

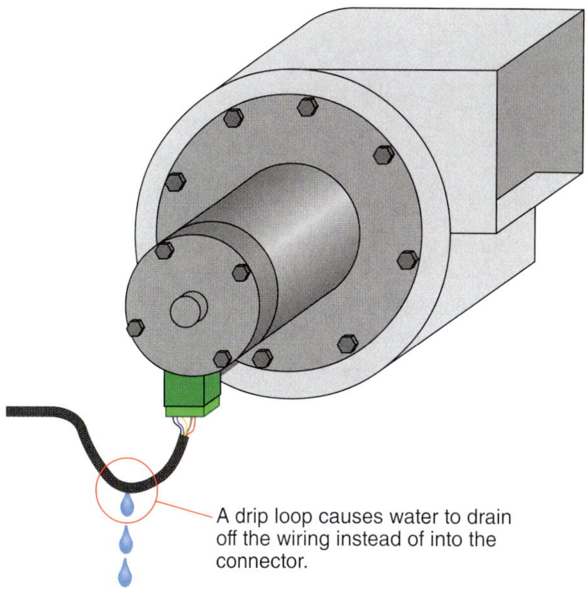

FIGURE 15-35 Wiring drip loop.

Secure the belly-band mount to the blower housing, then turn the housing over. The shaft of the motor should be protruding through the blower wheel hub. Rotate the motor shaft until the flat spot is aligned with the setscrew on the blower wheel, then lightly tighten the setscrew. Turn the blower cage by hand to ensure that it rotates freely and is not rubbing the sides of the housing at any point. It should be centered between the walls of the housing. Then tighten the setscrew to secure the wheel to the shaft.

Return the blower assembly to the unit and secure it. Then connect the wiring harnesses. Ensure that the harness forms a drip loop, to prevent moisture from entering the motor connectors. The final steps are to clear the work area of tools, spin the wheel a final time by hand to ensure it remains free to turn, and restore power to test the motor.

To practice how to replace an ECM, follow the steps in **SKILL DRILL 15-2**. Note that, other than the wiring connections, there is no significant difference in the process of replacing common AC blower motors.

SKILL DRILL 15-2 Practicing ECM Replacement

This drill focuses on the replacement of an ECM used in an indoor blower application, as outlined in the previous section.

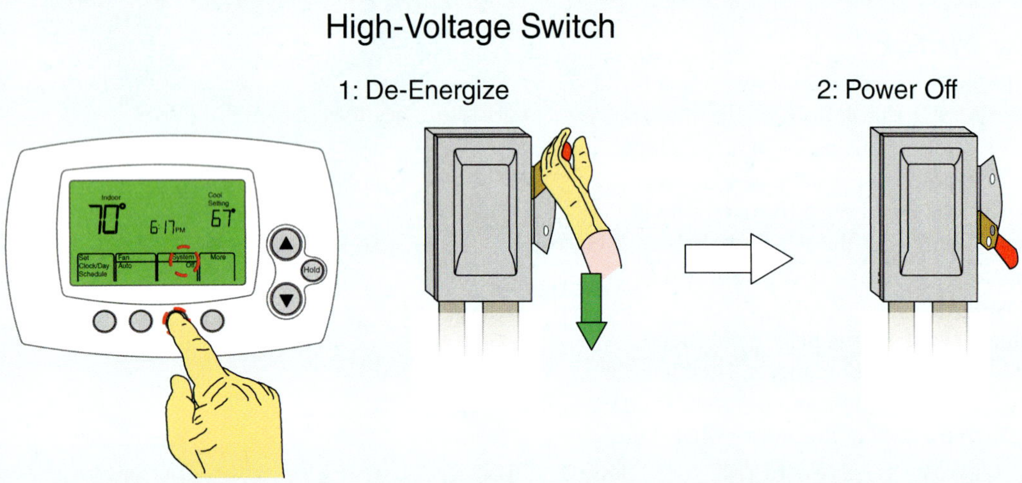

1. Ensure that all controls, including the thermostat, are set to the Off position. Deenergize, lock out, and tag the equipment being serviced. Be sure that the process deenergizes not only the line voltage feeding the equipment but also the low-voltage control power source that may be located in a separate piece of equipment.

SAFETY TIP

Working with electrically energized circuits is hazardous. Avoid contact with any surface that could potentially be energized. When working inside energized enclosures, maintain a stable position and avoid placing both hands inside the enclosure whenever possible. All work within energized equipment by unqualified individuals should be supervised by a qualified instructor or technician.

Continued

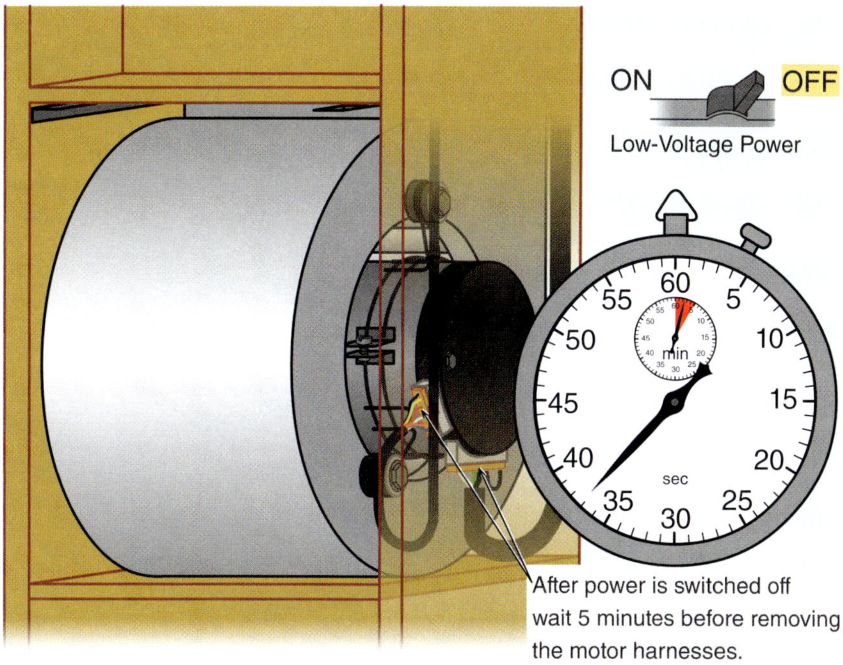

2. Wait five minutes to allow the internal capacitors to fully discharge. Carefully disconnect the wiring harnesses or leads from the motor module; remember that one or more harness connections likely have a locking feature. Also ensure that tubing or wires secured to the blower housing for vibration control have been disconnected.

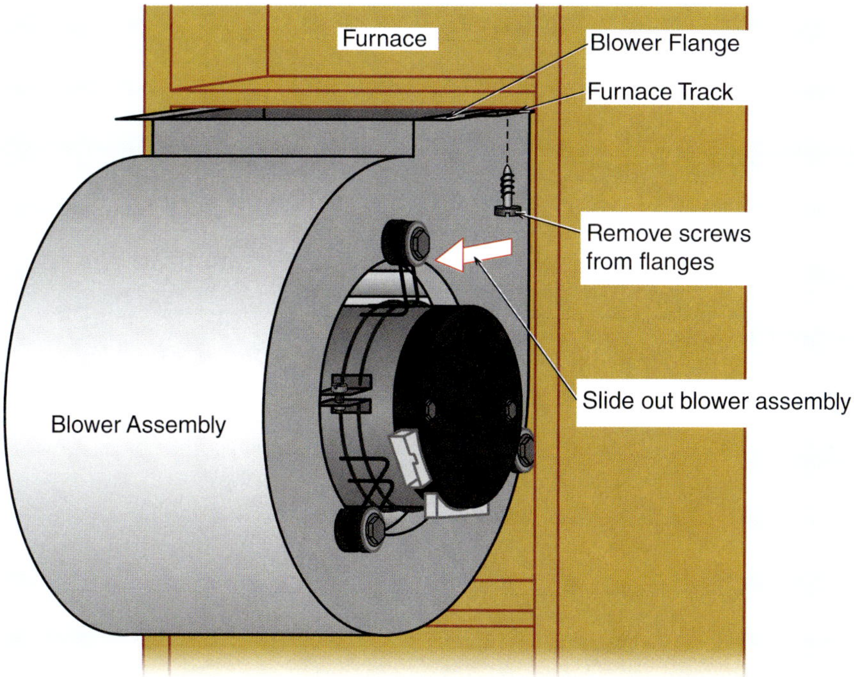

3. Remove the screws holding the blower assembly in place. Support the blower assembly with a block of wood or other material if the assembly is suspended from screws alone. Then slide the blower assembly out of the unit.

In the illustration shown here, the blower assembly slides in and out of a set of tracks in the unit cabinet. However, there are several different approaches to mounting and securing the blower assembly inside a furnace or fan coil unit. Examine the mounting approach and determine the appropriate way to safely remove the assembly, based on the equipment being serviced.

Continued

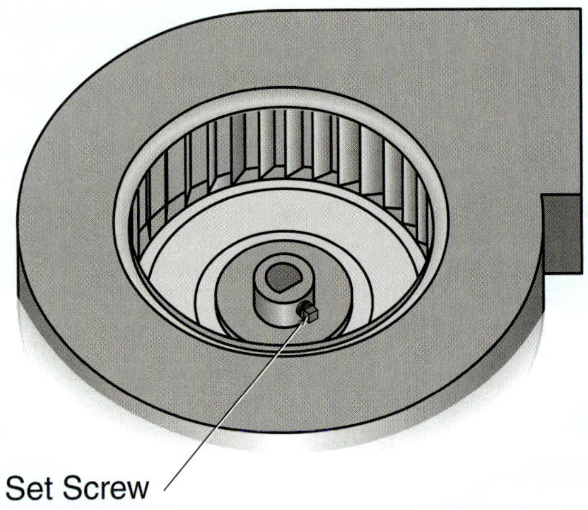

4. Place the assembly on its side with the blower wheel facing up. Examine the setscrew holding the wheel to the motor shaft and select the correct tool to loosen it (open-end wrench, adjustable wrench, or hex key). Use penetrating oil if the setscrew has suffered corrosion and is seized. Ensure that the wheel is free to move on the shaft after loosening the set screw.

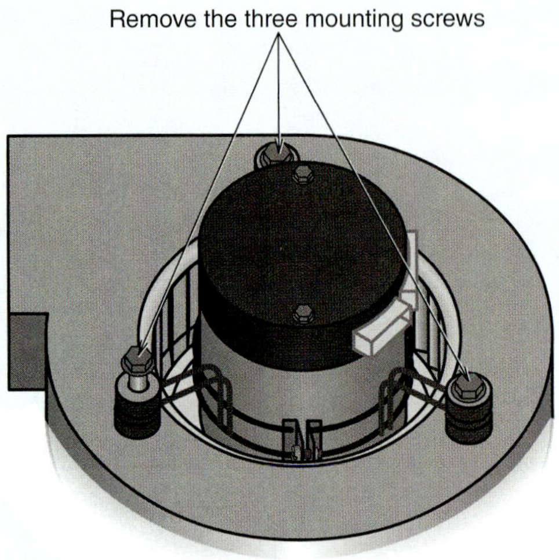

5. Turn the assembly over to access the motor side of the housing, and remove the screws or bolts holding the motor-mounting assembly to the housing.

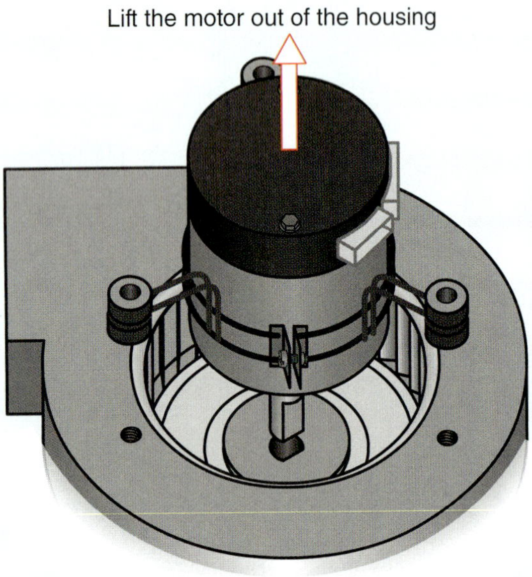

6. With the blower wheel already loose on the motor shaft, lift the motor and belly-band mounting bracket straight up and out of the housing. The wheel remains inside the housing.

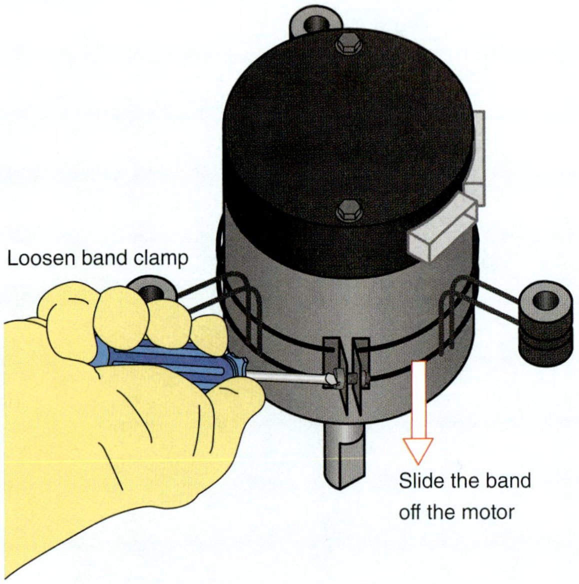

7. Examine the position of the belly band on the motor housing so that the new motor can be placed in a similar position inside the band. Loosen the bolt on the belly band that maintains tension on the motor housing, and then slide the bracket from the motor.

Continued

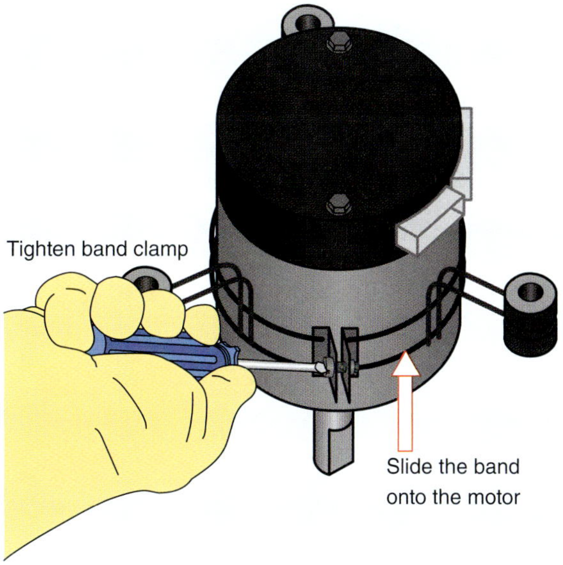

8. Place the new motor into the mounting bracket and tighten slightly. Then dry-fit the motor and bracket back to the housing. Envision the assembly in its position in the unit, and ensure that the motor wiring connections will be under the motor with the assembly in position. Then securely tighten the belly-band clamp around the motor.

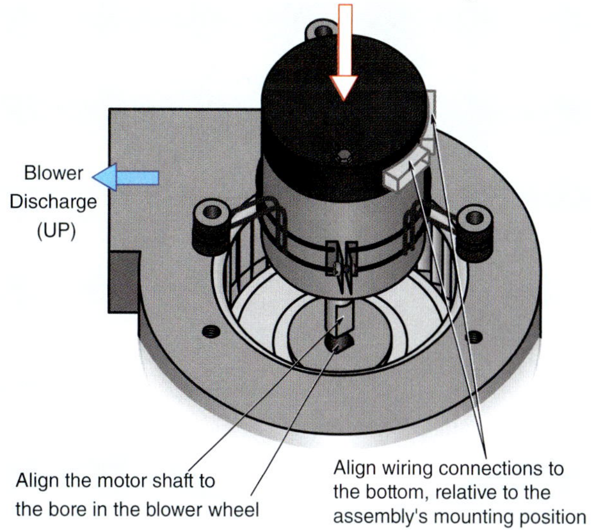

9. Position the motor and mount bracket into the housing by aligning the motor shaft with the bore of the blower wheel and sliding it through.

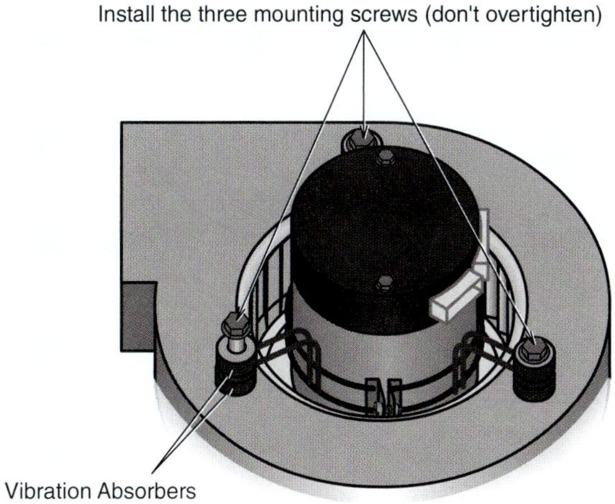

10. Install and tighten the hardware to secure the motor-mounting bracket to the blower housing. Make sure all vibration absorbers (if present) are also in position. Do not overtighten the hardware! In some cases, the hardware consists of large sheet metal screws, and the threads in the housing can be stripped easily. In other cases, there are small nut plates welded to the opposite side of the motor housing and a bolt is used. In either case, they are not substantial enough to avoid damage if an excessively large wrench is used or too much force is applied.

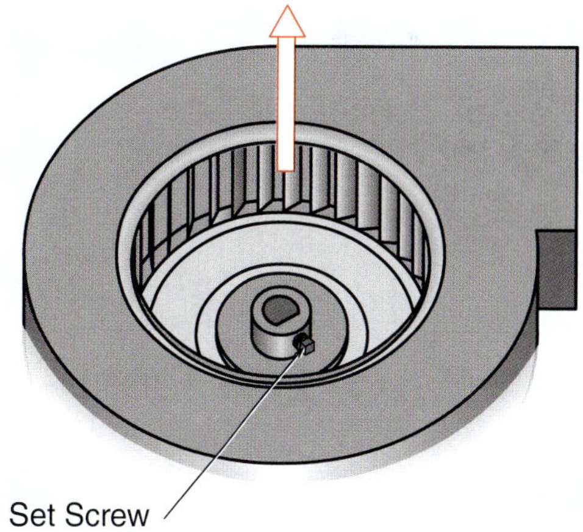

11. Turn the blower assembly over again, with the blower wheel now facing up. Rotate the shaft or wheel to align the setscrew with the flat side of the motor shaft. Then slide the wheel up or down on the shaft as necessary to center it between the walls of the blower housing and tighten the setscrew slightly. Rotate the wheel by hand to ensure that it does not make contact with the housing at any point. Adjust again if necessary, then fully tighten the setscrew. Ensure it is properly aligned with the flat side of the motor shaft.

Continued

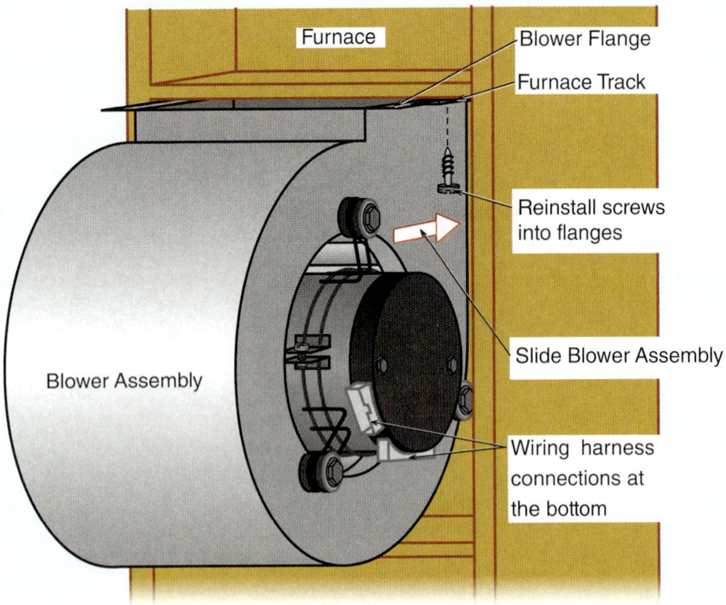

12. Now slide the completed assembly back into its position in the unit. Take care when choosing the positioning of the motor: the wiring connections toward the bottom of the motor but fully accessible for connection and testing. Then secure the blower assembly with the original hardware. Rotate the blower wheel again by hand after securing the blower assembly, to make sure it turns freely without touching the blower housing or wobbling.

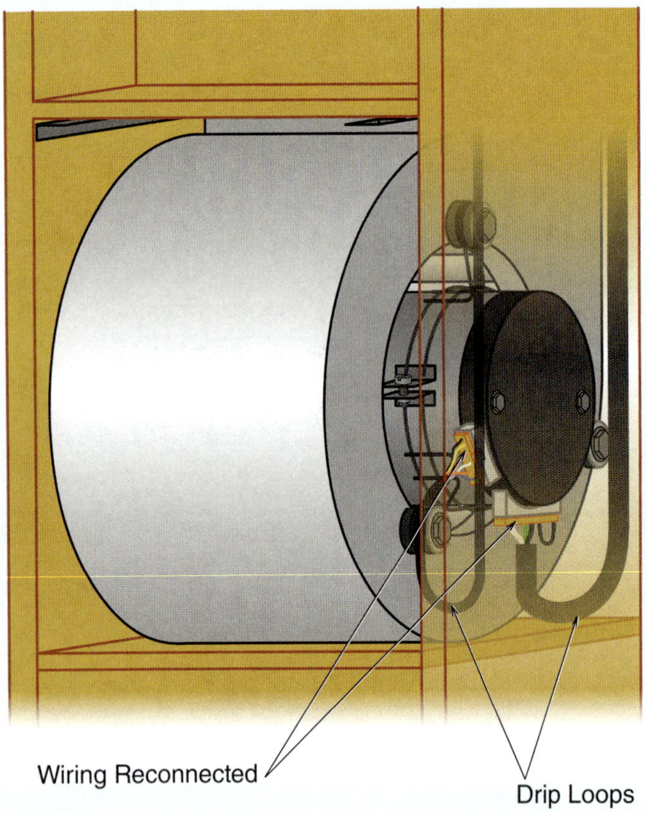

13. Reconnect the wiring harnesses or leads to the motor. Make sure a drip loop is formed in the wiring. Reinstall any other fasteners that might have been securing wire bundles or tubing to the blower housing.

Continued

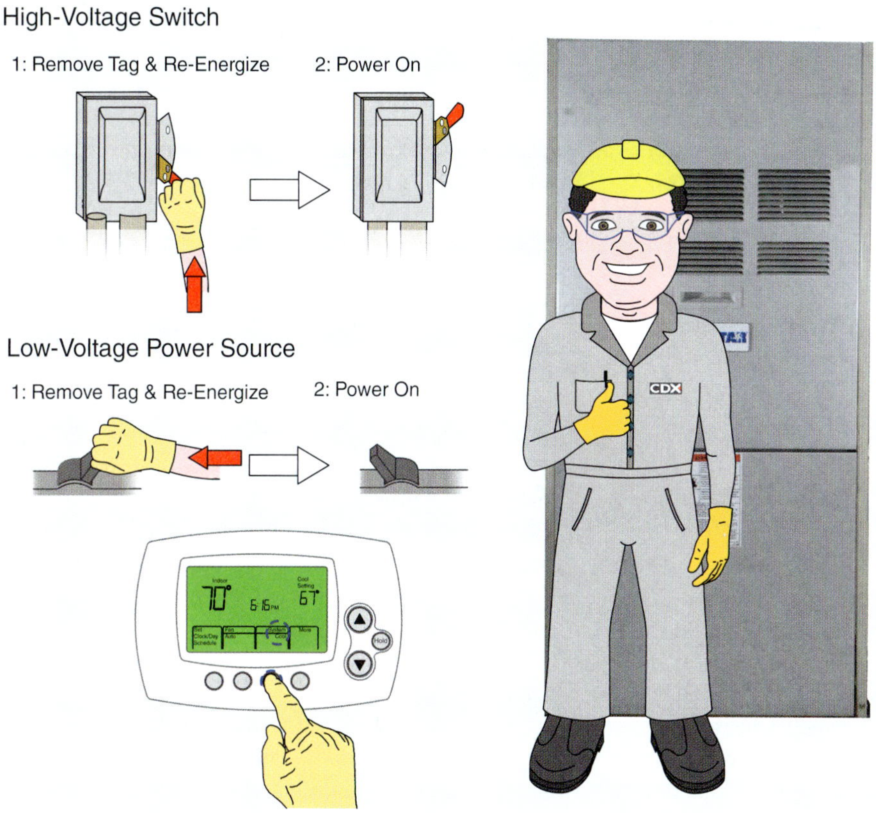

14. Ensure that all controls, including the thermostat, remain set to the Off position. Remove the locks and tags from the power source(s) and restore power to the system. Then test the new motor and module by manipulating the thermostat to operate the motor in all operating modes. Troubleshoot if necessary, as described in the previous Skill Drill. Once the new motor is confirmed to be operating properly, reinstall the access doors and return the system to service.

▶ Wrap-Up

Summary

The high operating efficiency and versatility of ECMs has enabled OEMs to build equipment that meets and exceeds the efficiency levels required of HVACR systems. ECMs are often used to power indoor blowers, condenser fans, and refrigeration evaporator fans, in place of traditional AC induction motors. ECMs differ from conventional AC in that they contain integral control circuits that allow them to be programmed by OEMs to provide consistent performance across a wide range of conditions and modes of operation.

ECMs combine some of the features of AC and DC motors. The main difference between DC motors and ECMs is that DC motors use mechanical commutators and brushes to create a fluctuating voltage that is used to excite the armature of the motor. ECMs are developed around what is essentially a three-phase DC motor package.

Constant-airflow and constant-torque ECMs have unique wiring connections that receive line-voltage power and control signals from the host equipment. Most ECMs have a detachable control module with an internal three-pin connector through which the motor is connected to the module. ECMs are preprogrammed by the OEMs to perform in a specific way. Although the motors themselves may be somewhat interchangeable, the programming between them is not. When a constant-airflow or constant-torque ECM blower motor must be replaced, the replacement should be provided by the OEM or its authorized distributor.

If an ECM appears to have failed, it must be tested to verify the failure before it is condemned. Diagnostic tools, such as the TECINspect device, are available to test the functions of some motors to determine whether it is functional. In many cases, the ECM control module can be replaced without replacing the motor. The motor can also be replaced separately from the module. Thorough testing is necessary to ensure that an ECM has actually failed, before condemning it.

Key Terms

armature Rotating component of a DC-powered electric motor.

brushes Carbon components used to transfer incoming DC voltage to the commutator of a DC motor.

brushless A descriptor for motor construction that eliminates the need for brushes.

commutator A split ring attached to the armature of a DC motor. It causes the DC voltage to reverse direction every 180 degrees, simulating an AC voltage.

constant-torque ECM An ECM that is designed to produce a given torque value on demand.

constant-airflow ECM An ECM with exceptional flexibility in its programming, allowing it to maintain a consistent volume of airflow by managing the torque applied to the blower shaft, thereby managing rotational speed and providing the same airflow even as the resistance to airflow changes.

constant-speed ECM ECMs that run at a predetermined and programmed speed. They are most commonly used in axial fans in various applications.

excitation voltage The voltage applied to motor windings to create the necessary magnetic field.

external static pressure (ESP) The total of static pressure losses created by ductwork, filters, coils, and other components in an air-distribution system.

modulate In this context, to vary the properties of a voltage waveform.

pulse-width modulation (PWM) A digital signal modulation approach that creates a square wave to send a signal through on and off pulses of the wave.

seasonal energy-efficiency ratio (SEER) A comparative measure of energy efficiency used for air conditioners and heat pumps, determined by dividing the total cooling output over the cooling season (measured in Btu or kJ) by the total power input (measured in watts) over the same period. The result provides the number of Btu (kJ) provided per watt of power consumption. The higher the SEER rating, the more efficient the equipment.

torque In the case of motors, the amount of force applied to the shaft that causes it to rotate.

volute The funnel-shaped structure closely positioned around most axial fan blades that captures air from the blade tips, helping to gather and guide the air in the desired direction.

Review Questions

1. The motor at the core of an ECM is also referred to as a ______.
   a. brushless DC motor
   b. AC induction motor
   c. DC induction motor
   d. PWM motor
2. The type of ECM that has a field-separable and replaceable control module is the ___________.
   a. X13
   b. ECM 3.0
3. Variable-frequency drives (VFDs) are most commonly used with ______.
   a. single-phase AC induction motors
   b. three-phase AC induction motors
   c. DC induction motors
   d. ECMs
4. The operating voltage ultimately supplied to the motor in an ECM is ______.
   a. single-phase AC
   b. three-phase AC
   c. single-phase DC
   d. three-phase DC
5. If a constant-airflow ECM has a wire jumper connected to Pins 1 and 2 of its power connector, it is configured for ____________ operation.
   a. 120 VAC
   b. 240 VAC
6. Unlike the standard DC motor, the excitation voltage for an ECM is applied to the __________.
   a. rotor
   b. stator
7. The type of ECM that uses preprogrammed charts of torque and speed to adjust itself during operation is the _____________.
   a. constant-airflow
   b. constant-torque
8. Which of the following is a true statement about ECMs?
   a. The ESP in an operating system is measured and signaled to a constant-airflow ECM.
   b. ECMs offer increased reliability when operated at torques and/or speeds near the top of their capacity.
   c. ECMs should not be considered a solution to poor air-distribution system design.
   d. Constant-speed ECMs are best used for indoor blower applications.
9. If an ECM rotates back and forth on start-up before rotating in the right direction consistently, __________.
   a. it demonstrates that the module is failing to receive the proper inputs
   b. the motor is orienting itself to the proper direction of rotation
   c. it indicates that the motor is nearing failure
   d. the PC board in the host equipment is sending mixed signals
10. During a continuity test of ECM windings, the resistance values should be within ______.
    a. ±5%
    b. ±10%
    c. 3 Ω of each other
    d. 20 Ω of each other
11. In which communicating mode is the operating-program information stored in the host equipment and not in the ECM itself?
    a. DSI mode
    b. Thermostat mode
    c. PWM mode
    d. Airflow mode
12. Which communication mode makes use of the most pins in the communication connection of a constant-airflow ECM?
    a. DSI mode
    b. Airflow mode
    c. PWM mode
    d. Thermostat mode
13. Which type of ECM typically has an electronic module that can be placed away from the motor itself?
    a. The constant-airflow ECM
    b. The constant-speed ECM
    c. The constant-torque ECM
    d. Only the X13 constant-torque ECM

14. Before removing the power-wiring harness or leads from an ECM, always _____.
   a. disconnect the communication harness or leads first
   b. rotate the motor shaft by hand first, to determine whether it is seized
   c. deenergize the line-voltage power source first
   d. check for any feedback voltage from the module first

15. In the absence of a diagnostic tool and with the proper documentation, communication with an ECM from the host equipment can be tested through ______________ ______________.
   a. a parallel connection to a laptop
   b. pin-out testing

16. The problem that exists when the power is initially turned off to an ECM is that _____.
   a. the motor can rotate at any time for a few minutes after power has been removed
   b. capacitors in the control module retain a high-voltage charge for minutes
   c. the start capacitor retains a high-voltage charge
   d. an incoming communication signal can permanently reverse the motor's rotation

17. Which of the following is a true statement about constant-torque ECMs?
   a. The speed taps can be energized by using either 24 VAC or 15–33 VDC.
   b. If the motor receives a signal on more than one-speed tap, it doesn't respond to either one.
   c. A programmed off delay can be eliminated by using a DIP switch in the motor module.
   d. All five-speed taps on the motor must receive a unique signal before the motor will run.

18. When the shaft of an ECM is turned by hand, why is consistent resistance felt at points through each revolution?

19. To prevent motor damage due to an obstruction that prevents the blower or fan from turning when it is energized, some ECMs use _____.
   a. overload relays
   b. auto-reverse programming
   c. stall-detection programming
   d. solid-state fuses

20. Which of the following steps should come first when replacing an ECM that powers an indoor blower?
   a. Loosen the setscrew that holds the blower wheel to the motor shaft.
   b. Remove the mounting bolts holding the belly-band mount to the blower housing.
   c. Loosen the belly-band clamping screw that maintains tension on the motor housing.
   d. Form a drip loop in the wiring harnesses or leads.

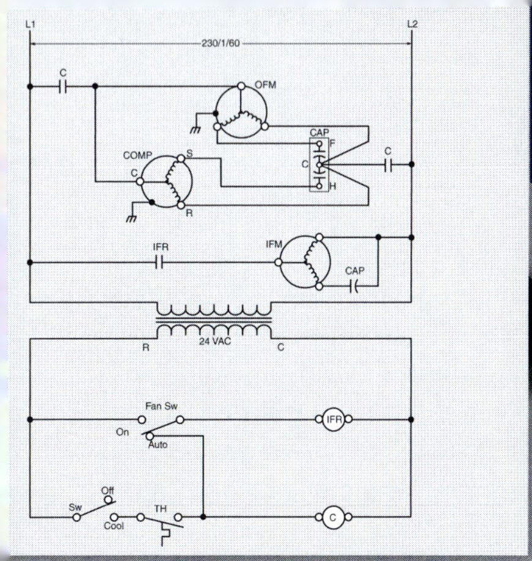

CHAPTER 16

Electrical Diagrams

Knowledge Objectives

After reading this chapter, you will be able to:

- **K16001** Describe the basic elements of an electrical diagram.
- **K16002** Identify common electrical diagram symbols.
- **K16003** Identify and interpret ladder diagrams.
- **K16004** Identify and interpret connection and installation diagrams.

▶ TECHNICIAN TIP

Diagram Diversity

Diagrams from different manufacturers (unless they are aligned in business) will never be the same. Although this chapter presents many common symbols and other drawing elements, people that conceive and generate electrical diagrams are not bound to create them in a highly specific way. This has always been the case with HVACR systems. However, in today's global economy, differences in the approach to creating diagrams have grown substantially. Asian and European cultures, for example, have long had their own approach to electrical diagrams, and the influence of those approaches is often seen in the diagrams that originate outside of the United States.

▶ Introduction

Electrical diagrams serve as the road map for the electrical circuits of HVACR systems. Long before any components are assembled to create equipment, wiring diagrams are developed around the desired sequence of operation. Electrical diagrams show technicians how the components interact and are physically connected, and that information can be translated into a sequence of operation. Although an experienced technician often needs to review a written sequence of operations for complex systems, the electrical diagrams provide the necessary information for technicians to comprehend the operating sequence of most residential and light-commercial systems.

It is essential that HVACR technicians be able to read and interpret electrical diagrams. It is a skill mastered by experienced and successful HVACR technicians. Although a complex wiring diagram may seem challenging to understand now, they will eventually become easy to comprehend. Understanding electrical diagrams begins with recognizing the elements that come together to create them and then seeing those elements applied to simple diagrams.

Before diving into electrical diagrams, remember that they all share one feature. Electrical diagrams, unless otherwise noted, always show the state of switches and relays in their deenergized position. In the case of HVACR equipment, also assume that a proper refrigerant charge exists in the unit. For example, when a low-pressure safety switch is in a circuit, it will be shown closed, as it would be with the unit deenergized but containing a proper refrigerant charge. If there were no refrigerant in the circuit, the low-pressure switch would be open.

▶ Elements of an Electrical Diagram

K16001 Describe the basic elements of an electrical diagram.

There are many possible elements that a wiring diagram might contain, but not all the elements are present on every diagram. Two elements that are likely to be a part of every diagram are lines and symbols. Symbols are a very important part of every diagram, and there are many of them. Symbols will be discussed separately in the next section of this chapter. Other elements that will be discussed in this section are **legends** and notes.

Lines

The lines on a wiring diagram represent electrical paths. Unlike mechanical and architectural drawings, though, only a few line types are needed. **FIGURE 16-1** shows the simple line legend found on some drawings. Some electrical drawings do not include a line legend. The legend shown here also does not represent the precise line types used on every electrical drawing. One of the first things to understand about electrical drawings is that no comprehensive standard is followed throughout the industry. Although most are certainly similar, there can be many subtle variations from one diagram to another. This is true of all diagram elements, including the symbols.

Not all electrical drawings will use different line weights for low- and high-voltage wiring. But most are consistent in their use of dashed or dotted lines to represent wiring that is outside of the unit cabinet (**FIGURE 16-2**). This wiring is usually referred to as field wiring or external wiring.

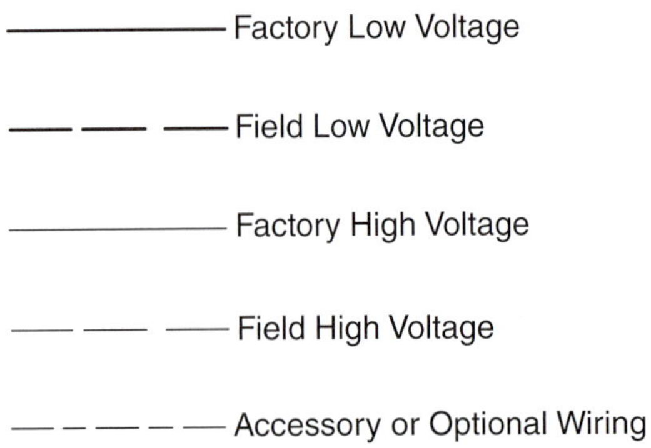

FIGURE 16-1 Electrical diagram line types.

Dashed or dotted lines may also represent other things. In **FIGURE 16-3**, a dashed line is shown connecting two poles of a switch. In this case, the line shows there is a mechanical connection between the two poles; what one pole does, the other will do. Dashed lines might also be used to show optional wiring, as shown in **FIGURE 16-4**.

It is important not to confuse lines representing wiring with other lines on a diagram. Solid or dashed lines are often drawn around portions of a wiring diagram to indicate the boundaries of a device. A solid line usually indicates that the device is inside the equipment, whereas a dashed line indicates it is external to the equipment. **FIGURE 16-5** shows how those lines might appear.

Even from the small portion of diagrams shown so far, it is evident that many lines cross each other. But are they actually connected to each other? **FIGURE 16-6** shows how many diagrams distinguish between wires that are connected and those that are not. When wires cross on a diagram and they are connected, a solid circle is used. The empty circle is used at device connections. Also shown in the figure is an alternate means of showing two wires crossing without being connected.

In **FIGURE 16-7**, part of a compressor contactor (C) control circuit is shown. A low-pressure switch (LPS) and a high-pressure switch (HPS) are also shown in the circuit. Many high- and low-pressure safety switches simply have two wire leads; there are no terminals. The solid black circles show that one wire from each switch is connected to the other, placing them in series. Which of the two leads are used for each connection doesn't matter, since it is merely a switch in series. Note,

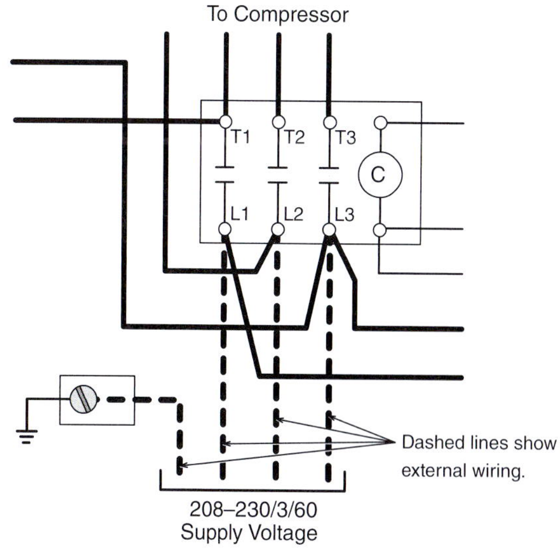

FIGURE 16-2 Dashed lines showing external wiring.

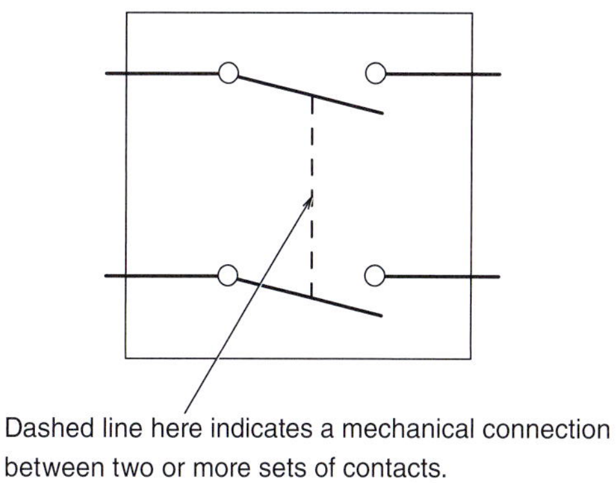

FIGURE 16-3 Dashed lines showing a mechanical connection.

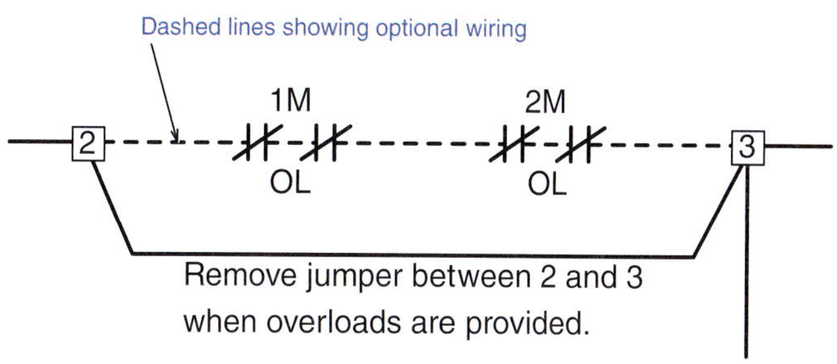

FIGURE 16-4 Dashed lines showing optional wiring.

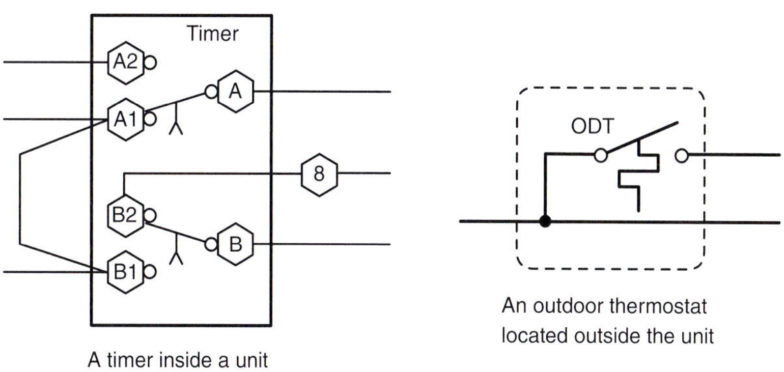

FIGURE 16-5 Lines outlining components.

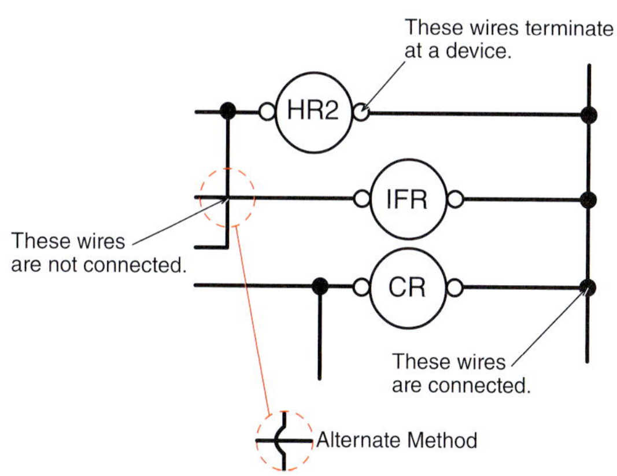

FIGURE 16-6 Examples of connected and unconnected wires.

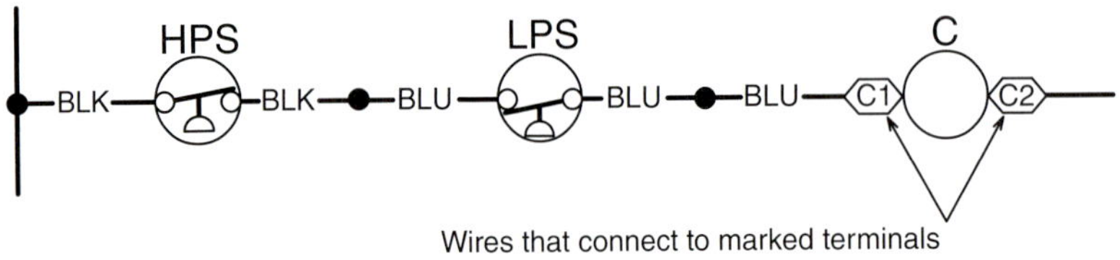

FIGURE 16-7 Wires that connect to marked terminals.

however, that the compressor contactor has marked terminals C1 and C2, and the labels are shown inside a hexagon. Some manufacturers use this hexagon shape to show that those terminals are marked in some way. The terminal identifiers may appear in another part of the diagram or in other diagrams. Note that the pressure switch connections are shown only as empty circles.

Also note that wire colors are shown in Figure 16-7. Showing wire colors, like many other aspects of electrical diagrams, is at the discretion of the provider. Some diagrams are small yet complex; they may become too cluttered when color tags are added. Some manufacturers may number wires instead, using only black wire throughout. In those cases, wire numbers may be shown on the diagram instead of color tags.

Legends

Legends provide the user with a translation of the many abbreviations shown on the diagram. **FIGURE 16-8** shows a typical legend for a residential unit. Although the abbreviations are not fully standardized across the industry, many of the same ones are used by different manufacturers. After some experience working with both wiring diagrams and the related equipment, most abbreviations will become familiar. Note that the example shown offers a legend for wire and connection types and for components.

	Field Splice	C............Contactor
	Terminal (marked)	CAP....... Capacitor
	Terminal (unmarked)	CCH...... Crankcase Heater
	Splice	COMP... Compressor
	Splice (marked)	CTD....... Compressor Time Relay
	Field Control Wiring	GND...... Ground
	FieldPower Wiring	HPS.......High Pressure Switch
	Accessory Wiring	HR.........Heat Relay
	Factory Low Voltage	IFB.........Indoor Fan Board
	Factory High Voltage	IFM........Indoor Fan Motor
		LPS....... Low Pressure Switch
		OFM...... Outdoor Fan Motor
		SOL.......Solenoid
		TRAN.....Transformer
		TSTAT... Thermostat

FIGURE 16-8 Typical legend.

Notes

Notes are sometimes tied to references found on an electrical diagram, but not always. Some notes may not be tied directly to a specific diagram feature. Notes and note references may appear on any type of diagram. They can address a variety of issues, as shown in **FIGURE 16-9**. They provide a way for the manufacturer to share essential information that would be challenging to place on the diagram itself.

With a sophisticated and versatile piece of equipment, quite a few notes may be required. Although they may not be found on a list along with other notes, information such as that shown in **FIGURE 16-10** is often found within the electrical diagram itself or in an adjacent area. The information shown here refers to the position of dual-inline package (DIP) switches. DIP switches are found on many printed circuit boards (PCBs), and they may also be referred to as microswitches. The settings shown here are related to the programming for an electronically commutated motor (ECM) to establish the proper air volumes for the application.

Whenever a reference to a note is encountered, it is important to read the note and determine whether it applies to the current situation. The best practice is to always begin reviewing an unfamiliar diagram by reading the notes.

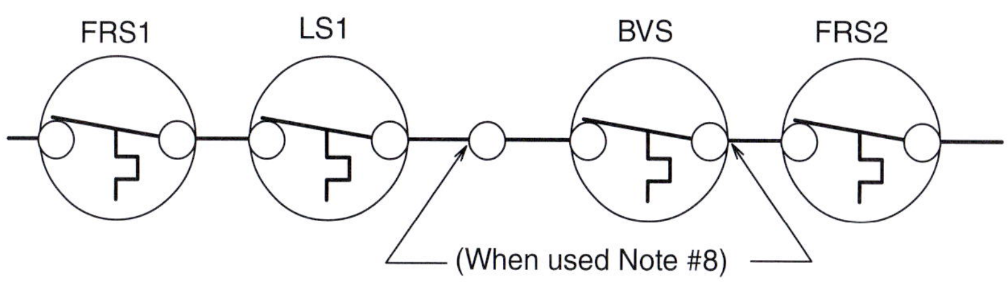

NOTES
1. Replace any original wiring with wire rated for 105°C (221°F).
2. This wire must be connected to a reliable ground for the flame-proving circuit to function.
3. Only copper conductors must be used between the disconnect switch and unit junction box.
4. Solid lines within the printed circuit board (PCB) are integral conductors and not included in legend.
5. Replace with 3.2-amp fuse only.
6. Jumper not present when LGPS is used.
7. For gas heating, blower off-delay selections are 120, 150, and 180 seconds.
8. LS1 and FRS2 are factory-connected when BVS option is not provided.

FIGURE 16-9 Example of electrical diagram notes.

BASED ON 400 CFM/TON (SETUP SWITCH SW1-6 0N, SW4-5 OFF)

MODEL SIZE	SETUP SWITCH SW3 OR SW4 POSITIONS							
	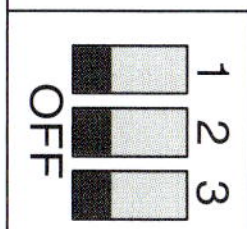	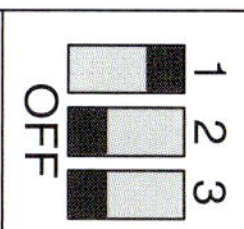	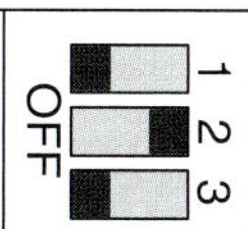	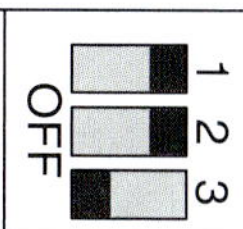	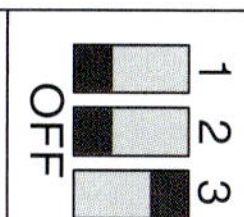	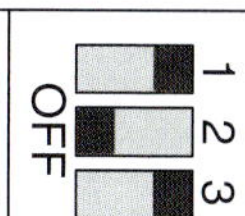	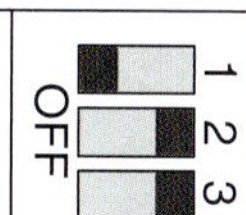	
075	DEF	600	800	1000	1200	1400	1400	1400
095	DEF	600	800	1000	1200	1400	1600	1600
115, 140	DEF	800	1000	1200	1400	1600	2000	2200

FIGURE 16-10 Equipment setup notes.

K16002 Identify common electrical diagram symbols.

▶ CORE CONNECTION

ANSI and Electrical Symbols

The purpose of the American National Standards Institute (ANSI) is to "enhance both the global competitiveness of U.S. business and the U.S. quality of life by promoting and facilitating voluntary consensus standards and conformity assessment systems, and safeguarding their integrity." Over the years, ANSI has developed standards in a large number of areas. Although ANSI standards are not legally enforceable in all cases, many federal organizations, such as the Occupational Safety and Health Administration (OSHA), have incorporated ANSI standards by reference into their enforceable requirements. In the world of electrical diagrams and their symbols, ANSI has published standards. In some cases, ANSI partners with industry organizations such as the Air Conditioning, Heating, and Refrigeration Institute (AHRI), the Institute of Electrical and Electronics Engineers (IEEE), and the National Electrical Contractors Association (NECA). Despite the availability of standards developed through these partnerships, there remains a great deal of diversity in symbol interpretation and diagram publication.

▶ Electrical Diagram Symbols

Symbols are used on all types of electrical diagrams. They are used to represent a variety of electrical switches, relays, load devices, and many other components. In a number of cases, the symbol not only identifies a component but also provides some important information about that device. To a certain degree, symbols are the alphabet of electrical diagrams.

There are electrical diagrams that depict the circuits in almost everything that operates using electricity: vacuum cleaners, automobiles, home lighting, etc. The subject of the diagram has everything to do with the symbols required. In this section, we will focus on the symbols that HVACR technicians are most likely to encounter. Although many of them are common to diagrams for other systems and devices, some may be unique to HVACR drawings.

Switches

If there is a single electrical device that dominates electrical diagrams, it is likely switches. Switches are control devices as opposed to loads, but loads do little or nothing until an electrical circuit is completed to them through a switch or set of contacts.

Basic Switches

Switches can be manually operated, like the handle of a power disconnect switch, or automated, like a pressure switch. Some of the basic symbols for switches are shown in **FIGURE 16-11**. Note the dashed line connecting some contacts, as was discussed in a previous section. This shows they are mechanically connected; when one moves, the other moves.

There a few unique switch types that a technician may also encounter. **FIGURE 16-12** shows how momentary switches and a rotary switch appear on a diagram. Momentary switches do not remain in position when actuated. A normally open (NO) momentary switch, for example, closes when pressed, but it returns to its normally open position as soon as it is released. Most of the time, a holding relay of some type is activated, which maintains the circuit after the momentary switch reopens. Normally open momentary switches are encountered more often than the normally closed (NC) type.

A rotary switch can have a number of different positions—something difficult to do with typical switch construction. A typical Hand-Off-Auto (HOA) switch on the face of a motor starter is a common example of a simple rotary switch.

Temperature Switches

Temperature switches have their own specific symbols. **FIGURE 16-13** shows four ways a temperature switch may appear. They must be drawn with care because the visual form tells a technician what to expect from the switch. Depending on how they are drawn, a technician interprets them as normally open or normally closed, like other switches. Further, a technician can tell whether the switch opens or closes on a rise or on a fall in temperature. This strategy also applies to pressure switches, flow switches, and time-delay switches.

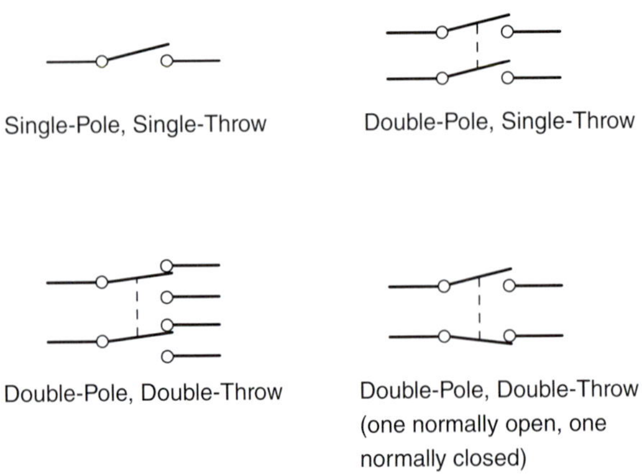

FIGURE 16-11 Examples of basic switches.

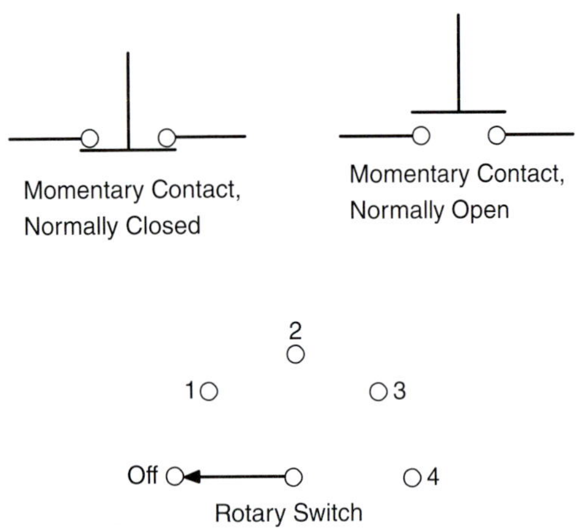

FIGURE 16-12 Momentary and rotary switch symbols.

Pressure Switches

Pressure switches have their own unique look, as shown in **FIGURE 16-14**. However, like temperature switches, the way they are drawn informs the technician of how the switch will react to changes in pressure. The same type of symbol is generally used regardless of whether the pressure range is positive or negative (vacuum).

Flow Switches

Flow switches are usually found on systems that either produce chilled water or are water cooled. They are rarely needed in most residential heating and cooling systems, except for water-source and geothermal heat-pump systems. The four possibilities for flow switch symbols are shown in **FIGURE 16-15**. Flow switches that close on a rise in flow rate are far more common than those that open when flow increases.

Time-Delay Relays

Time-delay relays are a little different. They can be confusing at first glance. Note that time-delay relays generally do not fall into the category of switches, but their contacts are shown with the same basic characteristics as switches. This allows the extra information about their operation to be added. As will be discussed in the next section, contacts are drawn differently from switches and do not support the added information. **FIGURE 16-16** shows four different states for time-delay switches.

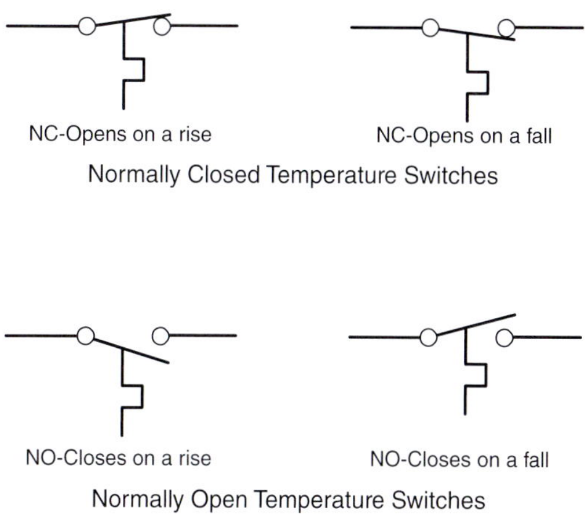

FIGURE 16-13 Temperature switch symbols.

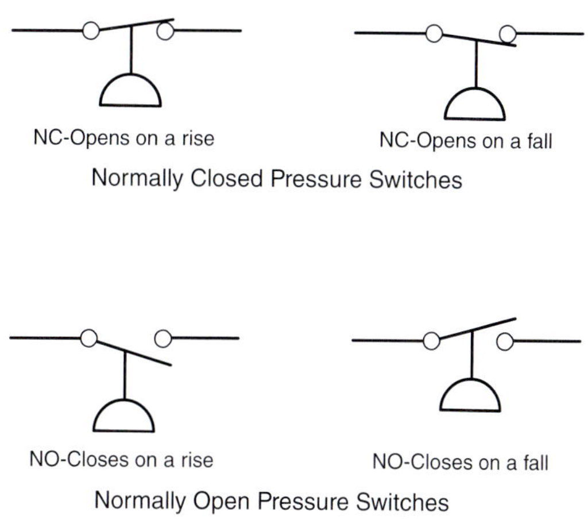

FIGURE 16-14 Pressure switch symbols.

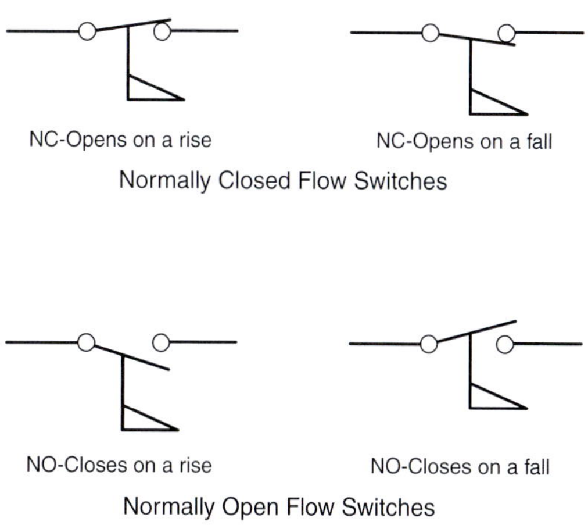

FIGURE 16-15 Flow switch symbols.

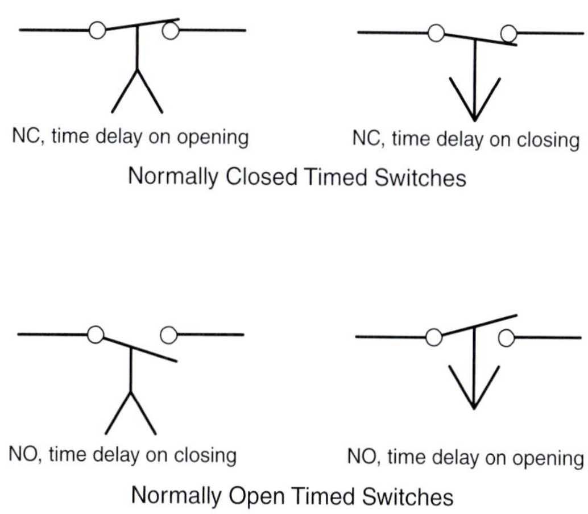

FIGURE 16-16 Time-delay relay contact symbols.

Time-delay relay contacts use arrows to indicate whether the position—open or closed—is delayed after the timer is energized or delayed after the timer is deenergized. Some time-delay relays, for example, initiate a timing period after power has been removed from the timer. This may seem unusual, but there are applications where they are used.

A normally closed, timed-open (NCTO) contact will open within a set period of time after the timer is energized. A normally open, timed-closed (NOTC) contact will close within a set period of time after the timer is energized. Both of these are shown with the arrow pointing upward.

A normally open, timed-open (NOTO) contact doesn't open until the timer has been deenergized for a set period of time. A normally closed, timed-closed (NCTC) contact doesn't close until the timer has been deenergized for a set period of time. For time-delay contacts that do not take their normal position until some period has elapsed after the timer is deenergized, the arrow points downward.

Again, the operation of time-delay relays can be confusing. When in doubt and seeking clarity, refer to the manufacturer's written sequence of operation.

Relays and Contactors

Contact sets are generally considered to be automated in some way. Although there are exceptions, most contacts are housed inside relays and contactors. Both normally open and normally closed contacts are shown in **FIGURE 16-17**. Contacts are typically shown this way regardless of whether they are part of a relay, contactor, or similar device.

On connection (or pictorial) diagrams, a contactor or relay is usually shown as it naturally appears: intact with the operating coil and contacts shown together inside a border. **FIGURE 16-18** shows an example of each in this form. On the contactor, note the L1/L2 and T1/T2 labels on the contacts. The L represents line voltage coming in. The T terminals are where the load conductors are connected.

On **schematic diagrams**, the coil and the contacts are separated from each other, shown in a place that is more representative of their place in the circuit. **FIGURE 16-19** shows a simple schematic diagram where the relay contacts and coil are separated. Note that the coil is in the low-voltage section of the diagram, while the contacts are in the line-voltage portion of the diagram. Both the contacts and the coil of the relay share the same label, providing a way to match the two. This simple diagram shows a normally open thermostat controlling 24 VAC power to the R1 relay coil. When the thermostat

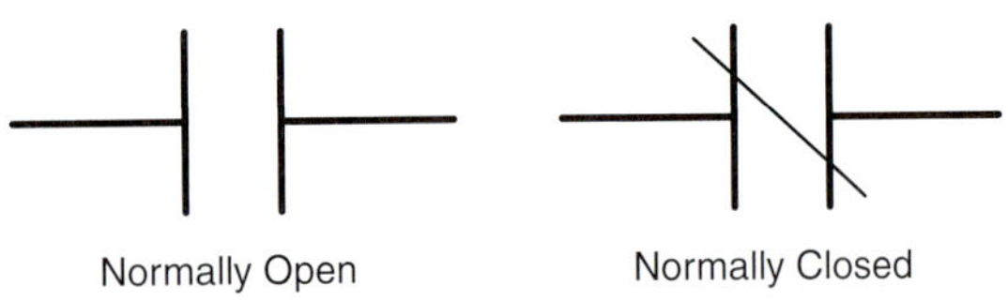

FIGURE 16-17 NO and NC contacts.

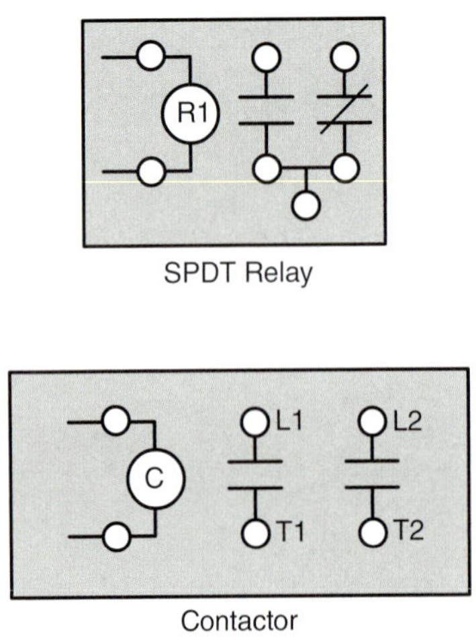

FIGURE 16-18 Relay and contactor connection diagrams.

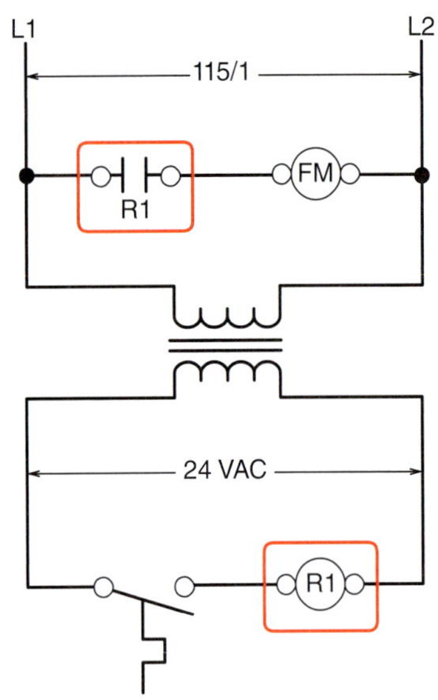

FIGURE 16-19 A relay coil and contacts on a schematic diagram.

closes, the relay coil is energized, its contacts close, and the small fan motor runs on the 115 VAC power supply.

Although the coil and contacts of a relay or contactor may be on separate sides of the drawing, they may also be shown on the same side. Relays and contactors often have contacts that control low-voltage functions as well as line voltage.

It is important to remember that relays and contactors, as well as some other controls, can have many sets of contacts. For example, a four-pole contactor will have a coil as well as four sets of contacts shown separately on the diagram.

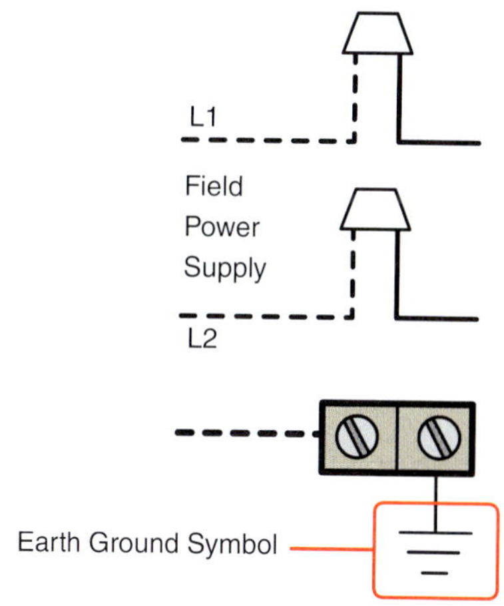

FIGURE 16-20 Line-voltage power supply connections.

Power Supplies and Protection

The line-voltage power supply for a system can be shown in several ways, depending on the type of diagram.

FIGURE 16-20 shows a common method. This figure represents a small portion of an **installation diagram**. Note the ground symbol at the bottom of the diagram. This is the point of connection for the ground wire routed with the power supply. Also note the use of the dotted and solid lines to separate external wiring from internal wiring. In this case, wire nuts are to be used to connect the power supply, meaning there are two conductors prepared inside the unit for this purpose. The power supply may also connect to a terminal board or connect directly to a contactor.

Transformers are very common on HVACR diagrams. **FIGURE 16-21** shows a common configuration for a transformer in a packaged unit or air handling unit. The transformer shown here can operate with 208 or 230 VAC connected to the primary windings, but it is designed to produce only one secondary voltage. As shown, the transformer is operating on a 230 VAC power supply; the 208 VAC lead remains unconnected. Remember, if the incorrect primary-winding lead is connected to the power supply, the secondary voltage will be incorrect as well.

On the secondary side, the common terminal is marked. This leg of power will remain unswitched, connected to each 24 VAC load or control device. The other lead will be switched through a variety of relays and controls.

Power Supply Protection

Circuit-breaker symbols are shown in **FIGURE 16-22**. On some diagrams, the symbol remains simple and does not identify the circuit-breaker type. In others, the type is identified with the additional details shown in the figure.

It is also common for fuses to be used to further protect HVACR equipment, in addition to circuit breakers. Two different symbols for fuses are shown in **FIGURE 16-23**. The image on the right also shows a power disconnect switch in the circuit.

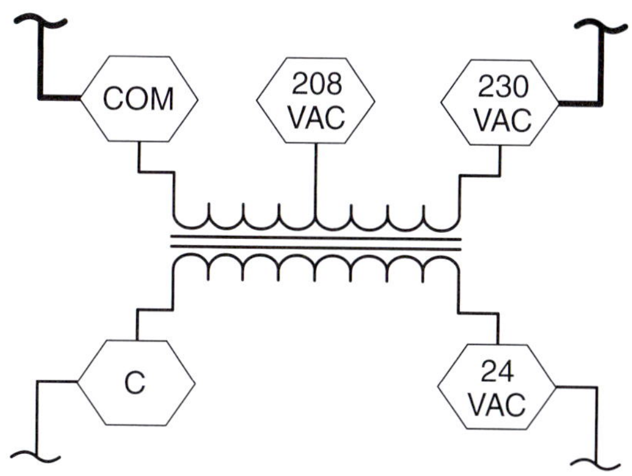

FIGURE 16-21 A control transformer.

Overload Devices

Both thermal and magnetic overload devices are used, primarily for compressor protection but also for larger motors. **FIGURE 16-24** shows an inline thermal overload. Also shown are the symbols for the sensing portion of pilot-duty

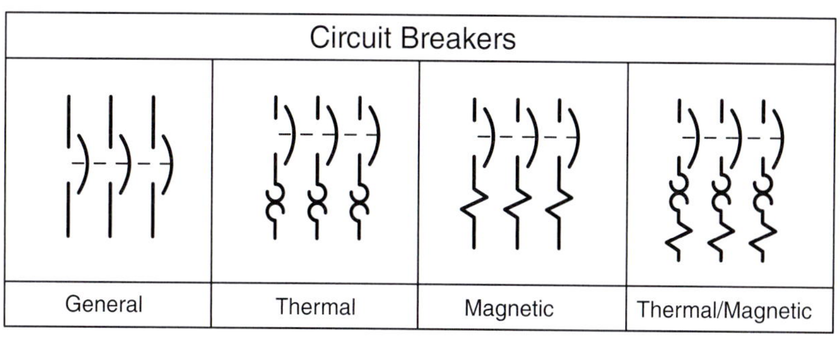

FIGURE 16-22 Circuit-breaker symbols.

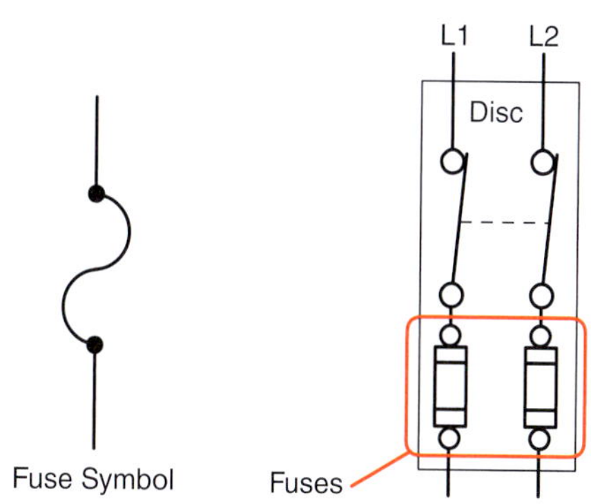

FIGURE 16-23 Fuse symbols.

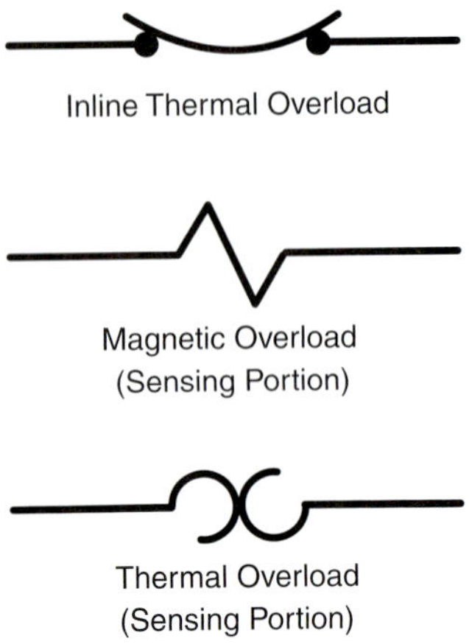

FIGURE 16-24 Thermal and magnetic overloads.

overloads. Pilot-duty devices have a sensing portion that monitors for an overload condition, but there will also be a set of normally closed contacts in the control circuit. Like relays, both portions of the device share the same label.

Loads

Loads include all devices that consume power and accomplish the primary work in an HVACR system. Some loads, of course, consume more power than others. The following is a list of typical loads:

- motors
- heating elements
- transformers
- capacitors
- relay, contactor, and solenoid coils.

Motors and Capacitors

There are many different motors used in HVACR systems, including those used in compressors. From an electrical point of view, and on electrical diagrams, compressors are represented in the same fashion as other motors.

Electrical diagrams may identify motors with very simple symbols, such as those shown in **FIGURE 16-25**. The leads may be positioned as shown, or they may extend from the central circle in different directions, depending on the layout of the diagram. However, most diagrams will show motors in a bit more detail.

FIGURE 16-26 shows some more complex and representative motor symbols. **FIGURE 16-27** shows how a single-phase compressor motor with a start capacitor and start relay might look on a diagram. Note the symbol for the capacitor. The curved line represents the cathode side of the capacitor, which will be at a lower voltage than the anode. Therefore, the flat line indicates the side of the capacitor that should be connected to the Start winding of the motor. Although this is the correct way to use the capacitor symbol, it will appear backwards (incorrect) on some drawings. Also note that the symbols for the start and run capacitor are the same. If there is a bleed resistor on the start capacitor, it may be shown by using the symbol for a resistor.

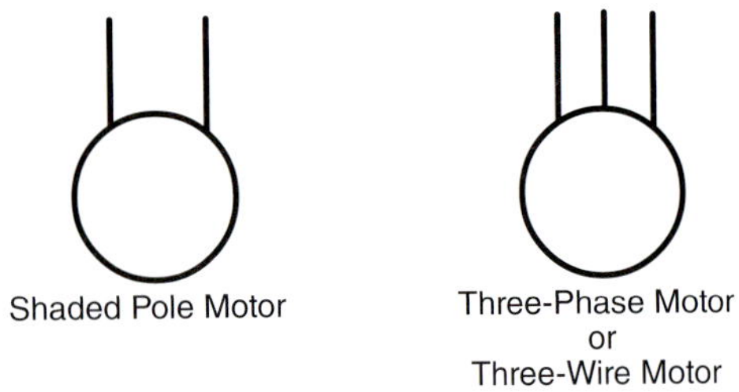

FIGURE 16-25 Simple motor symbols.

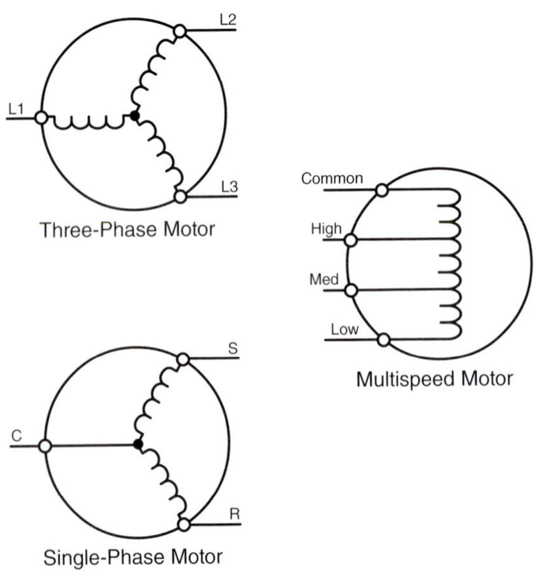

FIGURE 16-26 Motor symbols.

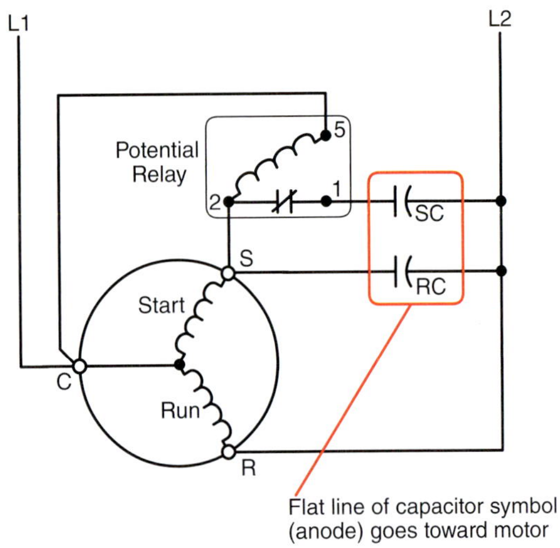

FIGURE 16-27 A motor with run capacitors and start components.

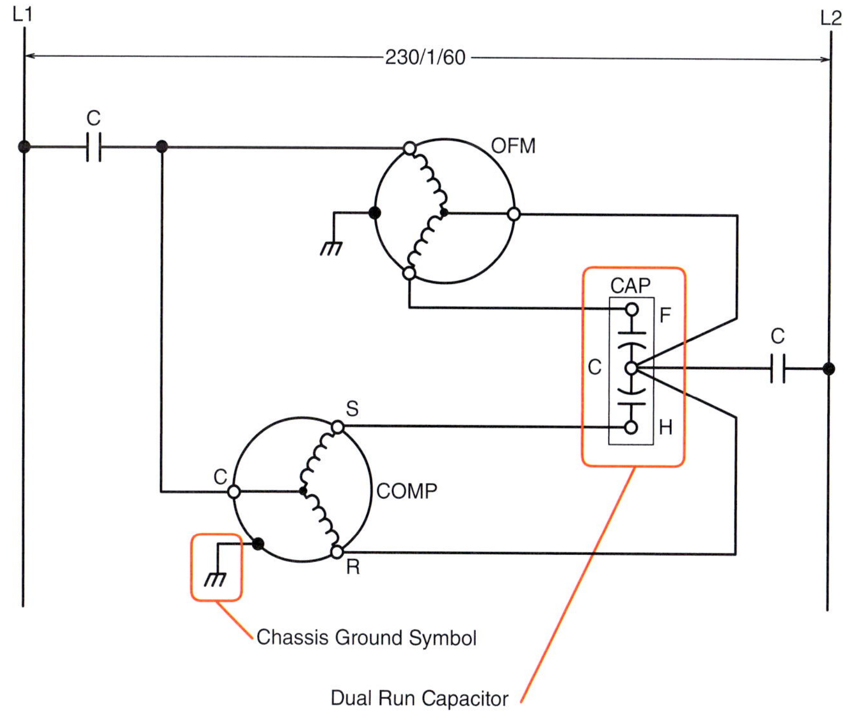

FIGURE 16-28 Dual run capacitor connection diagram.

▶ **TECHNICIAN TIP**

Capacitor Wiring

Most oil-filled run capacitors, as well as some start capacitors, identify the cathode side of the capacitor. However, don't expect the identification to be obvious. The marking is often very subtle, using a dot of paint or perhaps a raised spot in the stamped-metal casing. The cathode is the plate that is positioned closest to the outer casing. When a capacitor fails by becoming grounded, it is usually the cathodic plate that makes contact with the casing. Connecting the opposite terminal to the compressor, and the cathode to the power supply, helps protect the motor from damage and delays the response from a fuse or circuit breaker if the capacitor becomes grounded.

Dual run capacitors are popular in residential and light-commercial systems. **FIGURE 16-28** shows how a dual run capacitor, serving a compressor and outdoor fan motor, appears on a drawing.

Heating Elements

Heating elements are drawn by using one of two symbols, as shown in **FIGURE 16-29**. One of the two symbols is also used to identify a resistor. Since electric heating elements are resistive in nature, they share a symbol. However, accessible resistors in AC-powered HVACR circuits are rare. Resistors are found almost exclusively in DC circuits on PC boards, where the details of board connections are not shown. The rare exception is the bleed resistor often found on motor capacitors.

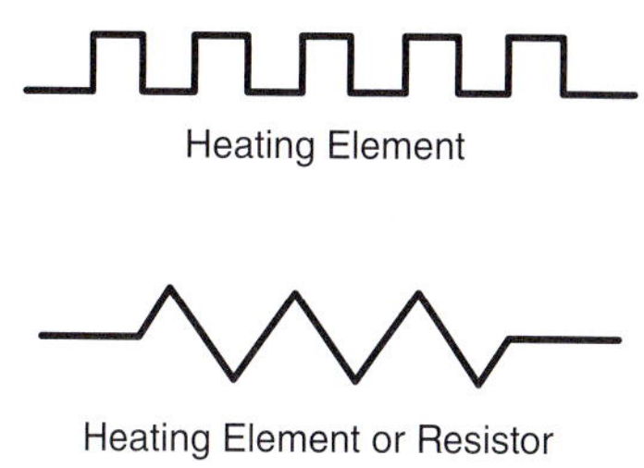

FIGURE 16-29 Symbols for heating elements.

▶ Ladder Diagrams

K16003 Identify and interpret ladder diagrams.

Now that the common symbols used on wiring diagrams have been presented, this section can focus on how they relate to each other on a working diagram. The most valuable type of diagram in troubleshooting is the **ladder diagram**. It is also referred to as a schematic diagram, or even more simply as a schematic. The three terms are synonymous in the HVACR trade.

FIGURE 16-30 shows a simple ladder diagram. Ladder diagrams get their name from the layout, which is similar to a ladder. The vertical legs represent the power source, with the electrical loads and controls drawn between the legs, like the rungs of a ladder.

The power of the ladder diagram is that it is representative of how the various circuits function, rather than representing how the circuit physically appears in the equipment. A technician can quickly locate an electrical component on the schematic that is not functioning as expected and see what other components are in the circuit with it.

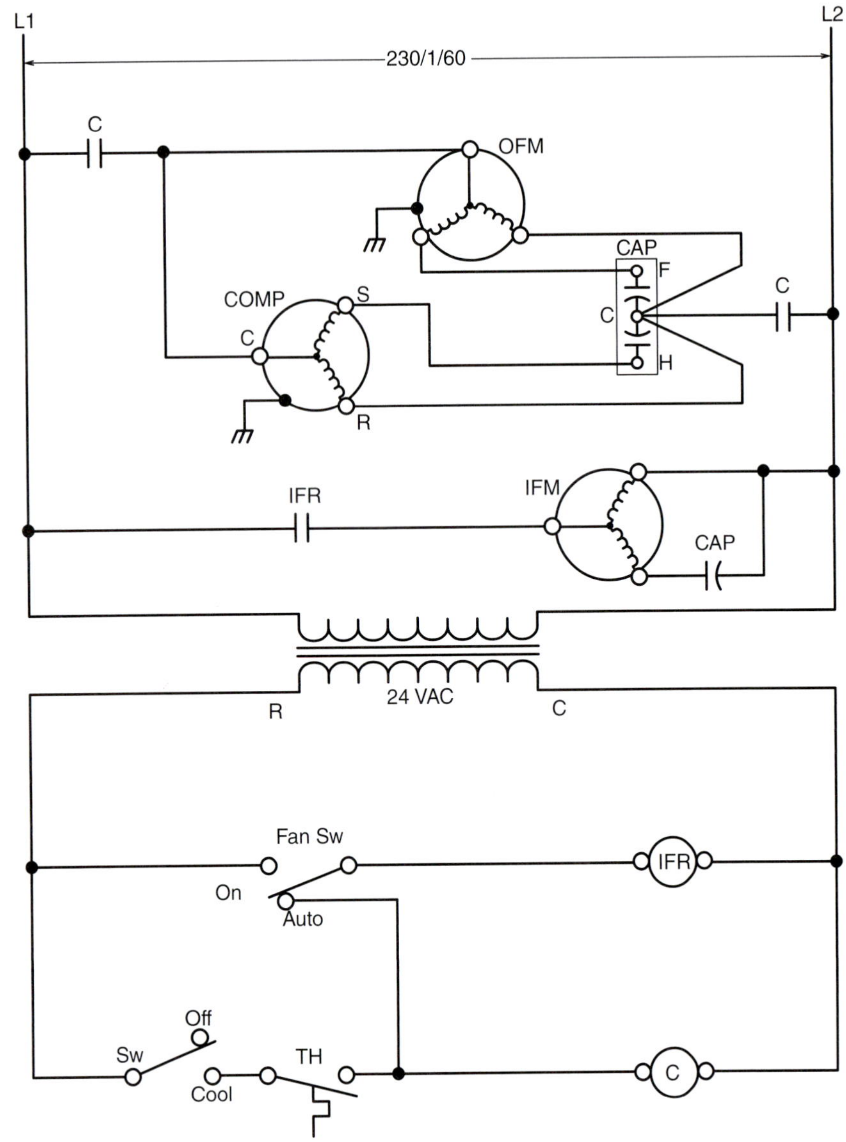

FIGURE 16-30 Simple ladder (schematic) diagram.

Notice how the diagram resembles a ladder. The power sources are shown as the vertical legs, with load and control circuits shown as the rungs between the legs. Note how the two power sources—line voltage and the control transformer—are separated from each other.

Low-Voltage Section

All wiring below the transformer represents the low-voltage section of the diagram. Ladder diagrams usually flow from the highest voltage at the top to the lowest at the bottom. However, some do choose to lay them out differently, including sideways. Examine the low-voltage wiring, beginning with the highlighted portion of the diagram shown in **FIGURE 16-31**.

The indoor fan control circuit operates on 24 VAC. The coil of the indoor fan relay (IFR) is in this rung of the ladder, along with the Fan switch of the thermostat. The Fan switch has two positions—On and Auto. In the Auto position as shown, the fan cannot run unless power is provided to the switch from the ladder rung below it. In the On position,

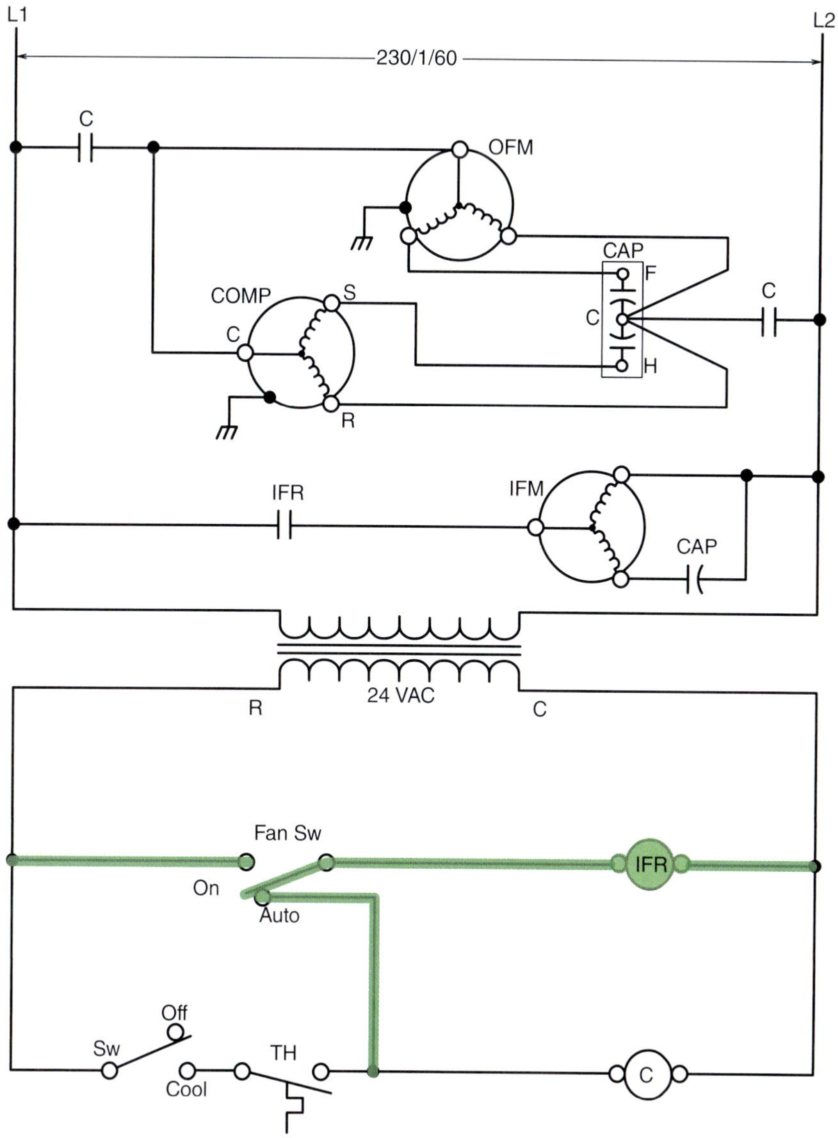

FIGURE 16-31 The indoor fan control circuit.

the switch completes the 24 VAC circuit to the IFR and thus the coil is energized. When it does, the contacts of the relay change their position.

Now examine the temperature control part of the circuit, shown in **FIGURE 16-32**. The coil of the contactor (C) is in this rung of the ladder. There are two circuit interruptions that prevent it from being energized. The first is the thermostat switch (SW). The switch must be closed for the compressor and other components needed for cooling to operate. The thermostat (TH) symbol indicates that it closes on a rise in temperature—appropriate for a cooling system. If the temperature rises above the set point of the thermostat, it will close. With the switch set to the Cool position and the thermostat closed, the closed circuit will energize the contactor coil. Its contacts, found in the line-voltage portion of the diagram, will then change position. In addition, when the Fan switch is in the Auto position, it will receive power from the thermostat and energize the IFR. When the room cools and the thermostat reopens, both the contactor and the IFR are deenergized. Note that there are no breaks in the circuit feeding the primary side of the transformer; it is connected directly to L1 and L2. As is common in HVACR systems, the transformer is energized at all times when the line-voltage power supply is present.

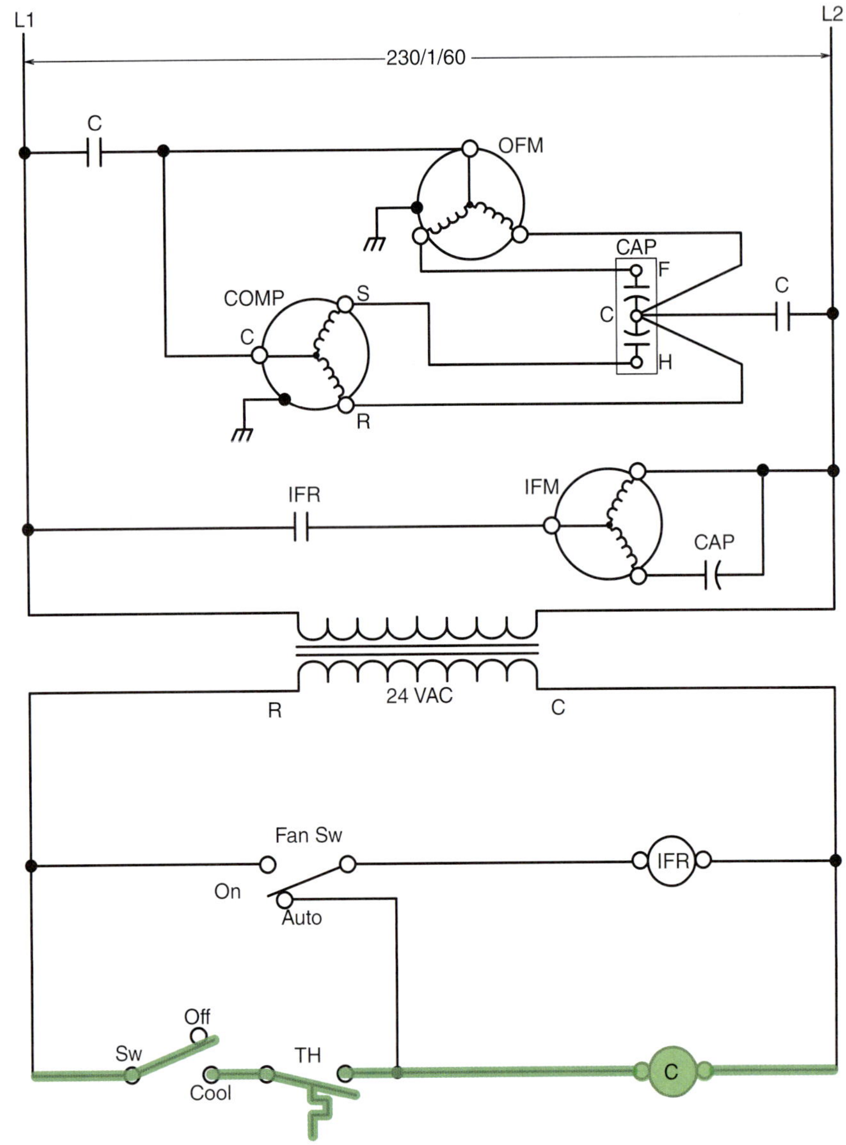

FIGURE 16-32 Temperature control.

Before moving on, notice that the Common (C) side of the control transformer is connected directly to each of the components in the control circuit, without interruption. All switching occurs in the other leg of power, labeled as R. This is the primary reason that one leg of power from the transformer is referred to as Common, and a very important lesson to remember.

Line-Voltage Section

The line-voltage portion of the ladder is examined in this section. In **FIGURE 16-33**, the line-voltage indoor fan circuit is highlighted. The motor is shown in the rung, and there is also a run capacitor in the circuit. There is only one interruption in the circuit, and that is the contacts of the IFR. Note that the IFR contacts share the same identifying label as the operating coil. Although, in reality, they are housed together as a single component, they are separated on the wiring diagram to better represent their independent functions in the line- and low-voltage circuits. All that is required for the indoor fan motor to run is the closure of the IFR contacts. Since there are no additional leads for the indoor fan motor shown, we can assume that this one is a single-speed motor.

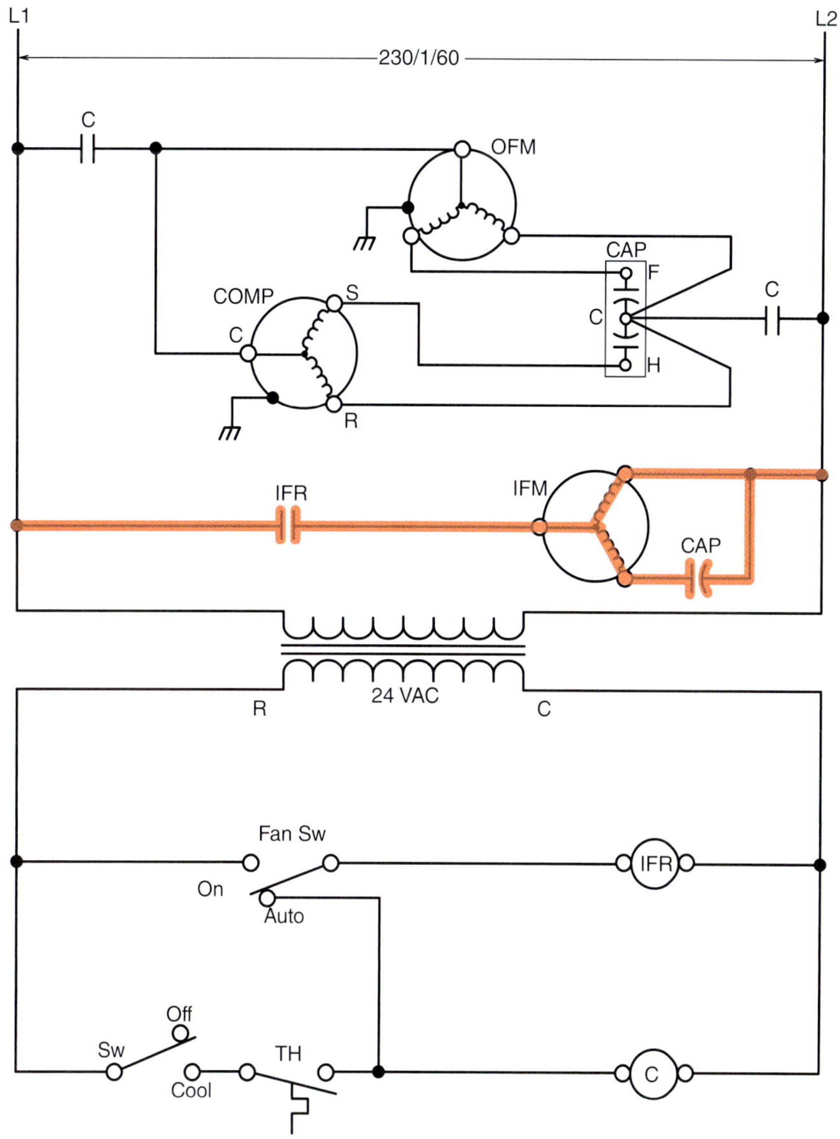

FIGURE 16-33 The indoor fan motor power circuit.

The compressor motor (COMP) is shown in the line-voltage portion of the circuit, as expected (**FIGURE 16-34**). However, this rung of the ladder has breaks in it on both sides—no portion of it is consistently connected to the power source. This contactor has two sets of contacts (double pole). Until the contactor is energized and closes, no portion of this circuit is energized. Again, the contacts share the same label as the contactor coil, located in the low-voltage portion of the diagram.

Since the two sets of contacts inside the contactor close together, both legs of power are provided to the compressor motor at the same time. One leg is connected to the Common terminal of the dual run capacitor. A second wire on this same capacitor terminal is routed to the compressor Run winding. The lead connected to the anode of the capacitor goes to the Start winding. The contact on the left side of the diagram provides power to the Common terminal of the compressor.

The final portion of the diagram is highlighted in **FIGURE 16-35**. The outdoor fan motor (OFM) is controlled by the same two sets of contacts as the compressor. It too is connected to the dual run capacitor. When both sets of contacts close, the condenser fan motor will run.

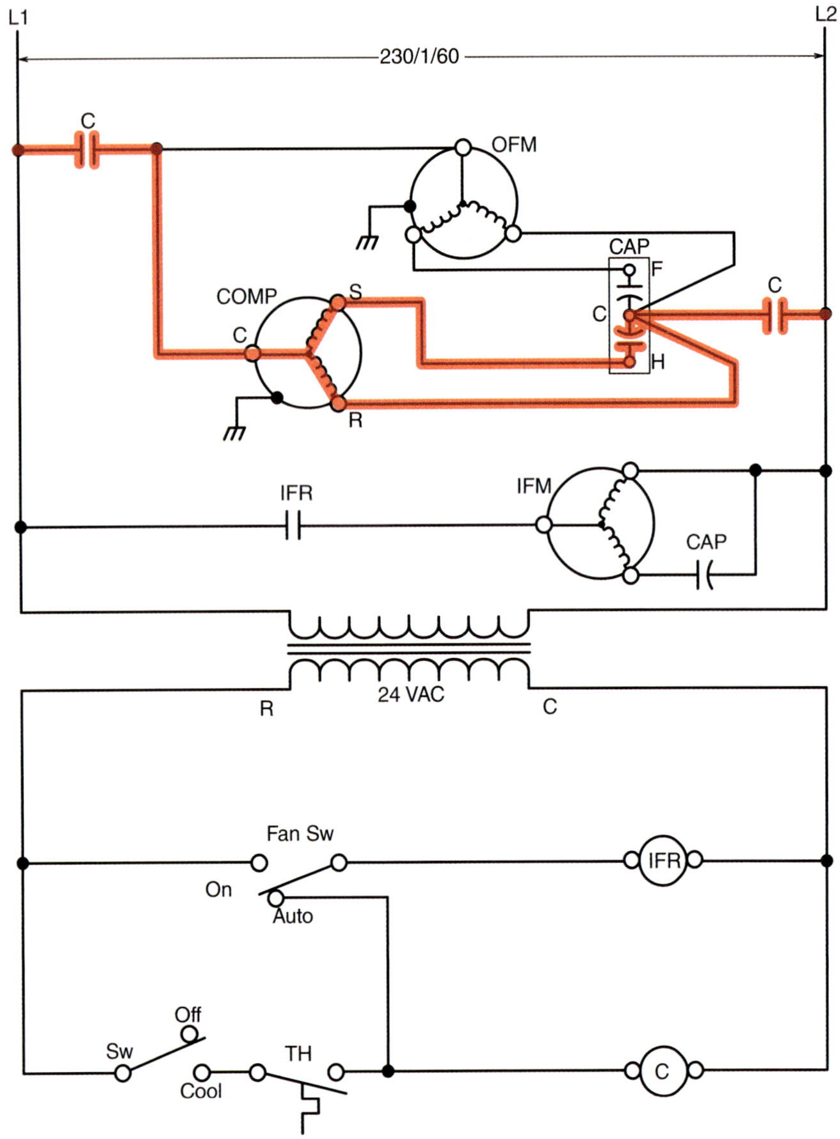

FIGURE 16-34 The compressor motor power circuit.

Line Numbers

The diagram examined is a very simple one; simpler than most. Diagrams can be much larger with many more rungs in the ladder. The line-voltage side may be laid out and drawn in various ways, but they basically show the same information. A standard residential or light-commercial heating/cooling system has a relatively predictable number of line-voltage loads (a compressor, fan and blower motors, and a transformer are the most common significant loads), with limited variations. The control side, however, is more likely to vary in complexity and control devices.

Large or complex ladder diagrams use a line numbering system to help users find their way around. The concept starts with adding a string of numbers down the left side of the diagram. Each number represents a row number. The row numbers are then shown on the right side of the diagram to identify where relevant parts of a device are.

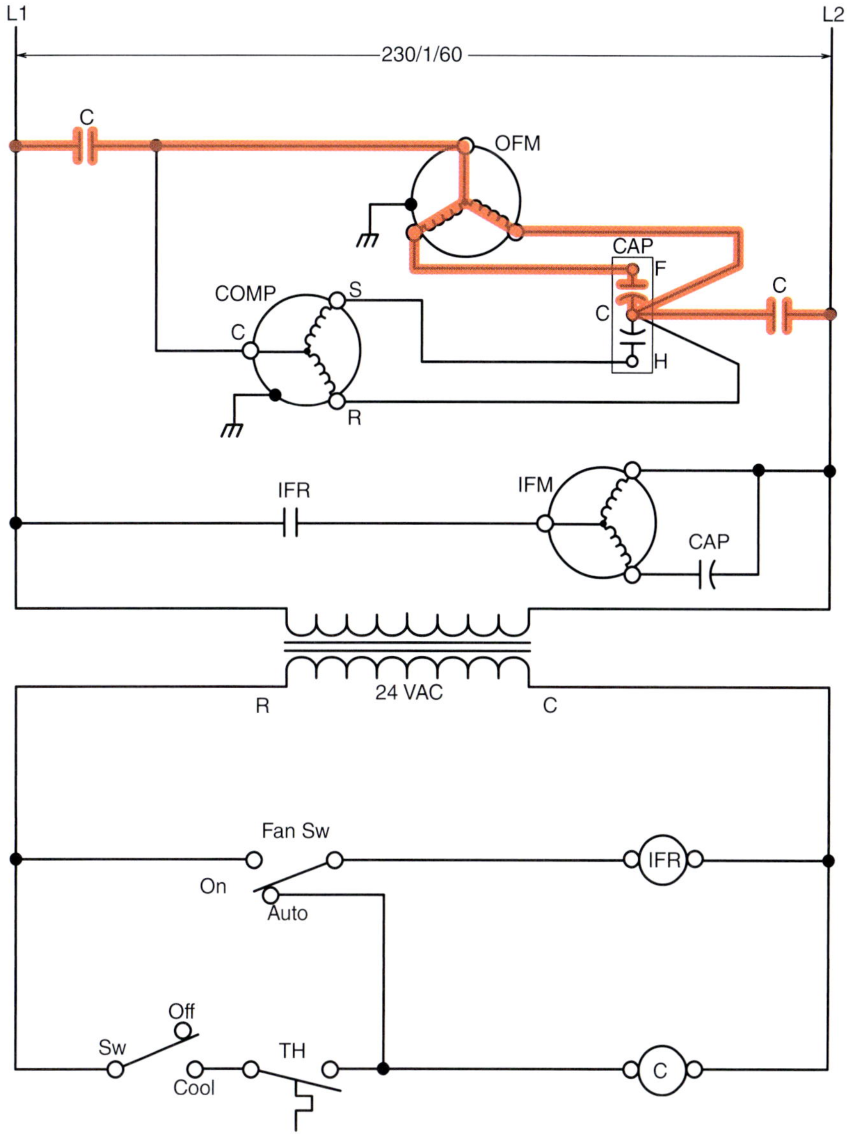

FIGURE 16-35 An OFM power circuit.

Refer to **FIGURE 16-36**. A column of numbers has been added to the left side, used as position indicators. When a device (not its label) is not precisely lined up with a number on the left, it is associated with the nearest number below. In Figure 16-36, Line 1 is well aligned with one contact from contactor C. The other contact is found near, but just above, Line 5.

Now look at the low-voltage section, and at the C contactor coil specifically. To the right are the numbers 1 and 5. When examining the diagram and following circuits, simply look at the numbers and locate all the contacts associated with C.

The IFR contacts can be located the same way. The IFR contact is located at Line 10. Note that if an associated contact is normally closed, the line number is underlined. In this example, none of the line numbers on the right side is underlined, indicating that all the associated contacts are normally open.

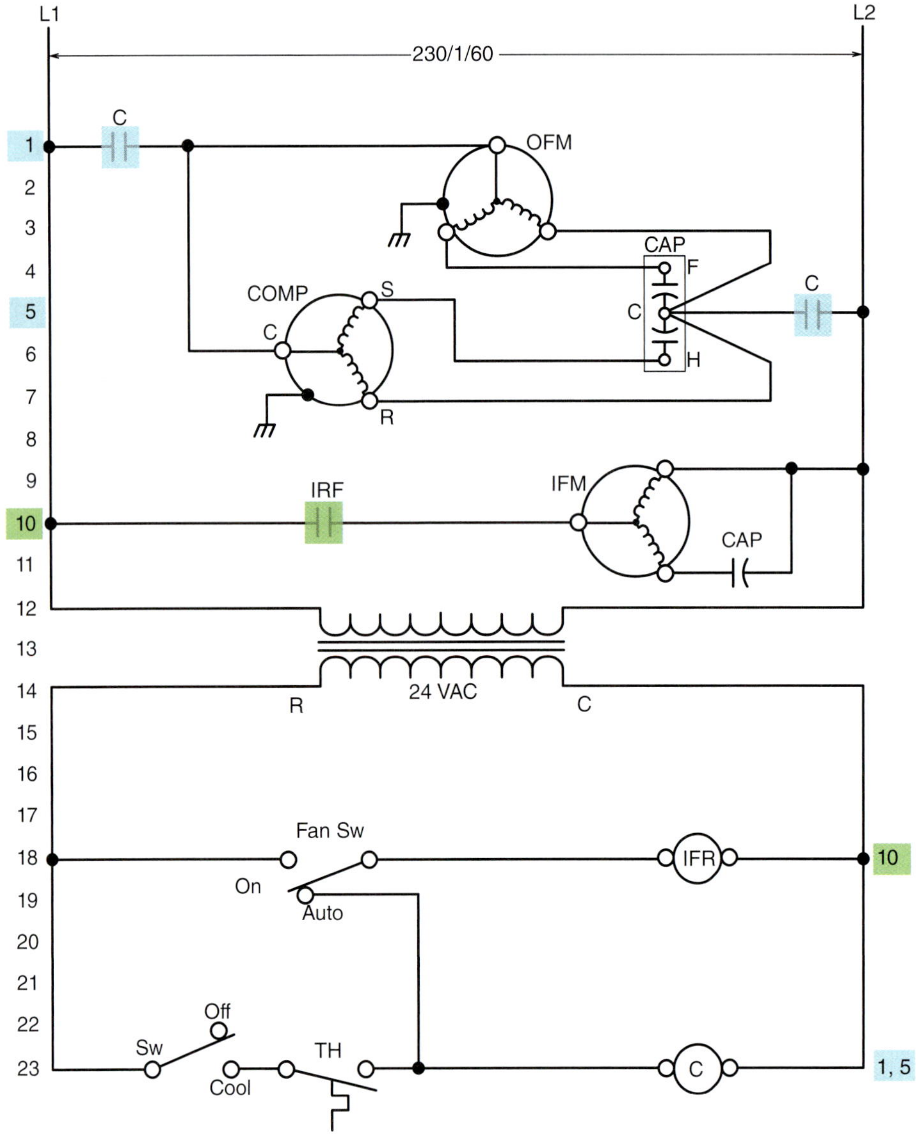

FIGURE 16-36 Using line numbers.

Although this system may not seem necessary on a diagram this simple, it is very helpful on larger and more complex diagrams. It is also helpful when trying to examine a diagram on a smartphone or tablet screen on a smaller scale.

▶ Connection and Installation Diagrams

K16004 Identify and interpret connection and installation diagrams.

Connection and installation diagrams are usually provided with HVACR equipment, along with schematics. Although schematics are far better to use for troubleshooting purposes, **connection diagrams** are more helpful when tracing wires from point to point, confirming that a unit is correctly wired. They are also helpful when replacing an electrical component, especially when that component is not identical to the one removed.

Connection Diagrams

Connection diagrams are also referred to as pictorial or line diagrams. The term "pictorial" stems from the fact that devices are displayed more realistically. For example, the coil and contact set(s) of a contactor will be shown as a single assembly, with a border around them to represent the enclosure. Connection diagrams focus on how the equipment is physically connected, whereas ladder/schematic diagrams focus on how the devices interact with each other.

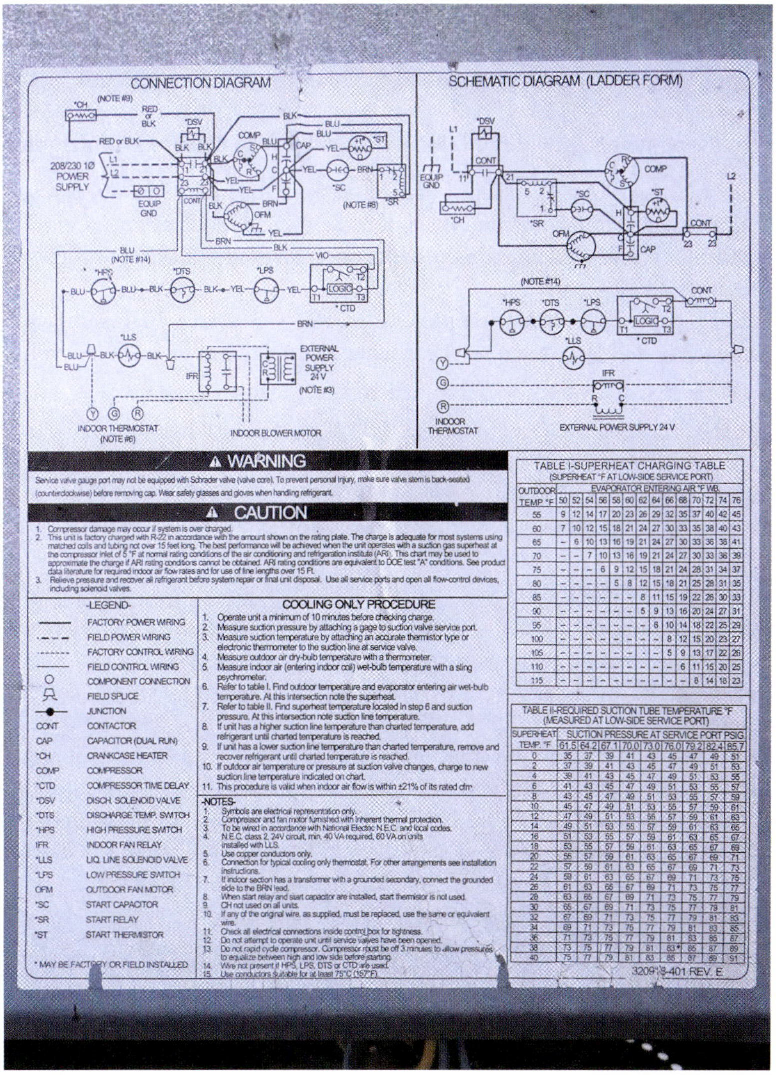

FIGURE 16-37 Typical label diagram.

A typical label diagram for a simple air conditioning condensing unit is shown in **FIGURE 16-37**. The label diagram shown includes the connection diagram with wire colors identified; a schematic diagram in the upper right corner; a legend for abbreviations; and a collection of notes. Many label diagrams also include refrigerant-charging information as shown here. A technician can expect to see diagrams like this on the access door of a unit. Their layout and appearance will, however, vary among manufacturers.

FIGURE 16-38 One of several available CoreSense modules.

▶ TECHNICIAN TIP

Air Conditioning Diagnostics

Technicians may encounter control wiring on a wiring diagram associated with a device labeled CSM—a CoreSense module (**FIGURE 16-38**). CoreSense modules are designed for use with Copeland scroll compressors. They monitor compressor conditions and provide diagnostic information to technicians. Information is provided through flashing LEDs on the control. Codes include, but are not limited to, excessive run time, shutdowns due to pressure safety switches, short-cycling, and compressor locked-rotor conditions. Conditions such as short-cycling are first alerted but will eventually result in the module taking control and shutting down the compressor to prevent damage. This prompts a visit from a technician. Look for similar, and ever more intelligent, devices to find their way into HVACR systems.

Note that the connection diagram usually requires more space for clarity than does a ladder diagram of the same equipment. More area is needed to avoid confusion in the many crossing wires. Wire colors are shown, facilitating a visual examination of the actual wiring.

The following points provide additional insight into the details of the diagram:

- A note on the left upper corner of the connection diagram (Note #9) is related to the crankcase heater, indicating that it is not present in all models.
- Note the double-dashed lines on the connection diagram. The legend shows that these represent field power wiring.
- Note #8 is related to starting components for the compressor. In this unit, a PTC start thermistor is standard, but a start capacitor and relay can be installed in its place as an option.

Again, it is important to remember that HVACR electrical diagrams can vary dramatically in the details. Although they all share the goal of depicting the electrical function and connections of a unit, they can vary significantly in the details. The diagrams of some manufacturers are easier to read and comprehend than others.

Installation Diagrams

Installation diagrams are provided to assist installers and clarify field-wiring requirements for a system. As a general rule, they provide little or no troubleshooting assistance to technicians. They can be used, however, to verify that field wiring was done correctly if the problem points in that direction.

Installation diagrams may be a component part of a label connection diagram, provided in the installation manual, or both. Since most systems require both low- and line-voltage field wiring, installation diagrams for both are generally needed. Installation diagrams are also found in the documents provided with thermostats and similar control devices.

FIGURE 16-39 shows an example of a line-voltage installation diagram for a single-phase unit. It is not unusual for references to notes on the label diagram to be added. The notes may refer the installer to tables for wiring sizing, specify a maximum wire size allowed due to the size of the screw terminals, and show similar information about the required power supply.

Similar installation diagrams are usually provided for low-voltage wiring. Refer to **FIGURE 16-40**. In some cases, the field-wiring terminals of the chosen thermostat do not match the terminals as labeled on the installation diagram. When that happens, the installer must determine the function of the wire and match it with the appropriate function of the

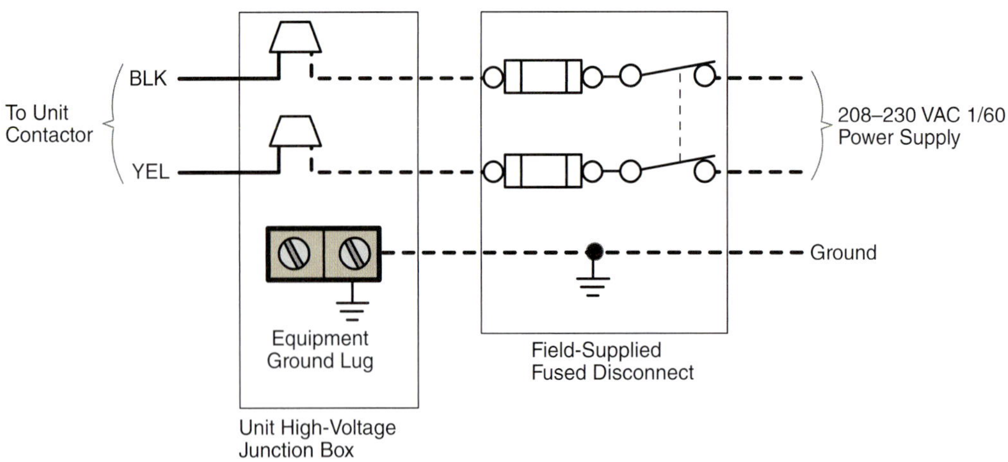

FIGURE 16-39 Line-voltage installation diagram.

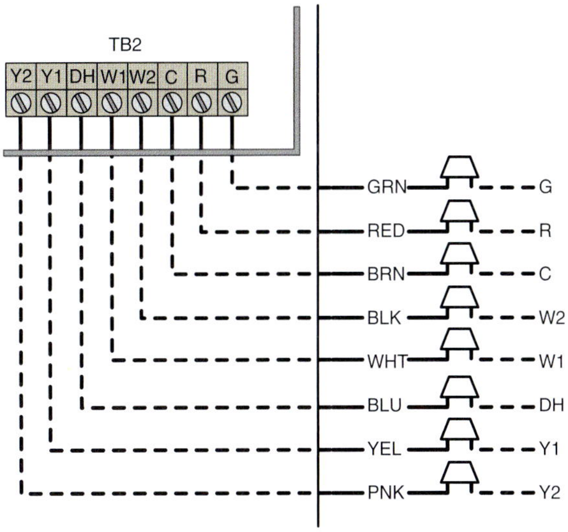

FIGURE 16-40 Low-voltage installation diagram.

thermostat. This is more likely to happen with unique terminals such as X, S, L, and Aux. Commonly used terminals such as C, R, Y, W, and G generally match most thermostats on the market.

In Figure 16-40, there is a note associated with the DH (dehumidification) terminal. This terminal is reserved for units that offer a specific dehumidification mode (usually cooling operation at a reduced fan speed). The related note simply indicates that the terminal is used only with specific unit models that offer the function. Units equipped with ECM blowers are more likely to offer this feature, due to their ability to modulate the blower speed.

▶ Wrap-Up

Summary

Electrical diagrams provide the roadmap for technicians to comprehend the electrical operation of HVACR systems. While most HVACR diagrams consistently provide the same basic information, they can vary significantly in the details.

Electrical diagrams consist of elements such as lines, legends, notes, and symbols. Symbols are used to represent electrical devices and, in many cases, represent how those devices respond to changes. The symbols for temperature and pressure switches fall into this latter category.

To best serve the needs of technicians and installers, electrical diagrams are intentionally drawn in various forms. Ladder, or schematic, diagrams, are best for troubleshooting and determining the sequence of operation. Connection, or pictorial, diagrams are best to see how the devices are connected and, in some cases, are physically related to each other. Label diagrams found on equipment access doors generally include ladder and connection diagrams, as well as the relevant notes and legends. Installation diagrams are usually small and relatively simple, yet they provide essential information regarding external wiring required for system operation.

One of the characteristics of successful technicians is their ability to interpret electrical diagrams quickly and accurately. These technicians are held in high regard and are often approached by others for guidance. Mastering this skill represents a sizable step toward mastering the HVACR trade in general.

Key Terms

connection diagrams Electrical diagrams that represent circuit components as intact devices and how the interconnecting wiring is arranged.

installation diagram An electrical diagram that shows wiring that must be completed in the field and essential details regarding equipment connection.

ladder diagram An electrical diagram that has the appearance of a ladder, with individual circuits forming the rungs. Ladder diagrams best represent how the equipment functions electrically and is used primarily for troubleshooting.

legends Sections of wiring diagrams that provide interpretation for abbreviations, line types and their meaning, and similar information.

schematic diagrams Electrical diagrams that represent how equipment functions electrically, typically drawn in the ladder-diagram format.

Review Questions

1. Unless otherwise noted, electrical diagrams show the state of switches in the __________ position.
   a. energized
   b. deenergized
2. Wiring that is external to a piece of equipment and field-installed is usually represented by using __________.
   a. solid, bold lines
   b. solid, thin lines
   c. dashed or dotted lines
   d. double lines
3. In the figure below, the dashed line represents __________.

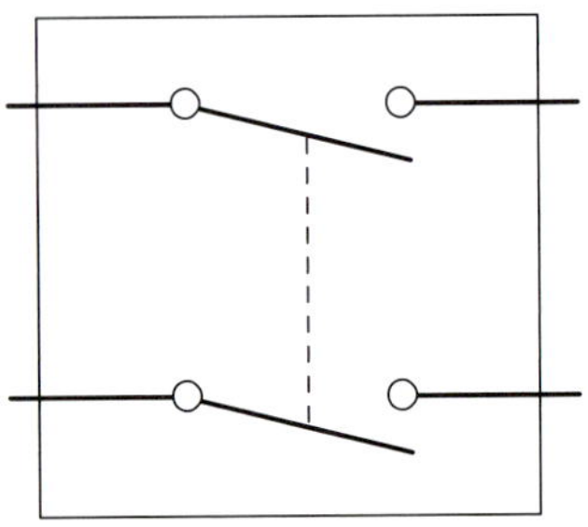

   a. field line-voltage wiring
   b. field low-voltage wiring
   c. a jumper wire
   d. a mechanical interlock between the two switch poles
4. The meaning of an unfamiliar component abbreviation can be found by consulting the__________.

Refer to the figure below to answer questions 5–14. Place the identifying letter of the items shown beside their matching description. *Note*: There are more symbols than descriptions—not every symbol is accounted for with a description.

5. Three-phase motor _____
6. Temperature switch, normally closed, opens on a fall _____
7. Flow switch, normally closed, opens on a fall _____
8. Temperature switch, normally open, closes on a rise _____
9. Contactor coil _____
10. Wires that do not connect _____
11. Capacitor _____
12. Double-pole, double-throw contacts___
13. Time-delay switch, normally open, timed closed (on delay) _____
14. Pressure switch, normally closed, opens on a fall _____
15. On schematic diagrams, the coil and contacts of a relay are __________.
    a. always shown adjacent to each other
    b. shown where they best represent their places in the circuit
    c. pictured together as a unit
    d. shown wired in series
16. In a ladder diagram, the vertical legs represent the __________.
    a. power source
    b. control circuits
17. Ladder diagrams that use a line numbering system also provide row numbers on the opposite side to identify the row where associated contacts can be found. If the row number is underlined, it means the contact is __________.
    a. normally open
    b. normally closed
18. Connection diagrams are also referred to as _____.
    a. pictorial diagrams
    b. schematic diagrams
    c. installation diagrams
    d. field-wiring diagrams
19. A feature more likely to be seen on connection diagrams than on ladder diagrams is _____.
    a. low-voltage wiring
    b. line-voltage wiring
    c. wire colors
    d. line numbers
20. To verify that field low-voltage wiring was installed and connected correctly, the best diagram to refer to is a(n) _____.
    a. ladder diagram
    b. installation diagram
    c. line diagram
    d. pictorial diagram

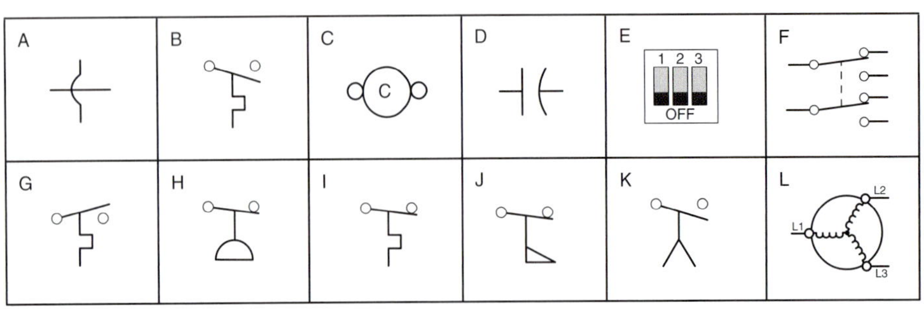

CHAPTER 17

Basic Troubleshooting Concepts

Knowledge Objectives

After reading this chapter, you will be able to:

- **K17001** Describe the five steps in strategy-based troubleshooting.
- **K17002** Explain how to use a multimeter in electrical troubleshooting.
- **K17003** Explain how to apply alternative troubleshooting measures and work with tech support.

Skill Objectives

Skills required to meet the objectives of this chapter:

- **S17001** Measure a power supply voltage and determine whether the voltage is within specifications.
- **S17002** Use the hopscotch method to troubleshoot an electrical problem.

▶ Introduction

Troubleshooting HVACR systems successfully requires careful thought and a well-conceived strategy. Technicians that demonstrate their ability to diagnose and solve problems set themselves apart quickly, and they generally become the go-to people when a problem proves difficult to diagnose.

Although there will certainly be challenging problems, troubleshooting most problems isn't that difficult. A sizable portion of the effort involves mental effort—analyzing a situation and determining what things can be eliminated from the list of possibilities. Physical troubleshooting should always follow mental troubleshooting.

This chapter will offer a simple five-step strategy for the troubleshooting and repair process. Although by this point in training, a technician will likely have completed many exercises with multimeters and similar instruments to check the condition of individual components, using the meter to quickly identify problems in circuits will be presented here. A discussion of alternatives to traditional troubleshooting will complete this introduction to basic electrical troubleshooting.

▶ Strategy-Based Troubleshooting

K17001 Describe the five steps in strategy-based troubleshooting.

As mentioned earlier, a great deal of troubleshooting is a mental exercise. When the situation is carefully analyzed, the physical portion of troubleshooting (**FIGURE 17-1**) to isolate the problem is more focused and effective. When the problem is not analyzed carefully at first, a lot of unnecessary testing often results.

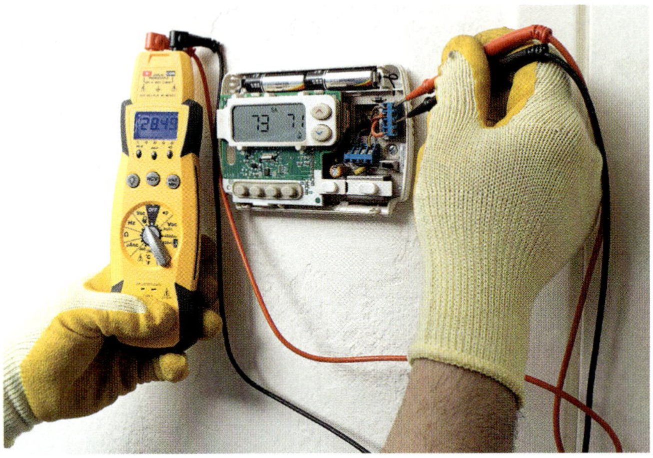

FIGURE 17-1 Troubleshooting a thermostat.

It is important to note that the five steps to be presented here are not just about troubleshooting. Troubleshooting only identifies the problem. But the job isn't done until the repair is done and the system is returned to service. This strategy will offer some assistance in approaching the repair after having proceeded with the diagnostic process.

The five-step strategy presented here includes the following:

1. verifying the problem
2. considering the possibilities
3. focused testing
4. performing the repair
5. final testing.

Each of these steps will be discussed as the section progresses.

Verifying the Problem

Upon arrival to the location of a service call, do not forget that the customer pays the bills—air conditioners and furnaces don't write checks or offer up credit cards. The task is to solve someone's problem, not just fix something. Greet the customer with a smile and a sympathetic ear.

The mental exercise begins with verifying the problem. All too often, technicians discount the input of the owner or system user. But the information they offer is extremely valuable. Yet there can be a great deal of disparity between a customer's explanation of what they are experiencing and the real problem. Today's savvy customers, with the Internet at their disposal, often know more about their system and its problem than the customer of yesteryear. Many owners research their problem and seek information before the service call is ever placed (**FIGURE 17-2**). But just as patients experiencing pain can discover all sorts of serious as well as not-so-serious causes on the Internet, so too can HVACR system owners. In some cases, they have read enough to

FIGURE 17-2 Owners often research their problem on the Internet.

make a decision of their own about the cause and what needs to be done. Other customers prefer to avoid that exercise altogether and rely solely on the expertise of a trusted technician. Regardless of the customer's approach to the problem, the information they offer is valuable and helps get the troubleshooting process started.

Verifying the problem also requires making a quick visual inspection of the system. Check to determine whether the system is doing (or not doing) what the customer says. A user can easily misinterpret what they see or hear. For example, a customer may step outside to look at the unit and find it running. However, it may be running intermittently for short periods with limited effect—a symptom of a unit cycling on a low-pressure switch due to a low refrigerant charge. They may also see a condenser fan running, but not realize the compressor isn't running. Indeed, the customer may not know what a compressor is. A quick visual inspection can provide additional information and help determine where to begin focused testing.

During the verification process, some simple actions can help determine the extent of the problem. Check the thermostat settings, for example. If it is an electronic model, consider accessing the configuration menu and checking the settings if the complaint indicates an issue that such settings could affect. A complaint that indicates it maintains comfort but cycles on and off a lot may simply indicate that the cycle-rate setting is incorrect. Note that this doesn't mean that a technician should react to the first thing said, make a change, and then leave. A system with that complaint needs to be checked more thoroughly. Making an assumption from a complaint and being less than thorough often leads to an embarrassing return visit.

Active Listening

Active listening is a valuable skill that every technician should work to develop (**FIGURE 17-3**). It is not complicated, but it does support clear communication with others. Active listening isn't just for troubleshooting—active listening skills can be applied to all aspects of life.

Active listening is simply one person making an obvious attempt to show that they are listening and interested in what another has to say, and doing so involves taking steps to ensure messages are communicated accurately. While listening to the customer, look at them and use positive body language, such as nodding of the head, to show they're being heard. An active listener uses other senses in addition to the ears, especially the eyes. Watch how the speaker is delivering the message (**FIGURE 17-4**). Do they seem unsure? Are they already annoyed or perhaps fearful about what might be wrong? Pay close attention to the words being said, the tone of voice, and the body language being displayed. Body language can say a lot and is typically expressed instinctively. That means it is usually genuine.

FIGURE 17-3 Active listening leads to improved communication.

FIGURE 17-4 Pay attention to body language when communicating with the customer.

One of the most important things to do when listening to the customer's description of a problem is to use the art of **paraphrasing**. It is an easy way to confirm information and shows that the conversation is being taken seriously. Paraphrasing refers to repeating what a customer has said using words similar to those of the speaker and in the form of a question, to confirm understanding. The following is an example of paraphrasing during a service call:

Mrs. Jordan: "It was hot in the house when I came home. Nothing seemed to be running. I did turn a switch on the thermostat to On, and I hear a fan running, but it's not getting cooler."

Technician: "So, you changed the position of one switch to On?"

Mrs. Jordan: "Yes."

Technician: "Let's look at the thermostat together. Ah—I see what's happened. The system switch is still set to Off; only the Fan switch has been moved to the On position. We'll turn the cooling on now and then check the system. I'll make sure it's running as it should."

This simple exchange shows how paraphrasing—saying what has been said in different words to confirm or prove understanding—supports problem solving. In this case, listening carefully to what was said and how it was said led to an instant solution. When someone actively listens and uses paraphrasing, it shows that they value what the customer is saying and that they are an interested participant in the conversation.

Troubleshooting Questions

The troubleshooting process truly begins when the conversation with the user begins. Each comment or complaint from the customer leads to a mental list of possibilities. As the conversation continues and more information is gathered, the list begins to shorten. For example, if the customer says it is hot inside and the system doesn't seem to be running, the mental list begins with loss of line-voltage power or control power. Knowing this, the follow-up question "Are you sure the indoor fan isn't running?" can reveal a great deal. If the answer is "Oh yes—the indoor fan is running, but the air isn't cold," a total loss of control power can usually be eliminated as a possibility. Any concern about the loss of line-voltage power to the indoor unit (or a packaged unit) can also be discarded.

In the case of a system that is fully offline, it would seem that there is little to discuss with the customer. However, even in these cases, the customer can offer valuable information about the circumstances of the failure. Ask the customer whichever of the following questions apply:

- When did you first notice it was not running?
- Was the failure sudden, or did you notice other problems before it failed completely?
- What was the weather like? Were there thunderstorms in the area?
- Did you experience a power outage before the system failed?
- Is there anything else in the building not working?
- What have you tried to get it restarted?

The answers to these questions will help make sense of what happened and what may have failed, if anything. If there was a thunderstorm and the problem started with a power outage, the problem may be a blown fuse or tripped circuit breaker. In commercial buildings with a three-phase power source, the loss of a single phase to the building will leave some things functional while others are not. Thus, if multiple items aren't working, the technician will often find that the HVACR system is not the only thing lacking an appropriate power supply.

Listen carefully to the answers and adjust the mental list of possibilities accordingly. Then consider and ask questions that can help shorten the list further and move the process along.

While gathering information, customers may ask questions as well. A good attitude and a willingness to answer these questions will serve a technician well. It is OK to say, "I don't know the answer to that right now, but I will find out." Then do exactly that, and let the customer know the answer once it has been ascertained. Always follow through with commitments made to the customer, without fail. When that's not possible, proactively reach out to the customer with an apology and explain why a commitment can't be honored.

Consider the Possibilities

After gathering as much information as possible, the mental list can be analyzed and narrowed to the most likely areas of concern. Rather than thinking of individual components (although that may be appropriate for small, simple systems), think in terms of systems or circuits. For example, if a compressor is trying to start but its internal overload is opening on every attempt, it should be obvious that the thermostat is calling for cooling and the control system is energizing the contactor. The place to begin in this scenario is at the contactor, checking for the appropriate power source and examining the wiring connections. If nothing is found there, another restart must be attempted while the compressor motor current is being monitored. It is likely that the compressor is mechanically seized or that a start capacitor or relay has failed (if present). Motors with open or grounded windings do not usually try to rotate; they either trip a breaker, blow a fuse, or do nothing at all.

While considering the possibilities, consider what does work as well as what does not. If the compressor is doing nothing but the condenser fan motor runs fine, then power is being provided to the unit. Simply approaching a split-system condensing unit in this condition allows the indoor unit, the power source, and the incoming control wiring to be eliminated from the mental list of possibilities. If some components in the unit are responding to the call for cooling as expected, then those three areas are not likely to be involved.

This step in the strategic approach to troubleshooting and repair is where experience will improve judgment and accuracy in diagnosing problems. Honing this skill will serve technicians well as their careers progress. One of the common responsibilities of service managers is to mentor technicians and provide guidance in this step of troubleshooting (**FIGURE 17-5**). Once focused testing begins, technicians must function on their own. But service managers often assist less-experienced technicians with guidance in analyzing a situation and providing direction. Those who master the art of analyzing information to guide the troubleshooting effort often advance much faster than their peers.

▶ TROUBLESHOOTING TIP

The If-Then-Else Statement

In computer programming, the If-Then-Else statement has long been used to execute diverse instructions based on conditional information. **FIGURE 17-6** shows the basic flow of the statement and how one conclusion leads to another set of conditional statements. But this statement also applies to troubleshooting. Some people use it mentally already but perhaps do not recognize it nor do so consciously. The If portion of the statement is usually a true-or-false condition: the answer leads to one of just two possible paths. If a condition is true, then another conclusion, or set of conclusions, is also true. If the condition is false, then conditions under the "else" portion of the statement are true. Think about the If-Then-Else statement the next time a situation calls for analyzing information to determine where to begin focused testing.

Focused Testing

Focused testing is when the technician begins to eliminate subsystems and components from the remaining list of possibilities. Instruments are used to test specific circuits and components (**FIGURE 17-7**), which are then identified as the problem or eliminated from the list of possibilities.

It is important during focused testing to know what to expect from every test. As testing circuit continuity begins, for example, know whether the circuit should be open or closed. If it is not as expected, stop and consider why. Has the component failed? Is control voltage present at the correct places? This often leads to more focused testing, but with the emphasis on a much smaller area or number of components. Eliminating possibilities until the real problem is identified is the purpose of focused testing.

FIGURE 17-5 Mentoring a technician.

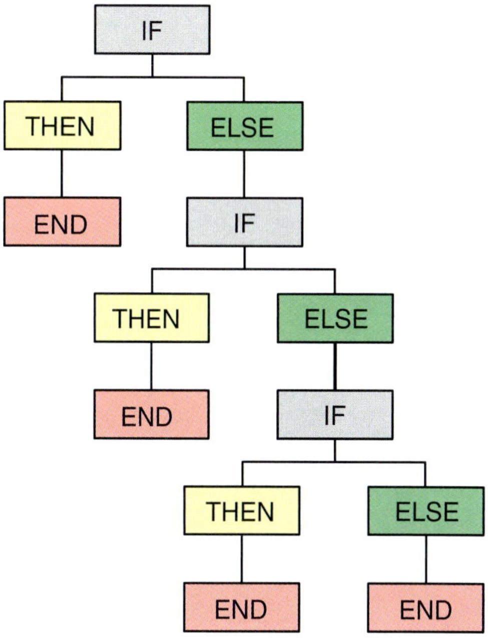

FIGURE 17-6 The If-Then-Else statement.

FIGURE 17-7 Troubleshooting with instruments.

It is important to remember the testing actions taken and to ensure that they are documented in some form. Every specific step taken doesn't need to be recorded, but a thorough outline of the process may be needed for warranty claims or for future reference. Customers also feel more comfortable with such information in hand, indicating that a strategic approach to the problem was taken. It is much better to clearly document some details of the troubleshooting effort than to simply say a defrost control is defective after being on the job for 1.5 hours.

A good place to begin focused testing is to start with the simplest actions. Some simple tests can confirm suspicions about the problem and, more importantly, reveal some additional information before more detailed testing begins. In some cases, problems reveal themselves very quickly through the simplest of tests. Look both at and beyond the obvious, and consider what each new test reveals and how it affects the list of possibilities.

Later in this chapter, a method of electrically testing an energized control circuit will be presented to support approaches to focused testing.

Making Repairs

Once the problem has been clearly identified, repairs can be made (**FIGURE 17-8**). However, it is very easy to think the problem has been identified when in fact it was only another symptom. A blown fuse is a good example of this. A blown fuse will certainly keep a system from running: it is a reason why a system fails to operate. But fuses do not simply fail without having been exposed to excessive current. A blown fuse should always stimulate an interest in cause and under what circumstances an overcurrent condition occurred. Once the fuse is replaced, the job is not done until major loads (primarily motors) are tested for grounds and the power wiring is inspected for poor connections, burned terminals, and damaged conductors. These are sources of problems, rather than mere symptoms.

It is important to follow the correct service procedures for any part being replaced. Manufacturers will often make changes in parts, especially solid-state components, after a product is released. These parts may require changing the wiring or using different terminals than the original part offered. With solid-state devices and PC boards, there may be jumpers or switches that must be positioned to suit the installation. Follow any packaged instructions carefully, and if there is a discrepancy, contact the manufacturer's technical support department for guidance.

Although customers are anxiously awaiting the completion of the work, do not rush through the process by eliminating important steps. Do it right the first time or redo it, to avoid regrets. A poor wiring connection made today will not get better with age. Be careful not to create new problems as the ones at hand are resolved. Whenever possible, monitor the system through a complete operating cycle. Callbacks on completed work destroy customer confidence, regardless of the reason.

There may be times when a temporary repair can be made to return a system to service while parts are acquired. For example, a defrost relay can perhaps be bypassed in a heat pump during the heat of the summer while a new relay is acquired. Technicians find themselves in a delicate position when a system serves people who are ill or in discomfort or when the system serves critical processes. Customers are often hoping for a miracle of some sort. When faced with these situations, carefully consider the implications of a temporary repair. A temporary repair may place the system in jeopardy for a different problem altogether. If that is the case, the customer must be informed and helped to understand the implications of their choices. On the other hand, the right

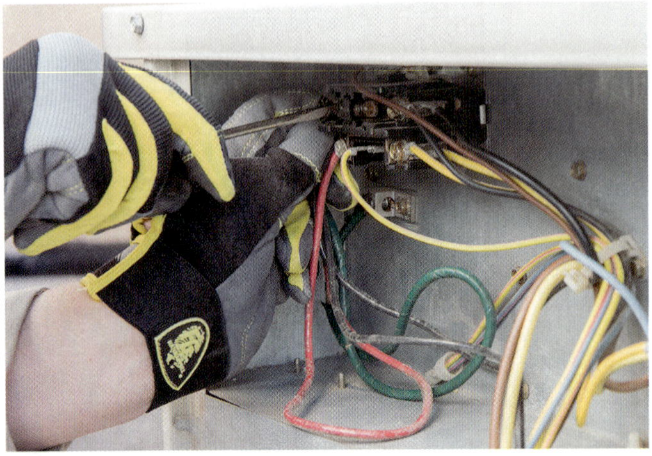

FIGURE 17-8 Making electrical repairs.

attitude and a sincere effort to minimize discomfort or process disruptions until permanent repairs can be completed will enhance a technician's reputation and value. However, poor judgment and taking steps that create more problems at someone's expense can quickly take a technician's reputation in the other direction. Consider the situation carefully while prioritizing the customer's needs.

Follow-Up Testing

Once a repair is made, don't allow any feeling of accomplishment prevent the final step in the process from being taken: follow-up testing. Each new part or repair needs to be tested to ensure that the problem has been resolved and that no new problems have been created. In many cases, a complete system diagnosis cannot be accomplished with a failed part in place.

FIGURE 17-9 A refrigerant-vapor acid test kit.

A failed compressor is a perfect example of this. On a service call, a technician may find that a compressor's motor windings are grounded. This failure could be the result of a compressor defect, or acid from a previous compressor failure could have contaminated the refrigerant circuit and destroyed the winding insulation. A major power surge from a nearby lightning strike could also have caused the failure. But there is little additional testing, beyond an acid test of the refrigerant (**FIGURE 17-9**), that can be done without a functional compressor. The compressor can be disconnected, and the control sequence that leads to compressor operation can be tested. But the rest will have to wait until the repairs are completed. When a problem like this is reported to a customer, it is important to clarify that all the needed tests cannot be completed at that moment and that more testing needs to be done once a new compressor is in place to determine the root cause. Keep the client informed during this process to avoid misunderstandings.

During follow-up testing, attention will naturally be focused on the replaced part and its function. But it is important to also watch the big picture and analyze all aspects of operation, in every practical mode of operation. If the sequence of events or circumstances that existed at the time of failure can be duplicated, it is best to do so. This helps prove the system can now operate as expected in those same circumstances. Doing this with the knowledge of the customer is a good idea too. The more comfortable they are with the results of the work and the thought put into it, the more likely they are to call again in a future time of need. A technician may be fully aware that the circumstances were not related to the failure, but the customer tends to associate those conditions with problems. Breaking that association with evidence through thorough testing goes a long way toward satisfying a customer's concerns.

The following exercise will help develop the mental skills needed to develop a troubleshooting approach. Read the scenario and determine what the most logical and practical steps to take would be. Note that there isn't necessarily a perfect answer, as there are legitimate reasons to support more than one approach. The intent is to practice analyzing the information available and determine how to proceed.

Imagine being inside the home of a customer that has no cooling. All the homeowner can say is that the system doesn't work and that something isn't right: it's very warm inside. When the customer is asked when the problem was first noticed, they indicate that it seemed fine at bedtime but woke to find it warmer than usual and getting worse. The owner then leaves hastily before any follow-up questions can be asked.

What should the first step be in this scenario?

Step 1:

____

____

____

The thermostat appears to be active and shows the current set point, which is well below the room temperature. The display also indicates that cooling is on, but the thermostat can't provide that information—it merely knows that it has sent a signal to start the

▶ TROUBLESHOOTING TIP

Using a Jumper Wire

A simple jumper wire can be a valuable test instrument. One place it can be used effectively is in testing thermostat functions (**FIGURE 17-10**). Working from the thermostat subbase, a jumper wire can be used to energize circuits in the same way that a thermostat does. If a function, such as connecting R to G at the wall plate to start the indoor blower, operates with the thermostat removed but not when it is installed, then the thermostat has been identified as the problem.

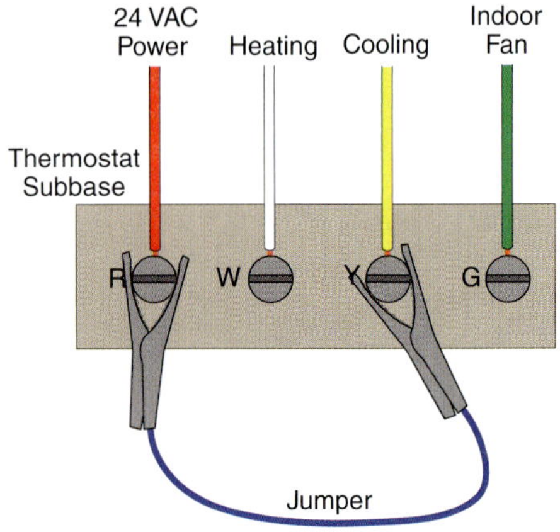

FIGURE 17-10 Using a jumper wire to test thermostat functions.

cooling system. Approaching the thermostat often means the system return-air grille is nearby, and now the indoor fan can be heard running. What helpful information has been discovered, and what would be the next step in the troubleshooting process?

Step 2:

____

____

____

It has now been confirmed that control power is available and that the contactor in the condensing unit is energized. Yet neither the compressor nor condenser fan motor, both connected to the contactor, is running. That means that the problem is common to both motors. Consider this and list as many potential problems as possible that could result in both motors failing to operate. To add to the challenge, arrange the list from the most likely to least likely problem.

Step 3:

____

____

____

▶ Troubleshooting with a Multimeter

K17002 Explain how to use a multimeter in electrical troubleshooting.

S17001 Measure a power supply voltage and determine whether the voltage is within specifications.

S17002 Use the hopscotch method to troubleshoot an electrical problem.

For electrical troubleshooting, the multimeter is the best choice. A good multimeter with the right accessories can handle most of the tests that need to be conducted. If it cannot test and report the microfarad value of capacitors, a separate instrument for that purpose is necessary. Reliable capacitor testing is often needed.

The Voltmeter

The voltmeter is used for energized testing. When a system with problems is first approached, energized testing is generally chosen first because it reveals the most relevant information. Energized testing typically leads to other types of focused testing, often with the ohmmeter. Depending on what control has opened the circuit, checking the refrigerant charge and oil pressure and performing similar exercises are often required as well.

Energized testing can be applied to both line-voltage power sources and control circuits.

Line-Voltage Power Sources

One of the first tests to perform on an inoperative system is a test of the line voltage that serves the system. In many cases, this step can be skipped because of some sort of evidence that the system is energized. For instance, a condensing unit with a running condenser fan is obviously being powered, regardless of what the compressor is doing. However, it is always good practice to check the line voltage and to ensure that it is correct, especially under load.

For most residential and light commercial split systems, the voltage can be tested at the contactor (**FIGURE 17-11**). Larger systems often have terminal blocks for line-voltage connection. If there is no voltage, begin checking upstream at the unit disconnect switch. If the disconnect is fused, check for power at the top first (incoming power) and then at the bottom. If there is no power to the disconnect switch, the system circuit breaker is next. Either the breaker has tripped, or it has suffered an internal failure (which would be rare). It is also possible that a conductor at the breaker or disconnect has become loose or damaged.

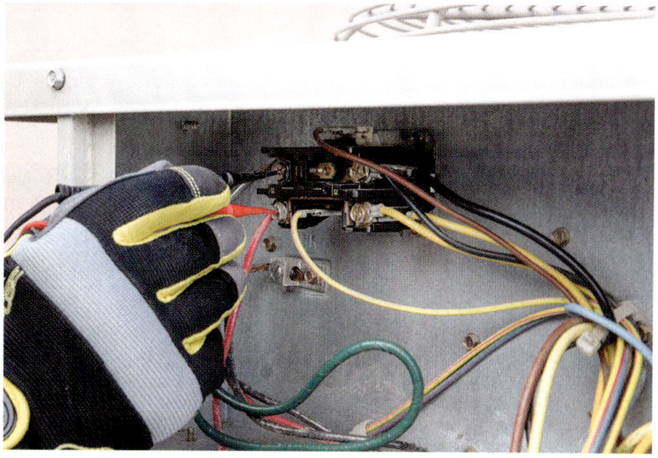

FIGURE 17-11 Checking the line voltage at a condensing unit contactor.

Major loads, such as motors, may be rated for a single voltage, such as 230 VAC, or for a range of voltages. Many motors are rated to operate at 208–230 VAC to increase their flexibility across the range of voltages distributed to utility customers. The input-voltage requirements for motors are as follows:

- For both single- and three-phase motors rated for a single voltage, the voltage must be within ±10% of the nameplate value. For instance, if a motor is rated at 230 volts, the input voltage must be within the range of 207–253 VAC (±23 volts, representing 10% of the nameplate rating).
- For both single- and three-phase motors with dual-voltage ratings, the input voltage must be within 10% of the lowest voltage and 10% of the highest voltage listed. For example, if a motor is rated for 208–230 VAC operation, then the minimum voltage is 187 VAC and the maximum voltage is 253 VAC.

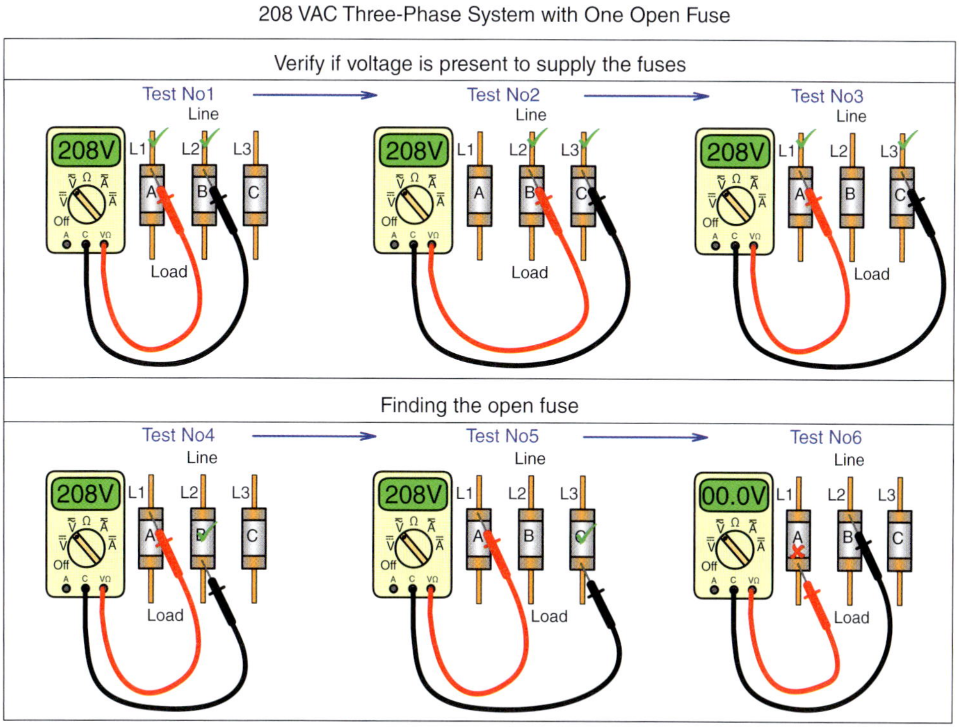

FIGURE 17-12 Testing fuses in an energized circuit.

▶ TROUBLESHOOTING TIP

Testing Fuses in an Energized Disconnect

Checking a fused power source is often initially done with the fuses in place and the disconnect energized. First, remember to practice the highest level of electrical safety, recognizing that an open, energized disconnect switch represents a significant hazard. Start at the top, where the incoming power supply to the fuses can be checked, as shown in **FIGURE 17-12**. Then, with one meter probe remaining at the top of a fuse, use the other lead to check the voltage at the bottom of the fuses. If the fuse is intact, the meter will display the line voltage. If the fuse is open, it will show 0 volts. Once the fuse is identified, deenergize the disconnect, remove the fuse, test it with an ohmmeter to confirm it is open, and replace it.

▶ CORE CONNECTION

Although motors can operate at voltages that are higher and lower than the nameplate value (usually with ±10%), it is not healthy for them. Operating a motor at or near the outside of its operating range often causes premature failure—especially for a motor operating at a voltage below the nameplate value, which increases its operating current and decreases torque.

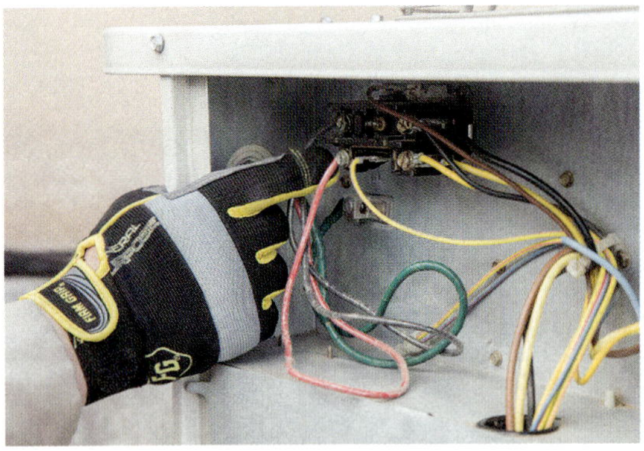

FIGURE 17-13 Checking conductor terminations.

When testing the power supply, apply the above calculation to ensure that the voltage is sufficient. It is also important to check the line voltage, both with and without a load. If there is a significant voltage drop due to a poor wiring termination or similar problem, it will only be evident when the voltage is tested while the circuit is under a load. Remember that voltage drop increases motor current (more accurately, the current of all inductive loads), which then leads to overheating. Inspect all the power-wiring terminations by deenergizing the system, locking/tagging the power source, and visually checking each termination. Gently tug each one and check the tightness of the lug or screw (**FIGURE 17-13**). An insufficient wire size will also cause the voltage to drop under load. Ensure that the power wiring is the proper size.

To practice how to measure a power supply voltage and determine whether the voltage is within the required range, follow the steps in **SKILL DRILL 17-1**.

SKILL DRILL 17-1 Practicing How to Measure a Power Supply Voltage

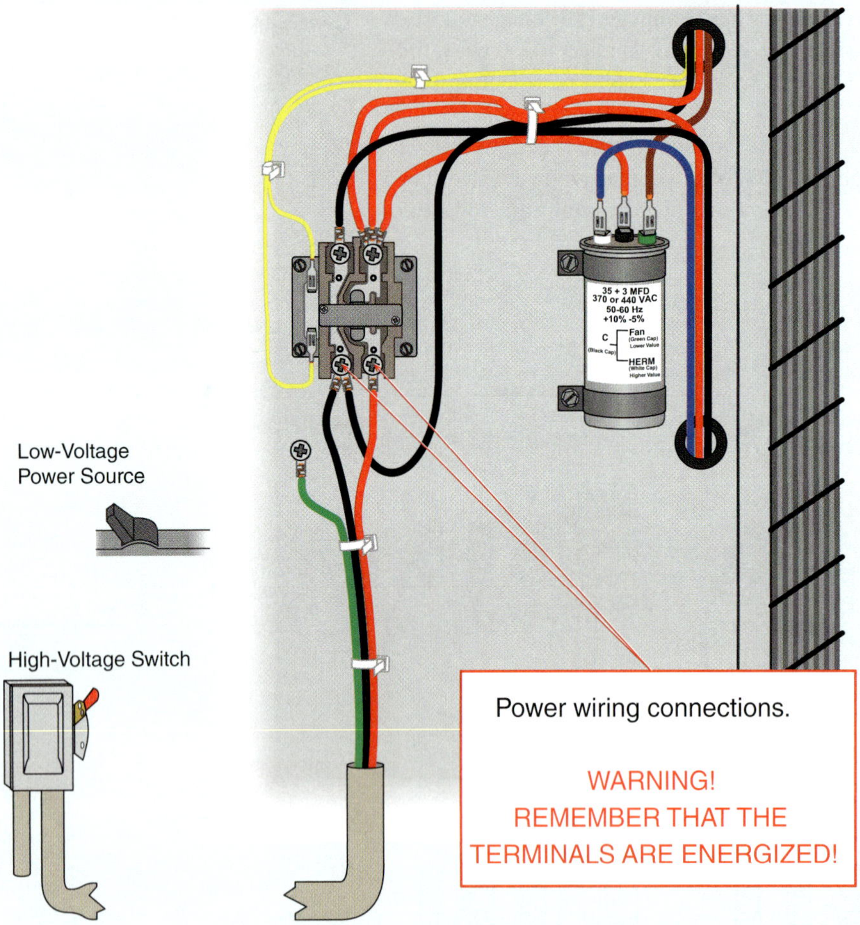

1. Ensure that all controls, including the thermostat, are set to the Off position. Remove the access door and locate the line-voltage power connections to the unit. If the wiring connections are made with wire nuts, locate the terminals of the contactor where they are connected.

SAFETY TIP

Working with electrically energized circuits is hazardous. Avoid contact with any surface that could potentially be energized. When working inside energized enclosures, maintain a stable position and, whenever possible, avoid placing both hands inside the enclosure. All work on energized equipment by unqualified individuals should be supervised by a qualified instructor or technician.

Continued

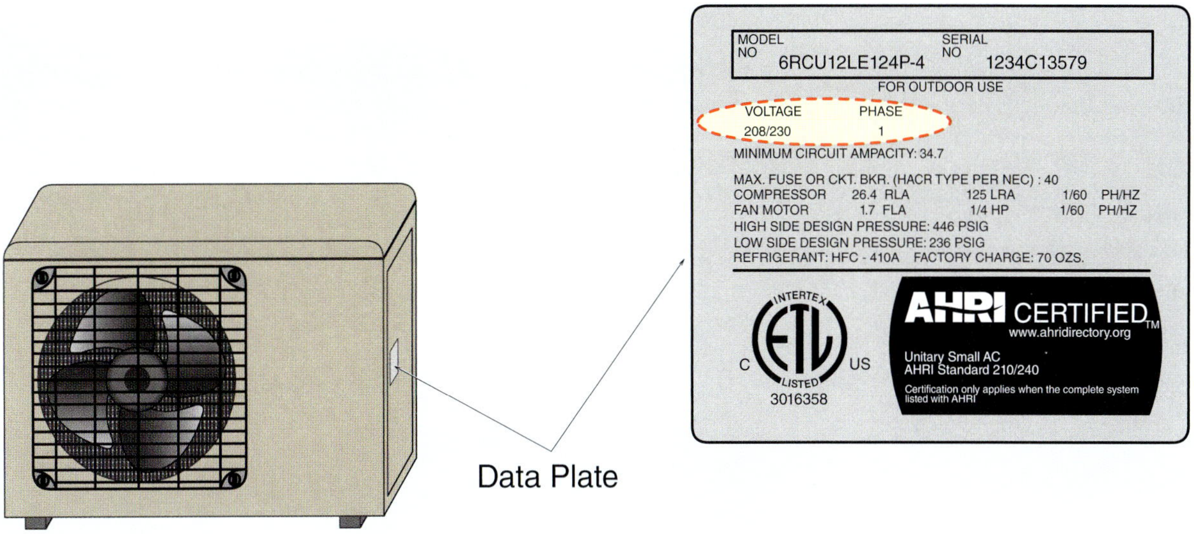

2. Check the unit data plate for information regarding the power supply requirements. Most single-phase condensing units will specify a voltage range rather than a single voltage.

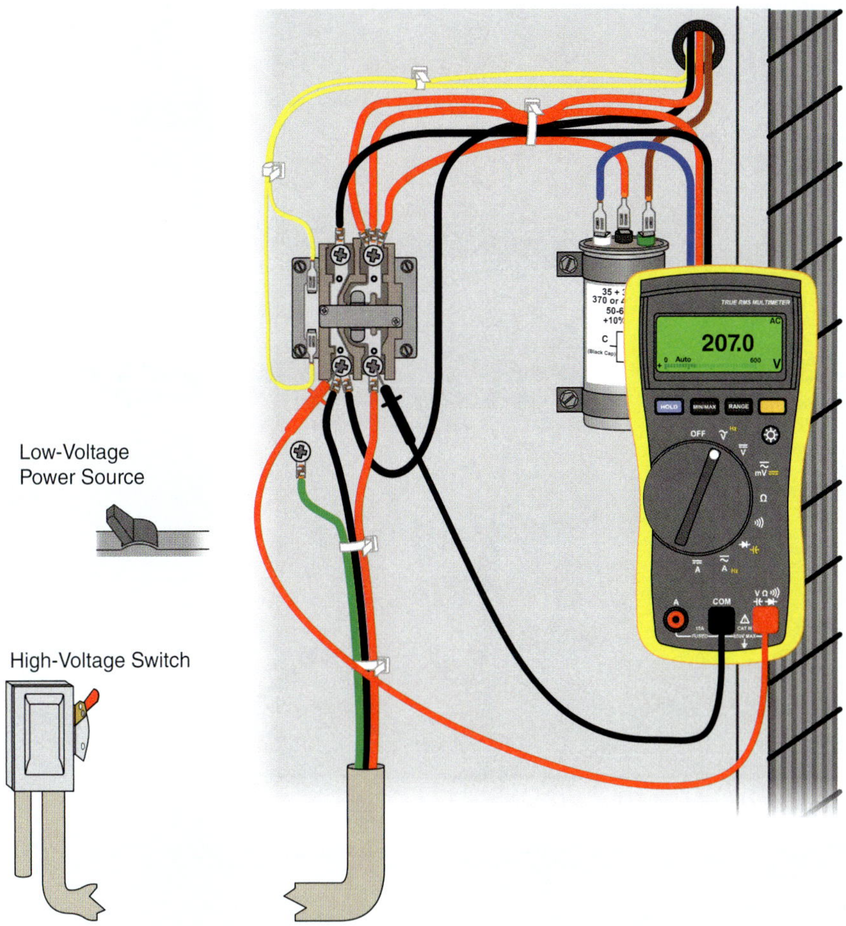

3. Set the multimeter for AC (alternating current) voltage testing. Check the line voltage at the terminals of the contactor while the unit remains off.

Continued

208 voltage required × 0.10 = 20.8 volts

Minimum allowable voltage = 208 volts - 20.8

Minimum allowable voltage = 187.2 volts

4. Apply the ±10% rule to the data plate voltage. In this example, we will assume a measured voltage of 207 volts on a system that requires a 208–230 VAC power supply. Since the reading of 207 volts is well below the upper limit of 230 VAC, the only need in this example is to determine whether it is too low. Complete the calculations as shown in the text to determine the minimum allowable voltage for the unit power supply, and then determine whether the measured voltage is within the required range.

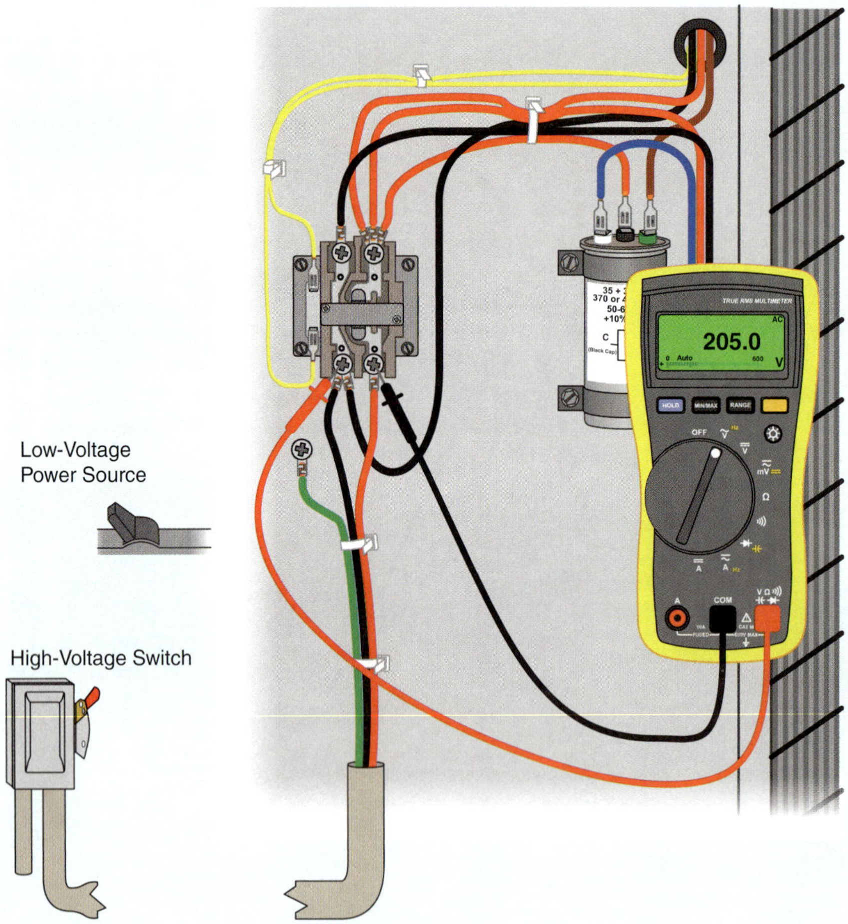

5. Prepare the unit for start-up. Start the unit and allow its operation to stabilize. Measure the voltage at the contactor with the unit in operation. Apply the same calculated minimum voltage to the measured voltage under load and determine whether it remains acceptable.

Continued

6. Once testing is complete, carefully reinstall equipment covers and access doors, and return the system to service.

Using the Hopscotch Method to Test Control Circuits

The hopscotch method refers to an approach often used in troubleshooting control circuits. It can be done with a voltmeter during energized troubleshooting, in search of an open condition in a circuit.

FIGURE 17-14 shows the 24 VAC portion of a simple control circuit, with several normally closed safety controls in series with the compressor contactor coil. Note that the safety control symbols show the high-pressure switch (HP) opens on a rise in pressure; the low-pressure (LP) switch opens on a fall; and the discharge line thermostat (DLT, found in scroll-compressor systems) opens on a rise in temperature. All three are normally closed (NC).

SAFETY TIP

Never disregard the unseen hazard of energized electrical circuits. Practice electrical safety at all times, and keep tools and other conductive objects away from exposed conductors and terminations.

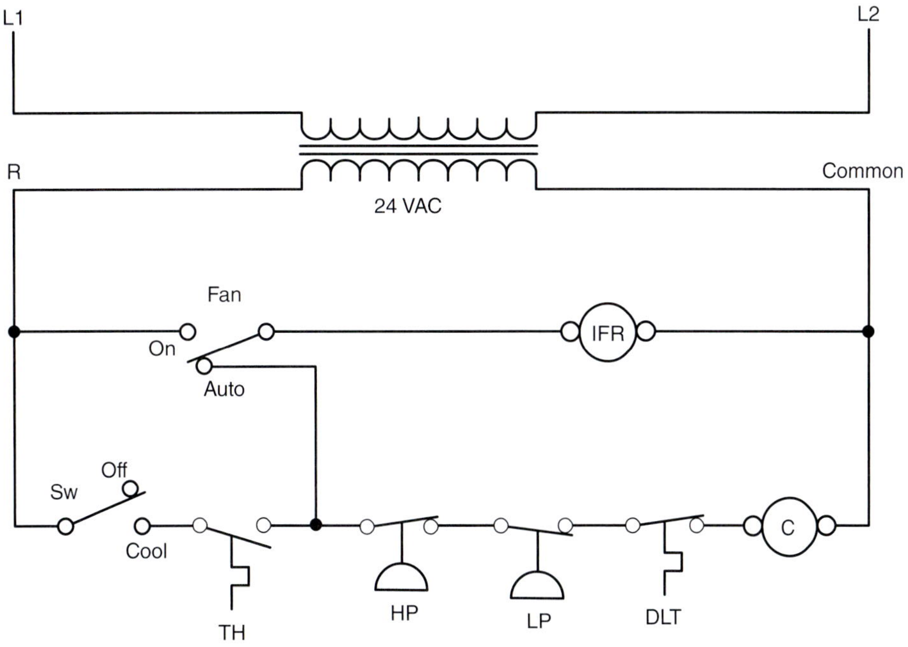

FIGURE 17-14 Example control circuit.

Apply voltage-testing techniques to the circuit as it is presently. Assume that the circuit is energized through the 24 VAC control transformer, the thermostat switch is set to Off, and there is no call for cooling, as shown in the figure.

Now refer to **FIGURE 17-15**. If the meter leads are applied across the open contact, as shown in the upper part of the figure, the meter will read 24 VAC. If they are applied as shown in the lower part of the diagram, 0 VAC or minor background voltage will be displayed. There is no closed path for the other leg of 24 VAC power to make it to the test-lead position.

Now assume that the thermostat switch has been placed in the Cool position and that the set point has been lowered, closing the thermostat (TH) contacts. A technician expects the indoor blower to start and the compressor contactor to be energized, followed immediately by the compressor starting. But assume that the compressor doesn't start, and yet the indoor blower does start as expected.

This portion of the example falls back to analyzing the available information and adjusting the list of potential problems. When the indoor blower starts and runs, what does that say about the circuit?

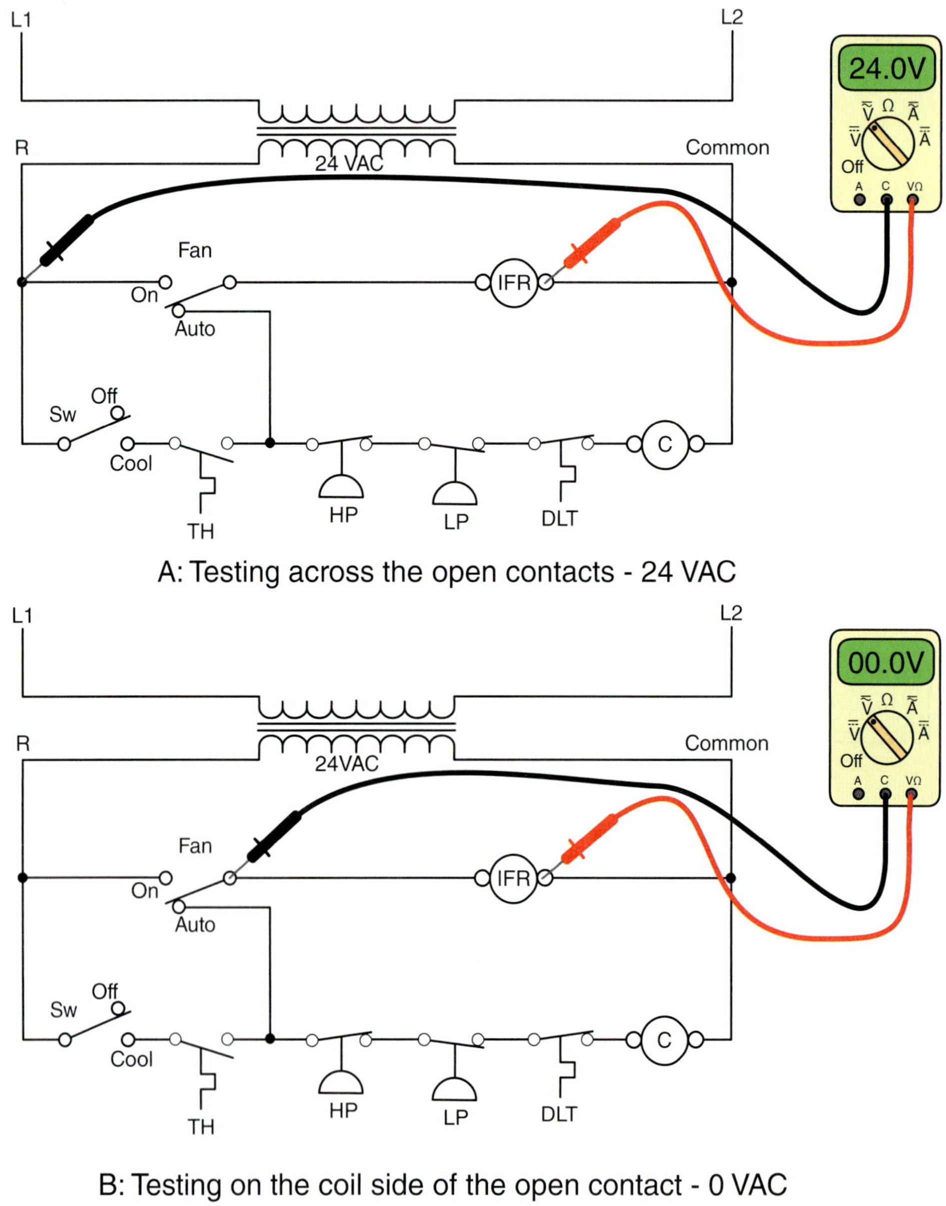

FIGURE 17-15 Testing for control power—system energized but idle.

With the FAN switch in the Auto position, the only source of power to energize the indoor fan relay (IFR) coil comes through the thermostat switch and contacts. Thus, they must both be closed and functioning properly; otherwise, the IFR coil would not be energized. They are quickly eliminated from the list of potential reasons that the compressor contactor did not energize.

The hopscotch method is used to test a series of controls in a circuit to determine which one is open. As we started the system, there were five sets of contacts in the circuit that could interfere with the compressor contactor. Two of them were eliminated since the indoor blower runs. However, they will be included in the hopscotch sequence here, for the sake of comparison.

Refer to **FIGURE 17-16**. One meter lead is placed on the contactor terminal where the Common wire is attached. The other is used to "hopscotch" from one point to the other. The meter will continue reading 24 VAC at each of the first four numbered points until the open contact is reached. Once the meter leads are placed as shown in the lower part of the figure, the meter will no longer read 24 VAC. This indicates that the LP switch (or the wiring

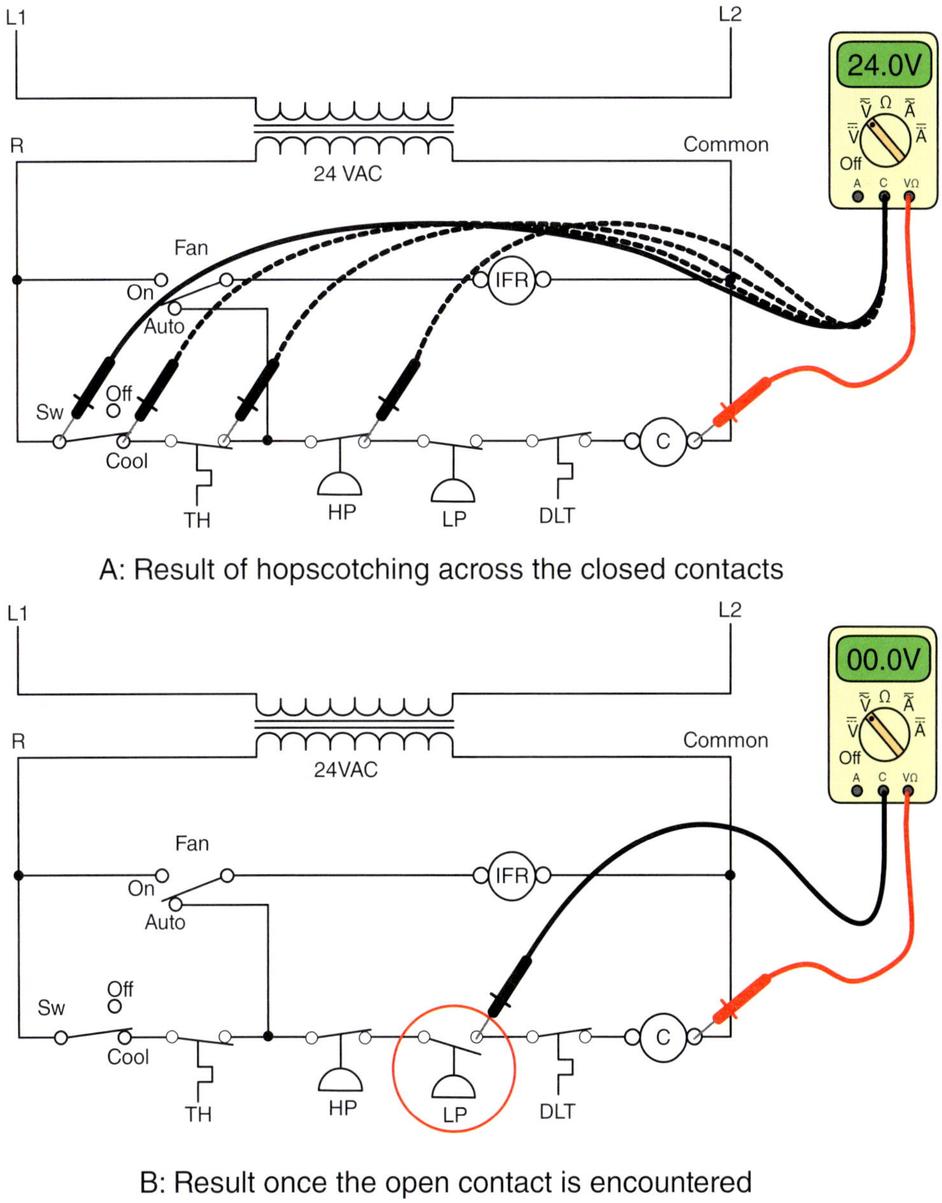

FIGURE 17-16 Hopscotch testing of the control circuit.

▶ **TROUBLESHOOTING TIP**

Condensate Pumps

Condensate pumps are used to collect condensate (water) from the indoor coil when the coil is located where gravity alone cannot support proper drainage. For example, a furnace with an evaporator coil may be in a basement, and the condensate needs to be pumped up several feet (meters) to reach ground level. There are a number of installation situations that require a condensate pump.

Most pumps are equipped with a safety switch that will shut down the cooling system if the pump fails and begins to overflow with condensate water draining into its sump. To prevent additional condensate from being produced, the outdoor unit/compressor is disabled. The switch is usually a simple float switch with two wires or terminals. The Y circuit that energizes the compressor is wired through this switch to provide protection. For condensate pumps that support high-efficiency condensing furnaces, the switch may be wired into the W, or heating, circuit.

In the cooling season, it is always a good idea to quickly check the condensate pump on systems that are equipped with one when investigating a sudden loss of cooling. The sump can overfill, initiating the switch, due to pump failure or a blockage in its discharge line, which is typically small plastic tubing. A quick look at the pump and its switch can often save a lot of troubleshooting effort.

beyond the HP switch) is open. Now it becomes a matter of determining if the refrigerant pressure is truly low, a wiring termination is flawed, or the switch itself has failed. Now the list of possibilities has been reduced and the ones that remain can be explored.

To practice how to use the hopscotch method to troubleshoot an electrical problem, follow the steps in **SKILL DRILL 17-2**.

SKILL DRILL 17-2 Practicing How to Use the Hopscotch Method to Troubleshoot an Electrical Problem

Conduct hopscotch testing at an outdoor condensing unit or heat pump, with the line-voltage power supply to the unit disabled for an added degree of safety. Working with a split system will allow for control power to be available from the indoor unit while the power to the outdoor unit remains off. To enhance the experience, instructors should create a circuit fault in the contactor control circuit that cannot be detected easily by visual inspection.

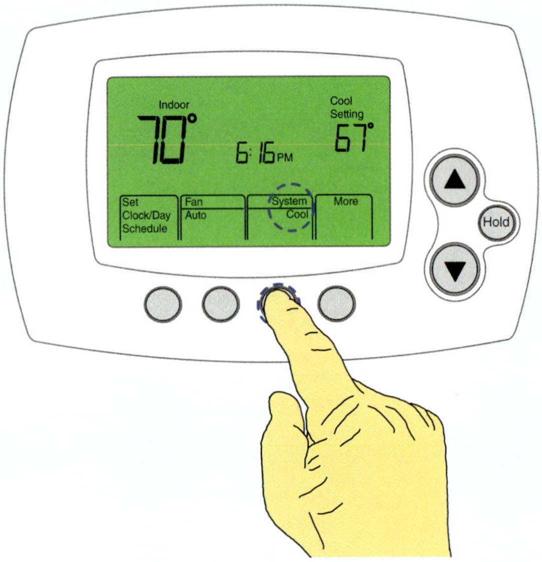

SAFETY TIP

Working with electrically energized circuits is hazardous. Avoid contact with any surface that could potentially be energized. When working inside energized enclosures, maintain a stable position and, whenever possible, avoid placing both hands inside the enclosure. All work on energized equipment by unqualified individuals should be supervised by a qualified instructor or technician.

1. Set the thermostat as required to initiate a call for cooling—switch the mode to Cool, and ensure the set point is several degrees below the room temperature. Since there is a known defect in the control circuit, the compressor contactor should not energize when called upon by the thermostat. Remember that the system is electrically energized, so practice the appropriate electrical safety techniques.

2. Remove the equipment access doors as necessary. Locate the unit wiring diagram, and identify the control switches and/or contacts that are in series with the compressor contactor coil. Then physically locate the control devices shown on the diagram inside the unit. Trace the interconnecting wiring until all the devices and the wiring terminations become familiar.

Continued

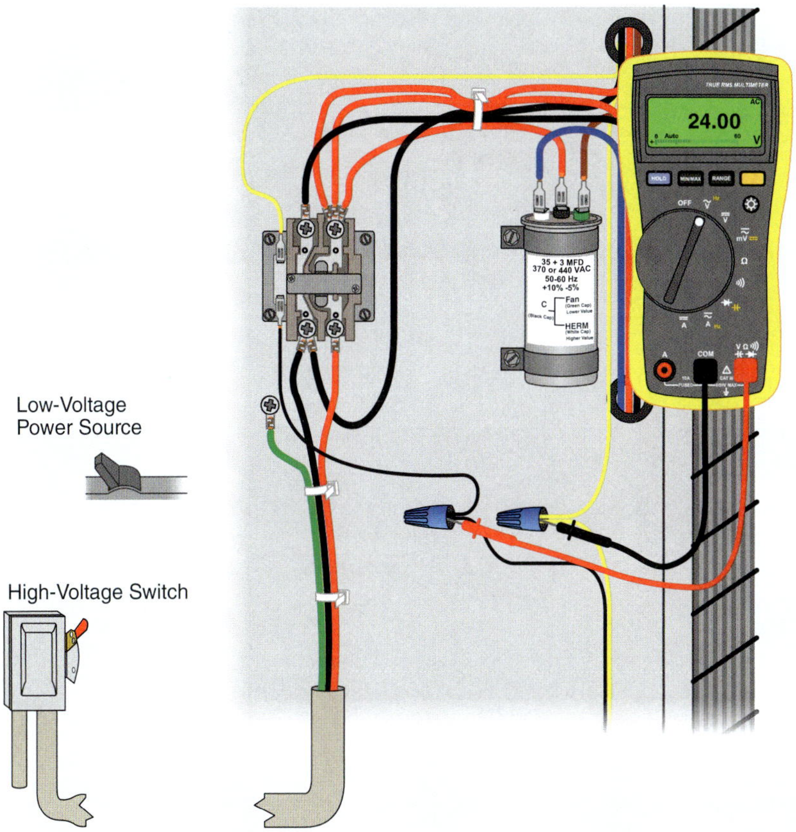

3. Begin testing by ensuring that the thermostat and interconnecting control wiring to the outdoor unit is calling for cooling. Using a voltmeter, test the incoming Common and Y wires at the connection to outdoor or packaged unit; 24 VAC should be present. If control voltage is not present there, the problem is at the thermostat, the interconnecting wiring, or a control device indoors.

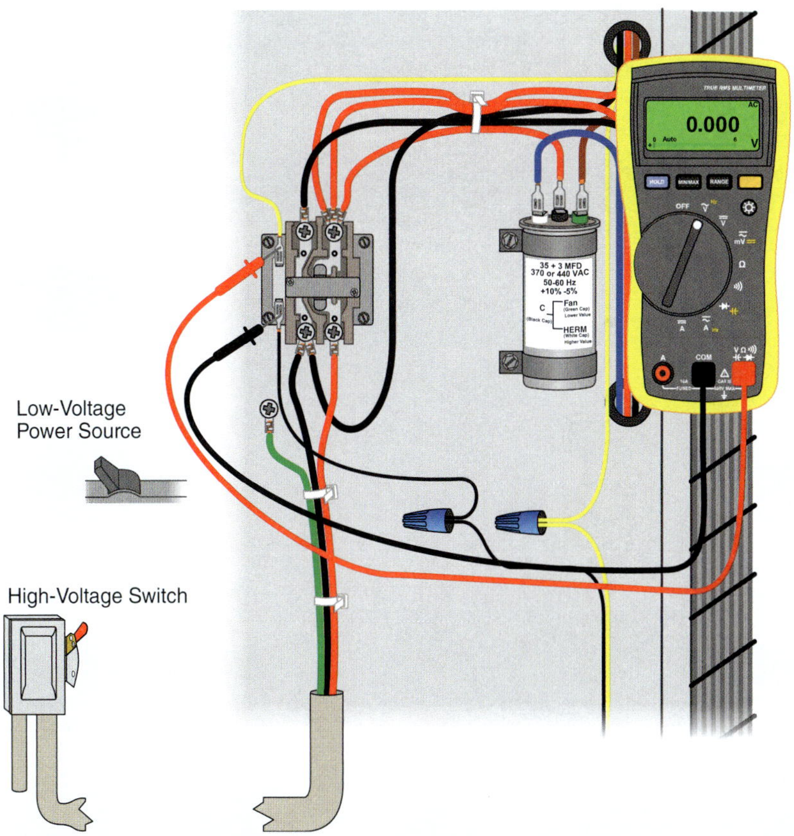

4. Next, confirm the presence or absence of control power at the contactor coil. Touch the meter leads to the coil terminals. Control voltage should not be present. If 24 VAC is present at the terminals and the contactor is not energized, the contactor coil is likely open. If this is the case, deenergize the control circuit, disconnect the wiring from the coil, and test the coil for continuity with an ohmmeter to confirm the failure.

Continued

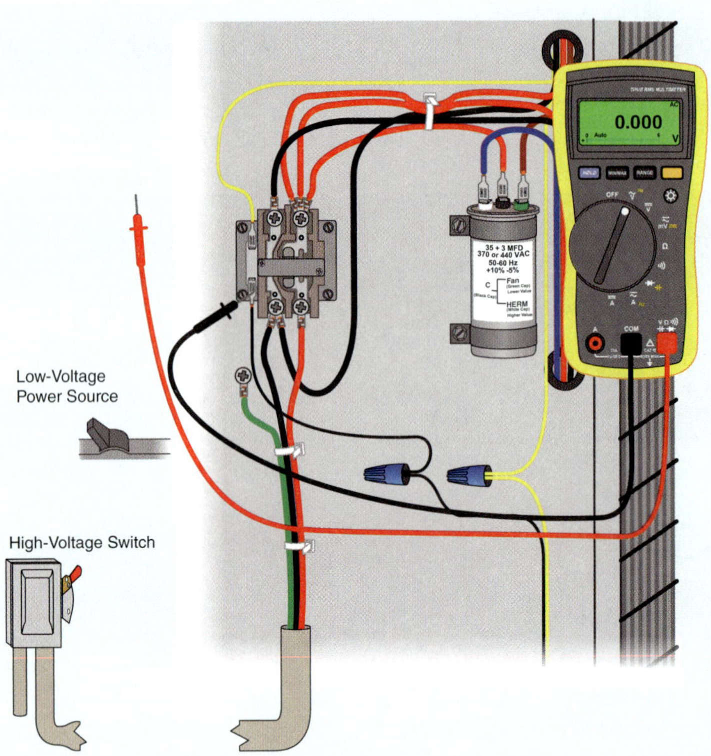

5. Determine which of the connections at the contactor coil represent the Common lead. This can be determined from the wiring diagram and confirmed by tracing the internal wiring of the unit from the point where the Common control lead enters the cabinet. One meter probe will be connected to this terminal of the coil, and it will remain there during hopscotch testing.

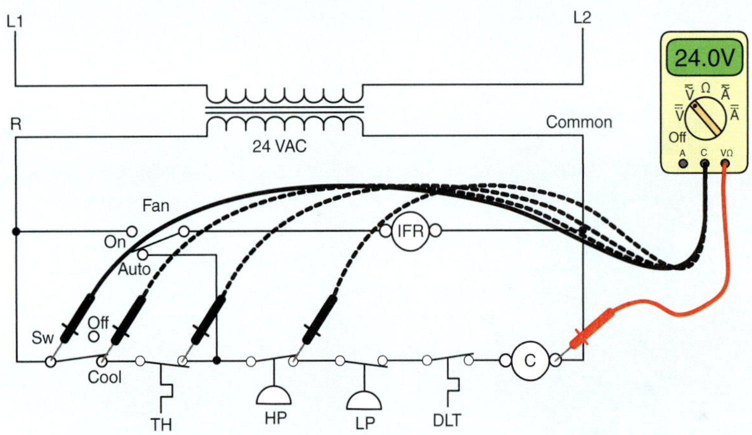

A: Result of hopscotching across the closed contacts

6. With one meter lead firmly connected to the Common terminal of the contactor, begin hopscotching from one control connection to another. This might include, but is not limited to, relays, anti-short-cycle timers, temperature safety switches, and pressure safety switches. It is best to touch the probe to each of the two leads connected to the contacts or switch in question. A reading of 24 VAC when the meter probe is touched to each test location confirms the control/switch is closed and passing power. Continue until a point is located where 24 VAC power is lost in the circuit. This confirms the open in the circuit is between this point and the previous test point. Either the control itself is open or there is a poor wiring connection interfering with unit operation.

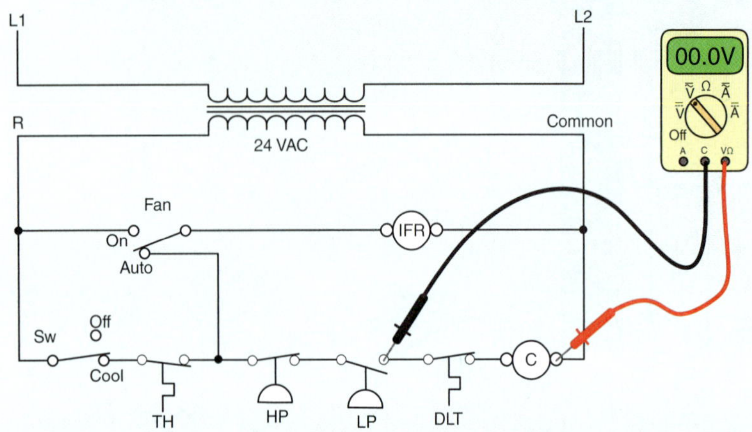

B: Result once the open contact is encountered

Continued

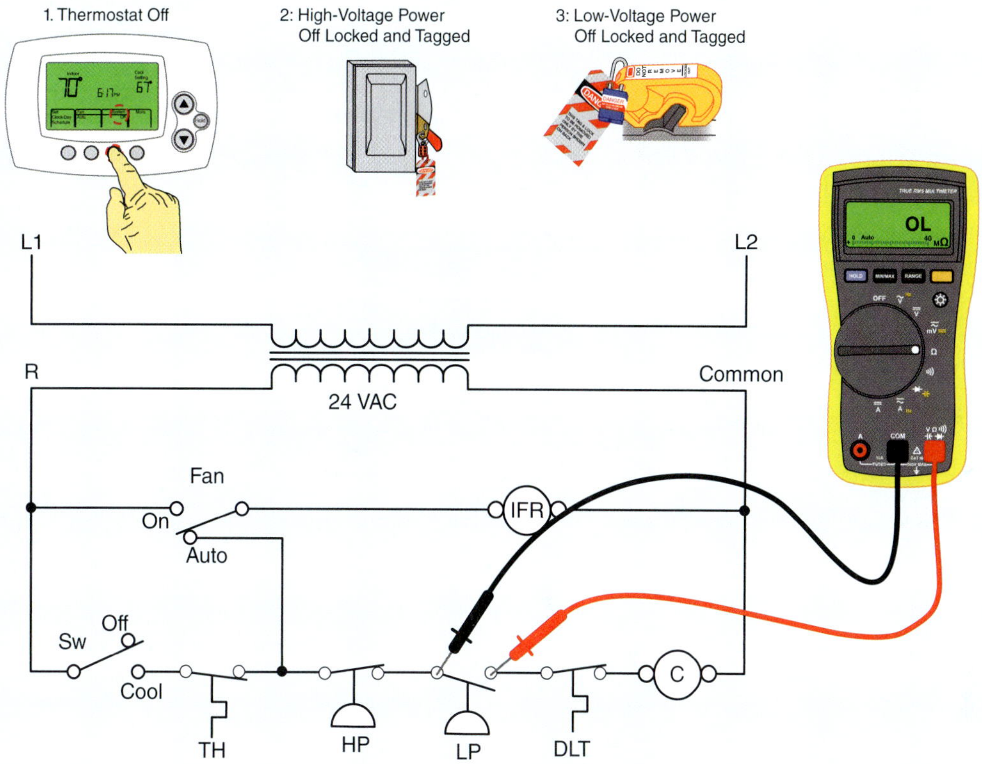

7. Once the open point in the circuit has been located, deenergize the control circuit by turning the thermostat to Off and disabling the power sources. Disconnect the suspected control from the wiring, taking note of the condition of the connections. Use an ohmmeter to check the contacts of the device.

If there is continuity through the switch or contacts, the device is functional and the problem is likely in the connections. Otherwise, the faulty control must be replaced to restore the unit to service.

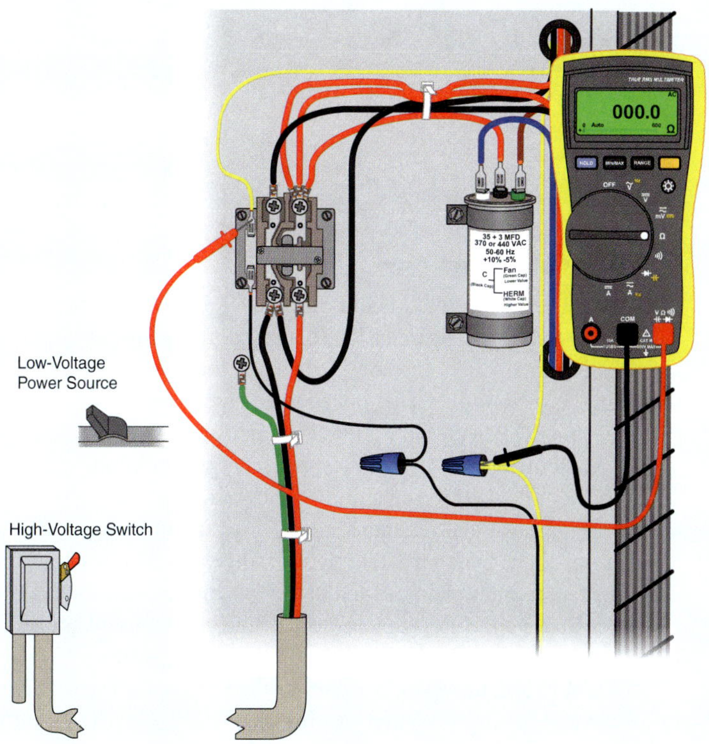

8. Inspect and reconnect the wiring to the device, making reliable connections. Perform a final test of the device with the ohmmeter, checking again for continuity through both the leads and the contacts. The ohmmeter can also be used to check for a complete circuit through all safety devices in the unit, as shown in the figure.

Continued

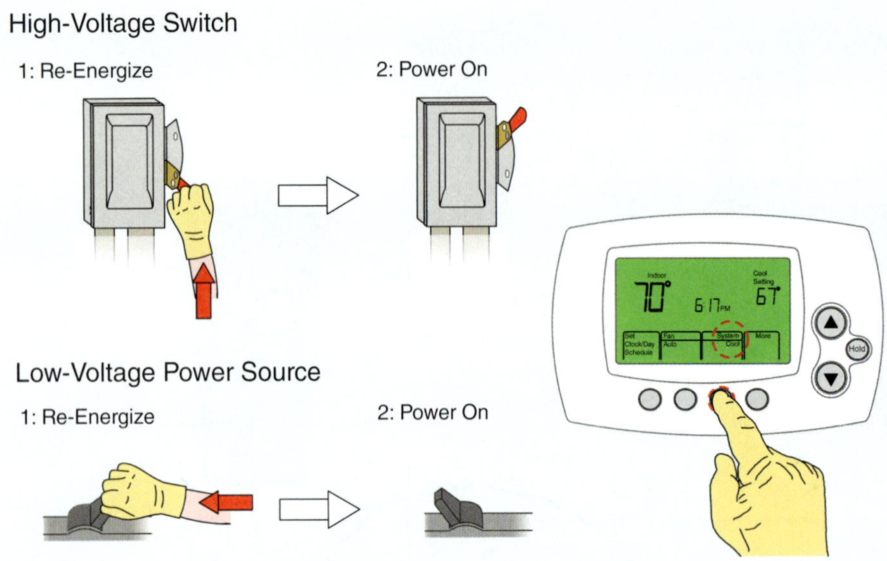

9. Perform a final inspection of the wiring to ensure that all tools are removed, wiring connections are sound, and the unit is ready for start-up. Restore power to the system. Set the thermostat as required to call for cooling. If the problem has been corrected, the contactor will energize and the compressor should start normally.

10. Once testing is complete, carefully reinstall the equipment covers and access doors and return the system to service.

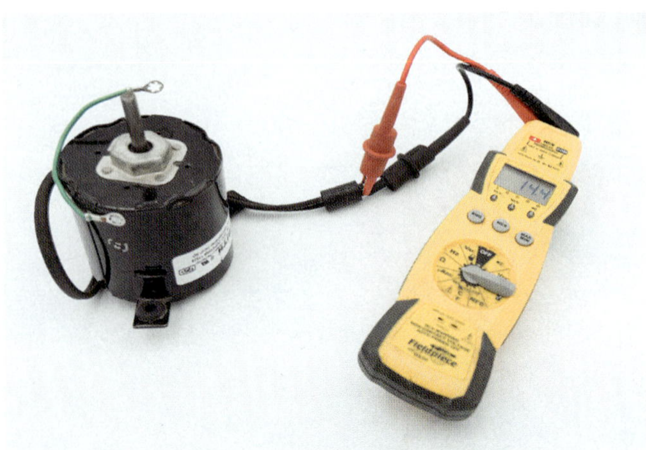

FIGURE 17-17 Using an ohmmeter on motor windings.

The Ohmmeter

The first lesson to recall about ohmmeters is that they cannot be used on an energized circuit. The battery in an ohmmeter is used to provide a small direct current (DC) voltage for resistance/continuity testing. If the ohmmeter is applied to an energized circuit, the protective fuse will likely blow and the meter can be damaged.

Ohmmeters are used to test a deenergized circuit, with any number of devices in the circuit, for continuity. During troubleshooting, the actual resistance reading is rarely important. What *is* important is whether a complete circuit exists. For example, the contacts of a relay can be tested to verify which sets are closed and which are open. Unless there is something very unusual going on, the resistance reading in ohms is of little consequence.

There are, however, occasions when the resistance reading is very important. When testing the windings of a motor (**FIGURE 17-17**), for example, measuring the precise resistance through each winding might be necessary. When there is no

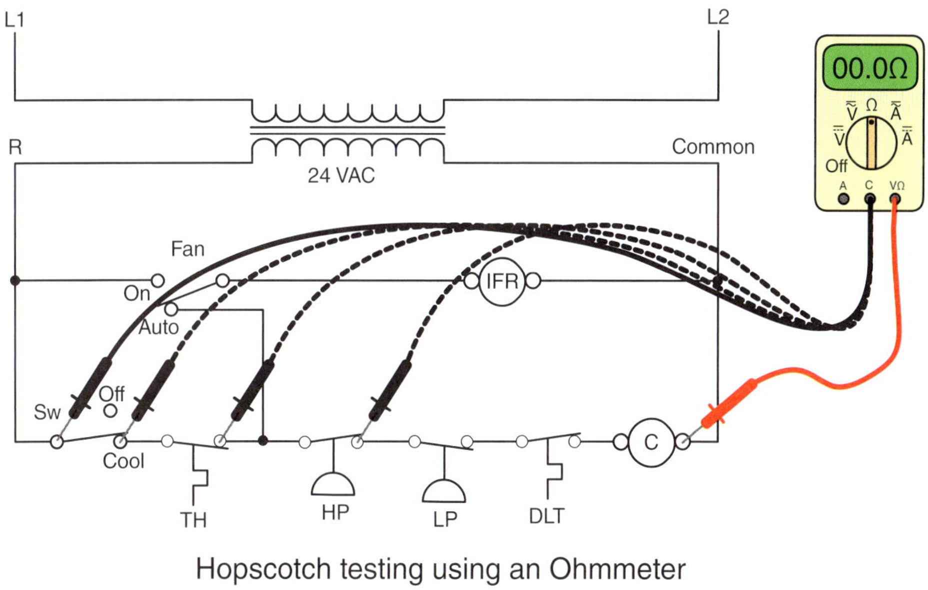

FIGURE 17-18 Hopscotch testing with an ohmmeter.

continuity, it is an indication that the winding is open. When the windings are intact, their resistance values can help identify the start and run windings in a single-phase motor. The resistance values of the windings in a three-phase motor should be very close to the same. Motor testing is covered in detail in Chapter 14.

Note that the hopscotch method of troubleshooting can be applied with an ohmmeter as well as with a voltmeter (**FIGURE 17-18**). The method is applied in the same way, but with the electrical circuit deenergized. One probe of the meter is placed at one end of a circuit, and the other probe is used to hopscotch from control to control, looking for open switches or contacts that are expected to be NC.

The Ammeter

Since there is no significant current flow in a circuit until a load is energized, there is little to test with an ammeter in a non-functioning system. The current of an energized load, however, provides helpful diagnostic information or confirms proper operation.

Most of the time, an ammeter will be used to determine the current consumed by a motor (**FIGURE 17-19**). Fan and blower motors that are drawing more current than their data plate allows may be suffering from a mechanical problem, such as failing bearings, or from an electrical problem, such as a failed run capacitor. The current consumption of a motor is also directly related to the load imposed on it. For example, if a motor that is too small (insufficient horsepower) is used to replace a fan or blower motor, the current will be excessive and can only be reduced by reducing the airflow. Compressors that start and operate on an extreme load, such as a very hot building, will typically use slightly less current as they gain control of the temperature (lowering the compression ratio), thus reducing the load imposed on them at the evaporator.

When measuring current with an ammeter, also remember that open access doors can have a significant effect on the current for blowers and fans. A condenser fan mounted on a split-system condensing unit will generally draw more current with the access door removed than with it in place. Likewise, removing the access door from a furnace and operating the blower will result in a higher-than-normal motor current. Again, the current of a motor is directly related to its load. When the access doors are removed (**FIGURE 17-20**), the resistance of drawing air through a return duct or across a finned condenser coil is removed. Thus, the fan or blower moves more air and the current rises. This is very important to remember when testing or troubleshooting motors.

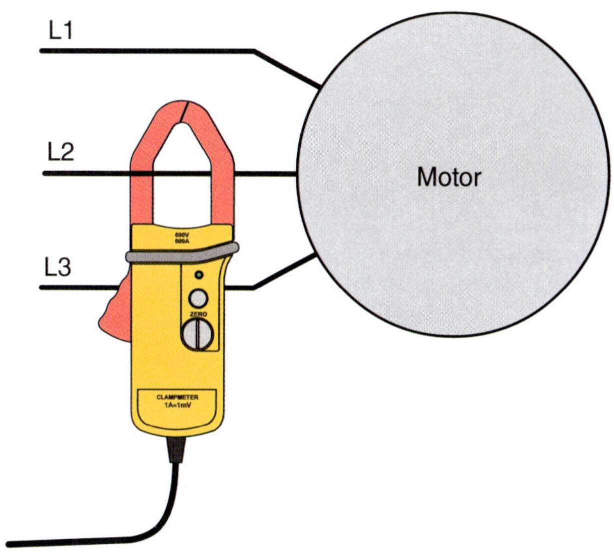

FIGURE 17-19 Checking motor current.

K17003 Explain how to apply alternative troubleshooting measures and work with tech support.

FIGURE 17-20 With the access door off, air freely enters through the opening, increasing motor current.

▶ Alternative Troubleshooting

Sometimes, despite best efforts having been made, the solution to a problem will be elusive. Regardless of how well a technician knows and understands the operation of a system, a problem can be challenging to identify with confidence. This is especially true of intermittent issues. When a unit fails to operate at times, yet always seems to be fine upon arrival at the site, new tactics may be necessary.

The first step when a unique challenge arises is to return to the strategy-based troubleshooting presented earlier in this chapter. Reconsider everything that the owner said, and ask additional questions if necessary. Reconsider the possibilities, perhaps reintroducing possibilities that were removed earlier because they seemed unlikely. Return to focused troubleshooting, retest as needed, and test some components that had been assumed to be functioning normally.

Another approach is to try to recreate the circumstances of the failure. Perhaps it was exceptionally hot or cold when the problem occurred but the temperatures were significantly different upon arrival at the site. Flawed circuit boards, motors, and controls may be affected at temperature extremes but operate normally in milder conditions. If the answer to a recurring and intermittent problem is elusive, consider trying to simulate the conditions that seem to accompany the problem.

It reflects poorly on the technician and their employer to give up on a problem too easily. There is also no knowledge or experience gained from walking away. In some cases, another technician with a different perspective on the issue can solve it quickly. Discuss intermittent problems freely with other technicians, and new ideas that lead to a solution will often surface. Always be prepared to try alternative approaches when the conventional approach isn't getting results.

▶ TROUBLESHOOTING TIP

Keeping a Set of Eyes on the Job

Controls that open and automatically reclose following an event can be very frustrating to identify after the fact. When an intermittent problem such as this occurs, having a technician stand by and watch for hours is expensive and often unproductive. However, there are other ways to monitor a control.

One reliable way to determine whether a set of contacts is opening at some point is to place a very small fuse in parallel with the contacts. An inline fuse holder for small glass fuses should be wired in parallel with the contacts and a fuse inserted in the housing. If the contacts open, the load of the control circuit is then placed solely on the fuse. The key is to use a fuse with a very tiny current rating. A rating of 1/10A or less is best, since the control circuit doesn't use much current. Many technicians choose 1/100A fuses for certainty. Although such fuses are not inexpensive, they can be invaluable in these situations. If the contacts open, the fuse will blow. Eventually the contacts will close again and the system will continue as it was, but on the next return visit, the blown fuse reveals that the control's contacts opened at some point. Alternatively, a device designed for reuse, such as the tool shown in **FIGURE 17-21**, can be used. It features a light that illuminates when the contacts open, and it remains lit until the technician arrives. Devices such as fuses and control monitors that are used to determine whether a switch or set of contacts is opening are often referred to as tattle tales, and the name is also a registered trademark for one such device.

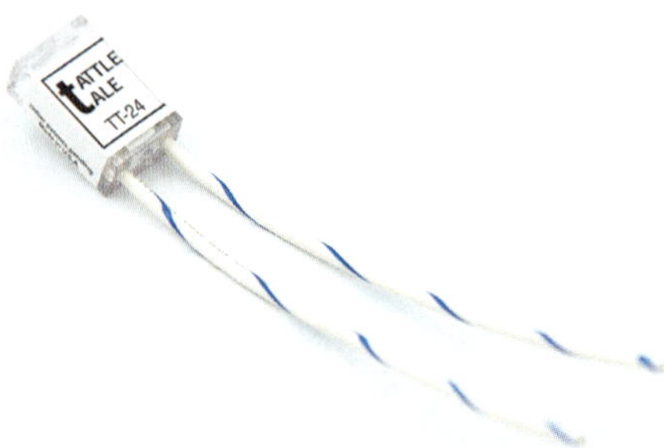

FIGURE 17-21 Control monitor.

Exchanging Parts

Although it should not be something a technician regularly resorts to, exchanging key parts can help solve some challenging problems. In some cases, there is financial risk. If, for example, an expensive solid-state control is purchased and installed, only to find it did not solve the problem, the part is probably unreturnable. Although a vendor may on occasion be willing to accept an electrical part back from a good customer, they should not be expected to do so. Electrical parts that have been placed in service for any period of time are rarely accepted for return, and rightfully so.

Common parts, such as relays and contactors, are easy to replace and are often taken from the normal inventory of a service truck. However, contactors and relays aren't known for intermittent operation; generally, they work or they don't. Intermittent problems are generally related to slightly more complex devices or to unreliable electrical connections.

Commercial installations where some identical units are in operation offer a different opportunity (**FIGURE 17-22**). Perhaps a defrost control board is suspected to be the source of an intermittent problem. If there are identical units serving the facility, the defrost boards from two units can be swapped. If the board has a problem, it will now manifest itself in the other unit. However, this should be done with the customer's knowledge. Explain the reason behind suggesting the swap and address any concerns. Always be honest with the customer, and avoid taking unusual actions with their equipment without their knowledge and acceptance. Being honest and avoiding hidden actions are practices that should become the backbone of any troubleshooting strategy. Before resorting to a parts exchange, though, exhaust all practical troubleshooting efforts first.

FIGURE 17-22 A group of similar units.

▶ FOCUS ON CUSTOMER SERVICE

Finishing Up Right

Once a troubleshooting and repair process has been completed, make it count by finishing up the job on the right note. Many callbacks result from carelessly executing the last bit of work. Here are a few tips to finish a job up right:

- Set the temperature controls at the preferred settings; don't leave them set at unusual settings that were used when testing. It is ideal to stay long enough to see the system cycle on its temperature controls normally. Allow ice machines and similar equipment to complete a full operating cycle without intervention.
- Make sure the system is clearly marked regarding the refrigerant and oil in use. If either has been changed, make it clear and obvious.
- Make sure all refrigerant ports are capped. Check for leaks around the stems of service valves, and carefully tighten the packing as necessary.
- Ensure that all access doors are properly reinstalled and that there are no rattles and squeaks that will cause a follow-up complaint.
- Last but not least, leave the work area cleaner than it was. Don't focus solely on bits and pieces that were left while working; pick up any other debris in the area as well. Leave the equipment clean too. A quick wipe down and a even a little bit of wax applied to the cabinet over the course of a few minutes can make a big difference in how a customer perceives the work.

Working with Tech Support

Believe it or not, there was a time when the Internet and the vast resources it has made available to the trade didn't exist. In the days before the Internet, fax machines, and mobile phones, the only way to transfer books, manuals, and wiring diagrams was by mail. The point is that those days are long gone, and there are few excuses now to work without the proper documentation. Those resources have also become more important than ever, because the complexity of HVACR equipment has risen exponentially over the years.

All major equipment manufacturers make installation, operation, and maintenance manuals (IOMs) available online, either to the general public or to registered members of their network. Specifications, application data, detailed capacity charts, troubleshooting charts, and wiring diagrams are all available to resourceful technicians. Most materials can be downloaded quickly and accessed on the job (**FIGURE 17-23**). Many manufacturers also offer short training videos online to familiarize technicians with unique features and service tips for their equipment.

FIGURE 17-23 Reviewing downloaded documentation at the jobsite.

When time permits, issues can be discussed with technical support via email. When the need for guidance is immediate, contact technical support directly by phone. All the necessary contact information is posted on manufacturer websites. Before making contact, always be sure to have considered the situation carefully and professionally, and ensure the call isn't being placed for something that can be solved independently. That can be embarrassing, as well as a waste of time.

Regardless of how technical support is contacted, it is essential to have the right information on hand. Having the make, model, and serial number of the equipment, and the same information for the other part of the system (split systems) is a given. If any questions are related to specific parts, have the part numbers and any other identifying information readily available. Be prepared to share the environmental conditions and operating voltage provided to the equipment.

Explain the scope of the problem carefully, and share each step that has already been taken. It's common for technical support personnel to direct a technician to take some action while they remain on the phone. Be prepared with any tools and equipment that might be needed; have them organized and readily accessible. Have reasonable expectations (technical support personnel aren't magicians, after all), and work with them in a professional manner to get the best results.

▶ Wrap-Up

Summary

Troubleshooting is a primary task for HVACR technicians. Successful troubleshooting relies on a strategic process that begins with gathering information from the customer and ends with the successful functional testing of the repaired system.

The multimeter and its accessories are vital in electrical troubleshooting. Practice will foster the ability to interpret meter readings of all types very quickly in order to identify problems. The hopscotch method of testing a circuit is often applied with a voltmeter to an energized circuit, but it can also be applied with an ohmmeter to a deenergized circuit.

Troubleshooting doesn't always go as expected, and some problems can elude even the best technician. Making use of the many resources available today via the Internet is an important skill to develop. In unique cases, alternative approaches may be necessary. When all else fails, making contact with the manufacturer's technical support department can often provide additional insight to a problem and quickly resolve it.

Key Terms

active listening A personal communication approach that requires a listener to display an obvious commitment to understanding and to clearly respond to the input of a speaker.

paraphrasing In personal communication, expressing the meaning of a speaker's words by restating the meaning in the listener's own words to ensure clarity.

Review Questions

1. Which of the following is not a part of the five-step troubleshooting and repair strategy presented?
   - a. Performing the repair.
   - b. Considering the possibilities.
   - c. Swapping parts.
   - d. Verifying the problem.
2. Today's savvy customer often knows more about their system than they would have in years past, thanks to the _______.
3. Embarrassing callbacks often result when a technician is _______.
4. A personal communication approach that requires a listener to display an obvious commitment to understanding and to respond to the input of a speaker refers to _______.
   - a. conscious listening
   - b. paraphrasing
   - c. passive listening
   - d. active listening
5. Restating what a speaker has said in the listener's own words is the definition of _______.
   - a. conscious listening
   - b. paraphrasing
   - c. passive listening
   - d. active listening
6. A split cooling system has been set at the thermostat to the cooling mode, and the set point is well below the room

temperature. The outdoor unit isn't running at all, but the indoor blower is operating. Which of the following problems can be eliminated from the list of possible causes?
   a. Failure of the control transformer.
   b. Blown fuse(s) at the outdoor unit disconnect switch.
   c. An open compressor contactor coil.
   d. An open low-pressure (LP) safety switch.

7. A compressor is starting repeatedly but runs only a minute or two before cycling off. Which of the following problems can be eliminated from the list of possible causes?
   a. Line voltage serving the unit is too low.
   b. Incorrect thermostat setup.
   c. Low refrigerant charge (low-pressure switch opening).
   d. Excessive indoor airflow.
8. Physical troubleshooting should always follow _____.
9. Making temporary repairs to quickly return a customer's system to service is an example of _____.
   a. poor judgment
   b. laziness
   c. prioritizing a customer's needs
   d. taking an unnecessary risk
10. A(n) _____ can be a valuable tool to quickly test specific functions, such as thermostat wiring.
    a. screwdriver
    b. ammeter
    c. jumper wire
    d. capacitor
11. For residential and light-commercial split systems, an ideal location to measure the line voltage for the condensing unit is at the _____.
    a. outdoor unit contactor
    b. furnace
    c. circuit breaker panel
    d. control transformer
12. Refer to the image below. If the L2 fuse is open on this 240 VAC energized power supply, the voltmeter shown in the figure will read _____.
    a. 0 volts
    b. 240 volts

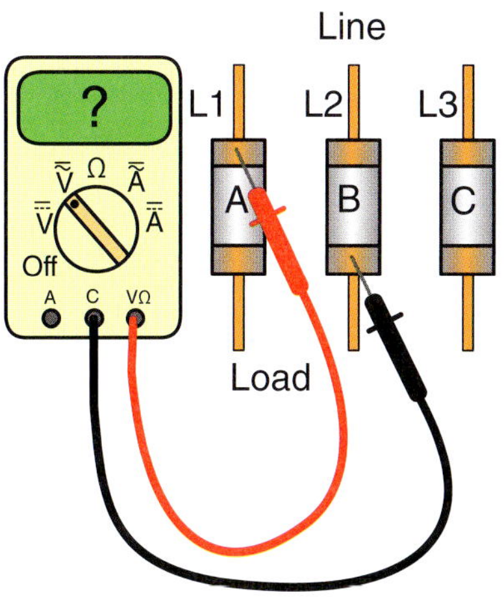

240 VAC, THREE-PHASE SERVICE

13. The line voltage serving a system must typically be within what percentage of the voltage stated on the data plate?
    a. 5%
    b. 10%
    c. 15%
    d. 20%
14. If poor wiring connections or undersized conductors are causing an excessive voltage drop, that drop will be present when the electrical system is _____.
    a. idle
    b. under a load
15. Keeping one lead of a voltmeter at a single point while the second lead is used to move from control to control, looking for a loss of voltage, is associated with _____.
    a. continuity testing
    b. simulating the conditions of a failure
    c. the hopscotch testing method
    d. unacceptable electrical-safety standards
16. Under what circumstances should an ohmmeter *not* be used in troubleshooting?
17. When testing the contacts of a relay with an ohmmeter, which piece of information is generally of the greatest importance to a technician—continuity or the precise resistance of the circuit?
18. On inductive loads like motors, a reduction in voltage causes a(n) _____.
    a. decrease in winding resistance
    b. decrease in motor current
    c. increase in winding resistance
    d. increase in current
19. When the access door of a furnace blower is removed while the blower is running, will the motor current likely increase or decrease?
20. Which of the following is *not* a true statement about alternative troubleshooting?
    a. When traditional troubleshooting doesn't reveal the source of a problem, reconsider the possibilities, perhaps reintroducing possibilities that seemed unlikely.
    b. When a problem seems to occur only under certain conditions, a technician should try to simulate the conditions whenever possible.
    c. Swapping parts from one unit to another on a commercial installation should only be done quietly, to avoid a complicated discussion with the customer.
    d. Technical support personnel will always need to know the make, model, and serial number of the equipment being serviced.

CHAPTER 18

Heating Systems: Sequence of Operation and Troubleshooting

Knowledge Objectives

After reading this chapter, you will be able to:

- **K18001** Describe common gas furnaces.
- **K18002** Describe common gas-furnace electrical controls.
- **K18003** Describe the sequence of operation for an induced-draft gas furnace.
- **K18004** Describe common gas-furnace malfunctions and troubleshooting procedures.
- **K18005** Describe common oil furnaces and burners.
- **K18006** Describe the sequence of operation for an oil burner.
- **K18007** Describe common malfunctions of and troubleshooting procedures for oil burners.

▶ Introduction

To be a successful HVACR technician, it is essential to understand a unit's sequence of operation and how to troubleshoot it when the sequence fails to execute as expected. A basic review of wiring diagrams and system troubleshooting was provided in previous chapters. To fully understand this chapter, those chapters should be reviewed first. Previous chapters related to the electrical components found in heating equipment should also be reviewed.

This chapter will present the sequence of operation and troubleshooting techniques of both gas and oil furnaces. Note that this refers to warm-air furnaces using gas or oil as a fuel, as opposed to boilers that heat water. For simplicity, they will be referred to simply as gas and oil furnaces throughout.

▶ Common Gas Furnaces

K18001 Describe common gas furnaces.

There are three common types of gas furnaces in operation today. Older furnaces rely on a natural draft through the combustion chamber to support combustion and carry away combustion byproducts. Older **natural-draft furnaces** (**FIGURE 18-1**) use a standing pilot flame to light the main burner. The pilot flame is manually ignited on the oldest versions, and it remains burning at all times. In newer versions, a spark ignites the pilot flame when heat is needed. A technician may still encounter natural-draft furnaces from time to time, but they are rare today. They are also the least efficient type of gas furnace—the primary reason for their departure from the market. Since the required efficiency levels for gas furnaces were raised beyond the capabilities of natural-draft models many years ago, few remain in service.

The most popular gas furnace today is the **induced-draft furnace** (**FIGURE 18-2**). Instead of relying on the heat of combustion to create a natural draft through the furnace, a small blower is used. One of the main advantages is that the furnace heat exchanger can be much more complex; the path through the heat exchanger can be longer and have twists and turns. This allows more heat to be extracted and delivered to the structure, significantly increasing the efficiency. It also provides greater heating capacity in a smaller package. The **annual fuel utilization efficiency (AFUE)** of an induced-draft furnace is typically around 80%. The HVACR industry considers these to be medium-efficiency units.

Note that an induced-draft furnace is so named because the inlet of the combustion blower is positioned at the outlet of the heat exchanger. This means that the heat exchanger will operate at a slight negative pressure, inducing (drawing) combustion air into the combustion chamber. Some furnaces are designed with the blower pressurizing the heat exchanger, forcing air through it rather than drawing it through. These furnaces

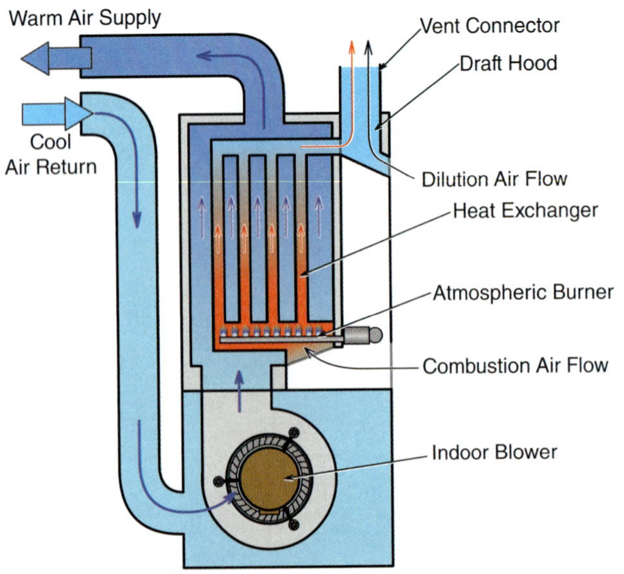

FIGURE 18-1 Natural-draft furnace.

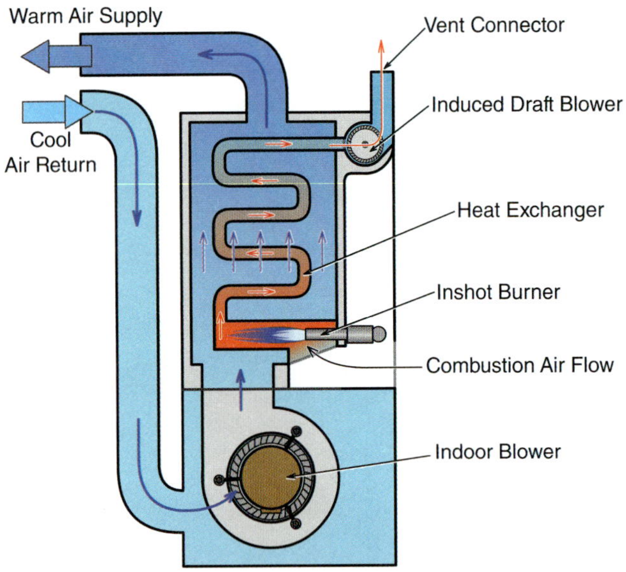

FIGURE 18-2 Induced-draft furnace.

are more accurately referred to as forced-draft furnaces, but the term "induced-draft furnace" is often applied by technicians to all furnaces of this design.

Gas furnaces with the highest efficiency values are known as **condensing furnaces** (**FIGURE 18-3**). They also use a blower to force air into the combustion chamber and heat exchanger. The combustion chamber is sealed, however, unlike less efficient models. This means that the combustion blower has a dedicated air intake routed outside of the furnace, rather than taking in air from the area around the burner. Combustion air for these furnaces is typically drawn from outside the building.

In a condensing furnace, the heat and byproducts of combustion also pass through a second heat exchanger, where additional heat is extracted. So much heat is extracted from the exhaust gases that water vapor condenses from the gases and must be collected and removed. The condensed water also contains some acid byproducts. For this reason and because the exhaust gases are relatively cool, PVC pipe, instead of metal, is used for venting. A second PVC pipe is used to provide the air source for the combustion process. These furnaces have the most complex sequence of operation, but they offer an AFUE of 90–98% using this additional technology.

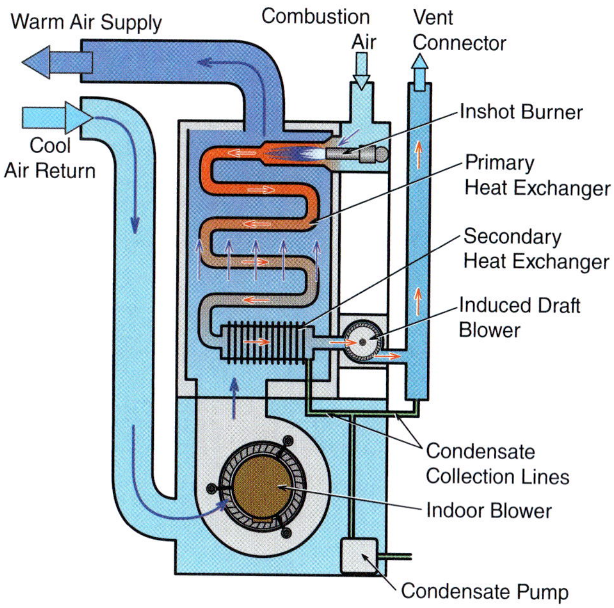

FIGURE 18-3 Condensing furnace.

In this chapter, there will be some limited discussion of natural-draft furnaces, but it will primarily focus on induced-draft and condensing furnaces since they dominate the market.

▶ Gas-Furnace Electrical Components

K18002 Describe common gas-furnace electrical controls.

Electrical components in gas furnaces (other than loads such as motors) can generally be categorized as either control components or safety components. These components vary depending on the age and complexity of the furnace. Control components initiate and monitor normal furnace operation in response to external temperature controls. Safety components protect the furnace, as well as buildings and occupants, from hazardous operating conditions.

Gas-Furnace Control Components

Thermostats are the primary external control device for furnaces. They may be simple two-wire devices or highly sophisticated models that include such features as occupancy scheduling, multistage control, alarm capability, and networking options for remote monitoring and control. A thorough understanding of the thermostat used on a particular system is essential to determine whether system failures are related to it or the furnace itself.

Controls within the furnace switch loads such as the main blower, draft inducer, and the gas valve(s). An optional electronic air cleaner and a humidifier may also be powered and controlled. Air cleaners are usually energized whenever the main blower is energized. How these devices are specifically controlled varies depending on the age and type of furnace.

Gas-Furnace Safety Controls

The type and number of safety components on a gas furnace can, again, vary based on the age and type of furnace. Very old furnaces may have only a pilot **thermocouple** and one or more high-temperature limit switches to protect the furnace. However, modern furnaces can have multiple safety and limit switches. The technician must know how to identify these switches and their intent.

The blower door switch (**FIGURE 18-4**) is mounted on the furnace cabinet at the edge of the blower compartment access door. This switch is spring-loaded to the open position, closing its contacts when the access door is properly positioned over the cabinet opening. It breaks the furnace power supply when the

FIGURE 18-4 Furnace blower door switch.

FIGURE 18-5 Furnace pressure switch.

door is removed. This switch prevents the blower and electrical devices from being energized when the door is off, for the safety of the user or technician.

On furnaces with induced-draft fans, pressure switches (FIGURE 18-5) are used to ensure there is no obstruction within the heat exchanger or venting system. These switches are used on both medium- and high-efficiency furnaces. Since the design pressure in the heat exchanger is relatively low, they must be very sensitive and factory calibrated for accuracy. As a rule, they should not be adjusted in the field. The proper way to test these switches is to use a squeeze bulb and pressure gauge to ensure that they close at the pressure set point, measured in inches of water column (in. wc) or millimeters of mercury (mmHg). These pressure switches may measure the difference in pressure compared to atmospheric pressure, or a differential pressure across the heat exchanger. Furnaces with multiple stages of heating capacity may have several pressure switches.

The primary high-temperature limit switch (FIGURE 18-6) is located in the heat exchanger area, where it can sense an overheat condition. This could occur, for example, if the main blower motor fails while combustion continues. It is nonadjustable and is designed to open when the temperature of the heat exchanger exceeds the set point. The gas valve closes immediately when the high-temperature limit switch opens. Note that most limit switches related to combustion safety are nonadjustable (fixed). The limit switch shown in Figure 18-6 also has a manual-reset button located between the contacts. For an important safety device like this, a manual-reset feature is common.

In the event that a furnace does not receive enough air to support a proper draft, or the venting system becomes partially blocked, there is the potential for the flame to "roll out" of the combustion chamber and into the furnace **vestibule**. The gas valve, wiring, and control devices are all in this area. To prevent a fire hazard, furnaces have rollout switches (FIGURE 18-7) just outside the combustion chamber. There may be more than one, depending on the size of the furnace. Most have a manual-reset button. When a technician needs to reset this switch, problems with the combustion airflow must be suspected and investigated.

Gas-Furnace Ignition Controls and Components

Modern gas furnaces typically use a solid-state device (PC board) to control the ignition sequence and other furnace operations. An electronic ignition control generally controls only the combustion process, while an integrated furnace control (IFC) (FIGURE 18-8) usually controls most or all aspects of furnace operation. Both types control power to ignition devices, induced-draft fans, gas valves, and other components essential to the combustion process.

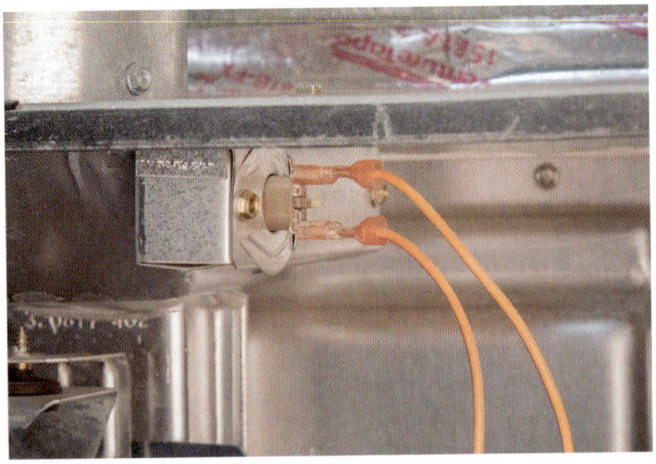

FIGURE 18-6 Furnace high-temperature limit switch.

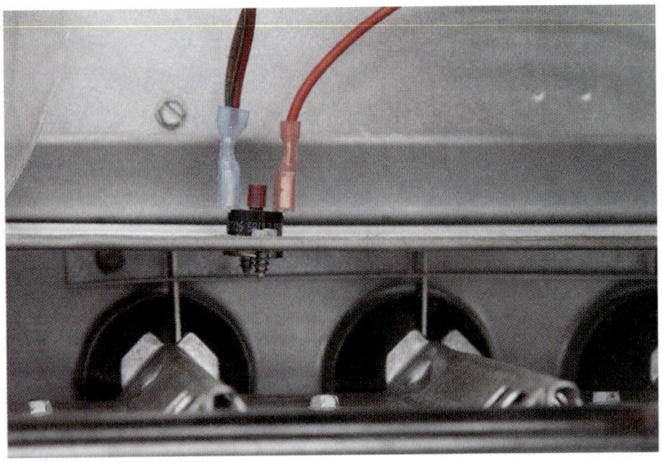

FIGURE 18-7 Rollout safety switch.

Electronic ignition modules and IFCs use several approaches to ignition. Older furnaces with a standing pilot flame do not require a source of pilot ignition, although over the years, many have been field retrofitted with electronic relight devices. Modern furnaces use either a spark-generating ignition module or a **hot-surface igniter (HSI)** to light a pilot flame or the main burner. When the spark or HSI lights the main burner directly, the system is referred to as a **direct-ignition system**. Some spark and HSI systems do, however, continue to ignite a pilot flame first. When a pilot flame is used, once it has been proven to exist, the main gas valve is opened and then the pilot flame ignites the escaping gas. When the main burner is being ignited directly, the ignition of the gas-air mixture must be quickly confirmed, since a great deal of gas is being admitted into the combustion chamber.

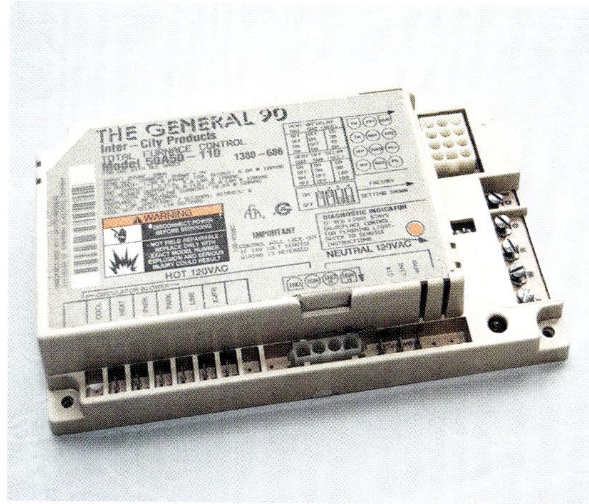

FIGURE 18-8 IFC.

To prevent gas from pouring into the combustion chamber when there is no source of ignition or when ignition does not take place as expected, modern furnaces use **flame rectification** (FIGURE 18-9). A special rod exposed to the pilot flame or main burner, called a **flame rod**, rectifies an alternating current (AC) to create a small, pulsating direct current (DC) when heated. In fact, the AC from the ignition control actually passes through the ionized flame. The flame rod is wired to the ignition control or IFC, where the rectified current is sensed. The current required to prove that a flame exists is very small—generally 2–5 microamps (μA). Note that because the current is tiny, the wiring connections and the flame rod's grounding point must be clean and sound. In addition, the polarity of the line voltage powering the furnace must be correct for the circuit to function properly.

Testing the rectified current requires placing an ammeter in series with the sensor wire (FIGURE 18-10). Note that the flame sensor is often located on the opposite side of the furnace from the igniter. If the flame does not light all the way across the series of burners, the sensor will not detect a flame, which will result in an ignition failure.

If no current is sensed from a flame rod on a furnace with a pilot flame, the main gas valve will not open. For direct-ignition furnaces, the main gas valve will open for only a brief period. If ignition does not occur quickly and/or the rectified current is not generated and sensed, the gas valve will close. The induced-draft fan continues to run to push (purge) any unburned gas out of the heat exchanger before ignition can be attempted again. Most ignition controls will allow several attempts before locking the combustion process out completely. At that point, intervention is required to correct the source of the problem and return the furnace to service.

▶ TECHNICIAN TIP

Measuring Current from a Flame Rod

Remember that measuring a very small current requires that an ammeter be placed in series with the device. Closing the jaws of a clamp-on ammeter around the wire will not work for measuring microamps accurately, if at all. It is also important to ensure that the ammeter used can read in the expected current range; not all ammeters can read such tiny current values accurately.

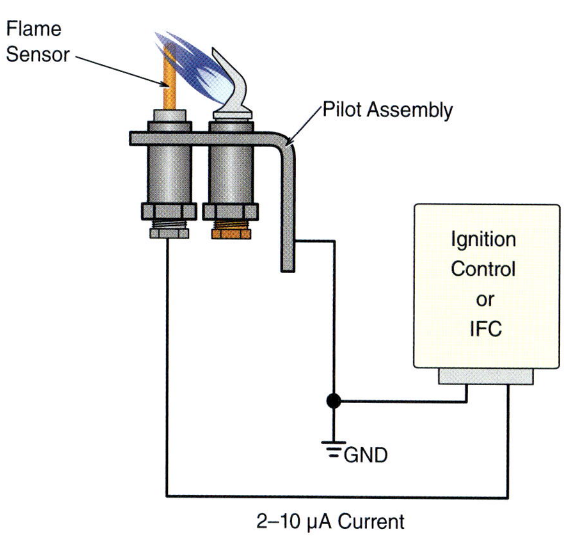

FIGURE 18-9 Flame rectification.

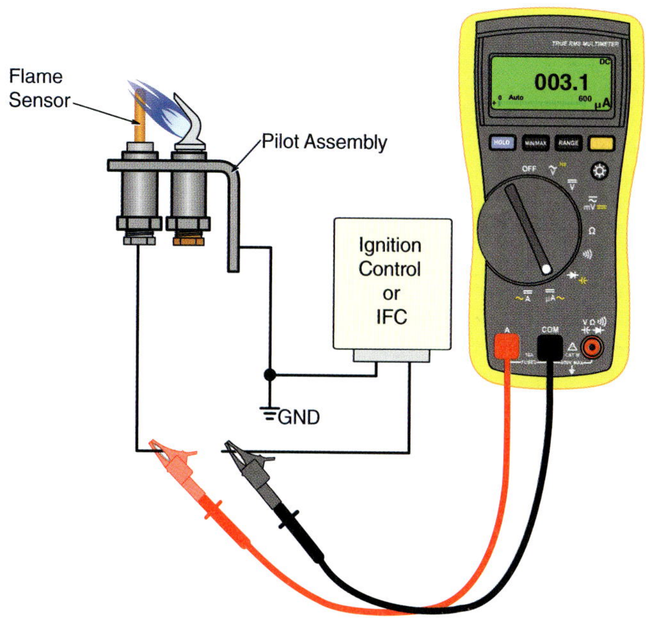

FIGURE 18-10 Testing the current flow of a flame sensor.

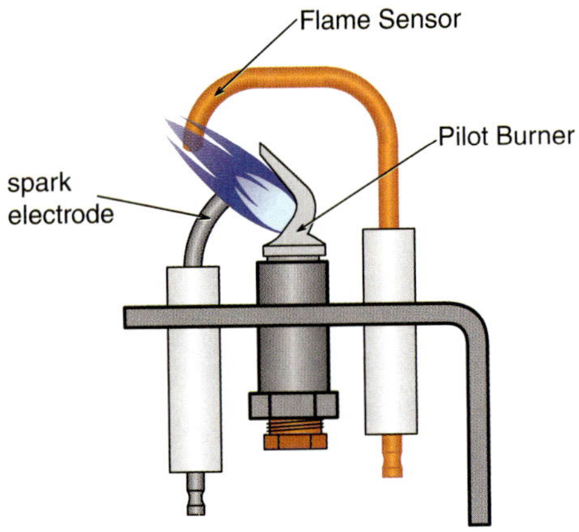

FIGURE 18-11 Pilot burner with spark igniter.

FIGURE 18-12 Silicon-carbide HSI.

> ▶ **TECHNICIAN TIP**
>
> **Handling HSIs**
>
> It has long been thought that handling an HSI with bare hands transfers oils to the material and that the oils would cause premature failure. However, manufacturers have largely debunked this myth. On the other hand, today's work gloves offer improved dexterity and feel, and there is no harm in handling them with gloves. What is most important is to prevent sharp impacts to the material.

K18003 Describe the sequence of operation for an induced- draft gas furnace.

Spark Igniters

Furnaces with spark ignition do not have a standing pilot flame. The pilot, if used, is ignited during each operating cycle. A high-voltage signal is sent to a spark **electrode** (**FIGURE 18-11**). Although the current is low, the voltage required for an arc to form is usually 10,000–20,000 volts. The high voltage creates an intermittent spark that jumps across an air gap above the pilot or main burner, igniting the escaping gas. Once the flame sensor detects a flame, the igniter will be deenergized.

Hot-Surface Igniters

HSIs (**FIGURE 18-12**) are very common in today's gas furnaces. As the name implies, a silica-based material is heated to a very high temperature that is sufficient to light the gas-air mixture. HSIs are constructed of silicon carbide or silicon nitride, the latter being the more recent material introduced and generally more durable.

HSI material is relatively brittle and can be damaged by rough handling. Cracks are generally identified by a white line in the material. Cracking can also occur from normal operation over time.

When they are energized and operating properly, the glow from the intense heat is quite obvious. However, a glowing HSI is no guarantee that it is reaching a sufficient temperature to ignite the burner. When an HSI glows but fails to ignite the gas-air mixture, it is often an indication that some of the resistance has been lost over many operating cycles. The resistance of a functional HSI is commonly 40–90 Ω, but it is important to know the manufacturer's specification for the specific type in use. When the resistance is too low, the HSI will not reach a sufficient temperature to support ignition. Resistance testing must be performed when the element is cold. Under no circumstances should an element be handled until it has cooled.

Note that the HSI is powered by the line voltage provided to the furnace, and not by the control voltage in most scenarios. A line-voltage test is what is usually performed when testing for power to an HSI that doesn't glow when it should. However, there are some specific gas valves that control the HSI directly and do power it with 24 VAC.

▶ Gas-Furnace Sequence of Operation

The sequence of operation for a gas-fired furnace will vary depending on its type. Although there are similarities in the operating sequence of induced-draft and condensing furnaces, there will always be some differences. Also remember that every wiring diagram is different as well.

The wiring diagram in **FIGURE 18-13** shows an example of a ladder diagram for an induced-draft furnace with spark ignition. A component legend is provided in **FIGURE 18-14**. This diagram is the basis for the operating sequence that follows.

The following sequence of operation describes furnace operation under normal conditions:

1. A call for heat is signaled by the thermostat by closing the circuit between R and W. Before any other action occurs, the IFC checks to ensure that power is *not* sensed through the draft pressure switch (DPS), for safety purposes.
2. Control power from the W terminal flows through the (IFC) board, through the normally closed manual reset limit (MRL), and through the automatic reset limit (ARL). Once the control voltage returns to the IFC through the ARL, proving that the limit switches are closed, the board energizes the induced-draft fan relay (IDR). The IDR contacts close, starting the induced-draft motor (IDM).
3. The IDM creates a negative pressure in the combustion chamber, closing the draft pressure switch (DPS). After a pre-purge cycle lasting 25–30 seconds, meant to ensure the combustion chamber is clear of gas, the IFC closes the gas-valve relay (GVR). This initiates the ignition sequence.
4. The ignition sequence, governed by the IFC, lasts for 10 seconds. The main gas valve (MGV) is opened and the igniter (IGN) is energized simultaneously.
5. Once the burner ignites, the flame sensor (FS) sends its rectified signal back to the IFC, proving ignition. The igniter is deenergized one second later, and burner operation continues.

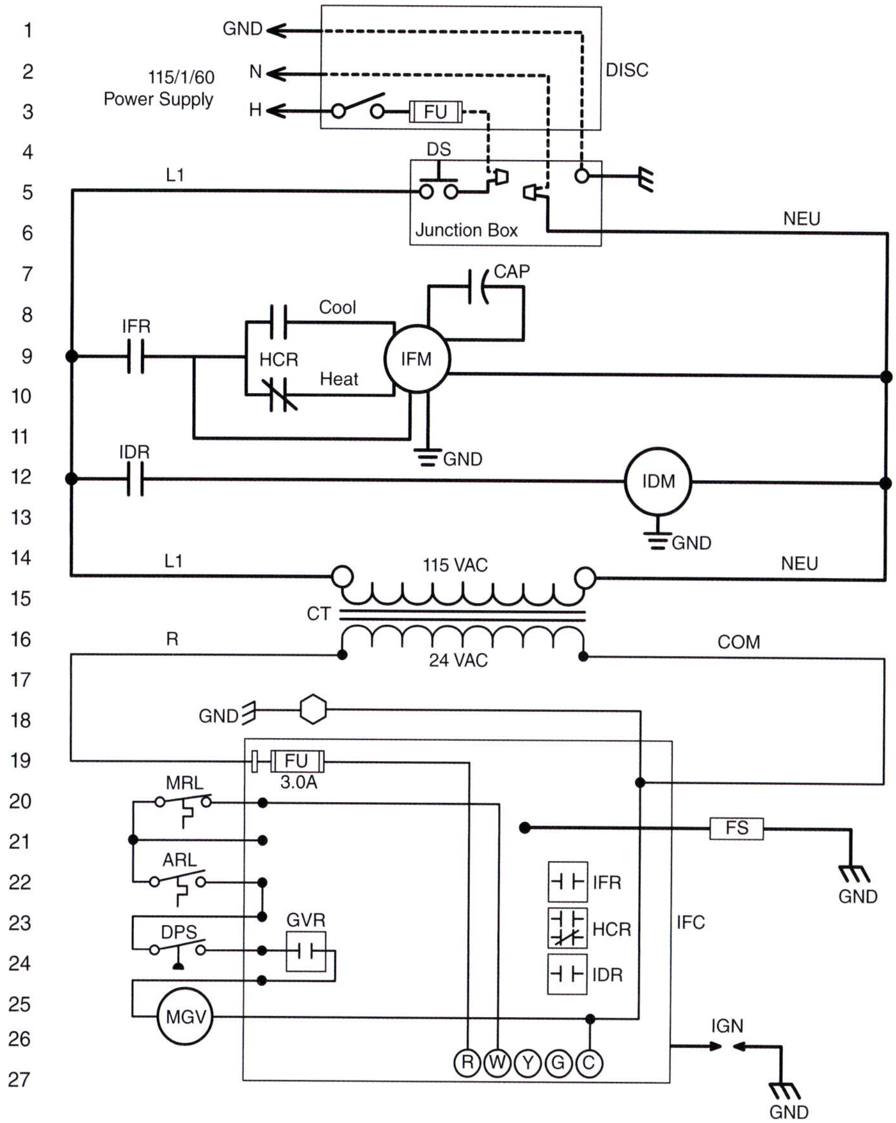

FIGURE 18-13 Example ladder diagram for a gas furnace.

COMPONENTS

ARL	AUTOMATIC RESET LIMIT CONTROL	IDM	INDUCED DRAFT MOTOR
C	COMMON	IDR	INDUCED DRAFT RELAY
CAP	RUN CAPACITOR	IFC	INTEGRATED FURNACE CONTROL
CT	CONTROL TRANSFORMER	IFM	INDOOR FAN MOTOR
DISC	DISCONNECT SWITCH	IGN	IGNITER
DPS	DRAFT PRESSURE SWITCH	IFR	INDOOR FAN RELAY
FS	FLAME SENSOR	MGV	MAIN GAS VALVE
FU	FUSE	MRL	MANUAL RESET LIMIT CONTROL
GND	GROUND	NEU	NEUTRAL
GVR	GAS VALVE RELAY	PBS	PUSH BUTTON SWITCH
HCR	HEAT/COOL RELAY		

FIGURE 18-14 Component legend for ladder diagram.

6. Approximately 15 seconds after ignition has been proven, the IFC energizes the indoor fan relay (IFR), closing its contacts. The heating/cooling relay remains deenergized, resulting in the IFM operating at the chosen heating-mode speed through its normally closed contacts.
7. Once the thermostat is satisfied, the circuit between R and W is broken. This deenergizes the GVR, closing the gas valve. The IDM continues to run for 10 seconds to purge the heat exchanger of combustion byproducts and any unburned gas. Once the IDM stops and the heat exchanger is no longer at a negative pressure, the DPS opens.

The IFR remains energized, and the blower continues to run for two minutes after the external heating-control circuit is broken, cooling the heat exchanger and distributing the residual heat to the structure.

It is important to note that this is simply an example of a furnace operating sequence. Although most induced-draft furnaces operate using a similar sequence, there are many variations.

▶ Common Gas-Furnace Malfunctions and Troubleshooting

K18004 Describe common gas-furnace malfunctions and troubleshooting procedures.

There are a number of problems that can occur during furnace operation. Remember that the first three steps in strategy-based troubleshooting are as follows:

1. Verify the problem.
2. Consider the possibilities.
3. Perform focused testing.

Modern furnaces controlled by an IFC, such as the furnace in this example, are usually equipped with some sort of diagnostic display. Diagnostic codes can be displayed by LED lights that flash in a sequence, or simple two- or three-digit displays can display specific codes. Example diagnostic codes are shown in **TABLE 18-1**. If more than one code is appropriate, the display will alternate between the active codes. Note that codes 0, C, H, and F are displayed when the furnace is in standby or when a normal operating mode is active.

Using diagnostic codes is a quick way to verify a problem. Although, on rare occasions, codes can be misleading, they typically direct the technician's attention in the proper direction. Without them, the furnace must be observed through an operating sequence to determine precisely where the sequence is broken. This step is still often required to verify a problem, but the codes can generally speed up the process and provide valuable information. Once a problem has been verified, the possibilities can be considered and mentally organized from most to least likely. Then focused testing can begin.

Many IFCs are designed to clear diagnostic codes after some period of time, such as a week, has passed. Many also provide a manual means of clearing the codes from memory. When the power is cycled off and back on, many IFCs will continue to display the most recent codes that have not been cleared from memory.

TABLE 18-1 Diagnostic Fault Codes

CODE	DESCRIPTION
0	STANDBY
C	COMPRESSOR ENERGIZED
H	GAS HEAT ON
F	FAN CIRCULATION MODE
10	HARD LOCKOUT
15	FAILED IGNITION
17	SOFT LOCKOUT
18	FLAME LOST MID-CYCLE
22	ARL CONTROL OPEN
26	LINE AND NEUTRAL REVERSED
30	MRL (MANUAL RESET LIMIT CONTROL, "ROLL OUT") OPEN
40	PRESSURE SWITCH STUCK CLOSED WHEN IT SHOULD BE OPEN
42	PRESSURE SWITCH OPEN WHEN IT SHOULD BE CLOSED
61	BLOWER FAULT – BLOWER UNABLE TO RUN
80	INTERNAL CONTROL FAULT

A brief discussion of diagnostic codes and their possible causes is provided in the following sections. As each fault is discussed, the possibilities and some specific focused-testing procedures will also be presented. Note that fault codes are not always presented in a specific order, as is the case in Table 18-1. In the discussion that follows, they will be presented in a different order than shown in the table to better facilitate understanding.

Code 26—Line and Neutral Reversed

In a 120-VAC furnace, the line voltage is the switched leg. The IFC is sensitive to the polarity of the power supply, as is the flame rectification circuit. For this reason, the incoming power supply must be wired to the correct leads in the junction box of the furnace. If the leads are switched, the furnace will not operate and this fault code will be displayed.

Troubleshooting this fault code is relatively simple. Check the incoming power to the furnace and ensure that the L1 (hot) leg of power is connected to the appropriate lead in the junction box. It is rare for such a code to be displayed once the furnace has been properly started up after installation. It is possible, however, that an error in the power wiring outside of the furnace itself could occur during an unrelated electrical repair, causing this problem.

Code 22—Automatic-Reset Limit Open

This limit switch monitors the temperature in the heat exchanger. When this limit opens, it is usually due to a lack of airflow from the main blower. Without a sufficient amount of airflow to move heat away from the heat exchanger, the furnace quickly overheats.

Although the focus is on electrical troubleshooting, some nonelectrical issues can also impede airflow, such as the following:

- a clogged air filter
- selecting the incorrect blower speed
- a loose or broken belt on belt-driven blowers
- closed dampers or another obstruction in the air distribution system.

This limit switch might also open due to the furnace being overfired. If the gas pressure to the furnace is excessive or if the gas orifices are larger than they should be, too much gas will be admitted and excessive heat results. If the airflow through the furnace is appropriate, the gas pressure should be checked. The size of the orifices should also be checked against the manufacturer's specifications.

Electrical issues that might cause this problem include the following:

- *Indoor fan motor failure.* The motor must be tested using the proper procedures for the motor type: AC induction or electronic control module (ECM). A run capacitor could prevent a permanent split capacitor (PSC) motor from running properly.
- *Failure of the indoor fan relay (IFR) or heating/cooling relay (HCR).* These failures are tested by monitoring the voltage outputs from the IFC, since the relays are mounted directly on the board. When the expected outputs do not occur when expected and all inputs are normal, the IFC must be replaced.
- *Limit-switch failure or out of calibration.* Measuring the actual temperature at the same location while the furnace operates can confirm that the switch is opening before it reaches the design set point.

Code 30—Manual-Reset Limit Control Open

This limit switch is positioned near the combustion area and protects against flame rollout into the control vestibule. If it is tripped due to a high-heat condition, it will need to be manually reset. The IFC will lock out all functions related to combustion and heat operation until the switch is manually reset, closing its contacts.

Overheating in the vestibule can be caused by a blockage in the heat exchanger, which might also prevent the furnace from attempting ignition (see Code 15—Failed Ignition). This limit switch might also open due to **delayed ignition**. The temperature at the switch location can be measured to determine whether it is opening when it shouldn't. Like the ARL, the switch itself can also fail.

Code 40—Pressure Switch Closed When It Should Be Open

The DPS is normally open. As shown on the wiring diagram, it closes on a fall in pressure. Thus, the heat exchanger is at a negative pressure when the furnace is operating normally, and the switch closes after the appropriate pressure has been reached. The IDM must be running for this to happen. Before the IFC begins the ignition sequence, it checks to ensure this switch is open. If it is not, the code will be displayed, and the heating sequence will not continue.

Testing the DPS with the furnace and power source off is simply a matter of checking through the contacts with an ohmmeter. In most cases, adjusting this pressure switch in the field is ill-advised. They are very sensitive, and an incorrect setting can allow combustion to occur when it is not safe. If the DPS is not operating properly, replace it with the proper, factory-calibrated part.

Code 42—Pressure Switch Open When It Should Be Closed

If the DPS does not close within a set period of time—one minute, for example—after the IDM has started, then this code will be displayed. Nonelectrical reasons for the switch to remain open include the following:

- A leaking pressure switch hose. The pressure switch is connected to the heat exchanger with flexible tubing and barbed fittings. The tubing can come loose, or the tubing can split at the end, causing a leak.
- A leaking induced-draft blower gasket.
- A blocked vent or extreme downdraft through the vent piping.

Electrical reasons for the failure include switch failure and a failure of the IDM. Note that the small blower wheel mounted on the motor shaft can also become loose or damaged. The motor would then run but would not produce the necessary negative pressure.

If the motor does not run, check for the appropriate voltage at the IFC terminals. If power is present, the motor will then be tested for an open winding, grounded condition, or failed bearings. Some IDMs have a run capacitor that could be the source of the problem and require testing. If power is not present at the IFC terminals and no other codes are displayed, the IDR or the IFC has likely failed. Either condition means that the IFC must be replaced.

Code 61—Blower Fault

In this example, the IFC has no direct knowledge of blower operation. If the ARL opens during a heating cycle and remains open for more than two and a half minutes, it will assume the blower is not running and displays this code. Other systems may use another means of determining whether the blower is running. If the blower does not start within the appropriate time after burner operation has been proven through the FS, test for the proper voltage at the IFC terminals that power the blower. If present, the blower motor must be checked for open windings, grounds, and bearing failure. Loose or flawed wiring terminations can also cause the problem.

The HCR for this furnace uses the normally closed contacts during the heating cycle; it is not energized in the heating mode. Therefore, it is not a likely source of the problem. The IFR, however, must be energized and must close its contacts. If it fails to do so, the IFC must be replaced since it is integrated into the board.

Code 15—Failed Ignition

They are several reasons why an ignition attempt might fail. Nonelectrical issues include a low, or complete loss of, gas pressure. It can generally be assumed that the limit switches and the DPS are closed, since a trial for ignition will not take place if any of them is open. Since the DPS is closed, it can also be assumed that a draft has been proven.

Assuming a gas supply at the required pressure is available, a spark to ignite the gas-air mixture may not be present. The IFC generates high-voltage, low-current pulses to the igniter rod, which must arc across a gap bathed in the gas-air mixture. The example furnace uses direct ignition, so there is no pilot assembly. If the arc cannot be heard or seen, check for the following problems:

- *A loose, disconnected, or damaged ignition lead to the igniter.* A flawed connection can occur at either end. The ignition lead can be tested with an ohmmeter (when the system is deenergized) from one end to the other. It must have continuity, and the resistance should be very low—in the range of 0.1 Ω.
- *The ceramic insulator around the igniter mount is cracked or broken.* This would allow the current to flow through the mount to ground, instead of arcing across the gap. With the system deenergized, check for continuity between the tip of the igniter rod to ground; there should not be continuity. If there is, the ceramic insulation around the rod is most likely cracked or broken, even if it is not visible.
- *The incorrect spark gap.* Like an automotive spark plug, the gap must be correct. If it is too wide, the arc cannot be established. If it is too narrow, it may not be sufficient to ignite the gas.
- *Poor grounding.* As noted on the wiring diagram, the current from the IFC passes across the gap and directly to ground. If the igniter assembly is not mounted properly or if corrosion is present, it can interfere with the path to ground. Ensure that all components are properly grounded and that all ground terminations are sound. Poor grounds can cause an intermittent problem as well.

If the igniter is not grounded and the ignition lead shows no sign of damage, the high-voltage pulse from the IFC or igniter can be tested, but it must be done with care. Do *not* use a common voltmeter to test for voltage here. The igniter voltage exceeds the capacity of many voltmeters. The most common method of testing is to disconnect the ignition lead from the board. Hold the ignition lead well away from the end of the wire by using a pair of well-insulated pliers. While the board is powering the igniter, slowly move the end of the lead toward its connection on the board. An arc should form with the end of the lead within ½" (1.3 cm) from the connector. If the arc occurs at this distance, it should easily be sufficient to jump the roughly ⅛" (3.2 mm) gap at the igniter assembly.

Gas-Valve Problems

Again, assuming the gas supply to the furnace is sufficient, the gas valve may not be opening, preventing ignition. Gas-valve relay failure is one possibility. The output from the relay to the gas valve can be checked at the appropriate IFC terminals and at the gas valve. The valve itself is operated by a 24 VAC solenoid coil. One leg of power comes directly from the Common terminal and the other from the relay output. If 24 VAC power is present at the valve, test the gas-valve coil for continuity with the power off. Although it is also possible that the coil is being energized and is functional but the valve has suffered an internal failure, this scenario is rare.

Many of today's gas valves are equipped with an On/Off switch mounted directly onto the valve. This switch must be in the On position for it to function.

The Flame Sensor

The IFC is informed of a successful ignition trial by the flame sensor (FS). If the rectified signal does not return to the IFC, it assumes that ignition has failed. The example furnace is equipped with an LED on the IFC that signals the status of the burner. If it is on and steady, an appropriate signal is being received. If it is blinking, the sensor signal is weak. If it is not illuminated at all, no signal is being received. Thus, the LED assists in the troubleshooting process.

Like the igniter, good ground connections are essential for the sensor to operate properly. All grounds should be checked. Flame sensors might also produce a weak signal due to oxidation buildup. They can be cleaned with an abrasive pad or fine steel wool.

The current from the sensor can be tested as shown in Figure 18-10. The required current should be determined from the manufacturer's documentation.

▶ CORE CONNECTION

Gas Pressure

Gas valves in modern gas furnaces are designed to reduce the pressure to that required by the manufacturer in the manifold. The manifold includes the distribution piping downstream of the valve that delivers the gas to the orifices. The most common manifold pressure for a natural-gas furnace is 3.5 in. wc (6.5 mmHg). However, for furnaces with more than one stage of heat, the low stage(s) of heat operate(s) at a lower pressure. Furnaces that use propane for fuel usually require a manifold pressure of 10–11 in. wc (18.7–20.6 mmHg).

To exert good control over the pressure in the manifold, the pressure entering the gas valve must be higher. Many gas valves require a minimum of 4.5–5 in. wc (8.4–9.3 mmHg) at the inlet for proper control of downstream natural-gas pressure. The inlet pressure of propane must also be higher than manifold pressure for the valve to effectively control the downstream pressure. But common gas valves also have an inlet pressure limitation: a maximum inlet pressure of ½ psig (25.9 mmHg). Converted to inches of water column, ½ psig = 13.9 in. wc. Therefore, the gas-valve inlet pressure for a furnace operating with propane must remain below that pressure.

It is also very important to remember the maximum inlet pressure of gas valves when pressure testing a new gas line. The gas valve must be isolated from the test pressure to avoid damage. A manual gas cock must be installed ahead of the gas valve and remain closed during pressure testing. For additional protection in case the gas cock leaks through, do not assemble the piping union between the gas cock and the gas valve during the test. That way, any pressure that leaks past the gas cock is harmlessly vented to atmosphere and won't damage the valve.

Testing Hot-Surface Igniters

Although the example furnace is equipped with spark ignition, HSIs are very common today. As mentioned previously, the resistance can be tested when it is cold (**FIGURE 18-15**). There must be continuity, and the required resistance is usually 40–90 Ω. However, it is important to know the required value for the igniter being tested. An HSI that glows weakly and does not get hot enough to ignite the burner is often characterized by a low resistance.

As shown in Figure 18-15, the HSI is typically connected electrically with a small plastic plug, with pins in one side and sockets in the other. While a furnace is sending power to the igniter during an ignition trial, the voltage can be tested at the power connector from the IFC or other ignition control (**FIGURE 18-16**). In most cases, line voltage should be displayed. Remember, however, that some furnaces do use 24 VAC–powered igniters.

Code 18—Flame Lost Mid Cycle

This fault can occur only after ignition has been proven. Otherwise, a Code 15 fault will occur. Since the IFC relies on the flame sensor to prove that a flame is present throughout

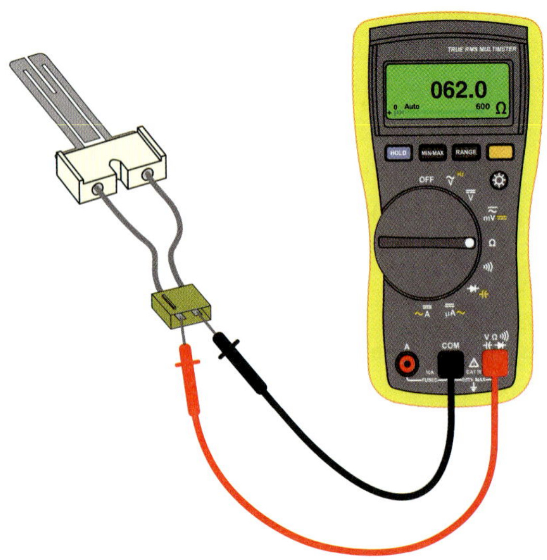

FIGURE 18-15 Testing HSI resistance.

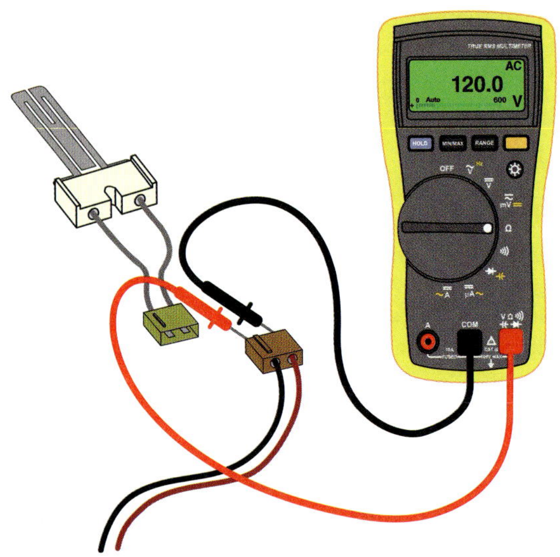

FIGURE 18-16 Testing for power to the HSI.

the heating cycle, a loss or significant reduction of current from the sensor would cause this fault. An unexpected loss of gas pressure could cause the flame to go out, leading to a loss of current from the flame sensor. Troubleshooting this fault typically requires careful observation through one or more heating cycles. If the IFC closes the gas valve and displays this code while combustion seems fine and the sensor is consistently bathed in the flame, monitor the current from the flame sensor through the cycle. Remember that good ground connections are essential. During the heating cycle, as the furnace components warm up, already-weak ground connections can be adversely affected.

Code 17—Soft Lockout

These codes can vary from unit to unit, but soft lockouts generally reset themselves automatically. In the example furnace, soft lockouts prevent the furnace from operating in the heating mode for one hour. At that point, the IFC resets and the furnace tries again. The one-hour wait is true for all the conditions below, except for the first one:

- The DPS fails to close (before an ignition trial) after the draft inducer has been operating for one minute. After five minutes, the lockout resets and the furnace tries again.
- No flame has been detected after five attempts at ignition. The first four failed attempts result in the system attempting ignition again immediately after sufficient time is allowed for the induced-draft motor to clear any unburned gas from the heat exchanger. No more attempts will be made for 1 hour after these five attempts.
- The flame has been lost five times during a thermostat cycle. The furnace will repeat the ignition sequence immediately after the first four losses. No more attempts will be made for one hour after the five occurrences.
- The ARL has opened and remained open for two and a half minutes. The furnace will initiate a soft lockout and then repeat the entire ignition sequence, provided that the ARL closes again during the lockout period. This cycle can repeat itself three times.

When this furnace is timing the soft lockout period, Code 17 will be displayed, followed by the code for the condition that caused it to lock out.

Soft lockouts offer a way for furnaces to experience something unusual yet continue to try to operate to maintain comfort. This is due to the accepted fact that providing heat is considered essential, especially in exceptionally cold climates.

Code 10—Hard Lockout

Once this code is displayed, the furnace will no longer attempt to operate in the heating mode. The following conditions lead to a hard lockout:

- After the fourth time that the ARL has remained open for more than two and a half minutes. Three soft lockouts have already occurred, and the ARL continues to open during the heating cycle for more than two and a half minutes. In this case, the IFC can be reset at the thermostat by turning it off for a few seconds and then back on again.
- When the circuits of the IFC detect an internal fault (Code 93). When this occurs, the line voltage serving the furnace must be cycled off and back on again, allowing the IFC to initiate a new self-test. If there is something wrong and the test fails, a hard lockout results again. Replacing the IFC is then typically required to restore heat.
- When the IFC detects that voltage is being sent to the gas valve when it should not be present. Code 93 is also displayed for this condition. Cycling the line voltage on and off will also initiate a new test in this case. If voltage is again detected at the contacts of the gas-valve relay before power is intentionally sent to it, Code 93 results again. The IFC must be replaced since the gas-valve relay is integrated into the board. The relay is not a field-replaceable component.

It should be clear that modern furnaces are quite sophisticated and require patience to troubleshoot. Remember that this is just one example of a set of fault codes and that each brand and model of gas furnace has its own unique features and operating characteristics. Not every furnace currently operating is equipped to report faults; those that aren't so equipped leave technicians to rely only on their knowledge of the operating sequence. Using all the available resources for troubleshooting, including the sequence of operation, troubleshooting chart, flow diagrams, and wiring diagrams, is essential for success. Use the strategy-based troubleshooting process, and begin with visual observations, followed by an evaluation of the possibilities to help determine where to begin focused testing.

▶ TROUBLESHOOTING TIP

IFC Fuses

If line voltage is present but there is no evidence of control voltage on an unresponsive furnace, check the transformer output to the IFC. Although an IFC often switches line-voltage power to furnace motors, the IFC itself is powered by the control voltage. Most IFCs are equipped with a control circuit fuse (**FIGURE 18-17**). This fuse is very similar to an automotive-type fuse (glass cartridge or blade type) and is simple to replace.

As is the case with any blown fuse, the technician must first try to determine the source of the problem. This fuse often protects the entire 24 VAC circuit, not just the board itself. Note that the 3-amp fuse shown on the wiring diagram in Figure 18-13 is in series with the 24 VAC output to the entire control circuit. Always replace the fuse with one of the correct type and current rating.

FIGURE 18-17 Control fuse on an IFC.

K18005 Describe common oil furnaces and burners.

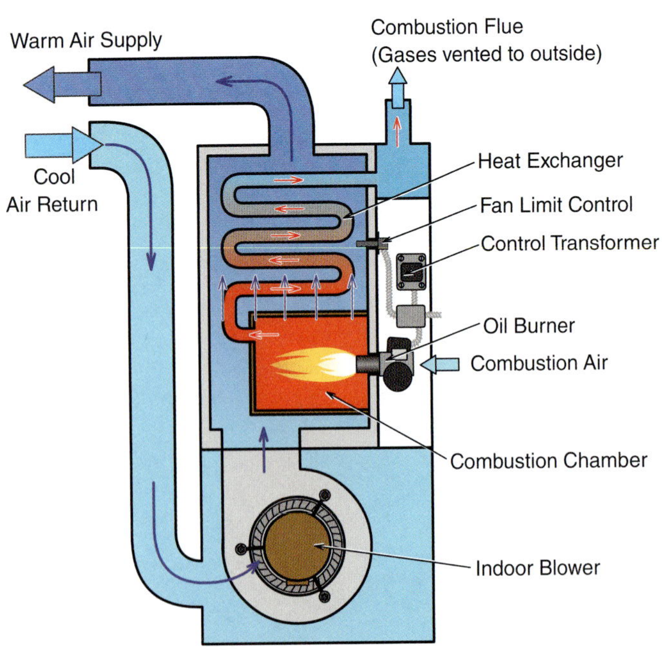

FIGURE 18-18 Oil furnace.

▶ Oil Furnaces

Common oil furnaces (**FIGURE 18-18**) burn fuel oil that is atomized into a fine mist and sprayed into the combustion chamber. The oil is atomized by being forced through a small, precisely machined nozzle that creates a specific spray pattern. As it exits the nozzle, the mist combines with combustion air to form an ignitable mixture.

Oil furnaces are generally simpler than modern gas furnaces. Modern oil furnaces, however, may be equipped with ECM motors to offer greater flexibility and electrical energy efficiency. They may be equipped with solid-state boards, much like gas furnaces, that manage blower operation and accessories such as electronic air cleaners and humidifiers.

Many oil furnaces are still used in remote locations where natural gas is not available. Propane is an option for these rural locations, but for many years, propane was a less accessible fuel with a higher operating cost. Although the use of fuel oil for comfort heating continues to shrink, it is still preferred in some areas. In many cases, oil is used in areas where summer cooling is not a necessity but where robust and reliable heating capacity throughout cold winters is essential.

Indoor Blower and Limit Controls

Basic oil furnaces use a very simple device to control the blower and provide an overtemperature limit function. A fan/limit control (**FIGURE 18-19**) uses a long, spiral-shaped bimetal element that twists as it is heated, due to expansion along its length. The probe is inserted into the furnace near (but not into) the heat exchanger. Three separate trip levers (**FIGURE 18-20**) are positioned to provide fan control and high-temperature protection.

FIGURE 18-19 Fan/limit control.

The lowest temperature setting deenergizes the indoor blower as the temperature of the heat exchanger falls below the setting. The second lever is set to energize the indoor blower as the heat exchanger warms up. The third lever sets the high-temperature limit, which deenergizes the burner. The twisting of the bimetal element causes the individual trip levers to contact the switch mechanisms at the appropriate temperature. This simple control is easily tested using skills previously learned in this book.

This high-temperature limit switch resets automatically after the bimetal element cools. A decrease of 25°F (14°C) is common. The differential is factory-set. Note that it is important to keep the fan-on and fan-off set points some distance apart, such as 40°F (22°C). If they are set too close together, the fan will cycle on and off multiple times during each heating cycle. Many are equipped with a switch that protrudes through the cover to manually operate the fan for air circulation. It is not a reset for the high-temperature limit switch.

FIGURE 18-20 Fan/limit control trip levers.

When cooling is added to basic oil furnaces, a basic transformer/relay package (**FIGURE 18-21**) can provide a control-power source and indoor blower control in the cooling mode. As a general rule, the blower will need to operate at a higher speed in the cooling mode, so the blower relay will energize a different winding of the indoor blower motor than the furnace fan/limit control. Oil furnaces used only for heating can operate using a small control transformer built into the **primary control**, since the control circuit consists of two wires to the thermostat. These devices are also simple to troubleshoot using previously learned skills, and are discussed later in this section.

Due to the simplicity of other devices in oil furnaces, this portion of the chapter will focus on the operation of the oil burner. Oil burners are significantly different from gas burners and require some specific knowledge to service them successfully.

Oil Burners

The oil burner (**FIGURE 18-22**) is a packaged assembly that includes the motor, oil pump, transformer, and primary control. The burner motor drives a centrifugal blower wheel inside the housing (**FIGURE 18-23**) that provides combustion air and creates turbulence

FIGURE 18-21 Transformer/relay package for control power and blower control when cooling is added.

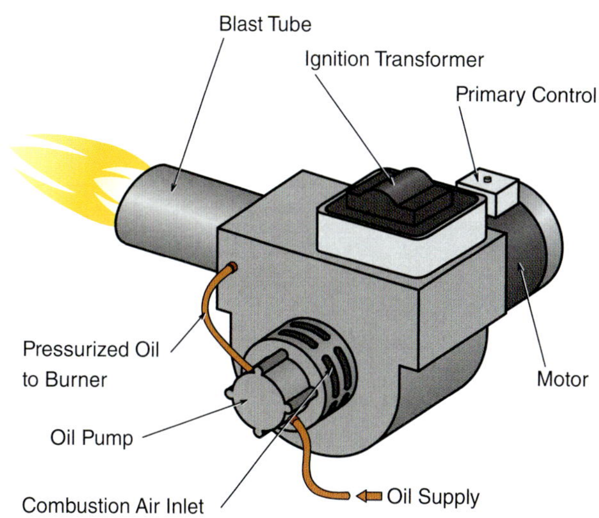

FIGURE 18-22 Oil-burner assembly.

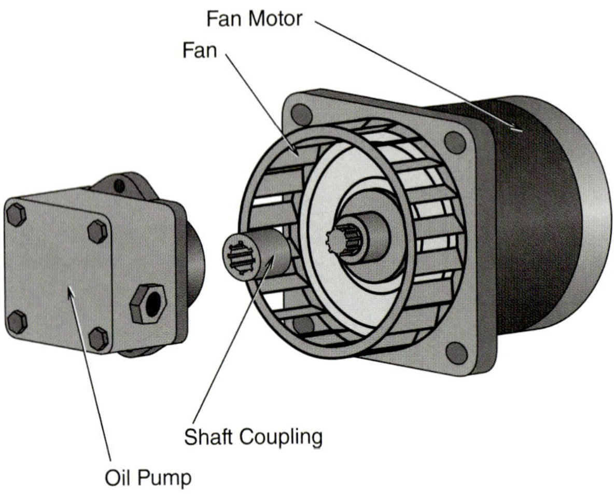

FIGURE 18-23 Burner motor, fan, and pump arrangement.

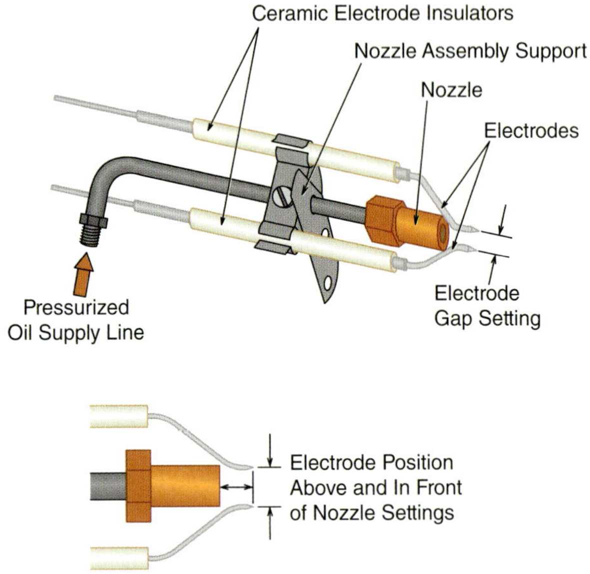

FIGURE 18-24 A nozzle assembly.

to thoroughly mix air with the oil spraying from the nozzle. In addition, the motor is mechanically coupled to the oil pump. Most oil pumps are designed to deliver oil at a pressure of 100 psig (690 kPa) to the nozzle.

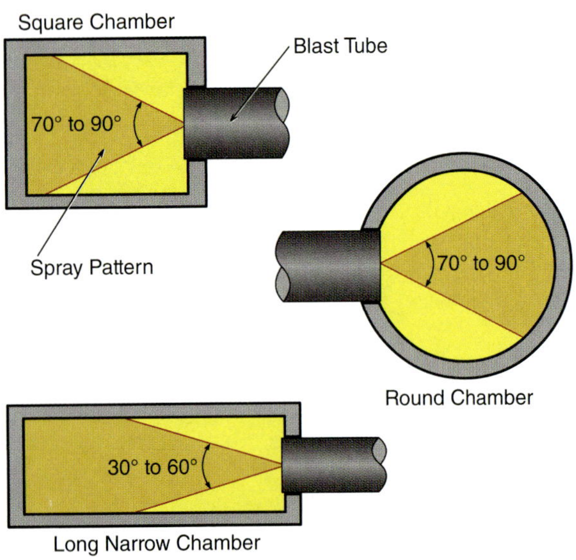

FIGURE 18-25 Nozzle spray angles.

The nozzle assembly (**FIGURE 18-24**) is enclosed in the burner blast tube and includes the oil nozzle and electrodes that establish the arc for ignition. The size of the nozzle determines the heating capacity. The primary fuel used for residential and commercial burners is #2 fuel oil with a heating capacity of approximately 140,000 Btuh per gallon (36,984 Btuh per liter). However, due to losses in the process and vent, the average oil furnace operates at an AFUE of roughly 85%. Therefore, each gallon delivers roughly 119,000 Btuh of usable heat.

Nozzles are designed to deliver oil in a specific spray pattern, determined by the nozzle design and angle. The round pattern can be hollow, with little or no atomized oil dispersed in the center of the cone; semi-hollow; or solid, where the oil spray is dispersed evenly throughout the cone. The spray angle determines the length and width of the cone, as shown in **FIGURE 18-25**.

One might ask how this information relates to the electrical operation of a burner. The answer is related to the ignition process. The type of nozzle used, especially the spray angle, determines the proper position of the electrode tips in front of the oil nozzle. The position of the electrode tips relative to the nozzle is critical for reliable performance. **FIGURE 18-26** shows an example of how the proper electrode position for

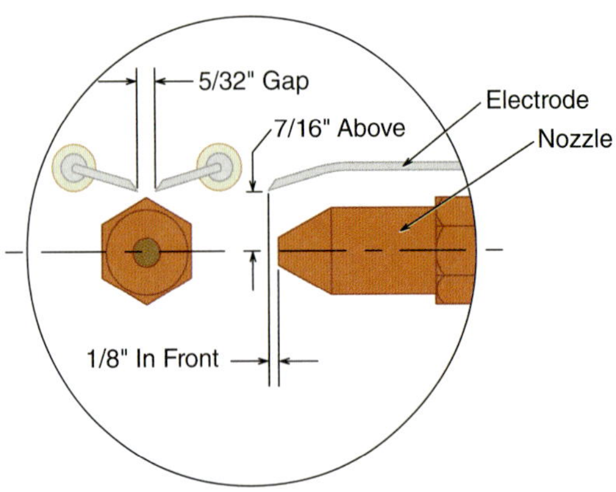

FIGURE 18-26 Diagrammatic example of the electrode position in relation to the nozzle.

a given burner and nozzle type is provided by a manufacturer. To make adjustments easier, the tool shown in **FIGURE 18-27** is used by most technicians who service oil burners. The tool is designed to properly position the electrodes for various spray angles, with four possible positions built into the tool. It is slipped over the nozzle, and then the electrode tips are positioned against the stops. The goal is to position the electrode tips at the edge of the oil spray cone.

As shown in Figure 18-24, each electrode has a conductive rod extending back toward the **ignition transformer** (**FIGURE 18-28**). The transformer has two contacts (contact styles vary) that, when the transformer is properly positioned on top of the burner housing, make contact with the conductive rods for the electrodes (**FIGURE 18-29**). Standard iron-core ignition transformers typically provide 10,000 volts at a low current, such as 20 milliamps (mA). Like gas ignition systems, this high voltage allows the current to jump across a gap, creating an arc that ignites the oil-air mixture. If the arc is not present or properly positioned in relation to the oil spray pattern, unburned fuel oil will be sprayed into the combustion chamber. An accumulation of unburned oil in the chamber can create a hazardous situation once a source of ignition returns. Therefore, the trial for ignition occurs for a limited period of time. If the mixture doesn't ignite, the primary control will stop the process.

Electronic igniters (**FIGURE 18-30**) are becoming more common on modern burners. The following are some of the benefits that they offer:

- smaller size and roughly 85% lighter in weight
- improved ignition of cold oil
- an arc that pulses at 15,000–30,000 cycles per second, compared to the 60-cycle frequency offered by common transformers
- epoxy-sealed, providing increased moisture resistance
- greatly reduced power consumption.

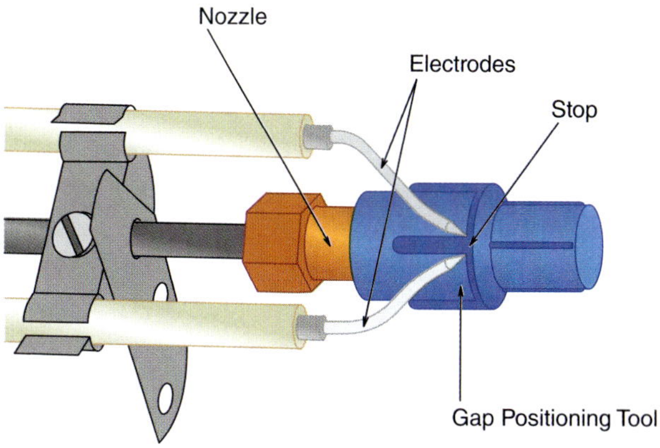

FIGURE 18-27 Electrode positioning tool.

FIGURE 18-28 An ignition transformer.

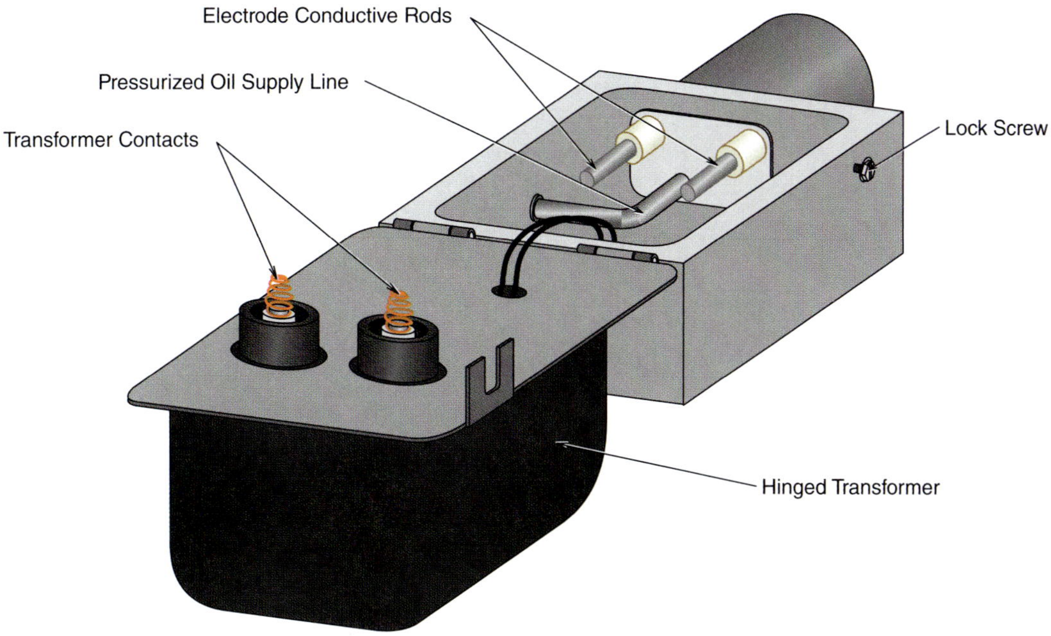

FIGURE 18-29 Electrodes extending into the burner housing.

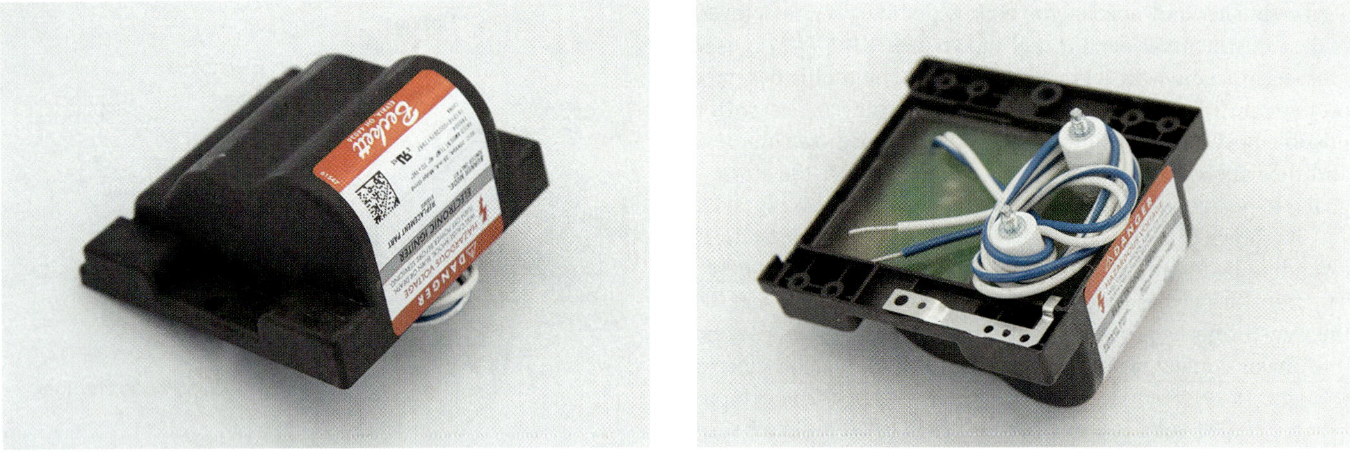

FIGURE 18-30 Electronic oil igniter.

FIGURE 18-31 Oil-burner primary control.

The Burner Primary Control

The primary control (**FIGURE 18-31**) accepts the signal from the thermostat or another external control on a call for heat. From that point, it controls burner operation and supervises the flame. Newer primary controls provide status lights for diagnostics, whereas older controls do not. Modern primary controls also offer some options to customize the sequence of operation and the ability to interact with it by using an available communication interface.

For the primary control shown, the wiring connections (other than the thermostat terminals) are made on the bottom of the control, which keeps the wiring protected inside the burner housing (**FIGURE 18-32**). Older controls often have exposed terminal boards. This control has terminals to accommodate connections to the ignition transformer, burner motor, limit switch(es), and oil solenoid valve. Some systems have a solenoid valve in the oil line that closes during the off cycle.

Line-voltage power is connected to the control at the limit-switch terminal (L1) and the L2 (neutral) terminal. In addition, there are two terminals designed to accept the input from a cadmium cell, typically referred to as a **cad cell** (**FIGURE 18-33**). Note that the cell is the small black device on the end that plugs into the wiring harness. It can be replaced by simply unplugging the old one and inserting a new one into the harness socket.

The cad cell is a light-sensitive device that changes its resistance in the presence of light. The primary control depends on changes in the resistance of the cad cell to determine whether

FIGURE 18-32 Primary-control wiring connections.

FIGURE 18-33 A cad cell.

FIGURE 18-34 An oil-burner motor.

the flame is present during the operating cycle. It must be positioned inside the burner housing where it has a clear line of sight to the burner flame. In the presence of the bright light from a proper burner flame, the resistance through the cad cell should be 300–1,000 Ω. Anything beyond 1,600 Ω when the cad cell is exposed to bright light typically represents a defective cell or possibly an obstruction in its line of sight. The primary control will not allow burner operation to continue if the resistance does not fall below the 1,600 Ω threshold.

Oil-burner motors (**FIGURE 18-34**) are designed to properly mount to the burner housing in a unique way. Beyond that feature, they are simple motors. Today's oil burners generally use 3,450 rpm motors, whereas older burners used 1,725 rpm motors. Although the older burners tended to be quieter due to the lower speed, higher efficiency has been achieved by using the higher speed.

▶ Oil-Burner Sequence of Operation

FIGURE 18-36 provides a typical wiring diagram for a primary control. In this diagram, an oil solenoid valve is not connected. Most systems do not need a solenoid valve in the oil line, but having one can help to prevent oil from draining back to a tank installed below the furnace during the off cycle.

The sequence of operation for most oil furnaces is fairly consistent, regardless of the age and efficiency of the system. As already mentioned, modern primary controls offer

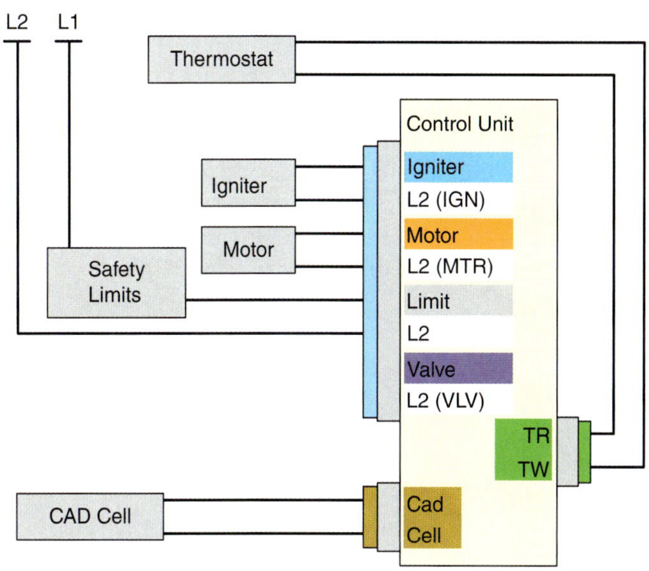

FIGURE 18-36 A wiring diagram of a primary control.

▶ TECHNICIAN TIP

Stack Switches

Older oil furnaces and boilers were often equipped with a stack switch (**FIGURE 18-35**) that served as the primary control. These controls are installed in the vent stack near the furnace or boiler—hence the name. Like primary controls, they accepted the signal from the thermostat and supervised burner operation. Instead a cad cell to detect the presence of a flame through light, a bimetal element extending into the vent piping reacted to the heat, acknowledging that ignition was successful. However, they were much slower to respond—75 seconds were commonly allowed for ignition to take place. Primary controls have largely replaced stack switches over the last 20 years.

FIGURE 18-35 Stack switch.

K18006 Describe the sequence of operation for an oil burner.

technician-accessible flexibility and programmable options that weren't available on earlier models. Certain aspects of the sequence can therefore be changed on modern primary controls.

1. On a call for heat, the thermostat closes its R_H-to-W contacts, completing the circuit from T_R to T_W at the primary control. Note that the primary control provides its own 24 VAC control voltage to the thermostat. At the thermostat, the R_H terminal must remain electrically isolated from the R/R_C terminal, which would be powered by a cooling-system control transformer.
2. The burner motor and ignition transformer are energized simultaneously. Since the oil pump is mechanically connected to the motor, it is in operation any time the motor is running. If an oil solenoid valve is connected, its opening is delayed for a predetermined time; a 15-second delay is typical. This is an option that can be disabled via programming on this primary control.
3. Once the oil solenoid valve is energized (or as soon as the transformer and motor are energized, if the solenoid-valve option is disabled), a 15-second ignition trial begins. With the motor running, the pump providing a pressurized oil supply to the nozzle, and the transformer energized and providing an arc, the oil-air mixture should ignite quickly.
4. The cad cell is exposed to the light of the burner flame, and thus its resistance drops, indicating successful ignition to the primary control. The burner continues to operate, and the ignition transformer remains energized for an additional 10 seconds to ensure the flame can support itself. Some burners keep the ignition transformer in operation throughout the entire burner cycle.
5. The fan/limit control, or furnace control board if present, starts the indoor blower as the heat exchanger warms to its normal operating temperature. Solid-state blower controls will likely operate on a timed basis. If the blower fails to operate, or the airflow through the furnace is insufficient, the temperature will continue to increase until the high-temperature limit control opens. This would immediately deenergize the primary control, motor, ignition transformer, and oil solenoid valve.
6. Once the call for heat has been satisfied, the thermostat circuit opens. The oil solenoid valve (if equipped) is closed immediately while the motor continues to run for a preset (programmable) period. When there is no oil solenoid valve, the motor is immediately deenergized to stop the flow of oil into the combustion chamber.
7. The indoor blower continues to run until the heat exchanger cools below the fan/limit control set point. For solid-state blower controls, the blower will shut down after a preset time delay.

This primary control can initiate both soft and hard lockouts, as well as an automated recycle sequence. A soft lockout is indicated by the Reset button's flashing red light, and a steady red light indicates a hard lockout. The primary control will enter a soft lockout under the following circumstances:

1. *The 15-second trial for ignition time elapses and the cad-cell resistance remains high (no flame established).* This initiates a soft lockout, and the Reset button must be pressed for at least one second to restart the burner. If the Reset button is used to try to restart the burner three times unsuccessfully, then a hard lockout will be initiated. This helps prevent the user from continuing to cycle the burner without ignition, allowing oil to accumulate in the combustion chamber and creating the potential for explosive ignition. To reset the burner from a hard lockout, the Reset button must be pressed and held for a full 15 seconds. Ideally, the user is not aware of this, prompting a visit by a qualified technician to resolve the problem.
2. *The cad cell detects light (resistance low) before the ignition sequence begins.* This can occur if the cad cell is inadvertently exposed to external light, its leads are damaged and shorted together, or it is defective. According to the primary control, this means that it has lost its ability to supervise the burner. Three unsuccessful uses of the Reset button to recycle the burner result in a hard lockout.
3. The flame is lost (cad-cell resistance increases above 1,600 Ω) while the burner is running—an indication that the flame has extinguished. The burner will shut down immediately. In this case, the primary control will enter an automatic Recycle mode. After

one minute, the burner will initiate a fresh start sequence. After three failed attempts to maintain a flame after ignition, the control will enter a soft lockout mode. Three unsuccessful uses of the Reset button to recycle the burner will result in a hard lockout.

Note that this control provides other functions for the Reset button. For example, if the technician presses the button while the burner is operating, the burner will stop. As long as the button is held down, the burner will remain off. This can be helpful during troubleshooting. The Reset button can also be used by a technician to enter a Pump Priming mode. Priming the pump (operating the pump to lift and stabilize the flow of oil to the burner from the supply tank) is often required for new installations, after the replacement of an oil filter or pump, and other similar circumstances. This may take longer than the normally allotted 15-second trial for ignition. The Reset button can be used to enter the Priming mode, temporarily disabling the cad cell while priming is underway. The burner will operate for up to four minutes without locking out, providing time to get the oil flowing.

▶ Oil-Burner Malfunctions and Troubleshooting

K18007 Describe common malfunctions of and troubleshooting procedures for oil burners.

Like gas furnaces, oil furnaces are much easier to troubleshoot with a clear understanding of the sequence of operation. Although the sequence of operation for a primary control and burner have been provided here, remember that there are variations in the sequence of other models and brands of primary controls and furnaces. However, the basic sequence of operation will be similar. The sequence is intended to provide safe, reliable supervision over the combustion process.

An Unresponsive Furnace

A visual examination of the thermostat and furnace is always the first step. Check them for any indication that the power supply is intact, such as icons on the thermostat (an icon on the display usually shows that control voltage has been lost), flashing or steady indicator lights on the primary control, or an operating blower. If there is no indication that the equipment is energized, check for line-voltage power at the junction box, nearby disconnect switch, and circuit breaker for the furnace, in that order. If there is power at the furnace junction box, check for line voltage at the primary-control terminals.

Remember that the high-temperature limit switch is often wired into the circuit that provides power to the primary control (Figure 18-36). If there is power where line voltage enters the furnace but not at the primary control, check the limit switch to determine whether it is open (**FIGURE 18-37**). With power supplied to the switch, a voltmeter

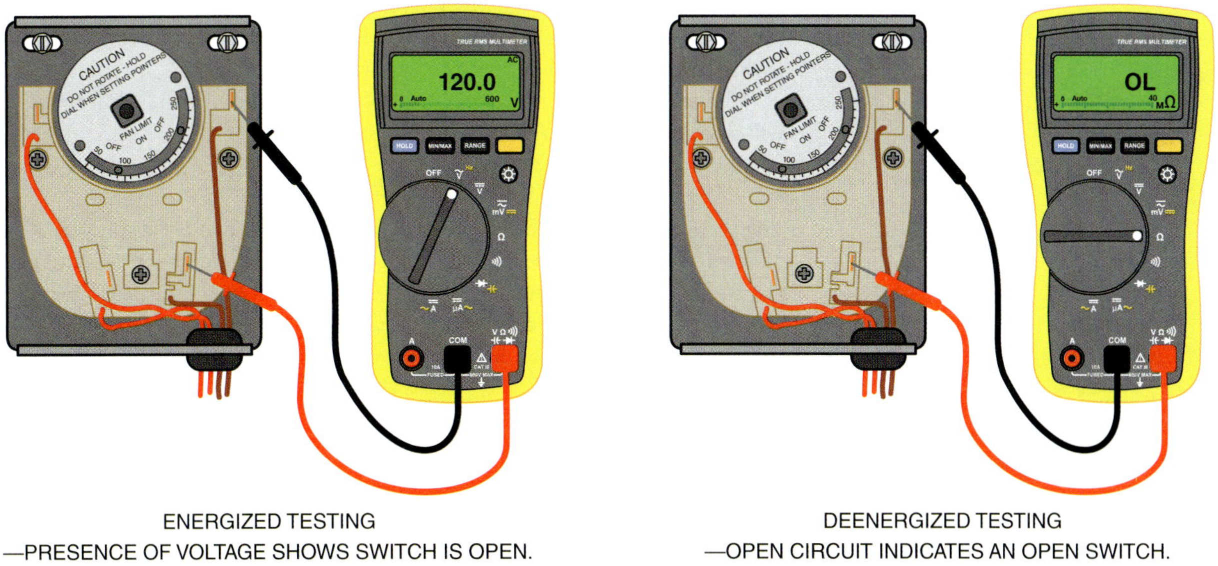

FIGURE 18-37 Checking the high-temperature limit switch.

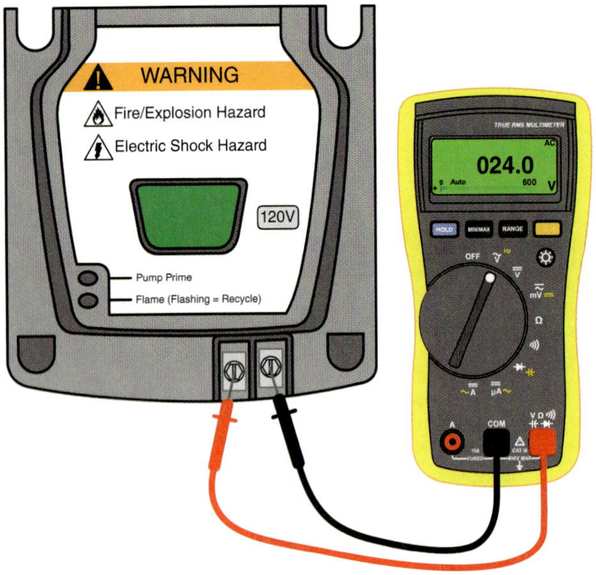

FIGURE 18-38 Checking for control power at the primary control.

is applied across the limits switch contacts. With the furnace deenergized, it can be checked with an ohmmeter. It may be open because the furnace has recently overheated or because the switch itself may have failed open.

Also remember that primary controls provide their own control voltage to and through the thermostat. The control power for cooling-system components, fan relays (if present), and other control functions will be provided by a separate control transformer. A failure of the transformer inside the primary control will leave the burner unable to operate, but the cooling system and indoor blower will remain unaffected.

If the primary control is powered and a lockout or recycle-timing condition does not exist, placing a jumper wire across the T terminals should cause the burner to start. If not, its internal transformer may have failed. transformer failure can be confirmed, when the device has line-voltage power applied, by checking for control voltage at its T terminals with the thermostat wires disconnected (**FIGURE 18-38**). If the control is not producing a control voltage, it must be replaced.

If line voltage and control power are present and the burner starts when a jumper is placed across the T terminals, the problem is likely at the thermostat or in the interconnecting control wiring. In some furnaces, there may be limit switches placed in the control circuit.

▶ TECHNICIAN TIP

Alarm Outputs

Many primary controls offer alarm outputs as an accessory or as a standard feature in a specific model. This is true of traditional controls as well as modern offerings. The alarm output is usually a simple set of contacts that close when the control enters a soft and/or hard lockout state. The contacts can be used to illuminate a light on a remote panel, initiate an audible alarm, or even initiate an auto-dialer to notify someone of the problem in critical applications.

Soft Lockouts

For this control, a soft lockout is indicated by a flashing red Reset button. A one-second press of the Reset button will reset the control and, assuming the thermostat is calling, it will begin a new ignition cycle.

Failures during the ignition process, often referred to as the trial for ignition, are more likely than failures that occur during the operating cycle. Ignition failures related to electrical issues most often occur due to the issues discussed in the sections that follow.

Motor Failures

If the burner motor fails to start, the oil pump will not provide oil flow, and there will be no blower to provide the required airflow through the burner. Most burner motors are split-phase motors equipped with a centrifugal switch to deenergize the start winding after start-up. Some permanent split capacitor motors are marketed for oil-burner use, and they offer significantly lower starting current and tend to operate at a lower current as well. Motor troubleshooting was covered thoroughly in Chapter 14. It is worth noting that oil-burner motors are quite durable.

On rare occasions, a mechanical failure of the pump, causing it to seize, will prevent the motor from rotating at start-up.

Improperly Positioned Electrodes

The arc must be established at the right position relative to the spray pattern of the air-fuel mixture. Ensure that the electrode gap, height above the nozzle, and forward position are appropriate for the nozzle type and angle in use (**FIGURE 18-39**). The use of the aforementioned tool shown in Figure 18-27 to set the electrodes is recommended.

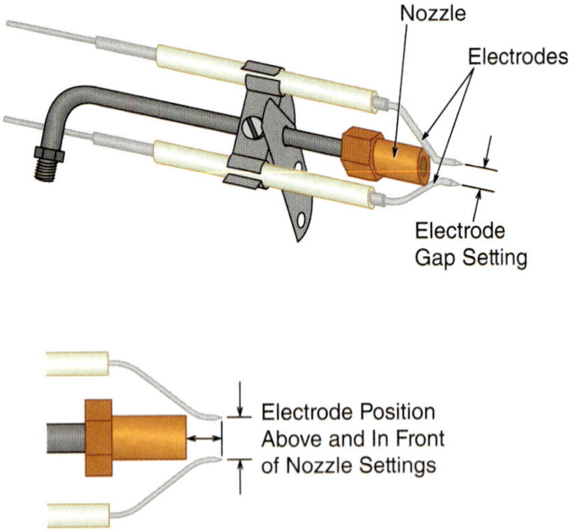

FIGURE 18-39 Required settings for the electrodes.

Electrode Contact and Ignition-Transformer Posts

When the transformer closes over the housing, the electrodes must be in firm contact with the transformer posts. Another problem that can occur is cracks in the ceramic insulators around the transformer posts. Cracks can allow the ignition current to short to the transformer housing.

Transformers should also be kept reasonably clean. Oily dirt and grime encourages carbon tracking, which is the development of conductive paths or tracks across the surface of the transformer.

Failed Ignition Transformer

The initial test of an ignition transformer begins with energizing it with the proper voltage and continues with listening for the sound of the arc produced at the electrodes. If there is no obvious sound, additional troubleshooting will be necessary to determine why.

Standard ignition transformers can be tested for grounds and continuity through the windings, like a control transformer. If the windings are intact and the transformer is not grounded, check for power at the ignition-transformer terminals on the primary control when the transformer should be energized. If there is no output from the primary control to the transformer, the primary control is at fault.

FIGURE 18-40 Ignition-transformer tester.

Standard iron-core transformers generate approximately 10,000 volts. Exercise extreme caution when testing the output voltage, and use a meter that is capable of measuring this voltage safely. Any measurement below 9,000 volts is generally cause for replacement. There are also special testers on the market, such as the one shown in **FIGURE 18-40**. The globe-shaped contacts are placed against the posts of the energized transformer, and the LEDs indicate whether the output is sufficient.

A traditional method of testing transformers is to use a large, well-insulated screwdriver to provide a conductive path across the terminals. The burner motor must first be disconnected completely so that it does not run during the test. The cad cell also must be disconnected, because the ambient light with the transformer tilted back will prevent the primary control from energizing the transformer.

Once the transformer is energized, place the screwdriver across the posts (**FIGURE 18-41**). Then, with the screwdriver remaining in contact with one post without touching the transformer housing, slowly lift the opposite end of the screwdriver away from the second post. A white arc should develop, and the arc should sustain itself across a gap of at least ½" (1.3 cm). Do not continue the test any longer than necessary, because arcing across a wide gap can shorten the life of the transformer. The normal gap at the electrodes is much smaller at 1/8"–5/32" (0.32–0.40 mm).

Electronic igniters can also be tested in this manner. However, they can deliver a peak output voltage up to 16,000 volts. Using the appropriate test equipment in lieu of this method is highly recommended.

> **SAFETY TIP**
>
> Ignition transformers produce extreme voltages. Using this method to test transformers must be done with extreme caution and a well-insulated tool in good condition. Testing in this manner is not recommended, in spite of its common practice in the field, due to the risk of electrical shock.

Failure to Detect a Flame

The cad cell must lower its resistance below 1,600 Ω for the primary control to detect the flame. In most cases, the resistance will be significantly below this threshold.

This primary control is equipped with an LED to indicate the presence of a flame, which is energized when the cad cell detects a flame. If the light is on when the burner is not running, the cause must be determined. Ensure that the transformer h properly seated against the burner housing and that no other source of light is reaching the cad cell. Some transformers have a gasket to help prevent ambient light from entering. Another possibility is that the resistance of the cad cell has dropped below normal in darkness, making it ineffective. If the wire leads have been pinched and damaged, allowing them to short to the housing, a low resistance path can be formed.

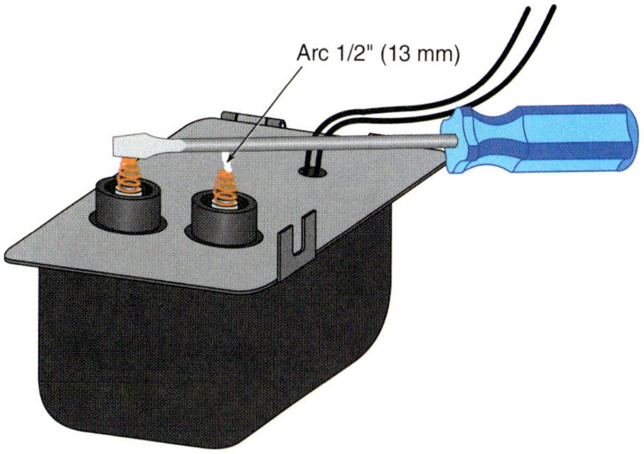

FIGURE 18-41 Testing an ignition transformer.

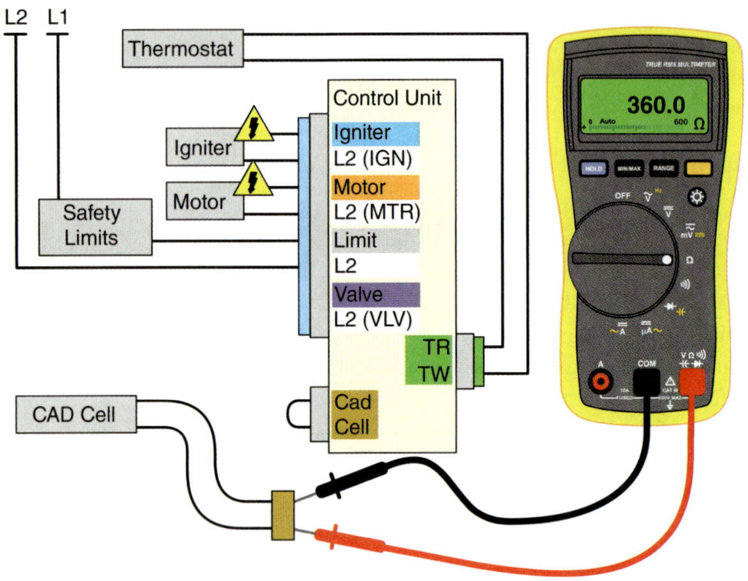

FIGURE 18-42 Testing cad-cell resistance while the burner runs.

Any of these conditions will prevent the primary control from energizing the burner, since a low resistance should not be sensed before the trial for ignition. If removing one of the cad-cell leads from the primary control allows the burner to start (creating an open circuit), one of the above conditions likely exists.

If the Flame light is not on and the burner starts normally but locks out after 15 seconds of run time, the primary control is not sensing the low resistance needed to confirm ignition. In this case, prepare a jumper wire for installation across the cad-cell terminals of the primary control. Start the burner, and before the 15-second period elapses, disconnect the cad-cell leads and install the jumper across the terminals. This will allow the burner to continue running while the cad-cell resistance is tested with an active flame (**FIGURE 18-42**).

The resistance should be something less than 1,600 Ω but well above zero. If it is above 1,600 Ω, then the cad-cell lens could be dirty; its line of sight to the burner may be obstructed; or the flame is unstable. If none of these conditions exist, the cad cell must be replaced. If there is an infinite resistance reading (open), the cad cell is likely defective. Before replacing it, though, make sure that the leads are undamaged and that there is no apparent damage to the socket.

Post-Ignition Failures

The most likely electrical issue to cause a flame failure after ignition is a high-resistance signal from the cad cell. Troubleshooting this issue is the same as described immediately above. The resistance reading of the cad cell must be tested while the burner is in operation. However, a fouled nozzle that creates a poor spray pattern can also cause flame instability after the ignition transformer is deenergized after a successful trial for ignition.

Hard Lockouts

As already noted, the primary control will enter a hard lockout after a soft lockout has been reset three times without successful ignition or after three attempts to automatically recycle the burner following a loss of flame during the operating cycle. When the Reset button is held down for more than 15 seconds, the control will reset and try again.

The primary control featured here is also equipped with welded-relay protection. If the control senses that the contacts of an internal relay, such as the one used to energize the burner motor, are closed when they should be open, it will not energize the burner, and therefore a hard lockout will occur. This is another example of how incorporating solid-state technology into common controls helps provide greater functionality and safety.

▶ Wrap-Up

Summary

Modern gas furnaces offer a great deal of flexibility and control options but can be challenging to troubleshoot without the proper information specific to each furnace. On the other hand, modern furnace controls also assist the troubleshooting process and generally guide a technician in the right direction.

Oil burners have changed little over the years, but their controls have become more sophisticated and versatile. When an oil furnace is used for cooling applications, additional controls, including solid-state devices, may be added. They may also be equipped with ECM motors to drive the indoor blower. Although this chapter provides a good foundation in troubleshooting electrical problems common to oil furnaces, it is also important for a technician to understand oil supply management to guarantee success in troubleshooting oil furnaces in general.

Key Terms

annual fuel utilization efficiency (AFUE) A measure of thermal efficiency in fossil-fuel furnaces and boilers that represents the season-long average efficiency of the equipment in delivering heat for the intended use. Heat lost through the vent or chimney, for example, reduces the AFUE rating since it does not contribute to the intended use.

cad cell A device containing cadmium sulfide that responds to the presence of light by significantly reducing the resistance of its circuitry; it is used to prove the presence of flame in an oil furnace.

condensing furnaces Furnaces offering exceptional efficiency through the use of a sealed combustion chamber and a secondary heat exchanger to extract the maximum amount of heat.

delayed ignition The result of excess gas buildup inside the burner compartment before successful ignition, causing a somewhat violent detonation.

direct-ignition system A gas furnace that ignites the main burner directly, rather than lighting a pilot flame that subsequently lights the main burner.

electrode A device that carries a high voltage to its tip, where an arc is initiated to ignite an air-fuel mixture. The conductive material used for the igniter is virtually unaffected by the arc, although it may experience wear after long usage.

flame rectification The process in which the flame in a furnace changes an alternating current (AC) to direct current (DC); it is used to verify the presence of a stable flame.

flame rod The rod bathed in the flame used on a gas furnace to direct the rectified current back to the operating controls.

hot-surface igniter (HSI) A device typically constructed of silicon carbide or silicon nitride used to ignite a gas furnace by being heated to an extreme temperature.

ignition transformer A transformer that provides a high voltage, generally in the range of 10,000 volts, which is used to create an intermittent arc across a gap and ignite an air-fuel mixture.

induced-draft furnace A furnace equipped with a fan or blower on the heat exchanger outlet to reduce the pressure in a heat exchanger and overcome the resistance to flow, helping combustion gases exit the heat exchanger and enter the vent system. Technically, forced-draft systems use a blower at the inlet of the heat exchanger to increase the pressure with the same intent. However, the term "induced-draft" is commonly applied to both types.

natural-draft furnaces Gas furnaces that rely on the heat generated from combustion to create a natural draft through the heat exchanger to move combustion byproducts into the venting system. Natural-draft furnaces cannot make use of complex heat exchangers, due to the resistance to flow that they impose, thus limiting their heat-transfer efficiency.

primary control A control used to manage and supervise the oil-burner combustion process and the related components.

thermocouple A device used to prove the existence of a pilot flame, generally used in natural-draft furnaces that are equipped with a standing pilot flame. Thermocouples generate a voltage of roughly 30 mV, which is used to hold a pilot gas valve open. Thermopiles, used in self-powered furnaces such as wall and floor models, generate roughly 750 mV, which is used as the power source for the entire furnace.

vestibule The section of a furnace immediately in front of the combustion chamber, where the gas valve, gas manifold, and various controls are housed.

Review Questions

1. The highest annual fuel utilization efficiency (AFUE) is provided by _____.
   a. condensing furnaces
   b. induced-draft furnaces
   c. natural-draft furnaces
   d. oil furnaces
2. Pressure switches are used on modern gas furnaces to _____.
   a. prove the burner has ignited
   b. ensure the heat exchanger is not obstructed
   c. prevent carbon monoxide from escaping
   d. delay the main blower until the heat exchanger is warm
3. Rollout switches are typically located _____.
   a. in the blower compartment
   b. in the exhaust vent hood
   c. on the gas valve
   d. inside the vestibule
4. A gas-furnace igniter that lights the main burner instead of a pilot flame is characteristic of a(n) _____.
5. In flame rectification, the alternating current (AC) (before rectification occurs) must pass to the flame rod _____.
   a. through the flame
   b. from the igniter
   c. through the gas piping
   d. from the integrated furnace control (IFC)

6. The difference between a forced-draft and an induced-draft furnace is _____.
   a. the length of the flue vent
   b. the efficiency of the furnace
   c. the location of the combustion blower
   d. the type of ignition system used
7. The typical resistance of a function hot-surface igniter (HSI) is _____.
   a. 5–10 Ω
   b. 15–25 Ω
   c. 40–90 Ω
   d. 150–200 Ω
8. An HSI is most often powered by _________.
   a. line voltage
   b. control voltage
9. A circuit between which two terminals at the thermostat must close to initiate a call for heat?
   a. R and G
   b. R and W
   c. C and W
   d. C and Y
10. When an IFC is used, which of the following would typically need to occur before the induced-draft blower is energized?
    a. The HSI or spark igniter must be energized.
    b. The indoor fan relay (IFR) must close.
    c. The high-temperature limit switches must be closed.
    d. The gas-valve relay (GVR) must be closed.
11. Which of the following is a true statement about gas-furnace operation?
    a. An electronic, or spark, ignition system must provide at least 50,000 volts to establish an arc.
    b. Failed HSIs are often identified by a significant increase in their resistance.
    c. A flame rectification circuit must generate at least 5 amps of current to prove a flame is present.
    d. The induced-draft motor (IDM) remains energized for a short time after the gas valve has closed.
12. How might reversing the polarity of the line-voltage power supply affect a gas furnace with an IFC?
13. Lockouts that are initiated by an IFC and that allow the furnace to automatically attempt ignition one or more times without intervention are referred to as _____.
14. If an IFC repeatedly detects that the GVR contacts are closed before ignition has been attempted, which will display the related fault code, repairing the problem will most likely require replacing _____.
    a. the IFC
    b. one or more high-temperature limit switches
    c. the gas valve
    d. the GVR
15. A typical fan/limit control on a basic oil furnace has _____.
    a. one setting lever
    b. two setting levers
    c. three setting levers
    d. four setting levers
16. In an oil burner, the position of the electrodes in relation to the nozzle is determined by the _____.
17. A cad cell in an oil burner _____.
    a. ignites the oil-air mixture
    b. changes its resistance in the presence of light
    c. generates a voltage in the presence of light
    d. rectifies the AC current from the ignition transformer
18. Which one of the following is a true statement about oil-burner primary controls?
    a. Primary controls typically provide their own control voltage for the thermostat circuit.
    b. If the cad-cell circuit is open (disconnected), the primary control will allow the burner to run indefinitely.
    c. Oil furnace limit switches are usually wired directly into the burner motor circuit from the primary control.
    d. The primary control will energize the oil solenoid valve before it energizes the burner motor.
19. Which one of the following is the approximate gap setting between the electrode tips on an oil burner?
    a. ½" (13 mm).
    b. ⅜" (9.5 mm).
    c. ⅛" (3.2 mm).
    d. 1⁄16" (1.6 mm).
20. The voltage threshold for replacing a standard iron-core ignition transformer is when it produces less than _____.

CHAPTER 19

Air Conditioning and Refrigeration System Operation and Troubleshooting

Knowledge Objectives

After reading this chapter, you will be able to:

- **K19001** Describe the basic refrigeration cycle and identify its major components.
- **K19002** Describe the sequence of operation for a typical packaged air conditioning unit.
- **K19003** Describe the sequence of operation for a typical refrigeration unit.
- **K19004** Explain how to analyze, troubleshoot, and diagnose common electrical problems.

▶ Introduction

To successfully troubleshoot air conditioning and refrigeration systems, a complete understanding of the sequence of operation is essential. Many problems with air conditioning and refrigeration systems do not result in complete shutdown, but failures of electrical load and control devices often do present this way. This is especially true for electrical loads, such as motors. Without operational motors to do the real work, cooling systems cannot function. Since loads are managed by electrical controls, the failure of a control device can lead to system shutdown as well. But there are occasions when the failure of an electrical device does not result in a total loss of cooling operation.

The previous chapters in this book presented the many electrical load and control devices found in common air conditioning and refrigeration systems. A review of basic wiring diagrams and troubleshooting strategy was presented in preceding chapters. This chapter will bring devices, wiring diagrams, and troubleshooting together by presenting sequences of operation and common problems that require troubleshooting. Since many hands-on troubleshooting skills have already been presented and practiced, this chapter will focus more heavily on critical thinking and analysis. Learning to analyze the available information to determine the possibilities is essential, because the analysis forms the basis of which tests to perform and when.

K19001 Describe the basic refrigeration cycle and identify its major components.

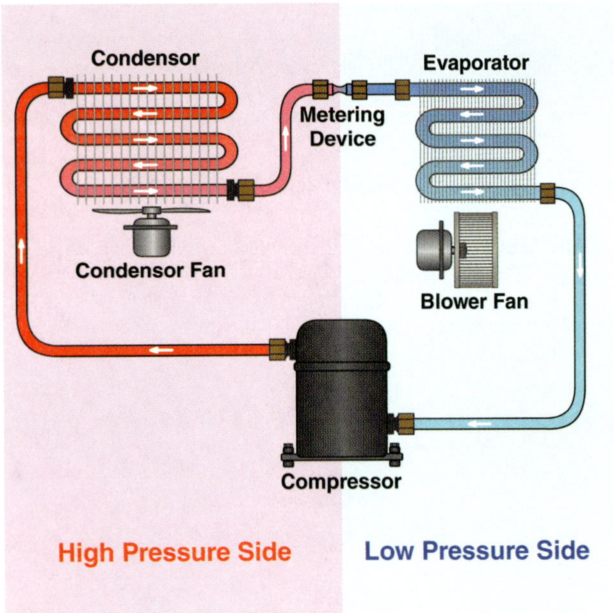

FIGURE 19-1 Basic refrigerant circuit.

▶ The Basic Refrigeration Cycle and Its Components

It could easily take an entire book to cover the direct-expansion, vapor-compression refrigerant cycle and its many possible variations. Only the most basic concepts will be presented here to support the sequence of operation presented later in this chapter.

For a common air conditioning and refrigeration system, there are four major components in the circuit to consider (**FIGURE 19-1**). The following four components and their respective functions will be presented in the sections that follow:

- compressors
- condensers
- metering devices
- evaporators.

The primary goal of air conditioning and refrigeration systems is to transfer heat from one point to another, and each of these components plays an important role in the process.

Compressors

Compressors are considered the heart of an air conditioning and refrigeration system. Their purpose is to compress a low-pressure refrigerant vapor into a high-pressure refrigerant vapor. The vapor being compressed typically enters at a low temperature and exits at a much higher temperature, due to the heat of compression. The temperature of the vapor entering the compressor is important since it serves the secondary function of cooling the compressor and its motor.

Most compressors are classified as either **hermetic**—permanently sealed with an enclosed motor (**FIGURE 19-2**)—or **semi-hermetic**, which allows them to be partially disassembled for field service (**FIGURE 19-3**). Larger compressors, generally used for industrial refrigeration applications, may be open-drive models (**FIGURE 19-4**). These compressors have exposed drive shafts that are coupled to an electric motor shaft. The compressor and motor are separate components, much like larger pumps.

There are several types of compressors used in air conditioning and refrigeration systems, including **reciprocating compressors**, **scroll compressors**, and even **rotary compressors**. Scroll compressors are very popular today for residential and commercial systems.

FIGURE 19-2 Hermetic compressor.

FIGURE 19-3 Semi-hermetic compressor.

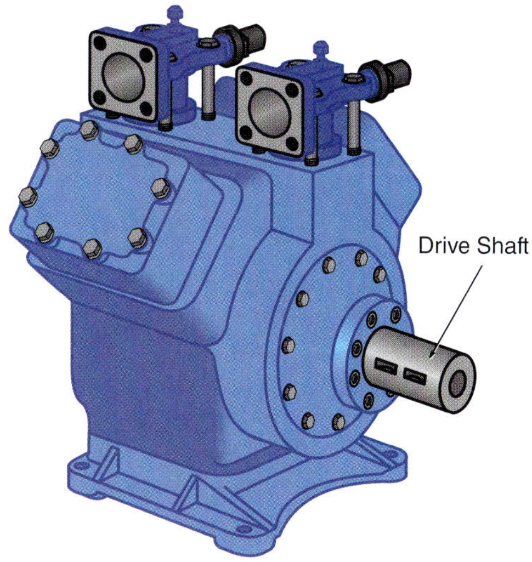

FIGURE 19-4 Open-drive compressor, to be coupled to an electric motor.

Groups of them are now found in large condensing units and **chillers** (FIGURE 19-5). This is due to their efficiency and reliability. Scroll compressors have far fewer moving parts than reciprocating models have and are generally hermetic by design. Semi-hermetic scroll compressors, however, are entering the market. They are primarily offered for commercial refrigeration applications.

What all these compressors have in common is an electric motor, regardless of whether it is field-accessible or inaccessible. Smaller compressors for residential and light-commercial use have single-phase motors. Larger compressors require three-phase motors. Again, their task is to compressor refrigerant vapor, increasing both the pressure and the temperature in the process.

Condensers

High-pressure, high-temperature vapor from the compressor is sent to the condenser (FIGURE 19-6). Most condensers for residential and commercial cooling, as well as those for commercial refrigeration, are air-cooled. This means that air is used to cool the refrigerant flowing within. Water-cooled condensers, however, are often used in commercial and industrial applications, as well as in some heat pump systems. Although water-cooled systems have more maintenance requirements, water is far more effective than air for transferring heat.

▶ CORE CONNECTION

Why Do We Need to Raise Refrigerant Pressure?

Compressors, as the name implies, compress gases to increase pressure. Air compressors, for example, increase pressure so that the air can be used to do work, such as turn an impact wrench or blow away debris.

In refrigeration, it's all about thermodynamics. To use outside air that might be 100°F (38°C) or higher to cool a gas (refrigerant vapor), the temperature of the gas must be increased above that temperature. Otherwise, the outside air blown across the coil would add heat to the gas, not remove it. Of course, there is much more to learn about thermodynamics and the vapor-compression refrigerant cycle, but the simple explanation is that it allows heat to be transferred to the outside air.

FIGURE 19-5 Chiller with scroll compressors.

FIGURE 19-6 Air-cooled condenser coil.

FIGURE 19-7 Condenser fan.

As the term "condenser" implies, the function of the condenser is to cool the refrigerant vapor to, and below, its **dewpoint** for the current pressure. Once cooled to its dewpoint while heat continues to be removed, the refrigerant condenses to its liquid state. As the condensation process takes place, the refrigerant temperature will remain the same, even though heat is being removed. Once the process is complete—all the vapor has condensed to a liquid—the refrigerant temperature can decrease as even more heat is removed. Ideally, the refrigerant leaves the condenser in a liquid state and has been cooled below its dewpoint temperature. When the refrigerant has been fully condensed and its temperature has been further reduced, it is said to be subcooled.

Theoretically, an air-cooled condenser can function without fans. In fact, the condenser coil in some domestic refrigerators, for example, has no condenser fan. However, a condenser coil that depends solely on natural air currents for comfort cooling must be much larger. To make the coil smaller yet more effective, condenser fans (**FIGURE 19-7**) are used to move large volumes of air across it. Therefore, the failure of a condenser fan or any electrical device that controls it results in little or no refrigerant condensation, and then the system fails to provide cooling to the user.

FIGURE 19-8 Capillary tube.

Metering Devices

Metering devices provide an essential function in the process, although some are very simple devices. Actually, all of them can be described in very simple terms: they are openings, or orifices, that are much smaller than the interconnecting piping. As refrigerant passes through, the pressure drops as a result. It is the design and functionality of the various types that sets each apart.

For many years, metering devices were purely mechanical devices, and many still are. A simple capillary tube (**FIGURE 19-8**) served the purpose for many years. The length and inside diameter of the tube is carefully engineered to provide a specific amount of pressure drop. Later, an even simpler device was developed to take the place of the capillary tube in many systems, especially heat pumps—the fixed-orifice metering device (**FIGURE 19-9**). It should be clear from these pictures that these devices are not complicated. Again, they simply represent a restriction in the refrigerant circuit that results in a significant pressure drop.

The problem with these two devices is that they have no means of responding to changes in the operating conditions. To make systems more effective in a variety of operating conditions, a metering device that can change the size of the orifice to accommodate the conditions is needed. Thermostatic expansion valves (TXVs or TEVs) offer this feature (**FIGURE 19-10**). These metering devices respond to the temperature of the refrigerant line downstream and change the size of the orifice inside. This affects the amount of pressure drop and the refrigerant flow rate.

FIGURE 19-9 Fixed-orifice metering device.

Today, an increasing number of metering devices are electronically controlled and positioned. Electronic expansion valves (EXVs or EEVs) gather inputs from the system, determine the ideal opening size by using **algorithms**, and adjust the size of the orifice accordingly (**FIGURE 19-11**). Although specific troubleshooting procedures for EXVs are not covered here, it is a skill that a technician must eventually develop to progress in the trade.

As already mentioned, all these devices provide a pressure drop. Once the liquid refrigerant passes through the metering device and emerges at the reduced pressure, its

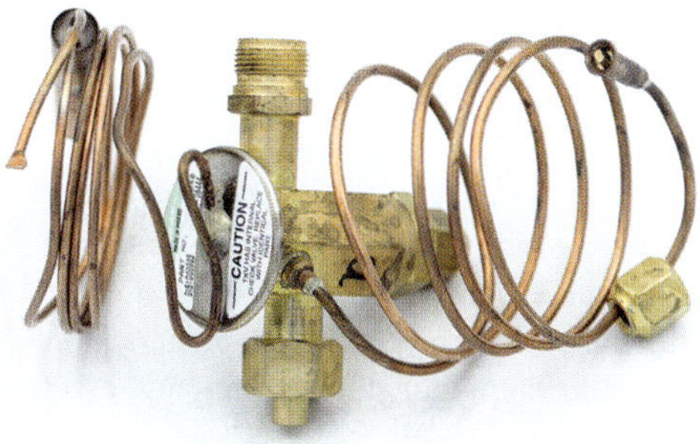

FIGURE 19-10 Thermostatic expansion valve (TXV or TEV).

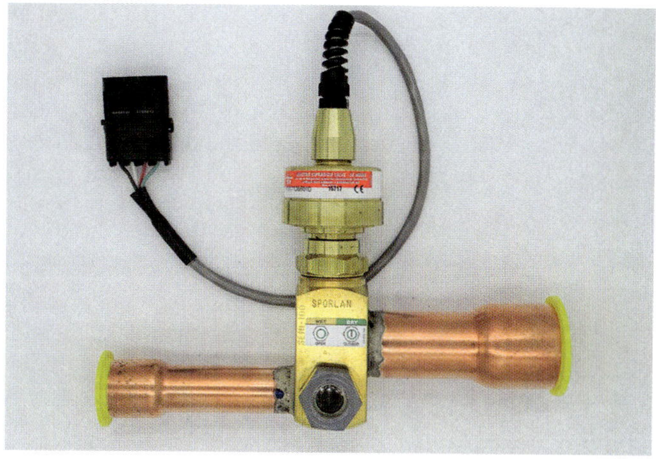

FIGURE 19-11 Electronic expansion valve (EXV).

temperature also drops significantly to bring it in line with the new pressure condition. Pressure-temperature (PT) charts for each unique refrigerant are used to determine the resulting change. But cooling the refrigerant rapidly—from 105°F to 40°F (41°C to 4°C), for example—isn't magic. A percentage of the liquid refrigerant must instantly convert into a vapor, absorbing heat from the remaining refrigerant in the process. Therefore, refrigerant typically leaves the metering device as a low-pressure, low-temperature mixture that is roughly 85% liquid and 15% vapor. Remember the following two points:

1. The cooler the refrigerant entering the metering device is, the better, since this reduces the amount of liquid refrigerant that must convert into a vapor as it passes through the metering device.
2. Any vapor leaving the metering device reduces the refrigerating effect of the evaporator, since vapor transfers very little heat. Ideally, all the refrigerant entering the evaporator is in the liquid state, but this rarely occurs in common systems due to the thermodynamics explained above.

Evaporators

The goal of most air conditioning and refrigeration systems—i.e., removing heat from a space—is achieved at the evaporator (FIGURE 19-12). The cold refrigerant liquid/vapor mixture enters the evaporator coil from the metering device. As air at room temperature is forced across the evaporator, it is cooled while the refrigerant inside is warmed. When the refrigerant reaches its boiling point for the current pressure, it boils and begins converting into its vapor state. During this process, the refrigerant temperature does not change; it will remain at the same temperature until all the liquid has completed the change of state. It is this conversion process that absorbs the vast majority of the heat from the air passing over the coil.

The blower responsible for moving air across the evaporator is driven by an electric motor. In many cases, for application flexibility, these motors offer multiple speeds or variable-speed operation. Like condensers, there are evaporators applied without a source of forced air, depending only on natural air currents. These are generally limited to certain refrigeration applications and not effective for comfort cooling. Of course, not all evaporators are designed to cool air. A chiller, for example, circulates water through its evaporator section.

FIGURE 19-12 Cased air conditioning evaporator coil with a TXV.

Refrigerant vapor leaving the evaporator returns to the compressor, where the cycle begins again. Maximum evaporator efficiency depends on exposing as much of its internal area as possible, and in turn the air passing over it, to boiling refrigerant. However, to protect the compressor from liquid, the vaporization process must be completed before the refrigerant leaves the evaporator. Compressors cannot compress liquid. Indeed, liquid refrigerant returning to a compressor is a common source of damage and failure. Therefore, it is important that the change of state from a liquid to a vapor be completed within the evaporator. To ensure only refrigerant vapor leaves the evaporator, most systems are adjusted so that the change of state is completed well before it exits. In the last portion of the evaporator, the vapor does absorb some heat and increases in temperature. The additional heating that occurs after the change of state is complete is referred to as superheat.

The process described in the preceding sections has many variations, and a number of details have been left unaddressed since they are beyond the scope of this book. As technicians advance in their career, they will encounter many of these variations. An interesting part of the trade is seeing how the basic principles remain the same but are applied in diverse ways to achieve a specific goal.

▶ The Sequence of Operation for a Common Air Conditioning System

K19002 Describe the sequence of operation for a typical packaged air conditioning unit.

Despite minor differences, the sequence of operation for most residential and light-commercial air conditioning systems is quite consistent. Some wiring diagrams appear more complex due to the addition of solid-state devices and electronically commutated motors (ECMs). Remember that connection diagrams are generally harder to read than ladder diagrams, since many circuits cross over each other. But the same basic events must take place, regardless of age or equipment efficiency level.

This section presents the sequence of operation for a typical packaged air conditioning system. The wiring diagram shown in **FIGURE 19-13** is the basis for the sequence that follows. A legend is provided in **FIGURE 19-14**.

1. The compressor **crankcase heater (CCH)** is energized whenever line-voltage power is applied to the unit. Line-voltage power is also available to the indoor fan motor (IFM) since it is an ECM.
2. When the room temperature rises above the set point, the thermostat contacts close to connect R to both Y and G.
3. Power from the Y terminal passes the normally closed high-pressure switch to one side of the compressor contactor (C). With Common connected to the opposite side of the coil, it is energized, and the contactor closes. Contacts L1-T1 and L2-T2 close, energizing the compressor (COM) and the outdoor fan motor (OFM).
4. Simultaneously, power from G energizes Terminal 1 of the IFM, starting the motor as programmed. Note that power from Y is also carried from the contactor terminal to Terminal 2 of the IFM. When both terminals are energized, the motor is programmed to operate at the higher of the two speeds.
5. The system will continue to operate until the thermostat is satisfied, opening the contacts connecting R to Y and G. The high-pressure switch can also open, breaking the circuit that is energizing the compressor contactor. In the latter case, the IFM will remain energized at Terminal 1. As a result, it will continue running but at a lower programmed speed.

The sequence of operation for the example cooling system is quite simple. Many systems will include other features, such as permanent split capacitor (PSC) motors for the indoor fan, fan relays, low-pressure switches, and additional solid-state controls that serve a variety of purposes. Again, the same basic events must take place for the unit to start the necessary loads and provide cooling. Keeping this simple fact in mind will help prevent confusion when evaluating more complex problems and wiring diagrams.

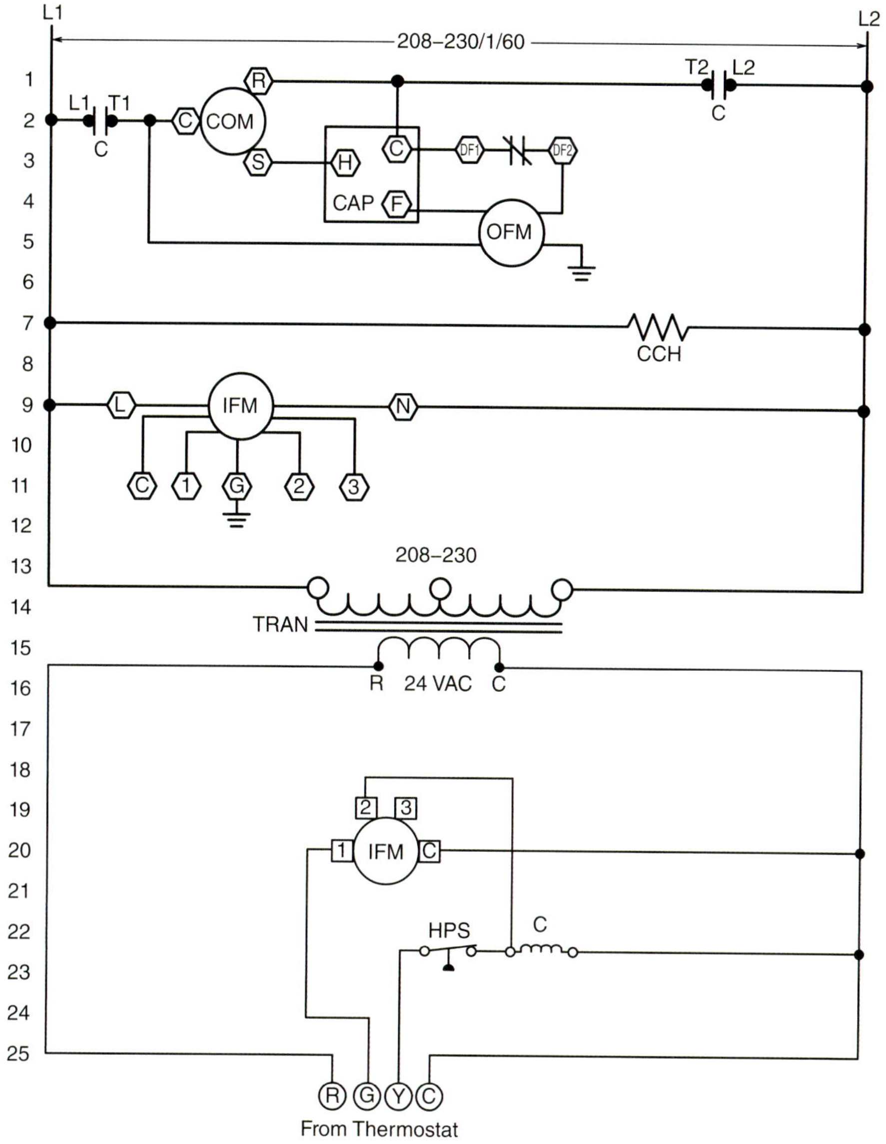

FIGURE 19-13 Cooling unit ladder diagram.

Legend	
C: Contactor	HPS: High Pressure Switch
CAP: Run Capacitor	IFM: Indoor Fan Motor
CCH: Crankcase Heater	OFM Outdoor Fan Motor
COM: Compressor	TRAN: Control Transformer

FIGURE 19-14 Legend for Figure 19-13 ladder diagram.

▶ The Sequence of Operation for Common Refrigeration Systems

Refrigeration systems differ from conventional air conditioning in that they operate at a variety of temperature ranges. The basic refrigeration cycle still applies, but the style of the control devices often differs. Many refrigeration systems require some form of defrost control for the evaporator, even if it is a simply a planned off-cycle to allow the evaporator temperature to rise above freezing. Of course, this is only effective for fixtures that maintain temperatures above freezing—usually 38°F (3°C) or higher. For a freezer, some form of heat is required to melt the frost on the evaporator, since the box temperature must remain

▶ TECHNICIAN TIP

Hot-Gas Defrost

Larger commercial and industrial refrigeration systems use hot refrigerant gas directly from the compressor to defrost the evaporator coil. The sequence of operation differs significantly from systems using electric defrost heaters, since the compressor must continue to run throughout the defrost cycle. Additional refrigerant flow controls are also required.

Hot-gas defrost systems are generally very effective and fast. Contrary to popular belief, the efficiency doesn't result from the high temperature of the it refrigerant. Instead, it primarily results from the massive amount of heat transferred as the refrigerant condenses. Most commercial cubed-ice machines also use hot refrigerant gas to release the cubes from the evaporator plate during the harvest cycle.

K19003 Describe the sequence of operation for a typical refrigeration unit.

below freezing. This section will present the sequence of operation for a simple commercial reach-in freezer with a time clock controlling the defrost cycle.

Refer to the wiring diagram shown in **FIGURE 19-15** for the following sequence of operation. Before beginning, however, it is important to understand the state of the system as it appears in the diagram, as outlined below:

- The defrost time clock is shown with its contacts in their normal positions, which is the refrigeration mode. This provides power to the N terminal, energizing the box temperature control. In this diagram, the condensing unit is considered a separate piece of equipment that is plugged into a receptacle on the freezer unit. The condensing unit includes the compressor, condenser coil, and condenser fan. Most of the compressors used for reach-in units use start capacitors and start relays, since they often need the additional starting torque. This is primarily due to the pressure difference between the high- and low-pressure sides of the system. The extra starting torque helps the compressor motor overcome this disadvantage.
- The temperature control is shown normally closed and opens on a fall in temperature. Since it is a freezer (assume that it is designed to maintain 0°F, or −18°C), the thermostat would be closed when it is first started at room temperature. In other words, the condensing unit is energized as soon as power has been applied to the system. In this example, assume the thermostat will open at −2°F (−19°C).
- The fan thermostat is shown normally open and closes on a fall in temperature. In freezers like this one, the evaporator fan is prevented from running until the evaporator coil has reached a temperature well below freezing. For this example, assume that it closes at 10°F (−12°C). This helps prevent the small compressor from overloading

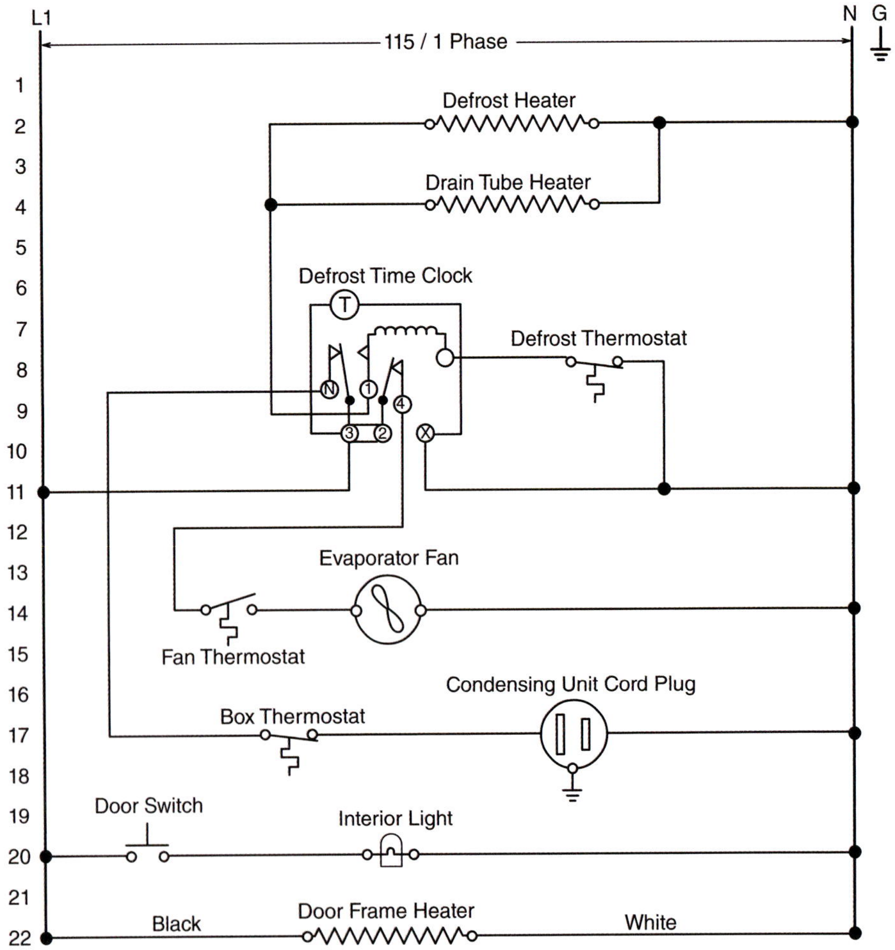

FIGURE 19-15 Example reach-in freezer wiring diagram.

when the box is started at room temperature, because it is designed for low-temperature operation. It also helps prevent the fan from blowing condensate off the coil while its surface is still above freezing. The fan switch will open when the box temperature rises above 35°F (2°C).

- The defrost thermostat is shown normally closed and opens on a fall in temperature. This thermostat is included to terminate the defrost cycle by energizing a solenoid coil in the defrost timer as shown. When the coil is energized, the contacts of the timer change back to their normal position (shown). This fixed-setting thermostat is mounted on one of the evaporator coil U-bends, opening at 40°F (4°C) and closing at 50°F (10°C). If the coil rises to that temperature during the defrost cycle, the contacts close, energizing the cycle-termination solenoid coil. Because the system is shown at room temperature, this thermostat is presently closed. The defrost timer contacts, however, are in their normal position, and there is no power to Terminal 1. Thus, the coil is not energized. The defrost thermostat can energize the cycle-termination coil only when the time clock is tripped to its defrost mode and power is made available to Terminal 1.
- The door switch is shown normally open since the door of the freezer is usually closed. The switch closes when the door is opened to light the interior of the box.

With the above details in mind, the sequence of operation will be easier to understand.

1. With power applied and the box at room temperature, the following devices are energized:
   i. The defrost timer motor is energized by Terminal 3 of the timer and Terminal X.
   ii. The condensing unit power receptacle is energized, providing power to the compressor and condenser fan. Both loads start once power has been applied.
   iii. The evaporator fan switch is powered by Terminal 4 of the defrost timer. Since the switch is open, the evaporator fan does not start.
   iv. The door frame, or mullion, heater is energized at all times. This low-wattage heater is routed through the door frame to ensure that the door gasket does not freeze to the frame.
2. The condensing unit starts, and cold refrigerant is provided to the evaporator coil through the metering device. Once the coil temperature falls to 40°F, the defrost thermostat opens. When the coil temperature falls to 10°F, the fan thermostat closes, energizing the evaporator fan.
3. The condensing unit and evaporator fan will continue to run until the box temperature reaches −2°F (−19°C) and the thermostat opens. The condensing unit compressor and OFM are then deenergized.
4. The defrost timer can initiate a defrost cycle at any time, regardless of whether the compressor is running, based on its settings. The number of defrost cycles set to occur each day depends on the equipment and fixture usage. When a defrost cycle is initiated, the two contacts shown change position and the following occurs:
   i. Terminal N is deenergized, which disables power to the box thermostat and condensing unit.
   ii. Terminal 1 is energized, applying power to the defrost heater (located on the face of, or within, the evaporator coil) and the drain tube heater.
   iii. Terminal 4 is deenergized, disabling the evaporator fan.
5. The defrost cycle can be terminated in one of two ways:
   i. The defrost timer reaches its maximum time setting for cycle duration.
   ii. The defrost thermostat closes when the coil temperature rises to 50°F (10°C), energizing the cycle-termination solenoid coil.
6. Once the defrost cycle terminates, the contacts return to their normal position, restoring power to Terminals N and 4. The defrost heaters are deenergized, and the condensing unit is energized through the closed box thermostat. The fan thermostat has opened during the defrost cycle, since the evaporator temperature rose, and it will reclose as the evaporator temperature falls below 10°F (−12°C). The evaporator fan will then start.

▶ TECHNICIAN TIP

What's Unusual About This Picture?

Closely examine the wiring diagram in Figure 19-15 and try to spot an unusual feature. As a general rule, the neutral of a 115-volt power source is routed to each load and control device in a circuit without being switched. In 230-volt systems as well, one of the two legs of power is often left unswitched. Some systems, however, such as the freezer in this example, do occasionally switch the neutral. The defrost thermostat is shown switching the neutral leg of power to the cycle-termination coil. Note that the neutral is unbroken to all other loads and controls in the diagram. Breaking the neutral is extremely rare and will not often be seen in the HVACR trade.

Remember that this is only one example for a simple reach-in freezer. There are many possible variations in the operating sequence and many variations in the style of controls used. Many modern commercial freezers, for example, rely on solid-state controls to manage both temperature control and defrost cycles. As is the case for all systems, technicians must have the correct documentation for each system serviced.

▶ System Electrical Diagnosis and Troubleshooting

K19004 Explain how to analyze, troubleshoot, and diagnose common electrical problems.

Any of the electrical devices involved in the sequences described above—both loads and controls—can fail and cause problems. This section will identify some potential problems, describe how they present themselves, and explain how they can be diagnosed with confidence. Remember to apply strategy-based troubleshooting principles throughout the troubleshooting process. In all cases, a visual examination of the system and temperature controls is always the best place to start.

Troubleshooting an Unresponsive System

When a system is not running at all, begin by examining the controlling thermostat. Ensure that the mode switch is in the proper position—Cool or Auto—and that the set point is below room or fixture temperature. If the thermostat is digital, check for any indication that the power supply has been lost. An icon will usually be displayed when the thermostat or temperature controller has lost its power source and is depending solely on its battery to maintain the display. For electronic refrigeration controllers with no battery backup, the display will be blank.

Since the transformer powers air conditioning thermostats directly through the R and Common wires, a power loss at the thermostat typically indicates that either the line-voltage power source to the transformer or the transformer itself has failed. If the icon is displayed, turn the thermostat mode to Off for the time being.

Continue the visual inspection by checking the circuit breaker position. If it has tripped, do not reset it until after an inspection and some testing of the equipment have been completed. First, use a voltmeter to confirm that the equipment is deenergized and also to ensure that the contacts are open. Then use an ohmmeter to check for grounds, first on the load side of the contactor and then at the line side of the contactor (**FIGURE 19-16**). If the contactor is open, a ground on the line side likely indicates that the power wiring to the unit has been damaged somewhere. A ground indicated on the load side means that one or more devices, such as the compressor or outdoor fan motor, are grounded. Disconnect each load from the contactor and test them individually to determine which load is grounded.

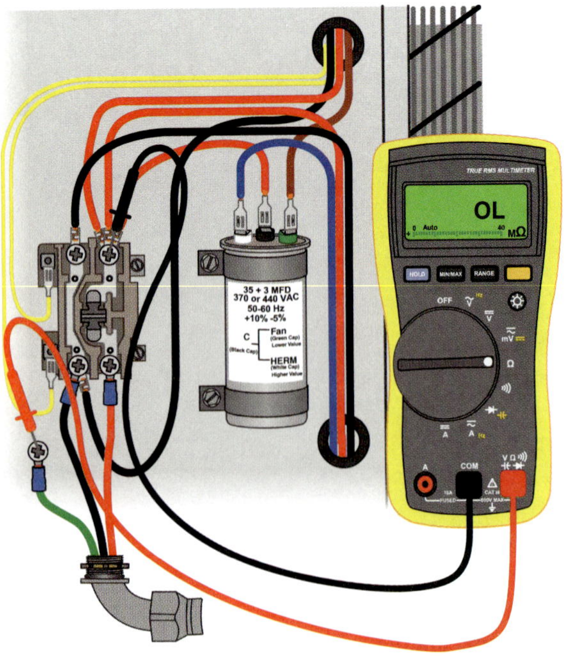

FIGURE 19-16 Checking for grounds from the load side of the contactor.

If no grounded loads or controls are identified, check any fuses before resetting the circuit breaker. Only after the system has been checked for grounds or obvious signs of electrical damage should the circuit breaker be reset. Also make sure that the thermostat has been turned off. It is always best to first reset the breaker and then restart the system from a strategic location. A technician should try to be in a position to observe the start-up of major loads.

If the circuit breaker trips immediately when reset, without any controls set to start the system, a grounded condition likely exists in the wiring or possibly at the transformer, since no other significant loads should be active. However, in the example air conditioning system, remember that the IFM is an ECM in this case, and it receives line voltage at all times. This is common for ECMs. The motor could be grounded, which would cause the circuit breaker to trip immediately. The compressor crankcase heater is also energized immediately when power is applied. Crankcase heaters can also become grounded in failure, though that situation is rare.

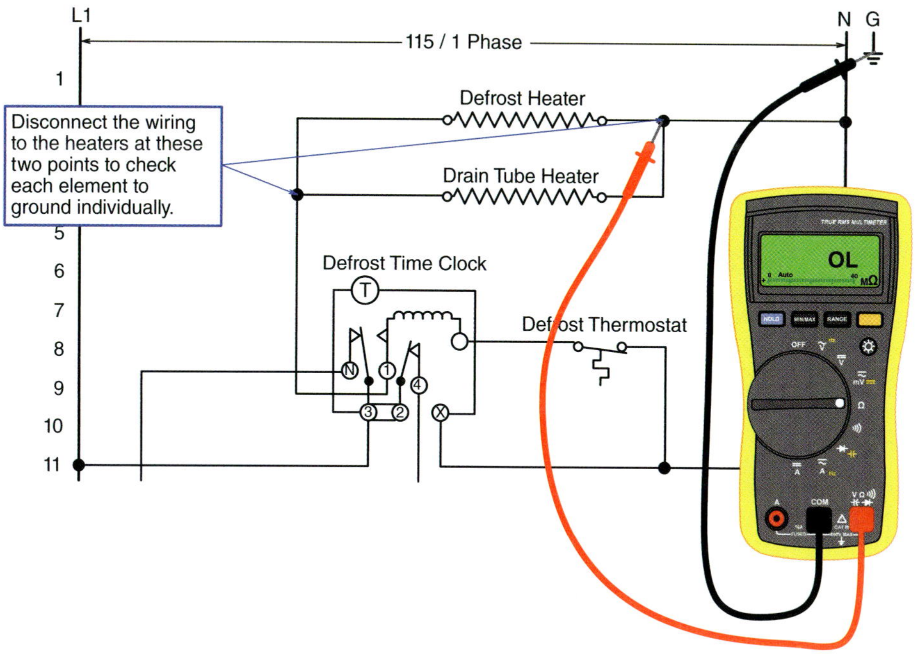

FIGURE 19-17 Checking the defrost heater for a grounded condition.

In the example refrigeration system, the time clock is energized immediately when power is applied. If power was lost while the system was in defrost, the defrost and door frame heaters will also be energized as soon as power is applied. These devices could also be grounded, causing the breaker to trip immediately. Electric heating elements are simple resistive devices and are easy to check for grounds. Connect one meter probe to either end of the element wiring and the other probe to ground (**FIGURE 19-17**). If a ground is noted, the heaters must be electrically disconnected from each other to determine which one is defective. Note, however, that it is much more common for electric heating elements, including crankcase heaters, to fail open than it is for them to become grounded.

If power is restored without incident, test the control transformer output to ensure it is providing the required secondary voltage. If everything appears to be normal, place a clamp-on ammeter onto the compressor Common lead to prepare for start-up, since the compressor is the largest and most important load (**FIGURE 19-18**). For the example air conditioning unit, the outdoor disconnect can be turned off before the thermostat is set for cooling. Then the disconnect can be turned on while the technician is at the unit to observe the starting sequence. This allows the compressor start-up current to be monitored firsthand, and the disconnect can be quickly turned off if something is amiss. If there is a persistent problem that caused the circuit breaker to trip, it will usually present itself as the equipment starts.

While the equipment is starting, use the ears as well as the eyes, listening for unusual sounds. If the compressor starts normally, use the ammeter to continue checking the current to other loads.

An Unresponsive System with a Power Source

When the power sources, including the control transformer, are found to be normal, a technician's attention turns to the controls first, and then to the loads.

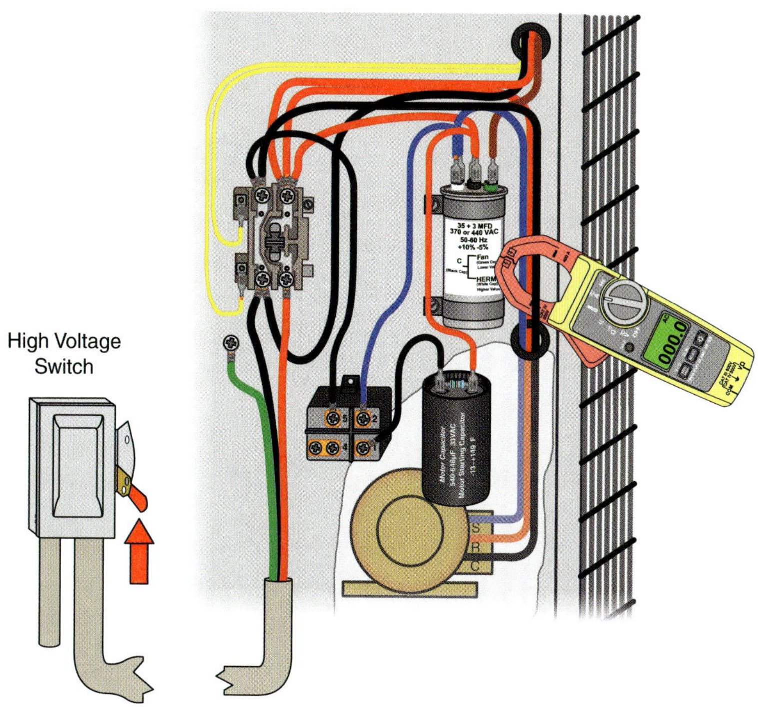

FIGURE 19-18 Ammeter ready to monitor compressor current.

The Example Air Conditioning System

When the line and control voltage power sources are intact and the thermostat is set to call for cooling, check for 24 VAC between C and Y. Testing for voltage at the contactor coil isn't conclusive here, since the high-pressure switch could be open; check where the control wiring enters the unit (**FIGURE 19-19**). If 24 VAC is not present, check for power at C and G as well, which will provide additional information.

If the control voltage is not present at one or both terminals, the problem is likely outside of the unit, either at the thermostat or in the interconnecting wiring. If 24 VAC is found in both tests, it indicates the thermostat is calling for cooling and the indoor fan and that the signals are reaching the unit. This confirms that the problem is inside the unit. Note that room thermostats transfer power from R to both Y and G when cooling is called for, to start the indoor fan. This would also be true in the example system. However, note that Y also energizes Terminal 2 of the ECM in this system. When two signals are received by the ECM (**FIGURE 19-20**), it is programmed to run in the higher of the two energized speed taps.

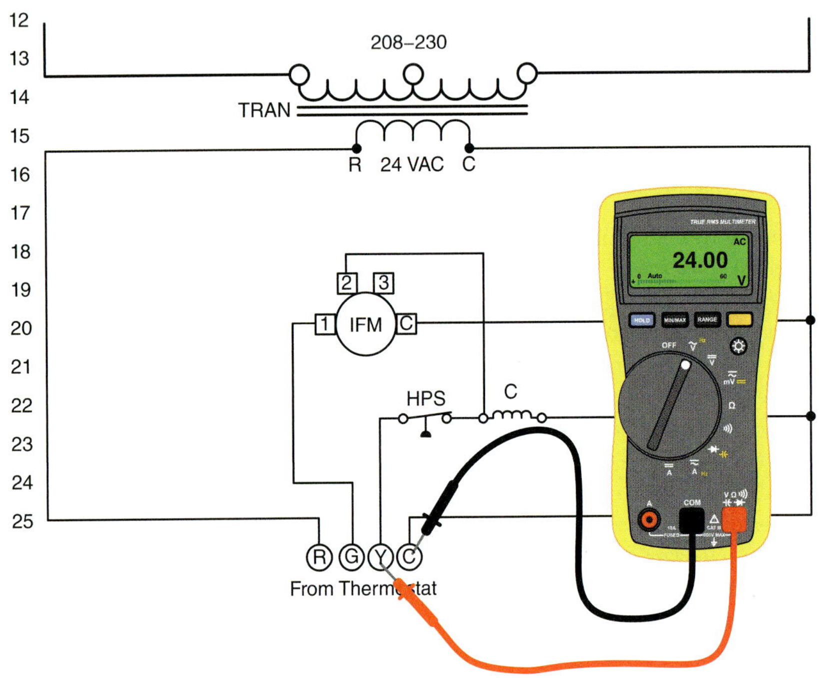

FIGURE 19-19 Checking for control voltage at the outdoor unit.

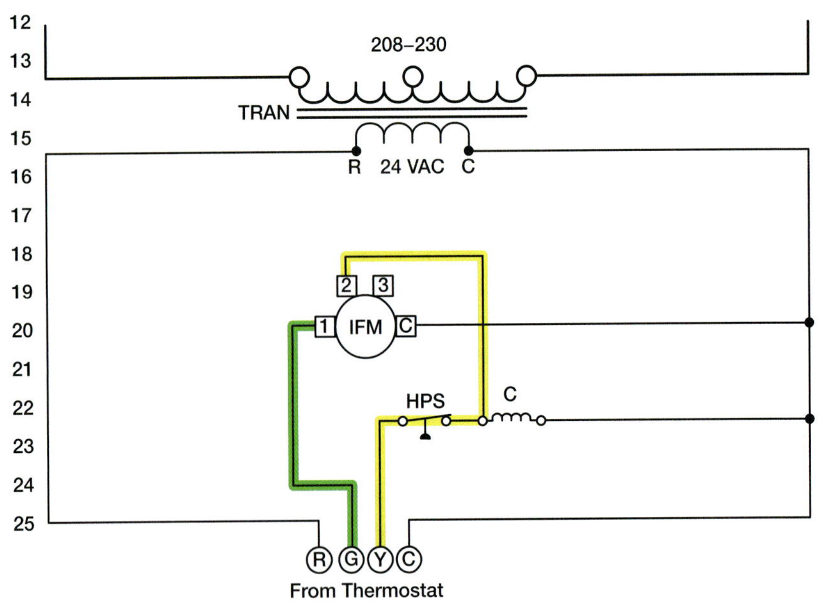

FIGURE 19-20 Two separate ECM control terminals energized on a call for cooling.

The example unit could also be found with the indoor blower running, but the compressor contactor is not energized. This would point to an open high-pressure switch (HPS). Since the thermostat is energizing the G terminal, the ECM can run, regardless of the position of the HPS. Although it would not be the correct speed for the cooling mode, the small difference would not likely be noticed by a technician. But the open HPS will prevent the compressor and condenser fan from starting while the indoor blower is running.

If the HPS is closed, indicated by 24 VAC measured at the contactor coil terminals, deenergize the system, disconnect the wires from the coil, and check the contactor coil for continuity (**FIGURE 19-21**). Replace the contactor if the coil is open. Also remember that, in rare cases, the contactor can be energized, but the contacts can be severely damaged or even burned off, preventing power from being applied to the compressor and outdoor fan motor.

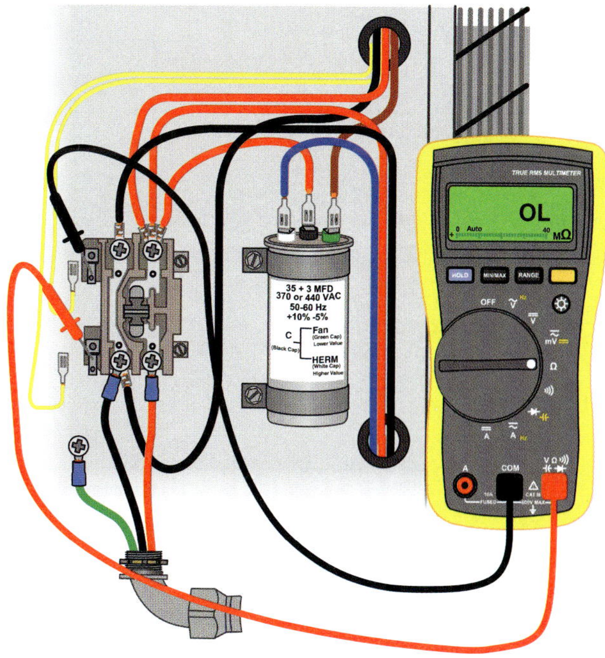

FIGURE 19-21 Checking the contactor coil for continuity.

The Example Freezer

If the power source to the freezer is intact without any obvious response from the equipment, first check the position of the defrost timer contacts. Use a voltmeter at the contacts as shown in **FIGURE 19-22**. Line voltage will be present as shown if the timer is in the defrost mode. If the unit is in defrost, there will be little or no sound, but the defrost heaters will be energized and should be functioning. The only possible sound may be melted ice dripping on a heating element, creating a sizzling sound.

If the time clock motor fails while it is in the defrost mode, the system will remain in defrost. Although the defrost thermostat should terminate the cycle, the timer failure, depending on its nature, may also affect the cycle-termination coil or even the mechanical linkage of the contacts. If so, the defrost cycle will simply continue indefinitely, and the interior of the box will become very warm after a while.

If the defrost timer is in the appropriate position, check to determine whether the box thermostat is closed. A simple way to begin on the example freezer is by unplugging the

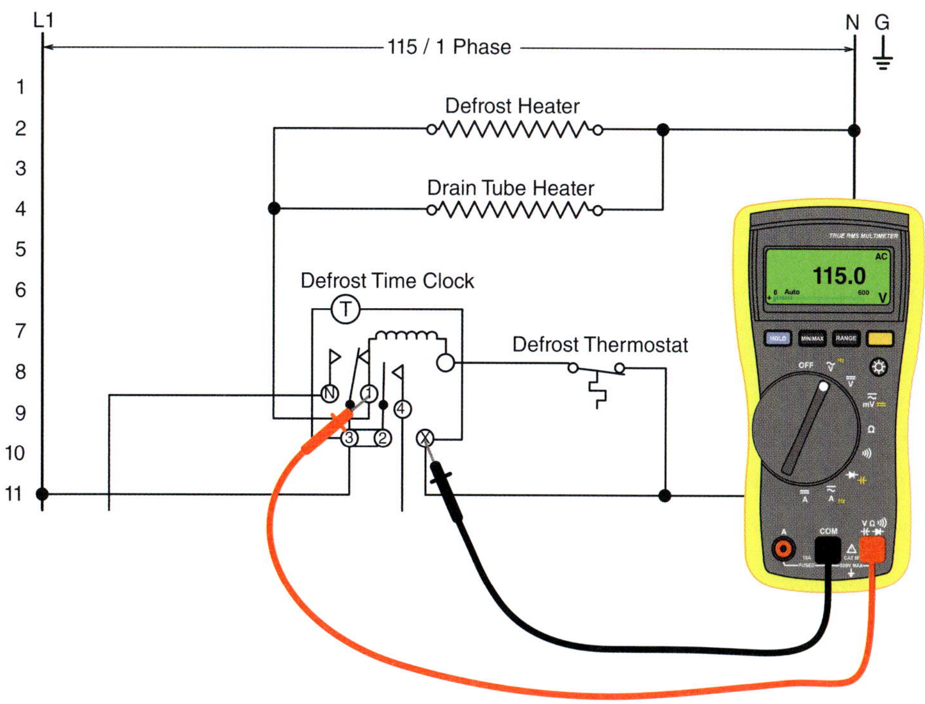

FIGURE 19-22 Checking the position of the defrost timer contacts with power applied.

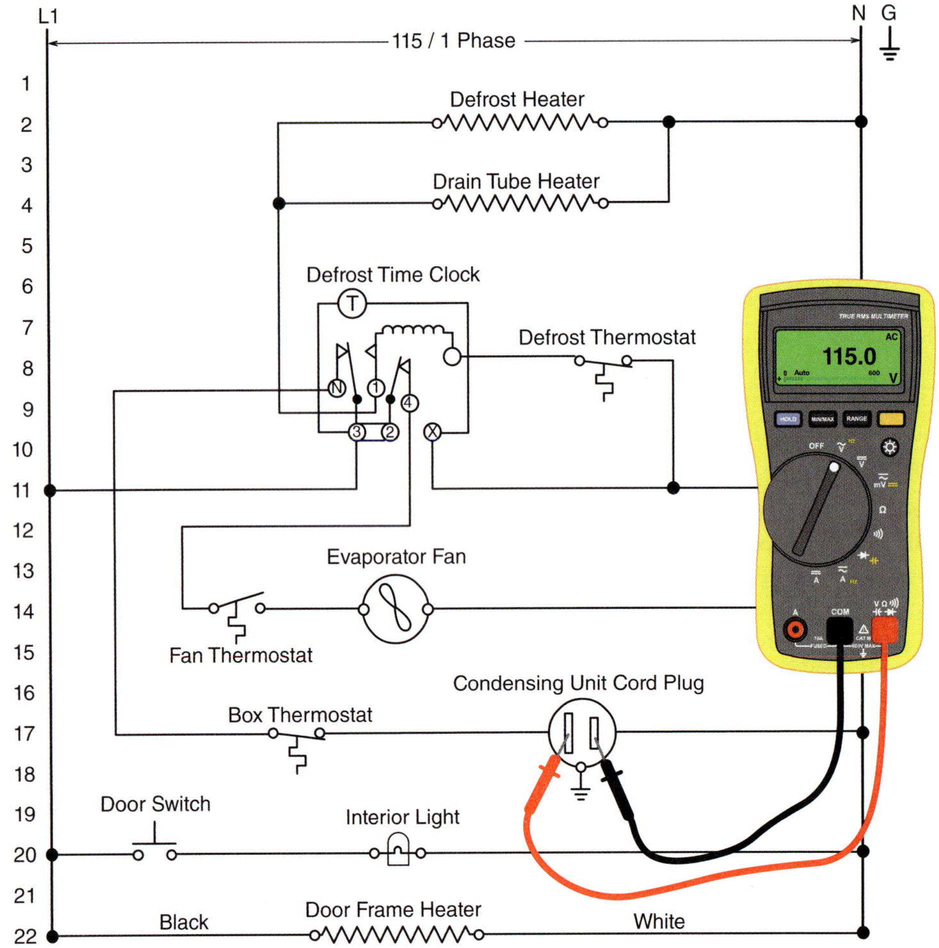

FIGURE 19-23 Checking the box thermostat position with power applied.

condensing unit and checking for power at the receptacle (**FIGURE 19-23**). If the box thermostat is closed, indicated by power at the receptacle, then the problem lies in the condensing unit. This would lead to troubleshooting of the compressor, condenser fan motor, start capacitor, and start relay, as demonstrated in previous chapters.

If there is no power to the receptacle, deenergize the system and test the box thermostat for continuity (**FIGURE 19-24**). If it is open while the box temperature is well above the set point, then the thermostat is defective and must be replaced.

Other Common Air Conditioning and Refrigeration Electrical Problems

Some electrical problems do not result in an unresponsive system. A system may run intermittently and/or fail to perform properly when it does run. In many cases, some components operate while others do not. This section will review problems that are not related to the power source and that present themselves in a variety of ways.

Air Conditioning System Problems

The following are some of the common electrical problems in air conditioning systems:

- *High-pressure switch opens.* Outdoor fan motors can fail completely, or they can overheat due to overload or bearing problems and may therefore operate intermittently. If the motor becomes grounded, it generally results in a blown fuse or tripped circuit breaker due to the excessive current. But when starting problems occur—e.g., seized bearings—the motor will draw its locked-rotor current only. For motors of this size,

▶ TECHNICIAN TIP

Outdoor and Indoor Fan Motors

For many years, the outdoor fan motors of air conditioning systems were referred to as condenser fan motors since they provided airflow for the condenser coil. The introduction of the air-cooled heat pump changed some of the traditional rules. When operating in heating mode, the outdoor coil serves as the evaporator coil, absorbing heat from the air rather than transferring heat to it. Since then, the condenser fan motor has been commonly referred to as the outdoor fan motor to more accurately address its dual responsibilities.

The same can be said of the indoor fan motor of a heat pump, although the term "indoor fan motor" was not uncommon prior to the introduction of heat pumps. Regardless, when discussing systems other than heat pumps, the terminology used is the choice of the speaker, and fellow craft professionals will understand the meaning.

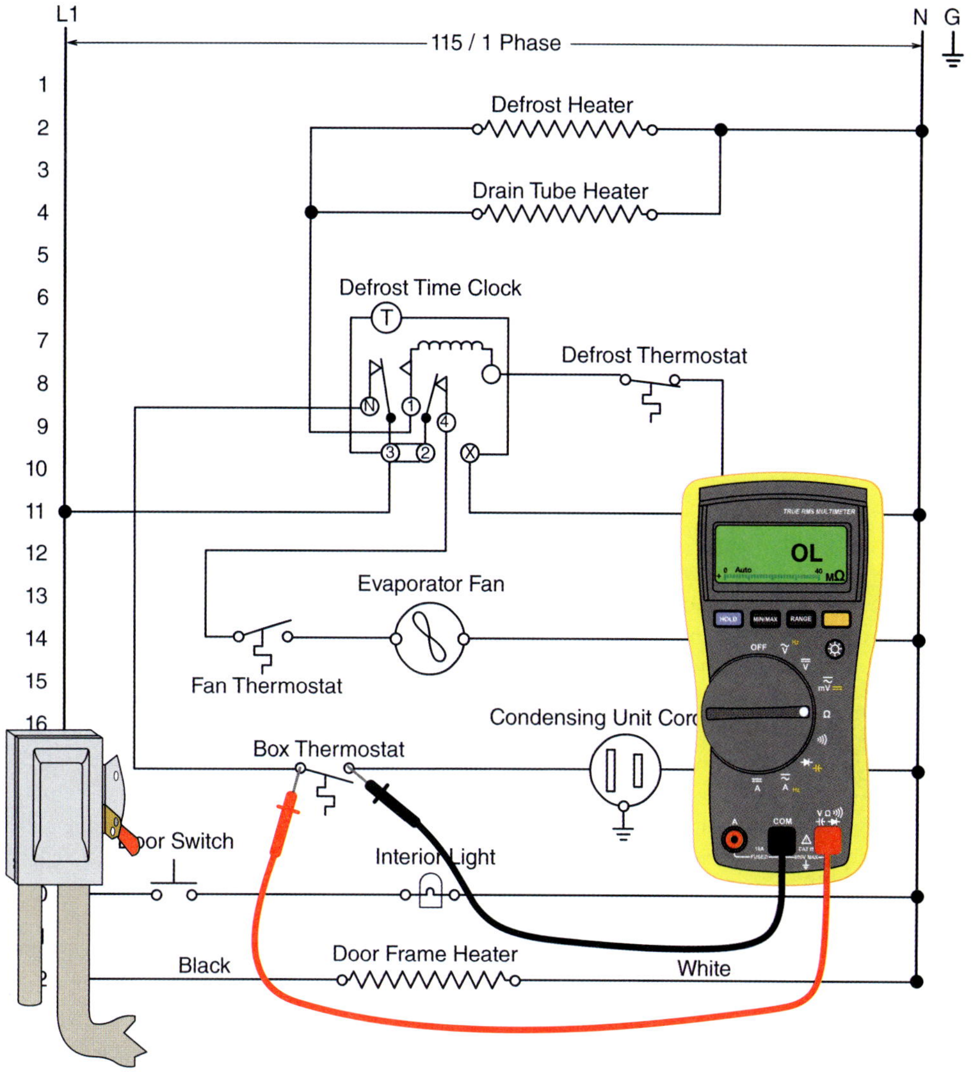

FIGURE 19-24 Checking the thermostat for continuity on a deenergized system.

the maximum current developed in the windings is not high enough to blow a fuse or trip a circuit breaker. Internal thermal protection usually stops the flow of current to the motor windings to prevent a fire, but once the windings cool, it can attempt to start again. This cycle can repeat itself any number of times unless a control device permanently opens the control circuit to the contactor.

An outdoor fan motor may also start and run for a short period with the bearings failing, but it will eventually overheat, then restart after it cools, and finally run again for a while. The high-pressure switch will likely open soon after the fan motor shuts down, but not instantly.

If the switch resets automatically when the pressure falls, this cycle can continue indefinitely, even after the motor fails completely. For this reason, most high-pressure switches require a manual reset to prevent this situation.

When the high-pressure switch is found open, deenergize the system and spin the motor shaft manually to determine whether it turns freely. Then check the motor for an open winding (**FIGURE 19-25**) or a grounded condition (**FIGURE 19-26**). Restart the system and check the motor current if the motor shows no obvious signs of failure.

If the motor operates as it should, start troubleshooting the refrigerant circuit. Air in the system, an excessive amount of refrigerant, and a blocked condenser coil are other common, nonelectrical reasons for a high-pressure switch to open.

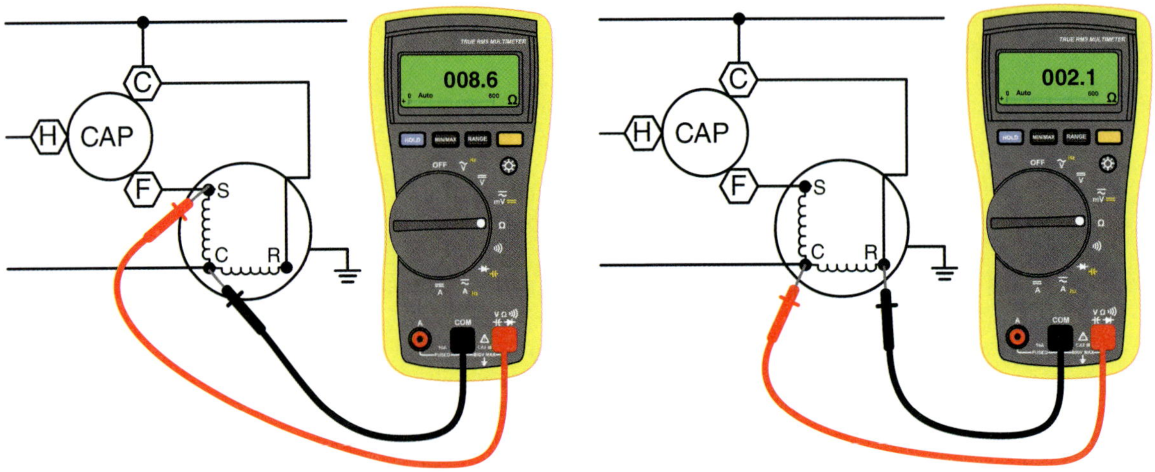

FIGURE 19-25 Testing motor windings for continuity.

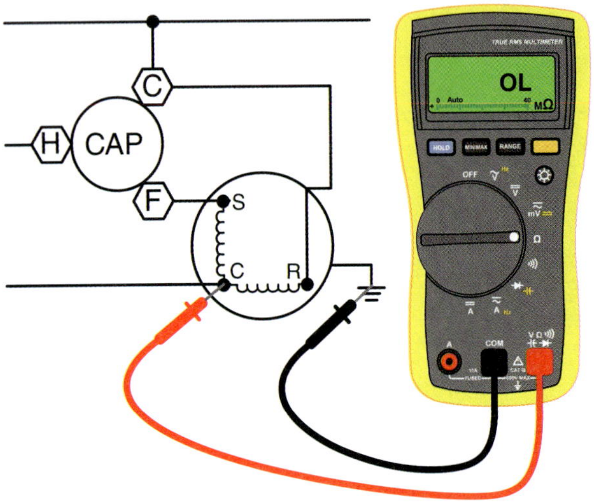

FIGURE 19-26 Testing motor windings for a grounded condition.

- *Low-pressure switch opens.* Many air conditioning systems do not have low-pressure protection. When a low-pressure switch is installed, it is often in series with the high-pressure switch (**FIGURE 19-27**). Most have a fixed pressure setting that protects only against a significantly low refrigerant charge. A failed blower motor, for example, will significantly reduce heat transfer at the coil, and the pressure will be lower. However, it often doesn't cause the pressure to fall far enough to open the low-pressure switch. Systems that are equipped with zone dampers can also obstruct airflow when the dampers are not working correctly. This too may not cause the pressure switch to open. The system continues to run as a result, and the evaporator will typically begin to freeze as time passes.

  When troubleshooting the problem, it may be necessary to bypass the low-pressure switch temporarily, allowing the system to operate for troubleshooting. Remember that an open low-pressure switch is usually a symptom of the real problem (loss of refrigerant; failed indoor blower motor) and not the cause. If the refrigerant pressures are normal and the switch will not close, it must be replaced.

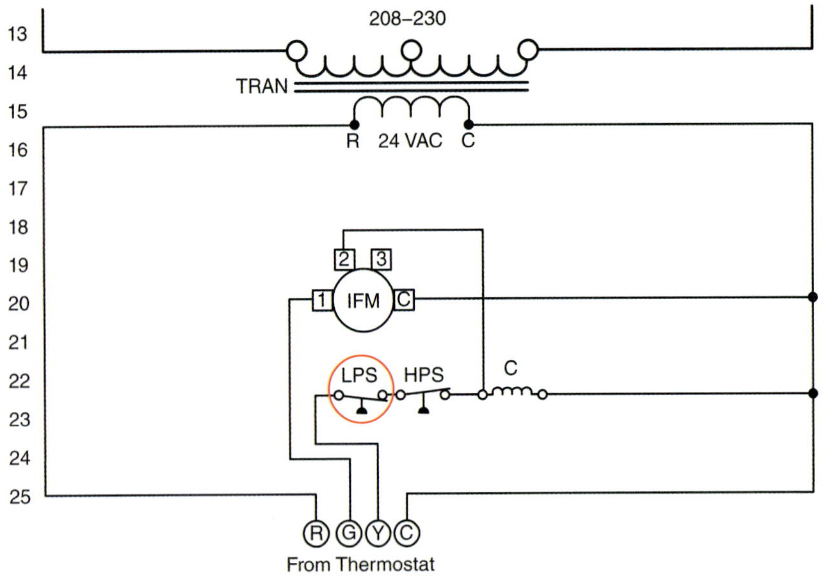

FIGURE 19-27 Low-pressure switch in the control circuit.

- *Anti-short-cycle protection failure.* When these generally reliable devices fail, they typically fail in the open position. In this case, the control circuit for the compressor contactor is not completed on a call for cooling and the user becomes aware of a problem rather quickly. The indoor blower would be running, but no cooling is provided.

  On the rare occasion that they fail closed, the only way to recognize the failure is to try to rapidly restart the system, noting that no time delay takes place. The failure can be confirmed by deenergizing the control and checking for continuity.

  Note that fewer anti-short-cycle devices are being used today, since most digital thermostats can provide this feature.

- *Crankcase heater failure.* This is a type of failure that often does not make itself known unless the technician happens to be in the right place at the right time. Crankcase heaters are designed to prevent refrigerant from condensing in the compressor crankcase during the off-cycle.

  For refrigerant to migrate and condense in the crankcase, that area must be cooler than the rest of the system. During the summer months, this is unlikely, even at night. It is much more likely to happen during the winter, but unless the system is operating in the winter due to unusual load conditions indoors (or it is a heat pump), there is no negative result, e.g. washing the oil out of the compressor crankcase at start-up. When this does occur, a rattling or similar noise may be heard as the compressor starts and attempts to compress oil and liquid refrigerant passing through.

  If these kinds of sounds are heard during start-up and the conditions are present for refrigerant migration and condensation in the crankcase to occur, check the operation of the heater. Use an ammeter to check the current (**FIGURE 19-28**), which will be relatively low but measurable. If there is no current, check to ensure that it is powered, then deenergize and check for continuity through the heating element.

Refrigeration System Problems

Since there are similarities in air conditioning and refrigeration systems, they share many of the same common problems. The following are some of the electrical failures that are somewhat unique to refrigeration systems:

- *Defrost timer failure.* Defrost timers (**FIGURE 19-29**) are relatively reliable. The contact structure and mechanical mechanisms can handle many cycles without a problem. Other than human error in the installation of the trippers, the mechanical clock mechanism is probably the most likely source of failure. This results in the unit failing to defrost. If the clock mechanism fails while the unit is in defrost, the cycle-termination switch will still terminate defrost at some point, placing the system back into its refrigeration mode.

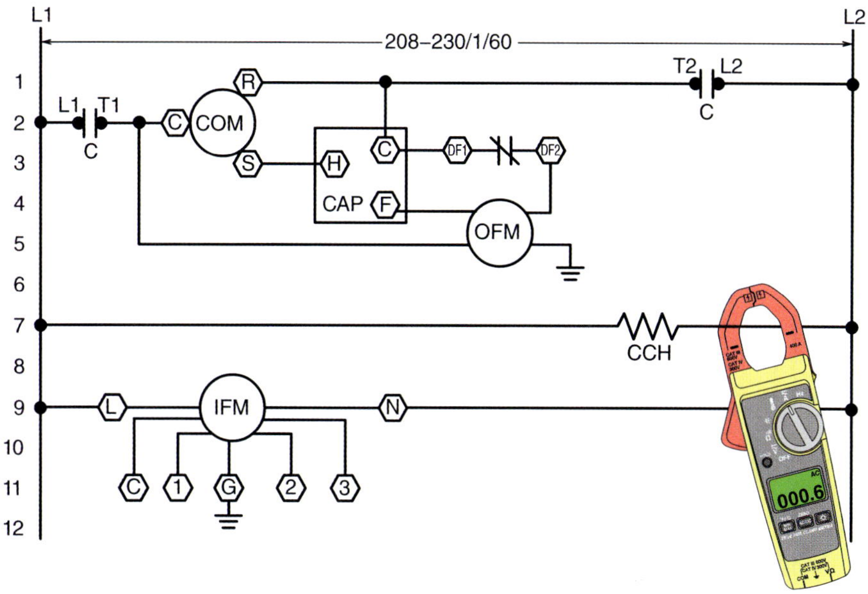

FIGURE 19-28 Checking the current to a crankcase heater.

▶ **TECHNICIAN TIP**

Fuses and Circuit Breakers

Remember that fuses and circuit breakers are not technically required or designed to protect equipment or loads, although they do provide that secondary function. They are designed to protect circuit wiring and wiring devices from conditions beyond their current-carrying capacity. This is why many motors require overload protection that will respond to currents that are often well within the capacity of the circuit and selected to only slightly exceed the expected normal running current of the motor.

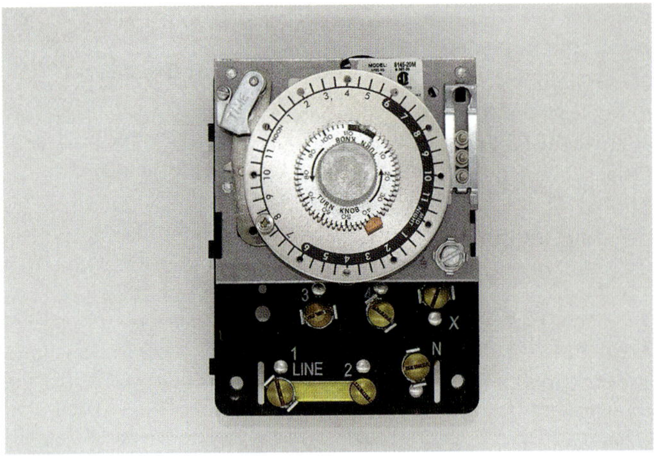

FIGURE 19-29 A refrigeration defrost timer.

The electrical diagram for a typical timer is shown in **FIGURE 19-30**. Most have this or a very similar electrical structure, regardless of whether the clock itself is electromechanical or digital. Terminal designations may vary slightly, especially for the cycle-termination terminal. There are two normally closed loads and one normally open load under control. One normally closed load is generally the refrigeration apparatus, and the other is the evaporator fan(s).

Both normally closed loads must be cycled off (contacts open) during the defrost cycle. The normally open load is the defrost heaters, which must be energized during defrost. Note that the cycle-termination switch, discussed earlier in this section, can be a thermostat or a pressure switch. The coil shown in Figure 19-30 represents a solenoid-operated mechanism that switches the contacts back to their normal position (refrigeration mode). When a defrost cycle is initiated, the two normally closed load contacts open, and the normally open load contacts close.

With a wiring diagram in hand, defrost timers are relatively simple to troubleshoot. If a load, such as the defrost heater, is not being energized in defrost, advance the timer to initiate a defrost cycle and check for power between Terminals 1 and X; line voltage should display on the meter (**FIGURE 19-31**). If not, check for power between Terminals N and X. If line voltage is shown, it indicates that the contacts are not changing position as they should. In the refrigeration mode, line voltage should be displayed when testing between Terminals N and X, but not between Terminals 1 and X. The clock contacts are subjected to significant loads during operation and can become welded closed or be damaged.

The cycle-termination solenoid coil can be tested for continuity with the power off (**FIGURE 19-32**). Apply an ohmmeter from Terminal 1 to the cycle-termination terminal (unlabeled in this diagram).

- *Defrost heater failure*. Electric defrost heaters are simply long heating elements that are placed on the evaporator coil. The size of the coil determines how they are placed to ensure even coverage. Large refrigeration coils may have several separate elements, and they are often specifically shaped to fit the coil. Like any resistance heating element, they are easy to test for continuity from the leads at each end (**FIGURE 19-33**).

  A defrost heater may also be placed along the bottom of the condensate pan under the coil, to ensure that defrost water doesn't freeze before it drains out. In addition, a defrost heater or heat cable may also be found attached to or inside a drain line, where it passes through the refrigerated space.

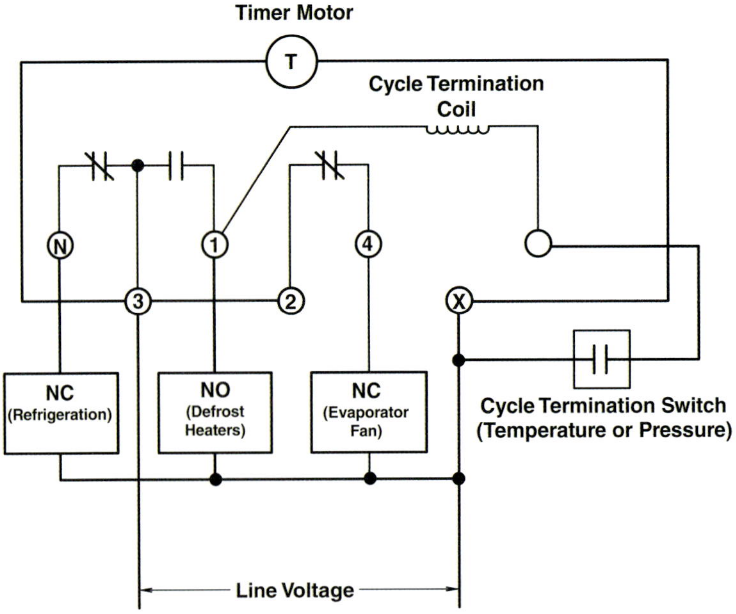

FIGURE 19-30 Defrost timer wiring.

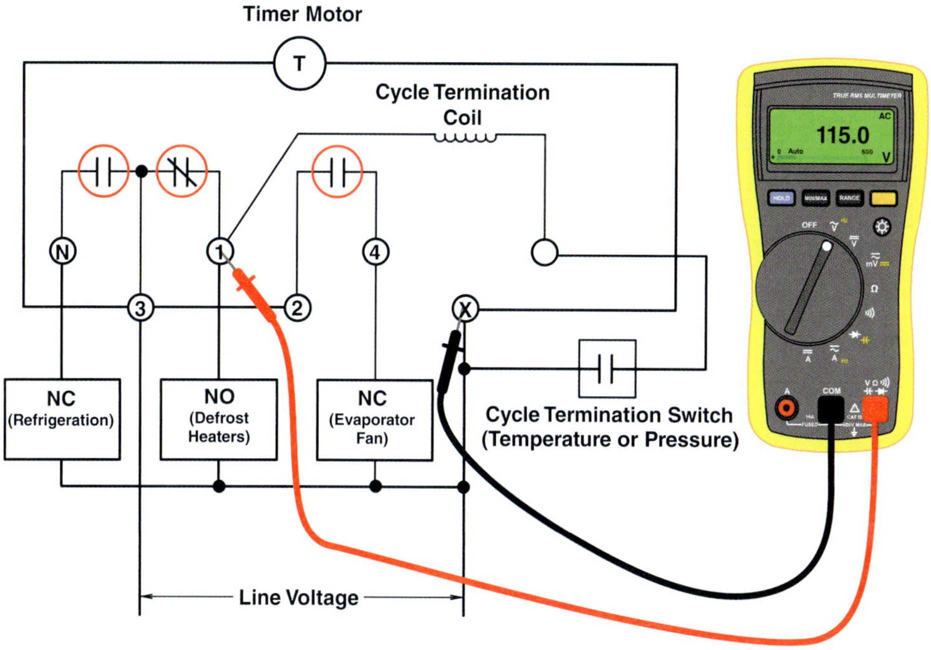

FIGURE 19-31 Checking for closed contacts in the defrost mode, with power applied.

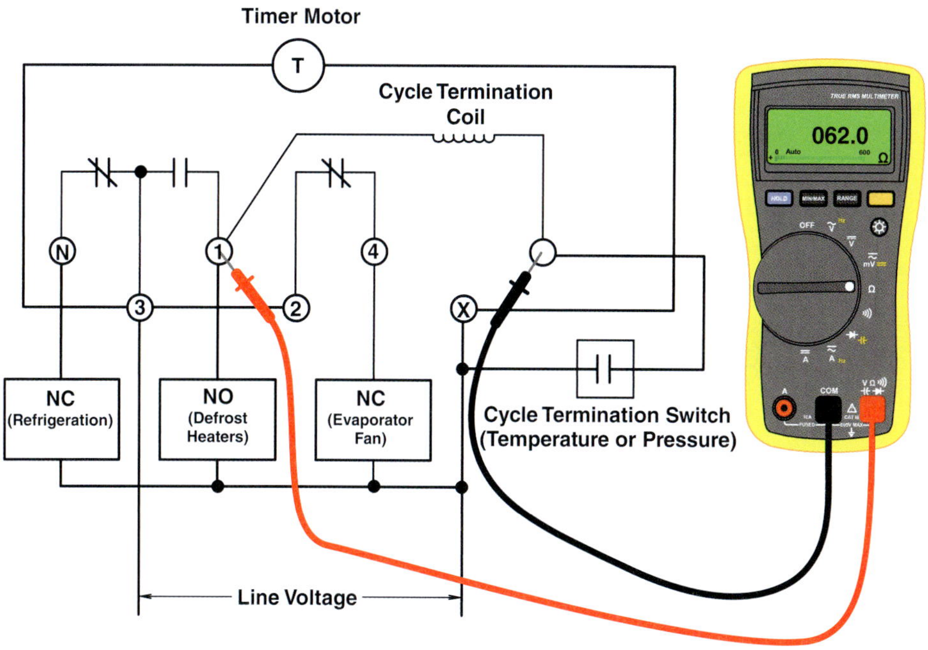

FIGURE 19-32 Testing the cycle-termination solenoid coil for continuity.

One obvious sign of heater failure is a frozen evaporator coil. If the defrost timer is working correctly, the heating elements must be tested next. When multiple heating elements are present, sections of the coil may defrost while others do not. This indicates that only one or more of the elements have failed. A coil that seems to defrost well while the pan beneath is filled with ice indicates a defective heating element in the pan.

- *Mullion (door frame) heater failure.* The mullion heater prevents frost around the door on reach-in and walk-in freezers. Coolers are usually equipped with them as well to prevent condensation around the door and gasket. Most mullion heaters are small-diameter, low-wattage heating cables (**FIGURE 19-34**) routed just beneath the door frame. The sign of a failed mullion heater is frost and/or condensate formation around the door gasket and frame. In severe cases, the door may freeze closed. Although it may not be physically difficult to separate and open the door with a tug, the gasket can be damaged and torn from repeated occurrences.

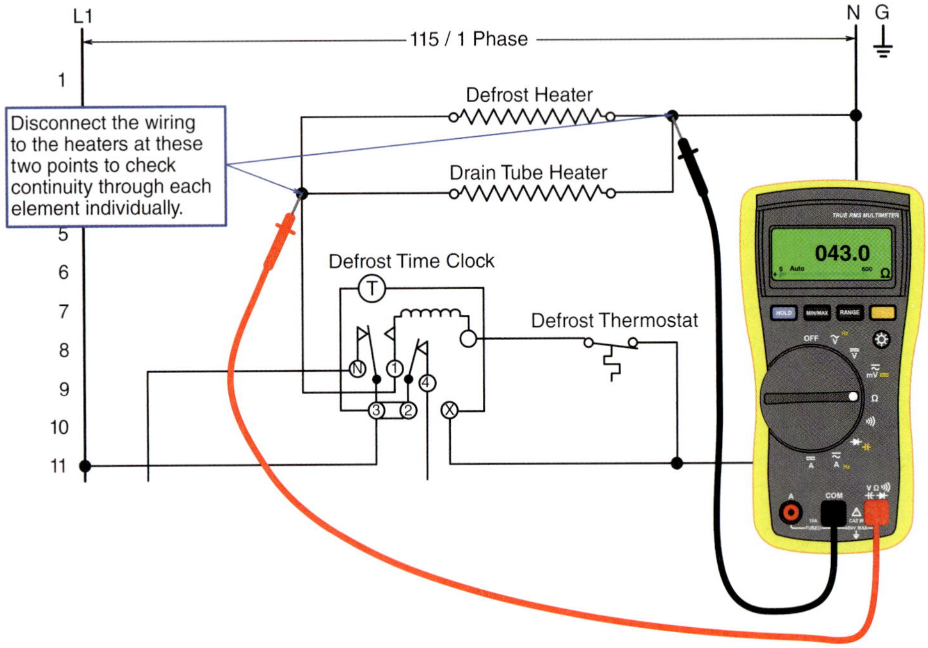

FIGURE 19-33 Testing a defrost heater for continuity.

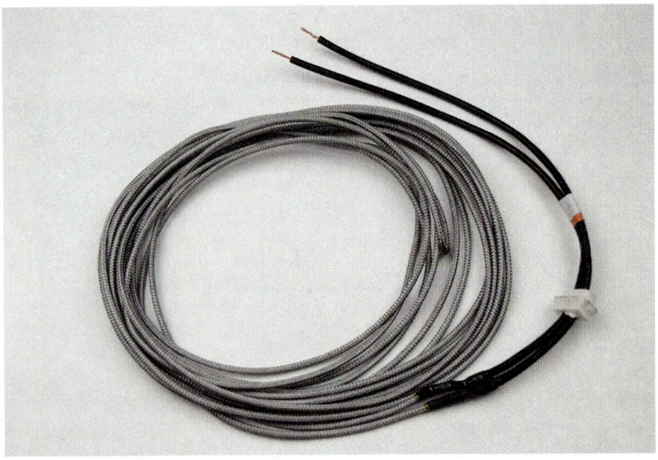

FIGURE 19-34 Heating cable.

▶ Wrap-Up

Summary

Troubleshooting and repairing air conditioning and refrigeration systems can be very challenging. While the source of some problems can be easily identified, others may take a great deal of time and patience to identify. But even the most challenging problems can be solved by focusing on the basic principles of strategy-based troubleshooting. Experienced technicians are not necessarily better than their less-experienced peers in the trade at understanding a multimeter display and deciding where to place the probes to complete a test. What sets them apart is their ability to analyze the information available, eliminate specific components and circuits from the realm of possibilities, and focus their attention on those areas that can produce a given set of symptoms.

This chapter provides some examples of how to analyze the information available in certain situations, but it is only the beginning. Each new troubleshooting challenge becomes a learning experience that will contribute to becoming a successful HVACR technician.

Key Terms

algorithms A set of programmed mathematical rules used in calculations by computers or processors.

chillers A refrigeration apparatus designed to chill water for cooling applications.

condensers Coils of tubing containing vapor condensing into a liquid, typically transferring heat to air or water.

crankcase heater (CCH) A relatively low-wattage, electric resistance heater used to warm a compressor to discourage the migration and condensation of refrigerant in the crankcase.

dewpoint The temperature representing 100% saturation for a vapor, below which the vapor begins to condense into its liquid state.

evaporators Coils of tubing containing liquid boiling to a vapor, typically absorbing heat from air or water.

hermetic In this context, a compressor that is sealed and unserviceable in the field.

metering devices Orifices, either fixed or variable in size, designed to create a pressure drop and meter the flow of refrigerant.

reciprocating compressors Positive-displacement compressors that use pistons driven by a crankshaft to compress refrigerant vapor.

rotary compressors There are two basic types of rotary compressors. Rotary-vane compressors use a rotor positioned off-center with respect to the casing and sliding vanes that compress vapor as they rotate. Rotary-screw compressors compress vapor between the flutes of two precisely matched helical-screw assemblies.

scroll compressors Compressors that use two scroll-like plates—one orbital and one stationary—to compress refrigerant vapor.

semi-hermetic In this context, a compressor that is partially field serviceable. As a general rule, only the oil pump and valve assemblies are reasonably serviceable under field conditions. The motor, crankshaft, pistons, and other moving parts can often be serviced and replaced in a shop with the proper equipment.

Review Questions

1. The temperature of the refrigerant vapor is important because its secondary function is to _____.
   a. keep the compressor quiet
   b. force oil out of the compressor
   c. lubricate the compressor
   d. cool the compressor
2. The type of compressor that has an enclosed motor but also has some parts that are accessible for field service is the _____ compressor.
3. The type of compressor known for its efficiency, its reliability, and having few moving parts is the _____ compressor.
4. Which of the following statements is true?
   a. Condensers are designed to increase the pressure of the refrigerant vapor.
   b. During the refrigerant condensation process, the refrigerant temperature does not change.
   c. Air transfers heat from refrigerant in a condenser more efficiently than water.
   d. No air-cooled condenser can function without fans to force air across it.
5. The metering device that requires a carefully engineered length and diameter is the _____.
6. The primary function of the metering device in the refrigerant circuit is to _____.
   a. cool the refrigerant
   b. increase the refrigerant pressure
   c. create a pressure drop
   d. separate vapor and liquid
7. The state of the refrigerant entering an evaporator coil is typically _____.
   a. both liquid and vapor but dominated by liquid
   b. both liquid and vapor but dominated by vapor
   c. liquid only
   d. vapor only
8. Any additional heat absorbed by refrigerant vapor after it has completely changed from its liquid state is referred to as _____.
9. Which of the following statements is true?
   a. Refrigerant must be subcooled before it leaves the evaporator.
   b. All evaporators are designed to transfer heat to air.
   c. Compressors cannot compress liquids.
   d. Only reciprocating compressors are capable of compressing liquids.
10. The easiest wiring diagram to follow for troubleshooting and understanding the operating sequence is the _____.
11. A refrigerated fixture can usually defrost satisfactorily during planned off-cycles when the box temperature is maintained at _____.
    a. 25°F (–3°C) and above
    b. 32°F (0°C) and above
    c. 38°F (3°C) and above
    d. 45°F (7°C) and above
12. When a freezer is initially started at room temperature, the evaporator fan is often designed to remain off until the coil is near or below freezing, __________.
    a. to prevent the evaporator fan motor from overloading
    b. to help reduce the initial load on the compressor
    c. to reduce the total current draw of the system
    d. to keep the low-pressure safety switch from opening
13. If a circuit breaker trips immediately when reset, __________.
    a. the compressor is likely grounded, since it is the largest load
    b. it is most likely undersized for the system
    c. it is defective and must be replaced
    d. a grounded condition exists somewhere in the circuit or system
14. To what terminal(s) does a thermostat typically transfer power from the R terminal on a call for cooling?
    a. Y and G.
    b. Y only.
    c. C and G.
    d. C only.

15. When a smaller motor, such as a condenser fan motor, draws locked-rotor current due to seized bearings, what is the most likely result?
    a. A fuse blows and shuts down the entire system.
    b. A circuit breaker trips and shuts down the entire system.
    c. Thermal protection inside the motor opens to stop current flow.
    d. The contactor opens and will not reclose until it has been reset.
16. When a low-pressure switch is provided in an air conditioning system, it is often wired in series with the ________.
    a. high-pressure switch
    b. crankcase heater
17. The setting of many low-pressure switches used in air conditioning systems is designed to protect against only ________.
    a. a slight reduction in airflow across the evaporator coil
    b. a failure of the outdoor fan motor (OFM)
    c. a significantly low refrigerant charge
    d. a failed crankcase heater
18. What is the primary function of a crankcase heater?
    a. It warms the crankcase to prevent refrigerant from migrating and condensing in the crankcase during the off-cycle.
    b. It keeps the compressor warm, allowing it to start more easily.
    c. It ensures that liquid refrigerant doesn't enter the compressor while it is running.
    d. It keeps the oil thinner so that it more easily flows out of the compressor with the refrigerant.
19. The contacts of a defrost timer that send power to (enable) the refrigeration system are ________.
    a. normally open
    b. normally closed
20. What prevents condensation and frost from accumulating around the door of a cooler or freezer?
    a. A mullion heater.
    b. A crankcase heater.
    c. The defrost timer.
    d. The door gasket.

CHAPTER 20

Heat Pump Sequence of Operation and Troubleshooting

Knowledge Objectives

After reading this chapter, you will be able to:

- **K20001** Describe the basic heat pump refrigeration cycle and its unique components.
- **K20002** Explain why supplemental heat is needed for heat pumps.
- **K20003** Describe the construction and operation of common electric heat packages.
- **K20004** Describe common approaches to defrost and typical defrost controls.
- **K20005** Describe the sequence of operation for a typical air-cooled heat pump.
- **K20006** Explain how to analyze, troubleshoot, and diagnose common heat pump electrical problems.
- **K20007** Explain how to analyze, troubleshoot, and diagnose common electric heat problems.

▶ Introduction

Heat pumps have been in use for many years now, and they offer some benefits over systems that depend solely on fossil fuels for heating. Heat pumps have gained popularity in rural areas, where oil and propane once represented the only practical choices, and they offer increased energy efficiency over electric resistance heating in all areas. They are especially popular in areas where the heating season is not as long or as brutally cold as the northernmost regions. However, heat pumps can also be paired with fossil-fuel systems to provide excellent comfort, efficiency, and flexibility in all climates.

Because heat pumps offer the ability to heat and cool using the same equipment, they possess a few unique components not found in conventional cooling systems. This chapter will focus on those components and the operating sequences that allow them to provide comfort in both heating and cooling modes.

▶ The Heat Pump Refrigeration Cycle

K20001 Describe the basic heat pump refrigeration cycle and its unique components.

A heat pump can be described as a reverse-cycle air conditioner. A conventional air conditioner uses the vapor-compression refrigeration cycle to transfer heat from the space to the outdoors. A heat pump can do the same thing, but it can also reverse the cycle so that heat can be transferred from outdoors into the space.

FIGURE 20-1 provides an example of refrigerant flow in a heat pump. In the cooling mode, refrigerant follows the same basic path as it does in a conventional air conditioning circuit. Hot refrigerant gas flows to the outdoor coil, where it is cooled and condenses into a liquid. From there, it passes through the metering device and into the indoor coil, where it absorbs heat and returns to its vapor state.

The primary component that sets the heat pump apart from a conventional system is the **reversing valve**. In Figure 20-1, the hot refrigerant gas leaves the compressor and flows directly into the reversing valve through the single port on one side. Reversing valves have two possible positions. Each of the two positions connects the center port with one of the ports positioned on either side of it. The center port is piped directly to the compressor inlet.

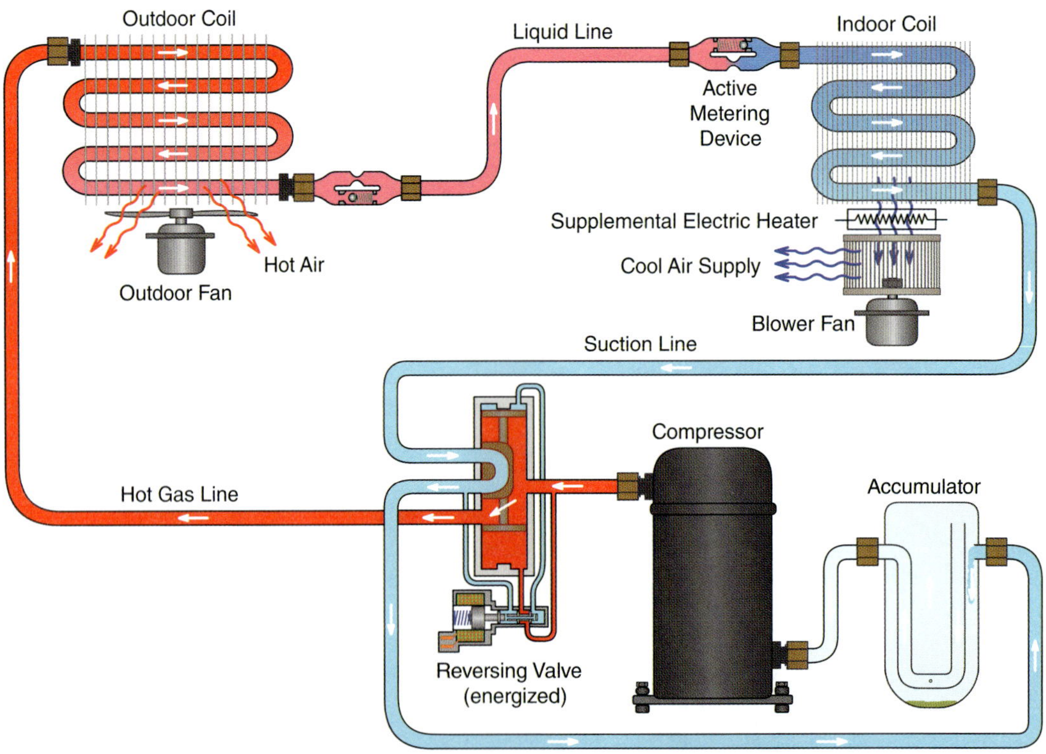

FIGURE 20-1 Heat pump refrigerant flow in the cooling mode.

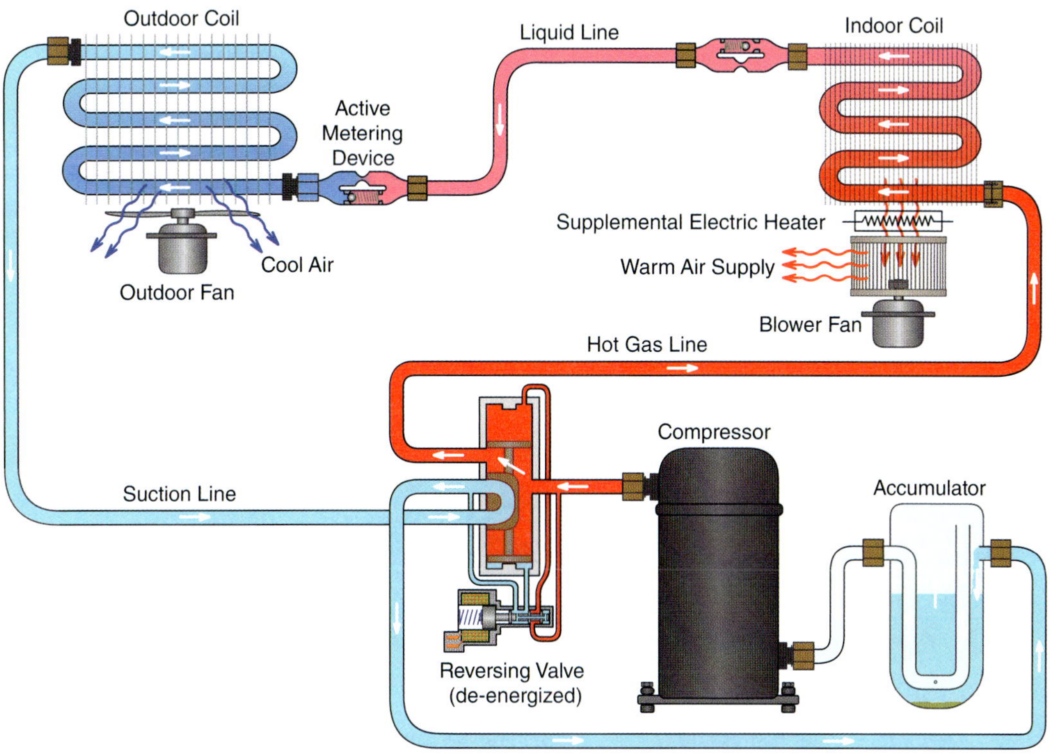

FIGURE 20-2 Heat pump refrigerant flow in the heating mode.

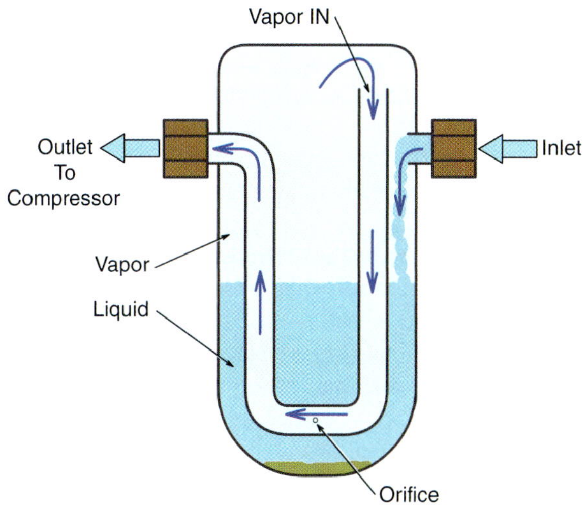

FIGURE 20-3 Inside an accumulator.

In the cooling mode, hot refrigerant gas from the compressor is directed to the port that leads to the outdoor coil. In this position, the center port is internally connected to the vapor line from the indoor coil.

In the heating mode, the reversing valve moves to its alternate position (**FIGURE 20-2**). The center port is now internally connected to the line that leads to the outdoor coil. Hot refrigerant gas flows from the compressor, through the valve, and out of the port connected to the line that leads to the indoor coil. Heat is then transferred to the space by the indoor coil.

A solenoid coil controls the position of the reversing valve (**FIGURE 20-4**). However, the solenoid coil does not move the valve directly, because its resistance to movement is too much for a small solenoid coil to move the valve on its own. Instead, it controls a miniature, two-position solenoid valve that uses high-pressure refrigerant from the compressor

▶ CORE CONNECTION

The Accumulator

In the heating mode, it is not unusual for some liquid refrigerant to remain after it passes through the outdoor coil and returns to the compressor. Not all of the refrigerant is vaporized, and both liquid and vapor are present. As a result, most heat pumps are equipped with a suction accumulator that captures the liquid before it reaches the compressor. The orifice shown at the bottom in **FIGURE 20-3** provides a way for small amounts of oil and liquid refrigerant to return to the compressor at a rate that prevents damage.

An accumulator may also be referred to by some manufacturers as a charge compensator. It is also needed because of the differences in internal volume of the two coils and the varying state of the refrigerant in the two coils during both operating modes. The liquid level in an accumulator during the heating mode is often indicated by a frost line on the outside.

FIGURE 20-4 Heat pump reversing valve.

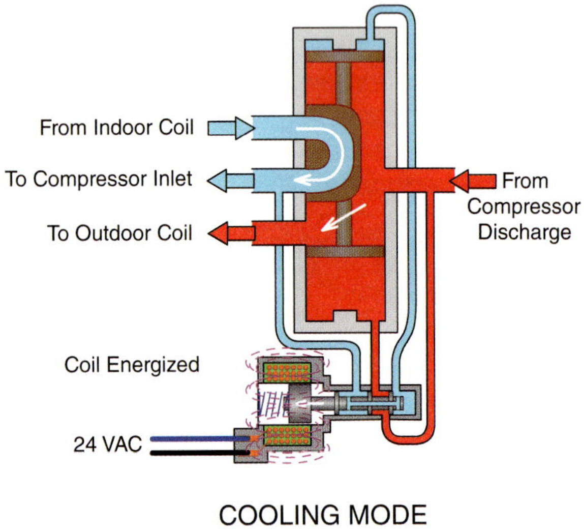

FIGURE 20-5 Reversing valve in the cooling mode.

to push the reversing valve shuttle, or slide, from one position to another. Technically then, a reversing valve is a **pilot-operated valve**.

FIGURE 20-5 shows the flow of refrigerant through the reversing valve in the cooling mode. High-pressure refrigerant pushes the shuttle from left to right, directing the hot refrigerant gas to the outdoor coil. Cool vapor from the indoor coil enters the valve and exits through the center port, returning to the compressor.

Note that the coil is shown energized. In most cases, heat pump reversing valves are energized by the thermostat in the cooling mode. The reason most often given is that the heating mode is the more critical operating mode. Thus, a failure of the reversing-valve solenoid coil results in the unit being able to operate in the heating mode, but not in the cooling mode.

However, this is not always the case. Some manufacturers have chosen to energize the valve in the heating mode over the years, so be aware of this detail when servicing a given heat pump.

The reversing valve is energized by the thermostat through either the B or O terminal. The thermostat energizes the B terminal when it is switched to the heating mode. The O terminal is energized when the thermostat is placed in cooling mode.

FIGURE 20-6 shows the position of the reversing valve in the heating mode (deenergized in this example). When the solenoid valve is deenergized, high-pressure refrigerant is directed to the right end of the valve, pushing the shuttle from right to left. Hot refrigerant gas is then directed to the indoor coil. Refrigerant vapor (often mixed with some liquid) from the outdoor coil enters the valve through the left port and exits the center port on its way to the compressor.

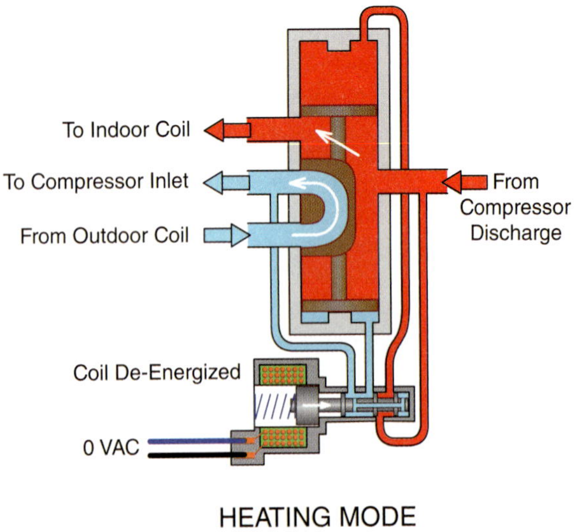

FIGURE 20-6 Reversing valve in the heating mode.

There is much to learn about the refrigerant cycle in a typical heat pump that goes beyond the scope of this book. However, operating characteristics that are relevant to the electrical operation of the unit will be presented as necessary.

The Defrost Cycle

Heat pumps must be able to provide heat even in subfreezing conditions. For the system to absorb heat from the outside air in these conditions, the refrigerant in the outdoor coil must be colder than the air. As a result, the coil temperature is often below freezing, and moisture that is in contact with the coil will certainly freeze (FIGURE 20-7). In very dry climates, frost development is rather limited. However, where it is both cold and wet, frost can develop on the coil surface rather quickly. This impedes airflow across the coil and interferes with heat transfer. Heat pumps, therefore, must periodically initiate a defrost cycle.

The defrost cycle of a heat pump is really quite simple. The basic approach is to allow it to return to its cooling mode for a short period of time, where hot refrigerant gas from the compressor is directed to the outdoor coil (**FIGURE 20-8**). The latent heat that it releases as it condenses to its liquid state, plus the sensible heat of the refrigerant gas, melts the accumulated frost. The only significant difference is that the outdoor fan is not energized during the defrost mode. Its goal is not to transfer heat to the surrounding air but rather to transfer as much heat as possible to the coil surface and melt the ice. Operating the outdoor fan would interfere with that process.

The problem, of course, is that operating in the cooling mode when heating is needed can obviously cause some discomfort for the user. To maintain reasonable comfort throughout the cycle, heat pumps with supplemental electric heat energize the heating elements to reheat the air as it leaves the evaporator. This does help maintain comfort, but it works against the energy efficiency of the system since both the unit and the electric resistance heaters are energized and are working against each other.

FIGURE 20-7 Frost on a heat pump outdoor coil.

▶ TECHNICIAN TIP

Water-Source Heat Pumps

Water-source heat pumps are generally more effective and efficient than air-source units, since the temperature of the water source (usually from, or circulated through, the ground) is more stable and has better heat-transfer characteristics. As it gets colder outdoors, the heating capacity of the system does not decrease. One distinct advantage is that a defrost cycle is not necessary. Defrost cycles work against the overall efficiency of a heat pump system.

▶ Heat Pumps and Supplemental Electric Heat

In addition to the need for heat to offset the cooling effect of the defrost cycle, heat pumps need supplemental heat for another reason. Although it is well-known that heat pumps can extract some heat from the air at very low outdoor temperatures, it is also understood that the heating needs of a space increase as the outside temperature decreases. The heat pump's ability to extract heat decreases as the outdoor temperature decreases. Eventually, the decreasing heating capacity of the heat pump and the increasing heating load of the space meet. When plotted on a graph, the two lines cross at some point (**FIGURE 20-9**). This point is referred to as the **balance point**: the temperature at which the heat load of the space equals the heating capacity of the heat pump.

K20002 Explain why supplemental heat is needed for heat pumps.

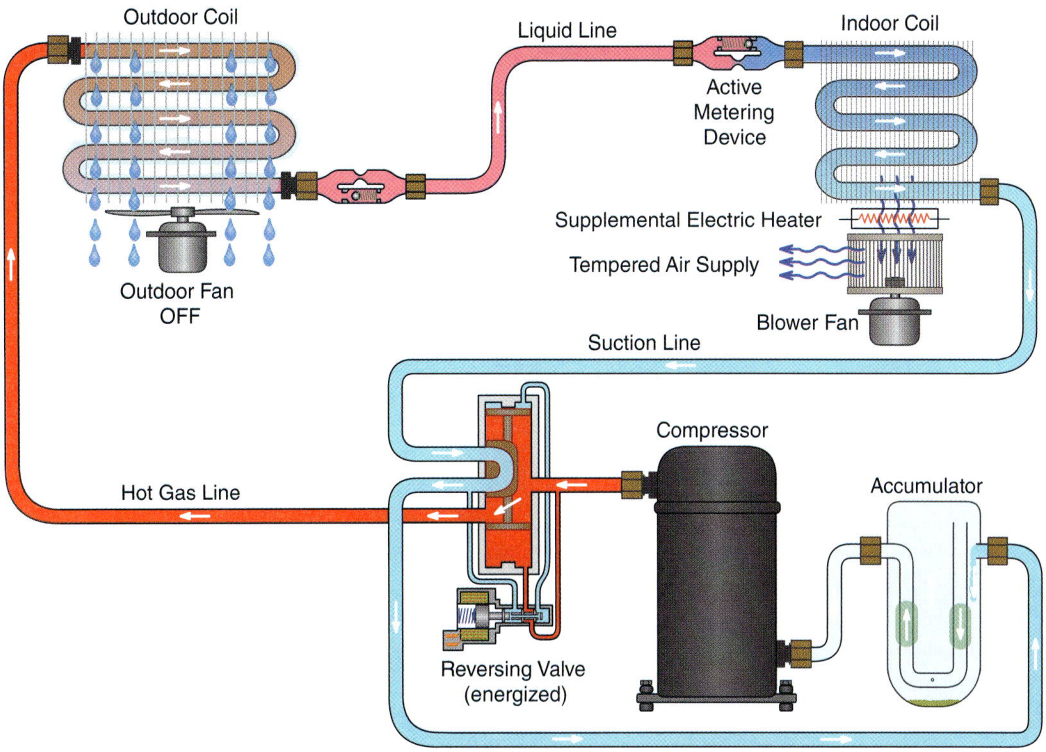

FIGURE 20-8 The defrost cycle.

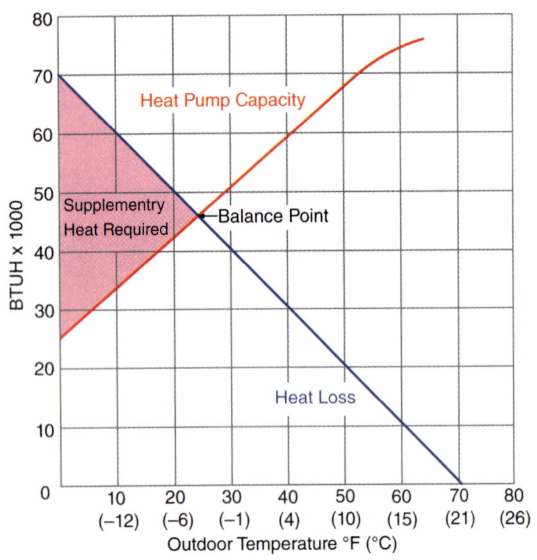

FIGURE 20-9 A typical balance point graph.

In theory, whenever the outdoor temperature is below the calculated balance point, the heat pump must run continuously to provide heat. As the outdoor temperature continues to drop, the heat pump requires either supplemental heat to maintain comfort or another source of heat to replace it altogether (**dual-fuel systems**). The source of supplemental heat is typically electric heating elements.

Note that the balance point is not a primary factor in the selection of heat pump size. Heat pumps are generally selected to match the cooling load, just like a conventional air conditioning system. The balance point is then determined by plotting the heat loss of the structure and the unit heating capacity, as shown in Figure 20-9. While it is not unusual for the balance point to be above freezing, it may also be as much as 10°F (6°C) below freezing.

Since the electric heating elements are positioned downstream of the heat pump indoor coil, they do not affect the load on the coil. Air leaving the coil is simply heated further, adding to the heating effect of the heat pump.

Electric heating elements provide one additional advantage to the heat pump system. If the heat pump itself fails during the cold days of winter, the heating elements provide a source of backup heat. Electric heat packages are simple and straightforward, making them very reliable. As long as the wiring connections are kept tight and maintained in good condition, they can provide trouble-free operation for years.

k20003 Describe the construction and operation of common electric heat packages.

▶ TECHNICIAN TIP

Dual-Fuel Heat Pump Systems

Dual-fuel heat pump systems do not use electric resistance heaters to supplement the heat pump when in the coldest weather. Instead, the heat pump is paired with a fossil-fuel furnace. However, since the indoor coil of the heat pump is downstream of the furnace, the heat pump and the furnace cannot operate at the same time. The controls are set to allow the heat pump to handle the heating load above the balance point while the furnace takes over the heating load exclusively below the balance point. Defrost cycles for the heat pump are usually unnecessary since the fossil-fuel system takes over at temperatures below the balance point.

▶ Electric Heating Packages and Controls

Electric heating elements (**FIGURE 20-10**) are most often furnished in 4–5 kW increments. An electric heat package rated at 15 kW, for example, will have three separate heating elements. Note that their actual wattage in operation depends on the voltage applied. A given heating element with an applied voltage of 230 VAC will provide more heating capacity than when 208 VAC is applied.

▶ TECHNICIAN TIP

Electric-Heat Current, Wattage, and Output

The calculations related to electric heat are relatively simple and are based on Ohm's law. One watt (W) of electric heat produces 3.413 Btu of heat. Therefore, a 5 kW (5,000 watt) electric heating element produces 17,065 Btu.

However, it is important to remember that the actual wattage of a given electric heater depends on the voltage supplied. For example, assume a heating element is rated to produce 5,000 watts at 240 VAC, producing 17,065 Btu. The current would be 20.8 amps. If the voltage provided is only 208 VAC, the wattage falls to 3,755 watts and the operating current falls to 18 amps. Only 12,818 Btu are produced, as a result. This does not damage the heating element in any way, but it significantly reduces the heat output. The available supply voltage should be considered when the size of the electric heat package for a job is selected.

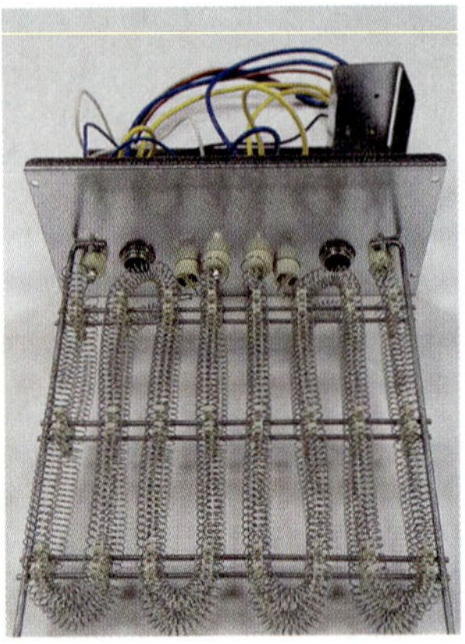

FIGURE 20-10 Electric heat assembly.

As shown in **FIGURE 20-11**, most electric heat assemblies have two different safety devices to prevent overheating. A normally closed, automatic-reset, fixed-temperature limit switch provides the primary protection. It is usually placed directly in the power circuit to the heating element rather than in the control circuit. As stated, it resets automatically after the temperature falls below a preset temperature. The fusible link (FUSL) shown is much like a fuse: if its rated temperature is reached, the internal link melts and the power circuit is broken. To return the heater to service, it must be replaced with a new link. Note in Figure 20-11 that each of the two independent heating elements (a 10 kW heating assembly) has a dedicated limit switch and fusible link.

Power Wiring

The opposite side of the electric heat package that was shown in Figure 20-10 is shown here in **FIGURE 20-12**. A comparable wiring diagram is shown in **FIGURE 20-13**. For residential and light-commercial systems, the power wiring for the air-handling unit (also known as the fan-coil unit) originates from the electric heat package line-voltage terminals. When no electric heat is provided (a cooling-only system), an accessory package from the manufacturer may be required to act as a point of power connection.

Incoming line voltage is connected to the terminal board lugs shown on the left side of the heat assembly. From there, the wires are routed to the air-handling unit through a plug connector to power the fan-coil unit transformer, blower motor, and any other electrical devices.

All the yellow and blue wires connected to the line-voltage terminals are used to power the electric heating elements. The electric heat relays (EHRs) are shown in their normally open state. Each relay controls one heating element, breaking both legs of power. One leg of power flows from the EHR through the normally closed limit switch (yellow wires). The other leg of power (blue wires) flows through the second set of EHR contacts through the fusible link. It is common for these safety devices to be directly in the power circuits for the heating elements rather than in the control circuits, because of how important they are to safety.

Control Wiring

The plug connector that carries power from the heat package to the air-handling unit also transfers control power from the air-handling unit back to the heat package. Power from the Common (C) side of the transformer is routed to one side of the operating coil on both EHRs (white wires). The smaller blue wires from W1 are wired to the other side of the operating coils. When W1 is energized by the thermostat, the EHRs are energized, their contacts close, and the heating elements are energized.

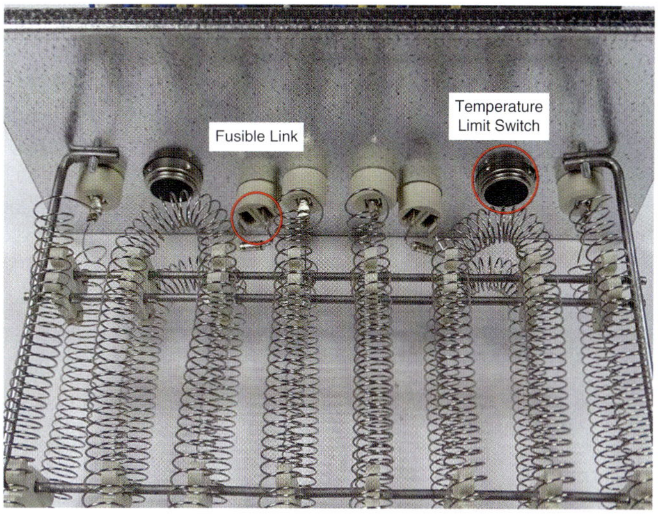

FIGURE 20-11 Electric heat safety devices.

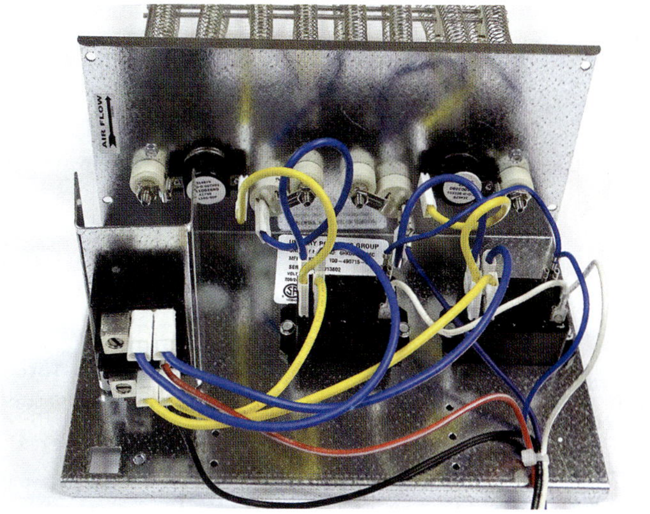

FIGURE 20-12 Electric heat package wiring.

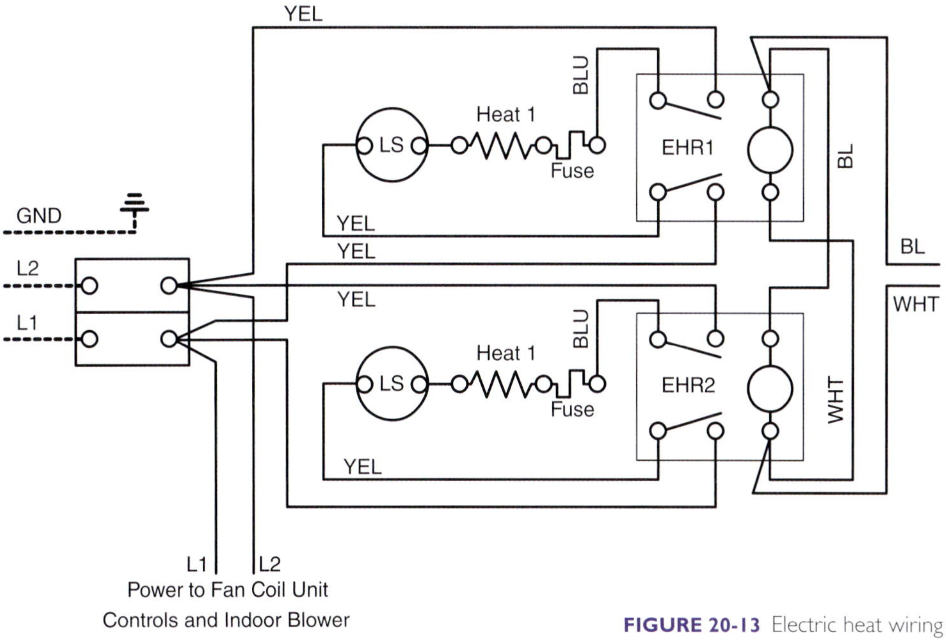

FIGURE 20-13 Electric heat wiring diagram.

▶ CORE CONNECTION

Think About the Electric Heat Needed During Defrost

If a 5-ton heat pump produces 60,000 Btuh of cooling when it is in defrost, how many watts of electric heat would be required to offset the cooling effect?

Note that W2 is not wired in this case. On larger heat packages (usually three or more heating elements), multiple stages of electric heat can be applied. When the outdoor temperature is not far below the balance point, for example, there is no need to energize the maximum amount of electric heat available. The first stage alone, perhaps 10 kW, is sufficient in milder weather. As it becomes colder, an outdoor thermostat or sensor can allow additional stages of electric heat to be energized as well. This same thermostat or sensor may also be used to disable all available electric heat when the outdoor temperature is above the balance point, preventing their use when the heat pump should handle the load alone. Switching the thermostat to the Emergency Heat mode will typically make all stages of electric heat available, regardless of outdoor temperature. The defrost-cycle controls may also be able call upon more stages of electric heat, usually depending upon the size of the system. A 5-ton heat pump, for example, can produce a great deal of cooling while in defrost, requiring more electric heat to offset the otherwise cold air temperatures.

Since the electric heat package shown in Figure 20-12 has a separate relay to control each of the two heating elements, it could easily be required to provide two-stage operation.

▶ Approaches to Defrost

K20004 Describe common approaches to defrost and typical defrost controls.

Over the years, many approaches to defrost have been developed. In the early days of heat pumps, for example, some manufacturers depended on precision air-pressure switches that monitored pressure on each side of the outdoor coil to detect frost buildup. Since the pressure differential was relatively small and since the environment is less than ideal for an air-pressure switch working at such precise levels, this approach did not prove to be very reliable. Much better defrost systems have replaced those early electromechanical setups, which also relied on mechanical timers.

A heat pump defrost cycle must enter and end the cycle while minimizing the time spent in this energy-wasting mode. A defrost system has three basic tasks, regardless of the approach:

1. Change the reversing valve position from its heating position to its cooling position.
2. Cycle the outdoor-fan motor off.
3. Energize the supplemental electric heat.

Remember that a heat pump in defrost is a heat pump that is working directly against its primary purpose. The defrost cycle also wastes energy. On the other hand, a badly frozen outdoor coil produces little to no heating effect. Thus, defrost-cycle controls are best designed to respond only when a defrost cycle is truly needed, and terminate the cycle as soon as the task of removing the ice is complete.

FIGURE 20-14 Defrost board time-selection jumper.

Time-Temperature Initiation and Termination

Many systems rely on compressor operating time to determine when a defrost cycle can be initiated. The operating time on many defrost control boards is selectable at 30, 60, or 90 minutes (**FIGURE 20-14**). In areas that are dry and not so cold, a 90-minute cycle is usually sufficient. In wetter, colder areas, a 30-minute cycle is often required.

At the end of the chosen interval, the timer will allow a defrost cycle to begin, if other conditions have been met. Note again that the timer runs only when the compressor is energized. The system will not initiate a defrost cycle when the compressor is off. Furthermore, if the thermostat stops calling for heat in the midst of a defrost cycle, the compressor will shut off, and the defrost cycle will stop with it. However, once a call for heat returns and the compressor starts, many systems will allow the defrost cycle to continue until it is complete.

Once the timer allows a cycle to occur, a defrost thermostat mounted in a specific location on the outdoor coil takes charge. The defrost thermostat closes at a predetermined, fixed setting—typically somewhere between 25°F and 35°F (−4°C and 2°C). This temperature varies by manufacturer and model. If it is closed, a defrost cycle will be initiated. The reversing valve changes position, the outdoor-fan motor is cycled off, and supplemental heat is energized. If the defrost thermostat is open, the system will return to heat and continue until the selected compressor runtime period elapses once again.

The defrost cycle is terminated in one of two ways. If the defrost thermostat reopens during the cycle, it is an indication that the coil is likely warm enough to have rid itself of ice, so the cycle terminates. The open-setting of a defrost thermostat generally varies between 55°F and 65°F (13°C and 18°C), again depending on the manufacturer and model. With the compressor still running, it returns to the heating mode: the reversing valve changes position, the outdoor fan starts, and control of the supplemental heat returns to the thermostat. If the defrost thermostat does not open, the defrost board will terminate the defrost cycle after a preset amount of time, usually around 10–12 minutes. This can happen when the ice buildup is especially heavy.

In some systems, a thermistor takes the place of the defrost thermostat. Rather than reacting to a switch closure to start the defrost cycle, these units respond to the resistance of the thermistor to determine the coil temperature.

Demand Defrost

Another approach is referred to as demand defrost. Demand defrost systems rely on two temperature sensors: one is mounted on the outdoor coil while the other monitors ambient temperature. The defrost board logic compares the two temperatures and responds based primarily on the difference in temperature between the two. A high temperature differential initiates a defrost cycle, whereas a low temperature differential indicates that a defrost cycle is not needed. The required differential between the two changes as the outdoor temperature changes. The cycle is usually terminated based on the temperature of the coil sensor rising to a predetermined temperature, again with a time limitation on the length of the cycle.

In some cases, this approach also relies on compressor runtime to determine when a cycle can be enabled. Whether this particular approach qualifies as demand defrost is debatable, but some manufacturers do define it as such.

There are many variations in the way that manufacturers approach these two primary defrost control options. Some possible variations are described just below:

- Some heat pumps that compile compressor operating time do not count the time unless the sensor on the outdoor coil is below a predetermined temperature, usually around the freezing mark. The reasoning is that there is no need to compile operating time unless frost has the potential to form.
- At the start of a heating cycle, a surge of cold refrigerant may enter the outdoor coil, but the coil then warms up some as the heating cycle continues. To accommodate this common occurrence, defrost cycles may be prevented if the coil temperature sensor has not been below the predetermined temperature for several minutes. This time requirement helps ensure that a defrost cycle is not falsely initiated by short-term or unusual operating conditions.
- Some systems initiate a defrost cycle after multiple hours of operation have passed and a defrost cycle has not occurred. The reasoning is that the compressor oil, which is constantly in circulation with the refrigerant, may be trapped in the accumulator and other parts of the system after hours of running in the heating mode. Initiating an occasional defrost cycle helps redistribute and move the oil back to the compressor.
- Some systems use a more sophisticated approach that allows the board to use some logic to determine the time between defrosts instead of monitoring only the compressor runtime. The defrost board monitors and graphs the length of time spent in defrost during each cycle. If the cycle times are short, the board logic will allow more time before the next cycle is enabled. If the cycle times are long and/or are terminated by the timer, the time between each defrost cycle will be shortened.

To successfully troubleshoot and repair any heat pump, a technician must know the details of its defrost-cycle operation. Many parts are needlessly replaced due to a lack of system knowledge.

▶ Heating Sequence of Operation

K20005 Describe the sequence of operation for a typical aircooled heat pump.

Other than the addition of the reversing valve, the electrical sequence of operation for a heat pump is much the same as any other air conditioning system. For that reason, only the heating and defrost sequences of operation will be presented here.

This sequence of operation is based on the packaged heat pump wiring diagram shown in **FIGURE 20-15**. The legend for the diagram is shown in **FIGURE 20-16**. This unit is equipped with an electronically commutated motor (ECM).

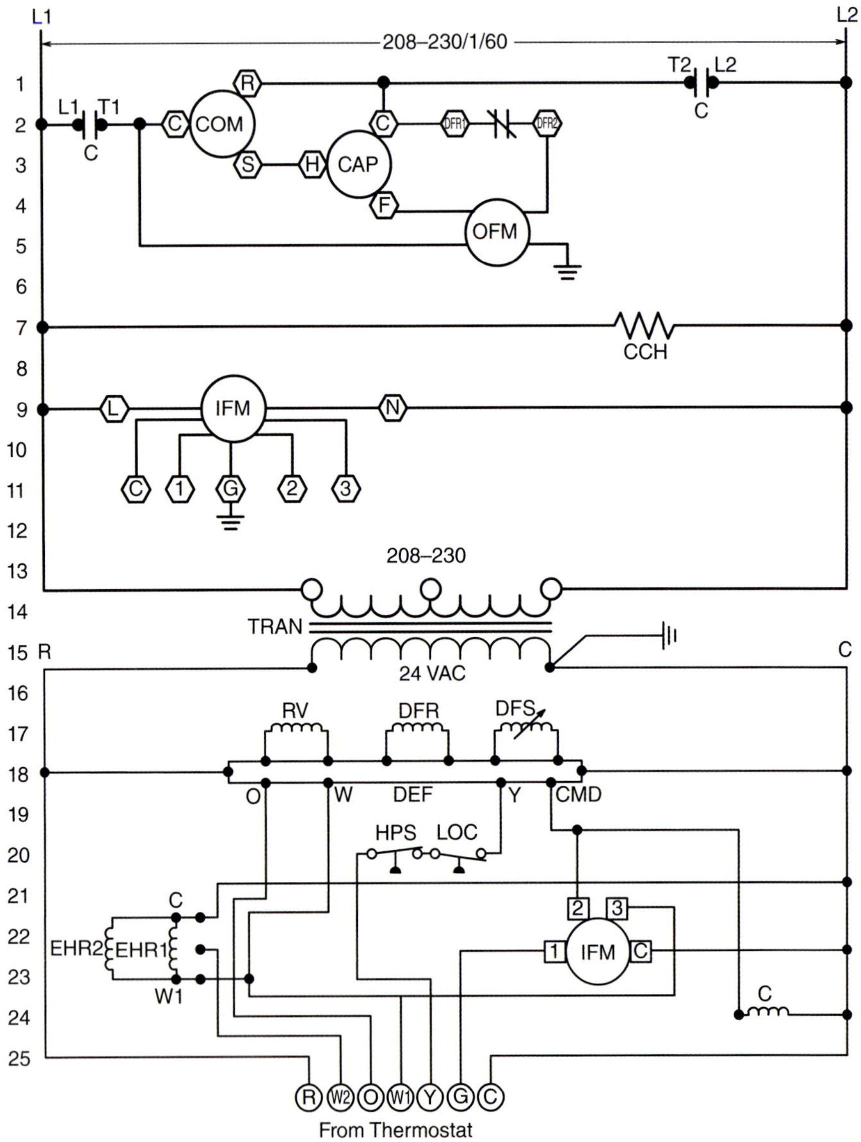

FIGURE 20-15 Heat pump wiring diagram.

Legend
C: Contactor
CAP: Run Capacitor
CCH: Crankcase Heater
COM: Compressor
DEF: Defrost Control Board
DFR: Defrost Relay
DFS: Defrost Sensor
EHR: Electric Heat Relay
HPS: High Pressure Switch
IFM: Indoor Fan Motor
LOC: Loss of Charge Switch
OFM Outdoor Fan Motor
RV: Reversing Valve
TRAN: Control Transformer

FIGURE 20-16 Legend for heat pump diagram.

Note the presence of the EHRs shown at the bottom left of the diagram. These two relays represent the EHRs shown in the earlier wiring diagram for an electric heat package.

This system energizes the reversing valve in the cooling mode. Any time the thermostat is switched to its cooling mode (regardless of the set point and room temperature), the thermostat transfers power from R to O, energizing the reversing valve. With the thermostat switched to the heating mode, the reversing valve remains deenergized.

With the above in mind, the following sequence of operation applies:

1. When the thermostat calls for the first stage of heat, power is transferred from R to both Y and G. Power flows from Y through the normally closed high-pressure switch (HPS) and the loss-of-charge (LOC) pressure switch to the defrost control board. Power is passed through the board to energize the compressor contactor (C), and to Terminal 2 of the indoor-fan motor (IFM).
2. Power on the G terminal from the thermostat energizes Terminal 1 of the ECM. With both Terminals 1 and 2 energized, the ECM responds by running in the higher of the selected speeds (Terminal 2).
3. The energized compressor contactor starts the compressor (COM) and the outdoor-fan motor (OFM).
4. If the space temperature continues to fall, the thermostat energizes the second stage of heat. Power from R is then transferred to W1—power is maintained on both Y and G. Terminal 3 of the IFM is then powered, increasing the speed of the IFM. In addition, electric heat relays 1 and 2 (EHR1 and EHR2) are energized, sending power to the electric heating elements.
5. The system continues to operate in this manner until the second stage of heat has been satisfied at the thermostat. The thermostat then deenergizes W1. The EHRs open, deenergizing the heaters, and Terminal 3 of the IFM is deenergized. This reduces the IFM speed to its previous level.
6. The IFM and OFMs, as well as the compressor, continue to run until the first stage of heat has been satisfied at the thermostat. The thermostat then deenergizes both the Y and G terminals in response. The contactor opens, deenergizing the compressor and OFM. Control power to the IFM is also removed, and the motor shuts down after a programmed two-minute delay.

Defrost Sequence of Operation

1. Whenever the compressor is energized in the heating mode, a solid-state timer on the defrost control board (DEF) compiles the elapsed time. Once the compressor has operated for a predetermined amount of time (30, 60, or 90 minutes, as selected by a jumper on the board), the board logic enables a defrost cycle.
2. If the defrost sensor located on the coil has a resistance value that indicates that the temperature is below 32°F (0°C), the defrost cycle will be initiated. If not, the timer of the defrost board will reset to zero and the system will continue to operate in the heating mode.
3. To initiate the cycle, the board energizes the W terminal, providing control power to the electric heat package and energizing the heaters. The reversing valve is energized, shifting the refrigeration circuit into the cooling mode. The defrost relay (DFR) is energized, opening the normally closed contacts and deenergizing the OFM. These three actions occur simultaneously.
4. When the defrost board senses that the coil temperature, based on the defrost sensor (DFS) resistance, has risen to 60°F (16°C), the defrost cycle terminates. The W terminal, reversing valve, and defrost relay are all deenergized. The board logic will terminate the cycle automatically after 12 minutes if the coil temperature does not rise to 60°F (16°C). The heat pump then returns to its previous operating status.

Note that this is just one example of a defrost-cycle sequence of operation. As mentioned earlier, there are many potential variations.

▶ Analyzing and Troubleshooting Electrical Problems in Heat Pumps

K20006 Explain how to analyze, troubleshoot, and diagnose common heat pump electrical problems.

Although heat pumps are generally reliable, they do experience electrical problems and failures, like all HVACR equipment. General electrical problems include loose, broken, or burned electrical connections, which can result in tripped circuit breakers and blown fuses. These types of problems have been addressed in previous chapters. This section will focus on problems that are unique to heat pumps and how to identify them through troubleshooting.

Troubleshooting Reversing Valves

Reversing valves have one specific function: to shift from one position to another on demand. Other than the potential for leaks that can develop on the valve assembly, and internal leaks that can sometimes occur, problems with reversing valves are generally limited to a failure to move from one position to the other.

The sliding mechanism inside the valve must be well sealed to prevent refrigerant on the high-pressure side from leaking into the low-pressure side. As mentioned earlier, the resistance to movement is too great for a solenoid coil to move the mechanism on its own. The solenoid coil controls a tiny, two-position valve to port high-pressure refrigerant vapor into one end of the valve, and it simultaneously allows vapor to escape from the opposite end. This is important to understand, because a significant pressure difference must exist for the valve to move.

When the compressor is not running to develop a pressure differential, the valve will not move, regardless of whether the solenoid coil is energized or not. Most valves require at least a 100 psi (690 kPa) pressure difference between the high- and low-pressure sides of the system for the valve to move. Although some valves will move with a small differential, most manufacturers do not consider the valve defective unless it refuses to change position at pressure differences greater than 100 psi.

When troubleshooting a reversing valve, the first thing to ensure is that an appropriate pressure difference exists. A low refrigerant charge or a compressor with failing valves, for example, can result in an insufficient pressure difference being developed. If there are no apparent problems with the refrigerant pressures, troubleshooting can proceed to the electrical system and controls.

▶ TECHNICIAN TIP

Reversing-Valve Internal Leakage

Reversing valves can be tested for internal leakage by taking temperature measurements on the two lines through which the cool refrigerant is passing. If the valve is leaking, it leaks hot refrigerant vapor from the compressor discharge into the low-pressure section of the valve, raising the temperature of the cooler vapor. As a rule, the temperature difference measured on the two lines, as shown in **FIGURE 20-17**, should not differ more than 3°F (1.7°C).

A temperature difference of just a few degrees is normal as the valve body is heated by the hot discharge vapor, and slight warming will take place by natural heat transfer. However, the temperature measurements should be taken 5"–6" (13–15 cm) away from the valve body to ensure heat from the valve body traveling through the copper refrigerant line isn't affecting the results. Note that a reversing valve that is leaking badly will mimic a compressor with damaged valves: The compressor will be unable to develop a significant pressure difference.

Reversing Valve Solenoid Coil

First, be certain of the mode—heating or cooling—in which the coil should be energized. Most, but not all, heat pumps energize the valve for cooling mode. This always means the valve must be energized during the defrost cycle. Remember that the valve will not move at all unless the system is running. If the compressor is not running, a faint "click" can usually be heard when the reversing-valve coil is energized, but it may or may not be audible.

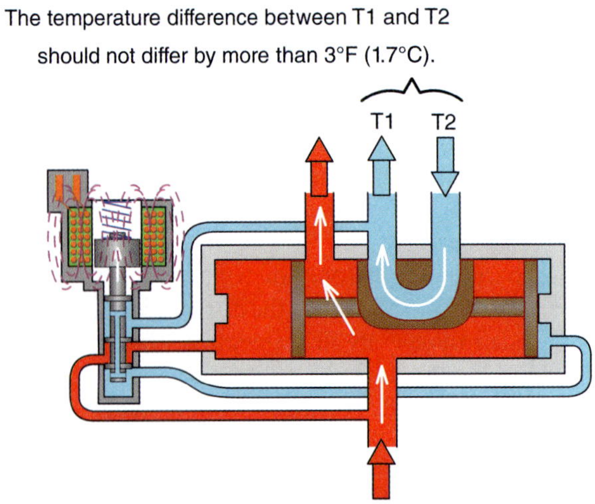

FIGURE 20-17 Temperature test of a reversing valve.

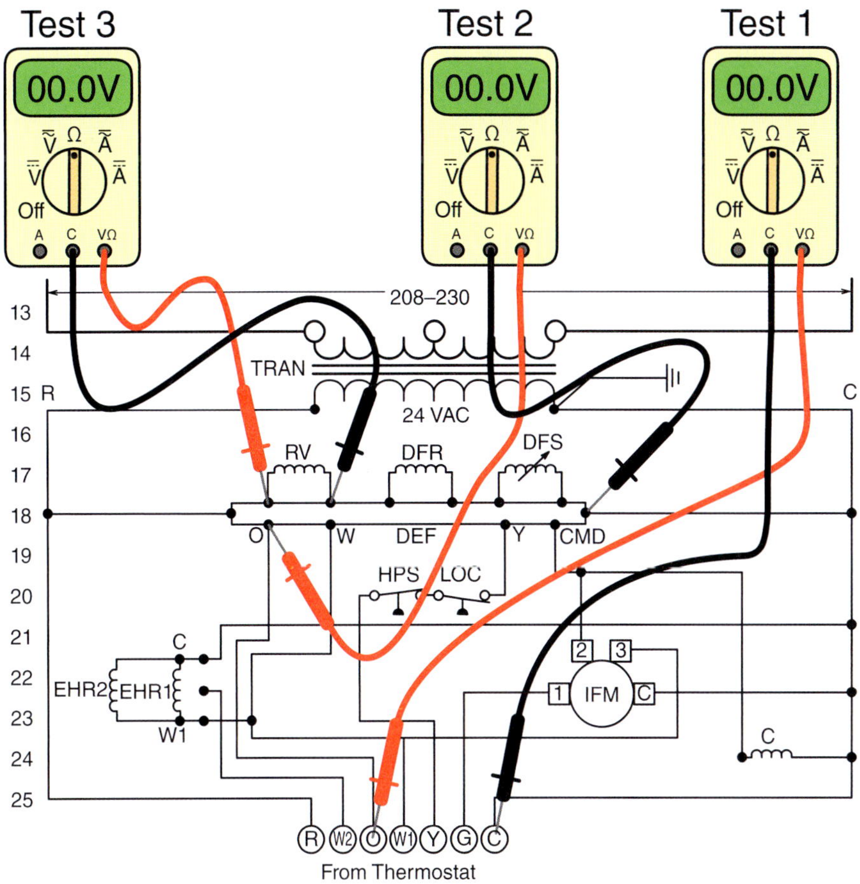

FIGURE 20-18 Checking for power to the coil.

Check for 24 VAC at the coil wiring connections or terminals from which it gets power, in the mode in which it should be energized. **FIGURE 20-18** shows a series of tests and test points, for units that energize the reversing-valve coil in the cooling mode (O for cooling; B for heating). If 24 VAC is not present during the first test, troubleshooting can proceed to the thermostat and the interconnecting wiring.

If power is available at the first set of test points but not the second set of points, there is likely a problem in the unit wiring connections, either in O or Common wiring to the board. Note, though, that if there is a problem with the Common connection to the board, it will be unable to operate at all.

If power is available at both the first and second test points but not the third set, the defrost board is the likely problem since it is not passing the power through. Always completely deenergize the system, including the control power, for several minutes, and test again. Take care to ensure the diagnosis is correct in all cases.

If the reversing-valve solenoid coil is being powered, the next step is to test the integrity of the coil. A quick test can be done by loosening the coil while it is energized. As it is pulled from the armature, a magnetic tug can usually be detected. It can also be removed completely, and a screwdriver can be slid through its center; again, the magnetic field pulling on the screwdriver shaft should be detectable. If the magnetic field is obvious, there is no need to test for continuity; the coil is intact. If not, deenergize the system completely, disconnect the two leads from the coil, and use an ohmmeter to check continuity through the coil (**FIGURE 20-19**).

If there is a sufficient pressure difference in the system and the coil is energized but the valve doesn't shift, try *lightly* tapping on the valve with

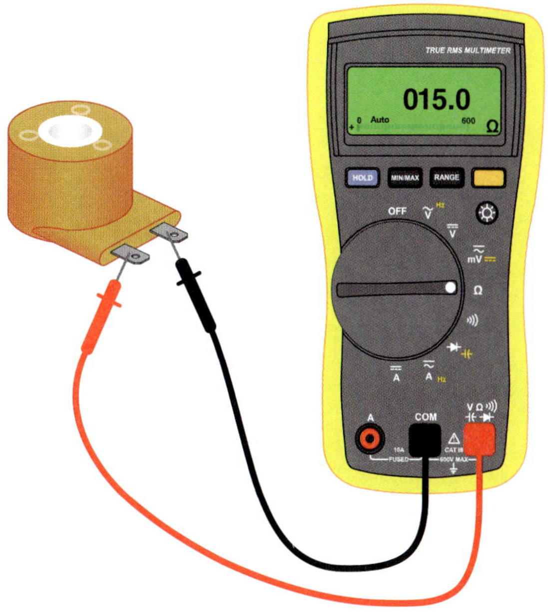

FIGURE 20-19 Testing the reversing-valve coil for continuity.

the insulated portion of a screwdriver. However, do not tap too hard or strike the small capillary tubes, because they are easily damaged and create a restriction. In the end, if the reversing valve refuses to shift and all the required elements are present, the valve must be replaced. Due to the challenging and expensive task of replacing the valve, make sure the diagnosis is correct and try deenergizing and energizing the coil a few times to determine whether the valve will shift. But it must shift reliably for it to be of any value to the user.

▶ TROUBLESHOOTING TIP

"Seeing" the Reversing Valve Move

One trick that some technicians have had success with is to place two small magnets on the side of the reversing valve body (**FIGURE 20-20**). At the two locations internally where the steel portion of the valve is against the brass casing, the magnets should stick to the side of the valve's nonferrous housing. As the valve shifts from one position to the other (which also produces an obvious "whooshing sound"), the two magnets will follow the steel assembly from one end to the other. While this provides an interesting visual, there is rarely any doubt that a reversing valve has moved. Temperature testing, described in a previous feature in this chapter, will show whether the valve hasn't moved fully into one position or the other.

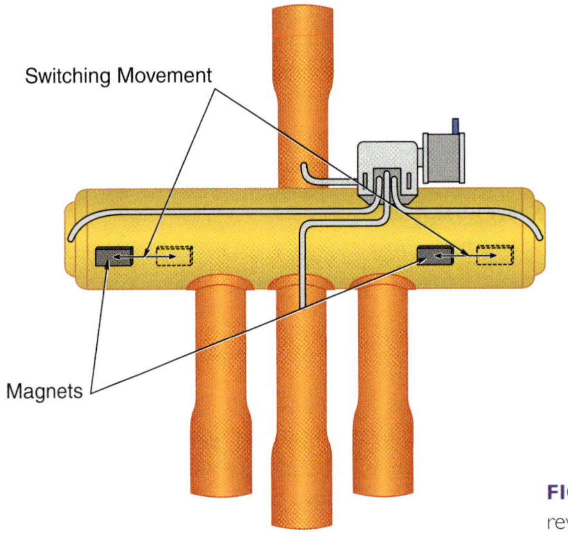

FIGURE 20-20 Magnet movement on a reversing valve body.

FIGURE 20-21 Photo of a Typical Defrost Thermostat.

Defrost Component Testing and Troubleshooting

One of the most beneficial features built into defrost control boards is its defrost-test pins. (**FIGURE 20-21**). Manufacturers provide these pins to enable technicians to quickly initiate a defrost cycle at any time. For many products, a small screwdriver or similar object is used to short the two pins together while the unit is running, usually for three to five seconds to ensure the contact is not accidental. The unit will then enter a defrost cycle, regardless of the coil and ambient temperature sensors or thermostats. Most units like this are designed to speed up the entire defrost cycle when in the test mode, so that the defrost cycle terminates faster than the normal 10–12 minutes.

If a required step is not occurring during the test mode, some troubleshooting can be done while it is in the test mode. For example, if the defrost cycle appears to initiate because the

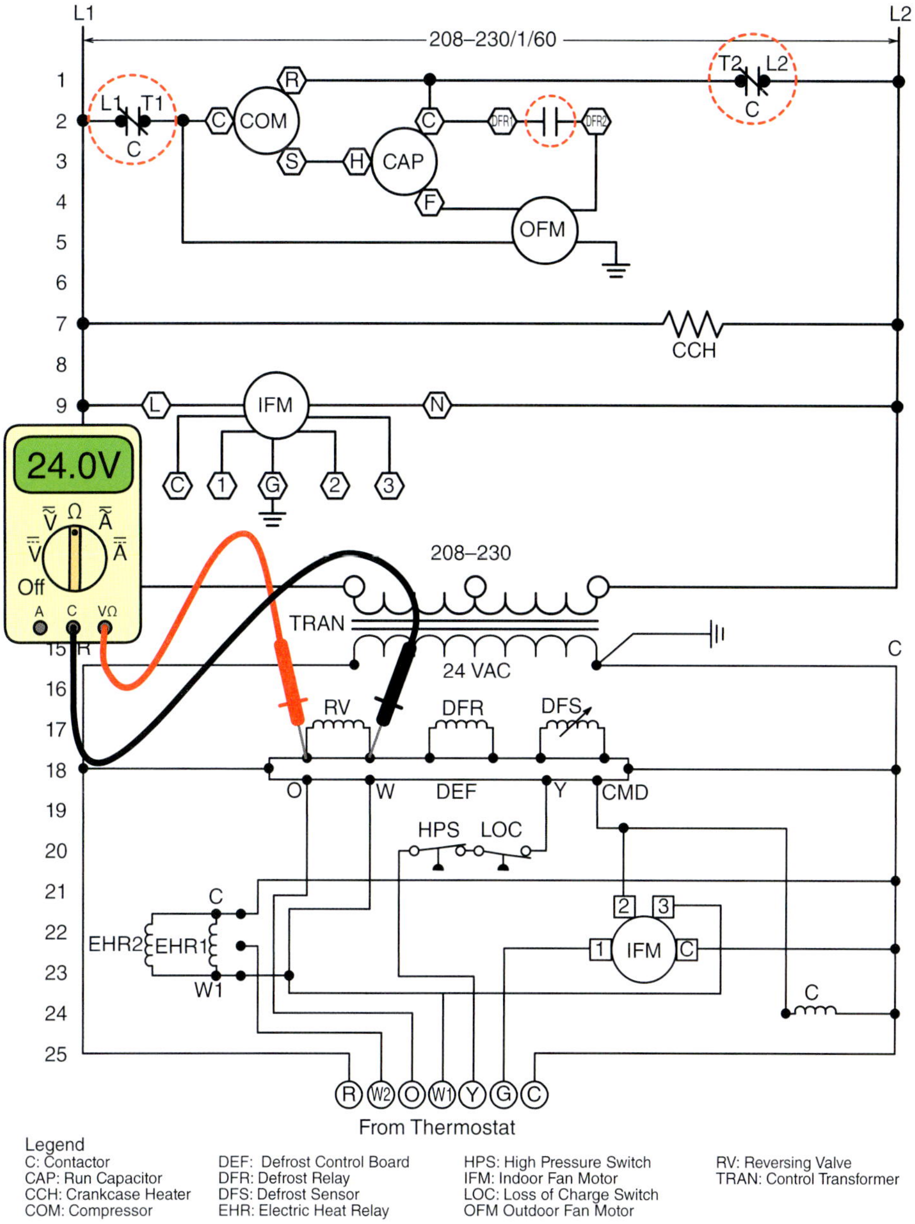

FIGURE 20-22 Checking reversing valve outputs while the unit is in defrost.

outdoor fan stops but the reversing valve does not shift, check for 24 VAC at the board outputs to the reversing-valve coil (or the lack of power, for units that energize the valve in the heating mode) as shown in **FIGURE 20-22**. If 24 VAC is available, troubleshooting proceeds to the reversing-valve coil itself.

▶ TROUBLESHOOTING TIP

Using a Meter at Board-Connecting Plugs

Many types of printed circuit (PC) boards, including defrost boards, have plugs to make factory wiring easier and more reliable (**FIGURE 20-23**). Many have open spaces on the side that expose a conductive surface for each pin. These spaces are used during testing, allowing access for meter probes.

FIGURE 20-23 Accessible board plugs.

FIGURE 20-24 Outdoor-fan relay on a defrost board.

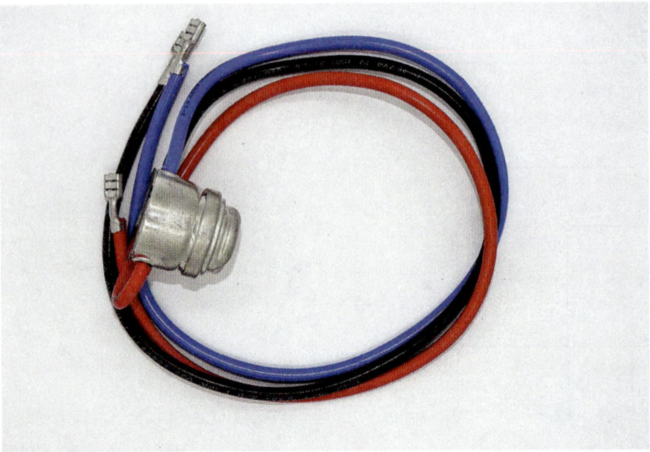

FIGURE 20-25 Typical defrost thermostat.

While in the test mode, the reversing valve may shift but the outdoor fan will continue to run. In this case, since the outdoor-fan relay is typically board-mounted (**FIGURE 20-24**), it is difficult or impossible to check for power to the relay coil. The board may be energizing it when it shouldn't, or the contacts may be stuck closed. If this continues through another test cycle, the board obviously has a defect and must be replaced—board-mounted relays cannot be replaced in the field. If the outdoor-fan relay does happen to be a separate component, the relay coil and contacts can easily be tested. A failed outdoor-fan relay can certainly be the source of this problem.

For units with a test mode that functions like the one just described, the process does not test the defrost-related thermostats or sensors. In most cases, shorting the defrost-test pins permits the system to ignore the defrost thermostat or sensor, allowing the technician to observe the defrost cycle in action regardless of its status.

Testing Defrost Thermostats

A defrost thermostat (**FIGURE 20-25**) is a simple switch that is either open or closed. It is clamped to an outdoor coil line or U-bend, allowing it to accurately sense the coil temperature. When the heat pump is running, it is primarily influenced by the temperature of the refrigerant flowing through the tube.

Most defrost thermostats for heat pumps close somewhere between 25°F–35°F (–4°C–2°C) and reopen at 55°F–65°F (13°C–18°C). They close on a rise and open on a fall in temperature. A closed thermostat allows a defrost cycle to initiate, while an open thermostat prevents or terminates the cycle. The characteristics of a particular thermostat can be determined from the equipment manufacturer's documentation or from the part itself. Many thermostats have the information printed on them, although it can be a little misleading. The pattern for its temperature characteristics is usually something like "L65-35F." These numbers mean that the switch opens at 65°F (18°C) and closes when the temperature differential is 35°F (1.7°C). Thus, this thermostat closes at 30°F (–1°C).

It is simple enough to test a defrost thermostat with the power off by using an ohmmeter to determine whether the switch is open or closed. If the temperature at its location

▶ TECHNICIAN TIP

A Bulging Defrost Thermostat

A technician may occasionally encounter a thermostat that is bulging (**FIGURE 20-26**). This occurs when the seal around the thermostat body fails, allowing water and/or water vapor to enter. As the water goes through freezing and thawing cycles, it pushes the potting compound up and out of the casing. A thermostat in this condition needs to be replaced, if it hasn't already failed.

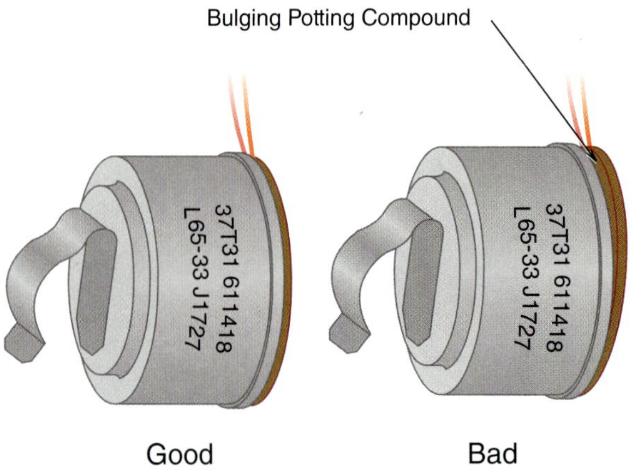

FIGURE 20-26 A bulging defrost thermostat.

is below the design closing temperature and it remains open, it should be replaced. However, that is not always the case when a technician is on-site. To determine whether it opens and closes at the appropriate temperatures, it must be exposed to those temperatures. Thus, thoroughly testing it usually requires that it be disconnected and removed from the unit.

To get an idea of the temperature at which it closes (the low temperature), the switch can be placed in a freezer. A solution of cracked ice and water could work, but remember that such a solution will not be colder than 32°F (0°C). In addition, although the switch should be waterproof, soaking it in water is not a great idea.

Attach a bead-style or similar thermocouple probe to the defrost thermostat as shown in **FIGURE 20-27**, ensuring that it makes good contact with the metal casing. Using a very small ball of soft putty, such as **duct seal**, is ideal. The bead must make good contact with the metal part of the thermostat casing where shown. Then plug it into the digital thermometer. Note that the thermocouple will likely react to temperature change faster than the bimetal switch mechanism inside the defrost thermostat. The result is that the temperature reading on the thermometer will likely read several degrees lower than the design closing temperature before the switch closes. Consider this during the test. But using duct seal as opposed to tape or a clamp, such as a clothespin, will help. Without laboratory-style test chambers and equipment, this is the best that can be done under field conditions.

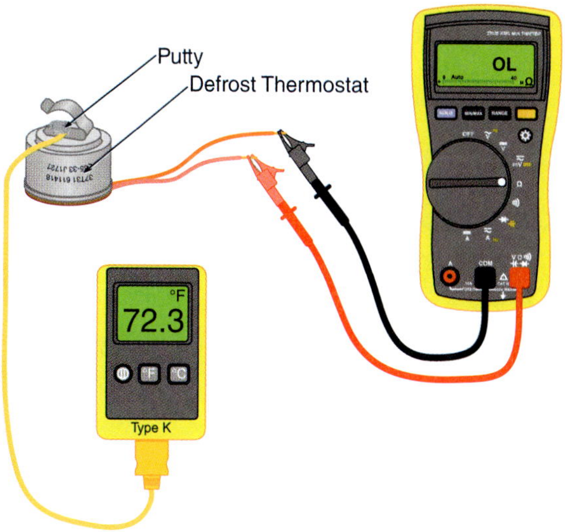

FIGURE 20-27 Thermocouple bead probe attached to a defrost thermostat.

Carefully place the assembly in the freezer to avoid disturbing the attached probe. An ohmmeter is also needed, with the probes attached to the two defrost thermostat leads with alligator clips. Watch the temperature fall on the thermometer while also monitoring the ohmmeter to see when the switch closes.

Once it has been closed, it will be a little less awkward to observe the temperature at which it opens. Remove it from the freezer, and place it in an area where it should warm up to the design opening temperature in a reasonable amount of time. Using a heat gun or a similar tool will usually result in false readings unless it's done with great care. Natural heat is best. When the ohmmeter shows an open circuit, the thermostat has opened.

Again, a defrost thermostat is a simple switch that is either open or closed. The defrost board, based on the accumulated runtime of the compressor and the time setting on the board, initiates or terminates a defrost cycle based on the switch position. When a defrost thermostat is found to be defective (will not close) in the cold of winter, it can be removed and a jumper temporarily placed across the terminals. This will allow the system to enter defrost each time the necessary compressor runtime has elapsed, but each defrost cycle will run the maximum length and be terminated by the timer since the defrost thermostat circuit will remain closed. Still, it is an option to provide heat pump operation to a customer until a new defrost thermostat is acquired and installed.

Testing Thermistors

Defrost boards may use one or more temperature thermistors, rather than a defrost thermostat, to initiate and terminate the defrost cycle. Thermistors were introduced in Chapter 12. They respond to temperature changes by changing their resistance in a consistent and predictable way. Thermistors are easier to test than defrost thermostats and are generally more reliable.

To test a thermistor, disconnect it from the defrost board. Use an accurate digital thermometer to measure the temperature at the location of the thermistor, and then measure the resistance through it. Compare the resistance to the appropriate graph for the thermistor in use. An example of a thermistor chart is shown in **FIGURE 20-28**. They may also be provided in chart form as opposed to a graph. If a thermistor is

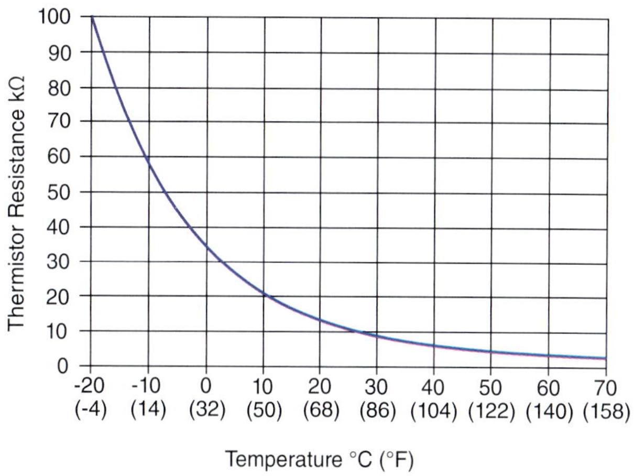

FIGURE 20-28 Example thermistor graph.

accurate at the tested temperature, it is likely accurate through the rest of its needed operating range.

▶ Analyzing and Troubleshooting Electric Heat Packages

K20007 Explain how to analyze, troubleshoot, and diagnose common electric heat problems.

Common electric heat packages are relatively simple to troubleshoot. Since the primary safety devices are typically in the line-voltage power circuit to the heating elements, the control circuit has few devices.

Troubleshooting Electric Heater Controls

For most heat pumps, the electric heat is enabled through the W/W1 circuit. Power to the circuit can originate from the thermostat when the second stage of heat is called upon (remember that Y starts the compressor and represents the first stage of heat) or from the defrost control board. Additional stages of electric heat are usually controlled by W2. One or more stages of electric heat may be interlocked with outdoor thermostats or thermistors that prohibit operation above a certain temperature to save energy.

When the electric heat doesn't appear to be working at all, first ensure that the thermostat is set to Heat mode and that the set point is well below room temperature—at least 3°F–4°F (1.6°C–2.2°C).

At the electric heat package, verify that line-voltage power is available (**FIGURE 20-29**). Remember, though, that power for the fan-coil unit control transformer and blower also originates from this point. If power was not available here, there would be no control power or system activity at all.

Next, check for the presence of 24 VAC power between Common and W1. For the heat package shown, if 24 VAC is present at the heat-relay coils, they should be energized, and their contacts should be closed. With power applied, carefully check the voltage across the two contacts (**FIGURE 20-30**). Since the heat relays are breaking both legs of power to the heating element, there should be line voltage at those two points. If not, one or both relay contacts are not closed.

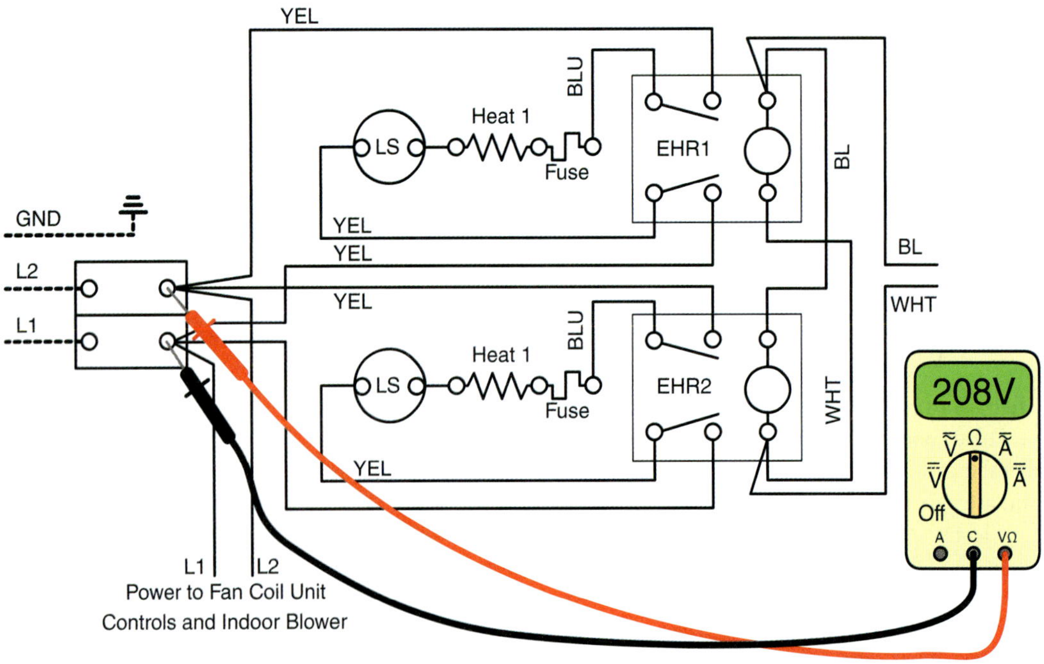

FIGURE 20-29 Checking for control power to the electric heat package.

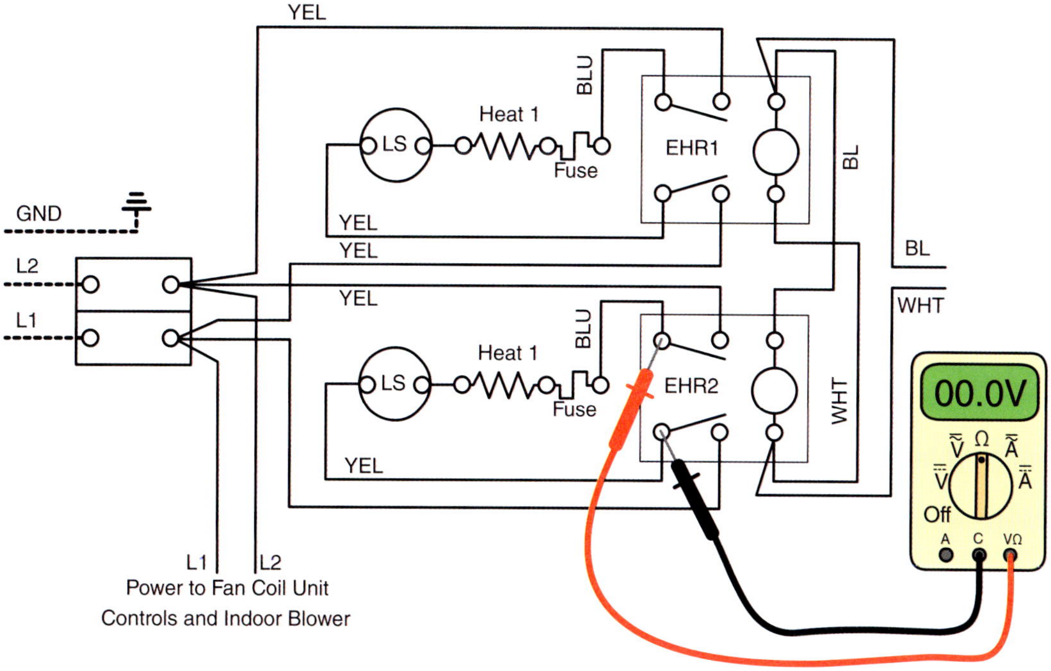

FIGURE 20-30 Checking the heat-relay contacts.

Once it is known that one or more heat-relay coils are energized with 24 VAC and their contacts are not closing, their failure can be confirmed by deenergizing all power and checking the coils for continuity. It is also possible that their coils are intact but that the contacts have been burned or otherwise damaged. Heat relays pass a considerable amount of current, although they do not experience increased inrush current when energized. The heating elements are a resistive load rather than an inductive load.

Troubleshooting Electric Heater Circuits

When the heat relays are found to be intact and functional, with line-voltage power found where indicated in Figure 20-30, troubleshooting can proceed to the heating element circuit. As previously noted, the two safety devices—the limit switch and fusible link—are wired into the power circuit rather than the control circuit.

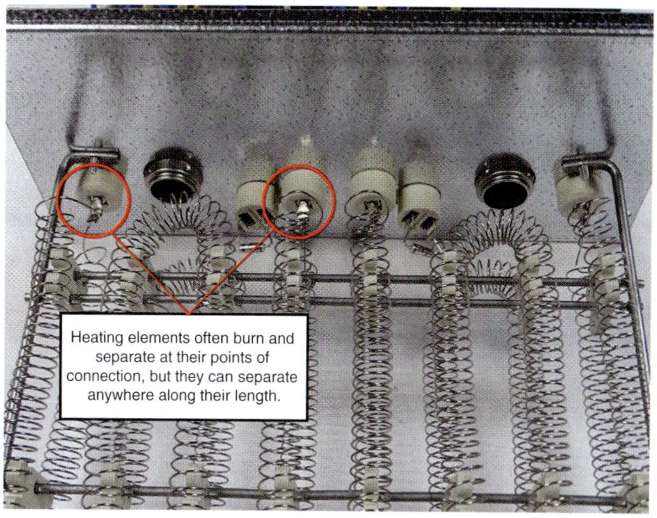

FIGURE 20-31 Failure points of heating elements.

When possible, visually examine the heating elements. Heating elements can fail and open at their points of connection or anywhere else along their length (**FIGURE 20-31**). However, the elements usually cannot be seen without removing the entire package from the unit or disconnecting the supply-air ductwork. For troubleshooting purposes, neither of these steps is necessary.

The best way to troubleshoot these circuits is by using the hopscotch method described in Chapter 17 (**FIGURE 20-32**). With the system deenergized, position and maintain one probe (the red probe shown here) at a single point while jumping from point to point with the other probe. If the ohmmeter displays an open circuit at any point, the open control or heating element has been identified.

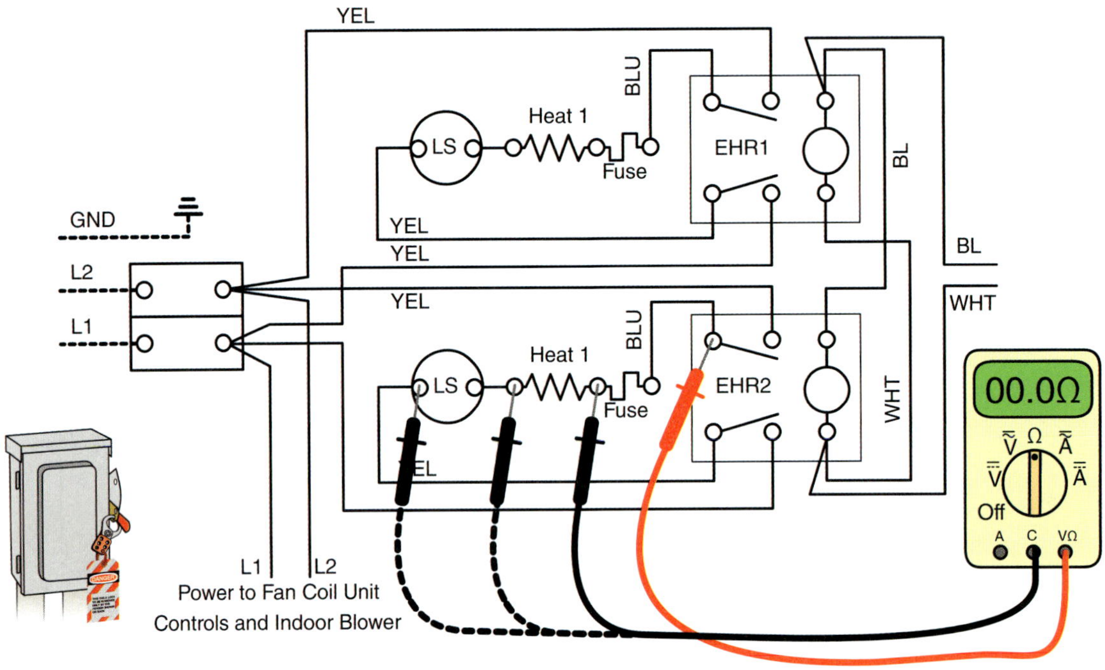

FIGURE 20-32 Hopscotch testing an electric heater circuit.

▶ TECHNICIAN TIP

Replacing Electric Heating Elements

Individual heating elements can be replaced when necessary. Replacement elements (**FIGURE 20-33**) can be acquired from most supply houses. Most new elements are sold to be a specific length in the close-coiled state, such as 24" (61 cm). First, measure the full extended length of the existing heating element. Then divide the length by three to find the ideal length of replacement element. For example, if the failed element measures 66" (168 cm), dividing by three results in 22" (56 cm). That would be the ideal length of the new, tightly coiled element. During the installation, they are carefully stretched to fit. In general, though, heating elements can replace existing elements that are two to five times their close-coiled state. Also choose a kit that provides the desired wattage at the voltage supplied to the system. Then follow the kit directions for installation, ensuring that no part of the element is too close to the metal frame. All the ceramic insulators, including those at the points of connection, must be in good condition as well. Broken insulators can allow the heating element to make contact with the metal part of the frame. Many, but not all, kits come with a replacement fusible link.

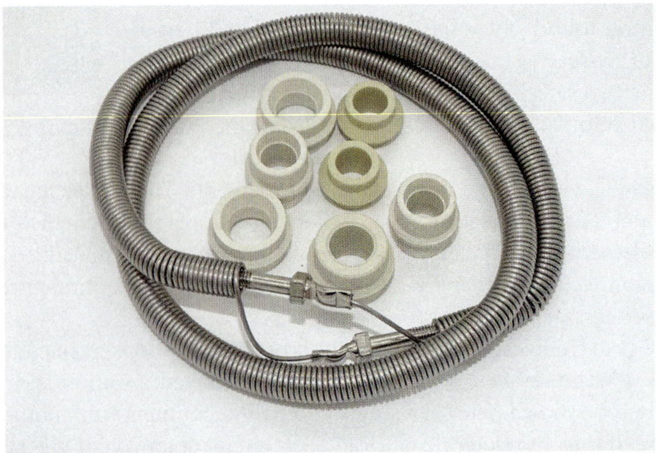

FIGURE 20-33 Replacement heating element.

▶ Wrap-Up

Summary

As with standard air conditioning units, it is important to understand the basic refrigeration cycle of heat pumps when analyzing and troubleshooting electrical problems.

Heat pumps also require some form of supplemental heat, provided by simple electric heat packages or by a fossil-fuel furnace in most cases. Electric heat systems and components are relatively simple, requiring only a few simple controls and safety devices to provide a reliable source of supplemental heat. In addition, electric heat packages can provide heat when the heat pump requires service, although it may not be enough to provide total comfort on the coldest days.

The primary difference between heat pumps and standard air conditioners is the reversing valve and its ability to reverse the flow of refrigerant. Heat pumps that provide heat in temperatures below freezing must also be able to defrost the outdoor coil. This requires them to operate in the cooling mode while the electric heat tempers the air inside. Since the defrost cycle works against the goal of providing heat in the winter, the number of cycles and their length must be minimized. Defrost controls are continually being improved and are much more effective than they were in the early days of heat pumps.

When analyzing and troubleshooting heat pumps and electric heating components, remember to use strategic troubleshooting techniques. Developing and implementing these techniques on each and every service call will help pave the way for success in troubleshooting larger and more complex systems in the future.

Key Terms

balance point The temperature at which the heat loss of a structure equals the heating capacity of a heat pump.

dual fuel systems Heat pump systems that are paired with fossil-fuel furnaces that provide heat when the temperature falls below the balance point and the heat pump itself is shut down.

duct seal A soft, gray putty, much like clay, that is used as a pliable sealant.

pilot-operated valve A valve that depends on a secondary valve to control a source of pressure used to position it.

reversing valve A multi-ported valve used to reverse the direction of refrigerant flow through the indoor and outdoor coils of a heat pump.

Review Questions

1. A heat pump can be described as a __________.
   a. capacitive heating system
   b. reverse-cycle air conditioner
   c. resistive heat source
   d. fossil-fuel heating system
2. The primary component that sets a heat pump apart from a conventional air conditioning unit is the _________.
   a. indoor coil
   b. thermostat
   c. reversing valve
   d. capillary tube
3. Some manufacturers may refer to a heat pump accumulator as a _____.
   a. receiver
   b. muffler
   c. heat exchanger
   d. charge compensator
4. A valve that relies on the operation of another, smaller valve to provide pressure for its movement is referred to as a __________.
   a. pilot-operated valve
   b. solenoid valve
   c. three-way valve
   d. mixing valve
5. In most cases, the reversing valve of a heat pump is energized in the ________ mode.
   a. heating
   b. cooling
6. When a typical heat pump thermostat is switched to the cooling mode, which of the following terminals related to the reversing valve is energized?
   a. Y2
   b. B
   c. O
   d. A
7. Which of the following statements about heat pumps is true?
   a. The reversing valve should never be energized while the compressor is running.
   b. Water-source heat pumps are always less effective than air-source models.
   c. The outdoor fan must continue to run during the defrost cycle.
   d. The defrost cycle works directly against the efficiency of a heat pump.
8. The heat pump alone has the capacity to handle the heating needs of the structure when the outdoor temperature is above the __________.
   a. ambient temperature
   b. balance point
   c. dew point
   d. indoor temperature
9. A 1,000-watt electric heating element produces _____.
   a. 3,140 Btu
   b. 3,413 Btu
   c. 12,818 Btu
   d. 17,065 Btu
10. In an electric heating package, a fusible link (FUSL) that melts at an excessive temperature is placed in the _________ circuit.
    a. power
    b. control

11. Which one of the following statements is true?
    a. A defrost cycle usually continues even if the room thermostat stops calling for heat, forcing the compressor to continue running.
    b. Defrost cycles usually last at least 30 minutes.
    c. Defrost thermostats typically close between 25°F and 35°F (−4°C and 2°C).
    d. Defrost thermostats typically open between 35°F and 45°F (2°C and 7°C).
12. Some systems also occasionally initiate a defrost cycle to ________________.
    a. help move the oil from other areas of the system and return it to the compressor
    b. make sure that ice never begins to form on the outdoor coil
    c. prevent the reversing valve from becoming stuck in one position
    d. recalibrate any sensors attached to the defrost board
13. In the heating mode, a typical heat pump thermostat energizes the compressor by sending control voltage out on the __________.
    a. B terminal
    b. W terminal
    c. Y terminal
    d. O terminal
14. During the heating mode with the heat pump running, if the room temperature continues to fall, the thermostat will ______________.
    a. stop the heat pump and energize the electric heat
    b. switch to Emergency Heat mode, since the heat pump is not keeping up
    c. disable the defrost function until the room temperature begins to rise
    d. energize the second stage of heat
15. When testing a reversing valve for internal leakage, the two temperature readings should be taken ________________.
    a. on the surface of the reversing valve body
    b. at the compressor
    c. on the body of the pilot valve
    d. about 5"–6" (13–15 cm) away from the reversing valve body
16. To move from one position to the other, most reversing valves require ____________.
    a. at least a 250 psi (1,724 kPa) refrigerant pressure difference
    b. at least a 100 psi (690 kPa) refrigerant pressure difference
    c. the compressor to be shut down first
    d. the electric heat to be energized first
17. The test pins on a defrost board allow a technician to ______________.
    a. initiate a defrost cycle at any time
    b. temporarily disable the compressor
    c. determine whether the defrost thermostat is open or closed
    d. determine whether the preset compressor runtime has elapsed
18. A defrost thermostat with the numbers "L60-35F" opens and closes at what temperatures?
    a. Closes at 60°F and opens at 35°F
    b. Closes at 35°F and opens at 60°F
    c. Opens at 60°F and closes at 25°F
    d. Opens at 25°F and closes at 60°F
19. Thermistors respond to temperature consistently and predictably by ______.
    a. changing the voltage applied
    b. changing the amount of power they consume
    c. becoming warmer as the temperature gets colder
    d. changing their resistance value
20. The best way to test for continuity through an electric heating element circuit that contains multiple safety switches is to use the __________ method.

GLOSSARY

absolute pressure The actual pressure exerted by a fluid with reference to a vacuum (zero pressure). Atmospheric pressure + gauge pressure = absolute pressure.

active listening A personal communication approach that requires a listener to display an obvious commitment to understanding and to clearly respond to the input of a speaker.

algorithms A set of programmed mathematical rules used in calculations by computers or processors.

alternating current (AC) Electric current that reverses its direction many times per second at a consistent interval. The interval is referred to as its frequency, which is typically 50 or 60 times per second. AC frequency is generally standardized by each country.

ambient Relating to the immediate surroundings or environment.

American Wire Gauge (AWG) system A system that assigns sizes to electrical wire, based on the cross-sectional area of the wire.

ampacity A measure of the maximum current that a conductor can carry without risking its immediate or progressive damage or failure.

ampere The unit of measure for electrical current, usually shortened to amp. One amp of current is equal to the current produced when 1 volt is applied to a resistance of 1 ohm. Amperes are typically represented in electrical formulas by the letter I, but by the capital letter A in other contexts.

amplifiers Electronic devices used to boost the amplitude of a voltage.

analog multimeters Multimeters that use a moving pointer or needle to display electrical values such as volts, amps, and ohms.

annual fuel utilization efficiency (AFUE) A measure of thermal efficiency in fossil-fuel furnaces and boilers that represents the season-long average efficiency of the equipment in delivering heat for the intended use. Heat lost through the vent or chimney, for example, reduces the AFUE rating since it does not contribute to the intended use.

arc An electric current that flows across an open space between two points, releasing heat and bright light in the process.

arc blast An intense explosion that occurs during an arc-flash incident.

arc fault A high-energy electrical discharge between two or more conductors.

arc flash A dangerous flash of light caused by the enormous release of thermal energy in an electric arc.

arc-flash boundary A boundary set at a distance from exposed energized electrical conductors or circuit parts; persons within the boundary could receive second-degree burns if an electrical arc flash were to occur.

arc-flash hazard analysis Investigation and documentation of a worker's potential exposure to arc-flash energy, which is conducted to prevent injury by determining safe work practices and appropriate levels of PPE.

armature Rotating component of a DC-powered electric motor.

auxiliary contacts Electrical contacts that are connected to and operate at the same time as the main contacts of a contactor or motor starter but provide switching for circuits other than the main power supply to the load.

balance point The temperature at which the heat loss of a structure equals the heating capacity of a heat pump.

base load The minimum level of electrical power demand required from a power system over a 24-hour period.

bellows In pressure-measuring instruments, a thin, enclosed, flexible, pleated cylinder that lengthens and shortens with changes of internal pressure.

bimetal strip A narrow strip composed of two layered dissimilar metals. When heated, the metals expand at different rates, so the strip will bend or straighten as the temperature changes.

biomass Plant materials, such as trees, vegetation, and animal waste, used as a fuel source.

Bourdon tubes C-shaped tubes closed on one end and attached to a source of pressure. Increased internal pressure causes the tube to straighten. The motion can be linked to an indicating or control mechanism.

breakover voltage The positive voltage at which a diode switches into the conductive state.

brushes Carbon components used to transfer incoming DC voltage to the commutator of a DC motor.

brushless A descriptor for motor construction that eliminates the need for brushes.

bus A thick strip of metal, often copper, that serves as a common distribution point for electricity; a bus is also referred to as a bus bar.

cad cell A device containing cadmium sulfide that responds to the presence of light by significantly reducing the resistance of its circuitry; it is used to prove the presence of flame in an oil furnace.

capacitance A measure, in farads, of the electric charge that a capacitor can store.

capacitor An electrical device designed to store electrical energy in the form of an electric field.

capillary tubes Small-diameter flexible tubing, usually made of copper, used for providing pressure inputs to instrumentation; named for the very tiny blood-carrying capillaries in humans and other animals.

centrifugal switch An electric switch that opens or closes due to the application of centrifugal force generated by a spinning motor or shaft.

chillers A refrigeration apparatus designed to chill water for cooling applications.

circuit A closed electrical path that includes one or more components or a portion of the circuit that consumes electric power.

clear language display A display that provides clearly understood language to guide the user, as opposed to codes or other indications that must be interpreted.

commutator A split ring attached to the armature of a DC motor. It causes the DC voltage to reverse direction every 180 degrees, simulating an AC voltage.

condensers Coils of tubing containing vapor condensing into a liquid, typically transferring heat to air or water.

condensing furnaces Furnaces offering exceptional efficiency through the use of a sealed combustion chamber and a secondary heat exchanger to extract the maximum amount of heat.

conductor A material through which it is relatively easy to maintain an electric current.

conduit Piping or tubing, usually steel, aluminum, or plastic, which is used to route and protect electrical wiring.

connection diagrams Electrical diagrams that represent circuit components as intact devices and how the interconnecting wiring is arranged.

constant-airflow ECM An ECM with exceptional flexibility in its programming, allowing it to maintain a consistent volume of airflow by managing the torque applied to the blower shaft, thereby managing rotational speed and providing the same airflow even as the resistance to airflow changes.

constant-speed ECM ECMs that run at a predetermined and programmed speed. They are most commonly used in axial fans in various applications.

constant-torque ECM An ECM that is designed to produce a given torque value on demand.

contactor An electrically controlled switch to control line voltage power flow to devices that represent a significant electrical load.

continuity As it pertains to electrical circuits, a state of continuity indicates a complete, unbroken electrical path. However, some amount of resistance to electrical current is typically present.

core In a transformer, the material around which a winding is wrapped. Most transformers contain laminated metal cores, but some use air or other materials.

coulomb An SI (International System) unit of electrical charge, equal to the power conveyed in one second by a current of 1 ampere. A capacitor with a capacitance of 1 farad can store 1 coulomb of charge on its plates when a potential of one volt is applied across its terminals.

counter-electromotive force (CEMF) An electromotive force, commonly called back voltage or back EMF, that is created in a motor and that opposes the electromotive force being applied by the circuit.

crankcase heater (CCH) A relatively low-wattage, electric resistance heater used to warm a compressor to discourage the migration and condensation of refrigerant in the crankcase.

current The flow of electrically charged electron particles.

current imbalance An imbalance in the current consumption of the three windings in a three-phase load. A current imbalance up to 10% is typically allowed before motor damage is likely to result.

current relay A relay with a low-resistance coil and a set of normally open (NO) contacts that responds to the passage of current through its coil.

current transformer (CT) A device that produces alternating current in a secondary winding that is proportional to the current in a primary winding.

cycle rate The number of heating or cooling cycles that occur per hour. Cycle-rate limitations can be established in most digital thermostats since they have no resistor for heating and cooling anticipation.

deadband A neutral zone between heating and cooling set points where no action occurs. For an auto-changeover mode setting, a deadband is needed to ensure the system does not rapidly alternate back and forth between modes to maintain a set point.

defibrillation The process of administering a controlled electrical shock to stop fibrillation and restore the normal heart rhythm.

defibrillator A device used to help the heart return to its normal rhythm by applying a controlled electric shock to the chest wall or heart.

definite-purpose contactor A magnetic contactor designed and built for use primarily in HVACR equipment, where frequent cycling of high-current loads is common.

delayed ignition The result of excess gas buildup inside the burner compartment before successful ignition, causing a somewhat violent detonation.

derated The quality of having the current-carrying capacity of a wire or wiring device reduced due to unusual circumstances.

dewpoint The temperature representing 100% saturation for a vapor, below which the vapor begins to condense into its liquid state.

diac A semiconductor diode that conducts electricity only after its breakover voltage has been reached. The term "diac" is an abbreviation of diode for alternating current.

diaphragm In pressure-measuring instruments, a thin, flexible disk that separates two sides of a chamber. Differences of pressure across the disk cause it to deflect in the direction of the lower pressure.

dielectric A non-conductive material positioned between the two conductive plates of a capacitor that store an electrostatic charge.

digital multimeters (DMMs) Multimeters that use a digital display of numbers and symbols to present electrical values such as volts, amps, and ohms.

DIN rail A metal rail constructed to one of several standards for size, designed to accommodate the mounting of electrical devices in an organized manner.

diodes Semiconductor devices containing N-type and P-type semiconductor material that allows current to pass only when a positive voltage is applied to the anode.

direct-ignition system A gas furnace that ignites the main burner directly, rather than lighting a pilot flame that subsequently lights the main burner.

disconnect A safety switching device used to enable and disable power to a load, usually located adjacent to the equipment. A disconnect means must be provided on, or within sight of, HVACR equipment.

doping The intentional introduction of impurities to a pure semiconductor to change its electrical properties.

dropout voltage The voltage value applied to a relay coil at which the contacts must return to their deenergized position.

dual fuel systems Heat pump systems that are paired with fossil-fuel furnaces that provide heat when the temperature falls below the balance point and the heat pump itself is shut down.

duct seal A soft, gray putty, much like clay, that is used as a pliable sealant.

eddy currents Circular electrical currents generated inside conductors due to the action of changing magnetic fields surrounding them. These currents sap energy from the magnetic field, reducing its strength.

electrically safe work condition (ESWC) A state in which the conductor or circuit part to be worked on or near has been disconnected from energized parts, locked/tagged in accordance with established standards, tested to ensure the absence of voltage, and grounded if necessary.

electrocution Death that results from being exposed to a lethal amount of electrical energy.

electrode A device that carries a high voltage to its tip, where an arc is initiated to ignite an air-fuel mixture. The conductive material used for the igniter is virtually unaffected by the arc, although it may experience wear after long usage.

electrostatic charge An electric charge that develops on the surface of an insulated body, such as the metal plates inside a capacitor.

electrostatic discharge (ESD) The release of accumulated static electricity that occurs when two objects touch. If not discharged to ground before handling solid-state components, ESD can damage or destroy electronic components.

encapsulated Relative to switches and electrical devices, an encapsulated device is enclosed or contained in a sealed housing.

evaporators Coils of tubing containing liquid boiling to a vapor, typically absorbing heat from air or water.

excitation voltage The voltage applied to motor windings to create the necessary magnetic field.

external static pressure (ESP) The total of static pressure losses created by ductwork, filters, coils, and other components in an air-distribution system.

farad A standard unit of electrical capacitance.

fault current Any abnormal electrical current, such as that developed in a short circuit.

ferromagnetic metal Any metal that can be easily magnetized, such as iron, due to the presence of unpaired electrons. Naturally ferromagnetic elements include iron, cobalt, and nickel.

ferrous metal Any metal containing iron, such as steel or cast iron. Aluminum is an example of a nonferrous metal.

fibrillation Very rapid irregular contractions of the muscle fibers of the heart that result in the heartbeat and pulse going out of rhythm with each other.

flame rectification The process in which the flame in a furnace changes an alternating current (AC) to direct current (DC); it is used to verify the presence of a stable flame.

flame rod The rod bathed in the flame used on a gas furnace to direct the rectified current back to the operating controls.

flame-retardant Substances that use a chemical reaction to resist or inhibit the spread of fire.

fossil fuels Fuels such as coal, oil, and natural gas, which are derived from the decomposition of plants and/or animals.

frequency In power generation, the frequency refers to the number of times per second that a generator's armature completes one full revolution. The unit of measure for frequency is hertz (Hz). In the United States, the standard frequency is 60 cycles per second, or 60 Hz.

full-load amps (FLA) The current carried by a motor or other inductive load when operating at its maximum designed load.

gauge pressure The pressure of a system measured in reference to atmospheric pressure. Gauge pressure = total pressure – atmospheric pressure.

ground fault circuit interrupter (GFCI) A safety device, often in the form of an electrical receptacle, that disables power flow when it detects that current is flowing along an unintended path, such as ground.

heat anticipator A resistor in the thermostat or subbase that generates a small amount of heat to warm the thermostat during a heat operating cycle.

helical In the shape of a helix—a three-dimensional spiral wrapped around a cylinder.

hermetic compressor A compressor that is factory sealed, making its internal components inaccessible without destroying the external shell.

hermetic In this context, a compressor that is sealed and unserviceable in the field.

holding circuit A circuit designed to allow an energized relay to remain energized until the power supply is deenergized, even though the original circuit/device that energized the relay has since returned to its normal position.

holdover voltage The voltage at which a diac ceases to continue conducting current. A diac will conduct current when the breakover voltage is reached, and cease to do so once the voltage falls below the holdover voltage.

horsepower (hp) A unit of power measurement that equals the power of 745.7 watts; it is also stated as 550 foot-pounds (ft-lb) per second.

hot-surface igniter (HSI) A device typically constructed of silicon carbide or silicon nitride used to ignite a gas furnace by being heated to an extreme temperature.

ignition transformer A transformer that provides a high voltage, generally in the range of 10,000 volts, which is used to create an intermittent arc across a gap and ignite an air-fuel mixture.

impedance The effective resistance of an AC circuit, resulting from the combination of simple resistance, inductive reactance, and capacitive reactance.

induced-draft furnace A furnace equipped with a fan or blower on the heat exchanger outlet to reduce the pressure in a heat exchanger and overcome the resistance to flow, helping combustion gases exit the heat exchanger and enter the vent system. Technically, forced-draft systems use a blower at the inlet of the heat exchanger to increase the pressure with the same intent. However, the term "induced-draft" is commonly applied to both types.

inductance A property of an electric circuit by which an electromotive force is induced in it through a variation of current, either in the circuit itself or in a nearby circuit.

induction The process in which a conductor becomes electrically charged by its proximity to a magnetic field or another electrically charged body.

inductive loads Electrical loads that operate based on the principles of electromagnetism. An electric motor, for example, rotates due to the formation of magnetic fields by energized electrical windings. Other examples of inductive loads include transformers and the operating coils of relays and similar control devices.

inrush current The instantaneous, high-volume current drawn by many electrical devices, especially those that operate through induction, such as motors and transformers, when they are first energized.

installation diagram An electrical diagram that shows wiring that must be completed in the field and essential details regarding equipment connection.

insulator A material that resists the flow of electricity and/or heat.

integrated circuit (IC) A semiconductor chip containing a variety of microminiature electronic circuits.

ladder diagram An electrical diagram that has the appearance of a ladder, with individual circuits forming the rungs. Ladder diagrams best represent how the equipment functions electrically and is used primarily for troubleshooting.

laminations The thin metal plates that make up the metal core of a transformer. They are required to minimize energy losses due to electrical eddy currents created within the metal by the alternating magnetic field.

legends Sections of wiring diagrams that provide interpretation for abbreviations, line types and their meaning, and similar information.

limited approach boundary A shock-protection boundary to be crossed only by qualified persons (at a distance from a live part); it is not to be crossed by electrically unqualified persons unless escorted by a qualified person, such as in a training situation.

line side The terminals of a device, such as a circuit breaker, connected to the voltage source.

load Any electrical device that performs work and consumes energy in the process. Common loads in DC circuits include resistors and capacitors.

load side The terminals of a device, such as a circuit breaker, connected to the load being powered.

locked rotor amps (LRA) The maximum current that is drawn by a given motor that is seized or so heavily loaded so that its rotation significantly slows or stops. The LRA may also be exhibited monetarily during motor start-up.

manometer A U-shaped tube containing a fluid such as water or mercury where one end is open to atmosphere and the other end is connected to a pressure source. Alternatively, both ends can be connected to a pressure source, for a differential reading. The pressure is measured by the difference in the displaced liquid height on the two sides of the U-shaped tube. Modern digital manometers, without using fluid, can supply and measure small air pressures for testing purposes.

megohmmeter A special type of ohmmeter used to measure very high resistance values (megohms, or MΩ), such as those associated with insulators. In HVACR applications, meggers, as they are often called, are used to evaluate the condition of motor-winding insulation.

metering devices Orifices, either fixed or variable in size, designed to create a pressure drop and meter the flow of refrigerant.

microamp (μA) 1 microamp equals 1/1,000,000th of an ampere.

microfarad (μF) A unit of measure for capacitance, equal to 1/1,000,000th of a farad.

microprocessor A user-programmable electronic device containing one or more integrated circuits (ICs).

milliamps (mA) A unit of measure for current equaling 1/1,000 of an amp or 0.001 A.

minimum approach distance (MAD) The closest distance at which a qualified worker is permitted to approach an exposed energized conductor or equipment. The MAD is the combination of two elements: the minimum air insulation distance (MAID) and the ergonomic factor.

modulate In this context, to vary the properties of a voltage waveform.

motor starter The combination of a contactor and one or more overload relays that protect the motor from damaging overload conditions.

multistage thermostat A thermostat that controls more than one stage of heating and/or cooling.

natural-draft furnaces Gas furnaces that rely on the heat generated from combustion to create a natural draft through the heat exchanger to move combustion byproducts into the venting system. Natural-draft furnaces cannot make use of complex heat exchangers, due to the resistance to flow that they impose, thus limiting their heat-transfer efficiency.

net oil pressure The difference between the crankcase pressure of a compressor and the measured oil pressure at the outlet of its pump.

noncondensable gases Gases that, under all operating pressures and temperatures of a given system, remain in the gaseous state. For systems using condensable fluids such as refrigerants, the presence of noncondensable gases such as air can significantly interfere with their operation.

ohm The unit of measure for electrical resistance. One ohm is equal to the resistance of a component or circuit when 1 volt is applied and result in 1 amp of current flow. Ohms are represented by the Greek letter omega (Ω) or by the capital letter R in formulas, such as Ohm's law.

Ohm's law The common name for an electrical concept that states that the strength of a direct current is directly proportional to the difference in electrical potential, and inversely proportional to the resistance of the circuit. It is written mathematically as E (volts) = I (amps) × R (resistance).

orifice An opening or passage. In the HVACR field, it commonly refers to a fixed refrigerant metering device that provides the pressure drop for refrigerant entering an evaporator coil.

overload relay A special type of relay that monitors current passing through and opens a control circuit when the current exceeds a predetermined value. Overload relays provide overload protection for large inductive loads, such as motors.

paraphrasing In personal communication, expressing the meaning of a speaker's words by restating the meaning in the listener's own words to ensure clarity.

Pascal's law A principle of physics stating that, for an enclosed liquid system, a pressure change at one point in the system is transmitted equally throughout the system.

photovoltaic cell An electronic device that converts sunlight (radiant energy) into electrical energy.

pickup voltage The voltage value applied to a relay coil at which the contacts of a relay must function and move to their energized position.

pilot-operated valve A valve that depends on a secondary valve to control a source of pressure used to position it.

plenum An air-filled space that in HVAC systems is commonly used as a path for air. For example, the area above a suspended ceiling is often used as a path for return air to the system, in which case it is referred to as a return-air plenum.

plenum-rated Wire and cable insulation that is approved for unprotected use in spaces used as a return-air path (plenums) for conditioned air. Such material must be flame-retardant and exhibit limited smoke production.

polarity A property of various points in an electrical circuit, related to their electrical potential. A negative pole has more electrons than a positive pole, and electrons tend to flow from negative poles to positive poles. Current, however, flows from a positive pole to a negative pole. For DC circuits, current flows in only one direction; one pole is always positive, and the other is always negative. In AC circuits, the polarity constantly changes, with current flowing in one direction and then the other and changing direction at a consistent frequency.

potential relay A relay with a coil and a set of normally closed (NC) contacts that responds to an increase of voltage applied to the coil.

power The energy, measured in watts, consumed by a load or loads in an electrical circuit.

primary control A control used to manage and supervise the oil-burner combustion process and the related components.

primary winding The coiled wires in a transformer that are connected to the supplied voltage source.

prohibited approach boundary A shock-protection boundary to be crossed only by qualified persons. Once a person crosses this boundary with a part of their body or with an object, the same PPE and equipment is required as if direct contact is to be made with a live part.

pulse-width modulation (PWM) A digital signal modulation approach that creates a square wave to send a signal through on and off pulses of the wave.

qualified person An individual who has demonstrated the skills and knowledge related to the construction and operation of electrical equipment and installations and who has received safety training to recognize and avoid the hazards involved. See *OSHA Standard 1910.332(b)(3)* for details on the required minimal training and skills.

raceway A track or channel used to route electrical wiring. Conduit is typically considered to be a type of raceway.

reactance The component(s) of impedance in an AC circuit that are not resistive in nature, resulting from the effect of inductance and/or capacitance.

reciprocal A mathematical value obtained by dividing the number one by another value.

reciprocating compressors Positive-displacement compressors that use pistons driven by a crankshaft to compress refrigerant vapor.

rectifier An electrical device that converts AC power to DC power.

relay An electrical device, often equipped with an electromagnetic solenoid coil, which is activated by one circuit to enable or disable another circuit(s).

resistance The characteristic of an electrical device that causes it to impede the flow of electrical current.

restricted approach boundary A shock-protection boundary set from an exposed energized electrical conductor or circuit part within which specific PPE and insulated equipment must be used. An electrical work permit is required prior to entry.

reversing valve A multi-ported valve used to reverse the direction of refrigerant flow through the indoor and outdoor coils of a heat pump.

ripple The small, undesirable periodic variation in a DC output that has been derived from an AC power source.

rotary compressors There are two basic types of rotary compressors. Rotary-vane compressors use a rotor positioned off-center with respect to the casing and sliding vanes that compress vapor as they rotate. Rotary-screw compressors compress vapor between the flutes of two precisely matched helical-screw assemblies.

rotor The rotating assembly of an electric motor, to which the motor shaft is usually attached.

rough-in The phase of construction in which electrical wiring, terminal boxes, and air conditioning ductwork are installed. Rough-in is done before walls and ceilings are covered with sheet rock or other finishing material.

schematic diagrams Electrical diagrams that represent how equipment functions electrically, typically drawn in the ladder-diagram format.

scroll compressors Compressors that use two scroll-like plates—one orbital and one stationary—to compress refrigerant vapor.

seasonal energy-efficiency ratio (SEER) A comparative measure of energy efficiency used for air conditioners and heat pumps, determined by dividing the total cooling output over the cooling season (measured in Btu or kJ) by the total power input (measured in watts) over the same period. The result provides the number of Btu (kJ) provided per watt of power consumption. The higher the SEER rating, the more efficient the equipment.

secondary winding The set of coiled wires in which the transformer's output voltage is electromagnetically induced.

semiconductor A material that is electrically neutral and serves as the building block for solid-state components. Its neutral status can be changed to positive (P-type) or negative (N-type) by doping.

semi-hermetic In this context, a compressor that is partially field serviceable. As a general rule, only the oil pump and valve assemblies are reasonably serviceable under field conditions. The motor, crankshaft, pistons, and other moving parts can often be serviced and replaced in a shop with the proper equipment.

shock hazard analysis Investigation and documentation of a worker's potential exposure to voltages from energized parts that pose electrical shock hazards, which is conducted to prevent injury by determining safe work practices and appropriate levels of personal protective equipment (PPE).

short circuit An accidental condition in which a low-resistance path around the intended load is created by a live conductor coming into contact with another conductive object.

slip rings Circular devices that transfer electrical energy between a rotating object and a stationary object.

slip The difference between the rotational speed of the rotor and the synchronous speed of the magnetic field.

solenoid coil An electrical mechanism that changes an electrical input into linear movement through magnetism.

solid-state components Electronic components made of semiconductor material that has been doped to make it positively or negatively charged.

stator The stationary portion of the motor that contains the motor windings that, when energized, generate electromagnetic fields.

substation A facility in which power from the power transmission system is reduced to voltage levels required by utility customers. Substations also serve as a point of protection for the downstream power distribution system, using circuit breakers, fuses, and power monitoring systems.

thermistors Resistors whose resistance varies predictably with the surrounding temperature.

thermistors Semiconductor devices in which the resistance varies in proportion to the temperature. Thermistors are used extensively in HVACR systems to sense temperature.

thermocouple A device used to prove the existence of a pilot flame, generally used in natural-draft furnaces that are equipped with a standing pilot flame. Thermocouples generate a voltage of roughly 30 mV, which is used to hold a pilot gas valve open. Thermopiles, used in self-powered furnaces such as wall and floor models, generate roughly 750 mV, which is used as the power source for the entire furnace.

three-phase power An electrical distribution system in which the three legs of the power supply are electrically out of phase with each other by one-third of a cycle.

torque A force applied to an object that causes it to rotate; it is commonly used to describe the force that must be overcome to start the rotor of a motor turning.

transformer An electrical device consisting of two or more coils of wire (windings) wrapped around an iron core in which voltage is transferred from one set of windings to the other(s) by induction.

transistors A semiconductor device capable of controlling a proportionally large amount of current with a small amount of voltage. Transistors are used in solid-state circuits as amplifiers and switches.

triac A diac with a gate circuit that must be triggered for the triac to conduct current. The term "triac" is short for tr*iode for alternating current.*

turns ratio The ratio of the number of primary-winding loops and the number of secondary-winding loops. The ratio determines the value of the secondary voltage in relation to the primary voltage.

VA An abbreviation for volt-amps, which serves as a common method of rating the power output of a transformer that is equivalent to watts. It is the product of the output voltage and maximum continuous design current for a transformer.

varistor A solid-state protective device in which the resistance varies in relation to the voltage applied to it.

vestibule The section of a furnace immediately in front of the combustion chamber, where the gas valve, gas manifold, and various controls are housed.

volt The unit of measure for electrical potential difference or electromotive force. One volt is equal to the electrical potential required to generate a current of 1 amp in a circuit with a resistance of 1 ohm. Voltage is electrical pressure used to push electrons in circuit. Voltage is represented by the capital letter V in most contexts but often by the capital letter E in power formulas.

voltage drop The amount of voltage loss that occurs in all or part of a circuit due to resistance.

voltage imbalance An imbalance in the voltage supplied on each of the three legs serving three-phase electrical loads. A voltage imbalance up to 2% is typically allowed. However, a lesser voltage imbalance can cause a current imbalance that is beyond an acceptable range.

volute The funnel-shaped structure closely positioned around most axial fan blades that captures air from the blade tips, helping to gather and guide the air in the desired direction.

watt The unit of measure for power. One watt represents the power produced by a current of one amp with a potential difference of 1 volt.

Watt's law A common name for the electrical formulas that state the relationship between power, voltage, and current. The various versions of the formula are also known as the power formulas.

whip An assembly of conductors and flexible conduit with connection fittings designed to connect a piece of equipment to a nearby source of power.

INDEX

D

E

I

L

M

N

O

T

U

V

W